Textbook of Veterinary Parasitology

S. C. Mandal

Textbook of Veterinary Parasitology

Springer

S. C. Mandal
Department of Veterinary Parasitology
Faculty of Veterinary and Animal Sciences
West Bengal University of Animal and Fishery Sciences
Kolkata, West Bengal, India

ISBN 978-981-99-2926-9   ISBN 978-981-99-2924-5 (eBook)
https://doi.org/10.1007/978-981-99-2924-5

Disclaimer: Every effort has been made to avoid errors or omissions in this publication. In spite of this, errors might have crept in. Any mistake, error, or discrepancy noted may be brought to our notice, which shall be taken care of in the next edition. The author, publisher, and seller cannot assume any responsibility for the validity of all the information or the consequences of their use. The doses of drugs and vaccines must be cross-checked before use. The author, publisher, and the seller cannot assume any responsibility for use of drugs and vaccines.

This Springer imprint is published by the registered company Springer Nature Singapore Pte Ltd.
The registered company address is: 152 Beach Road, #21-01/04 Gateway East, Singapore 189721, Singapore

Dedicated to my beloved parents

Preface

There are several textbooks of Veterinary Parasitology. Then, why one more new textbook? A robust question—indeed. Being a teacher in this subject having a stand of more three decades of experience, I closely observed the aspirations of different readers of this subject which created three motivations to write this book.

My first motivation to write this is to create wide readership of this book without increasing unnecessary volume. I have seen how the readers have to search different books for getting a particular course material. That's why I have tried to create a single window through which different type of course materials have been catered. Both auxiliary course materials, i.e., parasitic zoonosis, emerging and re- emerging parasitic diseases, parasites of wild animals, immunoparasitology, molecular parasitology along with core courses, i.e., veterinary cestodology, veterinary trematodology, veterinary nematodology, veterinary entomology and acarology, and veterinary protozoology have been catered. Few special chapters, i.e., remote sensing in veterinary parasitology, trends of parasite control incorporated with nanotechnology, etc. have also been provided in this book. Therefore, it is expected that all type of readers, i.e., students, teachers, researchers, and field veterinarians, will surely get benefit from this book.

My second motivation was generated to update the development of molecular parasitology in a simplified form. Last four decades have observed enormous genomic development in all biological sciences. It is very surprising that most of the textbooks are silent in this regard. The students use to get the tasks of molecular parasitology from the teachers, but they rarely find suitable textbook wherefrom they can get prepared. Obviously, some special books are available particularly on this aspect, but these are not for general utilization. I expect if the students incorporate simple touches of genomic or molecular matter in their answer, the answer will surely be more standard. Earlier, molecular matter in the biological sciences was considered as a very tough segment of a course. But today, this has been a routine work. Hardly we can find few laboratories of this discipline where PCR is not done. Then, why would the textbooks remain silent? This matter also motivated me to create this book.

Third motivation to create this book is related to proper presentation of the course matter. I don't want to create a book as an inert store house of information. I have described a particular matter under several headings. For example, biology of a parasite has been described under headings of fertilization, embryonation, egg passing, development in the environment, development in the intermediate host, infection of definitive host, and development up to adult in the definitive host. For pathology caused by parasite, it has been described under headings of anatomical pathology, hematopathology, chemical pathology, ectopic pathology, and secondary pathology. Every aspect of a parasite has been described in similar way. Essay-type description has been strictly avoided. Rendition of course material has been done in such a way that a reader would easily understand what they have to study putting a single glance.

S. C. Mandal

Acknowledgment

At first, I thank the Almighty, whose grace enabled me to accomplish this big project. I must thank my beloved students who are my inspiration, and I keenly observed their problems and aspirations during my long teaching period of more than three decades. My candid thanks are due to all the faculties of my department, Dr. S. Baidya, Professor; Dr. R. Jas, Asst. Professor; Dr. S. Pandit, Asst. Professor; Dr. A. Nandi, Asst. Professor, for their direct and/or indirect help during the creation of this book. I am indebted to my department, Dept. of Veterinary Parasitology, F/O VAS, West Bengal University of Animal & Fishery Sciences (WBUAFS), for taking some images of the parasites/parasitic specimens available therein. I must express thanks from the core of my heart to my beloved son, Dr. Srijan Mandal, for providing me different ideas and sharing his knowledge during the creation of this book. I must thank my wife, Mrs. Sudipta Mandal, for inspiring me with different ideas to beautifully distribute the content in this book. I am really seeking for a suitable word which can express my gratitude to the entire team of Springer Nature for assessing the proposal of this book, approval of the proposal, and bringing the manuscript to the final stage.

Introduction

Parasitology is a branch of science which deals with the phenomena of parasite and parasitism. But Veterinary Parasitology is not merely limited to the study of phenomena of parasites. Apart from the basic study of parasites and parasitism, the subject is focused more towards practical implication of basic knowledge i.e. diagnosis, treatment, control etc. Keeping the basic study little behind, the subject is marching towards development of different practical strategies related to diagnosis and control of parasites. Day by day, the scientists are exploring the knowledge and this knowledge is being exploited to develop strategies for better diagnosis and control of parasites. The subject has taken a new dimension with the rapid growth of immuno-parasitological and molecular parasitological field. However, without basic knowledge, it is not possible to develop any modern techniques related to diagnosis and control.

Different associations of organisms

To understand the definition of parasite and parasitism, clear knowledge of related definition of associations between different organisms is a pre-requisite. Any type association between two organisms is called as **symbiosis** (e.g. mutualism, parasitism etc.). **Mutualism** is an association between two organisms wherein both the partners benefit one another. The association between hermit crab and sea anemone is a good example of mutualism where sea anemone lives on the back of crab. The anemone get morsel of food torn off by the crab. On the other hand the crab is protected by the stinging tentacles of the anemone. **Commensalism** is an association between two organisms wherein one partner is benefitted and the other partner is neither harmed not benefitted. *Entamoeba coli* organisms remain in the intestine of the host and get food from the host. This is an example of commensalism. In **phoresis**, one organism is temporarily attached to other organism and simply carried from one place to another place. Sedentary protozoans, fungi etc. get attached on flies and are transported. In **parasitoidism**, one developmental stage of an organism attaches to other organism and gradually eat the organism. In this relationship, larvae or developmental stages feed tissue of other organism in such a gradual slow process that the later organism dies after complete development of former organism. In predation, one organism eat up other organism. In predation, one organism eat up another organism immediately. In both parasitoidism and **predation**, one organism is killed by another organism. But in parasitoidism, killing process is slow whereas in predation, one organism kills other organism immediately. We can understand the definition of parasite if we understand the definition of parasitism. **Parasitism** is a heterospecific association wherein one organism (parasite) is metabolically dependent on the other organism (host). It is usually stated that the parasites cause harm to the hosts. But the parasites may be harmless or even be beneficial to the hosts. *Toxocara canis* is an example of a parasite which lives in dog (host).

Different types of parasites

There are different types of parasites. Basing on seat of predilection, the parasites are either endoparasites or ectoparasites. Endoparasites occur within the hosts i.e. *Ascaris* spp. and those are ectoparasites which occur on the skin, within hairs, furs etc. (e.g. lice, mites ticks etc.). The parasites living in tissues of the hosts are histozoic parasites (e.g. *Trichinella spiralis*) and the parasites occurring in the hollow organs are called as coelozoic parasites (*Taenia solium*). Basing on duration of parasitism, they may be permanent parasites which occur on the hosts for their whole life (e.g. *Melophagus ovinus*) and temporary parasites which occur on the hosts for a short period for taking their food (e.g. mosquitoes). Obligatory parasites are those parasites which must require a host to pass a whole or part of their life (*Dipylidium caninum*). Basing on host range,they are stenoexenous (*Eimeria tenella* occuring only in birds) having narrow host range and euryexenous having broad host range (*Trypanosoma evansi* occurring in broad host range i.e. camel, horse, dog, cattle, buffalo etc.). Basing on involvement of intermediate hosts, they are monoexenous parasites which do not require any intermediate hosts (*Ascaris suum*) and heteroexenous parasites which require intermediate hosts (*Diphyllobothrium latum*). In case of autoheteroxenous parasites, the same host acts as definitive and intermediate hosts (*Trichinella spiralis*). Basing on reproductive ability, they are monocious parasite having both male and female reproductive systems in the same parasites (e.g. digenetic trematodes except schistosomes). The monoecious parasites are also called as bisexual parasites. Tape worms are good examples of monoecious parasites. Dioecious parasites are unisexual parasites and they have one reproductive system (either male or female). Round worms are examples of diocious parasites. Basing on food material taken by the parasites, they are haemophahus (hook worms), mucophagus (*Trichostrongylus*), biliphagus (*Fasciola*) etc. Two more parasites are yet to be mentioned. These are aberrant parasites and facultative parasites. Any parasite which affect unnatural hosts is called accidental parasites. *Hymenolepis diminuta* (natural host is rodent) is an accidental parasite of human beings. Erratic parasites are those parasites which invade unnatural organs or sites. Intestine is the normal site of *Entamoeba histolytica.* But this parasite may invade liver and lungs of human beings. Therefore, they may be erratic parasites of human beings if they occur in those organs. Facultative parasites are not parasitic but they behave as parasite if they are accidentally ingested or they penetrate through the opening of wounds of hosts. *Acanthamoeba* spp. occurring in human beings are facultative parasites.

Different types of hosts

Parasites have wide host range. Both terrestrial and aquatic animals may act as the hosts.Two types of hosts for parasites are commonly described i.e. primary hosts and secondary hosts. The parasites reach to the adult stage or attain sexual maturity in the primary hosts. The primary hosts are also called as final hosts or definitive hosts. Another type of hosts are the intermediate hosts in which, a part of biological development of parasite occurs. But in the intermediate hosts the parasites never attain sexual maturity. For example, dog is the definitive host of *Echinococcus granulosus* and sheep, goat, cattle etc. act as intermediate hosts of the parasites. Dog is the definitive host of *Dipylidium caninum* and fleas (*Ctenocephalides* sp.) serve as intermediate host of the same parasite. Paratenic or transport hosts are the hosts in which the parasites do not undergo any biological development. Only they take shelter in that host. The rodents act as the transport hosts of *Toxocara canis*. Reservoir hosts are the hosts which act as the source of infection and in these hosts they do not produce any pathology. A group of organisms are called as vectors which are arthropods and act as the source of infection. It is to be noted that the parasites attain sexual maturity in the vectors. Different blood sucking flies are the vectors of different *Trypanosoma* spp.

Seat of predilection of parasites

The parasites have wide range of predilection sites. Almost every organic system can be affected by the parasites. These are digestive system, respiratory system, nervous system, circulatory system, excretory system etc. There have site specificity of the parasites. They may either affect organ/organs of a particular system or the whole system may be affected. Not only the organ, but the parasites have even cell specificity. Some parasites may occur in some particular cells. Almost all cestodes are found in the small intestine. But *Stilesia hepatica* occurs in the bile duct. Trematodes affect bile duct and liver (*Fasciola, Dicrocoelium*), rumen and reticulum (amphistomes), oviduct and bursa of fabricious (*Prosthogonimus pellucidus*), lungs (*Paragonimus*). Likewise,the nematodes have variable predilection sites. They are mostly found in the GI tract (Ascarids, Strongyles), lymphatic and circulatory systems (Filarids), urinary system (*Dioctophyma* sp.), subcutaneous tissue (*Dracunculus* sp.). Protozoan parasites have also variable predilection sites. Trypanosomes occur in blood. *Leishmania* spp. occur in macrophages and endothelial cells. In many cases, the sexually matured stages of coccidian parasites occur in the GI tract and immature stages occur in muscles. *Tritrichomonas foetus* occurs in the genital tracts of cattle. *Babesia* and *Theileria* spp. occur in the blood.

Morphology of parasites

Study of morphology has taxonomic importance. In the earlier ages, when there was no microscope, study of gross morphology was the only parameter for identification of the parasites. After invention of microscope, morphological study has been more convenient. Different critical internal and external morphological features are easily examined. This method has been more enriched after invention of electronic microscope. In case of cestodes, morphometry is done for whole of the parasite, organ and part of organ of parasite. The head or scolex, rostellum, hooks, neck, segments, testes, ovary, vitelline glands are major areas of study. In case of trematodes, morphometry is done similar to cestodes. The body, tegument, suckers, pharynx, oesophagus, testes, ovary, cirrus, genital pores are studied for taxonomy. In case of round worms, body, mouth, lips, buccal cavity, buccalcapsule, pharynx, oesophagus, intestine, spicules, gubernaculum, bursa, bursal rays etc. are studied for morphological identification of the parasites. Head, proboscis, antena, arista, thorax, wings, legs, abdomen etc. are critically studied for morphological identification of insects and arachnids. In protozoan parasites, body, flagella, cilia, pseudopodea, nucleus, cyst, oocysts, sporocysts etc. are critically studied for taxonomical identification.

Biology of parasites

Life cycle of parasites are of two types i.e. direct and indirect. In case of direct life cycle, the parasites do not require any intermediate hosts. Biological development occurs in one host and environment (Trichostrongyles). In case indirect life cycle, the parasites require intermediate hosts to complete their life cycle (*Taenia* spp.). There may be requirement of two intermediate hosts by some parasites. Certain stage of biological development occurs in both these intermediate hosts (*Diphyllobothrium latum*). In the definitive hosts/ final hosts, the parasites reach to the adult stage attaining their sexual maturity. In case of trematodes, different developmental stages are egg, miracidium, sporocyst, redia, cercaria, metacercaria. In some cases, daughter sporocysts (*Dicrocoelium dendriticum*) and daughter rediae are produced. In case of *Schistosoma* spp., instead of metacercariae, schistosomula stage is produced. In some cases, instead of metacercaria, mesocercaria stage is produced.

Both sexual and asexual reproduction occur in cestodes. Fertilization of cestodes is either by self or cross fertilization. In the Eucestoda, life cycle of the members require one intermediate host. The larval stage is formed in the intermediate hosts. These larval stages are called as bladder worms. These are cysticercus, coenurus, hydatid cysts, strobilocercus etc. These are also called as metacestodes. The metacestodes are the infective stages for the final hosts. The final hosts get the infection by ingestion of metacestodes. In Cotyloda, two intermediate hosts are required for completion of life cycle.

In nematodes, the life cycle is completed either involving intermediate hosts (indirect life cycle) or without involving any intermediate hosts (direct life cycle). The nematodes are not hermaphrodite. Male and female parasites are different. The eggs are laid by the female parasites. The larvae hatach out from the eggs. There are five larval stages. The fifth larval stage is young adult.

Both sexual and asexual reproduction occurs in protozoa. Asexual reproduction include binary fission, budding (formation of small daughter organisms from parent organisms), Schizogony (multiple fission of organisms). Sexual reproduction is formation of zygote by male and female gametes and it also includes conjugation (exchange of genetic material). Both direct and indirect life cycles are found in protozoa. In some protozoa, same organism displays both sexual and asexual reproduction. For example, life cycle of *Eimeria* includes formation of sporozoites, schizonts, gametes etc.

Life cycle of arthropods mostly direct type. The eggs are laid by the female arthropods. The larvae hatch out. Other developmental stages are pupa and adult. This type of life cycle is holometabolous type. In some arthropods nymphs hatch out directly from the eggs. Adult develop from the nymph. This type of life cycle is hemimetabolous type.

Mode of infection of parasites

There are many ways by which the parasites enter the hosts. The common route of infection of parasites is ingestion. The infective stage of the parasites are ingested either with inanimate material (food, drinking water etc.). Or the infective stages are developed in the intermediate hosts and the final hosts get the infection by ingestion of intermediate hosts harboring the infective stages of the parasites. In some cases, the final hosts get the infection by ingestion of infected transport host. Skin penetration is other mode of infection. The infective stages of the parasites penetrate the skin of the hosts. Thereafter, they enter the internal organs and reach to maturity in their respective organs. The parasites can also enter the hosts by inoculation. The mosquitoes and other blood sucking insects transmit parasites during blood sucking. Some parasites (*Tritrichomonas foetus, Trypanosoma equiperdum*) are sexually transmitted. This type of transmission is called as venereal transmission. Congenital/transuterine, lactogenic/transcolostral are also some modes of infection. The parasites also can enter through ingestion of biomedicine, injection of medicinal agents, blood transfusion, biting etc.

Harm caused by parasites

The parasites cause harm to the hosts in many ways. The most common harm is utilization of host's nutrients by the parasites. Carbohydrate, protein, fat, vitamins, minerals etc. of host origin are utilized by the parasites. The parasites may utilize host's nutrients both in crude and simplified (digested) form. The cestodes take the nutrients by absorption through their body surface because they do not have digestive system. Trematodes have incomplete digestive system. They take host's material through their acetabulum and digest the material in the digestive system. The nematodes have complete digestive system. Different adult and larval stages of nematodes take host tissue, blood, body fluid and digest the food material. The arthropods have also complete digestive system. The protozoan parasites take food material in different ways i.e. pinocytosis, phagocytosis etc.

The parasites may cause direct injuries to the host tissues in different ways i.e. taking tissue materials as their food, biting the tissues by mouth. The parasites can cause mechanical damage to the hosts. There are many large parasites which can cause obstruction of the lumen of the G.I. tract thus causing functional disturbance of the gut. This is commonly found in some tape worm infection (*Diphyllobothrium latum, Dipylidium caninum*). The immature stages of parasites i.e. metacestode or bladder worms of cestodes may create pressure on the vital organs. For example, the hydatid cysts formed on the diaphragm may create pressure on the oesophagus. The excretory secretary substances released by the parasites cause damage to the hosts. The metabolic by product of many tape worms affect the nervous system of the hosts resulting in nervous disorders. The pathology caused by the parasites in different organs or organic systems cause complete or partial loss of function of that organ or organic system. For example, the digestive system is damaged by different parasites resulting in functional disturbance of digestion. Indirect harm caused by different parasites is a significant consequence of parasitic infections. The hosts suffer from debility, depression, anorexia etc. which result in loss of production, loss of body weight etc. In many cases, haemopoitic system is depressed and as a result of it, the affected animals suffer from anemia. Some parasites carry other parasitic pathogens. For example, the eggs of *Heterakis gallinarum* transmit harmful protozoan parasite, *Histomonas meleagridis* in turkey.

Diagnosis of parasitic diseases

There are different diagnostic approaches for diagnosis of parasitic diseases. These are history and symptoms-based diagnosis, parasitological diagnosis, immunological diagnosis, molecular diagnosis etc. There are some parasitic diseases which could be diagnosed basing on specific symptoms exhibited by the affected animals. For example, the dogs infected with tape worms drag their anus over ground to alleviate the irritation caused by gravid segments. Parasitological diagnosis is employed to directly observe the parasite or any parasitic stages by naked eyes or by aid of instruments. Examination of fecal sample reveals eggs, larvae, whole parasites during microscopical examination. Examination of blood smear reveals different haemoprotozoan parasites. Likewise, examination of urine, sputum also reveal different parasite or parasitic stages. Immunodianosis is done by different immunological tests i.e. IHA, CFT, RIA, ELISA etc. Molecular diagnosis is performed by different molecular tests i.e. PCR, northan blot, southern blot, using DNA probe etc.

Therapy of parasitic diseases

Broadly parasites are of five categories i.e. round worms, tape worms, flukes, arthropods, protozoans. First three categories are under helminthic parasites. The drugs which can kill the helminthic parasites are called as anthelmintics. The important anthelmintics are benzimidazole, imidazoles, tetrahydropyrimidines, piperazine, salicylanilides, avermectins etc. The drugs which can kill the protozoa are called anti-protozoals. There are different anti-protozoals effective against different protozoan parasites. Anti-trypanosomals are diamidine, antrycide salts, quinapyramine, suramin etc. Anticoccidials are amprolium, polyether antibiotics etc. Common anti-theilerial drugs are parvaquone, buparvaquone, menoctone etc. Anti- babesial drugs are trypan blue, acriflavin, imidocarb etc. Chloroquin, diodoquin etc. are antimalarial drugs. There are different arthropodicidal drugs. Lindane, chlordane, dieldrin, eldrin, DDT, BHC, HCH, dimethoate, dioxcarb, cypermethrin, permethrin are some arthropodicidal drugs. Drugs are administered in different ways. Oral administration constitutes drenching, feed- mixing, tablet- form, paste-form for licking, urea- block form etc. Different routes of injection of drugs include intramuscular injection, intravenous injection, intra- ruminal injection etc. Many drugs are applied topically and the methods of applications are dipping, spraying, pour- on, dusting powder etc. Drugs are also applied as collars, ear bands, tail bands, leg bands etc.

Control of parasites

Control of parasite is a difficult task as it has many limitations. These limitations include drug resistance constraints, problem of control of vector and intermediate hosts, problem in development of vaccine, problem of practical application of biological control etc. It is an established fact that not a single strategy can control parasites. There should have integrated approach for parasite management. The researchers are consistently working to unfurl the scientific reasons of drug resistance. They are also earmarking the resistant drugs against parasitic infection. Formulation of aniparasitic therapy is designed accordingly so that effective treatment can be done. Though it is very much difficult for mass control of vectors and intermediate hosts, but success has been achieved to certain level for controlling mosquitoes, snails and other vectors and intermediate hosts. A gamut of research works have already been carried out on development of vaccines, but success is limited. The reasons are stage variation of parasite, strain variation of parasites, complex nature of antigen, frequent change of antigenic characters. Many decades have passed but no potent vaccine against trypanosomosis has yet been produced because the trypanosomes frequently change their antigenic characters. Biological control of parasites is not effective and in many cases it is practically not applicable.

Contents

About the Author

S. C. Mandal is a Professor in the Department of Veterinary Parasitology, Faculty of Veterinary & Animal Sciences, West Bengal University of Animal and Fishery Sciences, Kolkata, India and former Head of the Department and former Dean-In-Charge of Faculty of Veterinary & Animal Sciences and former Controller of Examinations (Actg.) of the same university. He is a doctorate from a premier institution, Indian Veterinary Research Institute. He has more than three decades of teaching and research experience in the field of Veterinary Parasitology. He is the sole author of several books, including *Veterinary Parasitology at a Glance*, *Veterinary Helminthology*, *Question Bank of Veterinary Parasitology*, *Sure Success in Veterinary Parasitology*, *Encyclopedic Glimpses of Veterinary Parasitic Diseases*. Of these books, two books are amongst the best books of all time as selected by BookAuthority. He has contributed a chapter in a popular book, '*Pathology of Infectious Disease of Domestic Animals*'. He worked in six prestigious research projects (DBT-2,DST-1,ICAR-1,NATP-1,RKVY-1) in different capacities and acquired vast research experience. He has significant research contribution in the field of immuno-parasitology and molecular parasitology. He submitted forty one gene sequences of different parasites in the GenBank and got accession numbers of those. He was research editor of an international journal and presently he is a reviewer of different scientific journals. He has published many research articles of national and international repute. He got several awards for his significant contribution in his relevant field. He has been included in the National Academy of Veterinary Science, New Delhi, for his contribution to the field of Veterinary Parasitology.

1 Taxonomy of Parasites of Veterinary Importance

Taxonomy is a branch of science that is concerned with the classification of any living or extinct organisms. Taxonomy is the methodology and principle of systematic botany and zoology that creates a hierarchical stand from superior to inferior. There has been a clear difference between taxonomy and systematics. Taxonomy is the theory and practice for classifying organisms, whereas systematics is the science that deals with diversity of organisms.

Taxonomic Rank

In biological classification, taxonomic rank indicates the hierarchical position of a group of organisms in relation to the other group. Biologists pay equal attention for classification of the organisms from superior to subordinate.

A Brief History of Classification

The Aristotelian method was an old method of classification that dominated until the nineteenth century. It was based on the nature of the organism. In this method, only the constant characters are considered.

Carolus Linnaeus is known as the founder of modern taxonomy. His books are important for the basic foundation of zoological and botanical taxonomy and nomenclature.

Different aspects used for taxonomy are as follows:

- Gross morphology
- General microscopical features
- Specific morphology by scanning electron microscopy, transmission electron microscopy, atomic absorption microscopy, etc.
- Biochemical characterization
- Immunological characterization
- Genomic characterization

Taxonomic Character

Taxonomic characters are the attributes of one member of one taxon by which it can be differentiated from the member of the other taxon.

In general, taxonomic characters are as follows:

1. **Morphological Character**

   In the earlier period, gross and microscopic study of one organism was the prime area of attention for taxonomy. The term "morphology" is used till date to describe mostly the outer and inner structural features of any organism. The study of morphology was started to identify one organism and to place a particular organism in a phylogenetic tree. Later, this technique was exploited for many other purposes. In the case of bacteria and

S. C. Mandal, *Textbook of Veterinary Parasitology*, https://doi.org/10.1007/978-981-99-2924-5_1

virus, the placement of these organisms in a **phylogenetic tree** is mainly on the basis of outer and inner structural features, biochemical characters, immunological characters, genetic characters, etc. But in the field of Parasitology, biochemical, immunologic, and genetic characters are considered, but internal and external morphological features are the prime factors for phylogeny of parasites.

2. **Physiological Character**

   The physiology is different in different groups of animals. The carbohydrate, protein, and lipid contents are also different in different groups of organisms. Vitamin, mineral, enzyme, and other biochemical profiles are also different in different animals. Therefore, physiological activities are obviously different. Taxonomists very carefully analyze these physiological attributes for placing the organisms in particular taxon. Nutrition, excretion, and reproduction are differently performed in helminths, arthropods, and protozoa, and they are placed in different taxons.

3. **Ecological Character**

   Every group of organisms has its own niches. These different niches and other ecological factors are considered as attributes for the placement of one organism or a group of organisms. Different ecological factors include life process, adaptation, biomass, distribution of organisms, etc. All these things are taken into consideration during the study of taxonomy.

4. **Ethological Character**

   This character pertains to the behavior of one organism. The "behavior of one organism" is a broad term, which is dependent on different habits. Different habits are also considered for the taxonomy of organisms.

5. **Geographical Character**

   Biogeographic pattern is an important attribute of the organism. It depends on two important environmental factors, i.e., temperature and humidity. This creates different biogeographic pattern of organisms, which are also very important for determining the upper hierarchical positions of organisms.

Journey of Parasitic Taxonomy

Although parasites are also subject to biochemical characterization, immunological characterization, and genomic characterization, the morphology of a parasite is mostly used for identification. In the earlier period, before the invention of microscope, visual analysis of the external and internal structures of any organism was the sole resort for the identification and study of a parasite. At that time, analysis of gross morphometric data and different structures was the common trend for the study of any parasite. There have been many records of tapeworms, roundworms, etc. of the ancient period. The records tell us about the gross morphological features of those parasites. At that time, technique of staining of the parasites was not known to the people. The color, length of parasite, number of segments, size, shape, etc. were taken into account for morphological study. In the later period, the discovery of lens was done by Leeuwenhoek, and this lens later emerged as microscope. That was a great landmark of age, not only in the field of parasitology but also in other fields of science. The microscope did wonder at that time. There had been a vast study on microscopic morphology of different organisms. Day by day, many organisms were also discovered. Later, SEM (scanning electron microscope), TEM (transmission electron microscope), and atomic absorption microscope were invented. The images created by these microscopes based on the passing beam of electron through the sample or reflected electron or knocking off electron were extremely clear to visualize the minute structures of an object.

Molecular Taxonomy: A Trend

Advanced taxonomy is the trend of taxonomy for classification of organisms. This is a journey that started with using gross morphological features, and presently, it has reached to the state of analysis of genomic features. Time has come to pay attention toward more and more minute structures for proper placement of these organisms. In the recent past, phylogeny has taken a new dimension that has crossed the biochemical and immunological characterization of a parasite; rather, it has sharply entered into the field of molecular characterization and identification of parasite. The field of parasitology has been progressing toward the search for the genomic structure of different species of many parasites. The nucleotide sequences of many parasites have been derived. The primers of many parasites are now commercially available. For the study of phylogeny, the research is now directed toward the minute difference of genomic structure of the parasites. Even work on molecular epidemiology by using the PCR technology has been the present days' trend for the detection of different developmental stages of the parasites in a molecular way. The genes of parasitic stages are being detected in the intermediate hosts and vectors also. Molecular techniques are used to identify different molecular markers. The primers are constructed using different soft wares for these molecular markers.

The basic steps of genomic characterization are shown in the following figure.

Note

Ribosomal DNA sequences and mitochondrial DNA sequences are widely used for molecular characterization of parasites. The molecular markers of ribosomal DNA are internal transcribed spacer 1 (ITS 1), internal transcribed spacer 2 (ITS 2), etc. The mitochondrial DNA sequences mostly used as molecular markers are cytochrome oxidase subunit 1 (CO 1), NADH dehydrogenase subunit 1 (ND 1), etc.

Classification of Nematodes of Veterinary Importance

Phylum: Nemathelminthes

Class: Nematoda (cylindrical in shape; usually white or creamy white in color; unsegmented body; possessing body cavity; presence of digestive system)

Order: Ascaridida

(presence of three large lips; laterally placed caudal alae, absence of buccal capsule; lacking

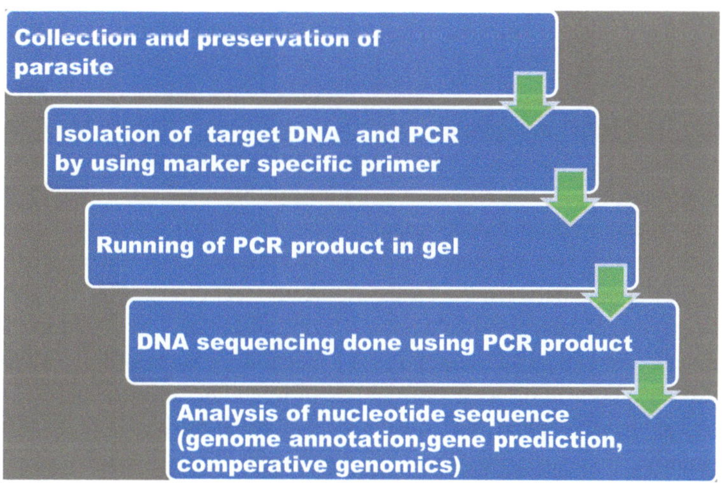

Fig. 1.1 Basic steps of genomic characterization

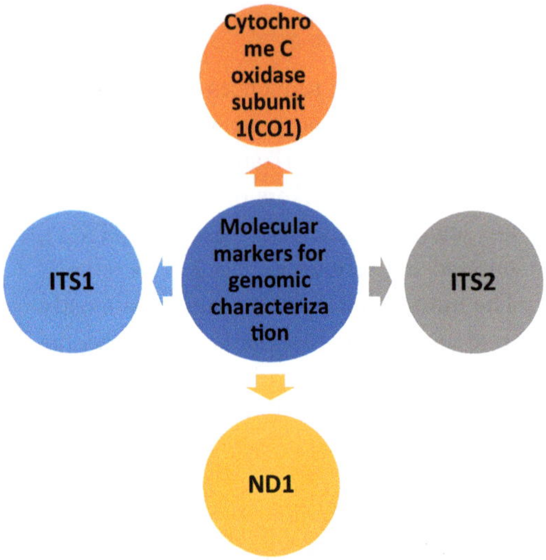

Fig. 1.2 Some molecular markers (schematically displayed)

> **Note**
> Molecular phylogeny is a branch of phylogeny in which molecular differences predominantly based on nucleotide sequences are determined and evolutionary status of the organism is also unfurled. By the accumulated data, phylogenetic position of any organism can also be derived, and these data become helpful for making molecular taxonomy.

esophageal bulb with some exceptions; two spicules)

Superfamily: Ascaridoidea

Family: Ascarididae (three well-developed large lips; each usually bearing two papillae; may be presence of interlabia in between the bases of lips; usually absence of buccal capsule and pharynx; club-shaped esophagus without posterior bulb; usually absence of well developed caudal alae in the male tail, two equal-sized spicule)

Family: Anisakidae (cylindrical and ventriculus esophagus, cecum present or absent)

Superfamily: Oxyuroidea

Family: Oxyuridae (presence of inconspicuous three lips, well-developed esophageal bulb; presence of cloacal papillae surrounding the cloaca; one spicule found in male; vulva found in the anterior end)

Family: Kathlaniidae (presence of esophageal bulb)

Superfamily: Subuluroidea

Family: Heterakidae(presence of three lips around the mouth; a small buccal cavity and pharynx, presence of preanal sucker with chitinous rim in male)

Family: Subuluridae (absence of or poorly developed lips; presence of buccal capsule frequently having teeth; bulbous esophagus; preanal sucker without chitinous rim)

Order: Rhabditida

Superfamily: Rhabditoidea

Family: Rhabditidae (small buccal cavity with three or six lips, rhabditiform esophagus)

Family: Strongyloididae (parasitic males not found; free living, parasitic, alternation of free-living and parasitic life cycle found; esophagus noticeably long in free living generation and valvular in parasitic generation)

Order: Strongylida (leaf crown or corona radiata may be present; club-shaped esophagus, bursate worms possessing large or small bursa supported by bursal rays)

Superfamily: Strongyloidea (mouth opening usually surrounded by leaf crown; teeth or cutting plates found in the buccal cavity; copulatory bursa well developed having a pair of lateral lobes and a dorsal lobe; bursa supported by bursal rays)

Family: Strongylidae (usually presence of leaf crown of two types, i.e., external leaf crown and internal leaf crown; called palisade worms as the leaf crown look like fence; teeth or cutting plates may be found in the buccal cavity; copulatory bursa well developed having a pair of large lateral lobes and a dorsal lobe; bursa supported by bursal rays; two equal spicules found)

Family: Trichonematidae (the buccal capsule typically cylindrical or annular; teeth absent; barbed spicule, vulva located near the anus)

Family: Amidostomidae (buccal capsule is short, shallow thick walled; absence of leaf crown)

Family: Stephanuridae (cup-shaped thick-walled buccal capsule with teeth at the base)

Family: Syngamidae (cup-shaped thick-walled buccal capsule with teeth at the base; absence of leaf crown; permanently copulated male and female in some members)

Superfamily: Ancylostomatoidea (subglobular buccal capsule; oral opening may be armed with teeth or cutting plate or unarmed; lip and leaf crown absent)

Family: Ancylostomatidae (head is dorsally bent giving hook-like appearance; absence of leaf crowns; well-developed subglobular buccal capsule; presence of teeth or chitinous cutting plates at the ventral margin)

Superfamily: Trichostrongyloidea (slender and thin bodied; rudimentary stoma; absence of leaf crown; 3, 6, or no lips)

Family: Trichostrongylidae (buccal capsule absent or rudimentary; teeth and lancets absent usually; well-developed bursa)

Family: Ollunanidae

Family: Dictyocaulidae (bursa with long bursal rays; the rays may be fused at the distal end in some members; short and stout spicule)

Superfamily: Metastrongyloidea

Family: Metastrongylidae (much reduced dorsal ray; long and filiform spicule)

Family: Protostrongylidae (bursa may be very small or absent; spicules may have membranous expansion, well developed gubernaculum and telamon)

Family: Filaroididae (bursa may be absent or rudimentary; ventral and lateral rays short and digitiform)

Family: Skrjabingylidae

Family: Crenosomatidae (ring formation in the anterior end; well-developed bursa)

Order: Spirurida (esophagus having anterior muscular and posterior glandular region; male tail end coiled usually bearing caudal papillae)

Superfamily: Spiruroidea (two lips each being subdivided, spicules unequal, hind end of male usually coiled spirally bearing lateral alae and papillae)

Family: Spiruridae (typical characters of superfamily)

Family: Thelaziidae (lips are absent; male tail end bearing pre- and post-anal papillae; unequal spicule)

Family: Acuariidae (ornamental structures called cordons are found at the anterior part of the body. The cordons may be recurrent or non-recurrent. These either may anastomose or do not anastomose)

Family: Tetrameridae (sexual dimorphism more; males slender and female globular or coiled)

Superfamily: Physalopteroidea

Family: Physalopteridae (a collar-like structure formed by cuticle of the worm at the head region; lips present with small teeth; pharynx absent)

Family: Gnathostomatidae (trilobed large lips; teeth like ridges present on the medial surface of cuticle)

Superfamily: Filarioidea (buccal capsule, lips, pharynx absent; esophagus having anterior muscular and posterior glandular part)

Family: Setariidae (head equipped with peribuccal ring bearing cuticular epaulette-like structure or small teeth)

Family: Filariidae (peribuccal ring or other structures are absent)

Family: Onchocercidae (striated cuticle, genital opening present anteriorly)

Superfamily: Dracunculoidea (rudimentary mouth; well-developed cephalic papillae)

Family: Dracunculidae (female member much larger than male; atrophied anal and genital openings)

Order: Enoplida

Superfamily: Trichuroidea (esophageal glands present outside the esophagus arranged in a row called stichosome)

Family: Trichinellidae (the hind end is thicker than the anterior end, male without spicule)

Family: Trichiuriidae (anterior part slender and posterior part thick called as whip worm; one spicule enclosed in spicule sheath; male tail end spirally coiled)

Family: Capillariidae (almost equal diameter through the entire length; esophageal part shorter than posterior part)

Superfamily: Dioctophymatoidea

Family: Dioctophymidae (male tail bears cup-shaped bursa without rays)

Family: Soboliphymidae (presence of cephalic sucker, campanulate bursal cup)

Superfamily: Mermithoidea (pharynx is exceptionally long reaching half or more length of the body)

Family: Mermithidae

Family: Tetradonematidae

Classification of Trematodes of Veterinary Importance

Phylum: **Platyhelminthes**

Class: Trematoda

Subclass: Monogenea (Parasitic primarily on cold-blooded aquatic vertebrates);

Subclass: Aspidogastrea (Parasitic in or on fish, turtles etc.);

Subclass: Digenea (Parasitic in domestic and wild animals)

Family: Fasciolidae (leaf shaped; closed oral and ventral suckers; presence of pharynx and short esophagus; branched or simple intestinal ceca and branched excretory bladder; testes tandem, branched, or lobed)

Genera

Fasciola

Fasciolopsis

Fascioloides

Parafasciolopsis

Family: Dicrocoeliidae (flattened body; internal organs visible from outside; pharynx and esophagus present; suckers located close; testes present posterior to ventral sucker; intestinal ceca not reaching up to the posterior end; uterus mostly in the hind part)

Genera

Dicrocoelium

Platynosomum

Eurytrema

Concinum

Athesmia

Family: Opisthorchiidae (flattened and characteristically translucent; suckers close; esophagus and pharynx present; intestinal ceca up to the posterior end; testes posterior in location and diagonal in position; excretory bladder having long stem and equipped with short branches)

Genera

Opisthorchis

Clonorchis

Metorchis

Parametorchis

Pseudoamphistomum

Family: Paramphistomatidae (thick, fleshy; suckers at both ends; anterior sucker sometimes with posterior pockets, absence of pharynx; frequently lobed testes usually anterior to the ovary; presence or absence of ventral pouch)

Genera

Paramphistomum

Cotylophoron

Gigantocotyle

Gastrothylax

Fischoederius

Gastrodiscus

Gastrodiscoides

Pseudodiscus

Calicophoron

Carmyerius

Family: Prosthogonimidae (spiny flukes; tapering anteriorly and rounded posteriorly; testes horizontal in position; ovary pretesticular; vitelline glands grape-like situated in the lateral field; Y-shaped excretory bladder)

Genus

Prosthogonimus

Family: Echinostomatidae (oral sucker small and surrounded by head collar equipped with single or double rows of spines; testes lobed or intact; tandem or diagonal in position; pretesticular ovary; vitellaria in the lateral field)

Genera

Echinostoma

Echinochasmus

Echinoparyphium

Hypoderaeum

Isthmiophora

Family: Heterophyidae (small-sized ovoid or pyriform-shaped fluke; pharynx and a long esophagus present; ventral sucker present almost in the middle of the body; genital pore frequently surrounded by a genital sucker (gonotyl) situated near to the ventral sucker; testes located horizontally or diagonally near to the posterior end of the body; cirrus sac absent; ovary pretesticular)

Genera

Heterophyes
Apophallus
Euryhelmis
Metagonimus
Cryptocotyle

Family: Nanophytidae (pear-shaped or elongated flukes; ventral sucker situated in the middle third of the body; ovary pretesticular; testes in the posterior part of the fluke)

Genus

Nanophyetus

Family: Troglotrematidae (flat-bodied fleshy flukes with convex dorsal surface and concave ventral surface; oral sucker subterminal, ventral sucker at mid-body region; vitellaria in the lateral field, genital pore situated posterior to the ventral sucker)

Genera

Troglotrema
Collyriclum

Family: Notocotylidae (commonly called monostomes, three to five rows of unicellular glands in groups situated in the ventral surface of the body; no ventral sucker, testes horizontal located at the posterior end of the body and lateral to the intestinal caeca; ovary is situated in between the testes)

Genus: *Notocotylus*

Family: Paragonimidae (spiny cuticle; ventroterminal oral sucker; lobed testes in the posterior half of the body; lobed ovary pretesticular; absence of cirrus pouch)

Genus

Paragonimus

Family: Clinostomidae

Genus

Clinostomum

Family: Strigeidae (have constriction creating anterior flattened or cup-shaped portion and cylindrical posterior part)

Genera

Apatemon
Cotylurus
Parastrigea

Family: Diplostomatidae

Genus

Alaria
Diplostomum
Posthodiplostomum
Neodiplostomum

Family: Allocreadiidae

Genus

Crepidostomum

Family: Sanguinicolidae

Genus

Sanguinicola

Family: Cyclocoelidae

Genus

Tracheophilus

Family: Schistosomatidae (cylindrical resembling roundworm; unisexual with marked sexual dimorphism; presence of gutter-like groove in male parasite used for carrying female parasite called gynaecophoric canal; intestinal ceca uniting posteriorly that further bifurcate and unite; eggs frequently equipped with terminal or lateral spines but have no operculum; metacercariae and rediae stages absent)

Genera

Schistosoma
Bilharziella
Ornithobilharzia
Bivitellobilharzia
Heterobilharzia
Gigantobilharzia
Austrobilharzia
Trichobilharzia

Classification of Cestodes of Veterinary Importance

Phylum: Platyhelminthes

Class: **Cotyloda**

Order: Diphyllidea

Family: Diphyllobothridae (distinct segmentation of the body; single or double set of reproductive organs; genital pore opening ventrally; ovary lying posterior to the segment)

Genus: *Diphyllobothrium*
Spirometra
Class: **Eucestoda**
Order: Dilepididea
Family: Dipylidiidae (rostellum armed with one or more rows of rose thorn-shaped hooks; suckers armed or unarmed; single or double sets of reproductive organs; gravid uterus filled up with egg capsule containing one or more eggs)
Genus
Dipylidium
Genus: Choanotaenia
Family: Dilepididae (rostellum armed with hook; sucker armed or unarmed, single or double sets of reproductive organs, sac-like gravid uterus)
Genus: *Amoebotaenia*
Order: Davaineidea (hammer-shaped hooks are present on the rostellum. They usually have single set of reproductive organs. The genital pore may be unilateral, regularly alternating, and irregularly alternating)
Family: Davaineidae (typical hammer-shaped hooks present on the rostellum; one or more circles of hooks usually present in the suckers; egg capsules present in gravid uterus)
Genus
Davainea
Raillietina
Cotugnia
Order: Anoplocephalidae (rostellum and hooks absent; broad segments; the uterus may persist without any replacement or may be replaced by egg capsule or par-uterine organ; eggs have three coverings comprising the innermost chitinous membrane, middle albuminous membrane, and outer vitelline membrane; innermost covering may form a specific structure with two hooked projections called pyriform apparatus)
Family: Anoplocephalidae (absence of rostellum and hooks; segments broader than long, regularly or irregularly alternating genital pore; the uterus not replaced and persists as a transverse tube or network of tubes; eggs have three layers, i.e., the outermost vitelline membrane, middle albuminous coat, and inner chitinous membrane, which is a pear-shaped structure having a pair of hooked projection called pyriform apparatus)
Genus
Anoplocephala
Paranoplocephala
Moniezia
Cittotaenia
Family: Thysanosomidae (the uterus is replaced by typical par-uterine organs)
Genus
Thysanosoma
Thysaniezia
Avitellina
Stilesia
Order: Hymenolepididea (rostellum bears single circle of hooks; genital pores unilaterally located)
Family: Hymenolepididae (rostellum armed; ovary transversely elongated; less number of testes, commonly four and rarely more than four typically arranged in a row or triangle)
Genus: *Hymenolepis*
Order: Taeniidea (gravid segments longer than broad; rostellum may be equipped with hooks; ovary usually bilobed; marginal genital pores irregularly alternating; gravid uterus having a median stem with lateral branches)
Family: Taeniidae (rostellum absent or present; when present, armed with two rows of hooks; irregularly alternating genital pores; ovary situated in the posterior part of the segment; uterus having a median longitudinal stem with lateral branch arising from either side of the longitudinal stem; eggs are initially covered by delicate membrane, which is lost, and the embryophore becomes exposed)
Genus
Taenia
Echinococcu0

s

Order: Mesocestoididae (rostellum absent; genital pore located midventrally)
Family: Mesocestoididae (rostellum absent; genital pore located midventrally)
Genus: *Mesocestoides*

Classification of Protozoa of Veterinary Importance

Subkingdom: Protozoa

Phylum: Sarcomastigophora (presence of single nucleus, flagella, pseudopodia, or both flagella and pseudopodia serving as locomotory organs; absence of spore formation; syngamy is a major mode of sexual reproduction if sexuality is present)

Subphylum: Mastigophora (reproduction basically by binary fission, tropozoite possessing one or more flagella)

Class: Zoomastigophorea (flagella number varying from one to more or may be absent in some members; absence of chromatophores; a sexual reproduction by binary fission; sexuality in some group)

Order: Kinetoplastida (mostly parasitic; one to four flagella found; kinetoplast present near kinetosome having mitochondrial material)

Family: Trypanosomatidae (typically leaf-shaped and rounded form organisms having single flagellum; flagellum may be attached to the body by a membrane called undulating membrane)

Genera

Trypanosoma

Leishmania

Family: Cryptobiidae (organisms look like trypanosomes but they are biflagellate)

Genus: Cryptobia

Order: Retortamonadida (two or four flagella found in the organisms; of these flagella, one flagellum turns posteriorly)

Family: Retortamonadidae (two or four flagella)

Genus: *Chilomastix*

Family: Cochlosomatidae (presence of a ventral sucker anteriorly; presence of a parabasal body)

Genera: *Cochlosoma*

Order: Diplomonadida (bilaterally symmetrical body; two similar nuclei; cyst formation found)

Family: Hexamitidae (bilaterally symmetrical body; two nuclei and six or eight flagella)

Genera

Giardia

Hexamita

Order: Trichomonadida (four to six flagella; one is recurrent flagellum attached to the undulating membrane; axostyle present)

Family: Trichomonadidae (pyriform in shape; single nucleus; costa and axostyle present; four to six flagella, one recurrent flagellum attached to the undulating membrane)

Genera

Tritrichomonas

Trichomonas

Tetratrichomonas

Pentatrichomonas

Family: Monocercomonadidae (three to five anterior flagella; a recurrent flagellum present usually free; absence of costa)

Genera

Histomonas

Parahistomonas

Order: Retortamonadida (presence of two or four flagella; one flagellum projecting posteriorly)

Family: Retortamonadidae

Genus

Chilomastix

Subphylum: Sarcodina (pseudopodia present; presence of flagella only during developmental stages; asexual reproduction by fission; sexual reproduction usually by flagellate gametes)

Superclass: Rhizopoda

Order: Amoebida (presence of naked protoplast at the trophic stage, usually uninucleate)

Family: Endamoebidae (Multiplication by binary fission; cysts formation common).

Genera

Entamoeba

Dientamoeba

Endamoeba

Iodamoeba

Endolimax

Family: Hartmannellidae (mainly free living, absence of flagellum.)

Genera

Acanthamoeba

Hartmanella

Phylum: Apicomplexa

Class: Sporozoea (Well developed apical complex; both sexual and asexual reproduction;

formation of oocysts; microgametes possessing flagella)

Subclass: Coccidia (formation of intracellular and dissimilar gamonts by anisogamy)

Order: Eucoccidiida

Suborder: Eimeriina (Independent development of macro and microgametes; sporozoites found in sporocysts; formation of large number of micro gametocytes; non-motile zygote)

Family: Eimeriidae (Members commonly intracellular; undergo both sexual and asexual reproduction; formation of sexual microgamont and macrogamonts in sexual reproduction, zygote formation by union of macro and macrogamont; oocysts with nil to several sporocysts each having one or more sporozoites, non-motile zygote; endodyogeny absent or present; formation of schizonts in hosts; sporogony in outside)

Genera

Eimeria

Isospora

Tyzzeria

Wenyonella

Family: Cryptosporidiidae (oocysts with four free sporozoites not enclosed in sporocysts; sporogony occurring inside the body)

Genus: *Cryptosporidium*

Family: Sarcocystidae (oocysts with two sporocysts; asexual stages occurring in intermediate hosts and gametogonous stages in definitive hosts; cysts and/or pseudocysts are produced in cells)

Subfamily: Toxoplasmatinae

Genera

Toxoplasma

Hammondia

Besnoitia

Cystoisospora

Subfamily: Sarcocystidae

Genera

Sarcocystis

Frenkelina

Suborder: Adeleina

Family: Klossiellidae (microgamete without flagella; inactive zygote; typical oocyst not found)

Genus

Klossiella

Family: Haemogragarinidae (Ookinete is formed)

Genus

Hepatozoon

Suborder: Haemosporina (there occurs an individual development of micro- and macrogamete; motile zygote; absence of conoid; pigment formation in host cells; schizogony in vertebrate and sporogony in invertebrate)

Family: Plasmodidae (heteroxenous life cycle found; zygote motile, schizogony stages in vertebrate and sporogony stages in blood-sucking insects)

Genera

Haemoproteus

Leucocytozoon

Plasmodium

Subclass: Piroplasmia (short apical complex; absence of conoid; pigment does not form in the host cell)

Order: Piroplasmida

Family: Babesiidae (pyriform, oval, round pleomorphic organisms; short apical complex; schizogony or binary fission in erythrocytes)

Genus: *Babesia*

Family: Theileriidae (apical complex short; small round, oval, dot-like, comma-shaped pleomorphic organisms; schizogony in the lymphocytes, histiocytes, etc.)

Genus: *Theileria*

Phylum: Ciliophora (presence of cilia or ciliary organelles at least in one stage of biological development; usually two types of nucleus; reproduction by binary fission)

Class: Kinetofragminophorea (often presence of cytostome; usual presence of prominent cytopharyngeal apparatus, presence of macro and micro nucleus)

Order: Trichostomatida

Family: Balantidiidae (a vestibulum is present near the anterior end of the body; the body is uniformly ciliated, which is long and holotrichous)

Genus: *Balantidium*

Classification of Arthropods of Veterinary Importance

Phylum: Arthropoda
Subphylum: Mandibulata (presence of mandible; absence of chelicerae; single or paired antennae)
Subphylum: Chelicerata (presence of a pair of chelicerae; absence of antennae)
Subphylum: Pentastomida (called tongue worm; body transversely striated; two pairs of strong hooked claws at the anterior end)
Class: Crustacea (presence of two pairs of antennae; presence of many limbs frequently biramous; presence of typical compound eyes)
Class: Myriapoda (segmented body)
Class: Insecta (body divided into head, thorax, and abdomen; presence of eye and antennae; presence of wings and legs; presence of digestive system, respiratory system, circulatory system, nervous system, excretory system, and reproductive system; trisegmented thorax, i.e., prothorax, mesothorax, and metathorax; segmented legs; two pairs of wings, the hind pair of wings reduced to halters or balancers)
Subclass: Pterygota (generally winged insects)
Division: Exopterygota (wings develop externally like buds; simple metamorphosis found)
Order: Orthoptera (long and filamentous antennae; two pairs of wing, i.e., mesothoracic and metathoracic pairs; mesothoracic pair thickened serving as cover called tegmina for membranous metathoracic pair of wings)
Order: Odonata (predatory nymph stage found called naiads)
Order: Hemiptera (mouth parts adopted for piercing and sucking; proboscis retracted in the head and thorax but protrudes out during feeding)
Family: Cimicidae (dorsoventrally flattened body; incomplete metamorphosis)
Genus: *Cimex*
Family: Reduviidae (cone-shaped head called cone-nosed bugs; presence of conspicuous eyes; four segmented antennae)
Genus: *Triatoma*
Order: Mallophaga (biting lice and mouth parts adopted for chewing; head rounded anteriorly; a pair of antennae having three to five segments; fused meso- and metathorax forming undifferentiated pterothorax; absence of paratergal plates)
Suborder: Amblycera (broad and rounded head; mandibles bite horizontally; presence of antennae in ventral groove; antenna four segmented of which the third segment is stalked; maxillary palps may be present)
Family: Menoponidae
Genera
Menopon
Menacanthus
Trinoton
Family: Gyropidae
Genus: *Gyropus*
Family: Boophidae
Genera
Heterodoxus
Boophia
Suborder: Ischnocera (three to five antennal segments; head possesses mandible, which bites vertically; absence of palpi; usually fused meso- and metathorax)
Family: Trichodectidae
Genera
Damalinia (Bovicola)
Trichodectes
Heterodoxus
Felicola
Family: Philopteridae
Genera
Cuclotogaster (Lipeurus)
Lipeurus
Goniodes
Goniocotes
Columbicola
Suborder: Rhynchophthirina
Family: Haematomyzidae
Genus: *Haematomyzus*
Order: Siphunculata (called sucking lice; mouth parts adopted for sucking purpose; conical and pointed head; antenna usually five-segmented; often presence of paratergal plates; absence of mandibles)
Family: Haematopinidae (absence of eyes; antenna having five segments; broad thorax; presence of paratergal plates; presence of tibial pad in claws)
Genus
Haematopinus

Family: Linognathidae (absence of eyes; elongated thorax; absence of paratergal plates; first pair of leg smaller than the two; absence of tibial pad in the claws)

Genus

Linognathus

Solenopotes

Family: Pediculidae (presence of eyes, presence of paratergal plates; legs are of the same size)

Genus: *Pediculus*

Division: Endopterygota (wings develop internally like buds; complete and complex metamorphosis)

Order: Siphonaptera (Laterally compressed body; three body parts, i.e., head, thorax, and abdomen; short clubbed trisegmented antennae deeply buried in antennal groove; presence of pronotal comb and genal comb; presence of sensilium or pygidium)

Family: Ceratophyllidae

Genera

Ceratophyllus, *Nosopsyllus*

Family: Pulicidae (conical head; well-developed coxae)

Genera

Pulex

Ctenocephalides

Xenopsylla

Family: Sarcopsyllidae (large and angular head; absence of pronotal comb or genal comb)

Genera

Echidnophaga

Sarcopsylla

Order: Diptera (presence of a single pair of wings; metathoracic pair of wings being modified to halters or balancers; dicoptic eyes in females and holoptic eyes in males; larval stage apodous; obtectate or coarctate type of pupa)

Suborder: Nematocera (long filamentous antennae having 6–16 segments; absence of arista; maxillary palp with 4–5 segments, discal cell absent; mandibles of larva bite horizontally; pupa obtectate type)

Family: Ceratopogonidae (mouth parts adopted for piercing or sucking; very short piercing type of proboscis present; antennae long and have 13–15 segments; thorax hairy without any scale)

Genera

Chiironomous

Culicoides

Family: Simuliidae (horn-like antennae; piercing proboscis short)

Genus: *Simulium*

Family: Psychodidae (small and hairy flies; long antennae having 16 segments; short and piercing proboscis; hairy wings without any scales)

Genus

Phlebotomus

Family: **Culicidae** (slender insects, filiform antennae long and have 14–15 segments, long proboscis of piercing and sucking type, thorax characteristically wedge shaped; the segments fused)

Genera

Anopheles

Culex

Aedes

Suborder: Brachycera (robust bodied flies have shorter antennae with three dissimilar segments; short and stout maxillary palp; arista may or may not be present; mandibles of larvae bite vertically)

Family: Tabanidae (arista present; antennae trisegmented; last segment with annulations; piercing and sucking type of mouth parts present)

Genera

Tabanus

Chrysops

Haematopota

Pangonia

Suborder: Cyclorrhapha (short antenna having three segments; arista dorsally present on the antenna; wings have closed cubital cell and discal cell)

Family: Gasterophilidae

Genus

Gasterophilus

Family: Muscidae (arista plumose bilaterally; R5 cell is closed or nearly closed; M1+2 vein curve forward distally)

Genera

Musca

Stomoxys

Haematobia

Hydrotaea

Fannia
Family: Glossinidae
Genera: *Glossina*
Family: Calliphoridae (hypopleural bristles are present. The flies are characteristically bright with metallic shine)
Subfamily: Calliphorinae
Genera
Calliphora
Lucilia
Callitroga
Phormia
Chrysomyia
Cordylobia
Subfamily: Sarcophaginae
Genus
Sarcophaga
Family: Oestridae
Genus
Oestrus
Family: Hippoboscidae
Genera
Hippobosca
Pseudolynchia
Melophagus
Subphylum: Chelicerata (chelicerae present; antennae absent)
Class: Arachnida (body divided into two parts, i.e., prosoma and opisthosoma; both antennae and wings absent; number of legs variable; adult with four pairs of legs and larva with three pairs of legs; eyes either may be present or may be absent)
Order: Acarina
Suborder: Mesostigmata (brown or dark brown coloured; visible by naked eye; body usually divided into two parts, i.e., gnathosoma bearing mouth parts with chelicerae and posterior idiosoma; tarsi of legs usually bear claws and caruncles)
Family: Dermanyssidae (pear shaped; truncated posterior margin)
Genera
Dermanyssus
Ornithonyssus
Pneumonyssus
Family: Gamasidae
Genus
Raillietia
Suborder: Trombidiformes
Family: Trombiculidae
Genus
Trombicula
Family: Cheyletidae (large-sized palpal claws; M-shaped gnathosomal peritremes; chelicerae blade like; tarsal appendages comb-like)
Genus: *Cheyletiella, Psorergates*
Family: Myobidae (small elongated body with striations)
Genus: Myobia
Family: Demodicidae (elongated crocodile-like tapered body; distinct division with head, thorax, abdomen, and four pairs of stumpy legs; elongated abdomen striated)
Genus: *Demodex*
Suborder: Sarcoptiformes
Family: Sarcoptidae (true burrowing mites; rounded or oval in shape; absence of eyes; presence of spines on the body; absence of genital apodeme)
Genera
Sarcoptes
Notoedres
Cnemidocoptes
Family: Psoroptidae (all legs cross the body margin; large sized oval in shape; jointed pedicels; first, second, and third pairs of legs of male and first, second, and fourth pair of legs of female possess suckers; stylet-like chelicerae)
Genera
Psoroptes
Chorioptes
Otodectes
Family: Cytoditidae
Genus
Cytodites
Suborder: Ixodoidea
Family: Argasidae (leathery mammillated body without any dorsal shield or scutum; ventrally situated mouth parts; eyes may or may not be present; when present, they are found in the supracoxal folds; marked sexual dimorphism not found)
Genera
Argas
Otobius
Ornithodoros

Family: Ixodidae (dorsoventrally flattened body; absence of antennae; presence of scutum; presence or absence of eyes, if present located at lateral margin of scutum; presence of different plates; presence of festoons)

Genera

Ixodes
Rhipicephalus (*Boophilus*)
Rhipicephalus
Hyalomma
Dermacentor
Haemaphysalis
Amblyomma

2 Bionomics of Parasites

The word “bionomic” has been originated from two Greek words, i.e., “bio” meaning “life” and “nomos” meaning “law.” In general, it is the comprehensive study of an organism along with its relationship with the environment. To study bionomics of parasites, we must concentrate on their physiology encompassing their nutrition, metabolism, circulation, excretion, reproduction, etc. In this chapter, all have been briefly described so that a reader can have an overview about some physiological and biochemical aspect.

Brief Description of Physiology of Protozoa

Nutrition of Protozoa

There are three major types of nutrition found in protozoa, i.e., holophytic, holozoic, and saprozoic. Other types of nutrition are also found. Different types of nutrition performed by protozoa have been described below:

Pinocytic Nutrition

This is called as cell drinking. Some amoebae and ciliates perform this nutrition. In this process, the liquid food is taken up by pinocytosis. At first, minute channels are formed, through which liquid food enter. The minute channels end into small minute vesicles, and in these vesicles, the food material enter. After entry of food, the vesicles get separated off and transform to vacuole.

Holophytic Nutrition

In this type of nutrition, the protozoa can produce carbohydrate by chlorophyll present in their body. This type of nutritional mode is also called as autotrophic nutrition.

Holozoic Nutrition

In this type of nutrition, the protozoa take up solid or liquid food material like higher animals. This type of nutrition is characterized by internalization and internal processing of food. The different steps are ingestion, digestion, absorption, assimilation etc.

Saprozoic Nutrition

In saprozoic nutrition, the organisms absorb food material from decomposed organic substances surrounding it.

Mixotrophic Nutrition

In this type of nutrition, some protozoa can perform more than one type of nutrition if it is needed.

Some Examples of Food Intake by Protozoa

Protozoan organisms have different mode of food intake. *Plasmodium* has numerous channels in their parasitophorous vacuole membranes. These

S. C. Mandal, *Textbook of Veterinary Parasitology*, https://doi.org/10.1007/978-981-99-2924-5_2

channels create permeability of organisms toward different organic solutes to allow the uptake of necessary nutrients. Transporters have been recorded for *P. falciparum* which are important for uptake of glucose, fructose, etc. The organic molecules reach to the parasite crossing three membranes, i.e., plasma membrane of the erythrocyte, parasitophorous vacuole membrane, and plasma membrane of the parasite. Catching and trapping of food particles by some protozoa is interesting. The protozoa use their cilia, flagella and pseudopodia for collection and filtering the food particles. All protozoa do not have these organs. Some protozoan organisms (e.g., ciliates) take food into the cell through a specific structure called as cytostome present at a particular region of cell surface. Others may absorb simplified food material. The beating flagellum of choanoflagellates makes the parasite enable for feeding. The beating movement creates water current. This water current causes water to move through the collar. The food particles present in the water are tapped in the collar and are subsequently ingested by pseudopodia. The ingested food particles are enclosed by membrane-bound digestive food vacuole. Ciliate protozoans are filter feeders. Ciliary structures are present in cytostome. The specific beating of these cilia pushes a stream of water against the membranelle of cilia. Food particles are trapped in between the cilia. Other ciliates do not have any cilia. They have complex cytostome and cytopharynx supported by basket like cytopharyngeal structures composed of microtubules. They take filament-like food particles which directly enter into the cytopharynx. Then, the food material is digested by digestive enzymes in food vacuoles. Suctorians have feeding tentacles and piercing tentacles by which they perform feeding. Amoeba do not have cell mouth. Some protozoa live in nutrient-rich medium of the body fluid (blood, tissue fluid, etc.). By the action of pinocytosis, food in the form of fluid is pinched off into digestive vacuole either in specific site, i.e., cytostome in case of ciliates or flagellar pockets in case of trypanosomes. Other protozoans engulf the portion of host's tissue by phagocytosis much like amoebae. In case of some hemoprotozoa, the hemoglobin present in cytoplasm of RBC is only partly digested. The protein part gets degraded and simplified to amino acids. Iron-containing portion is converted to hemozoin (insoluble iron-containing substance). Digestion process proceeds and the vacuoles get increased in size. The content becomes acidic initially and then transforms to alkaline near the end of the process. The products of digestion are then absorbed into the surrounding of the cytoplasm.

Brief Description of Metabolism and Other Reactions of Protozoa

Different enzymes of central metabolism, i.e., glycolysis, citric acid cycle, and oxidative phosphorylation, are found in protozoa. Glycolysis is found in protozoan parasites. Many primitive eukaryotes perform metabolism favoring glycolysis for generating ATP in glucose-rich environment where they reside. All the enzymes involved in hexose metabolism are available. TCA cycle is a universal evolutionarily ancient cycle which has been reported in protozoan parasites. Though less work has been done on TCA cycle in protozoa, there have been reports of tricarboxylic acid cycle in ciliate and other protozoa. In these cycles, different metabolites are oxidized. Carbohydrate is transformed to pyruvate which is further transformed to acetyl CoA. Breakdown of fatty acid results into production of acetyl CoA. The protein is transformed to alpha keto acids through deamination and this alpha keto acid enters the cycle. At each turn of this cycle, there occurs production of H_2O and CO_2.

Redox Metabolism

Redox metabolism has been detected in different protozoan parasites. Redox metabolism has been reported in apicomplexan protozoa. In this, different toxic substances, i.e., superoxide, hydrogen peroxide, etc., are removed by oxidation/reduction mediated by this redox metabolism. Isoprenoid pathway has been identified in protozoa.

Electron Transport

Oxidation of any organism is a series of reactions involving transfer of hydrogen step by step from one acceptor to another acceptor. In metabolism, a number of enzymes play a role in oxidation-reduction systems. In protozoa also, evidences are being available on occurrences of different oxidation-reduction system mediated by different enzymes. In aerobes, cytochrome system is involved in transferring hydrogen. Each cytochrome is iron-porphyrin-protein which always exists in either oxidized or reduced form. Different cytochromes have been reported from several protozoa. These cytochromes are cytochrome a, b, c, etc. Pyridine nucleotides are coenzymes of a number of important oxidative enzymes. Two important coenzymes are involved in performing metabolism of protozoa. These enzymes are diphosphopyridine nucleotide (coenzyme I) and triphosphopyridine nucleotide (coenzyme II). These enzymes are involved in metabolism of protozoa. Flavoprotein enzymes possess riboflavin in their prosthetic group as flavin dinucleotide and riboflavin phosphate. There are several examples of flavoprotein enzymes. These enzymes have role in catalyzing oxidation. There have been evidences that some flavoprotein enzymes are present in some protozoa.

Adenosine Phosphate System

The main components of adenosine phosphate system are adenosine monophosphate, adenosine diphosphate, and adenosine triphosphate. These contain adenosine (a riboside of adenine) and phosphoric acid groups. This system plays a great role in phosphorylation of metabolites and enzymes. These are responsible for transfer of high-energy phosphate bonds. Information is scanty in this aspect in protozoa. However, adenylic acid, ADP, and ATP have been found in some protozoa.

Special Note

- Peroxidase and catalase are enzymes of iron-porphyrin-protein complex. Catalase protects any cell from damage by the action of reactive oxygen species. Hydrogen peroxide is reduced to water and molecular oxygen. Catalases have the role of catalyzation of this reaction. This enzyme also catalyzes oxidation of the substrates like tyrosine, bilirubin, adrenaline, etc. These enzymes are present in protozoa.
- Reduced form of glutathione is a tripeptide having the amino acids—glutamic acid, glycine, cysteine, etc. The reduced glutathione has been detected in some protozoa.

Brief Description of Biochemistry of Protozoa

The present state of knowledge of biochemistry of protozoa is limited. Some biochemical components which are found in higher animals are also found in protozoa. In many of the cases, only the presence of some biochemical components has been proved, but functions of many of those biochemicals have not been known in those organisms. But their role in protozoa could be presumed if their function is known in the higher animals.

Vitamin

It is not completely known about the requirement of different vitamins for protozoa. However, knowledge of vitamin requirement of protozoan organisms is increasing due to development of newer techniques for culture of protozoa. Previously, knowledge of culture of protozoa was insufficient. But pure culturing is possible, and requirement of different components has been reported. It has been proved that some vitamins are also required for the protozoa for performing different biological functions. Thiamine and riboflavin are essentially required for the ciliates and parasitic flagellates. Some protozoa can grow without addition of thiamine. Some workers also reported the requirement of pantothenic acid for growth of some protozoa. Likewise, pyridoxine complex is also required for growth of many protozoa. Many protozoa require pyridoxamine, pyridoxine, etc. There are several reports on requirement of nicotinic acid by protozoa. Involvement of nicotinamide for oxidative metabolism of *Plasmodium gallinaceum* was reported. In addition to this, other vitamins, i.e., vitamin B12, are also required by the protozoa.

Mineral

There are different minerals which are required for the animals and plants. In case of protozoa, very little is known. It has been reported in different times that the protozoa also require different minerals for their life. In addition to the basic elements, i.e., carbon, hydrogen, oxygen, and nitrogen, they also require iron, magnesium, cobalt, calcium, magnesium, manganese, phosphorus, potassium, sodium, silicon, sulfur, molybdenum, zinc, etc.

Reproduction of Protozoa

Both sexual and asexual reproduction occur in protozoa.

Asexual Reproduction

In this type of reproduction, the protozoa divide into two daughter organisms along with the same genetic component as parent organisms.

Example: *Trypanosoma* spp. Budding, schizogony, sporogony are different phenomena of asexual reproduction.

Budding

In this type of reproduction, one or more buds (small daughter organisms) are produced from parent organisms. Buds may be produced externally or internally. It is called as endodyogeny, if two buds are produced, and endopolygeny, if more than two organisms are produced.

Schizogony

In this phenomenon of reproduction, single organism gives rise to a number of organisms by multiple fission. At first, the nucleus is divided into a number of bodies. Each nuclear body is surrounded by a part of cytoplasmic mass and gives rise to a generation of new organisms.

Example: *Eimeria* spp.

Sporogony

In this process of reproduction, a number of sporozoites are produced from a zygote.

Sexual Reproduction

In this phenomenon of reproduction, the parent organisms produce both male and female gametes. The gametes (male and female) unite to form a zygote. If two identical gametes are produced, it is called as isogamy. If two gametes are not identical, it is called as anisogamy.

General Life Cycle of Protozoa

Both direct and indirect life cycles are found in protozoa. For example, trypanosomes perform indirect life cycle, whereas *Eimeria* spp. perform direct life cycle. In some protozoa, same organism displays both sexual and asexual reproduction. For example, life cycle of *Eimeria* spp. includes formation of sporozoites, schizonts, gametes, etc. This clearly indicates combination of both sexual and asexual reproductive processes in the life cycle of same organism.

Brief Description of Physiology of Trematodes

Nutrition

While describing about nutrition of trematode, description of gut of these parasites is necessary.

The intestine of fluke is incomplete ending blind. The pharynx and esophagus are often found in trematodes. The intestinal wall is lined with digestive and absorptive cells. Thin layer of musculature causes peristalsis. In many trematodes, lateral intestinal branches or diverticula are found in the internal tissues. Undigested residue is expelled out through the mouth. Digestive hydrolases are found in different trematodes. The flukes take food through ingestion of tissue, blood, serum, etc., or the flukes absorb the food through their body surface. For example, schistosomes are in direct contact of host blood and absorb nutrients through their cuticle. They also ingest blood for their nutrition. They utilize amino acids after degradation of hemoglobin.

Metabolism

Detailed information on metabolism by trematodes is meager. Umezurike and Anya (1980) found malate dehydrogenase (MDH), NAD-linked malic enzyme (ME), phosphoenolpyruvate carboxykinase (PEPCK), and lactate dehydrogenase (LDH) in tissue homogenate of *Fasciola gigantica*. The activity was determined both in cytosolic and mitochondrial fraction. Fumarase activity is predominantly found in the mitochondria. They also found the activity of NADH-cytochrome C oxidoreductase, NADH oxidase, and some cytochrome C-oxygen oxidoreductase.

Most of the helminths use carbohydrate as substrate for energy conservation. Most adult helminths are not able to oxidize organic compounds to a significant extent to CO_2 and water. For energy production, catabolism of amino acid is almost of no importance. Adult helminths generate energy through anaerobic metabolism. TCA cycle and conventional type of respiratory chain are almost absent. Free living part can degrade carbohydrate to carbon dioxide, but the adult helminths perform energy metabolism by malate dismutation in which carbohydrate is degraded to phosphoenolpyruvate (PEP) via a conventional Embden-Meyerhof-Parnas pathway. PEP is then converted to oxaloacetate which then reduces to form malate. Some species, i.e., schistosome and filarial nematodes, perform anaerobic metabolism, and in this metabolism, carbohydrate is degraded to lactate or ethanol.

Excretion

The flukes perform excretion by flame cells. The flame cells collect the waste material from the surrounding and propel this waste substance to the excretory bladder. The flame cells are connected with many excretory ducts.

Brief Description of Biochemistry of Trematodes

Carbohydrate, Lipid, and Protein

Carbohydrate, lipid, and protein content are variable in trematodes. Of these three basic components, carbohydrate content is highest. Carbohydrate is the main energy source. In both cestodes and trematodes, the major proteins are keratin and sclerotin. Keratin protein is used for formation of cyst wall of certain immature stages of trematodes. Sclerotin is found in both cestode and trematode eggshells, particularly in those whose larval stages are associated with aquatic environments. The lipid is mainly phospholipid found in trematode.

Minerals

Calcium, magnesium, potassium, and sodium are found in trematodes. The presence of phosphorus, copper, and zinc in trematodes has also been reported.

Vitamin

Different vitamins are found in trematodes.

Enzymes and Other Important Substances

Biochemical study of the parasites has been done less compared to other aspects, i.e., control, diagnosis, etc. However, many works have been carried out on functions of different enzymes, hormones, and other biochemical components of trematodes. Many significant findings on biochemistry of different trematodes have been recorded by different workers. Studies show that enterochromaffin cells of the intestinal mucosa of many trematodes secrete 5-hydroxytryptamine. This 5-hydroxytryptamine is responsible for causing the rhythmical activity of *Fasciola hepatica*. This also increases glucose uptake and level of cyclic AMP. Cyclic AMP-dependent protein kinase activity occurs in the *Fasciola hepatica*. The enzyme, phosphofructokinase, is also stimulated by 5-HT. Different neurotransmitters have been detected in *Fasciola hepatica*. These are nonspecific cholinesterase, dopamine and noradrenaline, etc. These neurotransmitters have also been detected in other parasites also. In adult *Fasciola hepatica*, it has been reported that glucose fermentation occurs via glycolysis and fumarate reductase pathways. It has further been observed that aerobic metabolism also occurs in flukes. In *Fasciola hepatica*, the presence of **o-type cytochrome** and **cytochrome aa 3** has been reported. Works on cyclic AMP, different enzymes like glutathione S-transferase, alkaline phosphatase, etc. indicated different activity of these biochemical components. Study shows the availability of serine protease inhibitors in *Paragonimus westermani, Schistosoma, etc.* Work has also indicated the presence of cysteine protease inhibitor in *Clonorchis sinensis, Fasciola, etc.*

Reproduction of Trematodes

Reproduction occurs both sexually and asexually. Sexual reproduction occurs in the definitive hosts. Except *Schistosoma*, other trematodes are hermaphrodite. In those flukes, male and female reproductive systems are present in the same individual. Fertilization of eggs occurs in the organisms. The eggs are laid by the female. Embryonation of eggs occurs later. In the intermediate hosts, asexual multiplication occurs. Formation of sporocyst, daughter sporocyst, redia, daughter redia, and cercaria occurs by asexual reproduction.

Life Cycle of Trematodes

These parasites are heteroxenous involving one or more intermediate hosts. The developmental stages are egg, miracidium, sporocyst, redia, cercaria, metacercaria, and adult. Eggs are voided through the feces. Hatching of eggs occurs in the environment or hatching may occur after being ingested by the intermediate hosts. In some trematodes, two intermediate hosts are required for completion of life cycle. In the snails, development up to cercaria stages occurs. The cercariae get transformed to metacercariae either in second intermediate hosts or in the vegetation, water, etc. The final hosts get the infection by ingestion of metacercariae. In case of schistosomes, the metacercaria stage is not formed. In this case, the cercaria penetrates the skin of the final hosts and sets up infection.

Brief Description of Physiology of Cestodes

Nutrition

The physiology of cestodes in all aspects is not yet known clearly. For that reason, culture of cestodes was very difficult. In the recent time, culture of cestodes is possible to the higher extent as compared to previous situation. By laboratory culture, plerocercoid can develop up to sexually matured tapeworm. The outer cuticle of tapeworm made up of scleroprotein is permeable to water, electrolytes, some amino acids, etc. Tapeworms live in the GI tract, wherein they get the access to readily available rich sources of predigested nutrients. Tapeworms do not need any digestive system since they directly absorb nutrients through the tegument by diffusion. The body of tapeworm is characteristically flattened with numerous microtriches to maximize the surface area available for nutrient absorption. Different transporter molecules are present in the surface of tapeworms by which the nutrients are transported. Different enzymes are present (i.e., pancreatic alpha amylase) which help in digestion. One interesting thing is that the cuticle of the tapeworm is not digested by host enzymes. Some nutrients can enter through the surface of cestode. The nutrients enter through the cuticle of tapeworm either by active transport or by the aid of transporter molecules or by osmosis. Osmosis may be simple or mediated. The osmotic pressure of the parasites usually remains lower than that of the surrounding host tissue environment. The internal fluid found in cysticercoids is actually a transudate of the host. It indicates that larval stages of the parasite also absorb host material.

Metabolism

Information on metabolism of cestodes is very less. It has been found that cestodes can utilize glucose, and the absorbed glucose is stored in the form of glycogen. Metabolic rate of cestodes is higher in warm-blooded animals. Cestodes can accept oxygen if offered. But anaerobic respiration primarily ensues in cestodes. Carbohydrate metabolism occurs both in the cestodes and trematodes. Some flatworms utilize glucose, maltose, and sucrose. Some utilize galactose and others utilize fructose. Hence, utilization of carbohydrate varies from worms to worms. Most of the helminth use carbohydrate as substrate for energy conservation. Most adult helminths are not able to oxidize organic compounds to a significant extent to CO_2 and water. Like trematodes, adult cestodes generate energy through anaerobic metabolism. Like termatodes, TCA cycle and conventional type of respiratory chain are almost absent. As is found in case of trematode, free living part can degrade carbohydrate to carbon dioxide, but the adult helminths perform energy metabolism by malate dismutation in which carbohydrate is degraded to phosphoenolpyruvate (PEP) via a conventional Embden-Meyerhof pathway. PEP is then converted to oxaloacetate which then reduces to form malate.

Brief Description of Biochemistry of Cestodes

Carbohydrate, Lipid, Protein, and Enzymes

As that in trematodes, carbohydrate, lipid, and protein are three major biochemical components found in cestodes. The amount of carbohydrate content is the highest compared to protein and lipid. In cestodes, carbohydrate is the main energy source like other helminths. Protein, found in the cestode, is majorly scleroprotein. And this protein is low in comparison to lipid and carbohydrate. In both cestodes and trematodes, the major proteins are keratin and sclerotin. Keratin protein is used for formation of the hooks and part of the protective layers of the cestode eggs. Sclerotin is found in eggshell. Proteoglycans and glycoproteins are available in different cestodes. Amino acid composition of cestodes is closely similar to that of hosts with some exception (Smyth and Mcmanus 2007). Both simple and conjugated lipids have been reported in cestodes. Like trematodes, lipid is available in the cestodes in the form of phospholipid. Carbohydrate is present in the form of glycogen stored in the parenchyma.

Different research workers have reported presence of different enzymes in cestodes. Metacestode cysteine protease of *Taenia solium* has been characterized. Cathepsin L like cysteine protease from *Taenia solium* metacestode has also been identified and characterized. Protease and protease inhibitors have been identified in plerocercoid larvae of *Diphyllobothrium latum*. Trypsin and chymotrypsin inhibitor have been identified in *Taenia pisiformis* (adult/larva). Proteolytic enzyme inhibitors have been detected in *Hymenolepis diminuta*. Functional characterization of cathepsin B like peptidases from metacestodes has been done. Trypsin adsorption in *Hymenolepis* has been found.

Minerals

The presence of calcium, magnesium, potassium, sodium, etc. is found in cestodes.

Vitamin

In cestodes, the presence of vitamin B6 and vitamin C has been reported in *Hymenolepis diminuta*.

Reproduction of Cestodes

Both sexual and asexual reproductions occur in cestodes. Cestodes are hermaphrodite parasites. Both male and female reproductive systems are present in the same individual. Self-fertilization and cross-fertilization both occur in the same cestodes. Egg of one segment may be fertilized by sperm of other segments. Fertilization of egg of one segment may be fertilized by sperm of the same segment. Fertilization may occur between two individuals of the same species. In this case, sperm of one individual is absorbed by other individuals through their body surface.

Life Cycle of Cestodes

In Eucestoda

In the Eucestoda, life cycle usually requires one or more intermediate hosts. Direct life cycle also occurs in some member i.e. *Hymenolepis nana*. In general, eggs/gravid segments pass out with the feces. The eggs hatch after being ingested by the intermediate hosts. After hatching, the oncospheres come out. The oncospheres transform to one more larval stage called as bladder worms. In accordance with the species, different types of bladder worms are produced. These are cysticercus, coenurus, hydatid cysts, strobilocercus, etc. These are also called as metacestodes. The metacestodes are the infective stages. The final hosts get the infection by ingestion of metacestodes.

In Cotyloda

In cotyloda, life cycle is indirect and involves two intermediate hosts. The eggs are voided along with feces. The eggs hatch in the water and ciliated coracidium comes out. The coracidium swims in the water. The coracidium is ingested by the first intermediate hosts (copepods). In the first intermediate host, one larval stage is formed called as procercoid. The infected copepods are ingested by second intermediate hosts (fish, reptiles, etc.). In the fish, one more larval stage is formed which is called as plerocercoid. The final hosts get the infection by ingestion of second intermediate hosts harboring plerocercoids.

Brief Description of Physiology of Nematode

Nutrition

Nematodes have complete digestive system. The gut of the nematodes starts from the mouth which proceeds to the buccal cavity, pharynx, esophagus, intestine, rectum, and anus. The parasitic nematodes take up varieties of food material. These are liquid material available in the gut of the host, tissue material, body fluid of hosts, etc. Food is taken up in different ways depending on the nature of food and aperture of the mouth and buccal cavity. These activities are engulfing, piercing, biting, and sucking. After ingestion, the food is forced to the esophagus and then to the intestine. A number of enzymes are available in the intestine and food gets digested. The nutrient is absorbed there. After absorption, the waste material reaches to the rectum and then is voided through the anus.

Metabolism

As mentioned earlier, most of the helminths use carbohydrate as substrate for energy conservation. Adult nematodes generate energy through anaerobic metabolism as found in cestodes and trematodes. TCA cycle and conventional type of

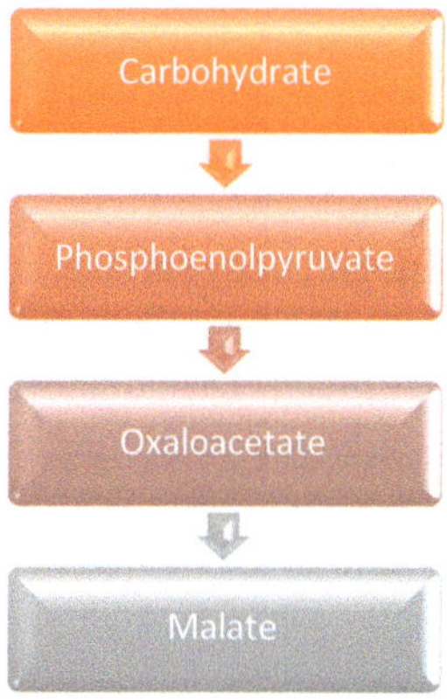

Fig. 2.1 Flowchart of formation of malate

respiratory chain are almost absent. Free living stages of nematodes can degrade carbohydrate to carbon dioxide. Like cestodes and trematodes, the adult nematodes perform energy metabolism by malate dismutation in which carbohydrate is degraded to phosphoenolpyruvate (PEP) via a conventional Embden-Meyerhof-Parnas pathway. PEP is then converted to oxaloacetate which then reduces to form malate.

Brief Description of Biochemistry of Nematode

Carbohydrate, Protein, and Lipid

Carbohydrate, lipid, and protein are basic biochemical components of nematodes. These are the structural and functional components as that found in higher animals. Different forms of carbohydrates have been identified. Most carbohydrate content is found in the form of glycogen. Protein has also been found in nematode. Total protein content varies widely in different nematodes. Different types of proteins have been characterized. The cuticle of nematode has collagenous and noncollagenous proteins. Noncollagenous protein is called as cutilin. Different enzyme proteins and other vital proteins have been characterized. These are peroxidase, transferase, protease, beta tubulin, calmodulin, etc.

Minerals

Different minerals have been detected in nematodes. These are sodium, potassium, calcium, phosphorus, zinc, copper, etc. The variation of these minerals in different nematodes and their developmental stages has been reported.

Vitamin

There have been research works on vitamin requirement for growth of nematodes. Different vitamins have also been identified and quantified in different nematodes.

Reproduction of Nematode

In nematodes, sexual reproduction occurs. The eggs of female are fertilized by the sperm of male. The fertilized female lays eggs or larvae.

Life Cycle

Life cycle is direct or indirect type. The hosts get the infection by ingestion of infective stage of the parasite. The infective stage may be free larvae or eggs in case of direct life cycle. In case of indirect life cycle, the hosts get the infection by ingestion of infected intermediate hosts harboring the infective larvae. There are also other modes of infections. Skin penetration by the infective larvae of nematode is another mode of infection. For example, the infective larvae of hookworm penetrate the skin of host, and in this way, the hosts get the infection. The filarial nematodes are transmitted to the hosts when the intermediate hosts (mosquitoes, blood-sucking insects) suck blood from the hosts. The intermediate hosts transmit infective stage of the parasites to the hosts.

Brief Description of Physiology of Arthropod

Nutrition

Nutrition is holozoic type. The arthropods take up their food material through their mouth. The digestion and absorption of nutrients occur in the digestive system. Egestion occurs through the anus. The digestive system is usually divided into three parts comprising stomodeum (foregut), mesenteron (midgut), and proctodeum (hindgut). Generally, the gut of arthropod includes the pharynx, esophagus, crop, proventriculus, intestine, and anus. The food is taken up by mouth which gradually reaches to the esophagus, crop, proventriculus, or gizzard. The muscular proventriculus serves as a valve which prevents regurgitation. The proventriculus may be equipped with teeth for grinding and disintegration of food material. The midgut secretes a number of enzymes which help for digestion. The hindgut includes the ileum and a dilated rectum. The rectum possesses several papillae which serve for reabsorption of water from the feces.

Excretion

Excretion is performed by different organs, i.e., Malpighian tubules, nephridia, and coxal glands. Coxal glands are found in the legs. The arachnids have both Malpighian tubules and coxal glands.

Circulation

The circulation is open type. The body cavity is called as hemocoel. Hemolymph is found instead of blood. A small tubelike structure is found called the heart. This is primitive tubular heart. In the heart, small openings are found called as ostia. The hemolymph enters the heart through the ostia. When the heart contracts, the ostia remain closed. The circulatory fluid is forced to reach in the tissue spaces.

Respiration

Respiration is performed by different respiratory organs found in arthropods. These are book lung, tracheal system, gill, body surface, etc. In insects, the tracheal system comprising a fine network of tracheal tubes is found. The minute tracheal tubes penetrate the cells. Air enters the body through some openings called spiracles. Two pairs of spiracles are found in the thorax region and eight pairs of spiracles are found in the abdominal regions.

Nervous System

The arthropods have complex nervous system which is more developed as compared to other invertebrate organisms. This consists of a central mass (brain) present on the stomodeum, nerve trunk, or commissure surrounding the esophagus. Ganglions are present in different segments. A ventral nerve trunk is present comprising nerve cords supplying segmental ganglions.

Brief Description of Biochemistry of Arthropods

Carbohydrate, protein, and lipid are present in the body of arthropod. Carbohydrate is stored as glycogen. The outer body cuticle of arthropod is formed by chitin proteins. In some internal organs, chitin proteins are also found. Other different proteins are also found. Lipids are also present in different forms in these organisms. Different fatty acids, i.e., oleic acid, linoleic acid, palmitic acid, and stearic acid are found in arthropods. Different minerals are found in arthropods. These are potassium, calcium, phosphorus, magnesium, manganese, copper, zinc, iron, molybdenum, etc. Vitamins found in arthropods are vitamins A, B1–12, C, D, E, and K. Permeases help to enable them to acquire nutrition from their hosts. Different permeases detected in ectoparasites import hexoses, carboxylases, amino acids, etc. Proteases reported in scabies mites is inactivated protease paralogues (SMIPPs).

Reproduction and Life Cycle of Arthropods

The arthropods are mostly dioecious. The male and female organisms are different. However, hermaphroditism and parthenogenesis are also found. The life cycles are of three types i.e. hemimetabolous, holometabolous, and ametabolous. Complete metamorphosis occurs in holometabolous life cycle (e.g., Diptera, Siphonaptera, etc.). In this life cycle, different stages, i.e., egg, larva, pupa, and adult are found. Three or more larval stages are found. In hemimetabolous life cycle, incomplete metamorphosis occurs (Hemiptera, Orthoptera). The developmental stages are egg, nymph, and adult. In ametabolous life cycle, metamorphosis is absent.

Juvenile hormone (JH) and methyl farnesoate (MF) function for development and reproduction of insects and crustaceans. The important enzymes which are responsible for catalyzing the biosynthesis of JH and MF are juvenile hormone acid O-methyltransferase (JHAMT) and farnesoic acid O-methyltransferase.

Molting of arthropods is triggered by important steroid hormone known as **ecdysteroids**. Molting and metamorphosis both are controlled by hormones. Ecdysteroids are produced primarily by the gonads of arthropods. Prothoracicotropic hormone secreted by specific neurosecretary cells of brain of arthropods activate prothoracic gland, an endocrine gland in prothorax, which in turn secretes a steroid hormone ecdysone. Young larval stages have a small gland behind the brain, called corpora allata which secretes juvenile hormone. This is also called as neotenin. In the last larval stage, the juvenile hormone is not produced. This undergoes maturation and adult stage.

References

Smyth JD, McManus DP (2007) The physiology and biochemistry of Cestodes. Cambridge University Press, Oxford

Umezurike GM, Anya AO (1980) Carbohydrate energy metabolism in Fasciola gigantica (Trematode). Int J Parasitol 10(3):175–180. https://doi.org/10.1016/0020-7519(80)90045-4

3 Host-Parasite Interactions

Host-parasite interaction is an important part of parasitology. This interaction determines many things related to survival of both host and parasite, grade of pathology caused by parasites, and immunity to the invading parasite. This is a continuous process which leads to host-parasite coevolution. When parasites invade a host, they compromise many aspects of their life and the parasites also adapt themselves in many ways to survive in the hosts. The compatibility of the hosts depends on many physiological and biochemical components of the parasite. These are size, virulence, developmental stages, biochemical composition, and many other things. A brief overview has been provided in this chapter about host-parasite interaction.

Distribution of Parasites on/in the Host

The parasites are found in different body parts of the hosts. The ectoparasites occur on the skin of hosts. The endoparasites occur within the body of the hosts. The parasites have organ specificity, cell specificity, etc. Distribution of different parasites in different organs and systems raises many questions. If it is the question, why do some parasites occur in the gut? Even, why do some parasites occur in particular part of the gut? The answer will be that these parasites are compatible in that particular host environment. They can easily survive in that particular location of the host. The internal host environment including the pH of that particular host environment is suitable for those parasites. This occurs as a result of adaptation of parasites. Different parasites have different predilection sites. This variation of predilection site has been so because of long-term evolution of the parasites for their survival. Many factors play a role for selection of different locations by the parasites. These factors are pH, presence of biochemical component in the location/site, etc. On the way of evolution, natural manipulation of genetic sequences (genetic mutation) is responsible for creating different identities of parasites in relation to their survival in different host's location. Ticks, mites, and lice are the ectoparasites which live on the skin. Infection by those ectoparasites on the skin causes pathological features there. Blood-sucking flies are the temporary parasites which intermittently suck blood for their survival. Some flies are called as myiasis-causing flies. These flies lay their eggs/larvae on the living tissues. The larvae mature in the vicinity or some other places in the hosts. These migratory larval parasites cause significant pathology in the hosts. Many parasites are endoparasites. In case of cestodes, it has been found that most of the cestodes occur in the small intestine of variable host range. Both in mammalian and avian hosts, cestodes are mostly found in the small intestine. Here, we are finding that despite these hosts have no phylogenetic relationship,

S. C. Mandal, *Textbook of Veterinary Parasitology*, https://doi.org/10.1007/978-981-99-2924-5_3

these phylogenetically unrelated hosts also harbor the tapeworms in the same location. Vivid research has to be directed toward this aspect so that the scientists can unfurl the evolutionary development in relation to selection of predilection site of the parasites. In case of nematodes, it is found that the parasites occur in the small intestine, large intestine, stomach, aorta, cecum, colon, trachea, gizzard, lungs, central nervous system, eye, nictitating membrane, etc. Much variation of predilection site is found in nematodes which is not found in cestodes. This variation of predilection sites in different groups of parasites is important. We should also consider the immunity status of a host. As for example, parasites occurring in the GI tract are immunocompatible to this location. The parasites which do not occur in the GI tract but occur in other organs other than GI tract; it could be presumed that those parasites fail to tolerate the local immune reaction mounted by the hosts in the GI tract. But research works should be carried out in this direction to delineate the matter scientifically.

Host-Parasite Interaction

The hosts get the parasitic infection in many ways. The route and activity during entry of parasites in the hosts are called as mode of infection. The main route of infection which is commonly found in case of parasites is ingestion of infective stages of parasites. The infective stages of a parasite may be eggs, larvae, cysts, oocysts, etc. In case of *Entamoeba* spp. and *Giardia* spp., the hosts get the infection by ingestion of cysts of the parasites. The sporulated oocysts are infective stages of *Eimeria* spp. In nematodes, the larval stages act as the infective stage and set up infection in the host. In case of trematodes, the metacercariae are the infective stages of the parasites. Metacercariae stages of *Fasciola* parasites develop on the vegetations, and the hosts get the infection by ingestion of the plants or vegetations harboring the metacercariae. In all these cases, the parasites/parasitic stages directly enter into the hosts through ingestion. In every case of parasitic infection, there are two possibilities—either the host's defense system directly can reject the establishment of the parasite in the hosts, or the parasites may fairly establish in the hosts. Host-parasite interaction depends on host-parasite coevolution. Through evolution, the hosts have developed immune mechanism which restrict the parasite from causing harm. Or the host's immune system directly kills the parasites. Through the process of evolution, the parasites have also developed some evasion mechanisms to combat the action of host's immune reaction. Therefore, evolution has role for determination of host-parasite interaction.

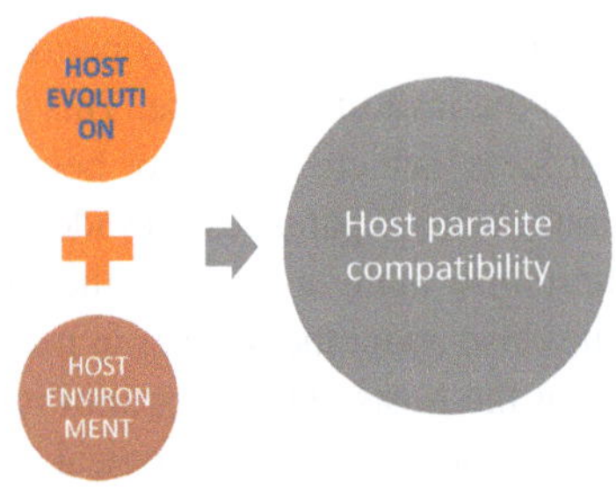

Fig. 3.1 Schematically displayed host-parasite compatibility

Morphological Specialization for Life of Parasite

Morphological specialization of parasites is markedly observed. It is thought that the shape of any parasite is not accidental. The morphological specialization has occurred as per the need of the parasites based on evolutionary process.

The cestodes have wide morphological variations. The specific structures enable the parasites to get firmly attached to the gut wall. For firm attachment, they have developed different types of apertures, i.e., hooks, rostellum, suckers, etc. The oncosphere has three pairs of hooks. These hooks help the oncosphere when they migrate in the tissues. Many adaptations are also found in the larval cestodes. The larval stages of cestode are bladder worms. Adaptation of larval cestode is specific, i.e., formation of cysts within which the larval cestodes remain and take nutrition from the cyst fluid. The adaptation phenomenon creates these structures, and the shape, composition, and size of different structures are responsible for host-parasite interaction.

In trematodes also, there were morphological specializations. The typical shape of *Fasciola* makes them enable to penetrate the liver tissues during migration by the immature stages. Male schistosomes have characteristic gynaecophoric canal by which the female parasite is carried during copulation.

There occurs morphological variation of nematodes. There are different cuticular structures, i.e., buccal capsule, buccal cavity, leaf crown, epaulettes, cordons, teeth, lancets, scales, spines, etc. All these structures are the output of evolution of nematodes. As per requirement, these structures have been developed. These morphological features contribute to the survival of the parasites.

The larvae of nematodes have typical morphology which made the larvae enable to penetrate the host tissue. The typical shape of bursa makes the male parasite enable to get attached firmly with the female worm during copulation.

The flagella of flagellate protozoa help for locomotion. The typical shape of trypanosomes is useful to move in liquid. Pseudopodia, cilia, etc. are also helpful for locomotion of protozoa.

In protozoa, many specialized morphological structures are found. These are flagellum, cilia, pseudopod, etc. These are locomotory organs. The shape of different protozoa is important for movement in their vicinity.

Enormous morphological variations are found in the arthropods. These morphological structures are useful for carrying out different functions. In different chapters, different morphological features have been described.

All these morphological specializations are attributed to host-parasite interactions. For example, the tapeworms possessing hooks in rostellum and suckers can cause more damage to their attachment site. The hookworms possess teeth or cutting plates in their buccal capsule, and their mouth parts are adopted for continuous sucking of blood from the hosts. These hookworms are more harmful compared to other worms. Hence, the morphological specialization of the parasites contributes to the status of host-parasite interaction resulting in different grades of pathology.

Behavioral Defense by Hosts

The animals perform different behavioral defense to protect themselves from any infection including parasites.

The following are some examples which indicate some general behavioral defense:

(a) The animals stand or graze in groups to have less attack by blood-sucking flies.
(b) The animals avoid to graze on forage adjacent to recently dropped feces. Parasite/parasitic stages/any other pathogen are expelled out through feces. The animals escape parasitic infection by avoiding recently defecated feces.
(c) Grooming of hairs of animals occurs and they eliminate ectoparasitic infection in this way.
(d) Licking the wound by animals creates a saliva coat which protects the area from any pathogen contamination. Licking of the penis by animals protects them from venereal diseases.
(e) The dogs go out and take grass or other vegetation which has anthelmintic activity. They learn this technique from their parents. This is a kind of self-medication by the animals to treat themselves in this way.
(f) All the members of the animal group take care of any sick animal.
(g) Mothers pass their young offsprings to the group members so that the young can develop immunity against pathogens.

Fig. 3.2 Schematically displayed different behavioral defenses

Host Immune Response: A Consequence of Host-Parasite Interaction

Immune Response to Protozoa

Nonimmunological defense mechanisms play a role for resisting any parasitic infection. There are many examples of nonimmunological immunity. N'Dama cattle are good examples of this. They are trypanotolerant animals resistant to *Trypanosoma* organisms. Continuous genetic selection of the hosts against the organisms resulted this resistance. Protozoa can initiate both humoral- and cell-mediated immunity. Though all protozoan parasite fail to elicit good immune response, many protozoans are able to confer good antibody response. Antibody-dependent cellular cytotoxicity and complement-mediated lysis are two major phenomena for inhibiting protozoan infection. In some cases, local antibody plays a great role in this regard. In *Tritrichomonas* infection, release of local antibody (IgE) occurs which binds to the antigen. In this situation, the Fc receptor present onto the basophil and mast cells binds to the Fc part of the antibody. Then, the effector cells get lysed and histamine, prostaglandin, and leukotriene are released. This initiates a type 1 hypersensitivity and this hypersensitivity reaction increases permeability of blood vessel by the action of histamine. Then, IgG molecules can easily pass through the blood vessel as the permeability has been increased and reach to the site of infection and kill the organisms. The immunity occurring in *Babesia* infection is interesting in which infected erythrocyte incorporates the antigen on their membrane and the antibody molecules get adhere to the surface of the RBC which is called as opsonization; thus, they are cleared along with RBC by the phagocytic cells. In case of *Toxoplasma* infection, both cell-mediated and humoral immune response have been found. Usually, the sensitized T lymphocytes release cytokine which activate the fusion of lysosome-phagosome system in the macrophage, and after fusion of phagosome with lysosome, lytic enzymes are released which kill the organisms. But *Toxoplasma* sp. has enormous potential to evade host immunity. They inhibit fusion of phagosome and lysosome. The organisms can easily survive in these cells. T cytotoxic cells play cell-mediated immunity in case of the endogenous antigen present within the cells. The cytotoxic T cells release toxic substance which kills the organisms. Cytotoxic T cells kill the infected lymphoblasts in *Theileria* infection.

Immune Response to Helminth

Nonimmunological defense mechanism is also found in case of helminthic infection. There occurs variation of susceptibility of the hosts to the parasites. Breed, sex, species, etc. are the factors which cause variation of infection along with pathological entity. All types of immunoglobulins are produced in helminthic infection, i.e.,

IgG, IgD, IgA, IgM, IgE, etc. However, IgE has a great role in helminthic immunity which is called as reaginic antibody secreted from the local mucosa-associated lymphoid tissue. A hypersensitivity 1 reaction occurs as described in protozoan infection by the immune effector cells which kill the organisms. The role of IgE antibody is markedly found in case of self-cure phenomena. In this phenomenon, helminthic antigen and mast cell-bound IgE cause release of vasoactive amines (i.e., histamine) as a consequence of degranulation of mast cell, basophil, etc. This causes high contraction of intestinal musculature. At the same time, histamine increases permeability of blood vessels causing efflux of large amount of body fluid. Hyperplasia of goblet cells occurs. All these result in expulsion of the parasites. As a result of ADCC, eosinophils release eosinophilic basic protein and eosinophilic cationic protein which kill the organisms. Cytotoxic T cells (Tc) release toxic substances which kill the parasites. Macrophages bind to antigen-bounded IgE through Fc receptor present on the cell surface get activated and releases interleukin 1, leukotrienes, prostaglandins, etc. which conjointly cause destruction of parasite.

Immune Response to Arthropod

Basic immunological phenomena are the same, but some minute variations are taken into account. The fleas release saliva which contain proteins of low molecular weight acting as hapten which bind to the collagen material present in the skin. This in turn causes a local type IV hypersensitivity reaction. As a consequence, infiltration of different mononuclear cells occurs in the localized area which is a local allergic reaction. In some arthropod infection, type 1 hypersensitivity occurs. Reaction between antigen and mast cell-bound IgE occurs. The activated mast cells get degranulated and release vasoactive amine, i.e., histamine which causes local inflammation manifested as allergic reaction. Both cell-mediated and humoral immunity occur in case of tick infection. In mite infection, type IV hypersensitivity occurs as is found in case of flea bite.

Genetic Resistance to Parasites

Many hosts are resistant to parasites. For example, Merino and Finn-Dorset sheep are resistant to *Haemonchus* parasite. N'Dama cattle are resistant to trypanosomes. This phenotypic manifestation can be unfurled genetically. Many works are being carried out to see the actual genetic reason for occurrence of resistance in the hosts. Different cytokine profiles of the animals indicate the resistance of the animals against parasites. For instance, studies are being carried out to see the nucleotide profile of IFN gamma genes, IL9 genes, etc. The difference of single nucleotide polymorphism in resistant and susceptible animals is analyzed. The expression level of the different cytokines is done by qPCR for determination of the genetic reason of the occurrence of resistance.

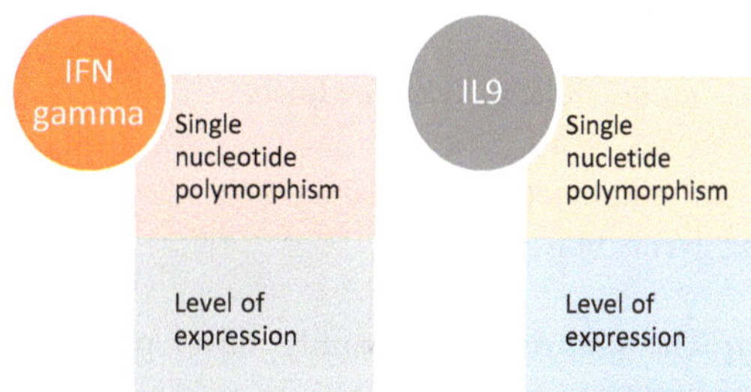

Fig. 3.3 Schematically displayed polymorphism for occurrence of genetic resistance

Establishment of Parasites in Different Hosts

There are many reasons how one host becomes immunocompetent. These are as follows:

1. Genetic makeup of the host
2. Proper nutrition of the host
3. Proper orchestration of immune system of the host

If immunological aspect is taken into consideration, the immunocompetent hosts perform the following functions:

(a) The parasites are rejected by host's immune system. The parasites are not able to set up any infection in the hosts.
(b) The parasites are killed by the host's immune system.
(c) The parasites may lose reproductive potential.
(d) Proper growth of the parasites is retarded in the immunocompetent hosts.
(e) Number of sexually matured parasites is less than the number of infective parasites.

Establishment of Parasite in Immunocompromised Host

There are many reasons how one host becomes immunocompromised. These are as follows:

1. Genetic makeup of the host
2. Nutrient deficiency of the host
3. Immunodeficiency of the hosts

In immunocompromised animals, the following events may occur:

(a) The parasites successfully develop in the hosts.
(b) Reproductive potential of the parasite is optimum.
(c) The parasites are not killed by the host's immune system.
(d) Growth of parasites is optimum.
(e) Number of sexually matured parasites is optimum.

Establishment of Parasite in Intermediate Host

In intermediate hosts, the following events occur:

(a) The parasites successfully develop up to certain stage in the intermediate hosts. If the life cycle of a parasite involves one intermediate host, infective stage of the parasite develops in that intermediate hosts. If life cycle involves two intermediate hosts, the infective stage of the parasites occurs in the second intermediate hosts.
(b) Growth of parasites is optimum.
(c) It is to be noted that the parasite never attains sexual maturity in intermediate hosts.

Establishment of Parasites in Unnatural Hosts

The following are observed if parasite enters in unnatural host:

1. The hosts which are not common for one parasite are called as unnatural hosts.
2. The internal biological environment of unnatural host is not suitable for parasite.
3. The parasites cannot develop up to sexual maturity in the unnatural hosts.
4. The parasites may perform aberrant migration.
5. The parasites affect unusual organs. For example, *Toxocara canis* are the parasites of dogs. But the larvae of the parasite migrate in different internal organs in the unnatural hosts, i.e., human (children).

Chronicity of Parasitic Infection

In case of helminthic infection, Th1 like cytokines are responsible for considerable immunopathology and occurrence of clinical signs. In spite of all Th2 cytokines in operation, the parasites survive in the hosts. What is the reason behind it and why do the parasitic infection enter into chronicity? Chronic infection of helminth is mediated by Th2-like response. A T cell subset is produced which are called as regulatory T cells or Treg cells. These cells produce IL10 which regulate immune reaction. This immunomodulatory Treg response or Treg activity is associated with Th2-like response, and anti-inflammatory cytokines are produced. This is called as modified Th2 response. Two important anti- inflammatory cytokines are IL10 and TGF. These anti-inflammatory cytokines downregulate the inflammatory reaction of the hosts. And the parasitic infection enters into chronic stage.

Host-Parasite Equilibrium

Host-parasite equilibrium is actually an evolutionary equilibrium between these two types of organisms (hosts and parasites). Host-parasite coevolution is a continuous process. Host-parasite interaction creates a selection pressure which results from many aspects related to parasitic infection. These are virulence of parasite, infectivity of parasite, and sustainability of a parasite in the host. All these are parasite-oriented adaptation. The host-oriented adaptations are susceptibility of hosts to a parasite, immunological modification, etc. In fact, host-parasite coevolution is a continuous journey for achieving equilibrium.

Immunoevasion Strategy of Parasite

Parasite can escape the host's immune attack by different phenomena. The following are some ways by which the parasites escape host's immune attack.

1. Trypanosomes frequently change their antigenic character of surface protein. Since the organisms are able to change their antigen frequently, the hosts fails to mount a strong immunity against this parasite.
2. Many helminthic parasites adsorb host's antigen on their surface. That's why the hosts do not mount any immunity as the parasites remain hidden within the host's molecule. In this way, the parasites remain safe. This phenomenon of immune evasion is called as molecular mimicry.
3. Some parasites release different enzymes (cathepsin L protease) which causes reduction of lymphoproliferation. Immunity is ineffective due to lack of lymphocytes which are the prime cells of immunity. These enzymes also cleave immunoglobulins of hosts.
4. Juvenile flukes absorb IgM on their surface. Since the parasites remain hidden within immunoglobulins, the eosinophils and other immune effector cells do not produce any Fc receptor causing nonoccurrence of ADCC or other types of immune reaction.

Pathology Caused by Parasites

Parasites cause damage to the hosts in different ways. They may cause mechanical damage to the hosts. Many tapeworms, *Ascaris*, and other worms obstruct the lumen of the GI tract resulting in obstruction of food passage and functional disturbances. Parasitic toxins and secretory proteins are the major components which cause pathological anomalies in the tissues. The toxins are usually protein in nature which cause allergic reaction. Allergic reaction due to endoparasitic and ectoparasitic infection is very common. A well-known parasite toxin (allergen) has been found in the body fluid of some worms. This allergen causes irritation of cornea and mucous membrane of the nasopharyngeal cavity, a commonly encountered clinical feature occurring in the parasitologists who work regularly with ascarid worm for a long period. Similarly, trematodes also produce some kind of toxin. Cercarial dermatitis is an example for the role of the toxin. There occurs severe inflammatory reaction in the human beings who was previously exposed with same antigen. Cercariae of nonhuman schistosome elicit an immunological reaction in the skin of human beings. This is a type of allergic reaction. Metabolic by-product of the worms of the parasites also causes damage to the hosts. It has been proved that the metabolic by-product of tapeworms affects the nervous system of the hosts. Similarly, the trematodes also release some kind of the metabolic by-product. This excretory-secretory material of parasites also causes pathological damage to the hosts. Very interesting example can be given. In case of the *Schistosoma* infection, there occurs formation of granuloma. The factors which predispose the formation of the granuloma is the release of soluble antigen from the miracidium present in the egg. The shell of the eggs is porous, and through the pores, soluble antigen gets released and elicits an immunological reaction leading to formation of the granuloma.

Different pathologies are caused by host-parasite interaction. The penetrating parasite (hookworm larvae, cercariae of *Schistosoma* spp., larvae of *Capillaria*, larvae of *Strongyloides* sp., etc.) causes different pathological damage during entrance and migration period. The hosts exert its defense against the parasite, and as a result, host immune response is triggered. Due to this, different cytokines and inflammation-mediating substances are released which cause inflammation and other pathological damage. Many parasites can evade the host immune response and live long duration in the host. In these cases, development of granuloma and fibrosis occurs as a result of immunomodulatory activity by different immune cells/substances.

Immunopathology and Molecular Pathology

Today's pathology concept is more toward the immunopathology and molecular pathology. The role of different immune cells and immune chemicals is becoming clearer at the molecular level. The role of T cells, B cells, complement, immunoglobulin, cytokine (interleukin), etc. is being unfurled in more detail. There is a balance of the functions of these cells and the immunochemical material. The cytokines which are released by any cell are obviously influenced by the genetic makeup of the parasite. Hence, the molecular pathology is more significant. In the field of medicine, cytokine therapy and immune cell therapy have been common practice. At present, the inflammation can be described in a more precise way. To describe molecular pathology, description of transcription of different genes encoding different cytokines or other proteins is prerequisite. For occurrence of any immunological event, signal transduction is also very important. Different protein molecules are expressed which cause pathology. Genes encoding those proteins are transcribed. It is the fact that all type of protein molecules are not expressed at a time. Expression of any protein molecule occurs whenever this protein molecule is required. Accordingly, respective genes are transcribed for translation of that particular protein.

Host-Parasite Relation Induced by Malnutrition and Micronutrient Metabolism

The host-parasite relation is dependent on nutrition. The parasites take upper hand if the nutritional status of the animal is not up to the level. If the nutritional status of the animals is up to the mark, then the animals show less susceptibility. Many nutrients play many roles. The following are some examples:

Nutrients	Mechanism
1. Vitamin A, D, E 2. Fatty acids 3. Vitamin B6, vitamin B12	Regulatory effects on cells of the immune system • Support immunity at cellular level • Help in antibody production
Protein	• Changing balance of hormone which regulates immunity • Help formation of immune cells and other cells
Zinc	• Help immunity at cellular level
Selenium	• Help antibody production

Amino acids present in the diet increase immunity. Young animals of poor health frequently pick up parasitic infection. Immune cells use some glucose. Experimentally, it has been proved that activated leukocytes express high level of nutrient transporters. When leukocytes are stimulated, they release IL1, TNF, and IL6 which go through the body and cause redistribution of the nutrient specially from the skeletal muscles.

Trace element nutrients, i.e., iron, copper, and zinc, are problematic because of their low concentration in the muscle. The body has a high need for these trace elements and it has been reported that dietary requirement for some trace nutrients may be higher for optimal immune function than it is for maximal growth or reproductive performance. Further work is needed for quantitative nutrient requirement.

What Is Nutritional Immunity?

We must understand host-parasite interaction in relation to nutrition. The body withholds essential nutrients to reduce the growth and multiplication of parasites. In birds, withholding nutrient can reduce bacterial colony in egg albumen. For instance, dietary ingredient, i.e., type of fat, can alter the proportion of prostaglandins and other eicosanoids which are released by leukocytes to coordinate their response during the process of disease challenges. Phospholipid is present in the cell membrane which is called as membrane phospholipid, a type of dietary fat. Eicosanoids are produced from phospholipid. The dietary fat has a role on eicosanoid which can modulate/change immunity. Fish oil possess high amount of eicosapentaenoic acid which is a predisposing factor to act on macrophages to release interleukins that drive T helper cells toward Th2 type response. Dietary manipulation cannot boost the entire immune system. Only this has a directional ability for particular immunity. For example, dietary fish oil can reduce those pathogens which are affected by Th2 type of response. Therefore, this oil has specificity toward immunity. Long-term dietary feed restriction causes elevated level of glucocorticoids which cause the T cells less functional resulting in decreased immune competence. Restrictive feeding to broiler chickens results in loss of glucagon, and this condition leads to affect mobilization of neutrophils.

Adequate dietary antioxidants can reduce the pathology caused by reactive oxygen intermediates released at the site of infection.

4 Veterinary Trematodology

The name "trematode" has been originated from the Greek word "trematos" meaning "pierced with holes." Digenetic trematodes are very important parasites in the field of veterinary science. These trematodes occur in different organs of a wide variety of domestic and wild animals. Some digenetic trematodes pose a great threat to the livestock industry. In this chapter, the description of trematodes of veterinary importance has been given encompassing different aspects of trematodes, i.e., classification, distribution, discovery, hosts, site, morphology, biology, pathogenesis, diagnosis, treatment, and control.

Classification of Trematodes of Veterinary Importance

Phylum: **Platyhelminthes**
Class: Trematoda
Subclass: Monogenia, **Subclass**: Aspidogastrea, **Subclass**: Digenea
Subclass: Digenea

Family: Fasciolidae
Genera
Fasciola
Fasciolopsis
Fascioloides
Parafasciolopsis

Family: Dicrocoeliidae
Genera
Dicrocoelium
Platynosomum
Eurytrema
Concinnum
Athesmia

Family: Opisthorchiidae
Genera
Opisthorchis
Clonorchis
Metorchis
Parametorchis
Pseudoamphistomum

Family: Nanophyetidae
Genus
Nanophyetus

Family: Echinostomatidae
Genera
Echinostoma
Echinochasmus
Echinoparyphium
Hypoderaeum
Isthmiophora

Family: Heterophyidae
Genera
Heterophyes
Apophallus
Euryhelmis

S. C. Mandal, *Textbook of Veterinary Parasitology*, https://doi.org/10.1007/978-981-99-2924-5_4

Metagonimus
Euryhelmis
Cryptocotyle

Family: Plagiorchiidae
Genus
Plagiorchis

Family: Troglotrematidae
Genera
Troglotrema
Collyriclum

Family: Paragonimidae
Genus
Paragonimus

Family: Prosthogonimidae
Genus
Prosthogonimus

Family: Notocotylidae
Genus
Notocotylus, Catatropis, Ogmocotyle

Family: Paramphistomatidae
Genera
Paramphistomum
Cotylophoron
Gigantocotyle
Gastrothylax
Fischoederius
Gastrodiscus
Gastrodiscoides
Pseudodiscus
Calicophoron
Carmyerius

Family: Schistosomatidae
Genera
Schistosoma
Bilharziella
Ornithobilharzia
Bivitellobilharzia
Heterobilharzia
Gigantobilharzia
Austrobilharzia
Trichobilharzia

Family: Clinostomatidae
Genus
Clinostomum

Family: Strigeidae
Genera
Apatemon
Cotylurus
Parastrigea

Family: Diplostomatidae
Genus
Alaria
Diplostomum
Posthodiplostomum
Neodiplostomum

Family: Allocreadiidae
Genus
Crepidostomum

Family: Sanguinicolidae
Genus
Sanguinicola

Family: Cyclocoelidae
Genus
Tracheophilus

General Morphology of Trematodes

Color

The color of digenetic flukes is widely variable. These may be pink, gray, dark gray, yellowish gray, etc. Some flukes have a combination of translucent and opaque area in the body.

Shape

The shape of flukes is also variable. Some are flat and leaf shaped (*Fasciola* spp.). Some are fleshy and pear shaped (*Cotylophoron* spp.). Some are elongated (*Fischoederius* spp.). *Dicrocoelium* spp. are lanceolate shaped. Some digenetic trematodes are completely different from others, viz., schistosomes are cylindrical in shape looking like roundworms. Sexual dimorphism is absent in digenetic trematodes except schistosomes. In this family, the male and female parasites are different.

Tegument

Tegument is the outer covering of trematodes, which may be either smooth or spiny. The tegument is a syncytial layer composed of fused cells. The cuticle is soft. The tegument of trematodes can actively absorb nutrients.

Digestive System

The digestive system is present in digenetic trematodes but it is a simple type. The different parts of the digestive system are the mouth, pharynx, esophagus, and intestine. The digestive system starts from the mouth, which is usually surrounded by oral sucker. The mouth is followed by pharynx. This is usually a short muscular structure. The shape of the pharynx is variable, which provides taxonomic clues for identification of trematodes. The next part is the esophagus, which is a tube-like structure. The intestine, in most of the digenetic trematodes, is bifurcated into two tube-like structures called as intestinal ceca. The intestinal ceca have taxonomic importance for identification. In some trematodes, these run a short distance, and in others, these may reach up to the posterior end. Most of the digenetic trematodes absorb nutrients through their tegument. Others take host tissue and body fluid through the acetabulum.

Nervous System

A very primitive type of the nervous system is found in the trematodes. A pair of ganglia representing the so-called brain is connected with the esophagus by a typical structure called circumesophageal commissure. From these ganglia, three pairs of nerves run anteriorly, and three backwardly. The digenetic trematodes usually lack any specialized sense organs. In the free-living forms, sense organs are found provided with pigmented areas (e.g., eye spots in cercariae and miracidia).

Excretory System

As a rule, flame cells are the excretory units of flatworms. There is a main canal structure called excretory canal that is surrounded by a number of tubules, and each of these is connected with flame cells. Both excretion and osmoregulatory functions are performed by the excretory system of trematodes. The excretory canal is opened through the excretory pore.

Reproductive System

Male Genital Organ

The different important organs/structures of male genital organ are as follows:

- Testes
- Vasa efferentia
- Vas differens
- Vesiculum seminalis

Description

Vasa efferentia unite to form vas deferens, which distally forms a sac-like structure called as vesicula seminalis. The vesicula seminalis is surrounded by prostate glands. This eventually terminates as cirrus. A cirrus sac may be present, which covers the vesicula seminalis, prostrate, and cirrus. The cirrus is a slender structure. The genital pore is situated anteroventrally or posterolaterally. The genital pore is surrounded by a genital sinus or genital atrium. In some trematodes (*Cotylophoron* spp., *Heterophyes* spp., etc.), it is surrounded by genital sucker.

Female Genital Organ

The different important organs/structures of the female genital system are as follows:

- Ovary
- Oviduct
- Ootype
- Uterus
- Melhis's glands or shell glands
- Vitelline glands

Description

The prime part of the female reproductive system is the ovary. From the ovary, a small tube-like structure arises called oviduct. The oviduct is connected with receptaculum seminis and Laurer's canal opening on the dorsal surface. The oviduct is further connected with a structure called ootype. A pair of vitelline glands consisting of many follicles is connected with vitelline duct or yolk duct. This vitelline duct joins the oviduct via a wide portion called as ootype. Vitelline glands are situated laterally on either side of the fluke. Numerous unicellular glands surround the ootype called Melhis's glands or shell glands. The secretion of this gland is used for the formation of egg shell.

Life Cycle

The digenetic trematodes are heteroxenous parasites. They always require one or two intermediate hosts. Some part of biological development occurs in the definitive host and some in the intermediate hosts and environment. The stages found in the life cycle of digenetic trematodes are egg, miracidium, sporocyst, redia, cercaria, metacercaria, and adult. In some trematodes, redia stage is not produced. In those trematodes, a generation of daughter sporocysts is produced. In the case of *Schistosoma* spp., metacercaria stages are not formed. The cercariae directly develop to a unique stage called as schistosomula.

The followings are the events of life cycle of digenetic trematodes:

Fertilization: Except Schistosomatidae, all digenetic trematodes are hermaphrodite. Both male and female reproductive systems are present in the same individual. Therefore, self-fertilization occurs within the body. External fertilization does not occur.

Egg passing: Adult parasites live in their predilection site and lay their eggs. Usually every predilection site is connected with the natural orifice through which the eggs are passed out. The eggs are usually passed with the feces, nasal discharge, urine, etc.

Egg hatching: Hatching of eggs occurs in the environment or in the gut of the intermediate hosts. Light and water are the main stimuli for hatching of the eggs in the environment. Enzymes, acid, etc. are the biochemical factors playing a role for hatching of eggs in the intermediate hosts. Light also stimulates development of miracidium and activity of the same in the environment.

Development in intermediate host: Different aquatic and land snails act as the intermediate hosts of digenetic trematodes. Miracidia enter the aquatic snails by penetrating the exposed soft tissues of the snail. The miracidia arrive at the hepatopancreas region where they start development. In the case of some trematodes, eggs do not hatch in the environment. The eggs are ingested by the intermediate hosts and hatch in their gut. Different developmental stages occurring in the snails are sporocyst, daughter sporocyst, redia, and cercaria.

Shedding of cercariae: The shedding of cercariae occurs when they are completely matured. There are several stimuli that act on the release of cercariae. Temperature variation and sudden flood are some factors that have been demonstrated by different workers.

Encystment and formation of metacercariae: The cercariae transform to metacercariae after casting off their tail appendages. The cercariae may encyst on the aquatic vegetation or in the second intermediate hosts (snail, ant, fly, tadpole, etc.) and transform to metacercariae. The metacercariae are quiescent form infective stage having tough outer wall.

Infection of definitive host: The final hosts get the infection by ingestion of infective metacercariae present in aquatic vegetation or drinking water. The final hosts may get infection by ingestion of second intermediate hosts carrying the metacercariae in their body. In case of schistosomes, the cercariae penetrate the skin and set up infection in the hosts.

Development in definitive hosts: The metacercariae get excysted in the small intestine. The juvenile flukes come out from the metacercarial cysts and develop to sexually matured parasites in their respective sites.

Family: Fasciolidae

Family Synopsis

This is an important family of trematodes. There are three genera in this family, i.e., *Fasciola*, *Fasciolopsis*, and *Fascioloides*. Of these three genera, *Fasciola* genus is the most important because the members of this genus cause significant loss of livestock economy in terms of loss of morbidity and mortality. The parasites are distributed all over the world. Moreover, *Fasciola hepatica*, *Fasciola gigantica*, and *Fasciolopsis buski* have zoonotic importance. Fasciolidae members are characterized by leaf-shaped body with spiny cuticle, close oral and ventral suckers, presence of pharynx and short esophagus, several branched internal organs. The life cycle is indirect and requires one aquatic snail as intermediate host. The developmental stages are egg, miracidium, sporocyst, redia, cercaria, and adult. Examination of feces is the gold standard diagnostic technique. Control of the parasites relies mostly on treatment and control of snails.

Genera

- *Fasciola*
- *Fasciolopsis*
- *Fascioloides*

Genus: *Fasciola*

History

Jehan de Brie first recorded *F. hepatica* in the year 1379. Redi first described *F. hepatica* with illustration in the year 1688. Thomas and Leuckart first established the life cycle during 1881–1883.

Distribution

The parasites are distributed in tropical and subtropical countries. They are commonly distributed in South East Asia, Africa, etc.

Etymology

Fasciola has been originated from Latin "fascia" meaning band.

Species

Fasciola gigantica
Fasciola hepatica

Site

Normal site	Rare site
Liver	Brain
Bile duct	Lung

Hosts

Primary hosts	Natural hosts	**Cattle, sheep, goat, pig**
	Unnatural hosts	Equids, lagomorph, rodents, man
	Domestic hosts	Cattle, sheep, goat, pig, etc.
	Wild hosts	Lagomorph, rodents, etc.
	Laboratory hosts	Rabbit
Secondary hosts	*Lymnaea*, *Pseudosuccinea*, *Galba*, etc.	

Disease caused: Fasciolosis

Morphology

Adult

Size: The size is 2–7.5 × 0.5–1.3 cm of *F. gigantica*. *F. hepatica* is smaller compared to *Fasciola gigantica*. The size is 1.6–5.0 × 0.4–1.3 cm.

Shape: Both the flukes are leaf shaped. Anteriorly, it has a cone-shaped prominence, which is called as anterior cone. A pair of shoulders are present. The shoulder is broader in *F. hepatica* compared to *Fasciola gigantica*.

Suckers: Two suckers are present. One is oral and another is ventral. Oral sucker is smaller than ventral.

Pharynx: Small.

Esophagus: Short.

Intestine: The intestinal diverticulum of *F. gigantica* is more branched in the median field, while it is less branched in the median field of *F. hepatica*.

Testes and ovary: Testes are branched occupying the median field. The testes are tandem in position. One ovary is also branched and pretesticular in position.

Cirrus: Scales like spines are present on the cirrus. The cirrus sac is partially or completely anterior to the ventral sucker.

Genital pore: The genital pore is located in between the intestinal bifurcation and the ventral sucker.

Vitelline glands: These are follicular and confluent occupying the lateral field from the anterior end to the posterior end.

Egg

The eggs are oval, yellowish, and operculated. The eggs are large in size (156–197 um × 90–104 um of *Fasciola gigantica* and 130–150 um × 63–90 um of *Fasciola hepatica*). The embryo appears as one-cell stage surrounded by a number of oval yolk cells. An umbilicus-like invagination is found at the posterior end (opposite to operculated end).

Larva

The larval stages are miracidium, redia, cercaria, and metacercaria. The morphology of these stages is as follows:

Miracidium: The miracidium is pear shaped. The outer epidermal covering is ciliated. Innumerable cilia are present. The anterior end of the miracidium is devoid of cilia, which is cone shaped. Digestive system is not found. However, one protonephridial system is present, which performs excretory function. As that of other trematodes, it has also apical papilla containing secretory glands. The secretion of these glands helps in penetration of host tissue. Slightly posterior to the apical papilla, paired pigmented eye spots are present. The epidermal cells or plates have definitive pattern. These cells are arranged in five rows: 6, 6, 3, 4, and 2.

Sporocyst: Sporocyst is a subsequent stage of miracidium. In sporocyst, some structures are found as in miracidium, i.e., apical papilla and secretory glands. The ciliated covering is lost, and the miracidium is transformed to sporocyst. The sporocyst has thin cuticle forming a sac-like structure. Sporocysts have brood chambers. Germ cells and germ balls are found here. Digestive system is absent. However, protonephridial excretory system is present. From the germ cells, rediae are formed.

Redia: Rediae are formed from the germ balls present in the brood chambers of sporocysts. The rediae have ambulatory buds (procruscula), which are antero- and posterolaterally projected giving a typical shape of rediae. The rediae have digestive system. Mouth, pharynx, and unbranched ceca are found. Excretory system is also found in redia.

Cercaria: Brood chamber is found in redia, and the germ balls in the brood chamber differentiate into cercariae. The basic internal structure is the same as the adult one. The mouth is surrounded by an oral sucker. The proportional location of ventral sucker is the same as that of the adult. The oral sucker is connected with the gut containing bifurcated ceca. The cercariae have penetration glands. They possess tail. Cystogenous glands are present in the cercariae. The secretory material of these glands forms the cyst wall. The excretory system is protonephridial type comprising flame cell system. The pattern of flame cells is the same as that of adult. The pattern of flame cells has taxonomic importance.

Metacercaria: A cyst is formed from the cercaria. The outer wall is acid-resistant, which resists the action of gastric juices. It is needless to mention that structural features of metacercariae are variable. In *Fasciola hepatica*, the metacercariae are enclosed by two cysts. These are outer cyst and inner cyst. The outer cyst is composed of several layers. The outer layer is composed of tanned protein. Underneath this layer, there is a fibrous layer of mucoprotein. The inner cyst has two layers. The outer layer is made up of mucopolysaccharide and mucoprotein. This layer has been subdivided into three layers further. The inner layer is made up of lamellae held in a protein-lipid complex.

Note

Electron microscopy study reveals that the tegument has foldings and spines. The spines are serrated with several points. Different sensory papillae are found, which are larger in *F. gigantica* as compared to *F. hepatica*.

Fig. 4.1 *Fasciola gigantica*

Fig. 4.2 *Fasciola gigantica* (gross)

Fig. 4.3 *Fasciola gigantica* (Microscopic)

Life Cycle

This life cycle is heteroxenous, which requires one intermediate host. All developmental stages are present. The stages are egg, miracidium, sporocyst, redia, cercaria, metacercaria, and adult.

Basic information of life cycle
• The definitive hosts are sheep, goat, cattle, etc. • The intermediate hosts are aquatic snails • Mode of infection of the intermediate hosts is penetration by cercariae • Mode of infection of the definitive hosts is ingestion of metacercariae • The infective stage of the parasite is metacercaria

Stages of Life Cycle Formed

In definitive host	In environment	In intermediate host
• Adult • Egg	• Miracidium • Metacercaria	• Sporocyst • Redia • Cercaria

The events of life cycle: a glimpse
• Voiding of eggs with feces of definitive host • Hatching of eggs with contact of water • Release of miracidium • Entrance of miracidium in aquatic snail hosts • Formation of sporocysts, rediae, and cercariae in snail • Transformation of cercariae to metacercariae in water surface, aquatic vegetation, etc. • Ingestion of metacercariae by definitive hosts • Development up to adult in definitive hosts

Detailed Description

The followings are different events of the life cycle described in detail:

Egg Passing

Adult parasites are hermaphrodite, which are located in the bile duct. Since the parasites are hermaphrodite, self-fertilization occurs. The eggs are laid by the adult parasites in the bile duct. The bile duct is connected with duodenum. Thus, the eggs are voided through the feces.

Embryonation of Eggs

Embryonation of eggs occurs in the environment. Water is important factor for embryonation of eggs.

Egg Hatching

Hatching of eggs occurs in the environment. The eggs hatch in water. Light is a good stimulus for hatching of the eggs. Light stimulates the development of miracidium and activity of the same. Water is also one important hatching stimulus. The miracidium causes alternation of the internal concave surface of viscous cushion. After hatching, the miracidia come out. The miracidia are ciliated and pear shaped.

Development in Intermediate Host

Lymnaeid snails serve as the intermediate hosts of the parasite. It has been found that *Fasciola hepatica* has species specificity of snails. *Lymnaea truncatula, L. bulimoides* and *L. luteola* serve as intermediate hosts of *Fasciola hepatica*, whereas *L. auricularia* var. *rufescens, L. rubiginosa* and *L. luteola* serve as the intermediate hosts of *Fasciola gigantica*. The snails are aquatic pulmonate snails. Miracidia enter the aquatic snail by penetration of the exposed soft tissue. The miracidia follow the snails by chemotaxis. The miracidia enter the route of lymphatic and circulatory system and arrive at the hepatopancreas region. The stages that are formed in snails are sporocyst, redia, and cercaria. The miracidium is transformed to a new stage of development called as sporocysts in the hepatopancreas. They are sac-like containing 5–8 germ balls. Each germ ball gives rise to the formation of rediae. The sporocysts do not have any digestive organs, and they perform their nutrition by absorption. A generation of rediae is produced from the sporocysts. The rediae have digestive organ and survive by taking snail tissues. The next developmental stage is cercaria. The cercariae have tail by which they can swim in water.

Release of Cercariae

The release of cercariae occurs when they are completely matured. There are several stimuli that act for the release of cercariae. Temperature variation and sudden flood are some factors that have been demonstrated by different workers.

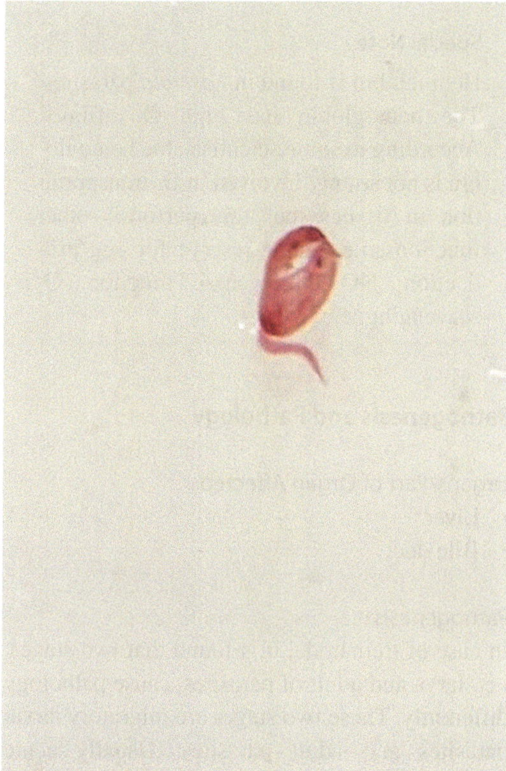

Fig. 4.4 Cercaria

Encystment and Formation of Metacercariae

The cercariae transform to metacercariae after casting off their tail appendages. The cercariae encyst on the aquatic vegetation or water surface and transform to metacercariae. The cystogenous glands secrete a specific substance that creates the outer covering of metacercariae. The metacercariae are quiescent form infective stage having tough outer wall. Within the cyst, the metacercaria remain alive for about 12 months. The final hosts get the infection by ingestion of metacercariae.

Infection of Definitive Host

The final hosts get the infection by ingestion of infective metacercariae. The metacercariae are formed on aquatic vegetation or on surface of water. The hosts pick up infection when they ingest aquatic vegetation or drink water contaminated with metacercariae.

Development in Definitive Hosts

The metacercariae get excysted in the small intestine. The juvenile flukes come out from the metacercarial cysts and penetrate the wall of the GI tract. They ultimately reach the liver and migrate in the liver tissues. They then reach to the bile duct where they get sexually matured. It takes about 12 weeks from infection for the formation of sexually matured parasite.

Special Note

- *The miracidia apply the papillae present at the broad anterior end and spin by using the cilia and penetrate the snail body.*
- *Excystation of metacercariae depends on many factors. These are temperature, gas, pH, oxidation-reduction potential, osmotic pressure, enzyme, and bile.*
- *Following excystment, newly excysted juvenile flukes (NEJ) utilize vital energy store and adjust metabolic pathway for their activity. The increase of temperature causes neoblast proliferation for their growth and development. The juvenile flukes release cysteine protease from their gastrodermal cells, which facilitate their migration. Aquaporin is released, which is called as tegumental transporter, which makes them enable for osmotic and other necessary changes of their penetration site.*
- *Molecular study depicts variation of gene expression during biological development of parasites. Different stages of parasites perform different functions. Therefore, different functional proteins are required in different biological stages of development. For example, the function of tissue-invading parasite is entirely different from stationary phase of parasite. Tissue-invading parasites require different enzymes for penetration. Therefore, gene encoding those proteins is tran-*

scribed as needed. Progressively differential gene expression is found from infection to subsequent developmental stage in the hosts. The data of gene expression from excystation to newly excysted juvenile (NEJ), traversal of the intestinal wall and migration in the liver, and transformation from immature to mature state are variable. Study of gene expression shows that the parasites switch from aerobic to anaerobic metabolism. The gene expression data also indicated that the production of vitelline and egg shell component is limited to the matured stage. Study shows upregulation of cell adhesion molecules, i.e., cadherins, integrins, and cytoskeletal proteins, which play an important role in sensing changes in the physiological environment. High degree of upregulation of genes associated with protein polymerization and microtubule-based movement, i.e., tubulin, dynein, and tegument antigen, was reported.

Special Note

Hemoglobin is found in *Fasciola* parasites. The hemoglobin has high O_2 affinity. According to some scientists, the hemoglobin is not simply involved in O_2 transportation in tissues, but this performs other functions, i.e., heme reserve for egg production, NO dioxygenase function, O_2 scavenging, etc.

Pathogenesis and Pathology

Organs/Part of Organ Affected

- Liver
- Bile duct

Pathogenesis

In case of trematode, it is found that two stages, i.e., larva and adult of parasites, cause pathology differently. These two stages are migratory larval parasites and adult parasites. Usually acute inflammatory reaction occurs due to massive invasion by the immature flukes. In chronic stage, fibrotic proliferation results in clay pipe stem

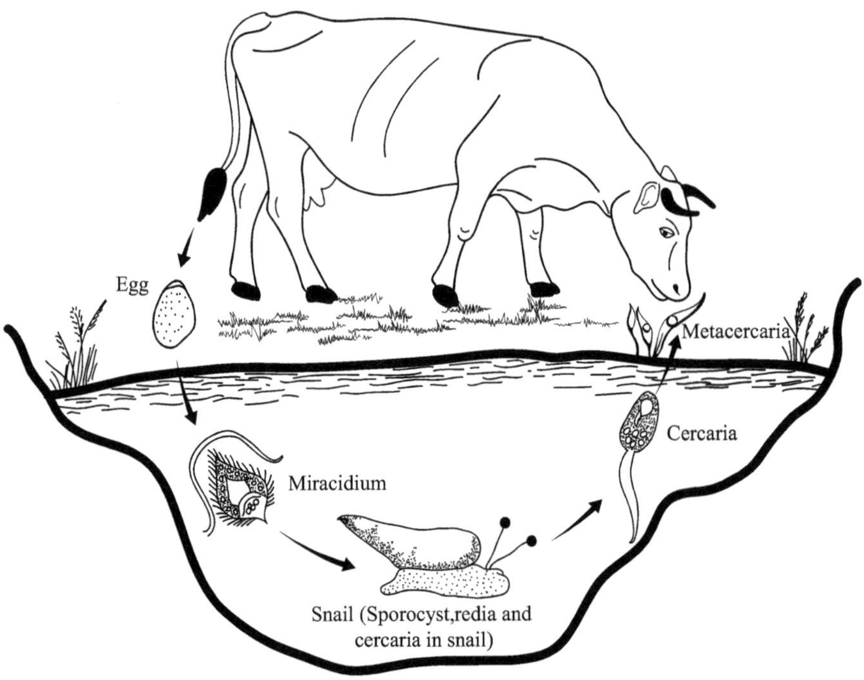

Fig. 4.5 Life cycle of *Fasciola* sp.

condition. Edema occurs due to massive loss of protein.

Pathogenesis in Chart

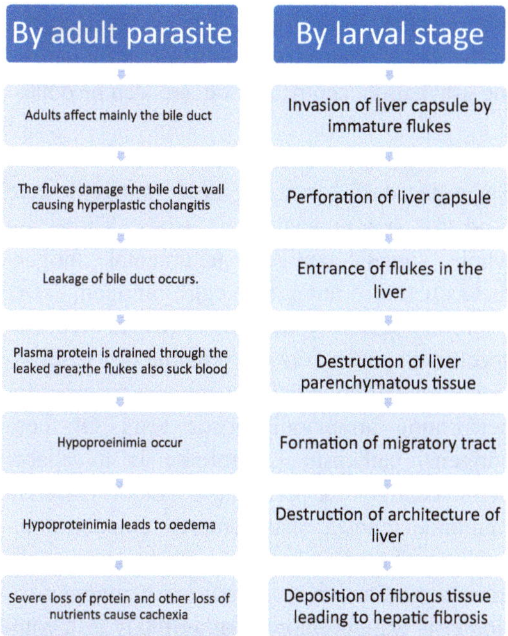

Fig. 4.6 Flow chart of pathogenesis caused by *Fasciola* sp.

Pathology

Anatomical Pathology

The migratory flukes cause damage to the liver capsule. Perforation of liver capsule occurs. Through these perforations, blood comes out. The flukes enter in the parenchymatous tissue of the liver and cause destruction of these tissues and form migratory tracts. The hepatic architecture is lost as a result of the healing process. The flukes reach to the bile duct where they take blood as food and get matured. In a long-standing case, fibrosis occurs.

Hematological Pathology

Hematological data reveals anemia and eosinophilia with Charcot-Leyden crystal. This anemia is due to massive blood loss when the immature flukes migrate in the peritoneal cavity, liver, and bile duct.

Chemical Pathology

Different studies indicate the rise of different vital enzymes, i.e., glutamate dehydrogenase, gamma-glutamyl transferase (GGT), etc. Massive loss of plasma protein has also been estimated, which results in edema (bottle jaw).

Immunopathology

For any parasitic infection, particularly the tissue-invading trematode, occurrence of inflammation is obvious, which appears as a primary pathological feature. Inflammation occurs mainly due to pro-inflammatory cytokines secreted mainly by macrophage. ADCC is another one important immunological phenomenon. In ADCC, different effector cells bind to Fc part of antibody already bound to antigen. Different effector cells are eosinophil, macrophage, NK cells, etc. These effector cells release toxic material. For example, eosinophil contains MBP (major basic protein), EPO (eosinophilic peroxidase), ECP (eosinophilic cationic protein), and EDN (eosinophil-derived neurotoxin). The toxic materials cause tissue damage. The role of NO (nitric oxide) secreted by activated macrophage is very important for killing helminthic parasite. Different studies indicate that this NO causes severe damage to the tissues. In this trematode infection, hyperplasia occurs. The pro-inflammatory cytokines have a role for causing hyperplasia.

Secondary Pathology

Secondary pathology is caused by anaerobic bacteria (*Clostridium oedematiens* (*novyi*)), which grow in anaerobic media of the affected area caused by migratory parasites. A complication of black disease and fasciolosis occurs simultaneously.

Molecular Pathology

Recent study has been directed toward different genetic markers encoding different protein molecules responsible for pathology. Study shows, in different laboratory animal models, upregulation of genes associated with fibrosis including genes in the JAK-STAT pathway. Study of tissue repair genes is also important for the assessment of pathology.

Pathology of Ectopic Origin

There have been records of aberrant migration or ectopic infection of *Fasciola*. The usual predilection sites are the bile duct and liver. But aberrant migration of the flukes occurs in lungs, kidney, spleen, etc. where the parasites cause different pathological features.

Symptoms

Acute Stage

The young animals suffer from acute stage of the disease. In acute infection, the sheep may die suddenly with similar symptoms of anthrax. Blood and froth come out from the mouth and nasal orifice. A large number of immature worms affect the liver. The animals show the symptoms similar to anthrax. Blood with froth comes out from the natural body orifices.

Chronic Stage

Cachexic condition is the characteristic feature of chronic fasciolosis. Alternatively, diarrhea and constipation are observed. The affected animals lose their vigor. They cannot resist slight thrust on their back and fall down easily. The animals become dull and depressed. The ribs become exposed due to loss of subcutaneous fat. The hairs are unhealthy and thin without any luster.

Diagnosis

Diagnosis of fasciolosis: a glimpse
• Examination of fecal sample for detection of eggs • Efforts made for the development of immunodiagnostic test by using crude and defined antigens (e.g., cathepsin proteins) • Molecular diagnosis by different molecular markers

Symptomatological Diagnosis

Diagnosis based on exhibited clinical signs is not possible. Alternation of diarrhea and constipation along with cachectic appearance might be suggestive of fasciolosis.

Parasitological Diagnosis

Examination of feces may reveal the presence of typical eggs. The eggs are yellowish in color and operculated. By direct smear technique, the eggs are visible. But, if the infection load is less and, as a result of which, if no egg is readily available in the smear, concentration techniques comprising gravitation, centrifugation, etc. can be done.

Immunological Diagnosis

Since decades, a panel of crude antigens were used for immunodiagnostic trials. These are whole somatic antigen, tegumental antigen, metacercarial antigen, egg antigen, etc. Excretory-secretory antigens were also used. But specific diagnosis was not possible due to cross-reactivity. Different defined antigens are used for performing immunodiagnostic tests. Of these antigens, cathepsin L protease is a reliable defined antigen for performing ELISA. Different immunodiagnostic tests comprise counterimmunoelectrophoresis, bentonite flocculation test, latex agglutination test, indirect hemagglutination test, indirect fluorescent antibody test, complement fixation test, ELISA, indirect ELISA, dot ELISA, Western blot, dipstick ELISA, etc. In this context, it is to be mentioned that two different steps are usually taken:

(a) Detection of antibody and determination of the level of antibody performing different immunodiagnostic tests using antigens
(b) Detection and determination of level of antigen by using monoclonal antibody

Molecular Diagnosis

Different molecular markers are used for the identification of this parasite. These markers are mainly of ribosomal DNA and RNA (ITS1, ITS2, 5.8 S rRNA, 18 S rRNA, etc.). Mitochondrial DNA (CO1, ND1, etc.) are also used as molecular markers for identification. Species-specific primer is available. Copro-PCR has also been described.

Treatment

Chemotherapy

Name of drug	Route of administration	Dose
Oxyclozanide	Oral	10–15 mg/kg (highly effective against adults) in cattle and buffalo 15–20 mg/kg (highly effective against adults) in sheep and goat 45 mg/kg (effective against immature worm in acute fasciolosis) in sheep
Rafoxanide	Oral	7.5 mg/kg (highly effective against both mature and immature worms in cattle, buffalo, goat and sheep)
Nitroxynil	Oral	10–15 mg/kg (highly effective against both mature and immature flukes in cattle. and in sheep 15–20 mg/kg)
Triclabendazole	Oral	10 mg/kg (highly effective in sheep and goat against all stages); 12 mg/kg (cattle and buffalo)

Nanotherapy

Studies indicate anti-*Fasciola* efficacy of silver nanoparticles and polylactic acid curcumin-nisin. Efficacy of gold nanoparticle against *Fasciola* has also been found. Gherbawy et al. (2013) produced silver nanoparticles using plant synthesis protocol for procurement of nanoparticles. They produced silver nanoparticles by challenging the fungus *Trichoderma harzianum* with silver nitrate solution. Efficacy of these nanoparticles was assessed by observation of hatching status of eggs of *Fasciola*. The percentage of unhatched eggs was 69.67 % (in treatment group of triclabendazole alone), while this percentage was 89.67% (treatment group in combination with drug and AgNPs).

Control

Do's	Don'ts
• The affected animals should be treated properly • The feces of the animals should be disposed properly • The animals should be grazed in high land area where there is no stagnant water • Bioclimatograph should be analyzed if available • Reports of remote sensing should be analyzed if available	• The animals should not be allowed to drink pond water or river water • The animals should not be allowed to graze in infected pasture

Description of Control in Detail

General Management

The affected animals should be treated properly. Since water is the prime environmental factor for the development of embryo in eggs of the parasite, the animals should not be allowed to graze in low land area where stagnant water is available. The animals should not be allowed to drink river water and pond water because the water may be contaminated with metacercariae. Underground water should be provided to the animals. The snails should be controlled by molluscicidal agents (copper sulfate at the concentration of 1:150 million parts of water, sodium pentachlorophenate, dinitro-o-cyclohexyl phenol at the concentration of 3–5 parts per million parts of water). The feces should be disposed in a particular fecal pit. Aquatic vegetations and weeds are the food material of snails. Therefore, though practical implication is too difficult; still drainage of vegetation can be done for reduction of snail population.

Immunological Control

An array of research works have been carried out for the development of vaccine using both crude and defined antigens. In many cases, satisfactory and encouraging results have been found. The main obstacle of commercialization of vaccine is the problem of immune modulation and immune evasion by the parasites by shifting of Th1 to Th2 for T regulatory response. In different experimental studies, vaccines were assessed by evaluation of different phenotypic characters comprising fluke number, fluke size, EPG, antibody profile, etc. Variation of vaccine efficacy was found in different breeds and different age groups. But surprisingly, variation of vaccine efficacy was found among the same breed, which led the scientists to know the genetic reason of this. Different genomic research revealed polymorphism of genes of these antigens, which have resulted in these phenotypic variations. But in some defined antigens, low allelic variation has been detected. This finding increases the hope for development of vaccine. The candidates of vaccines against fasciolosis are cathepsin L1 (FhCL1), cathepsin L2 (FhCL2), glutathione S-transferase (GST), fatty acid-binding protein (FABP), peroxiredoxin (Prx), gut-associated exopeptidase, and leucine aminopeptidase (LAP). Of these antigens, the lead candidates are cathepsin L1, cathepsin L2 (FhCL2), glutathione S-transferase (GST), fatty acid-binding protein (FABP), peroxiredoxin (Prx), paramyosin, gut-associated exopeptidase, and leucine aminopeptidase (LAP). The remaining key point is the selection of proper candidate antigen and selection of proper adjuvants. The adjuvants that have been used in different research trials are Montanide, Quil-A, etc. A number of recombinant antigens have been produced. But different trials have unfurled that native antigens were more effective than recombinant antigens. The main reason is related to the lack of post-translational modification of protein comprising glycosylation, phosphorylation, ubiquitination, nitrosylation, acetylation, methylation, lipidation, etc.

Molecular Control

Works are carried out to produce gene vaccine, but no molecular vaccine is available in the field.

Biological Control

There are different ways for biological control of this parasite.

(a) Studies indicate the use of different egg parasitic fungi for control of fasciolosis. But no such technology has been commercialized. Dias et al. (2011) worked on ovicidal effect of a fungus *Pochonia chlamydosporia* on eggs of *Fasciola hepatica*. In their works, the fungi were found effective preying upon the eggs. They suggested that the fungus *Pochonia chlamydosporia* could be used as biocontrol agents against helminthic parasites.
(b) Suhardono and Copeman (2006) investigated the control potential by exploiting the antagonism mechanism between the two larvae of *Fasciola gigantica* and *Echinostoma revolutum*. They found significantly less *Lymnaea rubiginosa* infected with *F. gigantica* in the rice fields where it get exposed to both duck and cattle dung.
(c) There have been reports in the earlier literature about biological control of *Fasciola* using ducks. But practical implication in mass scale is really very difficult.
(d) Prawns are observed to prey on snails. Practical implication of snail control by prawn is difficult.

Species: *Fasciola jacksoni*

The Indian elephants are affected by this fluke. The parasites affect the bile duct, liver, and duodenum. The fluke is pear shaped or round. The size is 1–1.4 cm × 1–1.2 cm. The whole tegument is spiny. The anterior cone is distinct. The ventral sucker is noticeably large reaching to double of the size of the oral sucker. The intestine is branched as that of *Fasciola gigantica* and *F. hepatica*, while the location of the testes is different, which is tandem, i.e., one behind another. The ovary is pretesticular and branched. The vitelline glands occupy the major part of the fluke. So far, the life cycle is not known. Clinical signs include edema (submandibular and abdominal), anemia, and hypoproteinemia.

Genus: *Fascioloides*

Etymology

This word "*Fascioloides*" has been originated from new Latin *Fasciola* + Latin oides.

Species: *Fascioloides magna*

Common name: Large American liver fluke, giant liver fluke

Distribution

The parasites are very commonly prevalent in Northern America. But these flukes are found in other parts of the United States. They are also found in South Africa, Europe, etc.

Hosts

Natural hosts: Primarily they are parasites of deer
Domestic hosts: Cattle, sheep, pig, and horse
Wild hosts: White-tailed deer, fallow deer and bison
Aberrant hosts: Yak, cattle, sheep, goat

Site

Common site: The liver is the common site.
Rare site: Lungs.

Morphology

Adult

This fluke is large, broad, and thick. The size is 10 × 2.5 cm. The fresh specimens are flesh-colored. The anterior cone is not distinct. They are roughly oval in shape. The vitellaria are restricted to the ventral region of cecal branches.

Egg

The eggs are large and distinctly operculated. Opposite to the operculum, a protoplasmic knob-like appendage is found.

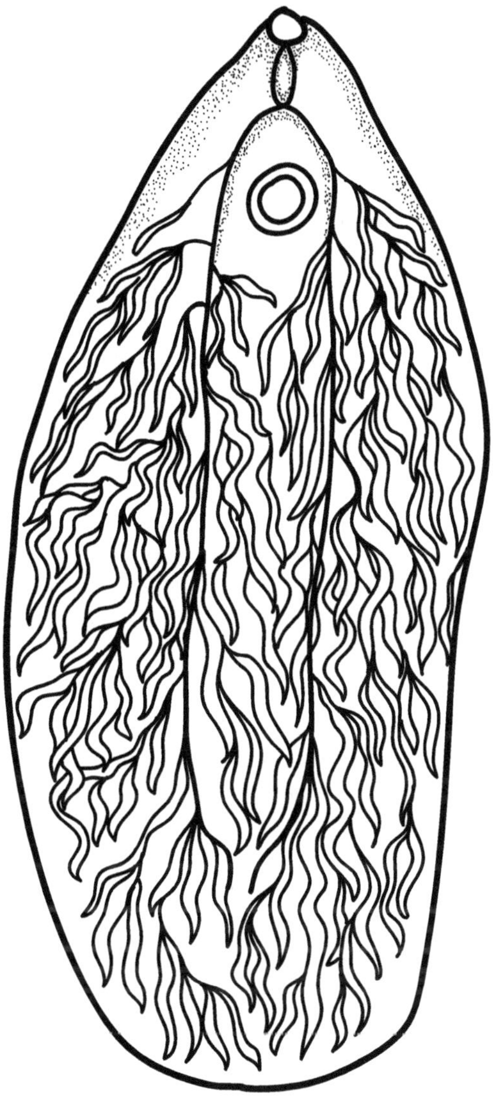

Fig. 4.7 *Fascioloides magna*

Life Cycle

The life cycle is indirect and requires one intermediate host for its accomplishment. The following events are observed during the life cycle of this parasite:

- The adult parasites live in the liver being present in the pseudocysts formed by them.

There, the adults lay eggs, which pass with the feces.
- In the aerated water, the eggs hatch. The ciliated miracidia are liberated. Hatching temperature ranges between 15 °C and 30 °C.
- The ciliated miracidia swim in the water and search for suitable snail hosts. *Lymnaea*, *Fossaria*, etc. act as the intermediate hosts. The miracidia enter the snail hosts and reach to the pulmonary sac where sporocysts are formed from the miracidia. Sporocysts are produced from miracidia. The next stage is redia, which is formed from the germ balls of sporocysts. The rediae migrate to the hepatopancreas whereby they start further development. Daughter rediae are produced from the germ balls of rediae. The cercariae are produced from daughter rediae. The cercariae are liberated from the snail after complete maturity.
- The cercariae encyst on grass blades or other vegetations and form infective metacercariae.
- The definitive hosts are infected when they ingest the grass or other vegetations on which the metacercariae are available.

Pathogenesis
The flukes remain encapsulated in fibrous cysts. Since the flukes are restricted in the cysts formed by the host's tissue, the pathogenicity is less. The cyst may be calcified later on.

Diagnosis

Symptomatological Diagnosis
Symptomatological diagnosis is not possible because no specific diagnostic symptoms are exhibited by the affected animals.

Parasitological Diagnosis
Examination of feces reveals eggs of *Fascioloides*. But it is not a confirmative diagnosis for identification of the parasite up to species level. Moreover, the eggs are rarely found in the feces as the flukes remain encapsulated in the liver.

Postmortem Diagnosis
Confirmatory diagnosis is possible during postmortem because the intact fresh worm can be identified easily. Or the flukes can be stained following standard protocol.

Immunological Diagnosis
Novobilský et al. (2007) demonstrated that the species-specific proteins 40, 120 kDa from FmESP (*Fascioloides magna* excretory-secretory product), and 80, 160 kDa from FhESP (*Fasciola hepatica* excretory-secretory product) was detected, by two-dimensional electrophoresis and immunoblot. No cross reaction was found. Those proteins could be potential immunodiagnostic markers.

Treatment
Rafoxanide, oxyclozanide, and albendazole are highly effective.

Genus: *Fasciolopsis*

Introduction
These are important flukes occurring in man and pig. Though the parasites are under the family Fasciolidae, they occur in the intestine, not in the liver. The parasite has zoonotic importance.

Species: *Fasciolopsis buski*

Etymology
The word *Fasciolopsis* has been originated from new Latin *Fasciola* + opsis (taxonomic suffix).
Common name: Giant intestinal fluke

Distribution
The parasites are commonly distributed in Asia.
Site: Small intestine
Hosts: Man, pig, and dog

Morphology

Adult
Size: 30–75 mm × 8–20 mm.
Shape: Oval and elongated.
Anterior cone/shoulder: Anterior cone and shoulder are absent.

Sucker: Ventral sucker is too large than the oral sucker and situated near the anterior extremity.

Intestinal ceca: Intestinal ceca is not branched and reach almost to the posterior end.

Pharynx: Small.

Esophagus: Short.

Testes: Highly branched and tandem in position located at the posterior half of the body.

Ovary: Pretesticular in position and branched.

Genital pore: Similar to *Fasciola*, genital opening is in between the intestinal bifurcation and ventral sucker.

Egg

The eggs are brown-colored and operculated. The size is 125–140 μm × 70–90 μm.

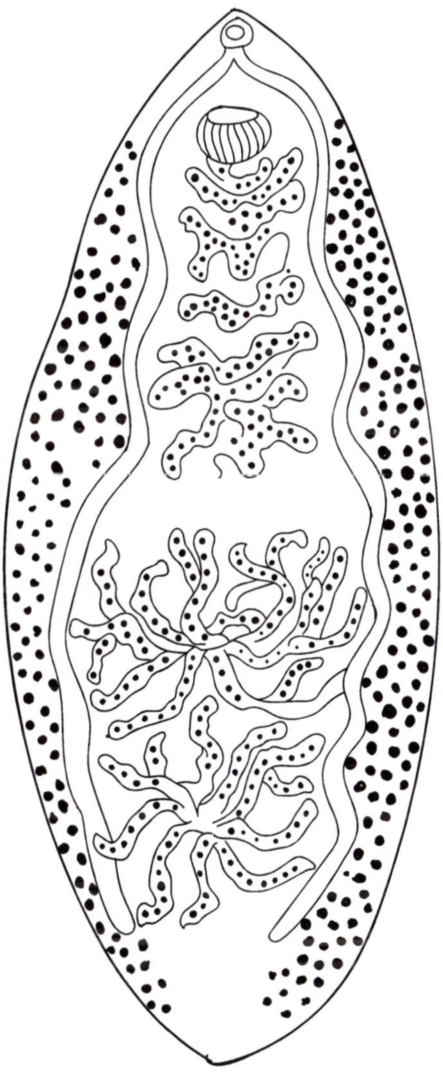

Fig. 4.8 *Fasciolopsis buski*

Fig. 4.9 *Fasciolopsis buski (Gross)*

Description of *Fasciolopsis buski* and *Fascioloides magna*

Fasciolopsis buski	*Fascioloides magna*
Common name: The common name is giant intestinal fluke	**Common name**: The common name is giant liver fluke and American liver fluke
Distribution: The flukes are mostly prevalent in East and Southern Asia	**Distribution**: The flukes are mostly prevalent in North America and Europe
Site: The flukes occur in the small intestine	**Site**: The predilection site is the liver
Morphology (a) The ventral sucker has a peculiar posterior cup-like evagination (b) The intestine is branched	**Morphology** (a) The ventral sucker has no posterior cup-like evagination (b) The intestine is not branched

Fasciolopsis buski	*Fascioloides magna*
Life cycle (a) Intermediate hosts are *Segmentina* and *Planorbis* spp. (b) The stages of life cycle are egg, miracidium, sporocyst, redia, cercaria, and adult (c) Since the parasites reside in the intestine, the eggs are laid in the intestine and are passed with the feces (d) The eggs undergo embryonation in the environment (e) The eggs hatch in the water (f) After hatching, miracidia come out (g) Miracidia enter the snail intermediate hosts (h) The parasites undergo three developmental stages, i.e., sporocyst, redia, and cercaria, in snail (i) The cercariae are released from the snail (j) The cercariae encyst on aquatic vegetation, i.e., water caltrop, lotus, etc. (k) The final hosts get the infection when the definitive hosts ingest aquatic vegetation (l) The metacercariae get excysted in the intestine and matured there	**Life cycle** (a) The intermediate hosts are *Fossaria*, *Stagnicola*, and *Lymnaea* spp. (b) The stages of life cycle are egg, miracidium, sporocyst, redia, daughter redia cercaria, and adult (c) The eggs are laid in the capsule formed by the parasite in the liver. The capsules are connected with the bile duct by connecting channel, and the eggs pass out with the feces (d) The eggs undergo embryonation in the environment (e) The eggs hatch in the water (f) After hatching, miracidia come out (g) Miracidia enter the snail intermediate hosts (h) The parasites undergo three developmental stages, i.e., sporocyst, redia, and cercaria, in snail (i) The cercariae are released from the snail (j) The cercariae encyst on aquatic vegetation (k) The final hosts get the infection by ingestion of aquatic vegetation (l) The metacercariae get excysted, and the juvenile flukes migrate through the abdominal cavity and subsequently penetrate the liver capsule and enter the tissue of the liver and remain there forming capsule

Pathogenesis

Pathogenesis caused by *Fasciolopsis buski* has been demonstrated in human beings only. They affect the intestine where they cause local inflammation with infiltration of inflammatory cells. Ulcer may be caused due to high load of infection. Symptoms in pig lack any clear report.

Symptoms

The affected persons exhibit clinical signs comprising abdominal pain, diarrhea, edema, ascites, etc. Symptoms in pig lacks any clear report.

Diagnosis

Symptomatological Diagnosis

Symptomatological diagnosis is not possible because no specific diagnostic symptoms are exhibited.

Parasitological Diagnosis

Examination of feces reveals eggs of the parasite.

Immunological Diagnosis

Study shows antigen-antibody reaction by complement fixation test using sera and extract of adult worm.

Treatment

Praziquantel 15–20 mg/kg (in pig).

Control

Eating raw and uncooked aquatic plants should be avoided.

Genus: *Parafasciolopsis*

Species name	*Parafasciolopsis fasciolaemorpha*
Hosts	Elk and wild goats. There have been reports of this fluke infection in moose
Site	Gallbladder and digestive system. There have been reports of infection in the liver of moose
Morphology	This fluke unlike other fasciolid members is too small ranging few mm only. The ventral sucker is roughly double of the oral sucker
Pathogenesis	In moose, septal fibrosis and liver cirrhosis have been reported
Diagnosis	Examination of feces reveals brownish eggs

Family: Dicrocoelidae

Introduction

Dicrocoelidae is undoubtedly an important family. The members of this family are small- to medium-sized worms; some are lancet shaped. They affect mammals, birds, reptiles, etc. Though the severity of pathogenesis is not comparable to the *Fasciola* parasites, *Dicrocoelium dendriticum* is important for causing marked loss of morbidity of sheep and goats leading to significant loss of economy. Dicrocoelidae members are characterized by flattened body, visible internal organs from outside, presence of pharynx and esophagus, closely located suckers, presence of testes, posterior to ventral sucker, intestinal ceca not reaching up to the posterior end, and uterus in the hind part. They are heteroxenous parasites and involve two intermediate hosts, i.e., land snail (first) and ants (second). Developmental stages include egg, miracidium, sporocyst, daughter sporocyst, and cercaria. Redia stage is absent. Diagnosis can be done by examination of feces. The control of the parasites mostly relies on proper treatment and control of land snail and ants.

History

Rudolphi discovered *D. dendriticum* in 1819. Loos discovered *D. hospes* in 1899. Krull and Mapes described the life cycle during 1951–1953.

Genera

- *Dicrocoelium*
- *Platynosomum*
- *Eurytrema*

Genus: *Dicrocoelium*

Etymology

The word *Dicrocoelium* has been derived from the Greek dikroos meaning twofold branching and koilia meaning body cavity.

Introduction

These are important flukes occurring in the bile duct, and gallbladder of the sheep, goat, etc. They also affect pig, deer, dog, etc. They are less pathogenic compared to *Fasciola* parasites.

Species

Dicrocoelium dendriticum
Dicrocoelium hospes

Species: *Dicrocoelium dendriticum*

Predilection Site

Normal site	Rare site
Liver	Adult flukes may aberrantly be found in lungs
Bile duct	
Gallbladder	

Hosts

Primary hosts	Natural hosts	Sheep, goat, pig, ox, donkey, dog, hare, rabbit, guinea pig, etc.
	Unnatural hosts	Human beings are accidental definitive hosts of *Dicrocoelium dendriticum*
	Domestic hosts	Sheep, goat, cattle, pig, ox, donkey, dog, hare, rabbit, guinea pig
	Wild hosts	Lagomorph, rodents, elk, coypu, etc.
	Laboratory hosts	Hamsters are good laboratory hosts
Secondary hosts	First intermediate hosts	*Cionella*, *Zebrina*, *Theba*, *Ena*, *Helicella*, *Planorbis*, *Arion*, etc.
	Second intermediate hosts	Ants: *Formica fusca*, *F. cunicularia*

Common name: Lancet fluke or lanceolate fluke

Distribution

The flukes are distributed in Asia, North Africa, Europe, and North America. In India, the flukes are commonly found in the Himalayan region affecting sheep, goat, etc.

Morphology

Adult

Size: The size of the fluke is 5–12 mm × 1–2.5 mm.

Shape: They are lancet shaped. That's why these are called as lancet flukes.

Pharynx: The pharynx is small.

Suckers: Two suckers are present. One is oral and another is ventral. Oral sucker is smaller than ventral.

Testes: Two testes are present immediately behind the ventral sucker. Testes are lobed and tandem in position.

Ovary: Ovary is round in shape and post-testicular in position.

Vitelline glands: Vitelline glands are situated in the lateral field.

Uterus: Typically arranged transverse uterine coils are present in the central part of the parasite.

Eggs

The eggs are oval, dark brown, and operculated. The size of eggs is 35–45 μm long and 20–30 μm width. The eggs are fully embryonated when they are laid.

Larva

The larval stages are miracidium, sporocyst, cercaria, and metacercaria. The description of different stages is as follows:

Miracidium: The general morphological features are almost the same as other digenetic flukes. The miracidium is pear shaped. This is covered with ciliated epidermal plates. Cytoplasmic ridges are present in between the epidermal plates. These ridges play a role during transformation of miracidium to other stage. The apical papilla is present at the anterior end consisting of secretory glands. The secretory material of the secretory glands helps during penetration of the host tissue. Eye spots are found in the miracidium. The brain of the miracidium lies in the parenchymatous tissues behind the apical region, which innervate the tissues. The excretory system is a protonephridial system comprising some flame cells, which collect the excretory material in liquid form. One excretory pore is found through which the excretory material is discharged. Later, the ciliated covering is lost, and the miracidium is transformed to sporocyst.

Sporocyst: The sporocysts are a sac-like body formed by the miracidium. Apical papilla, secretory glands, nervous tissue, and eye spots are absent. Digestive system is also absent. They have brood chambers. Daughter sporocysts are formed from the sporocysts.

Cercaria: Germ balls are present in the sporocysts, which differentiate into cercariae. The cercariae remain in the respiratory chamber. They are long-tailed stylet cercariae. In the respiratory chamber, the cercariae cause irritation, which results in secretion of mucus material. The cercariae remain in mucus balls (popularly called as slime balls).

Metacercaria: The metacercariae are the encysted form of cercariae. The outer wall of the metacercariae is resistant. The length of the metacercariae is variable, which may be more or less than 1 mm, and the width is 135–250 um.

Special Note

Cifrian and Garcia-Corrales (1988) demonstrated the tegument of adult *Dicrocoelium dendriticum*. They reported ridged network of the tegument. SEM of the body surface revealed high density of tegumental vesicles on the ridge wall, and valley floors of the ridges are found. Finger-like projections are found in some areas of tegument. The tegument is devoid of spines.

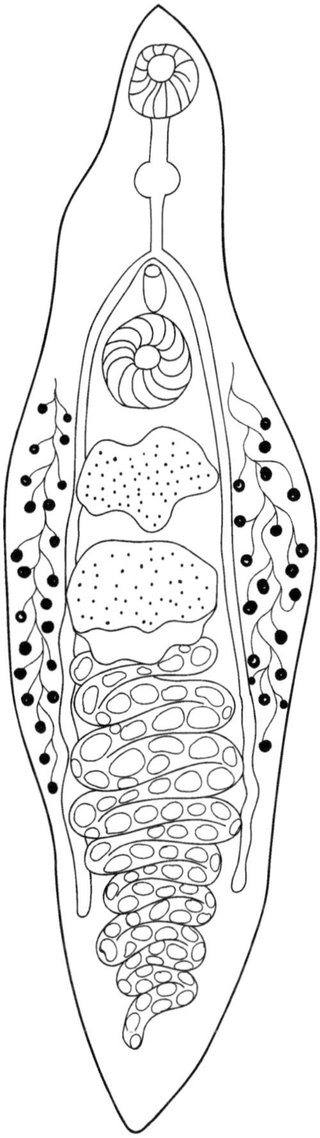

Fig. 4.10 *Dicrocoelium dendriticum*

Life Cycle

The life cycle is heteroxenous, which involves two intermediate hosts. The first intermediate hosts are land snails, and the second intermediate hosts are ants. All developmental stages are formed except redia stage. The stages are egg, miracidium, sporocyst, daughter sporocyst, cercaria, metacercaria, and adult.

Basic information
• Definitive hosts are sheep, goats, etc. • First intermediate hosts are land snails • Second intermediate hosts are ants • Mode of infection of the first intermediate host is ingestion • Mode of infection of the second intermediate host is ingestion • Mode of infection of definitive hosts is ingestion

Stages of Life Cycle Formed

In definitive host	In first intermediate host	In second intermediate host
• Adult • Egg	• Miracidium • Sporocyst • Daughter sporocyst • Cercaria	• Metacercaria

Events of life cycle: a glimpse
• Voiding of eggs with feces of definitive host • Ingestion of eggs by land snail • Hatching of eggs in the gut of snail • Formation of sporocysts, daughter sporocysts, and cercariae in snail • Release of cercariae remaining in slime balls • Ingestion of slime balls by ants • Transformation of cercariae to metacercariae in ants • Infection of definitive hosts by inadvertent ingestion of infected ants • Development of fluke up to adult stage in the definitive hosts

Detailed Description of Life Cycle

The following are different events of the life cycle discussed in detail:

Fertilization

These are hermaphrodite parasites and reside in the bile duct. Since the parasites are hermaphrodite, self-fertilization occurs.

Embryonation

Embryonation of eggs occurs within the parasite itself. The gravid fluke releases embryonated eggs.

Egg Passing

The adult parasites lay fertilized eggs in the bile duct. Bile duct is connected with the gut. These eggs are passed with the feces of the host.

Egg Hatching

It is to be noted that hatching of eggs of *Fasciola* occurs in the environment. The hatched out miracidium enters the snail hosts through the exposed soft tissues of the snail. But in case of *Dicrocoelium* spp., hatching of eggs occurs in the first intermediate hosts (land snails). The land snails ingest the embryonated eggs. In the gut of the snail, the eggs hatch.

Development in the First Intermediate Host

Different land snails (*Cionella* spp., *Zebrina* spp.) act as the intermediate hosts of the parasite. The stages that are formed in snails are sporocyst, daughter sporocyst, and cercaria. The miracidium is transformed to a new stage of development called as sporocyst. The sporocysts do not have any digestive organs. Daughter sporocysts are produced from the mother sporocyst. The next developmental stage is cercaria. The cercariae of *Dicrocoelium* spp. are xiphidocercaria type. They have long tails. The cercariae are not released freely as is found in other digenetic trematodes. In the case of *Dicrocoelium*, the cercariae are released being present in slime ball (a cyst-like structure). About 125–400 cercariae remain in a slime ball. Rainfall after a dry spell causes release of more slime balls.

Development in the Second Intermediate Host

The second intermediate hosts are different species of ants (*Formica fusca*, *F. cunicularia*). In the ants, metacercariae are formed in the abdominal cavity. Some metacercariae are formed in the esophageal nerve ganglion.

Infection

The final hosts get the infection by ingestion of infected ant.

Development in Definitive Hosts

The metacercariae get excysted in the small intestine of the definitive hosts. The juvenile flukes come out from the metacercarial cysts and penetrate the wall of the GI tract. They ultimately reach to the bile duct and liver via hepatoportal circulation and migrate in the liver tissues. They then reach to the bile duct where they get sexually matured.

Note

Numerous cercariae are developed from the second generation of sporocysts. They migrate from the hepatopancreas to pulmonary cavity of snail. This movement is mediated by tail, stylet, and enzyme. Here the cercariae are clustered and enveloped by mucilaginous slime ball. One slime ball contains many cercariae.

The ants Formica fusca, F. cunicularia, F. pratensis, and F. rufibarbis act as second intermediate hosts. After ingestion, cercariae (1–2 numbers) are called as "brain worm," which affects the brain of ants, and affected ants are compelled to climb on the top of vegetation. This event occurs every day until the ant is ingested. As the nervous system is affected by these brain worms, the movement of ant is stopped, and they remain on the top of herbage and grass and readily ingested by the animals.

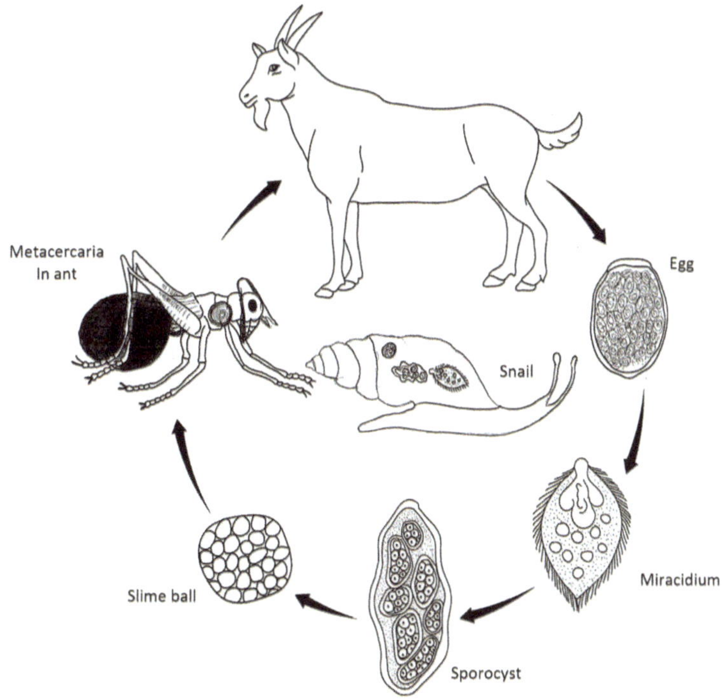

Fig. 4.11 Life cycle of *Dicrocoelium dendriticum*

Pathogenesis and Pathology

Pathogenesis

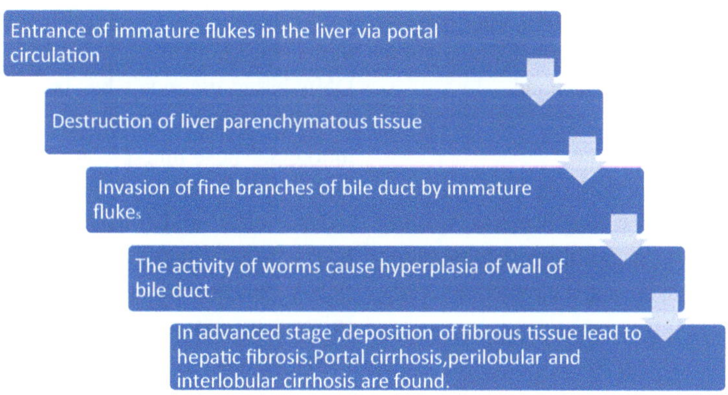

Fig. 4.12 Flow chart of pathogenesis caused by *Dicrocoelium dendriticum*

Pathology

Anatomical Pathology

The migratory flukes cause damage to the bile duct where they take blood as food and get matured. In a long-standing case, fibrosis occurs. Distension of bile duct is another pathological feature.

Hematological Pathology

Very little amount of pathology is caused due to dicrocoeliosis. However, in high load of infection, anemia has been reported.

Chemical Pathology

In high load of infection, loss of plasma protein has also been reported.

Secondary Pathology
There have been records of mixed infection of *Dicrocoelium* and *Fasciola*. *Dicrocoelium* sp. never causes significant pathology in animals. But remarkable pathology has been recorded by mixed infection with *Fasciola* parasites.

Immunopathology
The immature parasites cause inflammation of the affected area, which appears as a primary pathological feature. Inflammation occurs presumably due to pro-inflammatory cytokines secreted by macrophage. It could be presumed that other immunological phenomena, i.e., ADCC (antibody-dependent cellular cytotoxicity) and NO-mediated lysis, also occur. Different studies indicate that NO also causes damage to the tissue. In this trematode infection, hyperplasia also occurs. The pro-inflammatory cytokines might have a role for the occurrence of hyperplasia.

Molecular Pathology
Recent study has pinpointed toward different genes that express different protein molecules resulting in direct or indirect pathology. The literature is scanty on the study of molecular pathology of dicrocoeliosis. Study of multi-transcriptome may reveal upregulation of fibrosis-related gene. Upregulation of TGF-beta gene has also been found.

Clinical Signs
Clinical signs are not obvious. Colic pain, indigestion, diarrhea, etc. are some clinical features. Loss of body condition along with dullness, depression, and loss of body weight are some clinical features.

Diagnosis

Diagnosis of *Dicrocoelium*: a glimpse
• Examination of fecal sample for detection of eggs • Efforts made for the development of immunodiagnostic test using crude and defined antigens (e.g., 130 kDa) • Molecular diagnosis by different molecular markers

Diagnosis in Detail

Symptomatological Diagnosis
Diagnosis based on exhibited clinical signs is not possible. In fact, the infection does not cause marked pathogenesis. The symptoms displayed are vague and provide no clue for any diagnostic conclusion.

Parasitological Diagnosis
Examination of feces may reveal the presence of typical eggs. The eggs are smaller as compared to *Fasciola*. The eggs are dark brown colored and operculated.

Radiological Diagnosis
Literatures are almost silent about radiological diagnosis. However, some fragmented work shows the use of ultrasonography. Ultrasonography may reveal the areas of calcification of bile duct, which might provide some hints for *Dicrocoelium* infection.

Immunodiagnosis
Different crude and defined antigens have been tried for immunodiagnosis of *Dicrocoelium* infection. Study shows that 84 and 130 kDa of somatic antigen and 130 kDa of excretory and secretory protein have immunodiagnostic possibility. The immunodiagnostic tests that could be performed are indirect fluorescent antibody test, complement fixation test, ELISA, etc.

Molecular Diagnosis
DNA-based diagnosis has been a recent trend to identify parasite by using different molecular markers. These markers are mainly of ribosomal DNA and RNA (ITS1, ITS2, 5.8 S rRNA, 18 S rRNA, etc.) and mitochondrial DNA (CO1, ND1). Study shows the use of ITS2 and 18 S rRNA for the identification of *Dicrocoelium*. Primers are constructed for carrying out PCR. The PCR product is used for sequencing. The sequence is compared with reference sequence retrieved from NCBI database for confirmation. Species-specific primer of *D. dendriticum* is available.

Treatment

Name of drug	Route of administration	Dose
Fenbendazole	Oral	150 mg/kg bw
Albendazole	Oral	15 mg/kg bw
Cambendazole	Oral	25 mg/kg bw
Praziquantel	Oral	50 mg/kg bw

Control

General Control
Affected animals should be treated properly. It has been advised that land snails can be controlled by duck rearing. But practical implication of this technique is under question mark. It is also advised that snail can be controlled by using molluscicidal agents, i.e., copper sulfate, sodium pentachlorophenate, niclosamide, etc. But practical implication of this technique is still under question; because land snails remain in a vast area. Methaldehyde is a good molluscicidal agent and can be used with bran as bait. Therefore, it is practically very difficult to control land snail covering a large area using molluscicidal agents. Moreover, the molluscicidal agents are pollutants for the environment. Since the ants act as second intermediate hosts, they should be controlled by using insecticidal agent.

Immunological Control
Several attempts were made by different workers to develop vaccine against dicrocoeliosis. Somatic antigen of whole parasite and excretory-secretory proteins were subject to immunoprotection study. But no vaccine has been developed. Work on use of defined antigen was reported. Study shows immunoprotective potential of cathepsin protein against *Dicrocoelium dendriticum*. The feces should be properly disposed as the feces are the source of eggs.

Biological Control
Duck rearing can be done as a method of biological control as the snails are the food of birds. But this is possible in the case of organized farm only.

Species: *Dicrocoelium hospes*
Hosts: Ox
Site: Gallbladder

Distribution
The parasites are predominantly found in sub-Saharan West Africa. They are distributed in Senegal, Ghana, Mali, etc.

Life Cycle
The stages of life cycle are egg, miracidium, mother sporocyst, daughter sporocyst, cercaria, metacercaria, and adult. *Limicolaria* spp. act as 1st intermediate hosts. Ant (*Dolyrus* spp.) act as 2nd intermediate hosts.

Genus: *Platynosomum*

Etymology
The word *Platynosomum* has been derived from Greek platys (flat), nosos (disease), and soma (body).

Species: *Platynosomum fastosum*
Common name: Cat liver fluke
Disease caused: Lizard poisoning
Hosts: This is small (few mm) trematode that affects domestic and wild cat.
Site: Liver and bile duct.

Distribution
The flukes are distributed in Central and South America, West Africa, etc.

Morphology

Adult
The flukes are small in size ranging from 4 to 8 mm. The testes are typically horizontal in position. These testes are situated within the first half of the body. Both oral and ventral suckers are almost the same in size. As a rule, the vitelline glands are present on either side of the fluke.

Egg
The eggs are brown colored and operculated. They are oval in shape.

Life Cycle

There are two intermediate hosts of this parasite. The first intermediate host is aquatic snail (*Sublima octona*). Crustacean, lizard, etc. act as second intermediate hosts.

Some important information of life cycle
• Domestic and wild cat are definitive hosts • Aquatic snails are the first intermediate hosts • Lizards and crustaceans are the second intermediate hosts • Mode of infection of the first intermediate hosts is ingestion • Mode of infection of the second intermediate hosts is ingestion or penetration • Mode of infection of definitive hosts is also ingestion

Stages of Life Cycle Formed

In definitive host	In first intermediate host	In second intermediate host
• Adult • Egg	• Miracidium • Sporocyst • Cercaria	• Metacercaria

Events of life cycle: a glimpse
• Voiding of eggs with feces of definitive host • Ingestion of eggs by snails (*Sublima octona*) • Hatching of eggs in the gut of the snails • Release of miracidium from the eggs • Formation of sporocysts and cercariae in snail • Release of sporocysts containing cercariae from the snails • Second intermediate hosts (lizard, isopod crustacean) ingest the sporocysts containing the cercariae • Transformation of cercariae to metacercariae in the second intermediate hosts • Ingestion of the second intermediate hosts harboring metacercariae by definitive hosts • Further development in definitive hosts

Pathogenesis

The pathogenesis includes the common features comprising damage and lesions of parenchymatous tissues of the liver. There occurs denudation of bile duct epithelium. The bile duct may be dilated.

Clinical Signs

The clinical signs include diarrhea, vomiting, etc. The affected animal may die. This is commonly called as *lizard poisoning* as the animals suffer due to ingestion of infected lizard.

Diagnosis

Examination of feces can be done for detection of eggs. Necropsy finding for the presence of flukes in the bile and pancreatic duct is the most reliable diagnosis. Ramos et al. (2016) reported the use of FLOTAC technique for the detection of eggs of *Platynosomum fastosum* first time.

Treatment

Name of drug	Route of administration	Dose
Praziquantel	Oral	20 mg/kg

Genus: *Eurytrema*

Introduction

These are important flukes occurring in the pancreatic duct of sheep, goat, and cattle. Occasionally, the parasites may occur in bile duct and duodenum.

Etymology

The generic name "*Eurytrema*" was derived from two words, i.e., eurys (Greek) and trema (Ancient Greek). "Eurys" means wide and "trema" means hole.

Species: *Eurytrema pancreaticum*

Common name: Pancreatic fluke

Hosts: Sheep, goat, cattle, etc. are affected by this parasite.

Predilection Site

Common site: Pancreas and pancreatic duct are the common sites.

Rare site: Bile duct and duodenum are rare sites of occurrence.

Distribution: The flukes are predominantly distributed in Eastern Asia, Brazil, South America.

Morphology

Adult

The size of the fluke is 8–16 × 5–8.5 mm. They are flat and almost oval. Two suckers are present, of which the oral sucker is larger compared to ventral sucker. Pharynx and esophagus are short. The testes are horizontal in position. The testes are located just behind the level of ventral sucker. Ovary is post-testicular lying near the midline. The convoluted tubules are present mostly behind the level of testes and reach up to the posterior end.

Egg

The eggs are small in size and dark brown in color.

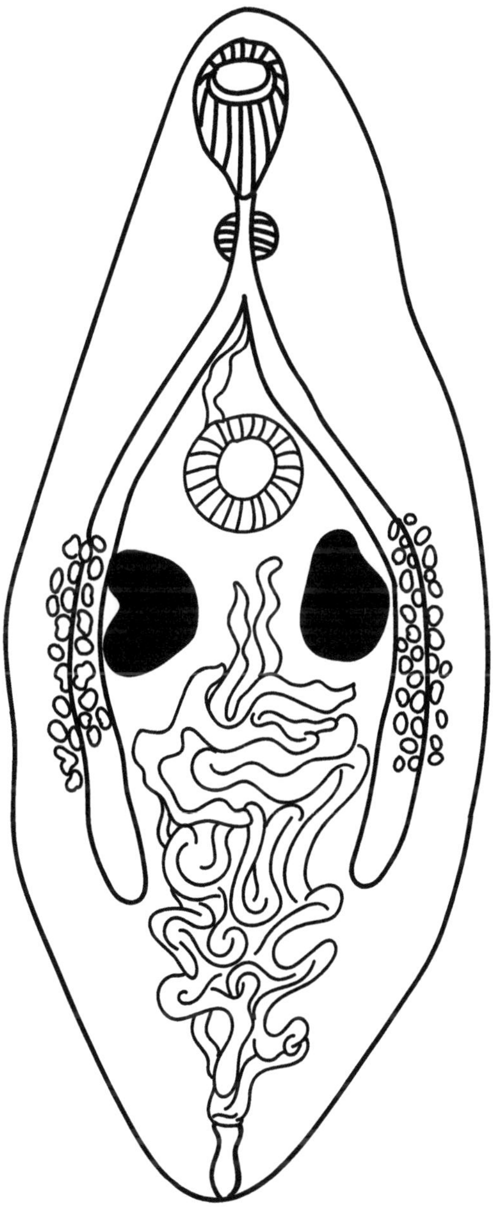

Fig. 4.13 *Eurytrema pancreaticum*

Life Cycle

Some basic information
• Definitive hosts are sheep, goat, cattle, buffalo, etc. • First intermediate hosts are *Bradybaena* spp. and *Cathaica* spp. • Second intermediate hosts are grasshoppers (*Conocephalus maculatus*) and tree cricket (*Oecanthus longicauda*) • The infective stage of the parasite is metacercaria • Mode of infection is ingestion of infected second intermediate hosts carrying metacercaria stages of the parasite by the definitive hosts

Stages of life Cycle Formed

In definitive host	In first intermediate host	In second intermediate host
• Adult • Egg	• Miracidium • Sporocyst • Daughter sporocyst • Cercaria	• Metacercaria

Quick look: events of life cycle
• Voiding of eggs with feces of definitive host • Ingestion of eggs by the first intermediate hosts • Hatching of eggs in the gut of the snails • Release of miracidia from the eggs • Formation of two generations of sporocysts and cercariae in snail • Cercariae are expelled by the snails • Ingestion of cercariae by second intermediate hosts or cercariae may penetrate second intermediate hosts • Transformation of cercariae to metacercariae in the second intermediate hosts • Infection of final hosts by ingestion of infected second intermediate hosts

Brief Description of Life Cycle

The life cycle is heteroxenous. Two intermediate hosts are required for completion of its life cycle. The first intermediate hosts are land snails, and the second intermediate hosts are grasshoppers. The eggs are passed with the feces of the hosts. The eggs hatch after being ingested by snail. Sporocysts, daughter sporocysts, and cercariae stages are produced in the snail host. As that of *Dicrocoelium* spp., redia stage is not produced. The cercariae are **microcercus** type having short tail. They encyst to metacercariae in the grasshopper. The final hosts get the infection by ingestion of infected grasshopper and other intermediate hosts inadvertently.

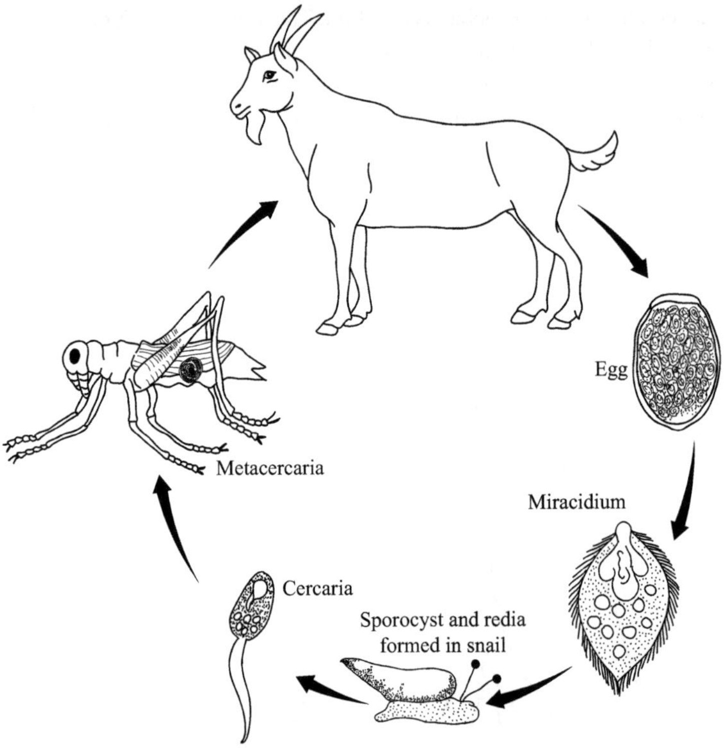

Fig. 4.14 Life cycle of *Eurytrema pancreaticum*

Pathogenesis

Acute Stage

In the acute stage, catarrhal inflammation with infiltration of cells occurs in predilection site. Erosion of duct epithelial cells occurs.

Chronic Stage

In the chronic stage of infection, the general inflammatory condition leads to deposition of fibrous tissue in the duct wall causing marked cholecystitis. There occurs occlusion and distension of pancreatic ducts. There is less or no production of digestive juice due to destruction of the duct wall.

Symptoms

Though different pathological features have been ascribed, no pathology-related symptoms are exhibited. Only the affected animals show poor body condition.

Diagnosis

Examination of feces reveals presence of eggs.

Clinical Signs

They cause digestive disturbances with clinical signs of diarrhea, constipation, nausea, abdominal pain, etc.

Treatment

Name of drug	Route of administration	Dose
Praziquantel	Oral	20 mg/kg bw for two consecutive days
Albendazole	Oral	15 mg/kg bw
Oxyclozanide	Oral	Two doses of 18.7 mg/kg bw at 2 days interval for cattle and buffalo

Species: *Eurytrema coelomaticum*

Common name: Pancreatic fluke.
Site: Pancreatic duct, bile duct, and liver.
Hosts: Cattle, sheep, goat, pig, man, etc.
Distribution: The flukes are distributed in Asia, Europe, and South America.

Morphology

They are 7.5–10 mm long. Two suckers (oral and ventral) are present, the oral being slightly larger as compared to ventral sucker. The pair of testes are slightly lobed. The ovary is also slightly lobed. The vitelline follicles are arranged in 6–8 groups. Behind the intestinal bifurcation, the genital pore opens.

Life Cycle

Life cycle is similar to that of other species of this family. Two intermediate hosts are required. Land snail (*Bradybaena* sp.) serves as the first intermediate host, and grasshopper of the genus *Conocephalus* or tree crickets serve as the second intermediate hosts in which the infective metacercariae are formed in about 3 weeks. The final hosts get the infection by ingestion of the infected second intermediate hosts.

Eurytrema dajii

These flukes occur in the liver and pancreas of cattle. They may also occur in bile duct, gall bladder, and intestine. Two suckers are present, the ventral being slightly larger than oral sucker. The pair of testes are oval and slightly lobed.

Genus: *Concinum*

Concinnum procyonis

The flukes occur in pancreatic duct, bile duct, and gallbladder of cat, fox, etc.

Family: Opisthorchiidae

Family Synopsis

The members affect the mammals, birds, reptiles, etc. They are small to medium sized occurring in the bile duct, gallbladder, etc. Opisthorchiidae members are characterized by flattened and translucent body, presence of esophagus and pharynx, intestinal ceca up to the posterior end, testes posterior in location and diagonal in position, and excretory bladder stem-like and equipped with short branches. Ovary is pre-testicular. They are heteroxenous parasites involving two intermediate hosts (first aquatic snail and second cyprinid fish). The developmental stages are egg, miracidium, sporocyst, redia, and cercaria. The parasites do not cause marked pathology except catarrhal inflammation in the wall of the bile duct. However, if heavy infection or long-term infection prevails, marked pathology is found in the affected area. Diagnosis can be done by finding the eggs in the feces. Control mostly can be done by proper treatment and avoiding uncooked fish.

History

Opisthorchis viverrini was first discovered by Arthur Looss in 1910. The parasite was isolated from cat in Thiland. *Clonorchis sinensis* was first recorded by McConnell in the year 1875. The parasite was detected in the bile duct of a Chinese carpenter.

Important Genera

- *Opisthorchis*
- *Clonorchis*
- *Pseudoamphistomum*
- *Metorchis*
- *Parametorchis*

Genus: *Opisthorchis*

Introduction

These are important flukes occurring in the bile duct of dog, cat, etc. The parasites have zoonotic importance.

Etymology

The generic name "*Opisthorchis*" is a Greek word derived from two words, i.e., "opisthen" and "orchis." "Opisthen" means "behind" and "orchis" means testicle.

Species

Opisthorchis tenuicollis
Opisthorchis viverrini

Distribution

O. tenuicollis is distributed in different parts of Asia. The flukes are also recorded in other parts of the world. *O. viverrini* are found in South East Asia.

Predilection Site

Normal site	Rare site
Bile duct	Intestine Pancreatic duct

Hosts

Primary hosts	Natural hosts	Dog, cat, pig, etc.
	Domestic hosts	Dog, cat, pig
	Wild hosts	Civet, wild cat
	Laboratory hosts	Hamsters are good laboratory hosts
Secondary hosts	First intermediate hosts	*Bithynia leachii*, *B. inflata*, etc.
	Second intermediate hosts	Cyprinid fish: *Tinca tinca*, *Abramis brama*, *Idus sp.*, *Barbus barbus*, *Leuciscus*, *Scardinius*, *Blicca etc.*

Morphology

Adult

Size: 5–12 mm × 1.5–3 mm.
Color: Fresh specimens are reddish colored.
Shape: Elongated oval, somewhat resembling a lance.
Cuticle: Smooth.
Pharynx: Small.
Esophagus: Small.
Intestinal ceca: The ceca terminate to the posterior end.
Suckers: Two suckers are present. One is oral and another is ventral. Oral sucker and ventral suckers are of the same size approximately.
Testes: Two lobed testes are present. Testes are oblique in position. These are located in the posterior region. Excretory bladder runs in between the testes.
Ovary: Approximately in the middle part, the ovary is situated. The ovary is pretesticular.
Vitelline glands: Vitelline glands are situated in the lateral field forming different clusters.
Uterine coils: Uterine coils are situated in the anterior part of the parasite, which spread from the ovary to the ventral suckers.

Egg

The eggs are oval, yellowish, or light brown and operculated. The embryo appears as a homogeneous mass. The size is 0.026–0.03 mm × 0.011–0.015 mm. The miracidium can be detected in the egg when laid.

Larva

The larval stages are miracidium, sporocyst, redia, cercaria, and metacercaria.

Miracidium: The miracidium is pear shaped. The outer covering is ciliated as of other digenetic flukes. Like miracidium of other flukes, eye spots are present. Unlike cercariae, the digestive system is absent in miracidium. A protonephridial system is present, which performs the function of excretion. Other structures present in the miracidium are apical papilla containing secretory glands.
Sporocyst: The next stage is sporocyst. The miracidium casts off the ciliated layer and is trans-

formed to a sac-like body called as sporocyst. The sporocyst has outer thin cuticular covering. A protonephridial system represents the excretory system comprising flame cells, excretory pores, etc. Eye spots are absent. Papilla and secretory glands are absent.

Redia: The germ balls present in the brood chambers of sporocysts give rise to rediae. Ambulatory buds projected antero- and posterolaterally provide a typical shape of redia. As a rule, redia has digestive system comprising the mouth, pharynx, unbranched ceca, etc. Excretory system is also found. Adam et al. (1995) reported that the tegument of redia is concentrically folded with dense arrangement of microvilli. The tegument is built up by typical syncytium with submerged tegumental cells.

Cercaria: Germ balls present in the brood chamber of rediae differentiate into cercariae. Both oral and ventral suckers are clearly visible. Eye spots, penetration glands, and excretory bladder are present. Dorsal and ventral fin folds of tail are detected. Adam et al. (1995) reported that the cercariae of *O. viverrini* possess two well-pigmented eye spots having rhabdomeric photoreceptors in the anterior part of body. They also reported that the cercariae of *O. viverrini* possess an extended protonephridial system, which is connected with a voluminous excretory bladder. The cercaria of *O. tenuicollis* is pleurolophocercous and oculate. Ten pairs of penetration gland cells are found. The flame cell formula is 2[(5)+(5+5+5+5)]=50.

Metacercaria: A cyst is formed from the cercaria. All species have similar morphological features. They are motile. The shape of the metacercariae is oval. Metacercarial body folds are observed as that of other trematodes. Both oral and ventral suckers are conspicuously visible. Ceca and excretory bladder are clearly visible and metacercaria can be identified.

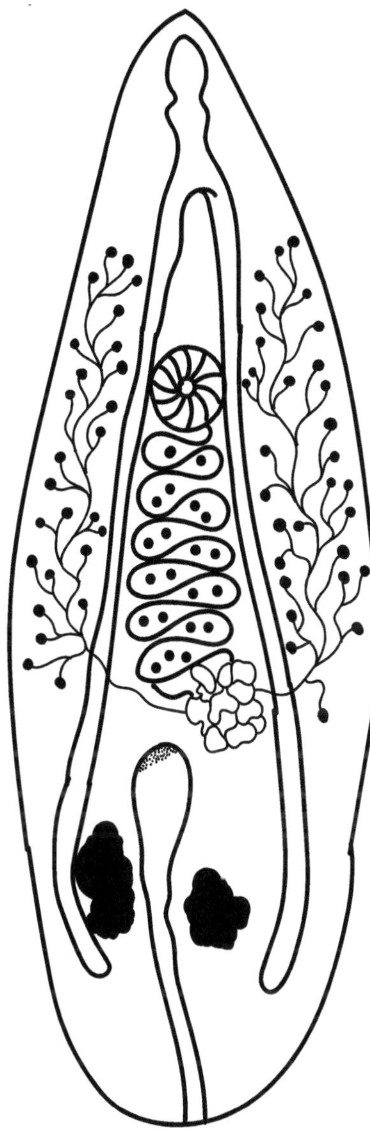

Fig. 4.15 *Opisthorchis tenuicollis*

Life Cycle

Some basic information of life cycle
• Definitive hosts are dog, cat, fox, and pig • First intermediate hosts are aquatic snails • Second intermediate hosts are cyprinid fish • The infective stage of the parasite is metacercariae • The mode of infection is ingestion of second intermediate host carrying metacercariae

Stages of Life Cycle Formed

In definitive host	In first intermediate	In second intermediate host
• Adult • Egg	• Miracidium • Sporocysts • Redia • Cercaria	• Metacercaria

Events of life cycle of *Opisthorchis*: a glimpse
• Passing of eggs with feces of definitive host • Ingestion of eggs by snail (first intermediate hosts) • Hatching of eggs in the gut of snail • Formation of sporocysts, rediae, and cercariae in snail • Release of cercariae • Entrance of cercariae in cyprinid fish • Transformation of cercariae to metacercariae in cyprinid fish • Ingestion of infected fish by definitive hosts • Development up to adult in definitive hosts

Life Cycle in Detail

The following are different events of the life cycle:

Fertilization

The seat of predilection of the parasite is the bile duct. The parasites are hermaphrodite. That means both completely functional male and female reproductive systems are present in the same individual. The adults self-fertilize in their predilection site.

Embryonation

Embryonation of eggs occurs within the parasite itself. The gravid parasites lay embryonated eggs.

Egg Passing

Embryonated eggs are passed with feces of the definitive hosts since the bile duct is connected with the intestine.

Egg Hatching

Hatching of eggs occurs after being ingested by snails (first intermediate host). Hatching does not occur in the environment as is found in the case of *Fasciola*, amphistomes, etc.

Development in Intermediate Host

Aquatic snails (*Bithynia leachii*, *B. infata*) act as the first intermediate hosts of the parasite. The stages that are formed in snails are sporocyst, redia, and cercaria. In some of the digenetic flukes, redia stage is not formed. But in the case of this parasite, redia stage is formed. The miracidium is transformed to a new stage of development called as sporocysts. The sporocysts do not have any digestive organs. A generation of rediae is produced from the sporocysts. Approximately 2 months is required from infection for formation of rediae. The rediae have digestive organs and survive by taking snail tissues. The next development stage is cercariae. They are equipped with tail by which cercariae can swim in water. The cercariae infect some fish (Cyprinidae members). They penetrate the skin of the fish and encyst as metacercariae there after shedding their tail appendages. The genera of fish are *Tinca*, *Idus*, *Blicca*, *Abramis*, etc. The metacercariae are quiescent form infective stage having tough outer wall. About 6 weeks is required for formation of metacercariae in fish muscles.

Infection of Definitive Host

The final hosts get the infection by ingestion of fish infected with metacercariae.

Development in Definitive Host

The metacercariae get excysted in the duodenum. The juvenile flukes are liberated from the metacercarial cysts and penetrate the wall of the GI tract. They migrate through the ampulla of Vater and then reach to the bile duct. They get sexually matured there within about 3–4 weeks.

Special Note

There are different common names of fish that transmit *Opisthorchis*. These are **tench**, **nase**, **common rudd**, **minnow**, **dace**, etc.

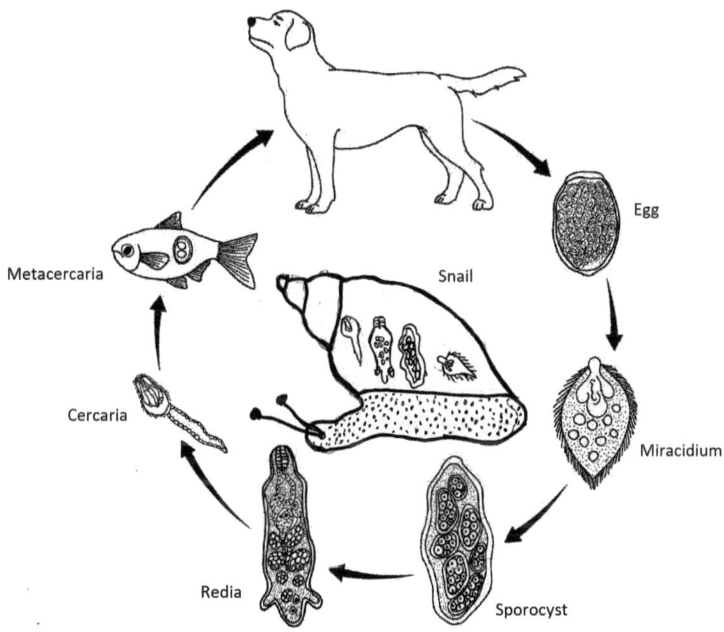

Fig. 4.16 Life cycle of *Opisthorchis tenuicollis*

Pathogenesis and Pathology

Cholangitis, catarrhal cholecystitis, and cholangiohepatitis are common pathological features found.

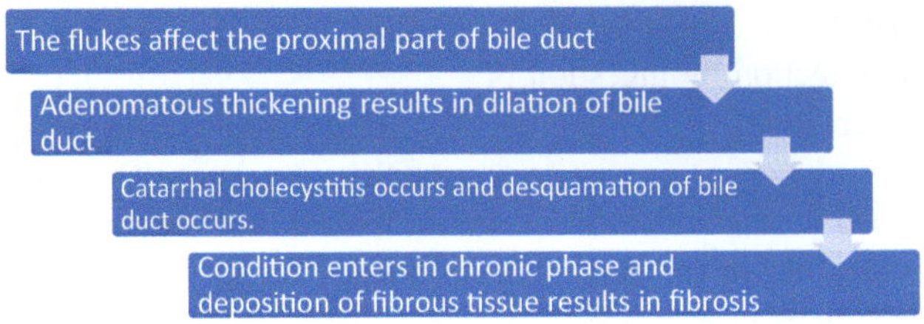

Fig. 4.17 Flow chart of pathogenesis caused by *Opisthorchis tenuicollis*

Pathology

Anatomical Pathology (Gross)
The parasites occur in the bile duct, pancreatic duct, and gallbladder. Inflammation and hyperplasia of bile duct epithelium commonly occur in the initial stage of pathology. In a long-standing case, fibrosis occurs resulting in portal cirrhosis. Distension of the bile duct and enlargement of the liver may occur in some cases. Sometimes, the parasite itself and the eggs act as nidus for formation of lodgment of the flukes in the bile duct causing biliary distention.

Histopathology
Histopathology of the affected tissue reveals infiltration of inflammatory cells. Histopathology also depicts fibrosis.

Hematological Pathology
Very little amount of pathology is caused due to opisthorchiasis. In most of the cases, the disease is asymptomatic. However, in man, anemia has been found.

Secondary Pathology
There have been records of mixed infection of *Opisthorchis* and different intestinal flukes, i.e., *Heterophyes* sp., *Metagonimus* sp., and *Echinostoma* sp.

Immunopathology
The parasites occur in the bile duct and liver. Inflammation and fibrosis are the main pathological features. There are certain pro-inflammatory cytokines, i.e., IL1 beta, TNF alpha, and IFN gamma, which are released by activated macrophages. The pro-inflammatory cytokines initiate inflammation. There also occurs cascade reaction of complement, and some complement fragments attract the inflammatory cells for infiltration. Fibrosis is another pathological feature. The alternatively activated macrophage secretes profibrotic factors, which activate the fibroblast cells to secrete fibrin. In *Opisthorchis* infection, different pro-inflammatory cytokines are responsible for causing pathology. Due to initiation of TLRs, two important cytokines, IL6 and IL8, are produced. These interleukins cause inflammation, anti-apoptosis, cell transformation, etc.

Molecular Pathology
Study on molecular pathology has been a new dimension to rule out the actual cause of pathology. Different genetic markers are being searched out to correlate the occurrences of pathological events. Initial pathology starts with inflammation mediated by pro-inflammatory cytokines, i.e., IL1 and IL6. Profibrotic cytokines and TGF beta induce fibrosis in *Opisthorchis* infection. Study shows upregulation of these cytokine genes.

Symptoms

Primary Symptom
The affected animals are apparently asymptomatic. However, high load of infection results in some clinical signs. Indigestion and diarrhea are the primary symptoms.

Secondary Symptoms
Icterus, ascites, and loss of body weight are some clinical sign.

Diagnosis

Symptomatological Diagnosis

Symptomatological diagnosis is not possible because no specific symptom is observed for confirmatory diagnosis of this parasitic infection.

Clinical Diagnosis

In case of opisthorchiasis and clonorchiasis in human beings, immediately after ingestion of infected fish, a discernible hypereosinophilia was observed in a report. Biochemical profile depicts elevated level of serum alkaline phosphatase and bilirubin level.

Radiographic Diagnosis

Ultrasonography, cholangiography, and computed tomography cannot provide confirmatory diagnosis, but these may provide supportive clues by observation of hepatobiliary dilatation, stone formation, neoplasia, etc. **Endoscopic retrograde chromatography** demonstrates shadow of the parasite.

Immunodiagnosis

Sirisinha et al. (1991) worked on the use of monoclonal antibody-based ELISA and DNA dot blot hybridization technique in terms of its potential for detection of *O. viverrini*. Teimoori et al. (2015) worked on development of immunodiagnostic tests by using a recombinant cathepsin protein (cathepsin F) of *Opisthorchis viverrini* in both indirect and sandwich ELISA assays. The rOv-CF was expressed in *E. coli* cells. The protein exhibited sensitivity and specificity of 62.1% and 84.05%, respectively, while tested with human sera.

Molecular Diagnosis

Different molecular markers are used for the identification of different parasites. These markers are of ribosomal DNA and RNA (ITS1, ITS2, 5.8 S rRNA, 18 S rRNA). Mitochondrial DNA (CO1, ND1, etc.) are also used as molecular markers for identification of parasites. Studies show use of ITS1 and ITS2 for molecular characterization of *Opisthorchis* sp.

Treatment

Chemotherapy

Niridazole, clioxamide, praziquantel, albendazole, and hexachlorophene are some effective drugs.

Nanotherapy

Study shows efficacy of nano-encapsulated curcumin and praziquantel to reduce periductal fibrosis and bile canalicular abnormalities caused by *Opisthorchis viverrini* in hamster. However, more works are required to carry out different nanotherapy in natural hosts.

Control

Do's	Don'ts
• Treat the affected animals properly • Efforts could be made to control aquatic snails because snails act as the first intermediate hosts • Public education about danger of eating raw and undercooked fish can be initiated	• Don't provide uncooked or undercooked fish to the animals

General Control

The affected animals should be treated properly. The animals should not be allowed to eat raw, uncooked, and undercooked fish as infective stages of the parasite may remain in the fish. The snails should be controlled by molluscicidal agents. The feces should be disposed properly because feces are the source of eggs of the parasites.

Immunological Control

Before understanding the scenario of vaccination, we should understand first the immunity occurring in natural infection of opisthorchiasis in the animals. Both innate and adaptive immunities play a role. This immunity provides only partial protection. Failure of host immunity to kill the parasite may be due to immune evasion by the parasites or the location of the parasite. Therefore, immunity that occurred due to natural infection indicates a challenging task to develop reliable immunocontrol strategies. Nevertheless, effort on immunological control by different researchers is done. Apart from using different crude antigens, different defined antigens have been used for vaccination trials. A number of works are in progress. In most of the cases, *Opisthorchis viverrini* or *O. felineus* are the model parasites. Generated data from the investigations depict only partial protection, and till date, no vaccine has yet been available. Different specific vaccine candidates so far used are glutathione S-transferase (GST), fatty acid-binding protein (FABP), cathepsin F, B1, ferritin, calreticulin, paramyosin, granulin-like growth factor (Ov-Grn-1), SH3 domain-containing protein, and Ly6-containing protein. Phung et al. (2019) carried out a work on recombinant *Opisthorchis viverrini* tetraspanin. Tetraspanin is a transmembrane that was expressed in *Pichia pastoris*. They investigated the potential of a recombinant protein encoding for large extracellular loop of *O. viverrini* tetraspanin-2 (rOv-LEL-TSP-2). Hamster model was used by them for vaccination trial using 50 and 100 µg of rOv-LEL-TSP-2 expressed from *Pichia pastoris* yeast combined with alum CpG adjuvant intraperitoneally. This vaccination significantly reduced the number of worms recovered from hamsters vaccinated with rOv-LEL-TSP compared to adjuvant control groups.

Molecular Control

No gene vaccine has yet been commercialized.

Biological Control

No biological control is available in the field.

Genus: *Clonorchis*

Etymology

The generic name "*Clonorchis*" was derived from two Greek words, i.e., "klon " and "orkhis." "Klon" means branched and "orkhis " means "testes."

Introduction

This genus comes under the family Opisthorchiidae. The important species is *Clonorchis sinensis*. This species is called as Chinese liver fluke or oriental liver fluke.

Distribution

The flukes are very common in the southeastern parts of Asia, China, and Japan.

Species: *Clonorchis sinensis*

Common name: Chinese liver fluke or oriental liver fluke.

Hosts: Man, dog, and cat are affected. The life cycle is similar to *Opisthorchis*.

Site: The seat of predilection of the parasite is the bile duct, pancreatic duct and, duodenum.

Morphology

Adult

Size: 2.5 cm × 0.5 cm.

Shape: Tapering anteriorly and wide posteriorly.

Color: Transparent pinkish.

Tegument: Spines are found in young flukes but adult's tegument appears smooth.

Testes: The testes are much branched and located posteriorly.

Egg

The eggs are brown colored. The size is 27–35 × 12–20 μm. The eggs are operculated having the miracidium within it. Opposite to the opercular pole, a hook-like structure is present.

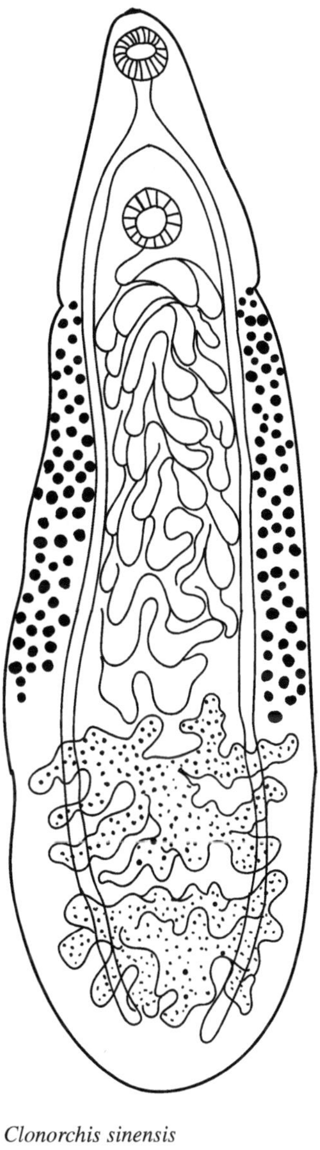

Fig. 4.18 *Clonorchis sinensis*

Life Cycle

The first intermediate host is aquatic snail like *Melania*, *Vivipara*, *Bithynia*, etc. The second intermediate host is cyprinid fish. The eggs are ingested by snail. Miracidia hatch out from the eggs. Sporocyst, redia, and cercaria are formed in the snail. The cercariae are released from the snails. The cercariae have noticeably long tail and pigmented eye spots. The cercariae get transformed to metacercariae in fish. The hosts are infected by ingestion of infected fish. Excystment of metacercariae occurs in the duodenum. The immature flukes reach the liver via bile duct where they mature.

Pathogenesis

In most of the cases, no marked pathology is observed except some mild pathology in the bile duct and liver. Marked pathology may occur in limited cases. There occurs desquamation of epithelial cells from bile duct wall. The main pathological features observed are biliary stasis and cholangiocarcinoma. Human beings get infection by ingestion of raw, undercooked, undersmoked, dried, and pickled freshwater fish. The immature flukes cause great damage in the bile duct comprising epithelial proliferation and neovascularization in peribiliary plexus. Intrahepatic calculi may be formed. There may be secondary bacterial infection.

Symptoms

The clinical features include persistent diarrhea, jaundice, hepatomegaly, etc. in man.

Diagnosis

Symptomatological Diagnosis

Symptomatological diagnosis is not possible as symptoms of diagnostic value are not found.

Parasitological Diagnosis

Examination of feces can be done for detection of eggs of the parasite.

Immunological Diagnosis

Cho et al. (2020) selected 18 candidate antigen proteins by protein array-based screening using *C. sinensis*-infected sera. Of these, CsAg17 antigen showed the highest sensitivity (77.1%) and specificity (71.2%). Therefore, CsAg17 antigen has immunodiagnostic potential.

Treatment

There is no satisfactory treatment.

Control

General Control

The affected individual should be treated properly. Eating of raw or undercooked fish should be avoided. Control of snail can be attempted. But practical implication of snail control is difficult.

Immunological Control

Bai et al. (2020) evaluated the molecular characteristics and immune responses of CsAg17 protein. Recombinant CsAg17 B-cell epitope protein and cDNA vaccines were produced. They evaluated immune response in FVB mice model. CsAg17 protein vaccine conferred 64% protection and cDNAvaccine provided 69% protection. No vaccine is available.

Molecular Control

No gene vaccine is commercially available. Research work on gene vaccination is in progress.

Genus: *Metorchis*

Introduction

These are minute flukes occurring in the bile duct and gallbladder of dog, cat, etc.

Etymology

The word metorchis has been derived from meta (Greek) meaning posterior and orchis (Ancient Greek) meaning testicle.

Species: *Metorchis albidus*

Hosts: This is small trematode, which affects dog, cat, fox, etc.

Site: Bile duct and gallbladder.

Distribution: The flukes are prevalent in different countries, i.e., Europe, Asia, USSR, North America, etc.

Morphology

Adult

The flukes are spatulate and small in size, ranging from 2 to 6 mm approximately. The body is spinose. The lobed testes are diagonally placed. The ovary is pretesticular. The testes are present in the posterior part of the fluke. The genital pore is located in front of the ventral sucker. As a rule, the vitelline glands are present on lateral side of the fluke.

Egg

The eggs are small in size (24–30 × 13–16 μm).

Life Cycle

There are two intermediate hosts. The first intermediate host is aquatic snail (*Bithynia*, *Amnicola*, etc.). The secondary intermediate hosts are different fish (*Carassius carassius*, *Leuciscus rutilus*, *Gobio gobio*, *Blicca bjoerkna*, etc.). The life cycle of *M. conjunctus* was studied which is as folloing. It is likely that life cycle of *M. albidus* will be same.

Some basic information of life cycle
• Definitive hosts are dog, cat, and pig • First intermediate hosts are aquatic snails (*Bithynia* spp., *Amnicola* spp.) • Second intermediate hosts are cyprinid fish • The infective stage of the parasite is metacercaria • The mode of infection is ingestion of metacercariae

Quick look: life cycle of *Metorchis conjunctus*
• Passing of eggs with feces of definitive host • Ingestion of eggs by aquatic snails (*Amnicola limosa porata*) • Hatching of eggs in the gut of the snails and liberation of miracidia from the eggs • Transformation of miracidia to sporocysts and further development of rediae and cercariae in snail • Escaping of cercariae from the snails • Penetration of the second intermediate hosts (cyprinid fish) by cercariae • Transformation of cercariae to metacercariae in the second intermediate hosts • Ingestion of the infected second intermediate hosts by definitive hosts • Development occurs up to adult in the definitive hosts

Pathogenesis

The pathogenesis includes the common pathological features comprising damage and lesions of parenchymatous tissues of the liver resulting in hepatitis.

Clinical Signs

The clinical signs include diarrhea, vomiting, etc.

Diagnosis

Nöckler et al. (2003) performed indirect ELISA for the detection of antibodies against

Opisthorchis felineus and *Metorchis bilis* in foxes. They demonstrated that ELISA could be suitable tool for detection of specific *O. felineus* and *M. bilis* antibodies in the fox by using excretory-secretory antigen of the parasites.Works are needed for use of ELISA as diagnostic test for *M. albidus*.

Treatment
Praziquantel is likely to be effective for eliminating this parasitic disease.

Genus: *Parametorchis*

Etymology
The word *Parametorchis* has been derived from three Greek words, i.e., para (near), meta (posterior), and orchis (testis).

Species: *Parametorchis complexus*
Distribution: They are prevalent in North America.
Hosts: These are small trematodes that affect cat, dog, etc.
Site: Bile duct.

Morphology
The flukes are small in size. As is found in other *Opisthorchis* members, testes are posterior in location. The testes are tandem in position. Uterus forms a rosette around the ventral sucker.

Life Cycle
Life cycle is not known but seems to be as other *Opisthorchis* spp.

Pathogenesis
The pathogenesis is as for other *Opisthorchis* spp.

Genus: *Amphimerus*

Etymology
The word "Amphimerus" has been originated from two Greek words. "Amphi" means "on both sides" and "meros" means "part".

Species: *Amphimerus noverca*
Site: The seats of predilections of this fluke are the liver and gallbladder.
Hosts: The hosts are dog, pig, human being, etc.

Morphology
They are 2.1–12 mm long. The oral sucker is doubly larger as compared to ventral sucker. The pair of testes are situated in the posterior third of the body. The testes are lobed or entire. These are tandem in position. The ovary is situated anterior to the testes.

Life Cycle

Events of life cycle
• Passing of eggs with feces of definitive host
• Ingestion of eggs by aquatic snails (*Melanoides lineatus*)
• Hatching of eggs in the gut of the snails and liberation of miracidia from the eggs
• The next developmental stages in the snails are sporocyst, redia and cercaria
• Discharge of pleurolophocercus cercariae from the snails
• Penetration of the second intermediate hosts by cercariae
• Transformation of cercariae to metacercariae in the second intermediate hosts (*Cirrhinus reba*, *Labeo bata*)
• Ingestion of infected second intermediate hosts by definitive hosts
• Development occurs up to adult in the definitive hosts

Pathogenesis
The parasites cause mechanical damage to the bile duct epithelium. Desquamation of bile duct epithelium occurs. Hyperplasia of bile duct is also found. In advance case, there occurs cirrhosis.

Diagnosis
The routine diagnostic method is examination of fecal sample. Cevallos et al. (2017) first developed ELISA for detection of IgG antibody against *Amphimerus* spp. in human sera.

Symptom
The affected animals do not show any marked clinical signs. But some common clinical signs include indigestion, diarrhea, loss of body weight, etc.

Treatment
Praziquantel is likely to be effective against this parasitic infection.

Genus: *Pseudoamphistomum*

Species: *Pseudoamphistomum truncatum*

Site: The seat of predilection of this fluke is the bile duct.

Host: The hosts are dog, cat, and human beings.

Morphology

They are 2.0–2.5 mm long. Two suckers (oral and ventral) are present, which are almost equal in size. The intestinal ceca is characteristically swollen posteriorly. The body has typical posterior truncation. The pair of testes are round to oval in shape situated at the posterior end. The ovary is pre-testicular. The uterine coils are mostly found in between the testes and ventral sucker.

Life Cycle

Not well known

Pathology

Not well known

Treatment

Not well known

Family: Nanophyetidae

Family Synopsis

The members affect the alimentary canal of mammals. The flukes are elongate or pyriform in shape. Ventral sucker is situated in the middle third of the body; ovary is pretesticular; testes are in posterior part of the fluke. They are heteroxenous parasites involving two intermediate hosts (first aquatic snail, and second salmonid fish). The developmental stages are egg, miracidium, redia, and cercaria. The parasites are important and transmit rickettsial agents causing salmon poisoning and Elokomin fluke fever. Diagnosis can be done by finding the eggs in the feces. Control can be done by proper treatment and avoiding uncooked fish.

Genus: *Nanophyetus*

The generic name "*Nanophyetus*" was derived from two words, i.e., nano (Greek) and likely Greek word "phyein". "Nano" means "dwarf" and "phyein" means " to grow".

Species: *Nanophyetus salmincola*

Hosts: These are small (few mm) trematodes, which affect dog, cat, fox, etc.

Site: Small intestine.

Distribution: The flukes are distributed in North America, eastern Russia, etc.

Morphology

The worms are white or creamy in color. The testes are situated at the posterior half of the body. The ovary is situated posterior and right to the ventral sucker. The vitellaria are situated laterally.

Life Cycle

There are two intermediate hosts. The first intermediate hosts are aquatic snails. The secondary intermediate hosts are fish specifically of Salmonidae family.

Some basic information of life cycle
• The definitive hosts are dog, cat, etc. • The first intermediate hosts are aquatic snails • The secondary intermediate hosts are salmonid fish • The infective stage of the parasite is metacercaria • The mode of infection of the definitive host is ingestion of metacercariae

Quick look: life cycle of *Nanophyetus*
• Voiding of eggs with feces of definitive host • Ingestion of eggs by stream snails (*Semisulcospira* sp., *Oxytrema* sp.) • Hatching of eggs in the gut of the snails • Release of miracidium from the eggs and infection of snail (*Semisulcospira, Oxystrema*) by them • Formation of rediae, daughter rediae, and cercariae in snail. Sporocysts are not found • Release of cercariae from the snails • Penetration of second intermediate hosts (fish) by cercariae • Transformation of cercariae to metacercariae in the second intermediate hosts (salmonid fish). The salmonid fish are *Salmo* spp., *Salvelinus* spp., etc. • Ingestion of the second intermediate hosts harboring metacercariae by definitive hosts • Development up to adult in definitive hosts

Pathogenesis

The parasites affect the upper part of duodenum. The duodenal wall is sloughed off resulting in hemorrhagic diarrhea. The parasites transmit a rickettsial pathogen, *Neorickettsia helminthoeca*, causing salmon poisoning in the definitive hosts. Salmon poisoning occurs primarily in canids. In the dogs, salmon poisoning is characterized by hemorrhagic diarrhea, vomiting, etc. Another complication of this fluke infection is "Elokomin fluke fever," which occurs in a wide host range, i.e., dog, fox, bear, raccoon, man, etc. Diarrhea, vomiting, fever, eye discharge, etc. are some clinical features.

Clinical Symptoms

Primary symptoms: The primary symptoms include diarrhea, vomiting, etc.

Secondary symptoms: The secondary symptoms are dullness, depression, anemia, etc.

Family: Echinostomatidae

Family Synopsis

The members affect the mammals, birds occurring in the intestine, bile duct, etc. The worms are characterized by small oral sucker surrounded by head collar equipped with single and double rows of spines, lobed or intact testes with tandem or diagonal position, pretesticular ovary, and vitellaria in the lateral field. They are heteroxenous parasites involving two intermediate hosts (first aquatic snail and second snail, fish, frogs). The developmental stages are egg, miracidium, sporocyst, redia, and cercaria. The parasites are generally regarded as harmless but may cause severe enteritis. Diagnosis can be done by finding the eggs in the feces. Control can be done by proper treatment and control of snails.

Genus: *Echinostoma*

Etymology

The generic name "*Echinostoma*" is derived from two Greek words, i.e., "echinos" and "stoma." "echinos" means "spined" and "stoma" means "mouth."

Species: *Echinostoma revolutum*

Distribution: The flukes are worldwide in distribution.

Hosts

Primary hosts	Natural hosts	Duck, goose, pigeon, etc.
	Domestic hosts	Duck, goose
	Wild hosts	Marsh birds, waterfowls
	Experimental hosts	Hamster, chicken
	Rare hosts	Human beings
Secondary hosts	*Stagnicola*, *Planorbis*, *Physa*, etc.	

Site: Cecum and rectum

Morphology

Adult

Size: 10–22 mm × 2–3 mm.

Shape: The worms are leaf shaped.

Color: The fresh parasites are reddish or pinkish in color.

Tegument: The tegument is spinose.

Head collar: 37 spines are found on the head collar. Of these spines, five larger spines are found on either side called as "corner spine."

Pharynx: They have prominent pharynx.

Esophagus: The esophagus is short.

Sucker: Ventral sucker is much larger than oral sucker.

Testes: The testes are lobed and tandem in position. These are located in the hind body.

Cirrus: The cirrus is located partly anterior to the ventral sucker.

Genital pore: Genital pore is situated near the bifurcation of intestinal ceca.

Ovary: The ovary is simple, round, and pretesticular in location.

Uterus: The uterus is extensively folded and present in the mid-field.

Vitelline glands: Vitelline glands are extended from the level of ventral sucker to the posterior end.

Egg

The eggs are 110 × 65 μm.

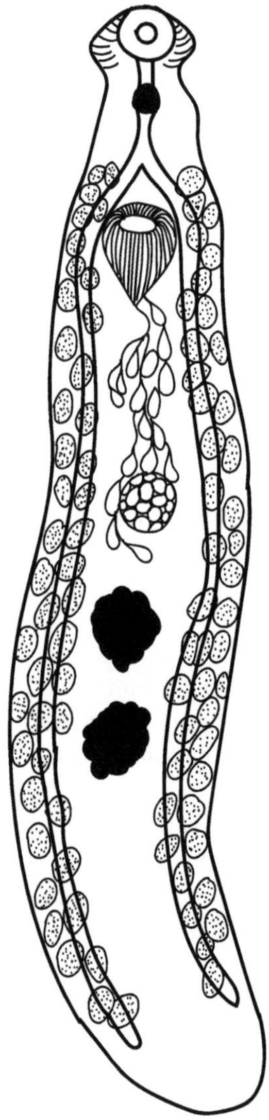

Fig. 4.19 *Echinostoma revolutum*

Life Cycle

Some basic information of life cycle
• The definitive hosts are domestic aquatic birds, i.e., duck, goose, fowl, partridge, etc. • The intermediate hosts are aquatic snails • Development of metacercariae occurs either in the same snail or different snail of the same species or in snail of different species • The mode of infection is ingestion of intermediate hosts carrying metacercariae

Stages of Life Cycle Formed

In definitive host	In environment	In first intermediate host	In second intermediate host
• Adult • Egg	• Miracidium	• Sporocyst • Redia (usually two generations) • Cercaria	• Metacercaria

Events of life cycle: a glimpse
• Voiding of eggs with feces of definitive host • Hatching of eggs in the environment and release of miracidia • Formation of next stages, i.e., sporocysts, rediae, and cercariae, in the intermediate hosts (snail). Two generations of rediae are produced. • Discharge of cercariae from the infected snails • Transformation of cercariae to metacercariae in the same snail or different snail of the same species or snail of different species. Tadpole may act as second intermediate hosts • Ingestion of infected second intermediate hosts by definitive hosts • Subsequent development in the definitive hosts up to adult stage

Detailed Description

Egg Laying

The adult parasites live in their predilection site where they lay eggs.

Egg Passing

The unembryonated eggs are voided with the feces of the host.

Embryonation

The eggs get embryonated in the presence of water.

Hatching

After full development of the embryo, the eggs hatch and the miracidia come out.

Development in the Intermediate Hosts

Life cycle involves two intermediate hosts. The first intermediate hosts are aquatic snails (*Bithynia*, *Lymnaea*, *Gyraulus*, *Physa*, *Valvata*, etc.), and the second intermediate hosts are snail, tadpole, etc. The miracidia actively enter the body of the first intermediate hosts. Development up to cercaria stage occurs in the snail hosts. The miracidia give rise to sporocysts. The sporocysts form 2–3 generations of rediae. The cercariae are known as *Cercaria echinata*. This type of cercariae possess collar of spines. The cercariae are released from the snail host. The cercariae either encyst on the same snail or on different member of the same species or on another snail species or tadpole. The cercariae actively enter the second intermediate hosts wherein they transform to metacercariae.

Infection to the Definitive Hosts

The final host gets the infection by ingestion of infected second intermediate hosts.

Pathogenesis

The flukes in less number do not produce marked pathogenesis. However, a large number of flukes may cause marked enteritis.

Clinical Signs

The affected animals display clinical signs of diarrhea.

Diagnosis

Examination of feces reveals the eggs of the parasite. Graczyk and Fried (1994) developed an ELISA using surface glycocalyx membrane crude antigen of adult *E. caproni*. Similar work may be carried out for diagnosis of *E. revolutum*.

Treatment

Praziquantel is an effective drug. Oxyclozanide can be tried.

Control

Control of snail of ponds can be done. The birds should be restricted from snail-contaminated water.

Species: *Echinostoma hortense*

The hosts are rat, dog, etc. Humans may also be affected. The parasite is slender. The head crown possess 27 spines. As that of other flukes, they have two suckers. The ventral sucker is much larger as compared to oral sucker. The testes are lobed and tandem in position.

Life Cycle

- Eggs are voided from the definitive host.
- Eggs hatch in water and miracidia are released.
- The miracidia enter snail.
- Different developmental stages are formed in the snails, i.e., sporocyst, mother redia, daughter redia, and cercaria.
- The cercariae are liberated from the snail.
- The cercariae penetrate in the second intermediate hosts. Frogs and loach may be the second intermediate hosts.
- Cercariae transform to metacercariae in the second intermediate hosts.
- Final hosts get the infection by ingestion of infected second intermediate hosts.

Pathogenesis

The parasite causes diarrhea, abdominal pain, etc. The parasite has zoonotic importance as they may affect human beings transmitted through eating infected frogs.

Control

Control of this parasite can be done by extermination of snails since the snails act as the first intermediate hosts.

Species: *Echinostoma ilocanum*

This parasite occurs in the intestine of man. Dogs are also affected. Snails act as both first and second intermediate hosts. Development up to cercariae stages occurs in the first intermediate hosts. Cercariae encyst on the second intermediate hosts. Pathogenesis is of general type comprising enteritis, which results in diarrhea as symptoms.

Species: *Echinostoma paraulum*

This parasite is found in the small intestine of duck and pigeon. However, this parasite may affect human beings also. Thirty-seven spines are present at the head collar, of which five corner spines are present on either side of the collar. Twenty-seven spines are arranged in double dorsolateral row. A pair of tandem testes are present in the third quarter of the body. The parasite may cause diarrhea.

Genus: *Echinochasmus*

Etymology

The generic name "*Echinochasmus*" is a word derived from two words, i.e., "echino" and "chasmus." "Echino" means "spined" and "chasmus" means "hiatus."

Species: *Echinochasmus perfoliatus*

Hosts: Dogs, cat, pig, and fox.

Site: The seat of predilection of this fluke is the intestine.

Morphology

They are 2–4 mm × 0.4–1 mm long. Two suckers (oral and ventral) are present. The oral sucker is subterminally located. The ventral sucker is almost twice as large as the oral sucker. The head crown is equipped with 24 spines. There are three corner spines which are small in size. The pair of testes are tandem and situated behind the middle part. The ovary is pretesticular. The vitelline glands are situated in the lateral field.

Life Cycle

Some basic information of life cycle
• The definitive hosts are dog, cat, and pig • The first intermediate hosts are aquatic snails • The second intermediate hosts are fish • The infective stage is metacercaria • The mode of infection is ingestion of the second intermediate hosts carrying metacercariae

Quick look of life cycle
• Passing of fluke eggs with feces of definitive host • Hatching of eggs with contact of water and release of miracidia • Entrance of miracidia in aquatic snail hosts (*Parafossarulus* spp., *Bulinus* spp.) • Formation of different subsequent stages of the parasites, i.e., sporocyst, redia, and cercaria, in snail • Penetration of fish by cercariae • Transformation of cercariae to metacercariae in fresh water fish (*Idus*, *Abramis*) • Ingestion of infected fish by definitive hosts • Development up to adult in definitive hosts

Pathogenesis

Enteritis is caused by this parasite.

Treatment

Oxyclozanide and rafoxanide can be tried.

Genus: *Echinoparyphium*

Species: *Echinoparyphium recurvatum*

Hosts: The hosts are domestic and wild ducks.

Site: Small intestine.

Morphology

The head crown bears 45 spines. Out of these, four are corner spines present on either side of the head crown. The oval testes are tandem and typically remain in contact with one another.

Life Cycle

First intermediate hosts: *Lymnaea* and *Planorbis* spp.

Second intermediate hosts: Frog (*Rana temporaria*) and snail (*Valvata* sp.)

Sohn (1998) experimentally studied the life cycle of this parasite, and they reported involvement of two intermediate hosts for completion of life cycle of parasite. Development up to cercaria occurs in the first intermediate hosts. Cercariae develop to metacercariae in the second intermediate hosts. Adults are formed in the definitive hosts.

Pathogenesis

The parasites affect the intestine and cause enteritis.

Genus: *Hypoderaeum*

Species: *Hypoderaeum conoideum*

They occur in the small intestine of duck, goose, and wild aquatic birds. The body is elongated. The head collar is poorly developed and equipped with 47–53 spines. Of these spines, two are "corner spines" present on either side. The ventral sucker is much larger. The testes are elongated, slightly lobed, and tandem in position. These are situated approximately behind the middle of the body. Two intermediate hosts are required for completion of the life cycle. The first intermediate hosts are *Lymnaea* snails, and the second intermediate hosts are tadpoles, in which the metacercariae are formed.

Family: Heterophyidae

Family Synopsis

The members of this family are commonly the parasites of mammals and birds occurring in the alimentary canal. The flukes are characterized by small size, spiny tegument, middle position of the ventral sucker, and presence of genital pore frequently surrounded by genital sucker near the ventral sucker. The parasites are heteroxenous and require two intermediate hosts for their life cycle (first aquatic snails and second fish or frog). The parasites are not harmful; but in some cases, enteritis or hemorrhagic enteritis may occur. Where possible, diagnosis can be done by finding the eggs in the feces. Control is mainly based on chemotherapy.

History

Heterophyes heterophyes was first discovered by Theodor Maximaillian Bilharz in the year 1851.

Genus: *Heterophyes*

Etymology

The generic name "*Heterophyes*" is derived from two Greek words, i.e., "heteros" and "phyes." "Heteros" means "different" and "phyes" means "form."

Species: *Heterophyes heterophyes*

Distribution: The flukes are distributed in Middle East, Far East, etc.

Hosts: Dog, cat, man, etc. are the hosts of this parasite.

Site: The seat of predilection of this fluke is the small intestine.

Morphology

They are 1–2 mm long. The body is covered with minute scale-like spines. Two suckers (oral and ventral) are present. The ventral sucker is much larger as compared to oral sucker. A characteristic genital sucker (gonotyl) bearing a circle of spines is located posterior and little lateral to the ventral sucker. The presence and location of the genital sucker is an important identifying feature. The pair of oval testes are situated horizontally. The testes are located just behind the tip of cecal ends. The pretesticular ovary is located medially.

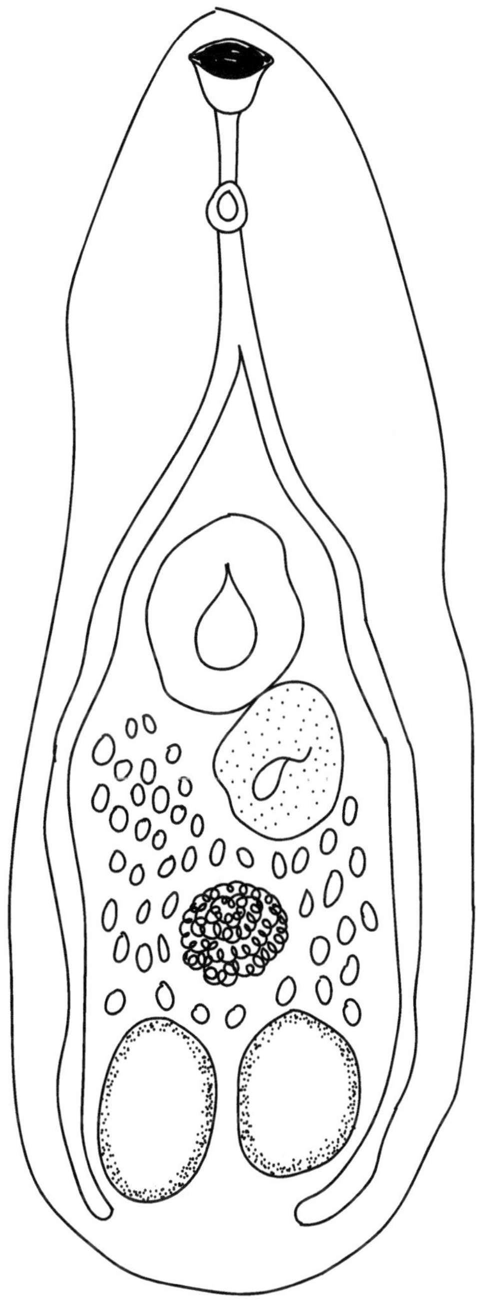

Fig. 4.20 *Heterophyes heterophyes*

Life Cycle

Some basic information of life cycle
• The definitive hosts are dog, cat, and man • The first intermediate hosts are aquatic snail (*Pirenella conica*, *Cerithidea cingulata*, *Tarebia* sp.) • The second intermediate hosts are fish (*Mugil cephalus*, *M. capito*, *Tilapia nilotica*, etc.) • The infective stage is metacercaria • The mode of infection is ingestion of second intermediate hosts carrying metacercariae

Events of life cycle: a glimpse
• Voiding of embryonated (formed miracidium) eggs with the feces of definitive hosts • Ingestion of eggs by snails (*Pirenella conica*, *Cerithidea cingulata*, *Tarebia* sp.) • Hatching of eggs in the gut of the snails • Release of miracidia from the eggs after hatching • Penetration of gut wall of the snail by the miracidia • Formation of sporocysts, rediae, and cercariae in snail. Daughter rediae may be produced. • Release of cercariae from the snails • Penetration of fish by cercariae (*Mugil cephalus*, *M. capito*, *Tilapia nilotica*, etc.). The cercariae are biocellate, monostome, and pleurolophocercous • Transformation of cercariae to metacercariae in the fish • Ingestion of fish harboring metacercariae by definitive hosts • Development up to adult in definitive hosts

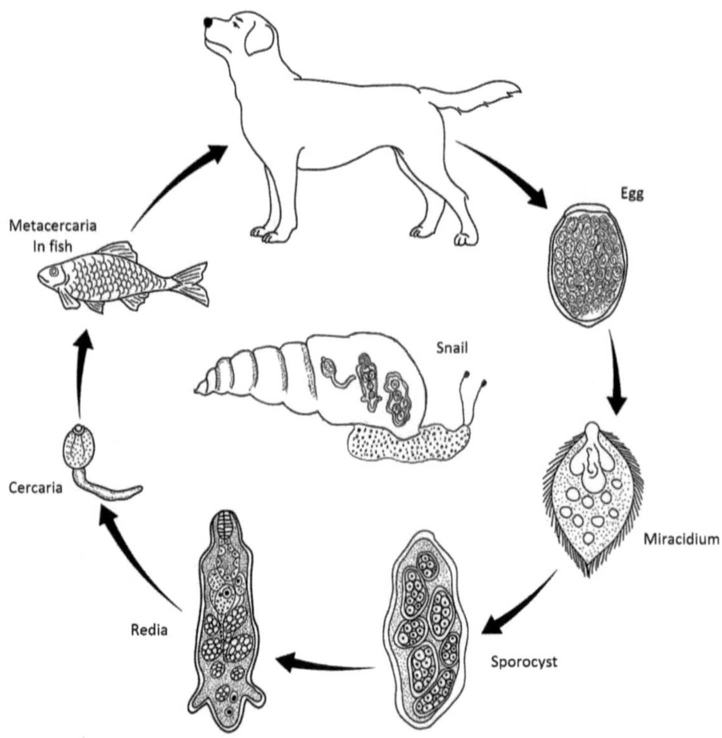

Fig. 4.21 Life cycle of *Heterophyes heterophyes*

Pathogenesis

The flukes do not cause marked pathogenesis, and they are usually regarded as nonpathogenic. But they may sometimes cause hemorrhagic enteritis in human beings.

Diagnosis

Examination of feces reveals eggs of the parasites. Different workers carried out different immunodiagnostic trials, i.e., counter-current Immunoelectrophoresis test, indirect fluorescent immunoassay, and intradermal test for diagnosis of heterophyiasis. But no immunodiagnostic test seems commercialized.

Clinical Signs

The affected animals show diarrhea. Hemorrhagic diarrhea is observed in human beings.

Treatment

Name of drug	Route of administration	Dose
Praziquantel	Oral	20 mg/kg is effective after repeating the dose for two consecutive days

Species: *Heterophyes persicus*

Site: The seat of predilection of this flukes is the intestine.

Host: The hosts are dogs.

Morphology

They are 1.4–2.1 mm long. Two suckers (oral and ventral) are present. The ventral sucker is noticeably much larger, which is roughly five times larger than oral sucker. The pair of testes are situated horizontally. The ovary is pre-testicular. The vitelline glands are distributed up to the ovary in 14–16 groups on each side.

Genus: *Metagonimus*

Species: *Metagonimus yokogawai*

Etymology

The generic name "*Metagonimus*" is a word derived from two Greek words, i.e., "meta" and "gone " "Meta" means behind and "gone" means "gonad."

Hosts: Dog, cat, pig, and man

Site: Small intestine

Common name: Yokogawa fluke

Morphology

The worms are spiny. The testes are diagonal in position. The ventral sucker is located right to the median line. Genital pore opens immediately anterior to it.

Life Cycle

Some basic information of life cycle
• The definitive hosts are dog, cat, and man • The first intermediate hosts are aquatic snails (*Semisulcospira* sp.) • The second intermediate hosts are cyprinid fish • The infective stage is metacercaria • The mode of infection is ingestion of the second intermediate hosts carrying metacercariae

Events of life cycle: a glimpse
• Embryonated eggs are voided with feces of definitive host • Eggs are ingested by snail (*Semisulcospira* spp.) • Hatching of eggs occurs in the gut of the snails • Miracidia are released from the eggs • Sporocysts, rediae, and cercariae are formed in the snail • Cercariae are released from the snails • Penetration of cyprinid fish by cercariae occurs • Cercariae get transformed to metacercariae in the fish • The final hosts become infected by ingestion of fish harboring metacercariae

Pathogenesis

The parasites cause enteritis.

Diagnosis

Examination of feces reveals eggs. Han et al. (2014) characterized 100 kDa somatic antigen of *Metagonimus yokogawai*. Performing Western blot using monoclonal antibodies, they found that it had less specificity to *Metagonimus yokogawai* and it cross-reacted with other trematodes but had not cross-reacted with nematodes and cestodes.

Clinical Signs

The clinical signs are abdominal pain, diarrhea, etc.

Treatment

Name of drug	Route of administration	Dose
Praziquantel	Oral	20 mg/kg is effective after repeating the dose for two consecutive days

Genus: *Euryhelmis*

Etymology

Euryhelmis word has been derived from two words, i.e., “Eury” meaning wide and “helmis” meaning worm.

Species: *Euryhelmis squamula*

Hosts: Mustelids, fox, polecat, mink, etc.
Site: Intestine

Morphology

They are very small in size and wide. The length is more or less 1 mm. The lobed testes are horizontal in position. The ovary is situated anterior to the right testes. The uterus is characteristically located in between the ventral sucker and left testes.

Life Cycle

Different events of life cycle

- Embryonated eggs are voided with feces of definitive host.
- Eggs are ingested by snails (*Bythinella*).
- Development up to cercaria occurs in the snail.
- Cercariae are released from the snails.
- Penetration of young frog (*Rana temporaria*) by cercariae occurs.
- Cercariae get transformed to metacercariae in the tadpole.
- The final hosts become infected by ingestion of the second intermediate hosts harboring metacercariae.

Pathogenesis

The parasites cause enteritis. Hemorrhagic enteritis may also occur in some hosts.

Clinical Signs

The clinical signs are abdominal pain, diarrhea, etc.

Genus: *Cryptocotyle*

Species: *Cryptocotyle lingua*

Distribution: The worms are distributed in the UK, Denmark, Japan, Canada, etc.

Etymology

The generic name “*Cryptocotyle*” is a word derived from two Greek words, i.e., “kryptos” and “cotyle.” “kryptos” means “hidden” and “cotyle” means “disk.”

Hosts: Different types of gulls are affected. Terns, kittiwake, mink, dog, and cat are also the final hosts of this parasite.
Site: Intestine.

Morphology

They are small in size. The body is roughly spatula shaped. The cuticle is spiny. The testes are present diagonally or horizontally. The ovary is trilobed and located anterior to the testes. Suckers are weak.

Life Cycle

Two intermediate hosts are required for completion of life cycle of this fluke. The first intermediate hosts are shore periwinkle (*Littorina* sp.). The eggs are passed with the feces of definitive hosts. The eggs hatch after being ingested by the first intermediate hosts. Development up to cercarial stage occurs in the first intermediate hosts. Encystment of metacercariae occurs in marine fish.

Pathogenesis

The flukes cause marked hemorrhagic enteritis.

Species: *C. concava*

These are parasites of sea birds.

Genus: *Apophallus*

Species: *Apophallus donicum*

Hosts: Cat, dog, fox, etc.
Site: Small intestine

Morphology

The cuticle of the fluke is spiny. Large-sized testes are present posteriorly.

Life Cycle

Stream snails serve as the first intermediate hosts, and cyprinid fish act as the second intermediate hosts. Development up to cercarial stage occurs in the first intermediate hosts. The released cercariae infect cyprinid fish in which the cercariae transform to metacercarial stage.

Family: Plagiorchiidae

Family Synopsis

The members of this family are commonly the parasites of birds occurring in the alimentary canal. The flukes are characterized by small spiny tegument, somewhat in lateral position of the genital pore in between two suckers. The entire or lobed testes are either horizontal or tandem in position. The ovary is pre-testicular. The parasites are heteroxenous and require two intermediate hosts for their life cycle (first aquatic snails and second larval stages of some insects).

Genus: *Plagiorchis*

Species: *Plagiorchis megalorchis*

The parasites affect different turkey poults. The life cycle involves two intermediate hosts. The first intermediate hosts are snails (*Lymnaea peregra*). The second intermediate hosts are larval stages of *Culicoides* spp.

Species: *Plagiorchis arcuatus*

The flukes are found in oviduct and bursa of Fabricius of fowl. The flukes are small in size reaching up to half cm. The tegument is spiny. Life cycle involves two intermediate hosts. The first intermediate host is *Lymnaea* spp. The second intermediate hosts are larval stages of *Culicoides* spp. Different crustacea also serve as the intermediate hosts.

Family: Troglotrematidae

Family Synopsis

They are parasites of the skin, frontal sinus, kidney, and intestine of carnivores and birds. The flukes are characterized by fleshy spiny body with convex dorsal surface and concave ventral surface, subterminal oral sucker, middle position of ventral sucker, lateral position of vitellaria, and genital pore situated posterior to the ventral sucker.

Genus: *Troglotrema*

Species: *Troglotrema acutum*

The parasites occur in the frontal and ethmoidal sinuses of fox. They cause damage to the skull. They may even cause perforation of the skull.

Morphology

The shape of the fluke is characteristic. They are anteriorly rounded and the posterior part has tail-like extremity.

Life Cycle

Life cycle requires two intermediate hosts. The first intermediate hosts are snails (*Bythinella*) and the second intermediate hosts are *Rana temporaria*. Extensive study on life cycle has not been done.

Pathogenesis

The parasite causes perforation of the skull as a result of decalcification.

Family: Paragonimidae

Family Synopsis

This is an important family of trematodes. The important genus of this family is *Paragonimus*. The parasites are distributed all over the world. The flukes are characterized by spiny cuticle, ventroterminal oral sucker, lobed testes in the posterior half of the body, pretesticular lobed ovary, and absence of cirrus pouch. The life cycle is indirect and requires two intermediate hosts (first aquatic snail and second crustacea). The developmental stages are egg, miracidium, sporocyst, redia, cercaria, metacercaria and adult. *Paragonimus westermani* is called lung fluke, which causes severe damage to the lungs. Control of the parasites relies mostly on treatment and control of snails.

History

Ringer in the year 1879 discovered *P. westermani* in the lung of a man.

Genus: *Paragonimus*

Species

Paragonimus westermani
Paragonimus kellicotti

Species: *Paragonimus westermani*

Distribution

The flukes are cosmopolitan in distribution. They are commonly prevalent in South-East Asia, China, etc.

Site (Predilection Site)

Normal site	Rare site
Bronchiole, lungs	Brain, spinal cord

Hosts

Primary hosts	Natural hosts	Dog, cat, man
	Domestic hosts	Dog, cat, etc.
	Wild hosts	Civet cat, tiger, mongoose, etc.
Secondary hosts	First intermediate hosts are aquatic snails and second intermediate hosts are crabs and crayfish (crustaceans)	
Experimental hosts	Wistar rat and other laboratory rats	

Morphology

Adult

Size: 7.5–16 mm × 4–8 mm.

Shape: They are oval in shape, somewhat resembling coffee bean shape.

Color: The color is reddish brown when fresh.

Tegument: The tegument is spiny. The spines are with bifid tips.

Pharynx: Small.

Esophagus: Short.

Suckers: Two suckers are present. One is dorsal and another is ventral. Ventral sucker is situated slightly above the middle of the body. Oral sucker is smaller than ventral.

Testes: Two testes are present posterior to the ovary. The testes are lobed.

Ovary: This is situated near the ventral sucker. The ovary is lobed.

Vitelline glands: Vitelline glands are situated in the lateral field.

Cirrus pouch: Cirrus pouch is absent.

Uterus: The uterus is short and coiled. The terminal uterus forms a metraterm.

Egg

The eggs are oval, yellowish to golden brown, and operculated. The egg size is 80–120 μm × 45–70 μm. The shell of the egg is thick. The abopercular end is more thick. The eggs are unembryonated when passed with sputum or feces.

Larva

The larval stages are miracidia, redia, cercaria, and metacercaria.

Miracidium: The miracidium is ciliated. It has 17 epidermal ciliated plates found in 4 rows. As that of other digenetic trematodes, the apical papillae are present, which are connected with penetration glands.

Sporocyst: The sporocysts are sac-like having a thin cuticular outer covering. Unlike cercariae and rediae, the digestive system is absent. The sporocysts have brood chambers in which germ cells are formed. Rediae are produced in the brood chamber. Digestive system is absent. Eye spots are absent.

Redia: Rediae are yellowish in color. They are formed from the germ balls present in the brood chambers. The redia has digestive system. Mouth, pharynx, and unbranched ceca are found. Excretory system is also found.

Cercaria: They are whitish in color. Brood chamber is found in the redia. Germ balls differentiate into cercariae. The mouth is surrounded by hairs. The skin is covered by spines. The ventral sucker is large and situated above the midline. Cercariae are microcercous type. They possess a stylet near the oral sucker.

Metacercaria: The metacercariae are oval or round in shape and whitish in color. A cyst is formed from the cercaria. Both oral and ventral suckers are clearly visible. The metacercaria is covered with spines.

Fig. 4.22 *Paragonimus westermani*

Note

- Waikagul (2007) reported a new species of *Paragonimus* from Western Thailand. Metacercariae were fed to cat. The adults were recovered from the dissected cat. The adult was morphologically similar to but smaller than *P. heterotremus*. The tegumental spines of the worm were singly spaced in arrangement, which were similar to but larger than *P. heterotremus*. They proposed its name as *P. pseudoheterotremus*.
- Donah et al. (2007) reported a new species of *Paragonimus* (*Paragonimus vietnamensis* sp. nov.) based on morphology of metacercariae and adult worms together with their molecular data of previous study.

Life Cycle

Some basic information of life cycle
• The definitive hosts are dog, cat, etc. • The first intermediate hosts are aquatic snail • The second intermediate hosts are crabs and crayfish • The infective stage is metacercaria • The mode of infection is ingestion of the second intermediate hosts carrying metacercariae

Events of life cycle: a glimpse
• Voiding of eggs with feces or nasal discharge of definitive host • Hatching of eggs with contact of water • Release of miracidium after hatching of eggs • Entrance of miracidium in aquatic snails • Formation of sporocysts, rediae (two generations), and cercariae in snail • Release of cercariae from aquatic snail • Penetration of fish by cercariae • Transformation of cercariae to metacercariae in fish • Ingestion of metacercariae by definitive hosts • Development up to adult in definitive hosts

Details of Life Cycle

Egg Laying

The parasites remain in the lungs forming nodules there. The parasites remain in these nodules. One thing should be noted that embryonation of eggs does not occur in the parasites. Unembryonated eggs are laid. The nodules have outlets like fine channels through which the eggs become free from the nodules. The nodules may be ruptured and the eggs are escaped.

Egg Passing

Eggs are coughed up and are passed from the hosts with the feces. The eggs may also be expelled out with the sputum or cough material also.

Egg Hatching

Eggs hatch after coming in contact with water. After hatching, the miracidia are released. About 3 weeks is required for hatching.

Development in First Intermediate Host

The ciliated miracidia penetrate the exposed soft tissue of the snails. *Melania*, *Thiara*, Asssiminea, and *Hua* spp. act as the first intermediate hosts of the parasites. Penetration is assisted by the secretion of penetration glands. The miracidia go through different developmental stages comprising sporocysts, rediae (two generations), and cercariae. The cercariae are microcercous type having short tail and stylet.

Release of Cercariae

The cercariae get released from the snails and swim in the water. They are microcercous type having a short tail. The cercariae soon seek for their second intermediate host (crab and crayfish). *Potamon*, *Eliocheir*, *Sesarma*, etc. act as the second intermediate hosts. The released cercariae come in contact with the fish and penetrate it. The cercariae decaudate their tails and are encysted as metacercariae in the second intermediate hosts in the soft body parts particularly in gills. As reported the crabs may also ingest the infected snails.

Development in Definitive Host

Infection of this parasite in the definitive hosts occurs by ingestion of uncooked or undercooked second intermediate hosts. Excystation of metacercariae occurs in the duodenum. The released juvenile flukes penetrate the wall of the GI tract and follow the abdominal cavity, diaphragm, and pleural cavity to reach their predilection site (lung).

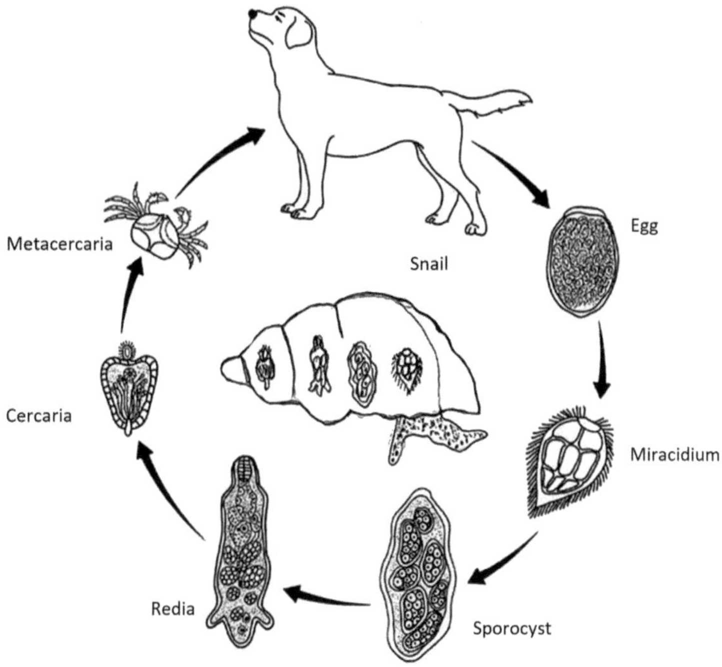

Fig. 4.23 Life cycle of *Paragonimus westermani*

Pathogenesis and Pathology

The immature *Paragonimus* flukes migrate into the lung parenchyma causing marked damage to the lung parenchymatous tissue. The adult parasites cause cyst formation in lungs. Hemorrhagic lesion, pneumonitis, and collection of exudates are some pathological features. The cysts are gradually filled up with purulent material.

> **Special Note**
>
> Cyst is formed by collection of infiltrated cells and exudate. The whole mass is covered up by fibrous tissue. Formation of fibrous tissue could be presumed to be as a result of normal immunological phenomena mediated by different cytokines. IL4 participates in fibrosis, and this cytokine is a potent profibrotic mediator. Other cytokines having a role for deposition of fibrous tissue are IL5, IL13, and IL21.

Pathogenesis is described in the flow chart:

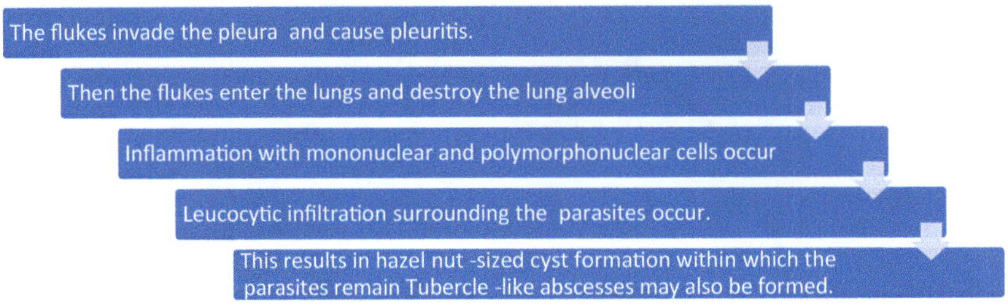

Fig. 4.24 Flow chart of pathogenesis caused by *Paragonimus westermani*

Pathology

Anatomical Pathology

The migratory flukes cause damage to the lung parenchyma. The migratory flukes cause damage in the course of their migration in the peritoneum, diaphragm, etc. Different natural infections and experimental infections reveal that the migratory flukes affect the pleura, which is filled up with hemorrhagic exudate. Then the parasites enter the lung alveoli causing significant damage in the vicinity. There occurs destruction of lung alveoli causing hemorrhagic pneumonitis. The parasite itself and their secretory material induce marked inflammation within the vicinity with infiltration of inflammatory cells predominantly with mononuclear cells, i.e., monocytes and macrophages. Later, the parasites are encapsulated by fibrotic and granulomatous tissues. Diffused fibrosis at the affected area of lungs is also found. Destruction and hyperplasia are marked in the alveolar linings. Congestion, consolidation, and emphysema are the characteristic pathological features.

Hematological Pathology

Demonstration of cellular picture is a part of determining pathology. Hematological data depict leukocytosis and eosinophilia in *Paragonimus* infection in human beings. In animals, leukocytosis has been found in some fragmented reports.

Chemical Pathology

Different studies indicate a rise of different vital enzymes, i.e., glutamate dehydrogenase, gamma-glutamyl transferase, etc. Massive loss of plasma protein has also been reported, which results in edema (bottle jaw).

Immunopathology

Inflammation invariably occurs in any parasitic infection, particularly in case of the tissue-invading trematodes. In this parasitic infection, occurrence of inflammation is obvious, which appears as a primary pathological feature. It is needless to mention that inflammation occurs mainly due to pro-inflammatory cytokines secreted by macrophage. ADCC is another one important immunological phenomenon. In ADCC, it is observed that effector cells (eosinophil, macrophage, NK cells) bind to Fc portion of antibody bound to antigen. These effector cells release toxic material. Particularly, in helminthic infection, NO (nitric oxide) secreted by activated macrophage is very important for killing helminthic parasite. Different studies show that this NO causes severe damage to the tissues while killing the pathogens.

Secondary Pathology

Secondary pathology is caused by secondary bacterial infection.

Molecular Pathology

Recent study has taken new dimension for analysis of pathology based on molecular methods. In the earlier days, pathology was described mainly by gross pathology and histopathology. Some cytokines play a role in inducing immunopathology. Due to this parasitic infection, upregulation of genes of pro-inflammatory cytokines and pro-fibrotic cytokines occurs.

Pathology of Ectopic Origin

As per reports, ectopic locations of *Paragonimus* are the brain and spinal cords. The parasites cause marked neuropathology there. This neuropathology results in different neurological signs.

Clinical Signs

The routes of migration of *Paragonimus* flukes are peritoneum, pleura, and pulmonary tissue. Different respiratory disorders are found. These are dyspnea, coughing, etc. Bilateral pneumothorax may cause death of the affected animals. Coughing with thick blood-tinged sputum material and eggs is observed. Sometimes, the parasites reach to the ectopic sites, i.e., intestine, liver, pancreas, etc., where they cause pathology. Diarrhea occurs if the intestine is affected.

Diagnosis

Symptomatological Diagnosis

It is not possible to diagnose this parasitic infection completely based on exhibited clinical signs. Respiratory signs are observed, which should be simulated with other diagnostic findings.

Parasitological Diagnosis

Parasitological diagnosis is possible by detection of eggs of *Paragonimus* in sputum/cough material. Gross examination of cough material reveals rusty color. The sputum should be mixed with 3% NaOH solution, which can disintegrate the mucus present in the sputum. If required, formalin-ethyl acetate concentration technique can be done. Examination of pleural aspirate can be done for detection of the eggs.

Immunological Diagnosis

Since decades, a panel of crude antigens was used for immunodiagnosis of this parasitic infection. These are whole somatic antigen, tegumental antigen, metacercarial antigen, egg antigen, etc. Excretory-secretory antigens were also used. But specific diagnosis was not possible due to cross-reactivity. Different defined antigens are used for performing immunodiagnostic test. Of these antigens, cathepsin L protease is a reliable defined antigen for performing ELISA. Different immunodiagnostic tests include counterimmunoelectrophoresis, bentonite flocculation test, latex agglutination test, indirect hemagglutination test, indirect fluorescent antibody test, complement fixation test, ELISA, indirect ELISA, dot ELISA, dipstick ELISA, Western blot, etc.

Molecular Diagnosis

The mainstay of molecular diagnosis is identification of conserved sequence of nucleotides and designing a specific primer. Different molecular markers are used for identification of this parasite. These markers are mainly of ribosomal DNA and RNA (ITS1, ITS2, 5.8 S rRNA, 18 S rRNA, etc.). Mitochondrial DNA (CO1, ND1, etc.) are also useful molecular markers for identification. Species-specific primer is available. Copro-PCR has also been described.

Treatment

Name of drug	Route of administration	Dose
Fenbendazole	Oral	20–50 mg/kg twice daily for consecutive 2 days in dogs
Albendazole	Oral	50–100 mg/kg daily for consecutive 2–3 weeks in cats

Control

Do's	Don'ts
• The affected animals should be treated properly • Measurements on snail control should be taken • Public education program for awareness of harm due to eating raw or uncooked crabs and crayfish should be initiated to control these parasites in human beings	The animals should not be allowed to eat raw and uncooked crabs, crayfish, etc.

Family: Prosthogonimidae

Family Synopsis

The members of this family are parasites of birds. The parasites are distributed all over the world. The flukes have spiny cuticle, pear-shaped body, elongated cirrus sac horizontal testes, pretesticular ovary, grape-like vitelline glands in the lateral field, Y-shaped excretory bladder. The life cycle is indirect and requires two intermediate hosts (first aquatic snail and second naiad stage of dragonfly). The developmental stages are egg, miracidium, sporocyst, cercaria, and adult. Redia stage is absent.

Genus: *Prosthogonimus*

Etymology

The generic name "*Prosthogonimus*" is a word derived from two Greek words, i.e., prosthenos meaning "at the tip" and "gonimos" meaning "fertile, productive."

Common name: Oviduct fluke

Distribution

Prosthogonimus ovatus are distributed in Asia, Africa, and Europe. *P. macrorchis* is predominantly available in North America.

Species

Prosthogonimus pellucidus
Prosthogonimus ovatus
Prosthogonimus macrorchis

Species: *Prosthogonimus pellucidus*

Distribution: The parasite is worldwide in distribution.

Common name: Oviduct fluke

Site

Normal site	Rare site
Oviduct	Intestine
Bursa of Fabricius	Peritoneum
Cloaca	

Hosts

Primary hosts	Natural hosts	Duck, goose, poultry birds
	Domestic hosts	Duck, goose, poultry birds
	Wild hosts	Turkey, pigeon etc.
Secondary hosts	First intermediate hosts: *Bithynia tentaculata, Amnicola limosa porata* Second intermediate hosts: dragonflies (*Leucorrhinia* sp., *Epicordulia* sp., *Tetragoneuria* sp., *Libellula* sp.	

Morphology

Adult

Shape: They are roughly pear shaped.

Size: 8–9 mm × 4–5 mm.

Color: The fresh specimens are pale reddish-yellow colored.

Suckers: Two suckers are found. Oral sucker is smaller than ventral sucker and elongated. The ventral sucker lies at about one third anterior part of the body.

Testes: Two testes are present. Testes are roughly oval in shape and horizontally placed behind the middle of the body.

Ovary: Ovary is pretesticular and substantially lobulated.

Vitelline glands: Vitelline glands are situated in the lateral field occupying approximately the middle third of the body.

Cirrus: Cirrus is small. The genital opening is near to the oral sucker. Cirrus sac is elongated.

Egg

The eggs are dark brown in color. They are operculated bearing a small spine opposite to the operculated pole.

Larva

The larval stages are miracidium, sporocyst, cercaria, and metacercaria.

Miracidium: The miracidium is pear shaped. The outer covering is ciliated.

Sporocyst: The miracidium is transformed to sporocyst after losing its ciliated covering. The sporocysts are roughly cigar shaped having a sac-like outer covering. They have brood chambers in which germ cells are present. Examination during initial stage reveals only germ ball but later developing cercariae are found. Rediae are not found in the sporocysts. Digestive system is absent.

Cercariae: A typical xiphidiocercariae are formed during the life cycle of *Prosthogonimus* spp. The cercariae are small and minute. The tail is simple. Two suckers are characteristically found. The oral sucker is well developed and present anteriorly. The ventral sucker is not markedly developed and present in the posterior half of the body. On either side of the ventral sucker, there are four penetration cells, which lead to anterior edge of oral sucker. Minute striations are found on the body of the cercariae.

Metacercariae: These are encysted form of cercariae. Two suckers are clearly visible. The ventral sucker is slightly larger than the oral sucker. A muscular pharynx is visible, which is followed by a short esophagus. The later part is bifurcated into two ceca.

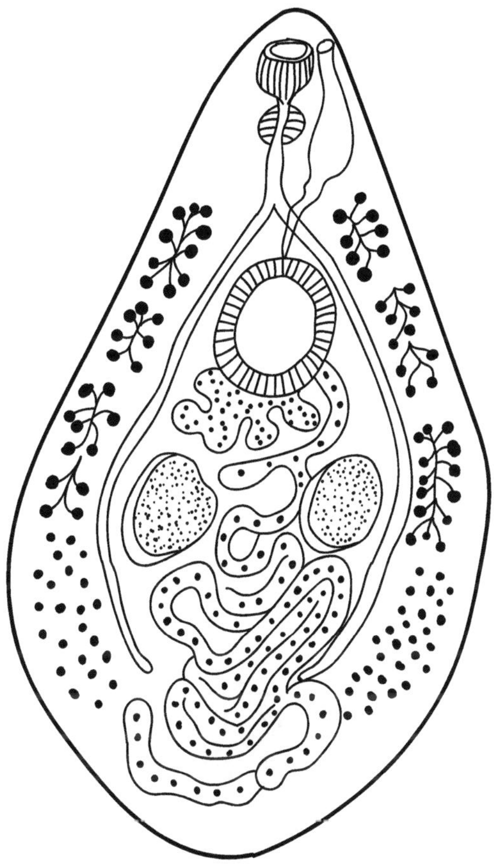

Fig. 4.25 *Prosthogonimus pellucidus*

Life Cycle

Some basic information of life cycle
• The definitive hosts are different birds • The first intermediate hosts are aquatic snails • The second intermediate hosts are naiad stage of dragonflies • The infective stage is metacercaria • The mode of infection is ingestion of the second intermediate hosts carrying metacercariae

Events of life cycle: a glimpse
• Voiding of eggs with feces of definitive host • Ingestion of eggs by snail • Hatching of eggs in gut of snail • Formation of sporocysts and cercariae in snail • Release of cercariae • Entrance of cercariae in naiad stage of dragonfly • Transformation of cercariae to metacercariae in naiad stage of dragonfly • Ingestion of nymph stage or adult stage of dragonfly harboring metacercariae by definitive hosts • Development up to adult in definitive hosts

Description

The parasites exhibit heteroxenous life cycle, which requires two intermediate hosts in their life cycle. Redia stage is absent. The stages are egg, miracidium, sporocyst, cercaria, metacercaria, and adult. Eggs are voided with the feces. Hatching of eggs occurs in the first intermediate host after being ingested. Different aquatic snails, i.e., *Bithynia tentaculata*, *B. leachii*, *Gyraulus albus*, etc., act as intermediate hosts. The stages that are formed in snails are sporocysts and cercariae. The cercariae come out from the snails. The cercariae are taken up by the second nymph stage of dragonfly. Different species of dragonfly, i.e., *Epicordulia*, *Leucorrhinia*, *Tetragoneuria*, etc., act as the second intermediate hosts. The uptake of cercariae by the nymph is typical. The breathing movement of the naiad stage directs the cercariae to enter through anal opening. The route of entrance of the cercariae is rectal respiratory chamber. The cercariae transform to metacercariae after losing its tail and remain encysted in the muscle of the nymph. Metacercariae are formed in the second intermediate host. The final hosts get the infection by ingestion of infected nymph or adult stage of dragonfly.

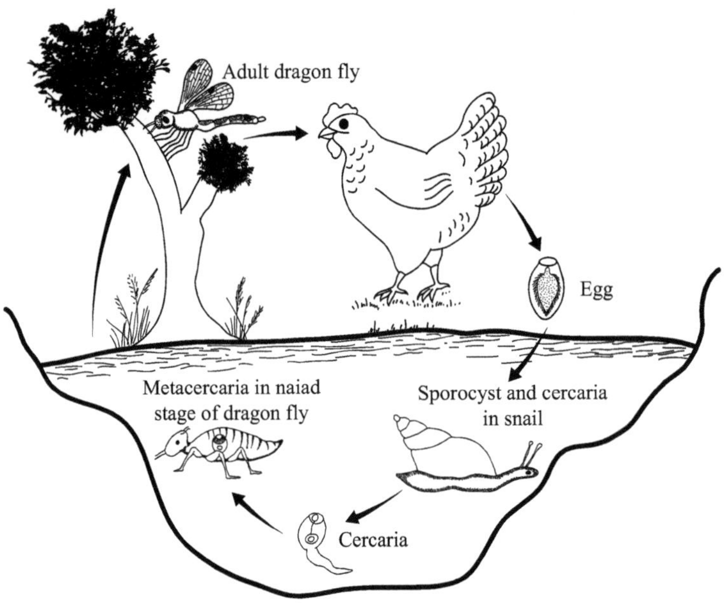

Fig. 4.26 Life cycle of *Prosthogonimus pellucidus*

Pathogenesis

The seat of predilection of the parasite is oviduct and bursa of Fabricius. The parasites cause irritation of oviduct using their spines, which cause severe inflammation. Yellow-white secretion is accumulated in the oviductal lumen. The irritation of the oviduct causes fast passing of eggs, and these are expelled out without formation of egg shell. The irritated oviduct also causes reversal of peristalsis, which causes entrance of parasitic material along with broken egg component in the peritoneum resulting in peritonitis.

Clinical Signs

Initially the birds do not show any clinical signs except some birds start laying eggs with soft shells or without any shells. The birds are weak and depressed and show the tendency to sit in the nest. As the shell is not formed, lime material in the form of milky fluid comes out separately. The cloaca and adjacent area are soiled with albumen, milky lime material. Peritonitis results in cyanotic combs and wattle. The birds die soon in this stage.

Diagnosis

Examination of discharged material reveals a large number of eggs.

Treatment

No satisfactory treatment is available. Praziquantel, fenbendazole and albendazole can be tried.

Control

Since the snails act as the first intermediate hosts, control of *Mollusca* by molluscicidal agent is important. Dragonflies act as the second intermediate host. Therefore, care should be taken about the population of dragonflies.

Family: Notocotylidae

Family Synopsis

The members of this family are parasites of birds and mammals. The flukes are characterized by presence of 3 or 5 rows of unicellular glands in groups situated in the ventral surface of the body, absence of ventral sucker, absence of pharynx, horizontal testes located at the posterior end of the body and lateral to the intestinal caeca, and presence of ovary in between the testes. They are heteroxenous parasite and require snail as intermediate hosts.

Genus: *Notocotylus*

Species: *Notocotylus attenuatus*

Distribution: The worms are distributed worldwide.

Hosts: The parasites occur in the ducks, goose, and other aquatic birds.

Site: Cecum and rectum.

Morphology

Adult

Size: They are 2–5 mm long.

Tegument: Spinosed ventrally and anteriorly.

Sucker: Only oral sucker is present. The ventral sucker is absent.

Pharynx: Absent.

Esophagus: Short.

Intestinal ceca: Long.

Testes: The pair of testes are lobed and situated lateral to the intestinal ceca near the posterior end.

Ovary: The lobed ovary is typically situated in between the testes.

Genital pore: Genital pore is located near the intestinal bifurcation.

Ventral glands: Three rows of ventral glands are present.

Vitelline glands: The vitelline glands are situated in the lateral field of posterior part of the body.

Egg

Eggs are small in size (20 × 10 μm). The eggs possess polar filaments at both ends.

Life Cycle

The events of life cycle are as follows:

- Eggs are passed with the feces.
- Eggs hatch after being ingested by aquatic snails (*Lymnaea* sp., *Planorbis* sp., *Bulinus* sp.).
- The stages are miracidium, sporocyst, redia, and cercaria. Cercariae encyst in the land snail.
- Final hosts are infected by ingestion of the infected snail.

Pathogenesis
The parasites are not harmful. They may cause enteritis resulting in diarrhea.

Diagnosis
Diagnosis can be done by examination of fecal sample for detection of the eggs. The eggs are typical, which have polar filaments.

Treatment

Name of drug	Route of administration	Dose
Oxyclozanide	Oral	15–30 mg/kg bw

Genus: *Catatropis*

Species: *Catatropis verrucosa*
The seat of predilection of this fluke is ceca and intestine. The hosts are mainly the ducks, goose, and other aquatic birds. They are 1–6 mm long. The oral sucker is small, whereas the ventral sucker is absent. The pair of testes are lobed and located at the posterior end. The lobed ovary is situated in between the testes. The vitelline glands are situated anterior to the testes. Three rows of ventral glands are present.

Species: *Catatropis indicus*
This is a species of trematode parasite under the family Notocotylidae. The seat of predilection of this fluke is ceca. The hosts are domestic fowls. They are 4.2–4.6 mm long. The ventral surface of the parasite is spiny and has three rows of ventral glands. A flask-shaped cirrus pouch is present. The oral sucker is round. Two intestinal ceca run up to the posterior end. The pair of testes are deeply lobed.

Genus: *Ogmocotyle*

Species: *Ogmocotyle indica*
Hosts: The hosts are cattle, sheep, and goat.
Site: Intestine and bile duct.

Morphology
Size: They are 0.8–2.72 mm long.
Color: The anterior part is elongated translucent. The posterior part is opaque.
Sucker: Only the oral sucker is present, which is round in shape.
Pharynx: Absent.
Esophagus: Short.
Testes: The testes are irregularly lobed and situated in the posterior end of the parasite.
Ovary: The ovary is lobed and situated at the posterior end.
Cirrus: Large papillae are present on cirrus. The cirrus pouch is markedly developed and present at the middle of the body.
Vitelline glands: The vitelline glands are characteristically arranged in a shape of horseshoe.
Life cycle: Not reported.

Pathology
In general, the parasites under this family are rarely associated with any pathology. But in some cases, pathogenic effects have been reported. The anterior part of the intestine is commonly affected. The intestinal wall is damaged by initial inflammatory reaction, and then necrosis occurs. The inflammation includes cellular infiltration with eosinophils and other leukocytes. Hemorrhages occur at the wall of the intestine resulting in hemorrhagic enteritis.

Symptoms
The affected animals exhibit different symptoms, which are diarrhea, dullness, debility, etc. There may be edema formation.

Family: Paramphistomatidae

Family Synopsis

The members of this family are usually the parasites of mammals. They are worldwide in distribution. The flukes are characterized by thick and fleshy body, suckers at both ends, anterior sucker sometimes with posterior pockets, absence of pharynx, lobed testes anterior to the ovary, and presence or absence of ventral pouch. Life cycle involves one snail host. The developmental stages are egg, miracidium, sporocyst, redia, cercaria, metacercaria, and adult. They cause less harm to the hosts. But immature stages cause marked pathogenesis. The gold standard diagnosis is examination of feces for finding the eggs of the parasite. Control mainly relies on proper treatment and snail control.

Morphology

Adult

The flukes are conical shaped. That's why they are called as conical flukes. The tegument is spineless. Two suckers are present. One is oral and another is ventral. Oral sucker is smaller than ventral. Two testes are present. Testes are located in different position. Ovary is usually post-testicular. Vitelline glands are situated in the lateral field.

Eggs

The eggs are oval, transparent, and operculated. Internally, it has mosaic appearance. The eggs of *Gigantocotyle* are yellowish.

Larva

The larval stages are miracidium, sporocyst, redia, cercaria, and metacercaria.

Miracidium: The miracidium is pear shaped. The outer epidermal covering is ciliated. Other structures found are similar to other digenean trematodes, which are eye spots, apical papilla containing sensory glands, and protonephridial type of excretory system comprising flame cells. Circulatory system and respiratory system are absent. Superficial argentophilic structures are found. Epidermal cells are 20 in number and arranged in 4 rows (6, 8, 4, 2 = 20).

Sporocyst: The ciliated covering of miracidium is lost and the miracidium is transformed to sporocyst. The sporocysts have brood chamber. Digestive system is absent. The body of the sporocyst is sac-like.

Redia: Fully matured rediae are sausage shaped. Both ends of rediae are rounded. Immature rediae display narrow posterior end. Features of rediae are similar to other digenetic trematodes. Rediae are formed from the germ balls present in the brood chambers of sporocyst. They have ambulatory buds, which are antero- and posterolaterally projected giving a typical shape of rediae. The redia has digestive system. Mouth, spherical pharynx, narrow esophagus, and saccular unbranched intestine are found in the digestive system. Excretory system is also found. Rediae usually contain 1–8 cercariae.

Cercaria: Brood chamber is found in the redia. Germ balls differentiate into cercariae.

Metacercaria: A cyst is formed from the cercaria.

Fig. 4.27 *Indoplanorbis* snail

Genus: *Paramphistomum*

Species: *Paramphistomum cervi*

Etymology

The word "*Paramphistomum*" has been derived from three Greek words, i.e., para (besides), amphi (on both sides), and stoma (mouth).

Distribution: The parasites are world wide in distribution. Prevalent in tropical and subtropical countries.

Hosts: Sheep, goat, cattle, and buffalo

Site: Rumen and reticulum

Morphology

The flukes are pear shaped and maggot-like. The tegument is without any spine. The body is dorsally convex and ventrally concave. The fresh specimens are pink (immature) to light red (adult). They are 6–7.5 mm long. Two suckers (oral and ventral) are present. The oral sucker is smaller as compared to ventral sucker. Both the suckers are large. The pair of testes are irregularly outlined and cauliflower-like in section. They are tandem in position.The ovary is post-testicular. The vitelline glands are situated in the lateral field.

Fig. 4.28 *Paramphistomum cervi*

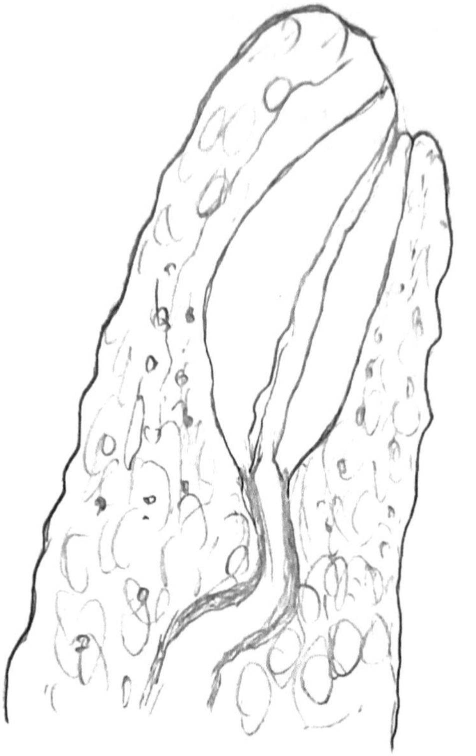

Fig. 4.29 Liorchis type of pharynx of *Paramphistomum cervi*

Species: *Paramphistomum cauliorchis*

This is a species of trematode parasite under the family Paramphistomatidae. The seat of predilection of this fluke is rumen and the hosts are cattle. They are 6–7.5 mm long. Two suckers (oral and ventral) are present. The oral sucker is smaller as compared to ventral sucker. The pair of testes are irregularly outlined and cauliflower-like in section. The vitelline glands are sparsely scatterred and situated in the lateral field.

Species: *Paramphistomum crassum*

The seat of predilection of this fluke is rumen and the hosts are cattle. They are 6.12–6.5 mm long and pyriform in shape. Two suckers (oral and ventral) are present. Both the suckers are large. The ventral sucker has a cavity, which opens on the ventral side. The pair of testes are characteristically cauliflower-like and situated diagonally and overlaps each other. The small ovary is situated right side near the testes. The vitelline glands are situated in the lateral field.

Species: *Paramphistomum epiclitum*

The seat of predilection of this fluke is rumen and the hosts are cattle and buffalo. They are 5–9 mm long. Two suckers (oral and ventral) are present. The ventral sucker is doubly larger as compared to the oral. The pair of testes are oval in shape and deeply lobed. The testes are tandem in position.

Species: *Paramphistomum gotoi*

The seat of predilection of this fluke is rumen and the hosts are buffalo. They are 5–10 mm long. Two suckers (oral and ventral) are present. The oral sucker is slightly smaller as compared to ventral sucker. The oral sucker is characteristically hourglass shaped possessing spine and papillae. The pair of testes are lobed. The ovary is situated posterior to the testes. The vitelline glands are situated in the lateral field.

Species: *Paramphistomum indicum*

The seat of predilection of this fluke is rumen and the hosts are sheep. They are 5.25–9.5 mm long. Two suckers (oral and ventral) are present. The ventral sucker possesses a cavity, which opens at the ventral side. The pair of testes are deeply lobed and cauliflower-like and either tandem or diagonal in position. The ovary is situated either left or right. The vitelline glands are situated in the lateral field.

Species: *Paramphistomum papillosum*

The seat of predilection of this fluke is rumen and the hosts are cattle. They are 5.12–6 mm long. Two suckers (oral and ventral) are large in size. Though the ventral sucker is large, its opening is small. The pair of testes are deeply lobed and tandem in position. The vitelline glands are situated in the lateral field.

Species: *Paramphistomum orthocoelium*

The seat of predilection of this fluke is rumen and the hosts are cattle and buffalo. They are 6.6 mm long. Two suckers (oral and ventral) are present. The oral sucker is smaller as compared to ventral sucker. The pair of testes are lobed and tandem in position. The ovary is situated in between the ventral sucker and posterior testes. The vitelline glands are situated in the lateral field.

Species: *Cotylophoron cotylophorum*

They occur in the rumen and reticulum of sheep, goat, cattle. The morphology is similar to *P. cervi*. But it has genital sucker surrounding the genital pore.

Fig. 4.30 *Cotylophoron cotylophorum*

Genus: *Gastrodiscoides*

Species: *Gastrodiscoides hominis*

The flukes occur in cecum of man and colon of pig. They are pear shaped. Fresh specimen is bright red in color. They are 4–14 mm long. Oral sucker possesses two pouches on either side. The ventral sucker has a deep cleft, which is an identifying feature. The two testes are lobed and placed diagonally. The round ovary is post-testicular in position. The vitelline glands are situated in the lateral field.

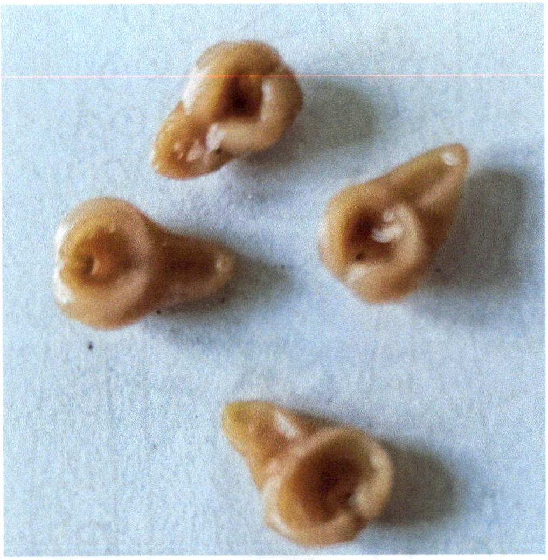

Fig. 4.31 *Gastrodiscoides hominis*

Genus: *Gastrodiscus*

Species: *Gastrodiscus aegyptiacus*

They are found in the intestine of elephant, horse, mule, etc. The body length is 9–17 mm. The oral sucker is smaller than the ventral sucker and equipped with well-developed pouches. The two testes are lobed and located diagonally. The ovary is oval in shape and post-testicular in position. The intestinal ceca reach to the poster end.

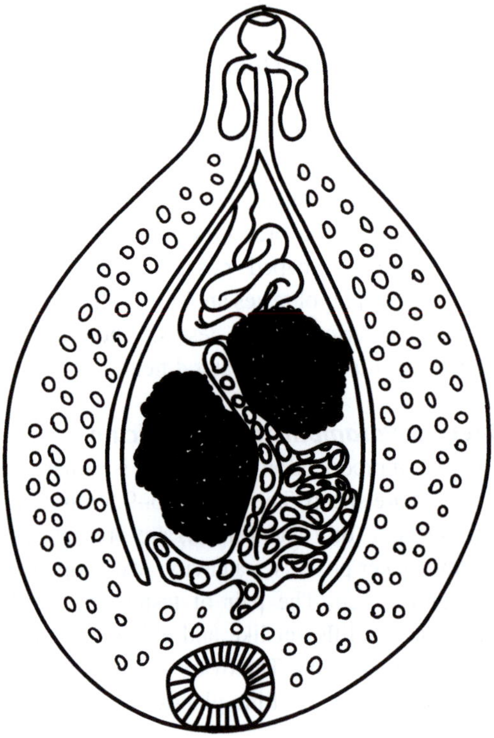

Fig. 4.32 *Gastrodiscus aegyptiacus*

Species: *Gastrodiscus secundus*

They are found in the intestine of elephant, horse, etc. The body length is 8 mm. The oral sucker is smaller than the ventral sucker. The two testes are lobed and located diagonally. The ovary is oval in shape and post-testicular in position. Fresh specimens pink in color. Anterior part is cylindrical and posterior part is saucer shaped. The intestinal ceca reach to the poster end. The vitelline glands are situated laterally. The genital pore is situated near the intestinal bifurcation. *Cleopatra, Indoplanorbis* snail act as the intermediate hosts of the parasite.

Genus: *Fischoederius*

Species: *Fischoederius elongatus*

The seat of predilection of this fluke is rumen and the hosts are cattle, buffalo, sheep, and goat. They are 10–20 mm long. A well-developed ventral pouch is present. The intestinal ceca are short and end behind the middle of the body. The pair of testes are deeply lobed and situated dorsoventrally. The ovary is situated in between the testes. The uterus runs straight in the median field. The vitelline glands are situated in the lateral and ventral sides, which extend from a little posterior to the intestinal bifurcation up to the end of posterior testes. The intermediate hosts are *Lymnaea luteola*. In case of Fischoederius cobboldi, the intestinal ceca are long and reach to up to the posterior end. This is the difference between these two species. In case of Fischoederius cobboldi, the intestinal ceca are long and reach to up to the posterior end. This is the difference between these two species.

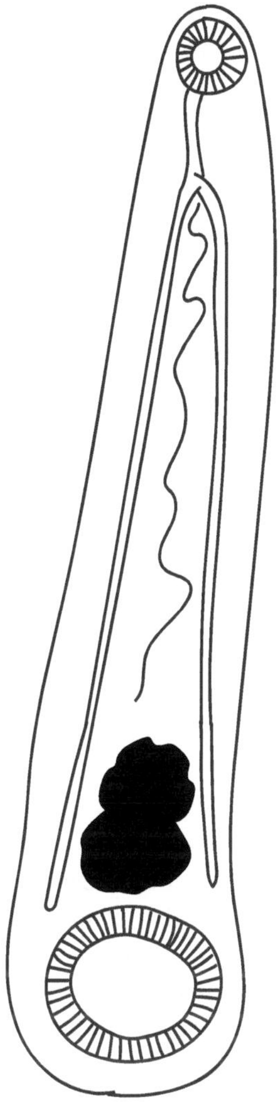

Fig. 4.33 *Fischoederius cobboldi*

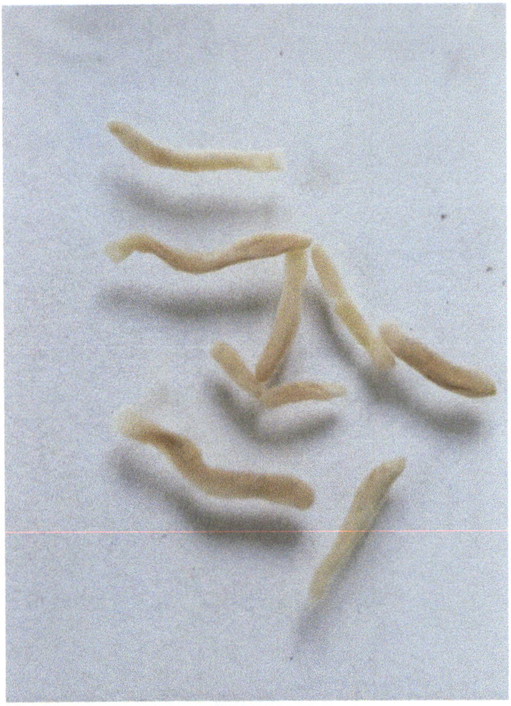

Fig. 4.34 *Fischoederius* sp.

Species: *Fischoederius cobboldi*

The seat of predilection of this fluke is rumen. The hosts are cattle, buffalo, goat, and sheep. They are 8–15.5 mm long. A deep ventral pouch is present. One testis is situated dorsal to another testis. The ovary is situated posterior to the testes. The intestinal ceca of this fluke end at the level of the posterior border of the posterior testes whereas the intestinal ceca of *F.elongatus* run a short distance and end behind the middle of the body.

Genus: *Gigantocotyle*

Species: *Gigantocotyle explanatum*

The important species of this genus is *G. explanatum*. They occur in the liver and duct. The size of the parasite is 8–13 mm × 3.6–4.3 mm. The ventral sucker is too large, which is about five times the oral sucker. The testes are pre-ovarian and placed diagonally. The cirrus pouch is absent. The uterus runs in the median field and has a convoluted terminal part. The intermediate host is freshwater snail, *Gyraulus convexiusculus*.

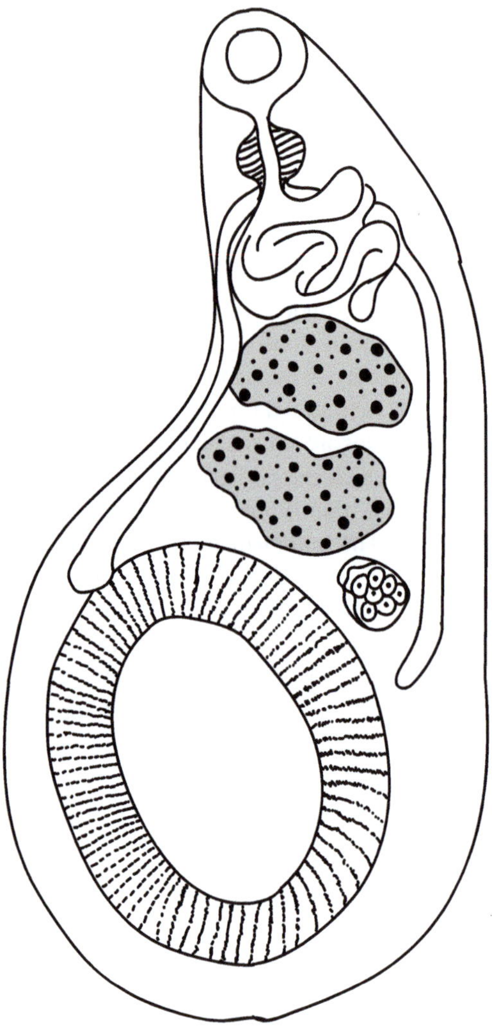

Fig. 4.35 *Gigantocotyle explanatum*

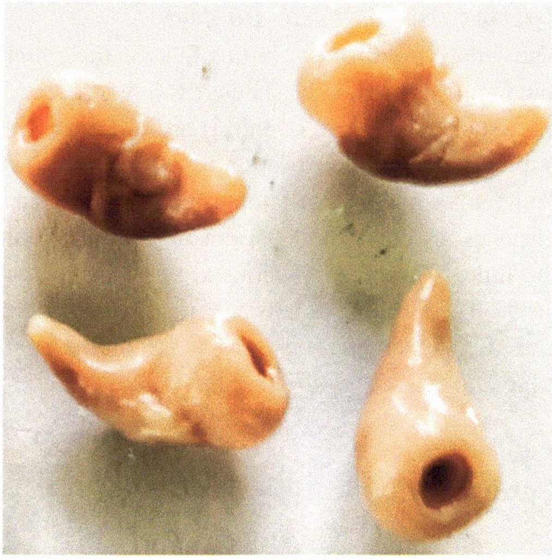

Fig. 4.36 *Gigantocotyle explanatum (Gross)*

Genus: *Gastrothylax*

Species: *Gastrothylax crumenifer*

The size of the fluke is 9–18 mm × 5 mm. They possess ventral pouch, which opens near the anterior end and extends up to the ventral sucker. The esophagus is characteristically "S" shaped. The testes are lobed and present just anterior to the ventral sucker. The ovary is post-testicular. *Indoplanorbis exustus*, *Gyraulus convexiusculus* are intermediate hosts.

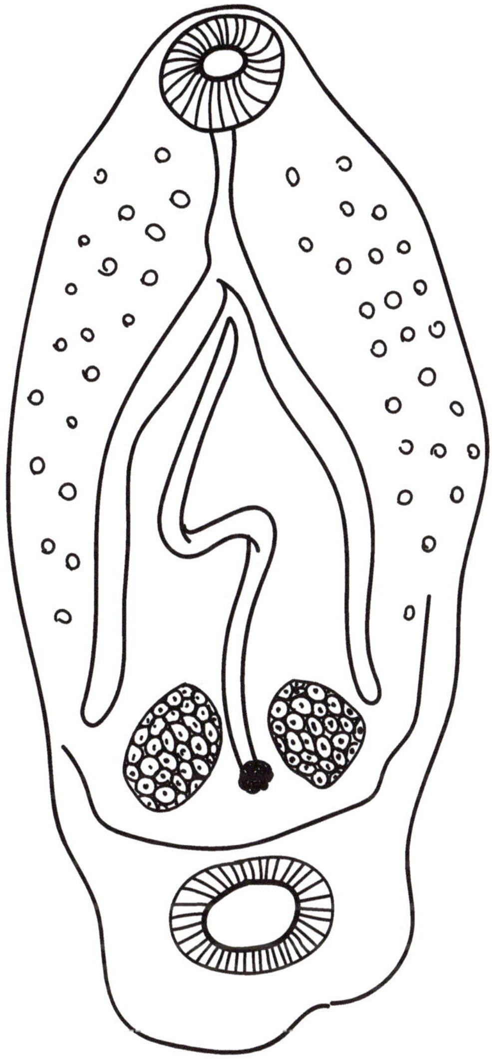

Fig. 4.37 *Gastrothylax crumenifer*

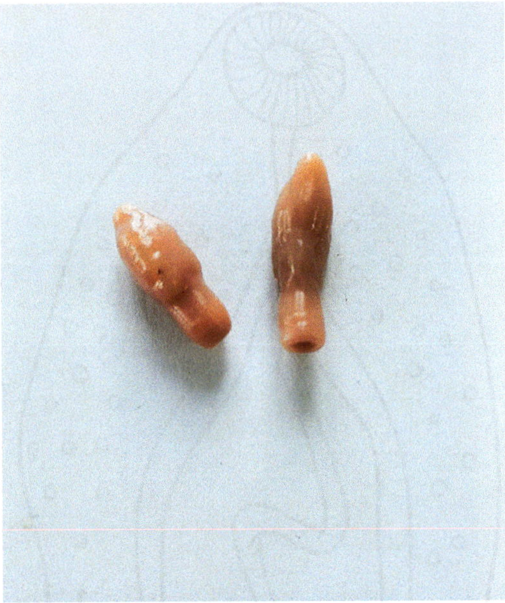

Fig. 4.38 *Gastrothylax crumenifer (Gross)*

Genus: *Pseudodiscus*

Species: *Pseudodiscus collinsi*

The seat of predilection of this fluke is the stomach, colon, and intestine. The hosts are horse, donkey, cattle, etc. This amphistome has also been reported from elephants in India. They are 4.85–12.5 mm long. Two suckers (oral and ventral) are present. The oral sucker is larger as compared to ventral sucker. Oral sucker is equipped with a pair of lateral pouches. The pair of testes are deeply lobed and located side by side. The ovary is situated posterior to the testes. The vitelline glands are situated in the lateral field. The intermediate hosts are *Planorbis*, *Indoplanorbis* etc.

Species: *Pseudodiscus hawkesi*

The seat of predilection of this fluke is the intestine and colon, and the hosts are Indian elephants. They are 3.6–11 mm long. Two suckers (oral and ventral) are present. The oral sucker has a pair of lateral pouches. The pair of testes are lobed and oblique in position and slightly overlapping. The testes may also be present one behind another. The ovary is situated immediately behind the posterior testes.

Species: *Pseudodiscus stanleyi*

The seat of predilection of this fluke is the colon and the hosts are horse. They are 8.6–9 mm long. Two suckers (oral and ventral) are present. The oral sucker has a constriction in the middle. The oral sucker has pair of lateral pouches. The pair of testes are lobed and present in the middle of the fluke. The ovary is situated posterior to the testes. The vitelline glands are situated in the lateral field extending from the base of the oral pouch up to the end of the ceca.

Genus: *Ceylonocotyle*

Species: *Ceylonocotyle scoliocoelium*

The seat of predilection of this fluke is the rumen and the hosts are cattle and buffalo. They are 2.7–7.86 mm long. The oral sucker is smaller as compared to ventral sucker. The pair of testes are lobed and tandem in position. The ovary is situated in between the ventral sucker and the posterior testes. The vitelline glands are situated in the lateral field.

Genus: *Homalogaster*

Species: *Homalogaster paloniae*

The seat of predilection of this fluke is the cecum, colon, and bile duct. The hosts are buffalo, cattle, goat, etc. They are 10 mm long. The pair of testes are situated obliquely in the anterior half of the body. The ovary is situated just anterior to the ventral sucker. The vitelline glands are situated in the lateral field.

Genus: *Carmyerius*

Species: *Carmyerius gregarius*

The seat of predilection of this fluke is the stomach and the hosts are cattle and buffalo. They are 7–10 mm long. Two suckers (oral and ventral) are present. The ventral sucker is slightly larger as compared to oral sucker. A well-developed ventral pouch is present, which extends up to the posterior end. The pair of testes are situated anterior to the ventral sucker. The ovary is situated posterior to the testes. The intestine runs a short distance and ends behind the middle of the body.

Species: *Carmyerius spatiosus*

The seat of predilection of this fluke is the rumen and reticulum, and the hosts are cattle and buffalo. They are 6–12 mm long. Two suckers (oral and ventral) are present. The oral sucker is smaller as compared to ventral sucker. The pair of testes are deeply lobed and situated at the posterior end symmetrically. The ovary is situated in between the testes. The vitelline glands are situated in the lateral field. The intestinal caeca run up to the end of second third of the body.

Genus: *Olveria*

Species: *Olveria indica*

The seat of predilection of this fluke is the rumen and the hosts are cattle and buffalo. They are 6.14–7.07 mm long. Two suckers (oral and ventral) are present, and the oral bears a pair of lateral pouches. The ventral sucker is larger than oral sucker. The esophagus is characteristically J shaped. The pair of testes are slightly lobed and situated diagonally. The vitelline glands are situated in the lateral field. The cuticular papillae are found at the region of oral sucker and genital sucker. The intermediate host of the fluke is *Gyraulus convexiusculus*.

Species: *Olveria bosi*

The seat of predilection of this fluke is the rumen and the hosts are cattle and buffalo. They are 7.6–9.04 mm. The cuticular papillae are present in the anterior region. The subterminal ventral sucker is much larger as compared to oral sucker. The oral sucker possesses a pair of pouches. The pair of testes are lobed and tandem in position. The ovary is situated immediately behind the posterior testes.

Life Cycle of Amphistomes

Some basic information of life cycle
• The definitive hosts are sheep, goat, cattle, etc.
• The intermediate hosts are aquatic snails
• The infective stage is metacercaria
• The mode of infection is ingestion of metacercariae

This parasite exhibits heteroxenous type of life cycle, which involves one intermediate host for completion of the life cycle. All developmental stages are found during life cycle. The stages are egg, miracidium, sporocyst, redia, cercaria, metacercaria, and adult.

Events of life cycle: a glimpse
• Passing of fluke eggs with feces of definitive host
• Hatching of eggs with contact of water and release of miracidia
• Entrance of miracidium in aquatic snail hosts
• Formation of different subsequent stages of the parasites, i.e., sporocyst, redia, and cercaria, in snail
• Release of cercariae from the snail
• Transformation of cercariae to metacercariae on aquatic vegetation and water surface
• Infection of definitive hosts by ingestion of metacercariae
• Development up to adult in definitive hosts

The following are different events of the life cycle:

Egg Passing

Adult parasites are hermaphrodite, which remain in different predilection sites, i.e., rumen, reticulum, etc. ***Gigantocotyle*** have different predilection sites (bile duct). The parasite self-fertilizes there. Eggs are voided through the feces.

Embryonation

Unembryonated eggs are passed with the feces. In the environment, the eggs get embryonated.

Egg Hatching

Eggs hatch when they come in contact with water. Water is an important stimuli for hatching. The eggs hatch after 2–3 weeks. After hatching, the miracidia come out. The miracidia are ciliated and pear shaped. They swim in water to detect suitable snail hosts. Within few hours, they penetrate the exposed soft tissue of the snails.

Development in Intermediate Host

The snails (*Planorbis*, *Indoplanorbis*) act as the intermediate hosts of the parasite. The snails are aquatic pulmonate snails. Miracidia penetrate the exposed soft tissues of the snail and enter the same. They follow the snail by the phenomenon of chemotaxis. The miracidia undergo different stages comprising sporocyst, redia, and cercaria. Immediately, the next stage of miracidium is sporocyst. The miracidia transform to sporocyst after losing its ciliated covering. The sporocysts get matured in 6–9 days approximately in summer season. More duration is required in winter season. The sporocysts do not have any digestive organs, and they perform their nutrition by absorption. A generation of rediae are produced from the sporocysts. Individual germ ball found in the sporocysts gives rise to formation of individual redia. The rediae have digestive organs comprising the mouth, pharynx, esophagus, and intestine. They survive by taking snail tissues as their food material. The rediae also have germ balls from which the cercariae develop. The rediae may produce daughter rediae. The next development stage is cercaria. The cercariae have tail by which they can swim in water. The cercariae decaudate their tails and encyst as metacercariae after shedding their tail appendages. The cercariae encyst on the aquatic vegetation and transform to metacercariae. The metacercariae are quiescent form infective stage having tough outer wall. The final hosts get the infection by ingestion of metacercariae.

Infection

The final hosts get the infection by ingestion of infective metacercariae.

Development in Definitive Hosts

The metacercariae get excysted in the small intestine (duodenum). The juvenile flukes come out from the metacercarial cysts and attach to the duodenal wall and feed there for about 6 weeks. The biochemical factors that play a role in excystation of metacercariae are pepsin, HCL present in abomasum, and trypsin and bile in small intestine. The prepatent period is about 7–10 weeks.

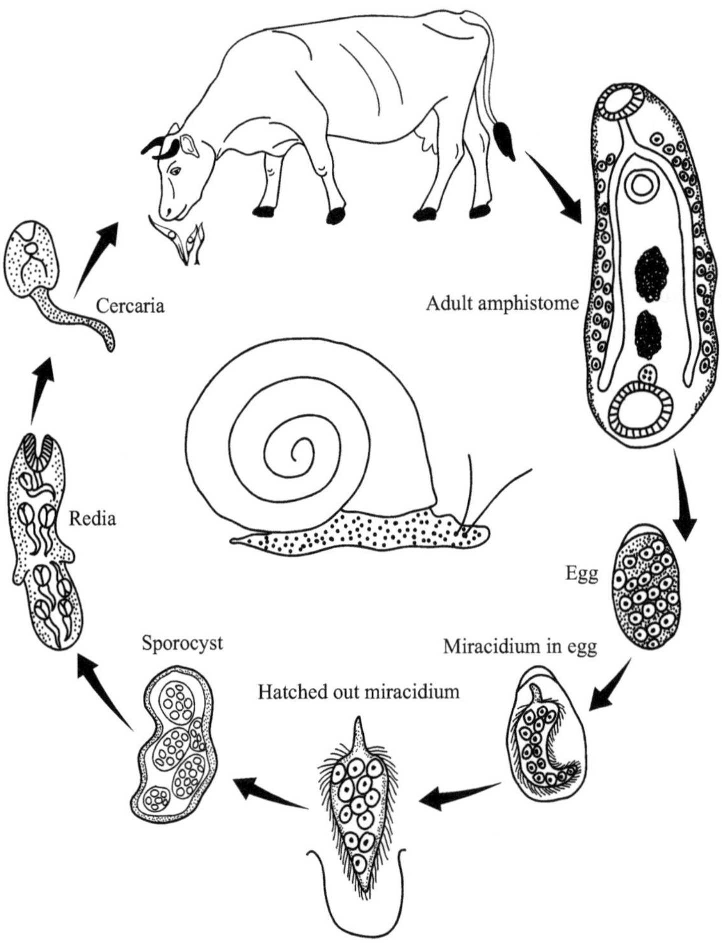

Fig. 4.39 Life cycle of amphistome

Pathogenesis

By Adult

The adult parasites do not produce any marked pathogenesis. Many adult parasites are found attached to the wall of the rumen and reticulum. During autopsy, sometimes, much area of the wall of rumen and reticulum is covered by these flukes. However, no marked pathogenesis is detected. Loss of ruminal papillae occurs. The less pathology occurs due to antigenic sharing of host and the parasite.

By Juvenile

Immature amphistomes cause massive pathogenesis than adults. Serious pathogenesis occurs in the young animals. Juvenile parasites are plug feeders and take out plug of mucosa from the wall of duodenum and place in their acetabulum. A large number of small immature amphistomes remain embedded on the wall of the abomasum and duodenum. Hemorrhage and erosion of epithelial cells are common features. Degeneration of cells results in marked necrosis. Erosion may even reach up to the muscle. There occurs destruction of glandular cells of intestinal wall. Continuous drainage of plasma protein from the affected wall of the intestine causes hypoproteinemia resulting in development of edema. Other features are hydropericardium, hydrothorax, pulmonary edema, etc. Edema also occur in the intermandibular space, abomasum, and mesenterium.

Pathogenesis and Pathology

Flow chart of pathogenesis

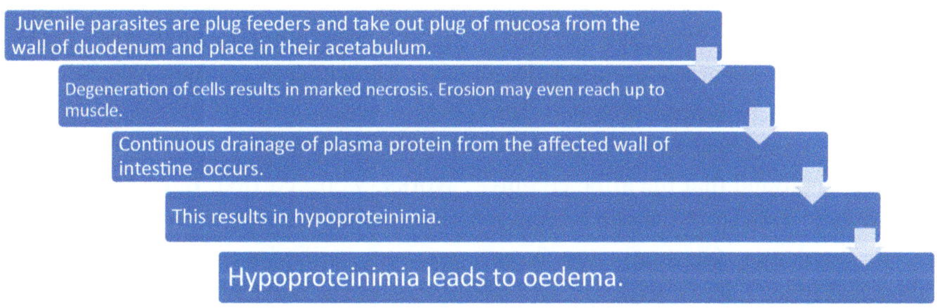

Fig. 4.40 Flow chart of pathogenesis caused by amphistomes

Pathology

Anatomical Pathology (Gross)

The immature flukes severely damage the wall of the intestine. The intestinal wall becomes ragged. Through this ragged surface of the intestine, plasma protein is drained out.

Histopathology

Histopathology of affected tissue of intestinal wall reveals cellular infiltration with neutrophil, eosinophil, etc. Congestion and hemorrhagic spot are common features. The flukes are found embedded deeply in the mucosa. The pathological features are also found in longitudinal muscle layer.

Hematological Pathology

Blood picture reveals decreased RBC count, eosinophilia, and neutrophilia. Decreased PCV and Hb% have been recorded in immature amphistomosis.

Chemical Pathology

Hypoalbuminemia is found due to severe loss of plasma protein.

Secondary Pathology

There have been records of mixed infection of amphistomes and different other intestinal flukes, i.e., *Heterophyes* sp., *Metagonimus* sp., and *Echinostoma* sp.

Immunopathology

Inflammation and fibrosis are the main pathological features. Both these two pathological features could be presumed in the light of immunopathology. Pro-inflammatory cytokines, i.e., IL1 beta, TNF alpha, and IFN gamma, are released by activated macrophage, and these cytokines play a major role for tissue damage and inflammation. There also occurs cascade reaction of complement, and some complement fragments attract the inflammatory cells for infiltration. Fibrosis is another pathological feature, which occurs in *Gigantocotyle* infection. The alternatively activated macrophages secrete profibrotic factors, which activate the fibroblast cells to secrete fibrin.

Clinical Signs

Marked pathogenesis is not caused by these flukes. Thousands of flukes fail to produce marked pathogenesis. However, significant pathogenesis is caused by immature amphistomes. Anorexia and diarrhea are common features in case of immature amphistomosis. The affected animals do not drink water or drink very less. Sometimes, it is found that the affected animals simply sip their jaws in the water. Initially, semisolid feces come out. Gradually the feces become watery and fetid. Edema occurs in different parts. But edema mainly occurs in the intermandibular area. Later anemia and emaciation prevail. Animals may die.

Diagnosis

Quick look: diagnosis of amphistomes
• Clinical signs, i.e., fetid diarrhea and edema, provide clue for speculating immature amphistomosis • Examination of fecal sample for detection of eggs and sometimes detection of pink-colored immature flukes in the diarrheic feces • Efforts made for the development of immunodiagnostic tests by using crude and defined antigens • Molecular diagnosis by different molecular markers

Diagnosis in Detail

History-based Diagnosis

History of grazing around the highly populated snail habitats during dry season provides clues for diagnosis. But this does not provide confirmation.

Symptomatological Diagnosis

No diagnostic symptoms are exhibited by the affected animals. However, some clinical signs, i.e., fetid diarrhea and edema, provide clue for speculating immature amphistomosis.

Postmortem Diagnosis

After death of the animals, the clusters of brownish pink immature flukes are found attached to the duodenal wall.

Parasitological Diagnosis

Sometimes, pinkish-colored small immature amphistomes come out with the feces. Examination of fecal sample may reveal the presence of characteristic amphistome eggs. The amphistome eggs are transparent and have mosaic appearance within the eggs. The eggs are large in size. They have prominent operculum at one end.

Immunodiagnosis

Works were carried out to perform different immunodiagnostic tests using crude somatic antigen and purified antigen. But no specific immunodiagnostic test has been commercialized due to the problem of cross-reaction. The antigens used for immunological characterization and immunodiagnostic trials are whole somatic antigen, excretory-secretory antigen, egg antigen, etc. Whole somatic antigen is prepared following different steps comprising collection of adult worms, washing by PBS, homogenization at 14,000 rpm in refrigerated centrifuge, and separation of supernatant. The supernatant is used as antigen. For preparation of excretory-secretory antigen, the adult flukes are kept in PBS in aseptic condition using antiprotease. The material is centrifuged and supernatant is used as antigen. For preparation of egg antigen, the eggs are washed properly for removal of host origin material. The eggs are sonicated. The sonicated material is centrifuged. The supernatant is used as egg antigen. Study shows immunodiagnostic potential of 52 kDa and 90 kDa protein of *Paramphistomum cervi* and 16 kDa protein of *P. gracile*. Different immunological tests are performed. The tests are agar gel precipitation test, immunoelectrophoresis, counterimmunoelectrophoresis, complement fixation test, indirect fluorescent antibody test, ELISA, indirect ELISA, sandwich ELISA, etc.

Molecular Diagnosis

Molecular diagnosis is possible using genus-specific or species-specific different molecular markers (ribosomal DNA, ribosomal RNA, mitochondrial DNA).

Control

Do's	Don'ts
• The affected animals should be treated properly • The feces of the animals should be disposed properly • The animals should be grazed in high land area where there is no stagnant water • Bioclimatograph should be analyzed if available • Reports of remote sensing should be analyzed if available	• The animals should not be allowed to drink pond water or river water • The animals should not be allowed to graze in infected pasture

General Control

The affected animals should be treated properly. Since water is the prime environmental factor for hatching of eggs, the animals should not be allowed to graze in low land area where stagnant water is available. The animals should not be allowed to drink river water and pond water because there is a chance of metacercariae contamination in this water. Underground water should be provided. The snails should be controlled by molluscicidal agents. There are different molluscicidal agents. These are copper sulfate, sodium pentachlorophenate, and di-nitro-o-cyclohexyl-phenol. Extermination of aquatic plants or weeds is another way to control snails because the aquatic plants are food materials of snails.

Immunological Control

Works were carried out to produce vaccine using different defined somatic antigen and purified antigens. But no vaccine is available in the field.

Molecular Control

No molecular vaccine is available in the field.

Biological Control

Studies indicate ovicidal effect of some fungi but no such technology has been commercialized.

Treatment of Amphistomes

Name of drug	Route of administration	Dose
Oxyclozanide	Oral	18.7 mg/kg (should be repeated after 2 days) in cattle and buffalo
Niclosamide	Oral	50–100 mg/kg for cattle, buffalo, and sheep

Family: Schistosomatidae

Family Synopsis

The members of this family are parasites of birds and mammals. They are worldwide in distribution. The flukes are characterized by cylindrical body resembling roundworm, unisexual with marked sexual dimorphism, and presence of gutter-like groove in male parasite used for carrying female parasite called gynecophoric canal. The intestinal ceca unite posteriorly, which further bifurcate and unite. The eggs are frequently equipped with terminal or lateral spines. They are heteroxenous parasite and require snails as intermediate hosts. The developmental stages are egg, miracidium, sporocyst, cercaria, and adult. They cause variable pathogenesis in the hosts. The gold standard diagnosis is examination of feces for finding the eggs of the parasite. Control of the parasites mainly relies on proper treatment and snail control.

Genera

- *Schistosoma*
- *Ornithobilharzia*
- *Heterobilharzia*
- *Gigantobilharzia*
- *Trichobilharzia*
- *Bivitellobilharzia*

Genus: *Schistosoma*

Distribution

The flukes are distributed in tropical and subtropical countries. Some species occur in Europe and America.

History

Calcified Schistsoma eggs were found in Egyptian mummies dating back to 3000 BC.

Site

Normal site	Rare site
• Mesenteric veins drained to both small and large intestine (*S. mansoni*) • Venous plexus draining to urinary bladder (*S. haematobium*) • *S. nasalis* occurs in nasal vein • *S. bovis* occurs in mesenteric and portal veins • *S. spindale* occurs in mesenteric veins • *S. indicum* occurs in portal and mesenteric veins • *S. incognitum* occurs in mesenteric vein	Brain and spinal cord (*S. mansoni, S. japonicum, S. haematobium etc.*)

Hosts

Primary hosts	Natural hosts	Cattle, sheep, goat, horse (*S. nasalis*) Cattle, sheep, goat (*S. bovis*) Pig, dog (*S. incognitum*) Ruminants and dogs (*S. spindale*) Ruminants, equines (*S. indicum*) Man and animals (*S. japonicum*) Man and animals (*S. haematobium*)
	Unnatural hosts	Lagomorph, rodents (*S. nasalis*)
	Domestic hosts	Cattle, sheep, goat, etc. (*S. nasalis*)
	Wild hosts	Lagomorph, rodents, etc. (*S. nasalis*)
	Experimental hosts	Mice, guinea pig, etc. (*S. nasalis*)
Secondary hosts	*Biomphalaria* (*S. mansoni, S. haematobium, S. nasale etc.*) *Indoplanorbis* (*S. nasalis, S. spindale etc.*) *Planorbis* (*S. nasalis, S. spindale etc.*) *Bulinus (S. haematobium, S. mattheei, S. bovis etc.)*	

Species: *Schistosoma nasalis*

Morphology

Adult

Shape: They are elongated cylindrical and look like roundworm.

Size: The length is approximately 6.5–11 mm.

Sex status: They are unisexual flukes. Male and female are different.

Tegument/cuticle: The tegument is tuberculated and spined.

Special feature: Male parasite has a gynecophoric canal by which the female parasite is carried during copulation.

Suckers: Two suckers are present. One is oral and another is ventral. Oral sucker is smaller than ventral sucker. Ventral sucker is definitively pedunculate.

Intestinal ceca: The bifurcated intestinal ceca reunite in the posterior half of the body which may again bifurcate and unite.

Testes: 3–6 testes are present in a row.

Ovary: The ovary is oval in shape and located just anterior to the site of reunion of the intestinal ceca.

Egg

The eggs of *Schistosoma nasalis* can be differentiated from eggs of other species. The eggs of *Schistosoma nasalis* are typically Napoleon's hat shaped or boomerang shaped or palanquin shaped. The eggs of *S. mecongi* have small lateral spine. The eggs of *Schistosoma mansoni* are oval in shape having a sharp lateral spine at the posterior end. The eggs of *S. haematobium* are also oval in shape, but the egg has terminal spine. The eggs of *S. japonicum* have rudimentary spine placed laterally. The eggs of *S. intercalatum* have prominent terminal spine.

Larva

The larval stages are miracidium, sporocyst, cercaria, and schistosomula.

Miracidium: The morphological features are similar to other digenetic trematodes. The miracidium is pear shaped. The body of miracidium is covered by ciliated flattened epidermal plates. These are present in four rows (6, 8, 4, 3 = 21). Beneath the covering, round and longitudinal muscle cells are present. At the anterior end, apical papillae along with secretory glands are present. Laterally, on either side of miracidium, papillae are present. Behind the apical region, a brain is present from which nerve fibers arise. Eye spots are present in miracidium. A protonephridial system is also present.

Sporocyst: The ciliated covering of miracidium is lost and the miracidium is transformed to sporocyst. The sporocysts have brood chambers. Digestive system is absent.

Cercaria: Cercariae are elongated. The body is broadly divided into two parts, i.e., head and tail. The head is tapering. Prominent penetration glands are found. The tail is forked. The outer covering is a glycocalyx, which is armed with spines and papillae. Several acetabular glands are present, which are used for adhesion with host and penetration of host tissue. A pair of head glands are found, which provide membrane material for development of schistosomula. A pair of escape glands are present near the mouth, the secretion of which enables the cercariae for escaping from the sporocysts. Brood chamber is found in the sporocyst. Germ balls differentiate into cercariae. The flame cell formula is 2[2+1+(1)]=6+(2).

Schistosomula: During penetration by cercariae, different proteolytic enzymes are secreted by pre-acetabular and post-acetabular glands. The tail is cast off. The trilaminate plasma membrane along with sugar layer glycocalyx is lost and actual tegument of schistosomula is formed.

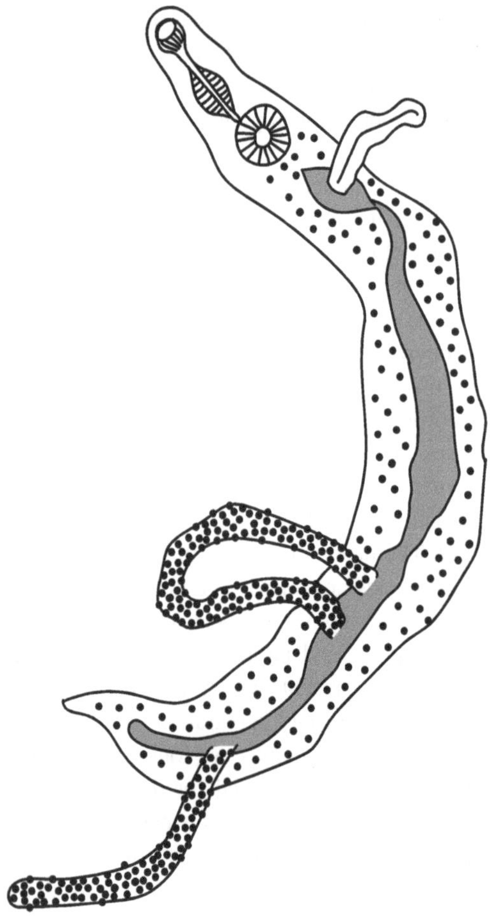

Fig. 4.41 *Schistosoma* sp.

Life Cycle

This parasite performs heteroxenous life cycle, which involves one intermediate host. All developmental stages are not found. Redia and metacercariae stage are absent. The stages are egg, miracidium, sporocyst, cercaria, schistosomula, and adult.

Some basic information of life cycle
• The definitive hosts are a wide variety of animals, i.e., cattle, sheep, goat, pig, etc. • The intermediate host is aquatic snail • The infective stage of the parasite is cercaria • The mode of infection is skin penetration by cercaria

Events of life cycle: a glimpse
• Voiding of eggs with feces of definitive host • Hatching of eggs in contact with water • Release of miracidium • Entrance of miracidium in aquatic snail hosts (*Indoplanorbis exustus*, *Lymnaea luteola*). • Formation of sporocyst, daughter sporocyst, and cercaria in snail • Penetration of the skin of the definitive hosts by the cercariae • Transformation of cercariae to schistosomula stage in the definitive hosts • Development up to adult in definitive hosts

Life Cycle in Detail

Egg Passing

Adult parasites are unisexual. Male and females are different. They occur in the nasal cavity. The eggs are laid by the adult parasites in nasal cavity. Eggs are voided through the nasal discharge or feces.

Egg Hatching

Hatching of eggs occurs in the environment. The eggs hatch in the water in minutes. Water, temperature, and salinity of the media are some stimuli required for hatching of the eggs. After hatching, the miracidia come out. The miracidia are ciliated and pear shaped.

Development in Intermediate Host

The snails (*Indoplanorbis exustus*, *Lymnaea luteola*) act as intermediate hosts of the parasite. The snails are aquatic pulmonate snails. Miracidia enter these snails penetrating the exposed soft tissue. This activity is mediated by chemotaxis. The stages that are formed in the snails are sporocyst and cercaria. The sporocysts are formed from miracidia. The sporocysts do not have any digestive organs, and they perform their nutrition by absorption. A generation of cercariae are produced from the sporocysts. The cercariae are non-ocellate, apharyngeate, brevifurcate, and furcocercous.

Infection to Definitive Hosts

The final hosts get the infection through skin penetration by the cercariae or by ingestion of the cercariae. The percutaneous route is more common.

Development in Definitive Hosts

During penetration, the tail is lost and transforms to a new stage called as schistosomula.

Special Note

High-density oligonucleotide array was conducted using all gene models from *Schistosoma mansoni* by Parker-Manual et al. (2011). cDNA derived from germ ball, cercariae, and day 3 schistosomula was hybridized to the array without amplification. Upregulation of different sets of developmental genes were found at each stage, which is depicted by upregulation of genes encoding elastase and other enzymes, which were found in germ ball. Upregulation of genes encoding cysteine protease and aspartic proteases was found in the cercariae and schistosomula. Micro-exon genes encoding variant secreted proteins were found to be highly upregulated in schistosomula in which upregulation of tegument and gut-associated genes was also found. Therefore, it is clear that different proteins are expressed in different developmental stages of parasite. The present information is related to *S. mansoni*, but this information provides a concept of expression of genes in different developmental stages of *Schistosoma*.

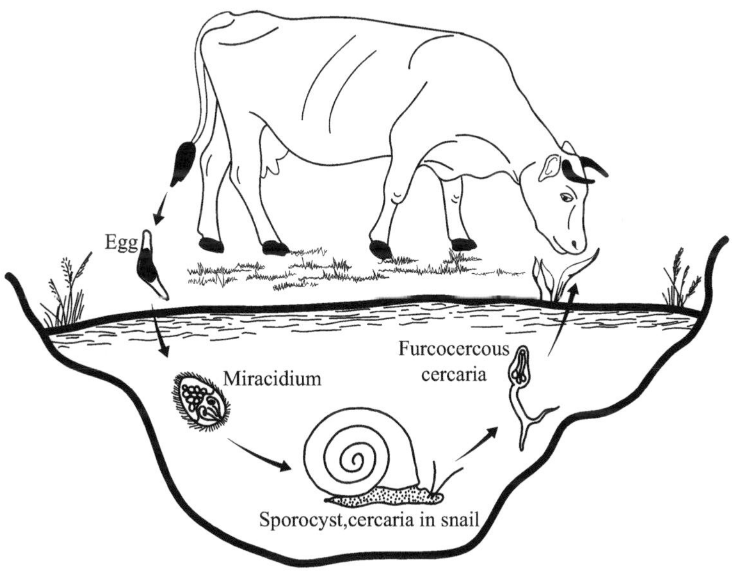

Fig. 4.42 Life cycle of *Schistosoma nasalis*

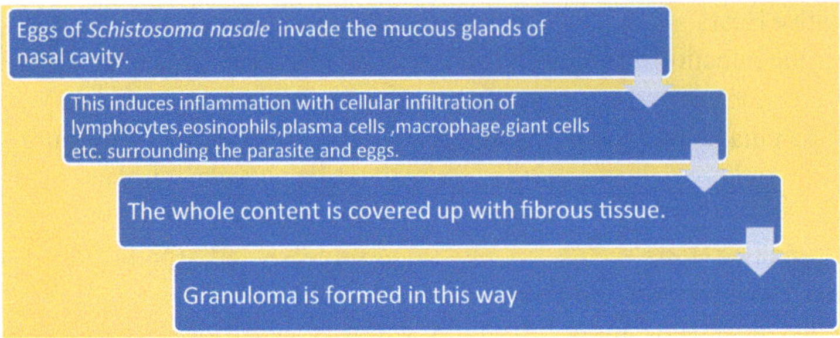

Fig. 4.43 Flow chart of pathogenesis caused by *Schistosoma* sp.

Pathogenesis and Pathology

Pathogenesis

Granuloma development occurs in case of *Schistosoma* infection. *Schistosoma nasalis* causes cauliflower-like growth in nasal cavity, which is called nasal granuloma or snoring disease. The eggs are more immunogenic and induce granuloma formation. The eggs of *Schistosoma nasalis* invade the mucosal glands of nasal cavity. The glands are markedly damaged. Many miliary abscesses are formed. The abscess bursts and the content of the abscess causes formation of another abscess. There occurs infiltration of cells comprising lymphocytes, eosinophils, plasma cells, macrophages, giant cells, etc. The area along with these cells is covered up by granuloma tissues. The condition of nasal granuloma is well marked in cattle than buffalo.

Pathology

Anatomical Pathology (Gross)

The flukes severely damage the wall of nasal cavity. Granuloma formation is a common consequence of schistosomosis. Nasal granuloma appears like cauliflower growth.

Histopathology

Histopathology of affected tissue reveals marked fibrosis and granulomatous cellular arrangement.

Hematological Pathology

Blood picture reveals a decrease of RBC count, eosinophilia, and neutrophilia. Decreased PCV and Hb% have been recorded in immature amphistomosis.

Chemical Pathology

Hypoalbuminemia is found due to severe loss of plasma protein.

Immunopathology

Granuloma formation occurs surrounding the schistosome and their eggs. Macrophage, eosinophils, and neutrophils are collected, which are finally covered up by fibrous tissue. IL4 participates in fibrosis, and this cytokine has long been considered a potent profibrotic mediator. Other cytokines having a role of deposition of fibrous tissue are IL5, IL13, and IL21. It could be presumed that these cytokines have role in deposition of fibrous tissue. It has been found that the efficacy of IL4 is nearly twice as TGFβ. Now it is a proven fact that synthesis of the extracellular matrix comprising types I and III collagen and fibronectin occurs as a result of IL4 stimulation. Study shows that anti-IL4 antibody causes significant reduction of hepatic fibrosis. This study indicates that IL4 has role in deposition of fibrous tissue.

Molecular Pathology

Study on molecular pathology mainly focuses on demonstration of expression level of genes of immune substances (cytokines, chemokines) and receptors related to immunity. Study shows upregulation of profibrotic cytokine genes in schistosomosis. Upregulation of genes of collagen type I, type II, and type IV has been observed. Upregulation of genes of three fibrogenic cytokines, i.e., TNF, TGF beta, and IL4, has also been reported. Research studies show elevated level of gene expression of fibrogenic cytokines.

Symptoms

Coryza is an important sign of this disease. The animals suffer from difficulty in breathing. Mucopurulent and sanguinopurulent discharge are observed. The affected animals show respiratory distress comprising sneezing, snoring, etc.

Diagnosis

Quick look: diagnosis of *Schistosoma*
• Examination of fecal sample or nasal discharge for detection of eggs • Efforts made for development of immunodiagnostic test by using crude and defined antigens. These antigens are cathepsin L, cathepsin L1, cathepsin B, cathepsin C, cathepsin C1, and cysteine protease • Molecular diagnosis by different molecular markers

Symptomatological Diagnosis

Some clinical signs, i.e., cauliflower-like growth in the nasal orifice, along with respiratory symptoms like nasal discharge and snoring sound confirm the case as nasal granuloma (NG) caused by *Schistosoma nasalis*.

Parasitological Diagnosis

Examination of nasal washing and nasal scraping may reveal the presence of characteristic eggs of *Schistosoma*. The eggs of *Schistosoma nasalis* are Napoleon's hat or boomerang shaped. 1 ml of nasal mucus material is mixed with 9 ml of 10% KOH solution in a test tube. This test tube containing the material is incubated in room temperature for 10–20 min. The material at the 1 ml of bottom is examined for detection of egg. The material should be boiled in restricted way and examined after centrifugation. The applied heat causes removal of the mucus.

Immunodiagnosis

Works were carried out to perform different immunodiagnostic tests using crude somatic antigen and purified antigen. But no specific immunodiagnostic test has been commercialized due to the problem of cross-reaction. In the earlier period, crude antigens used for the diagnosis of schistosomosis were whole somatic antigen, tegumental antigen, egg antigen, etc. Subsequently several defined antigens were used for immunodiagnosis of schistosomosis with variable success. These antigens are cathepsin L, cathepsin L1, cathepsin B, cathepsin C, and cysteine protease. The trials were done for different species, i.e., *S. haematobium*, *S. mansoni*, and *S. japonicum*. The different immunological tests comprise counterimmunoelectrophoresis, indirect fluorescent antibody test, ELISA, sandwich ELISA, Western blot, dot ELISA, etc. Similar immunodiagnostic works can be carried out in case *S. nasalis*.

Molecular Diagnosis

Molecular diagnosis is possible using genus-specific or species-specific primer. Different molecular markers are used for identification of parasites. These markers are mainly of ribosomal DNA and RNA (ITS1, ITS2, 5.8 S rRNA, 18 S rRNA, etc.). Mitochondrial DNA (CO1, ND1, etc.) are also used as molecular markers for identification of parasite.

Control

Do's	Don'ts
• The affected animals should be treated properly • The feces of the animals should be disposed properly • The animals should be grazed in high land area where there is no stagnant water • Bioclimatograph should be analyzed if available • Reports of remote sensing should be analyzed if available	• The animals should not be allowed to drink pond water or river water

General Control

The affected animals should be treated properly. Since water is the prime environmental factor for hatching of eggs, the animals should not be allowed to graze in low land area where stagnant water is available. The animals should not be allowed to drink river water and pond water because there is a chance of metacercariae contamination in this water. Underground water should be provided. The snails should be controlled by molluscicidal agent.

Immunological Control

Works were carried out to produce vaccine using different crude somatic antigens and purified antigens. Studies show the use of glutathione S-transferase with considerable success. But no vaccine is available in the field. The lead antigens that have been used for vaccination trials are 28 kDa GST (glutathione S-transferase) of *Schistosoma haematobium*, 14 kDa fatty acid-binding protein (FABP) of *Schistosoma mansoni*, tetraspanin protein of *Schistosoma mansoni*, and 80 kDa protein of *Schistosoma mansoni*. The short forms of these defined antigens are Sh28 GST, Sm14, Sm-TSP-2, and Smp80. Similar works can be carried out for *S. nasalis*.

Molecular Control

Works are carried out to produce gene vaccine but no molecular vaccine is available in the field.

Biological Control

No biological control has yet been commercialized. Genus: ***Schistosoma***

Other Species of ***Schistosoma***

Species: *Schistosoma indicum*

This species is prevalent in India. They occur in portal, pancreatic, mesenteric veins, etc. of cattle, sheep, goat, camel, etc. The male is 5–19 mm long and female is 6–28 mm long. The body cuticle is tuberculate and spined. The bifurcated intestine reunites posteriorly and continues further as a medium caecum. Just posterior to the ventral sucker, 5–12 testes are present forming a row. The parasites cause hepatointestinal schistosomosis in sheep, goat, cattle, buffalo, horse etc. Hepatointestinal schistosomosis includes acute intestinal form and chronic hepatic form. In intestinal form, there occur severe inflammatory conditions with hemorrhagic lesions. The wall of the intestine also becomes edematous. Hepatic schistosomosis involves the liver, which starts with inflammation followed by granulation and fibrosis. The eggs play a role for causing the pathology. The soluble antigen is released by the eggs, which induce immunopathology. Inflammatory cells gather around the eggs. Granuloma is formed there. Subsequently, deposition of fibrous tissue occurs. Clay pipe fibrosis is a characteristic pathological feature. Periportal inflammation leads to portal fibrosis.

Species: *Schistosoma mattheei*

The species occur in portal, mesenteric, and bladder veins of cattle, sheep, goat, camel, and man. They are prevalent in the Middle East and South and Central Africa. The body of the male is spiny behind the suckers. The dorsal surface of the male bears cuticular tubercles. The eggs are spindle shaped and not equipped with operculum. The size of the eggs is 170–280 × 72–84 mm.

Species: *Schistosoma incognitum*

The flukes are found in the mesenteric veins of pig and dog. They are prevalent in Indian subcontinent. The cuticle is tuberculated and spined. Four to seven testes are present. *Lymnaea luteola var australis* serve as the intermediate host. Yellowish brown eggs have prominent spine attached to the flattened side.

Species: *Schistosoma spindale*

The flukes occur in the mesenteric veins of cattle, buffalo, sheep, goat, equines, and dogs. The anterior part of the fluke is cylindrical and posterior part is flat. Two suckers are spined. The cuticle is tuberculated and spined. *Indoplanorbis exustus*, *Lymnaea* spp. are the intermediate hosts. The eggs are spindle shaped possessing a terminal spine. They are distributed in Far East and some parts of Asia.

Species: *Schistosoma haematobium*

Host: Human beings. Baboons, monkeys may also be affected.

Site: The adults are found in posterior mesenteric arteries. The parasites lay eggs in walls of bladder, ureters, and urethra.

Morphology

There are several identifying features of this worm. Males are 1–1.5 cm long having tuberculated tegument. The male has 4–5 testes. Ovary is situated behind the middle of the body. Intestine reunites in the middle of the body. The eggs have terminal spine.

Life Cycle

Eggs are laid by the female being seated in the gynecophoric canal in the veins. Through the vessels, the eggs reach to the mucosa of the urinary bladder and finally reach in the cavity wherefrom they pass out through the urine. The miracidia develop in the egg in the presence of water and hatch out later on. Then the miracidia pass in search of the intermediate hosts (snails). They penetrate through the exposed soft tissues of the snail and locate in the liver of the snail. Within a week or more, the miracidia transform into sporocysts. The sporocysts give rise to a generation of daughter sporocysts. The daughter sporocysts further transform into cercariae. The redia stage is absent. Then the cercariae are discharged from the snail in suitable climatic condition. Temperature, light, humidity, and other factors play a role in the release of the cercariae. Cercariae, coming in contact of the skin of the human being, start penetrating the skin and reach to the subcutaneous tissues at first. The cercariae transform into schistosomulae after shedding their tails. They reach to the pulmonary capillaries via the right heart. They then come to the left heart and enter into the systemic circulation. They then enter into the mesenteric artery, and from this area, they enter into the portal circulation. In the portal blood, they transform into the adult. The adult worms then reach to the rectal venous plexus to attain full maturity. They ultimately reach to the vesical plexus. The females lay eggs in the walls of bladder, ureter, urethra etc.

Pathogenesis

S. haematobium causes urinary or vesical schistosomosis. The eggs cause severe pathology as compared to other stage of this parasite. The eggs possess a terminal spine by which the eggs irritate and erode the affected area. There occur immunopathological reactions resulting in cellular infiltration. The area becomes hyperplastic. Deposition of granulomatous tissues leads to cause formation of the pseudotubercles, which are also called as egg granuloma. In the central portion of granuloma, there occurs degeneration of tissues, and later on calcification ensues. The disease is also called schistosomiasis hematobia. There are other associated complications. The metabolic by-product of the schistosomulae in the hepatoportal circulation may cause anaphylactic reaction, which exhibits the pathological features of fever, hepatosplenomegaly, etc.

Diagnosis

Diagnosis can be done by demonstration of the eggs in the urine. Biopsy of vesical mucosa can be done. The eggs can be seen in the tissues. Different clinical features i.e. eosinophilia, allergic manifestation may provide the guidance of diagnosis. Several immunological tests can be performed. These are circumoval precipitatin test, fluorescent antibody test, etc.

Treatment

Tartar emetic, anthiomaline, hycanthone, and niridazole are the drugs of choice.

Species: *Schistosoma mansoni*

Host: Man is the definitive host of this parasite. But animals are also affected.

Site: The worms are found in the portal veins. However, these worms are also found in mesenteric veins of sigmoidorectal area.

Morphology

There are several identifying features of this worm. The males are about 1 cm long having characteristic coarsely tuberculated tegument. The male has 8–9 testes arranged in zigzag fashion. Ovary is situated anterior to the middle of the body. The ovary is elongated and slightly lobulated. Intestine reunites in the anterior half of the body. Eggs have lateral spines.

Life Cycle

The eggs are expelled from the definitive hosts with the feces. The miracidium develops in the eggs in the presence of water and hatches out later on. Then the miracidia swim in water in search of the intermediate hosts (snails). They infect the snail and locate in the liver of the snail. The miracidia transform into the sporocysts. The sporocysts give rise to a generation of daughter sporocysts. The daughter sporocysts further transform to cercariae. Then the cercariae are discharged from the snail in suitable climatic condition. Cercariae come in contact with the skin of the definitive hosts. They start penetrating the skin and reach to the subcutaneous tissues at first. The cercariae transform into schistosomulae after shedding the tails. The schistosomulae are then carried to the liver. There the worms become adult. They migrate against the blood stream and reach to mesenteric veins and capillaries of sigmoidorectal region.

Pathogenesis

The disease caused by this parasite is called as schistosomiasis mansoni, schistosomal amoebiasis, or intestinal bilharziasis. The localized pathology comprises of dysentery. The lesions are hepatomegaly, portal hypertension, periportal fibrosis, etc.

Diagnosis

Diagnosis can be done by demonstration of the eggs in the feces. Biopsy of rectal tissue can be done for detection of eggs. The eggs can be seen in the tissues. Different clinical features, i.e., eosinophilia and allergic manifestation, may be the guidance of diagnosis. Different immunological tests can be performed. These are circumoval precipitatin test, fluorescent antibody test, etc. Tartar emetic, anthiomaline, hycanthone, and niridazole are the drugs for treatment.

Species: *Schistosoma japonicum*

The worms are distributed in China, the Philippines, Japan etc. Man is the common host of this parasite. The reservoir hosts are dog, cat, pig, etc. The worms are found in the portal vein, mesenteric vein, and rectal or hemorrhoidal plexus of vein. The males are about 1.2–2 cm long and the females are 1–2 cm long. The cuticle is not tuberculated as that of *Schistosoma mansoni* and *S. haematobium*. The testes has 6–8 lobes arranged in a single row. Ovary is situated in the middle portion of the body. Intestine reunites in the posterior fourth of the body. The penetration of the body by the cercariae results in dermatitis. Arrival of the immature flukes in the lung and liver results in pneumonia and hepatitis, respectively. The most serious damage is caused by the eggs of the parasite on the intestinal wall where they cause inflammation; masses of eggs of the parasite are surrounded by infiltrative cells predominantly by the eosinophils. Abscess is formed there. The flukes cause inflammation and granuloma in the affected area. Occasional embolic egg granuloma occurs in the brain and spinal cord.

Other Genera of Schistosomatidae

Genus: *Bivitellobilharzia*

Species: *Bivitellobilharzia nairi*

The seat of predilection of this fluke is the portal vein and the hosts are Indian elephants. The male is 9.4 mm long and the female is 10.5 mm long. The worms are stout and tuberculated. The male possesses a well-developed gynecophoric canal. The male parasite possesses 52 testes, which are arranged in a row. The ovary is situated at the anterior fourth of the body.

Genus: *Orientobilharzia*

Species: *Orientobilharzia dattai*

The seat of predilection of this fluke is the portal and mesenteric veins, and the hosts are cattle, buffalo, sheep, and goat. The male is 2.8–7.7 mm long and the female is 2.14–7.53 mm long. Thirty-seven to seventy testes are present in male in an alternating double row. *Lymnaea luteola* act as the intermediate hosts of this parasite.

Cercarial Dermatitis

This is an immunopathological complication, which occurs in human beings due to skin penetration by non-human schistosomes. Cercariae of avian species are very important in causing this condition. The initial infection results in mild erythema and edema formation. But repeated exposure to cercariae causes severe immunopathology comprising severe pruritus, development of papule, pustules, etc.

Treatment of Schistosomosis

Name of drug	Route of administration	Dose
Praziquantel	Oral	40 mg/kg for cattle, buffalo, sheep, and goat (single dose) or 25 mg/kg can be repeated after 14–21 days

Family: Strigeidae

Family Synopsis

The members of this family are commonly the parasites of different types of birds occurring in the alimentary canal. Some may occur in the mammals also. The flukes are characterized by a typical constriction, which divides the body into anterior flattened cup-shaped part containing the suckers and the posterior cylindrical part containing reproductive organs. The parasites are heteroxenous and require two intermediate hosts for their life cycle (first aquatic snails and second fish or leech). The parasites are not harmful and usually asymptomatic. Where possible, diagnosis can be done by finding the eggs in the feces.

Genus: *Apatemon*

Introduction

The members of this family are small flukes of different anatid birds. They cause hemorrhagic enteritis in the affected ducks, geese, etc.

Distribution

The worms are distributed in Europe, America, etc. Recently, it has also been reported from China.

Species: *Apatemon gracilis*

Morphology

The anterior part is typically cup shaped having an adhesive organ. Cirrus and cirrus pouch are absent. The vitellaria is found in the posterior part of the body.

Hosts: These trematodes occur in duck, goose, pigeon, etc.
Site: Intestine

Life Cycle

Events of life cycle: a glimpse
• Voiding of eggs with the feces of definitive host • Hatching of eggs in the environment • Release of miracidia from the eggs • Entrance of miracidia in the snails • Formation of sporocysts and cercariae in the snails (no formation of redia) • Release of cercariae from the snails • Penetration of the second intermediate hosts (rainbow trout, three-spined sticklebacks) by the cercariae • Transformation of cercariae to tetracotyliform metacercariae in the second intermediate hosts • Ingestion of the second intermediate hosts harboring metacercariae by definitive hosts • Development up to adult in definitive hosts

Note

Study shows that the first intermediate hosts are aquatic snails and members of Lymnaeidae family and the second intermediate hosts are fish (rainbow trout, stickleback), leech, etc.

Clinical Signs

The affected animals seem asymptomatic. However, the parasites may cause hemorrhagic anemia as a result of ingestion of blood after rupturing blood vessels of their predilection site (small intestine).

Diagnosis

Symptomatological Diagnosis

Diagnosis based on symptoms is not possible because specific symptoms are not exhibited by the affected birds.

Parasitological Diagnosis

Examination of fecal sample may reveal the eggs of the parasites. The adult parasite can be detected at the predilection site after necropsy.

Immunological Diagnosis

No immunodiagnostic test is available in the field.

Molecular Diagnosis

Study shows molecular identification of this parasite by using ITS 1 as molecular marker.

Treatment

Albendazole, oxibendazole, and praziquantel are some effective drugs.

Genus: *Parastrigea*

Introduction

These are small flukes occurring in different anatid birds.

Species: *Parastrigea robusta*

Hosts: Domestic ducks
Site: Intestine

Morphology

The parasite is 2–2.5 mm long. The body is roughly divided into two parts, i.e., anterior cup-shaped structure and cylindrical posterior part bearing reproductive system. Cirrus pouch is absent. The anterior part has two large-sized lateral expansions.

Life Cycle

Not known

Pathogenesis

The worms live in the intestine and cause marked damage to the intestinal villi. They firmly attach to the intestinal villi by their cup-shaped anterior structure. The blood vessels of intestinal villi are highly damaged and result in hemorrhage. Anemia is also a pathological consequence.

Diagnosis

Examination of fecal sample may reveal the eggs of the parasites. Necropsy of birds may reveal adult parasites at the predilection site.

Treatment

Albendazole, oxibendazole, and praziquantel are some effective drugs.

Genus: *Cotylurus*

Introduction

These are small flukes of different anseriform birds. They cause marked damage to the hosts by producing hemorrhagic enteritis.

Distribution

The worms (*Cotylurus cornutus*, *C. flabelliformis*, *C. platycephalus*) are distributed in Asia, Europe, and America.

Species: *Cotylurus cornutus*

Hosts: Domestic and wild anseriform birds
Site: Intestine

Morphology

The body is roughly divided into two parts, i.e., anterior cup-shaped or spoon-shaped structure with two lateral lobes and a poorly developed sucker. The posterior part is cylindrical, which contains the reproductive system. A strong copulatory organ is found in the bursa of the parasite.

Life Cycle

The life cycle is an indirect type, which involves two intermediate hosts. Lymnaeid snails act as both first and second intermediate hosts. The leeches may also act as the second intermediate hosts. The events occurring during the life cycle are as follows:

- The eggs are voided along with feces.
- The eggs hatch within few weeks. After hatching, the miracidia are liberated.
- The miracidia then penetrate the first intermediate hosts (*Lymnaea* spp., *Valvata* spp.).
- Sporocysts and cercariae are formed in the first intermediate hosts.
- The cercariae are furcocercous type and they enter in another lymnaeid snails of same species or other planorbids. The cer-

cariae transform to a tetracotyle type of metacercariae.
- The definitive hosts get the infection by ingestion of infective metacercariae being present in the second intermediate hosts.
- The metacercariae develop to adult stage in the definitive hosts.

Pathogenesis and Clinical Signs
The parasites may cause hemorrhagic enteritis. Clinical signs include diarrhea, debility, etc.

Diagnosis
Examination of fecal sample may reveal the eggs of the parasites. Adult parasite is found during necropsy.

Treatment
Praziquantel has been found effective.

Family: Diplostomatidae

Family Synopsis

The members of this family are found in the birds and mammals occurring in the alimentary canal. The flukes are similar to strigeids. But the anterior part is more flattened compared to strigeids. The parasites are heteroxenous and require two intermediate hosts for their life cycle (first- aquatic snails and second- frogs, fish). The parasites are not harmful and usually asymptomatic. Diagnosis can be done by finding the eggs in the feces.

Genus: *Diplostomum*

Introduction
These are small flukes of different gulls.

Distribution
The parasites are distributed in Europe, North America, etc.

Etymology
The generic name "*Diplostomum*" was derived from two words, i.e., "diplo-" and "stoma" "Diplo-" means "two" and "stoma" means "mouth."

Species: *Diplostomum spathaceum*

Distribution
The parasites are distributed in Africa, Europe, Asia, America, etc.
Hosts: Gulls
Site: Intestine
Common name: Eye worm of fish, eye fluke

Morphology
The suckers are small. The characteristic morphological feature is the presence of ventral sucker within the accessory adhesive organ.

Life Cycle
The following events occur during the life cycle of this parasite:

- Eggs are voided with the feces.
- The eggs are hatched in the environment, and liberated miracidia infect lymnaeid snails (*Lymnaea peregra*).
- Development up to cercarial stages occurs in the snails. The stages formed in the snails are sporocyst and cercaria.
- The cercariae are released from the snails.
- The fish are the second intermediate hosts (rainbow trout, cyprinid, salmonids, etc.). The cercariae get encysted in eyes of fish and transform into metacercariae.
- The definitive hosts get the infection by ingestion of infected fish.

Pathogenesis
The parasites may cause mortality of the young gulls. A marked pathogenesis occurs in fish when the cercariae enter the eyes. The metacercariae are formed in the eye, which appear as white spots. Subsequently, the whole eye may be damaged and colored white. The fish become blind.

Genus: *Posthodiplostomum*

Species: *Posthodiplostomum minimum*

Hosts: Heron, kingfisher, etc.
Site: Intestine

Life Cycle

The snails act as the first intermediate hosts, whereas fish act as the second intermediate hosts.

The following events are observed during the life cycle of this parasite:

- The eggs are passed with the feces.
- The eggs hatch in the environment and the miracidia are released.
- Development up to cercarial stages occurs in the first intermediate hosts.
- The cercariae are liberated from the snails.
- After leaving the snails, the cercariae encyst in the skin, fins, and different musculatures of fish (cyprinids).
- The final hosts get the infection by ingestion of the second intermediate hosts harboring the metacercariae.

Pathogenesis

The parasites do not cause severe pathology. However, they may cause enteritis in the herons. Marked pathogenesis occurs in the intermediate hosts (fish). They have location specificity for causing pathology. The cercariae transform to metacercariae on the fin, skin, cornea, etc. of fish. The metacercariae remain in light colored cyst. On the cysts, pigments get adhered and thus the cysts become brown or black colored. This is called "black spot disease."

Diagnosis

Examination of feces is the common diagnostic method, which reveals the eggs of these parasites.

Treatment

Praziquantel is an effective drug.

Control

The parasites can be controlled by effective treatment. Snail control has been described but practical implication doesn't seem effective.

Genus: *Neodiplostomum*

Introduction

These are parasites of different piscivorous birds. They cause gray pearl disease in fish.

Etymology

"*Neodiplostomum*" is derived from three Greek words, i.e., neos (new), diplo (two), and stoma (mouth).

Species: *Neodiplostomum perlatum*

Hosts: Different piscivorous birds (heron, kingfisher, etc.)
Site: Intestine

Life Cycle

The first intermediate hosts are snails and the second intermediate hosts are fish.

The following events are observed in the life cycle of this parasites:

- The eggs are passed with the feces of birds.
- The eggs are hatched in the environment.
- Development up to cercariae stages occurs in the first intermediate hosts.
- The cercariae are released from the snails.
- After leaving the snails, the cercariae encyst in cyprinid fish.
- The final hosts get the infection by ingestion of the second intermediate hosts harboring the metacercariae.

Pathogenesis

They may cause enteritis in the definitive hosts. Metacercariae are formed on the fin, skin, etc. and also in the internal organs of fish. These encysted metacercariae appear as pearl-like structures. That's why the disease caused by this parasite is called "gray pearl disease."

Symptoms

The presence of metacercariae gives a pearl-like structure on the skin and fin of fish.

Genus: *Alaria*

Introduction

These are parasites of different dog, cat, etc. They cause enteritis in the affected animals.

Distribution

The parasites are widely distributed in Europe. But they are also distributed in North and South America, Australia, etc.

Species: *Alaria alata*

Hosts: Dog, cat, fox, etc.
Site: Intestine

Morphology

The anterior part is flat, which is followed by posterior cylindrical region. Two ear-like tentacles are present in the anterior part. The posterior cylindrical part is too shorter than the anterior part. The two suckers are small. Testes are transversely elongated and located in the hind body. Cirrus pouch is absent.

Life Cycle

The following are the events occurring in the life cycle of this parasite:

- The eggs are passed with the feces of definitive hosts.
- The eggs hatch in the environment and the miracidia are released.
- Up to cercarial stages develop in the first intermediate host (*Planorbis* sp.). No rediae are formed. The next stage is mesocercariae (equivalent to metacercariae).
- The cercariae are released from the snails.
- The cercariae encyst in the second intermediate host (frog, tadpole, etc.).
- The final hosts become infected by ingestion of the second intermediate hosts harboring the mesocercariae. The birds, mice, and rat may act as paratenic hosts.

Pathogenesis

The parasites do not cause severe pathology. However, the adults may cause enteritis, and immature stage may cause pulmonary damage as a result of invasion of tissues during migration by these immature stages mostly found in man. Not only the lungs but also other organs like the kidney, brain, and heart are also affected by this parasite. Anaphylactic reaction and neuroretinitis are reported in man. Man acts as paratenic host.

Diagnosis

Examination of feces is the common diagnostic method, which reveals the eggs of these parasites.

Treatment

Praziquantel is an effective drug. Benzimidazole compounds comprising albendazole and fenbendazoles are also effective drugs.

Control

The parasites can be controlled by effective treatment. Snail control has been described but practical implication doesn't seem effective.

Family: Sanguinicolidae

Family Synopsis

The members of this family are the parasites of fish. The flukes are characterized by absence of suckers. They are heteroxenous and require one intermediate hosts (snails). The parasites are very harmful to the fish and cause mortality and usually asymptomatic.

Genus: *Sanguinicola*

Species: *Sanguinicola inermis*

Hosts: Fish
Site: Gill and other tissue

Morphology

The flukes are lanceolate shaped. Typical triangular eggs are found. The uterus contains only one egg.

Life Cycle

The following events occur in the life cycle of this parasite:

- The eggs are laid by the parasites in gills of fish.
- The eggs hatch in the gills.
- From the eggs, the miracidia come out.
- The miracidia leave the gills and enter lymnaeid snails.
- Development up to cercarial stage occurs in the snail. Two generations of sporocysts occur. No redia stage is found.
- The cercariae are discharged from the snails.
- The cercariae attack the gills of fish and develop to adult stages.

Pathogenesis

The fish suffer a lot due to this parasitic infection. The eggs block the brachial arteries of the gills. The miracidium causes severe damage. Thrombosis and hemorrhage in the blood vessels lead to mortality of the fish.

Treatment

Oxyclozanide is an effective drug.

Special Part of Trematodes

Immunity to Trematodes

The immunity against trematode is mainly associated with Th2 type of immune response. Tissue-dwelling migratory parasites are mainly controlled by ADCC and eosinophils, neutrophils, macrophages, and platelets that act as effector cells. IgE and IgG antibodies are associated in this reaction. The effector cells express Fc receptor (RFc), and by this Fc receptor, the effector cells bind to the Fc part of the antibody molecules. In this state (antigen+antibody+effector cell), the effector cells release different toxic materials. For instance, eosinophilic toxic products, i.e., eosinophilic basic protein, eosinophilic cationic protein, eosinophil-derived neurotoxin, etc., are released. Migratory immature *Fasciola* spp. are killed by the NO (nitric oxide) released by classically activated macrophage. Immunoregulation occurs by the cytokines produced by Treg cells, which are a subset of T cells. Treg cells produce IL10, which control the acute phase of inflammation. The Treg cells also produce TGF beta, which inhibit Th1 cytokines. Lastly, AAM (alternatively activated macrophages) are produced. In chronic *Fasciola* infection, alternatively activated macrophages have a great role, which produce substances toxic to the flukes. These alternatively activated macrophages release some cytokines, which participate in tissue repairing by depositing fibrous tissue.

In case of *Schistosoma* infection, granuloma occurs surrounding the parasite, which inhibits worm migration. The granuloma contains different cells comprising eosinophils, macrophages, and lymphocytes, which form a specific structure. The whole structure is covered by fibrotic extracellular matrix. Granuloma also occurs surrounding the eggs of *Schistosoma* spp.

In *Dicrocoelium* and *Opisthorchis* infection, inflammation, hyperplasia, and fibrosis occur. Pro-inflammatory cytokines, ADCC, and alternatively activated macrophage presumably play these roles. Nodule formation occurs in *Paragonimus* infection. A delayed type of hypersensitivity occurs in case of *Paragonimus* infection. Different cytokines play a role for deposition of fibrous tissue/nodule formation.

Many of the luminal flukes fail to produce any marked pathology. Mild enteritis and mild intestinal discomfort are found in these cases. Eosinophils and neutrophils infiltrate at the affected area. Goblet cell hyperplasia occurs, which might cause expulsion of intestinal trematodes. CD4+ T cells play a role for goblet cell hyperplasia. It is an established fact that adult rumen flukes cannot produce any marked pathology due to having sharing antigens between host and parasite.

It has been observed that Th1-like response is more protective than Th2-like response because the parasites are able to develop many evasion mechanisms against Th2 type of immune response. Treg cells regulate the balance by suppressing the action of Th1 cells and downregulate any excessive Th2-like response during granuloma formation.

In case of helminth infection, Th1-like cytokines are responsible for considerable immunopathology and occurrence of clinical signs. Chronic infection of helminth is mediated by Th2-like response. Treg activity is called as modified Th2 response. Anti-inflammatory cytokines are produced by these cells. These cytokines are IL10 and TGF beta. This immunomodulatory Treg response causes survival of parasites by downregulating the host's inflammatory and host's immunopathological response.

Immune Evasion by Trematode Parasites

Immune evasion by trematodes occurs by the following ways:

1. *Fasciola* parasites produce **superoxide dismutase**, which neutralizes superoxide radicals, extremely toxic to the juvenile flukes.
2. Juvenile flukes produce a vital enzyme, **cathepsin L protease**, which cleaves IgE and IgG. There occurs inhibition of ADCC as these two antibodies play a role in ADCC.
3. Different immune activities are inhibited by excretory-secretory (ES) substances released by the flukes. ES product induces inhibition of lymphocyte proliferation.
4. **Peroxiredoxin** is found in ES (excretory-secretory) product of *Fasciola hepatica*, which inactivates reactive oxygen species (ROS) of the host.
5. *Fasciola hepatica* and other helminths produce cathelicidin-like peptide. This peptide inhibits NLRP3 inflammasone which reduces secretion of IL-1 beta by macrophage.

> Note
> **Different evasion mechanisms performed by flukes**
>
> - Molecular mimicry
> - Destruction of host immune substances
> - Antigenic variation of parasites

Immunoregulation

A series of interleukins are produced due to helminthic infection, and there occurs a harmony of correlation among the cytokines. Th2 cytokines, i.e., IL4 and IL10, may play a crucial role in reducing the severity of acute disease and allowing survival of the host. Many experiments were carried out by different workers regarding the role of IL4 and IL10. It has been proved that absence of these cytokines can lead to cause severe pathology due to different trematode infection. A subset of T cells are called as Treg cells. These Treg cells play an important role of immunoregulation. Treg cells produce IL10 and TGF beta, which suppress action of Th1 response.

Fibrosis and Role of Cytokines

Fibrosis occurs in different helminthic infections. Lymphocytes release different cytokines, which further activate the macrophages. IL4, IL5, IL13, and IL21 have direct and indirect role in fibrous tissue deposition. When wound healing is required, different genes get activated to produce relevant cytokines and other important proteins. IL4 is a potent profibrotic mediator. IL5, IL13, and IL21 have also a role in deposition of fibrous tissue. It has been found that the efficacy of IL4 is nearly twice as TGF-β, and this cytokine plays a strong role in deposition of fibrous tissue. IL4 activation results in synthesis of the extracellular matrix comprising types I and III collagen and fibronectin. In schistosomosis, it has been found that deposition of fibrous tissue occurs eventually to encapsulate the granuloma structure. For this activity, IL4 plays a role. In this way, different cytokines play role in deposition of fibrous tissue. As mentioned earlier, IL13 has also a role in fibrosis. Both IL4 and IL13 exploit IL4Rα/Stat6 signaling pathway. All these works are in the research level, and many things are yet to be unfurled. While discussing about deposition of fibrous tissue, the role of eosinophils must be taken into consideration. IL5 plays a significant role in differentiation, activation, and recruitment of eosinophils. The eosinophils possess different fibrogenic cytokines. Alternatively activated macrophages have a role in deposition of fibrous tissue. IL4 has a role to produce alternatively activated macrophage.

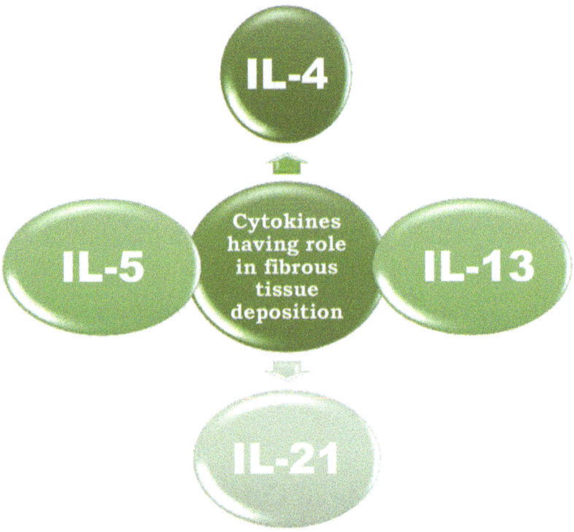

Fig. 4.44 Schematically displayed different cytokines having role in fibrosis

> **Note**
>
> Different immune phenomena triggering immunity are:
>
> (a) Nitric oxide produced by macrophages
> (b) Antibody-dependent cellular cytotoxicity
> (c) Complement-mediated lysis
> (d) Granuloma formation and deposition of fibrous tissue

Immunodiagnosis of Trematodal Diseases

An array of immunodiagnostic tests have already been developed. These tests either detect antibodies or circulating antigens in serum, plasma, urine, sputum, etc. Serological study is very important for determination of disease surveillance. The constraint for study of seroprevalence is that circulating antibodies against any particular pathogen remain even in the absence of parasites. There are several tests that are performed for detection of trematode infections. These are as mentioned in the table:

Different immunodiagnostic tests performed in different trematode infection
• Intradermal tests
• Circumoval precipitin test (COPT)
• Indirect hemagglutination test (IHT)
• Indirect immunofluorescence assay (IFA)
• Complement fixation test (CFT)
• Indirect or direct immunofluorescence antibody (IFA or DFA) tests
• Immunoblotting
• ELISA
• Falcon assay screening test-ELISA (FAST-ELISA)
• Dot ELISA

Brief Description of Different Immunological Tests

COPT (Circumoval Precipitation Test)

This test is based on production of a precipitate by exposing suspected serum to lyophilized schistosome eggs. The test has high sensitivity and specificity but it has some drawbacks. To achieve the result, the test arrangement is required to keep for long time. This test was used for seroevaluation of praziquantel—treatment for control of schistosomiasis.

CHR (Cercarien-Hullen Reaction)

This test is interesting. In this test, the live cercariae are allowed to the suspected serum. Immobilization of movement of the cercariae is visibly detected. Immobilization of cercariae occurs due to formation of precipitate. This test is labor intensive. This is done for diagnosis of *Schistosoma* infection.

IHT (Indirect Hemagglutination Test)

In this test, reactivity between antibodies in the serum of infected individual and antigen-coated red blood cells is detected. This test is simple and sensitive. The drawback of this test is its false-positive reaction. In many trematode infections, this test is done, i.e., *Fasciola* spp., *Schistosoma* spp., *Opisthorchis* spp., paramphistomes, *Dicrocoelium* spp., etc.

IFA (Indirect Fluorescence Antibody Test)

The test is based on the detection of reaction between parasite antigen and antibody. This test is done in different trematode infections, i.e., *Fasciola* spp., *Schistosoma* spp., *Opisthorchis* spp., paramphistomes, *Dicrocoelium* spp., etc.

ELISA (Enzyme-Linked Immunosorbent Assay)

ELISA is done to determine the reactivity between antibodies in suspected serum and extracted antigens from parasites. ELISA can be done by using different types of antigens, i.e., egg antigen, cercarial antigen, metacercarial antigen, excretor-secretory antigen, somatic antigen, etc. But unfortunately, there is every possibility of cross-reactivity in this test also. This test is tried for diagnosis of many trematode infections, i.e., *Fasciola* spp., *Schistosoma* spp., *Opisthorchis* spp., paramphistomes, *Dicrocoelium* spp., *Paragonimus* spp., etc.

Falcon Assay Screening Test-ELISA (FAST-ELISA)

FAST-ELISA is based on antigen-antibody reaction. Many experiments were carried out on comparative efficacy between ELISA and FAST-ELISA. FAST-ELISA is more sensitive and less time consuming. There had been works on FAST-ELISA for diagnosis of fasciolosis, schistosomiasis, etc. But constraint of cross-reaction is found in this test also.

Dot ELISA

This is a rapid test and result can be found within 15–20 min. In this test, nitrocellulose membrane is coated by antigen of trematodes. The antigen-coated NCP paper is incubated in primary antibody and then the same is further treated with secondary antibody. The secondary antibody is conjugated with enzyme, and this secondary antibody can be prepared or be procured commercially. This is further treated with substrate, which produces color. Detection of color is the positive result. However, in this test, selection of result is possible but no titer can be found. This test is tried for diagnosis of many trematode parasites, i.e., *Fasciola* spp., *Schistosoma* spp., *Dicrocoelium* spp., paramphistomes, etc.

Magnetic Affinity Enzyme-Linked Immunoassay

This is a rapid and easy test compared to ELISA. Works were carried out for diagnosis of different trematodes. Different defined antigens have been used for performing this test for diagnosis of different trematode infections. It has been found that this test is more specific than ELISA. Therefore, this test could be used as a good alternative to ELISA.

Different Source of Antigens of Trematodes

Different antigens are used by different workers for immunodiagnosis of different trematode infections. Whole somatic antigens are prepared and the same are used for raising the hyperimmune sera in rabbit. The sera are used for detection of the immunodominant proteins. After SDS-PAGE, the fractionated proteins are transferred onto the nitrocellulose paper. The same is then subject to Western blot technique. The Western blot technique reveals immunodominant proteins.

Some Studies on Different Immunodiagnostic Antigens

Studies show that **130 kDa** protein could be used for specific diagnosis of *Dicrocoelium dendriticum.* Martinez-Ibeas et al. (2013) constructed cDNA library by mRNA extracted from *D. dendriticum* adults. Among 173 new EST (expression sequence tag), **8 kDa** protein gene was cloned and expressed in *Escherichia coli*. This protein was evaluated for its diagnostic value using ELISA and Western blot. This recombinant protein displayed its value as an immunodiagnostic antigen. Sripa et al. (2012) carried out a work to assess immunodiagnostic potential of cathepsin B-like protein of *Opisthorchis* sp. This enzyme was actually detected during in vitro culture of the parasites in ES products. In this study, they evaluated the performance of recombinant *O. viverrini* protein **(rOV-CV-1)** for immunodiagnosis. The full-length Ov-CB-1 cDNA was cloned, and recombinant protein was produced in catalytically active form in *Pichia pastoris*. This protein was purified by affinity chromatography and tested in enzyme-linked immunosorbent assay (ELISA) with human sera. The sensitivity and specificity were 67% and 81%, respectively. Therefore, recombinant Ov-CB-1 could be used for serodiagnosis. Different antigens, i.e., whole somatic antigen, metacercarial antigen, etc., were tried for immunodiagnosis of *Schistosoma* infection. Tegumental antigen was also used for immunodiagnosis purpose. Zhang et al. (2015) reported immunodiagnostic potential of recombinant proteins of tegumental antigen of *Schistosoma japonicum* **(rSjPGM)**. These are recombinant *Schistosoma japonicum* phosphoglycerate mutase (rSjPGM) and *S. japonicum* recombinant RAD 23 homolog B **(SjRAD23)**. Both these proteins showed higher sensitivity and specificity. These tegumental proteins displayed satisfactory immunodiagnostic potential. Pothong et al. (2018) worked on immunodiagnostic potential of a recombinant antigen of *Paragonimus heterotremus*. A DNA expression library was constructed from mRNA of adult parasite. An immunoreactive protein encoded by CE3 gene sequence was selected for use in the construction of a recombinant *P. heterotremus* protein. For recombinant protein expression, the CE 3 gene sequence was isolated and inserted into the plasmid vector pRset, and the resulting product had the molecular weight of 35 kDa. The study displayed 88.89% sensitivity and 95.51% specificity of the aforesaid protein in ELISA. Kim et al. (2002) carried out one work on recombinant *Paragonimus westermani* yolk ferritin for serodiagnosis. This protein was found to be very much useful for serodiagnosis. This **recombinant yolk ferritin** protein of *P. westermani* was bacterially produced from a previously cloned complementary DNA. The work displayed 88.2% sensitivity and 100% specificity. In *Paragonimus kellicotti*, the abundant amount of immunoreactive proteins are **cysteine proteases**, **MFP6**, **myoglobin**, etc. Mirzadeh et al. (2018) carried out work on immunodiagnostic potential *of* three recombinant proteins for *F. hepatica*. These were ***F. hepatica* saponin-like protein 2 (SAP2)**, **ferritin protein (Ftn-1)**, and **leucine aminopeptidase (LAP)**. Recombinant proteins were recovered from *E. coli* by immobilized metal affinity chromatography (IMAC).This was purified by dialysis using chemical additive and reduced/oxidized glutathione. Sensitivity and specificity of these proteins were encouraging (95% sensitivity and 98% specificity for rtFhSAP-2, 80% and 83.3% for rtFhFtn-1,84%-88% for FhLAP) individually and in combination. Other native and recombinant proteins used for serodiagnosis of fasciolosis are cathepsin L, hemoglobin, paramyosin, etc.

Immunoprotection Against Trematodal Diseases

Different types of antigens are used for immunoprotection studies. These antigens are described in the following table:

Antigens used for immunoprotection studies
Crude antigen
(a) Soluble extract of whole parasite
(b) Soluble extract of larvae of parasite
(c) Soluble extract of eggs of parasite
(d) Excretory-secretory antigen of adult parasites
(e) Irradiation-attenuated metacercariae
Defined antigens
• 32 kDa protein isolated from newly excysted juvenile fluke of *Fasciola hepatica*
• Glutathione S-transferases of *Fasciola*, *Schistosoma*, and other flukes
• Cathepsin L of *Fasciola*, *Schistosoma*, and other flukes
• Cathepsin B of *Fasciola*, *Schistosoma*, and other flukes
• Paramyosin of *Fasciola* and other flukes
• Myosin of *Fasciola* and other flukes
• Sm26 and Sm30 of *Schistosoma* spp.

Description of Some Immunoprotective Antigens

FABP (Fatty Acid-Binding Protein)

This is an important protein found in many organisms. The full form is fatty acid-binding protein. This protein has attraction to the fatty acid. The trematodes cannot synthesize their sterol in their own body. Fatty acid is required for the lipid layer of cells. Therefore, the flukes are dependent for lipid material from their hosts. This protein plays a role in this regard. Looking into the importance of this protein, an array of research works were carried out for development of vaccine using the native purified FABP and recombinant FABP using a series of adjuvants comprising **Freund's**, **Quil A**, **Montanide 80**, etc. Certain level of efficacy was found, which was reflected by reduction of fluke number and fluke weight in case of vaccination trial using native FABP (Estuningsih et al. 1997). Rabia Aly et al. (2012) found heterologous protection against *Schistosoma mansoni* using native FABP from *Fasciola gigantica*. Kumar et al.

(2012) also carried out work on immunopotential capability of FABP and found certain level of protection.

Cysteine Proteases

This is a class of thiol enzymes. A number of enzymes are found under this class. The important enzymes having both diagnostic and immunoprotection potentials are cathepsin L, cathepsin B, etc. Cysteine proteases are present in a wide range of organisms, from lower group (bacteria, protozoa, rickettsia, etc.) to higher group of organisms (mammals, reptiles, etc.). This class of enzymes performs different biological functions. Cysteine proteases of parasites also perform different vital functions comprising digestion, tissue invasion, immune evasion, etc. Several works were carried out on immunodiagnosis and immunoprotection for parasitic diseases. Wijffels et al. (1994) carried out work on immunoprotection studies against *Fasciola* sp. using cysteine protease as vaccine candidate and got satisfactory result.

GST (Glutathione S-transferase)

Glutathione S-transferases are a family of **isoenzymes** that perform detoxification and excretion of xenobiotic agents. GSTs occur as multiple functional enzymes in most of life forms comprising aerobic eukaryotes and prokaryotes. These enzymes are essential for life of most organisms. The hypothesis is that anti-GST antibody can neutralize the function of GST of parasite and the parasite will be damaged/harmed. Based on this hypothesis, a series of works were carried out to envisage the immunoprotection potential of GST against *Fasciola* parasite in different animals (Estuningsih et al. 1997; Sexton et al. 1994; Morrison et al. 1996; Paykari et al. 2002). Native form of GST is isolated usually by glutathione-agarose affinity chromatography technique. Different adjuvants were used to potentiate the GST protein for conferring good immunoprotection. In different studies, a number of adjuvants, i.e., Quil A, Montanide 80, DEAE, Freund's, etc., were used in different combinations in different experimental works. In these works, animal models were mice, rabbit, sheep, cattle, buffalo, etc. Efficacy parameters were antibody titer, number of flukes recovered, weight of flukes, EPG, etc. Encouraging results were recorded in these experimental works. In some trials, efficacy reached 70% and above. Works were also carried out to produce recombinant GSTs. This protein was expressed in *E. coli* bacteria, and the same recombinant protein was used as a vaccine candidate along with different adjuvants as mentioned earlier. Immunoprotective capability of this enzyme was found against other flukes also.

Cathepsin

Cathepsin is a vital enzyme found in many life forms. This enzyme is under the **cysteine proteinase enzyme family**. Cathepsin L and cathepsin B are important. Cathepsin L has different forms, i.e. cathepsin L1, cathepsin L2, etc. Immunoprotective potential of this protein (cathepsisn L) was assessed against *Fasciola gigantica* with encouraging results (Estuningsih et al. 1997).

Hemoglobin

Hemoglobin of *Fasciola* parasite was tested for evaluating its immunoprotection potential. Several works were carried on its immunoprotection efficacy. The studies show low protection level as compared to GST and cathepsin. Dalton et al. (1996) carried out experiment on immunoprotection studies using hemoglobin as vaccine candidate. Wijffels et al. (1994) carried out similar work and found certain level of immune response.

> **Note**
> Some promising candidate proteins of *Fasciola* are glutathione S-transferase, cathepsin L, fatty acid-binding protein, paramyosin, hemoglobin, etc.

Molecular Trematodology

Molecular Taxonomy of Trematodes

Advanced taxonomy is a recent trend of taxonomy for classification of organisms. This is a journey that started with using gross morphological features, and presently, it has reached to the state of analysis of genomic features. In the recent past, phylogeny has taken a new dimension, which has crossed the level of biochemical and immunological characterization of a parasite. Many works have already been carried out on molecular or genomic characterization of different trematodes. Their evolutionary status is being unfold by analysis of phylogenetic tree. The output of analysis of phylogenetic tree is used for development of molecular taxonomy. Work on molecular epidemiology of different trematodes has been done. The work is also progressing further. Detection of different stages of trematode in the snails is possible.

Some Molecular Phylogenetic Study

Different researchers worked on molecular characterization of different trematode parasites. They used genomic DNA, ribosomal DNA, and mitochondrial DNA for genomic characterization of trematodes. They performed different types of PCR in accordance with their need of their relevant research program. Different molecular markers were used by them. These markers are ribosomal DNA (ITS1, ITS2, etc.), ribosomal RNA (5 S, 5.8 S, 18 S, 28 S), and mitochondrial DNA (ND1, CO1, etc.). Bogh et al. (1999) employed a mutation scanning approach for determination of genetic variability in mitochondrial DNA (mtDNA) of *Schistosoma japonicum* (within and among population) from people's Republic China. They used NADH dehydrogenase 1 gene (ND1) and cytochrome C oxidase subunit 1 (CO1) as molecular markers. Single-strand conformation polymorphism (SSCP) was done. This work reported genetic variation of the aforesaid parasite and concluded that SSCP was useful for determination of low-level genetic variation of mitochondrial DNA of *Schistosoma japonicum*. A number of works have been carried out to see the genomic structure of different species of *Paragonimus*, and remarkable findings have been obtained. Intapam et al. (2004) did characterization of five species of *Paragonimus*, i.e., *Paragonimus heterotremus*, *P. siamensis*, *P. harinasutai*, *P. westermani*, and *P. bangkokensis*, by RAPD markers. In this study, DNA was isolated and amplified using arbitrary primers and the RAPD pattern was examined. RAPD revealed polymorphism of those species. The species were distinctly different genomically. They constructed a relationship dendrogram based on the polymorphism. The relationship dendrogram depicted close relationship between strains of *P. heterotremus* and *P. harinasutai*, which was followed by *P. siamensis*, *P. bangkokensis*, and *P. westermani*. There have been several studies on genomic aspect of *Fasciola*. RAPD variation in two trematode species (*F. hepatica* and *Dicrocoelium dendriticum*) from a single cattle population was recorded taking the arbitrary primers by Morozova et al. (2002). The random amplification of DNA by standard RAPD-PCR was used to see the variation of genomic structure in *Fasciola hepatica* and *Dicrocoelium dendriticum*. Different indices like genetic similarity, genetic diversity, polymorphism level, and population subdivision were studied by them. From the indices of similarity in pairs (S), they constructed a dendrogram and this dendrogram reflected genetic relationship between the representatives of two species and between *F. hepatica* individuals isolated from the same or different host individuals. Genetic polymorphism was detected between the two strains (European and Egyptian) of *Fasciola hepatica* by Abdeltawab et al. (2000). The primer targeted 600 nucleotide sequences in length corresponding to 18 S rRNA. The findings revealed polymorphisms between the two strains. Egyptian strain had identical genetic sequence. Likewise, European strain also had identical genomic sequence. Molecular evidence of natural hybridization between *F. hepatica* and *F. gigantica* was carried out by Agatsuma et al. (2000). Two regions of mitochondrial DNA, cytochrome C oxidase subunit 1 (CO1) and NADH dehydrogenase subunit 1 (ND1), and two regions of the nuclear DNA (internal transcribed spacer 2 (ITS2) and the D2 region in the 28 S ribosomal DNA (28 S)) were taken for this study to elucidate taxonomic status. The study revealed interspecific cross hybridization between Fasciola hepatica and *F. gigantica*. Many such works are being carried out taking different molecular markers for genomic characterization.

Note

In the recent past, ribosomal DNA sequences and mitochondrial DNA sequences are vastly used for molecular characterization of parasites. The molecular markers of ribosomal DNA are internal transcribed spacer 1 (ITS1), internal transcribed spacer 2 (ITS2), etc. The mitochondrial DNA sequences mostly used as molecular markers are cytochrome oxidase subunit 1 (CO1), NADH dehydrogenase subunit 1 (ND1), etc.

Molecular Study of Biology of Trematode

Background

Many research works have been done by different scientists on life cycle of different species of trematode parasites. These works are done to record the time duration required for the development and complete maturity of different stages of the trematodes. Different stages of trematodes are cultured following standard protocol. In most of the digenetic trematodes, aquatic snails and land snails act as intermediate hosts. Different snails are collected, and the collected snails are identified. Natural infection of the snails by different trematode species can be confirmed. In vitro hatching mechanism of different eggs of trematodes has been unfurled. Epidemiological studies on the effect of geoclimatological factors on prevalence of trematodes have also been determined. The infected snails are maintained in the laboratory and pattern of metacercariae discharge is studied. Snail colony is developed in the laboratory, and the snails are artificially infected with hatched-out miracidium. Locations of development of the sporocysts, rediae, and cercariae in the snail intermediate hosts have been identified following standard protocol. Many workers identified many species of land snails and aquatic snails that act as intermediate hosts of different trematodes. Studies indicate geoclimatological role on biological development of trematodes in the environment. An array of works have been over by different workers on biology of different species of many trematodes. The work trend on life cycle includes identification of different species of snails and other intermediate hosts. Epidemiological studies on effect of geoclimatological factors on prevalence of the parasites are also done. Findings of these basic studies are prerequisites for study on molecular biology.

Molecular Study of Biology

Molecular study depicts variation of gene expression during different developmental stages of flukes. Different stages of parasites perform different functions. Therefore, different functional proteins are required in different biological stages of development. For example, activities of tissue-invading fluke are different from stationary phase of this parasite. Tissue-invading parasites require different enzymes for penetration of host tissues. Therefore, gene encoding those enzyme proteins is transcribed before/during this activity so that the parasites can perform their specific functions during different developmental stages. Different proteases (**cathepsin L**, **cathepsin B**, **cysteine proteases,** etc.) are such proteins that make the parasite enable for invading the host tissue. Therefore, understandably, these enzymes must be produced whenever these are required. mRNA encoding these proteins are transcribed. Therefore, mRNA profiles of different stages of parasites are presumably different. Gene expressions can be analyzed using RNA-seq data.

Some Molecular Study on Biology of Trematodes

Developmental time course of gene expression using RNA-seq was done for *Fasciola hepatica* by Cwiklinski et al. (2015). Progressively differential gene expression was detected from the infection to subsequent developmental stage in the hosts. The data was analyzed from excystation to newly excysted juvenile (NEJ), traversal of the intestinal wall and migration in the liver, and transformation from immature to mature state. As the parasites grow, they turn from aerobic to anaerobic stage, which was fairly reflected by gene expression indicating switching from aerobic to anaerobic metabolism. The gene expression data also indicated that the production of vitelline and egg shell component had limited to the matured stage. Upregulation of cell adhesion molecules, i.e., cadherins, integrins, and cytoskeletal proteins, which play an important role in sensing changes in the physiological environment, was noted. High degree of upregulation of genes associated with protein polymerization and microtubule-based movement, i.e. tubulin, dynein, and surface tegumental genes, was found in immature and adult flukes. It is needless to mention that peptidases play a vital role for the migratory or invasive parasites. These enzymes play a role for tissue migration. Cysteine peptidases are found in *F. hepatica* and many other trematodes so far studied. A concerted pattern of expression of these enzymes (cathepsin L, cathepsin B) was found, which could be connected with migration of parasite in the tissues where they encounter different profile of macromolecular substances. A high-density oligonucleotide array was conducted using all gene models from *Schistosoma mansoni* by Parker-Manual et al. (2011). cDNA derived from germ ball, cercariae, and day 3 schistosomulae were hybridized to the array without amplification. Upregulation of different sets of developmental genes was found at each stage, which was depicted by the picture of upregulation of genes encoding elastase and other enzymes found in germ ball. Upregulation of genes encoding cysteine proteinase and aspartic proteases was found in the cercariae and schistosomulae. Micro-exon genes encoding different secreted proteins were found to be highly upregulated in schistosomulae. Upregulation of tegument and gut-associated genes was also found in cercaria and/or day 3 schistosomulae. Therefore, it is clear that different proteins are expressed in different developmental stages of parasite, and transcription of genes encoding those proteins also ensues accordingly. Vitelline proteins are important for egg shell formation. Different vitelline proteins are expressed differently in different stages of biological development as per biological need of these proteins. A study of Bae et al. (2007) showed that the vitelline protein (PwVit-20) was restricted to the eggs and adults only, whereas PwVit-36 was constitutively expressed in metacercariae and juvenile stages. This finding suggests critical involvement of PwVit-20 in parasite's fecundity activity. This indicated the prepositioning of stem or primordial vitelline cells in juveniles before sexual maturity. Pw-Vit 36 might acquire a unique or additional function relevant to the maturation and/or development of the vitlline cells/follicles of *Paragonimus westermani* during evolutionary period.

Note

We know that different stages of parasites perform different biological functions. Therefore, different protein molecules might be required by different stages. Understandably, the protein molecules required for invading parasite will be somewhat different from stationary phase of parasite. We know that transcription of mRNA occurs as per the need of protein by the organism. Therefore, during a particular period of development, mRNA profile may differ in different stages of biological development of the same parasite.

Molecular Pathogenesis by Trematodes

General Pathological Background

Before entering into molecular pathology, we should know a general structural pathology caused by important trematodes. In case of trematodes, it is found that two stages of parasites, i.e., migratory larva and adult, cause pathology differently. Usually, acute inflammatory reaction occurs due to migrating tissue-invading flukes. For example, migration of juvenile *Fasciola* parasites causes damage in the liver parenchymatous tissues. These immature flukes cause peritonitis, hepatitis, etc. When chronic pathology starts, there occurs hyperplastic cholangitis. Two types of hyperplasia occur, which are glandular and papillary types. Thickening of bile duct occurs due to massive fibrotic proliferation resulting in clay pipe stem condition. In *Dicrocoelium* infection, hyperplastic cholangitis, portal cirrhosis, etc. are the common pathological features found in chronic phase of infection. In *Opisthorchis* and *Clonorchis* infection, cholangitis, catarrhal cholecystitis, and cholangiohepatitis are observed. Immature amphistomes cause massive pathogenesis than the adults. Juvenile parasites are plug feeders and take out plug of mucosa from the wall of duodenum. Granuloma development occurs in case of *Schistosoma* infection. *Schistosoma nasalis* cause cauliflower-like growth in the nasal cavity. The immature *Paragonimus* flukes migrate into the lung parenchyma causing marked damage to the lung parenchymatous tissue. The adult parasites cause cyst formation in the lungs. The cysts are gradually filled up with purulent material.

Description of Molecular Pathology

In case of tissue-invading trematodes, occurrence of inflammation is obvious, which appears as a primary pathological feature. Inflammation occurs mainly due to pro-inflammatory cytokines secreted mostly by macrophages. These **pro-inflammatory cytokines are IL1 beta, IL6, TNF alpha, IFN gamma,** etc. In this context, it is felt to describe the general events related to pro-inflammatory cytokines. The pro-inflammatory cytokines act on endothelial cells, and genes of different adhesion molecules comprising ICAM1, ICAM2, VCAM1, and PECAM1 are expressed. The adhesion molecules act as trafficking molecules of leukocytes. The leukocytes adhere to these adhesion molecules and pass out from the vascular system to tissue space. The genes encoding pro-inflammatory cytokines are transcribed by the influence of NFkB (nuclear factor kappa-light-chain-enhancer of B cells). In a canonical pathway, IKK complex phosphorylate IKB. Thereafter, NFkB is dissociated from the IKB-containing complex. The pro-inflammatory cytokines lead to extravasation of leukocytes from vascular system. In case of tissue-invading larval parasite of trematode, ADCC (antibody-dependent cellular cytotoxicity) is another one important immunological phenomena. In ADCC, it is observed that the Fc portions of antibody (already bound to antigen) bind to different effector cells (eosinophil, macrophage, NK cells, etc.). These effector cells release toxic materials. For example, eosinophil contains **MBP (major basic protein), EPO (eosinophilic peroxidase, ECP (eosinophilic cationic protein), and EDN (eosinophil-derived neurotoxin)**. The toxic material released by different effector cells cause the change of microenvironment of the tissue where the parasite is migrating. Likewise, the **NK cells also release perforin, granzyme,** etc. Different lesions and other pathological features occur due to the effect of these substances.

The role of NO (nitric oxide) secreted by activated macrophage is very important for killing helminthic parasites. Different studies indicate that this **NO causes severe damage to the tissues** affected by *Schistosoma* parasite. In this context, it is felt to describe the general events of NO production. This NO is derived from the amino acid, L-arginine, by the action of an important enzyme, nitric oxide synthase (iNOS). As a result, nitric oxide and L-citrulline are produced. NO gets oxygen from L-arginine. Molecular oxygen is taken up by an enzymatic process, which utilizes electrons donated by NADPH. The nitrogen atom of nitric oxide is derived from the terminal guanidino group of the arginine. iNOS is encoded by specific genes, which are transcribed by the action of some cytokines. Activation of iNOS gene promoter is an important mode of iNOS regulation by cytokines. NFkB is an important transcription factor for transcription of mRNA of iNOS.

While discussing about the immunopathology by immature flukes (*Fasciola* spp., *Schistosoma* spp., etc.), the role of complement should be described. The complements are a group of proteins that cause a series of cascade reactions resulting in inflammation and other pathology in the reactive process. These complement reactions occur in three pathways, i.e., classical, alternative, and lectin. In case of complement activity, IgM and IgG play a role. The key component is C1q. The IgM is a planner polymeric protein molecule in which the binding site of C1q remains hidden. When the IgM comes in contact with antigen, there occur conformational changes of IgM. Then C1q binding site of IgM gets exposed. Then the C1q binds to the IgM and complement reaction starts. The IgG is a monomeric protein, also having C1q binding site; but the binding capacity is less strong as compared to IgM. In the process of complement reaction, different complement fragments are produced, which attract the phagocytic cells, and at the same time, they enhance the activity of these cells. In this way, an inflammatory reaction occurs. In some trematodal infections, hyperplasia occurs. The pro-inflammatory cytokines have a role for occurrence of hyperplasia. Granuloma formation occurs in case of *Schistosoma* infection. Different cells gather around the pathogen. The aggregated cells form a specific structure. The whole structure is encapsulated by fibrous tissue. Alternatively activated macrophages cause deposition of fibrous tissue by releasing some cytokines.

> **Note**
>
> Molecular pathology is described by different features comprising gene expression of different molecules, which directly or indirectly cause pathology. For example, when cytokine or other protein molecules are required for eliciting any immune response, transcription of mRNA encoding those cytokines and other proteins occurs. Therefore, apparently the pathology observed by us is due to the complex molecular mechanism.

Trematodal Zoonosis

Opisthorchiasis

Causative Agents

The important zoonotic parasite is *O. tenuicollis*.

Mode of Infection

Two intermediate hosts are required for completion of the life cycle of the parasite. The first intermediate host is freshwater snail, and the second one is cyprinid fish. Parasites develop in both intermediate hosts. The final hosts get the infection by ingestion of cyprinid fish in which metacercariae are formed.

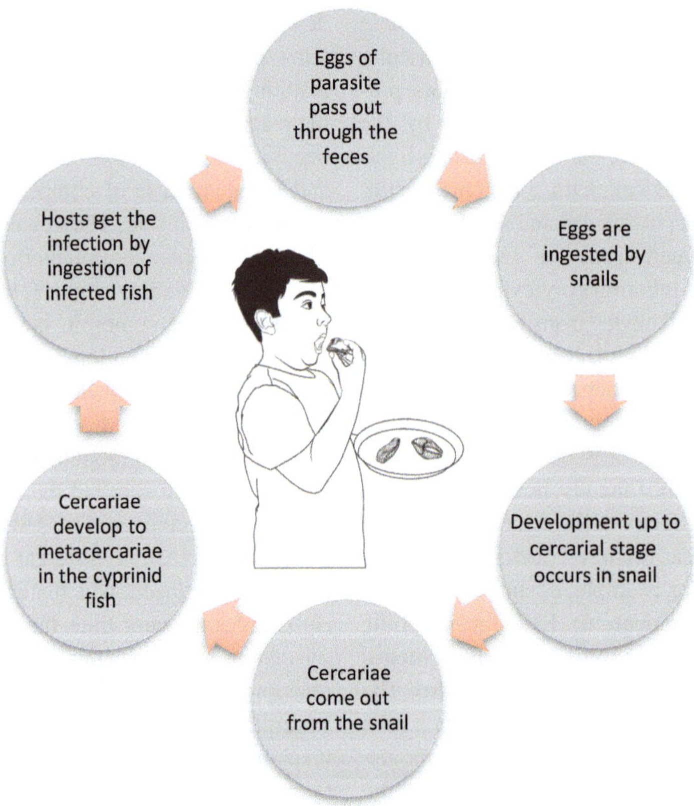

Fig. 4.45 Life cycle of *Opisthorchis* spp.

Pathological Features

In animals, cholangitis starts initially but it may turn sometimes to cholangiocarcinoma as a result of chronic inflammation. Dilation and obstruction of bile duct may occur. The parasites may cause mechanical obstruction of the bile duct, which may result in cholangiectasis, hepatomegaly, jaundice, etc. Abdominal discomfort, splenomegaly, edema, jaundice, etc. are associated clinicopathological features.

In man, cholecystitis, cholelithiasis, hepatomegaly, jaundice, etc. are some clinicopathological features. The liver and gallbladder may be enlarged. Cholangiocarcinoma may occur in some cases.

Diagnosis

Examination of fecal sample can be done for detection of parasitic egg.

Control

(a) Affected individuals should be treated properly.
(b) Feces should be disposed properly.
(c) Snails should be controlled using molluscicidal agents.
(d) Undercooked fish should be avoided.
(e) Sanitation improvement and proper health education.

Clonorchiasis

Causative Agents

The important causative agent is *Clonorchis sinensis.* The common name of this parasite is oriental fluke and Chinese liver fluke. The parasites affect the bile duct, pancreatic duct, and duodenum of man, dog, cat, pig, etc.

Mode of Infection

Two intermediate hosts are required for completion of the life cycle of the parasite. Different aquatic snails, i.e., *Bithynia* spp., *Bulinus* spp., *Parafossarulus* spp., *Melania* spp., etc., act as the first intermediate hosts. Up to cercarial development occurs in the first intermediate hosts. These cercariae get released from the snails and encyst on the fish of the family Cyprinidae. The final hosts ingest the second intermediate hosts and get infection with the metacercariae (infective stage).

Pathological Features

Cholangitis is an early pathological feature, found both in humans and animals, which may turn to cholangiocarcinoma as a result of chronic inflammation. There may be dilation of bile duct. The parasites mechanically obstruct the bile duct resulting in cholangiectasis. Hepatomegaly, jaundice, abdominal discomfort, splenomegaly, edema, jaundice, etc. are associated clinicopathological features.

Diagnosis

Symptomatological Diagnosis

The exhibited clinical signs are not so specific for confirmation of this parasitic infection.

Parasitological Diagnosis

- Diagnosis of this infection can be done by examination of stool for detection of the eggs. The eggs are typically **flask shaped**. If direct smear technique displays negative finding, then concentration technique should be done. The eggs should be differentially identified from *Opisthorchis*, *Heterakis*, etc.
- Examination of duodenal aspirate can be done.

Immunological Diagnosis

- Many works have been carried out using crude and purified antigens. ELISA test using recombinant cathepsin L proteinase is available for detection of specific antibody.
- Antigen detection is possible and ELISA test is available. Recent infection is detected by detection of egg antigen.

Molecular Diagnosis

- Molecular diagnosis is possible by using different molecular markers. Different ribosomal DNA, ribosomal RNA, and mitochondrial DNA could be used as molecular markers. Studies indicate the use of ribosomal DNA (ITS1, ITS2) and mitochondrial DNA (CO1) for characterization/identification of this parasite.

- **Multiplex PCR** can be done for detection of *Clonorchis* and *Opisthorchis* simultaneously. The test has good sensitivity and specificity.
- Molecular test is also available by **real-time PCR using mitochondrial NADH dehydrogenase subunit 2 (NAD 2)**.

Control

General Control

- Proper treatment should be done using specific drug available so far.
- Snails can be controlled by molluscicidal agents.
- Raw or undercooked fish should not be eaten.

Immunological Control

Different crude and defined antigens have been tried but no vaccine has yet been commercialized.

Molecular Control

Studies show protection against clonorchiasis in mice by use of DNA vaccine encoding FABP, cysteine proteinase, etc. But, so far, no DNA vaccine against clonorchiasis has been commercialized.

Fasciolopsiasis

Causative Agents

The zoonotic parasite is *Fasciolopsis buski.* The common name of the parasite is "giant intestinal fluke." These parasites affect the small intestine of man and pig.

Mode of Infection

Life cycle includes one intermediate host. The intermediate hosts are some snails, i.e., *Planorbis* spp., *Segmentina* spp., and *Hippeutis, Polyptis* spp. Development up to cercarial stages occurs in these snails. Cercariae are released from the snails and get attached on the tubers, water caltrop, water chestnut, on which they transform to metacercariae stages.

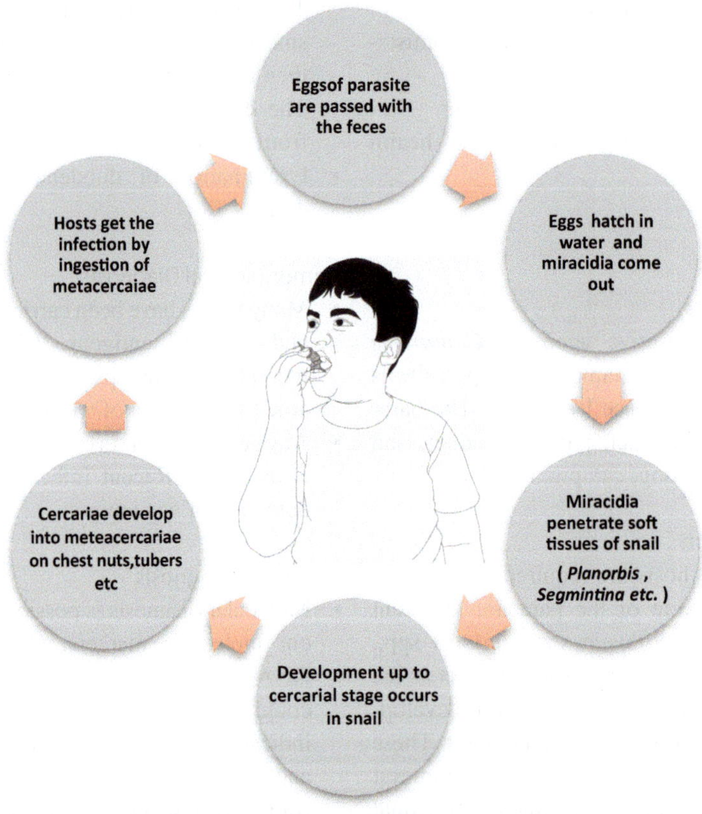

Fig. 4.46 Life cycle of *Fasciolopsis buski*

Pathological Features

In pig, initial infection causes inflammation. However, deep ulcerative lesions may be caused in heavy infection. Long-standing cases also result in ulcerative condition.

The human beings exhibit the clinical signs of nausea, epigastric pain, diarrhea, ascites, edema, etc. Blockage of intestinal lumen is another clinical feature.

Diagnosis

Examination of fecal sample reveals the presence of parasitic egg. Sedimentation method can be applied for better availability in the feces. Differential identification of eggs can be done from *F. hepatica*, *Echinostoma* spp., etc.

Control

Proper treatment should be done. The snails should be controlled using molluscicidal agents. But practical implication is difficult. Untreated and unfiltered water should not be taken.

Fascioliasis

Causative Agents

The zoonotic species is *Fasciola hepatica.*

Hosts Affected

Fasciola parasites affect the bile duct and liver of sheep, goat, cattle, etc. Human beings are the accidental hosts.

Mode of Infection

The eggs of the parasites are voided and the eggs hatch in the water. The miracidia enter the aquatic snail in which up to cercarial development occurs. The hosts get infected with metacercariae on ingestion. The cercariae are formed in the snail intermediate host. The cercariae are discharged from the snail hosts. They later get encysted on the vegetation as metacercariae. The hosts get the infection by ingestion of vegetation harboring metacercariae stages of the parasites present on vegetation. *In human beings, the parasite behaves like larva

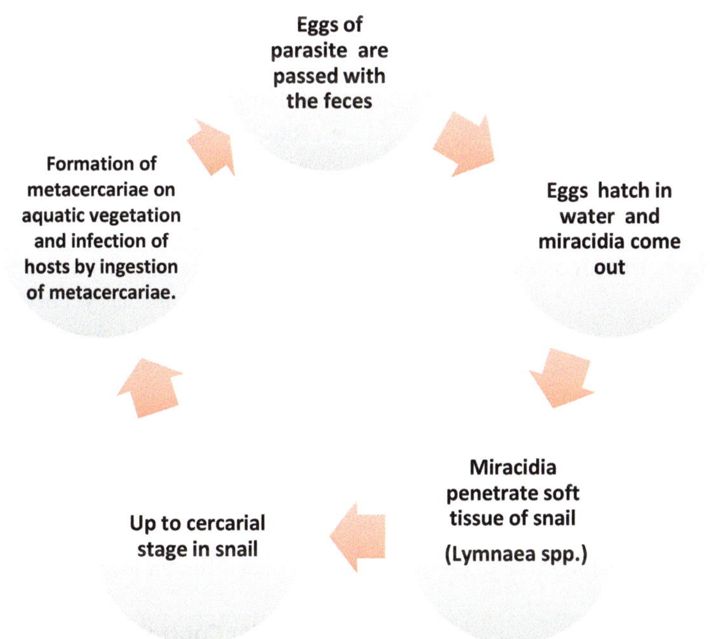

Fig. 4.47 Life cycle of *Fasciola* spp.

migrans. The juvenile flukes may not reach to the liver. The juvenile flukes reach to the larynx and lungs where they form **hazel nut-sized cyst.** *Adult *Fasciola* in colon was reported. The human beings get the infection by ingestion of raw watercress or other contaminated vegetations.

Pathological Features

In animals, hepatitis in acute cases occurs. Cirrhosis occurs in long-standing chronic case.

In human beings, this fluke can also cause disease. The larval stages wander in unusual sites. The parasites may reach to the larynx, lungs, etc. Epigastric pain, upper abdominal pain, and malaise are common clinical signs. In acute phase, fever and arthralgia may be observed. In early stage of infection, eosinophilia is readily detected. In this acute stage, urticarial eruption is also observed. In chronic phase of infection, cholelithiasis and empyema of the gallbladder are observed.

Diagnosis

Symptomatological Diagnosis

The exhibited clinical signs are not so specific for confirmation of this parasitic infection.

Parasitological Diagnosis

Diagnosis of this parasitic infection can be done by examination of the stool for detection of the eggs.

Immunological Diagnosis

Many works have been carried out using crude and purified antigens. ELISA test is available for *F. hepatica*.

Image-based Diagnosis

- **USG** (ultrasonography) can be done for detection of liver lesion.
- **MRI** (magnetic resonance imaging) is also useful for detection of liver pathology or detection of cysts caused by the parasite in any abnormal location, i.e., lungs, larynx, etc. CT scan is more sensitive.

Molecular Diagnosis

- Molecular diagnosis is possible by using different molecular markers. Different ribosomal DNA, ribosomal RNA, and mitochondrial DNA could be used as molecular markers. Studies indicate the use of ribosomal DNA (ITS1, ITS2) and mitochondrial DNA (CO1) for characterization/identification of this parasite. Copro-PCR is available.

Control

General Control

- Proper treatment should be done using specific drugs available so far.
- Underground water should be used for drinking.
- Feces are the source of eggs; therefore, feces should be disposed properly.
- Systematic treatment should be done for the whole herd of animals as the animals are the main source of infection.
- The animals should be allowed to graze in the high land area where stagnation of water is absent.
- Systematic control of snails should be done by using molluscicidal agents. Since the snails act as the intermediate hosts of the parasite, they should be controlled using copper sulfate solution, sodium pentachlorophenate, etc.
- Raw or undercooked vegetations should be avoided by the human beings.

Immunological Control

Different crude and defined antigens have been tried but no vaccine has yet been commercialized. In the veterinary field, a series of works were carried out using attenuated live vaccine comprising attenuated metacercariae, crude somatic protein of whole parasitic extract, ES protein (excretory and secretory protein), etc. Different purified proteins, i.e., cysteine protease, glutathione S-transferase, FABP, and paramyosin, were also used with different level of success.

Molecular Control

So far no naked DNA vaccine or gene vaccine is available.

Heterophyiasis

Causative Agent

Heterophyes heterophyes affects the small intestine of dog, cat, and fox. This parasite occurs in human beings also.

Mode of Infection

Two intermediate hosts are required for completion of the life cycle. The first intermediate hosts are aquatic snails, i.e., *Pirenella conica*. The second intermediate hosts are *Mugil cephalus*, *M. capito*, *Tilapia nilotica*, etc. The final hosts get the infection on ingestion of undercooked and unprepared infected fish. A salted fish preparation named 'fessikh' is the source of human infection in Egypt.

Pathological Features

In man, the parasites cause hemorrhagic diarrhea. Diarrhea, abdominal pain, etc. are common clinical signs. In unusual cases, the parasitic stages may affect the brain, spinal cord, heart, etc.

Diagnosis

Examination of fecal sample may reveal the presence of eggs of the parasites.

Control

- Affected individuals should be treated properly.
- Feces should be disposed properly.
- Human beings must avoid raw or undercooked fish.

Paragonimiasis

Causative Agents

Some species of *Paragonimus* are *P. westermani*, *P. africanus*, *P. kellicotti*, and *P. heterotremus*. Of

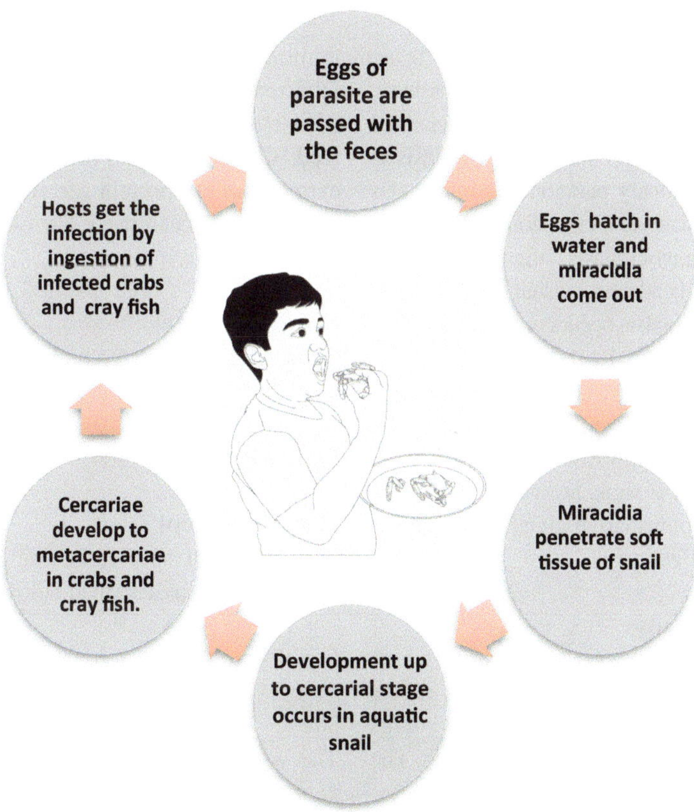

Fig. 4.48 Life cycle of *Paragonimus westermani*

these species, *Paragonimus westermani* is an important zoonotic parasite.

Mode of Infection

Two intermediate hosts are required for completion of the life cycle. The first intermediate hosts are aquatic snails in which up to cercariae stages are formed. The cercariae are liberated from the snails and get transformed into metacercariae and are encysted in the crabs and crayfish. The definitive hosts get the infection by ingestion of infected **crabs** and **crayfish.**

Pathological Features

The parasites affect the bronchiole and develop thereby. Flukes lay eggs in bronchi. Pathological lesions along with infiltration of neutrophils and eosinophils occur in the affected areas. The eggs of the flukes get encapsulated resulting in formation of cysts or abscess, which is further surrounded by fibrous capsule. Granuloma formation may occur due to immunological reaction mediated by degenerating eggs. **Microabscess** is formed elsewhere in the affected organ.

In human beings, the clinical signs are fever, blood in sputum, hemoptysis, anorexia, and malaise. There is evidence of cerebral paralysis characterized by epilepsy in man. Pulmonary paragonimiasis is common. Extrapulmonary paragonimiasis includes cerebral paragonimiasis and cutaneous paragonimiasis. Cerebral paragonimiasis affects the brain resulting seizure, headache, visual disturbances. Cutaneous paragonimiasis occurs in children.

Diagnosis

Symptomatological Diagnosis

The exhibited clinical signs are vague, which do not provide any confirmative clues for diagnosis of this parasitic infection.

Parasitological Diagnosis

(a) Feces should be examined properly, which may reveal the presence of the eggs of the parasite. There is similarity of the eggs between *Fasciola* and *Paragonimus* spp. The differential identification should be done.
(b) Image-based diagnosis by X-ray analysis may reveal the presence of the parasitic cysts.

Immunological Diagnosis

Many works have been carried out using crude and purified antigen. Dot immunogold filtration assay is a rapid test having high sensitivity and specificity where crude extract of the parasite is used. Antigen detection for diagnosis of early infection can be done by dot ELISA.

Molecular Diagnosis

Molecular diagnosis is possible by using different molecular markers. Different ribosomal DNA, ribosomal RNA, and mitochondrial DNA could be used as molecular markers. Studies show copro-DNA detection of *Paragonimus* sp. using different molecular markers, i.e., ITS2, CO1, etc.

Control

General Control

Proper treatment should be done using specific drug available so far. Undercooked crabs and crayfish should not be eaten.

Immunological Control

Different crude and defined antigens have been tried but no vaccine has yet been commercialized. Crude somatic protein of whole parasitic extract and ES protein (excretory and secretory protein) have been tried. No vaccine is commercially available.

Molecular Control

So far no naked DNA vaccine or gene vaccine is available.

Schistosomiasis

Causative Agent

The important species having zoonotic importance have been described below:

Schistosoma haematobium Infection

Site

The adult worms are found in the mesenteric veins and the eggs are laid in vesical, uterine, and prostatic plexus of veins. They are found in portal veins and cecum of cattle.

Mode of Infection

Eggs are laid by the female being seated in the gynecophoric canal in the veins. Through the vessels, the eggs reach to the mucosa of the urinary bladder and finally reach in the cavity wherefrom they pass out with the urine. The miracidia come out after hatching of the eggs. They penetrate through the exposed soft tissues of the snail and locate in the liver of the snail. The development up to cercaria occurs in the snails. The final hosts get the infection through skin penetration by the cercariae.

Pathogenesis

The eggs cause severe pathology in man. The eggs possess a terminal spine by which the eggs irritate and erode the affected area. There occurs immunopathological reaction resulting in cellular infiltration. The affected area becomes hyperplastic. Deposition of granulomatous tissues leads to cause formation of the **pseudotubercles**, which are also called as egg granuloma. The disease is also called as schistosomiasis hematobia. The metabolic by-product of the schistosomulae in the hepatoportal circulation may cause anaphylactic reaction, which exhibits the pathological features of hepatosplenomegaly, etc. This fluke causes urinary bilharziasis. Carcinoma of urinary bladder is also found very often.

Diagnosis

Symptomatological Diagnosis

The exhibited clinical signs are vague and do not have any diagnostic value.

Parasitological Diagnosis

Diagnosis can be done by demonstration of the eggs in the urine. **Biopsy of bladder** mucosa is done for demonstration of eggs in the biopsy material. The eggs have terminal spines.

Immunological Diagnosis

Many works have been carried out using crude and purified antigens.

Antibody detection: HAMA FAST-ELISA. HAMA stands for *S. haematobium* adult worm microsomal antigen. This test is reliable with high specificity and sensitivity. HAMA-EITB is also done for the same.

Antigen detection: Antigen detection is done by use of **monoclonal antibody** for detection of soluble egg antigen.

Molecular Diagnosis

Molecular diagnosis is possible by using different molecular markers. Different ribosomal DNA, ribosomal RNA, and mitochondrial DNA could be used as molecular markers.

Control

General Control

Proper treatment should be done using specific drug available so far.

Immunological Control

Different crude and defined antigens have been tried but no vaccine has yet been commercialized.

Molecular Control

So far as no naked DNA vaccine or gene vaccine is available.

Schistosoma mansoni Infection

The worms are found in the portal veins. However, these worms are also found in mesenteric veins of sigmoidorectal area. Man is the definitive host of this parasite. But animals are also affected.

Mode of Infection

Their eggs are expelled out of the definitive hosts through the feces. After hatching, miracidia come out. The miracidia penetrate through the exposed soft tissues of the snail and enter the liver of the snail. Development up to cercaria occurs in the snail. Cercariae come in contact of the skin of the human being. They start penetrating the skin and reach to the subcutaneous tissues at first. The cercariae transform into schistosomulae after shedding off the tails. The schistosomulae are then carried to the liver. There, the worms become adult. They migrate against the blood stream and reach to mesenteric veins and capillaries of sigmoidorectal region.

Pathogenesis

The disease caused by this parasite is called as schistosomiasis mansoni, schistosomal amoebiasis, or intestinal bilharziasis. The localized pathology comprises of dysentery. The ectopic lesions are hepatomegaly, portal hypertension, and periportal fibrosis.

Diagnosis

(a) Diagnosis can be done by demonstration of the eggs in the feces. The egg shell is acid fast, which could be stained by **Ziehl-Neelsen** stain. Biopsy of rectal tissue can be done for demonstration of eggs.
(b) **Rectal biopsy** material can be pressed in between two slides and stained. The eggs can be seen in the tissues.
(c) Different clinical features, i.e., eosinophilia and allergic manifestation, may be the guidance for the diagnosis.
(d) Several immunological tests can be performed. These are **circumoval precipitatin test, IFA, IHA, ELISA, EITB,** etc.

Schistosoma japonicum Infection

Hosts and Site

The worms are found in the portal vein, mesenteric vein, and rectal or hemorrhoidal plexus of vein. The definitive hosts are dog, cat, pig, man, etc.

Mode of Infection

Farmers, fisherman, and children are frequently affected by this parasitic infection. The eggs are passed from the host through feces. Development up to cercarial stage occurs in the aquatic snails. The cercariae are discharged from the snail. They infect the hosts through skin penetration.

Pathogenesis

The eggs cause severe pathology. The affected area becomes hyperplastic. Deposition of granulomatous tissues occurs, which are called as **egg granuloma**. Calcification of granuloma may occur in long-standing cases. It causes Symmers' fibrosis. This is also called clay pipe stem fibrosis. This condition occurs due to intrahepatic portal vein calcification. Among all the schistosomes, this species is extremely pathogenic. *Schistosoma japonicum* causes Katayama fever, portal hypertension, cirrhosis, etc. Some eggs may pass to the lungs and cause pathology.

Diagnosis

Diagnosis can be done by demonstration of the eggs in feces. Different clinical features, i.e., eosinophilia and allergic manifestation, may be guidance of diagnosis. Several immunological tests can be performed. These are circumoval precipitatin test, fluorescent antibody test, ELISA, etc.

Control

General Control

(a) Proper treatment of affected individual.
(b) Proper disposal of feces.
(c) Swimming in uncleaned and untreated pond and lake should be avoided.

Immunological Control

(a) Vaccine development using 28 kDa glutathione S-transferase (GST) of *S. haematobium* is in progress.
(b) **Tetraspanin** (Sm-TSP-2) has been used for vaccination trial in mice.
(c) Smp80 is a large subunit of **calpain**, which has been tried for cross protection against *S. haematobium* in baboon.
(d) Use of recombinant **paramyosin** muscle protein is in progress.

Research work can be done on use of those above-mentioned proteins for immunoprotection study of *S. japonicum* also.

Culture of Trematodes

Different works were carried out for culture of trematodes. Up to certain level of growth was found in different commercially available media along with some supplement. Of these media, **RPMI 1640** with human serum has been used successfully. In this medium, excysted juvenile flukes of *Fasciola* were cultured up to certain level. NCTC 135 plus heat inactivated chick serum and sheep RBC were also used for the same purpose. Different common supplements, i.e., diphasic medium, liver extract, bile, yeast extract, egg products, and embryo extract, were also tried to enhance the growth further but no enhancement of growth was found.

For culture of adult *Fasciola*, different research workers used Earle's salt solution with 100% survival for 5 days. A multi-component tissue culture medium is better than simple balanced salt solution. A series of experiments were carried out to maintain juvenile flukes from metacercarial stages. Addition of serum, blood, etc. increases the duration of survival of the flukes. A continuous flow culture apparatus is more suitable for culture of the adult parasites for considerably long period. Efforts were also made to culture *Paragonimus* using different types of culture media comprising simple balanced salt solution. Growth can be enhanced using blood, serum, etc. NCTC 109 supplemented with different substances, i.e., animal serum, Earle's solution, yeast extract, etc., showed good growth of this fluke up to development of reproductive system. Different culture media are used for culture of schistosomes comprising Earle's solution, Earle's solution supplemented with inactivate rabbit serum, rabbit red cells, lactalbumin hydrolysate, glucose, antibiotics, etc. Study shows investigation of effect of enrichment of media with chick embryo extract, rabbit liver extract, etc. Continuous flow culture apparatus is also used for culture of different flukes including schistosomes. For production of schistosomules from cercariae, RPMI 1640 medium containing inactivated serum, penicillin, and streptomycin is used

Emerging and Re-emerging Trematodal Diseases

Fascioliasis

Heydarian et al. (2021) reported emerging cases of fasciolosis in Lorestan Province, Western Iran. They reported *Fasciola hepatica* infection. Human fasciolosis is currently re-emerging in several regions mostly due to rapid changes of human activities. Until 1990, human fasciolosis was given secondary importance. But 1990 onwards, increased number of human fasciolosis are reported both for *Fasciola hepatica* and *F.gigantica*. There are different underlying reasons comprising changes of food habit, new settlements of humans, etc.

Paramphistomosis

Paramphistomosis is caused by different species of Paramphistomatidae. This disease is worldwide in distribution. In some parts, the prevalence is less important. But the incidences are increasing in those parts. Evidences indicate more prevalence of paramphistomosis as compared to fasciolosis in different parts of the UK and Ireland (Huson et al. 2017).

Schistosoma Infection

Schistosomiasis is emerging in man in different countries. The reason is environmental and sociopolitical. Systematic study of human cases is required to confirm emerging human cases in particular areas. The recent recurrent outbreaks of urogenital schistosomiasis in the south of Europe were unanticipated (Kincaid-Smith et al. 2017).

Opisthorchis Infection

Opisthorchis spp. are trematode parasites. They perform heteroxenous life cycle and require two intermediate hosts. The first intermediate hosts are aquatic snails and the second intermediate hosts are fish. A change of food habit causes introduction of parasite infection in human beings. Eating raw marinated tench fillets of fish has been increasingly popular in Italy. Pozio et al. (2013) reviewed 211 cases of opisthorchiasis in human in Italy which is due to consumption of trench fillets. More than 50 years ago, cases of opisthorchiasis was diagnosed in man in European Union countries i.e., Germany, Greece, Lithuania, Poland, Romania; but, in the last decade, the infection was documented in a few persons in Germany and greece (Reviewed, Pozio et al. 2013).

Dicrocoelium Infection

This is an important trematode parasite. There are two important species of this genus, i.e., *D. dendriticum* and *D. hospes*. These parasites commonly occur in animals. Cabeza-Barrera reported emerging spurious *Dicrocoelium* infection in Almeria of southern Spain. The infection was found in the foreigners because of their food habit of taking raw or undercooked liver.

Clinical Trematodology

(For Clinical Trematodology the readers must study relevant aspects i.e. etiology, pathogenesis, clinical signs, diagnosis, treatment, control etc. from the earlier description of different trematodes in this chapter. Here some additional information have been given in this context.)

Collection and Dispatch of Material to Laboratory for Diagnosis of Trematode Infection

Feces Collection

The feces should be collected afresh because in many parasitic cases, the eggs hatch out and the distinct eggs cannot be visible. For collection of feces from large animal, the hand after wearing a plastic glove is inserted to take out some amount of feces per rectum. Sufficient amount of feces should be collected. If possible, 100–200 g of feces should be collected so that gross examination, direct examination, and concentration techniques can be performed at a time. The specimen is kept in wide-mouthed glass containers with the arrangement of screw cap. However, many use different commercially available containers, i.e., plastic specimen tubes. The polythene bags are also used, which are feasible and disposable.

Preservation

As such, there is no need of preservation if the feces is examined immediately. But if the feces are examined late within a day, it should be kept in refrigerator. If the feces have to be transported to distant places, which would take few days, in these circumstances, the feces should be mixed with normal saline, and a homogenous mixture is produced. This mixture is mixed with 10% formalin ten times of the original mixture.

Collection of Blood

Blood is collected from the ear vein of dog, cattle, rabbit, etc. In case of trematode infection, for performing different immunodiagnostic tests, blood is collected for serum preparation. And for serum preparation, the blood should not be mixed with anticoagulant. Serum is essential for performing immunodiagnostic tests both for antibody detection and antigen detection. Vacutainer tube can be used for easy collection of blood. It is very easy using blood collection tube made up of glass or plastic test tube with a rubber stopper creating a vacuum seal within the tube facilitating for drawing of a predetermined volume of liquid.

Collection of Urine

Urine is directly collected into conical flask or beaker. Urine collection is essential for detection of eggs of some schistosome species.

Lacrimal Secretion

Collection of lacrimal secretion is not required for detection of trematode infection.

Nasal Discharge

Laboratory camel brush is used for collection of nasal discharge and a smear is prepared on the glass slide. Collection of nasal discharge is essential for detection of eggs of *Schistosoma nasalis*.

Animal Inoculation Test

Animal inoculation test is useful for diagnosis of many protozoan parasites. But it has no importance for diagnosis of trematode infection.

Blood and Biopsy Smear Examination

Blood examination and biopsy examination have some role for diagnosis of trematode infection. For example, rectal and bladder biopsy are done for detection of eggs of some schistosome spp.

Histopathology

Examination of histopathological section may reveal migratory trematode larvae, eggs of trematodes, etc. Hematoxylin and eosin is a very famous staining technique for staining of tissues. This is performed for visualizing microscopically the pathological status or detection of any pathogen therein. There are many ways to prepare the hematoxylin solutions (formulation). There are many protocols also for performing H&E staining. The nuclei are stained blue, and the cytoplasm and extracellular matrix are stained pink. The tissue is collected as per relevant method. The tissue is fixed in fixative agent. Dehydration of the tissue is done then. The tissue is embedded in melted paraffin wax, which is called as paraffin blocking of tissues. The block is mounted on a microtome to cut into thin slices. Different types of sections are done according to need. This may be lateral, medial, sagittal, mediosagittal, etc. The staining is done involving application of hematoxylin mixed with a metallic salt (mordant). The same is rinsed in a weak acid solution to remove excess staining. The tissue is placed in mildly alkaline water. After the application of hematoxylin, the tissue is countered with eosin.

Egg Identification

Egg identification is the gold standard technique for diagnosis of helminthic infection including trematode. Apart from this, different immunodiagnostic tests are done. But cross-reactivity is the major constraint for practical implication. Molecular diagnosis is reliable if genus and species-specific primers are available. In case of some trematodes, species-specific primers are available. Using these primers, specific identification of some trematode is possible.

Sheep, Goat, and Cattle

Parasite	Egg identification
Dicrocoelium dendriticum	The eggs possess an operculum (similar to hinged hatch) on one pole. The operculum is indistinct. They are dark brown, embryonated when laid. The size is 35–45 um × 20–30 um.
Fasciola	The eggs are oval and yellowish brown in color. Distinct operculum is present. The eggs are unembryonated when laid. The size is 156–197 um × 90–104 um (*F. gigantica*) and 130–150um × 63–90um (*F. hepatica*).

Dogs, Cats and Birds

Parasite	Clinical findings
Nanophyetus salmincola	The eggs are yellowish brown. They have indistinct operculum with a small blunt point at the end opposite to the operculum
Prosthogonimus pellucidus	The eggs are dark brown and operculated. The size is around 29 × 13 um having a small spine at abopercular pole.
Paragonimus westermanii	The eggs are found in cystic space of the lung. These are yellowish brown in color. The eggs have distinct operculum. The size is 80–120 um × 45–70 um.
Heterobilharzia americana	These are thin shelled eggs containing miracidium

Trematodes of Wild Animals

Parasites	Wild animals	Sites
Dicrocoelium dendriticum	Coypu, elk, donkey	Bile duct
Platynosomum fastosum	Wild cats	Bile duct, liver, gallbladder
Concinnum procyonis	Red and gray fox, raccoon	Pancreatic duct, gallbladder, bile duct
Heterophyes heterophyes	Fox	Small intestine
Metagonimus yokogawai	Pelican	Small intestine
Euryhelmis squamula	Fox, pole cat, mink, weasel	Intestine
Cryptocotyle lingua	Greater and lesser black backed gull, razorbill, common tern, kittiwake	intestine
Apophallus donicum	Fox	Small intestine
Apophallus muhlingi	Gulls	Intestine
Prosthogonimus macrorchis	Wild birds	Bursa of Fabricius and oviduct
Prosthogonimus ovatus	Wild birds	Bursa of Fabricius and oviduct

Parasites	Wild animals	Sites
Opisthorchis tenuicollis	Fox, wild pig	Bile duct; sometimes in intestine and pancreatic duct
Opisthorchis viverrini	Civet, wild cat	Bile duct, liver
Clonorchis sinensis	Pig, weasel, mink, badger	Bile duct, pancreatic duct
Pseudoamphistomum truncatum	Fox, glutton	Bile duct
Metorchis conjunctus	Mink, fox, raccoon	Bile duct
Metorchis albidus	Fox	Gallbladder, bile duct
Nanophyetus salmincola	Fox, coyote, mink, otter, opossum lynx, raccoon	Small intestine
Fasciola hepatica	Pig, hare, beaver, kangaroo, elephant, rabbit	Bile duct

Dicrocoelium Infection

The species of *Dicrocoelium* are important trematodes. They are found in bile duct and gallbladder of cattle, sheep, goat, etc. But this parasite also occurs in wild ruminants including red deer, roe deer, fallow deer, and mouflon. Adults are found in bile duct and gallbladder of a wide range of domestic and wild mammals, mainly sheep and mouflon. However, camel, horse, rodents, pig, lagomorphs, and primates may also be affected. The parasites are usually asymptomatic. Weight loss and loss of production are some vague clinical features exhibited by the affected animals.

Fasciola Infection

This is an important trematode parasite, which usually affects bile duct and liver of different domestic animals comprising cattle, sheep, goats, etc. But they are also found in wild animals where they serve as reservoir hosts. Red deer, rabbit, and wild boar are affected by this parasite. Wild animals act as the reservoirs of this parasite. Therefore, clinical findings are absent. Diagnosis can be done by examination of fecal sample. Molecular methods could be applied for identification of the parasitic eggs by copro-PCR. Coprological detection of *Fasciola* antigen may be possible by different types of ELISA and Western blot.

Schistosomes Infection

There are different species of *Schistosoma*. Of these species, *S. nasalis* and *S. suis* are important species occurring in animals. *S. nasalis* occurs in cattle, sheep, goat, etc. *S. suis* occurs in pigs. But schistosomes also occur in wild animals. S. *bovis* may affect camel, wild ruminants, rodent etc. *S. mattheei* may occur in wild ruminants,baboons and rodents etc. *S. haematobium* may affect baboon, monkeys etc. *S. rodhaini* occurs in rodents. Occurrence of *Bivitellobilharzia nairi* in Asian or greater one-horned rhinoceros (*Rhinoceros unicornis*) was reported without any clinical feature. Clinical form of schistosomosis in elephant comprising granuloma formation in the liver, uropathy, gastrointestinal bleeding, etc. was reported.

Paragonimus Infection

Paragonimosis has been reported from tiger, leopard, clouded leopard, rusty-spotted cat, fishing cat, leopard cat, golden cat, palm cat, jungle cat, small Indian civet, large Indian civet, Himalayan pine marten, etc. *P. westermani* is the most common species occurring in Indian wild carnivores, though *P. compactus*, *P. edwardsi*, and *P. kellicotti* have been reported from small number of wild hosts. The flukes do not produce marked pathology and remain in the cysts of lung parenchyma. Diagnosis is done by examination of fecal sample and isolation and identification of the worms during postmortem.

Heterophyes Infection

Infection of big cats by this parasite has been reported.

References

Abdeltawab AH, Shahat SA, Abdou M, Metwally KM, Woodward SR (2000) DNA sequence polymorphisms in two strains of *Fasciola hepatica* from Egypt and Europe. J Egypt Soc Parasitol 30(3):821–828. PMID: 11198380

Adam R, Arnold H, Hinz E, Storch V (1995) Morphology and ultrastructure of the redia and pre-emergent cercaria of *Opisthorchis viverrini* (Trematoda: Digenea) in the intermediate host *Bithynia siamensis goniomphalus* (Prosobranchia: Bithyniidae). Appl Parasitol 36(2):136–154. PMID: 7550441

Agatsuma T, Arakawa T, Iwagami M, Honzako Y, Cahyaningsih U, Kang SY, Hong SJ (2000) Molecular evidence of natural hybridization between Fasciola hepatica and F. gigantica. Parasitol Int 49(3):231–238.

https://doi.org/10.1016/s1383-5769(00)00051-9. PMID:11426578

Bae YA, Kim SH, Cai GB, Lee EG, Kim TS et al (2007) Differential expression of *Paragonimus westermani* eggshell proteins during the developmental stages. Int J Parasitol 37(3–4):295–305. https://doi.org/10.1016/j.ijpara.2006.10.006. PMID: 17112524

Bai X, Song JH, Dai F, Lee JY, Hong SJ (2020) Clonorchis sinensis secretory protein CsAg17 vaccine induces immune protection. Parasit Vectors 13(1):215. https://doi.org/10.1186/s13071-020-04083-5. PMID:32334611; PMCID:PMC7183723

Bogh HO, Zhu XQ, Qian BZ, Gasser RB (1999) Scanning for nucleotide variations in mitochondrial DNA fragments of *Schistosoma japonicum* by single-strand conformation polymorphism. Parasitology 118(pt1):73–82. https://doi.org/10.1017/s0031182098003527. PMID: 10070664

Cevallos W, Calvopiña M, Nipáz V, Vicente-Santiago B, López-Albán J, Fernández-Soto P, Guevara Á, Muro A (2017) Enzyme-linked immunosorbent assay for diagnosis of *Amphimerus* spp. liver fluke infection in Humans. Mem Inst Oswaldo Cruz 112(5):364–369. https://doi.org/10.1590/0074-02760160426. Epub 2017 Mar 27. PMID: 28443982; PMCID: PMC5398163

Cho PY, Lee J-Y, Kim TI, Song J-H, Hong S-J, Yoo WG et al (2020) Serodiagnostic antigens of Clonorchis sinensis identified and evaluated by high-throughput proteogenomics. PloS Negl Trop Dis 14(12):e0008998. https://doi.org/10.1371/journal.pntd.0008998

Cifrian B, Garcia-Corrales P (1988) Scanning electron microscopy of adult *Dicrocoelium dendriticum*. Parasitol Res 74(3):235–242. https://doi.org/10.1007/BF00539571. PMID: 3362829

Cwiklinski K, Dalton JP, Dufresne PJ, Lacourse J et al (2015) The *Fasciola hepatica* genome: gene duplication and polymorphism reveals adaptation to the host environment and the capacity for rapid evolution. Genome Biol 16(1):71. https://doi.org/10.1186/s13059-015-0632-2. PMID: 25887684

Dalton JP, McGonigle S, Rolph TP, Andrews SJ (1996) Induction of protective immunity in cattle against infection with *Fasciola hepatica* by vaccination with cathepsin L proteinases and with hemoglobin. Infect Immun 64(12):5066–5074. https://doi.org/10.1128/IAI.64.12.5966-5074.1996. PMID: 8945548

Dias AS, Araújo JV, Braga FR, Araujo JM, Puppin AC, Fernandes FM, Ramos RF, Bertonceli RM, da Silva RG, Perboni WR (2011) Biological control of *Fasciola hepatica* eggs with the *Pochonia chlamydosporia* fungus after passing through the cattle gastrointestinal tract. Parasitol Res 110(2):663–667. https://doi.org/10.1007/s00436-011-2538-6. Epub 2011 Jul 20. PMID: 21773773

Donah PN, Shinohara A, Horii Y, Habe S, Nawa Y, Le NT (2007) Description of a new lung fluke species, Paragonimus vietnamensis sp. nov. (Trematoda, Paragonimidae), found in northern Vietnam. Parasitol Res 101(6):1495–1501. https://doi.org/10.1007/s00436-077-0666-9. PMID:17674049

Estuningsih SE, Smooker PM, Wiedosari E, Widjajanti S, Vaiano S et al (1997) Evaluation of antigens of *Fasciola gigantica* as vaccines against tropical fasciolosis in cattle. Int J Parasitol 27(11):1419–1428. https://doi.org/10.1016/s0020-7519(97)00096-9. PMID: 9421734

Gherbawy YA, Shalaby IM, El-Sadek MS, Elhariry HM, Abdelilah BA (2013) The anti-fasciolіasis properties of silver nanoparticles produced by *Trichoderma harzianum* and their improvement of the anti-fascioliasis drug triclabendazole. Int J Mol Sci 14(11):21887–21898. https://doi.org/10.3390/ijms141121887. PMID: 24196355; PMCID: PMC3856041

Graczyk TK, Fried B (1994) ELISA method for detecting anti-*Echinostoma caproni* (Trematoda: Echinostomatidae) immunoglobulins in experimentally infected ICR mice. J Parasitol 80(4):544–549. PMID: 8064521

Han ET, Yang HJ, Park YJ, Park JH, Chai JY (2014) *Metagonimus yokogawai*: a 100-kDa somatic antigen commonly reacting with other trematodes. Korean J Parasitol 52(2):201–204. https://doi.org/10.3347/kjp.2014.52.2.201. Epub 2014 Apr 18. PMID: 24850966; PMCID: PMC4028460

Heydarian P, Ashrafi K, Rahimi Esboei B, Mohebali M et al (2021) Emerging cases of Fasciolosis in Lorestan Province, Western Iran: Case Series Report. Iran J Public Health 50(1):195–200. https://doi.org/10.18502/ijph.v50i1.5087

Huson KM, Oliver NAM, Robinson MW (2017) Paramphistomosis of ruminants: an emerging parasitic disease in Europe. Trends Parasitol 33(11):836–844. https://doi.org/10.1016/j.pt.2017.07.002. PMID: 28754416

Intapam PM, Kosuwan T, Wongkham C, Maleewong W (2004) Genomic characterization of lung fluke, *Paragonimus heterotremus*, *P. sinensis*, *P. harinasutai*, *P. westermani* and *P. bangkokensis* by RAPD markers. Vet Parasitol 124(1–2):55–64. https://doi.org/10.1016/j.vetpar.2004.05.023. PMID: 15350661

Kim TY, Joo I, Kang S, Cho S et al (2002) Recombinant *Paragonimus westermani* yolk ferritin is a useful serodiagnostic antigen. J Infect Dis 185(9):1373–1375. https://doi.org/10.1086/339880. PMID: 12001061

Kincaid-Smith J, Rey O, Toulza E, Berry A, Boissier J (2017) Emerging schistosomiasis in Europe: a need to quantify the risks. Trends Parasitol 33(8):600–609. https://doi.org/10.1016/j.pt..2017.04.009. PMID:28539255

Kumar N, Varghese A, Gaurav N, Dinesh C, Samanta S et al (2012) Vaccination of buffaloes with Fasciola gigantica recombinant glutathione s-transferase and fatty acid binding protein. Prasitol Res 110:419–426. https://800.gl/pkp.pmm

Martinez-Ibeas AM, Perteguer MJ, Gonzalez-Lazza C, Garate T, Manga-Gonzalez MY (2013) Analysis of an expressed sequence taq library from *Dicrocoelium dendriticum*. Exp Parasitol 135(2):287–296. https://doi.org/10.1016/j.exppara.2013.07.005. PMID: 23872427

Mirzadeh A, Yoosefy A, Kazemirad E, Barati Z, Golkar M et al (2018) Evaluation of a set of refolded recombinant antigens for serodiagnosis of human fasciolosis. PLoS One 13(10):e0203490. https://doi.org/10.1371/journal.pone.0203490. PMID: 30281608

Morozova EV, Ryskov AP, Semenova SK (2002) RAPD variation in two trematode species (*Fasciola hepatica* and *Dicrocoelium dendriticum*) from a single cattle population. Genetika 38(8):1155–1162. PMID: 12244701

Morrison CA, Colin T, Sexton JL, Bowen F, Wicker J et al (1996) Protection of cattle against *Fasciola hepatica* infection by vaccination with glutathione S-transferase. Vaccine 14(17–18):1603–1612. https://doi.org/10.1016/s0264-410x(96)00147-8. PMID: 9032888

Nöckler K, Dell K, Schuster R, Voigt WP (2003) Indirect ELISA for the detection of antibodies against *Opisthorchis felineus* (Rivolta, 1884) and *Metorchis bilis* (Braun, 1790) in foxes. Vet Parasitol 110(3–4):207–215. https://doi.org/10.1016/s0304-4017(02)00324-2. PMID: 12482649

Novobilský A, Kasny M, Mikes L, Kovarcik K, Koudela B (2007) Humoral immune responses during experimental infection with *Fascioloides magna* and *Fasciola hepatica* in goats and comparison of their excretory/secretory products. Parasitol Res 101(2):357–364. https://doi.org/10.1007/s00436-007-0463-5. Epub2007. PMID: 17265088

Parker-Manual SJ, Ivens AC, Dillon GP, Wilson RA (2011) Gene expression patterns in larval *Schistosoma mansoni* associated with infection of the mammalian host. PLoS Negl Trop Dis 5(8):e1274. https://doi.org/10.1371/journal.pntd.0001274. PMID: 21912711

Paykari H, Dalini A, Madani R (2002) Immunization of sheep against *Fasciola gigantica* with glutathione S-transferase. Vet Parasitol 105(2):153–159. https://doi.org/10.1016/s0304-4017(01)00546-5. PMID: 11900929

Phung LT, Chaiyadet S, Hongsrichan N, Sotillo J, Dieu HDT, Tran CQ, Brindley PJ, Loukas A, Laha T (2019) Recombinant *Opisthorchis viverrini* tetraspanin expressed in Pichia pastoris as a potential vaccine candidate for opisthorchiasis. Parasitol Res 118(12):3419–3427. https://doi.org/10.1007/s00436-019-06488-3. Epub 2019 Nov 14. PMID: 31724067; PMCID: PMC6899162

Pothong K, Komalamisra C, Kalambaheti T, Watthanakulpanich D, Yoshino TP (2018) ELISA based on a recombinant *Paragonimus heterotremus* protein for serodiagnosis of human paragonimiasis in Thailand. Parasit Vectors 11(1):322. https://doi.org/10.1186/s13071-018-2878-5. PMID: 29843786

Pozio E, Armignacco O, Ferri F, Gomez Morales MA (2013) Opisthorchis felineus, an emerging infection in Italy and its implication for the European Union. Acta Trop 126(1):54–62. https://doi.org/10.1016/J.actatropica.2013.01.005. PMID:23337391

Rabia Aly I, Diab M, AM EI-A, Hendawy M, Kadry S (2012) Fasciola gigantica Fatty binding protein (FABP) as a prophylactic agent against Schistosoma mansoni infection in CD1 mice. Korean J Parasitol 50(1):37–43. https://doi.org/10.3347/kjp.2012.50.1.37. PMID22451732; PMCID:PMC3309049

Ramos RA, Lima VF, Monteiro MF, Santana Mde A, Lepold R, Faustino MA, Rinaldi L, Cringoli G, Alves LC (2016) New insights into diagnosis of *Platynosomum fastosum* (Trematoda: Dicrocoeliidae) in cats. Parasitol Res 115(2):479–482. https://doi.org/10.1007/s00436-015-4763-x. Epub 2015 Oct 6. PMID: 26437646

Sexton JL, Wilce MC, Colin T, Wijffels GL et al (1994) Vaccination of sheep against *Fasciola hepatica* with glutathione S-transferase: identification and mapping of antibody epitopes on a three dimensional model of the antigen. J Immunol 152(4):1861–1872. PMID: 7509829

Sirisinha S, Chawengkirttikul R, Sermswan R (1991) Immunodiagnosis of opisthorchiasis. Southeast Asian J Trop Med Public Health 22(Suppl):179–183. PMID: 1822882

Sohn WM (1998) Life history of *Echinoparyphium recurvatum* (Trematoda: Echinostomatidae) in Korea. Korean J Parasitol 36(2):91–98. https://doi.org/10.3347/kjp.1998.36.2.91. PMID: 9637826

Sripa J, Brindley PJ, Sripa B, Loukas A et al (2012) Evaluation of liver fluke recombinant cathepsin B-1 protease as a serodiagnostic antigen for human opisthorchiasis. Parasitol Int 61(1):191–195. https://doi.org/10.1016/j.parint.2011.05.009. PMID: 21704728

Suhardono RJA, Copeman DB (2006) Biological control of *Fasciola gigantica* with *Echinostoma revolutum*. Vet Parasitol 140(1–2):166–170. https://doi.org/10.1016/j.vetpar.2006.02.028. Epub 2006 Jul 18. PMID: 16854529

Teimoori S, Arimatsu Y, Laha T, Kaewkes S, Sereerak P, Tangkawattana S, Brindley PJ, Sripa B (2015) Immunodiagnosis of opisthorchiasis using parasite cathepsin F. Parasitol Res 114(12):4571–8. https://doi.org/10.1007/s00436-015-4703-9. Epub 2015 Sep7. PMID:10831381.PUBMED

van Milligen FJ, Cornelissen JB, Bokhout BA (2000) *Fasciola hepatica*: an antigen fraction derived from newly excysted juveniles, containing an immunoreactive 32-kDa protein, induces strong protective immunity in rats. Exp Parasitol 94(3):163–171. https://doi.org/10.1006/expr.1999.4476. PMID: 10831381

Waikagul J (2007) A new species of *Paragonimus* (Trematoda: Troglotrematidae) from a cat infected with metacercariae from mountain crabs *Larnaudia larnaudii*. J Parasitol 93(6):1496–1500. https://doi.org/10.1645/GE-1054.1. PMID: 18314698

Wijffels GL, Salvatore L, Dosen M, Waddington J et al (1994) Vaccination of sheep with purified cysteine proteinases of *Fasciola hepatica* decreases worm fecundity. Exp Parasitol 78(2):132–148. https://doi.org/10.1006/expr.1994.1014. PMID: 8119370

Zhang M, Fu Z, Li C, Han Y et al (2015) Screening diagnostic candidates for schistosomiasis from tegument proteins of adult *Schistosoma japonicum* using an immunoproteomic approach. PLoS Negl Trop Dis 9(2):e0003454. https://doi.org/10.1371/journal.pntd.0003454. PMID: 25706299

5 Veterinary Cestodology

Cestodes are hermaphrodite, endoparasitic worms. Many tapeworms have veterinary importance. They occur in a wide variety of animals. Some tapeworms have zoonotic importance which occur both in man and animals. In this chapter, different aspects of tapeworms of veterinary importance have been described comprising history, distribution, host, predilection site morphology, pathogenesis, diagnosis, treatment, and control.

General Characteristic Features of Cestodes

Shape

The cestodes are tape-like or ribbonlike organisms.

Size

The size of cestodes is widely variable. The length may range from few mm to meters.

Color

The color of tapeworms is white or creamy white.

Body Structure

The body is dorsoventrally flattened and segmented.

Body Divisions

The cestode is broadly divided into three parts, i.e., head, neck, and body proper.

Head: The head is also called as scolex. Suckers are found in the scolex of a cestode. The scolex possesses different structures which have taxonomic importance. These are as follows:

- The scolex possesses four **suckers**. The suckers may be armed with hooks, or the suckers may be unarmed.
- The scolex may possess one protruded retractable part called as **rostellum**. Rostellum may be armed with hooks. Shape and size of hooks have taxonomic importance.

Neck: Next to the scolex is the neck. The neck does not have any segment. Segments are produced in this region. The process of proglottid formation is called as proglottidation.

Body proper: The body proper contains three types of segments. These are immature, mature, and gravid segments. Length and number of

S. C. Mandal, *Textbook of Veterinary Parasitology*, https://doi.org/10.1007/978-981-99-2924-5_5

segments of cestodes are widely variable. The immature segments do not contain any reproductive system or eggs. Completely developed reproductive system is found in the mature segment. The reproductive system is replaced by eggs in the gravid segment. The eggs may be contained in different uterine envelopes. These are capsule, paruterine organ, etc.

Important Systems

They do not have any body cavity. The different important systems are as follows:

Reproductive System

The cestodes are hermaphrodite. The same organism contains both male and female reproductive systems. Complete male and female reproductive systems are found in the mature segment. The gravid segment is packed up with eggs. In some cestodes, two sets of reproductive systems are present.

Male Reproductive System

The male reproductive system consists of one of more testes, vas deferens, and seminal vesicles.

Female Reproductive Systems

Female reproductive system consists of unlobed or lobed ovary, oviduct, ootype, uterus, etc.

> **Note**
> Digestive system, respiratory system, and circulatory system are absent in cestodes. The cestodes absorb nutrients through their body surface (cuticle).

General Biology

Biology varies between the members of Eucestoda and Cotyloda. In Eucestoda, except *Hymenolepis*, life cycle is of indirect type which involves two hosts. Both direct and indirect types of life cycles are found in *Hymenolepis nana*. In the intermediate host, infective stage of larva is formed. These are called as metacestodes/cysts/bladder worms. The names of metacestodes are cysticercoid, cysticercus, hydatid cyst, strobilocercus, coenurus, etc. The final hosts get the infection by ingestion of metacestodes. In case of Cotyloda, two intermediate hosts are required for completion of their life cycle. Procercoid is formed in the first intermediate host and plerocercoid is formed in the second intermediate host. The final hosts get the infection by ingestion of infected second intermediate hosts.

Classification of Cestodes of Veterinary importance

Phylum: Platyhelminthes
Class: **Eucestoda**

Order: Davaineidae
Family: Davaineidae
Genus: *Davainea*
Genus: *Raillietina*
Genus: *Cotugnia*

Order: Dilepididea
Family: Dipylidiidae
Genus: *Dipylidium*
Genus: *Choanotaenia*
Family: Dilepididae
Genus: *Amoebotaenia*

Order: Anoplocephalidea
Family: Anoplocephalidae
Genus: *Anoplocephala*
Genus: *Paranoplocephala*
Genus: *Moniezia*
Genus: *Cittotaenia*
Family: Thysanosomidae
Genus: *Thysanosoma*
Genus: *Thysaniezia*
Genus: *Avitellina*
Genus: *Stilesia*

Order: Hymenolepididea
Family: Hymenolepididae
Genus: *Hymenolepis*

Order: Taeniidea
Family: Taeniidae
Genus: *Taenia*
Genus: *Echinococcus*

Order: Mesocestoididea
Family: Mesocestoididae
Genus: *Mesocestoides*

Class: **Cotyloda**
Order: Diphyllidea
Family: Diphyllobothridae
Genus: *Diphyllobothrium*

Class: Eucestoda—Order: Davaineidea—Family: Davaineidae

Family Synopsis

The members of this family are small- to medium-sized tapeworms. The worms are found mostly in the birds and in some mammals. Hammer-shaped hooks are present on the rostellum. They usually have single set of reproductive organs. The genital pore may be unilateral, regularly alternating, or irregularly alternating. The intermediate hosts are different invertebrates, i.e., ant, fly, and slugs. Cysticercoid is the infective stage. Some of the members are important because they cause significant pathology. Heavy infection for long period causes incessant loss of productivity.

Species

Davainea proglottina
Davainea meleagridis
Davainea andrei

Species: *Davainea proglottina*

Prevalence

The worms are prevalent worldwide.

Common Name

- Minute tapeworm
- Small chicken worm

Hosts

Domestic hosts: Different types of birds are affected by these parasites, i.e., fowl, guinea fowl, peafowl, pigeon, and turkey.

Wild hosts: Wild gallinaceous birds are affected by the parasites.

Site

Common site: The tapeworms occur in the small intestine.

Aberrant site: There seems no report of aberrant site other than the small intestine.

Morphology

Gross Morphology

Size: The length is 0.5–3 mm and breadth is 0.18–0.6 mm.

Shape: They look as small tiny worms. Segments are not apparently detectable by gross examination.

Color: The color is white or translucent as that of other tapeworms.

Microscopic Morphology

Adult

Scolex: The scolex possesses one rostellum which is armed with hammer-shaped hooks. The suckers are armed with thorn-shaped hooks.

Segments: 4–9 segments are present, of which last segment is the gravid segment. The gravid segment contains egg capsules. Egg capsule contains single egg.

Testes: About 19 testes are present. They are located posterior to and by the side of the ovary.

Ovary: The ovary is bilobed and located in the middle of the segment.

Uterus: The uterus is normal and breaks into several pouches called egg capsules.

Vagina: It is found posterior to the cirrus pouch.

Genital pore: Genital pores alternate regularly.

Larva

Oncospheres: These are round and ciliated. Six hooks are present in the embryo (hexacanth embryo).

Cysticercoid: The cysticercoids are pear shaped.

Egg

Size: The eggs are 35 um in diameter.

Shape: The eggs are usually round in shape.

Structure: Egg is present in egg capsule. Egg capsule contains single egg. Six hooks are present in the egg.

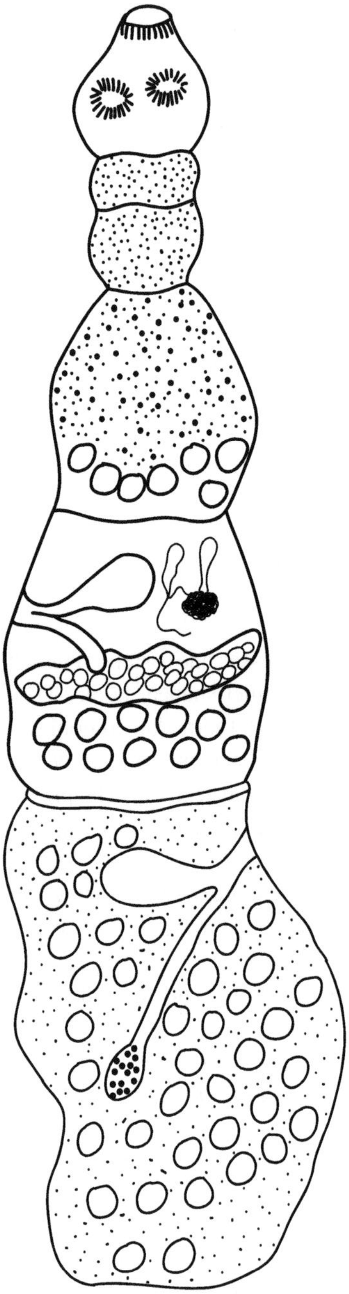

Fig. 5.1 *Davainea proglottina*

Life Cycle

The parasite is heteroxenous which requires one intermediate host for completion of life cycle.

Stages of Life Cycle Formed

In definitive host	In intermediate host
• Adult • Egg	• Oncosphere • Metacestode

Some basic information of life cycle
• The definitive hosts are birds • The intermediate hosts are slugs • The infective stage of intermediate host is metacestode (cysticercoid) • The mode of infection of definitive host is ingestion • The mode of infection of intermediate host is ingestion

Events of life cycle: a glimpse
• Voiding of gravid segments or eggs with feces of definitive host • Ingestion of eggs by the intermediate hosts • Hatching of eggs in the intermediate hosts • Release of oncospheres after hatching • Oncospheres develop to cysticercoids in intermediate hosts • Ingestion of intermediate hosts by definitive hosts • Development up to adult in definitive hosts

Detailed Description of Life Cycle

Embryonation

Embryonation of eggs occurs in the worm itself. After embryonation, segments become gravid. The gravid segment contains embryonated eggs.

Detachment of Gravid Segment

The integrity of junction of segments is lost. Then, the gravid segments are detached from the worm.

Discharge of Gravid Segment

The gravid segments are passed with the feces. Usually, one gravid segment is passed in every 24 h. Different factors play on discharge of gravid segment. The peak time of discharge of gravid segment is afternoon or night as regulated by the feeding time of the birds.

Liberation of Eggs from the Gravid Segment

From the gravid segment, the eggs are released. The discharged proglottids possess capability for locomotion. They are also positively phototactic in nature. They can crawl grass blades and other vegetation. During this crawling activity, the eggs are liberated from the gravid segment.

Development in Intermediate Host

The slugs act as the intermediate hosts. These slugs are *Arion*, *Agriolimax*, *Milax*, *Zonitoides*, *Gasterodonta*, *Polygyra*, etc. Eggs are ingested by the intermediate host. Hatching of eggs occurs in the gut of the snail. The hexacanth oncospheres are released after hatching which migrate in the body cavity of slugs. The oncospheres transform to infective stage (cysticercoid) within 3 weeks. The cysticercoid is pear shaped and somewhat solid-bodied structure.

Infection of Definitive Host

The definitive hosts get the infection by ingestion of intermediate host containing cysticercoids.

Development in Definitive Hosts

In the gut of the definitive host, the slug is digested and the cysticercoid is released. The scolex of cysticercoid gets attached to the wall of the intestine and gradually develops to become adult within approximately 2 weeks.

Note

Abdou (1956) carried out an experiment on biology of *Davainea proglottina* in Britain. They reported that gray snail (*Agriolimax reticulatus*) and garden snail (*Arion hortensis*) of Britain seemed to be important intermediate hosts of this tapeworm. They also reported slugs of nearly every age exposed to experimental infection showed 100% infectivity. Cysticercoids produced in the life cycle of *Davainea* can be identified by observing specific morphological features. In the stained slides of cysticercoids, distinct hammer-shaped hooks were found. The outer surface of cysticercoids is corrugated and wrinkled

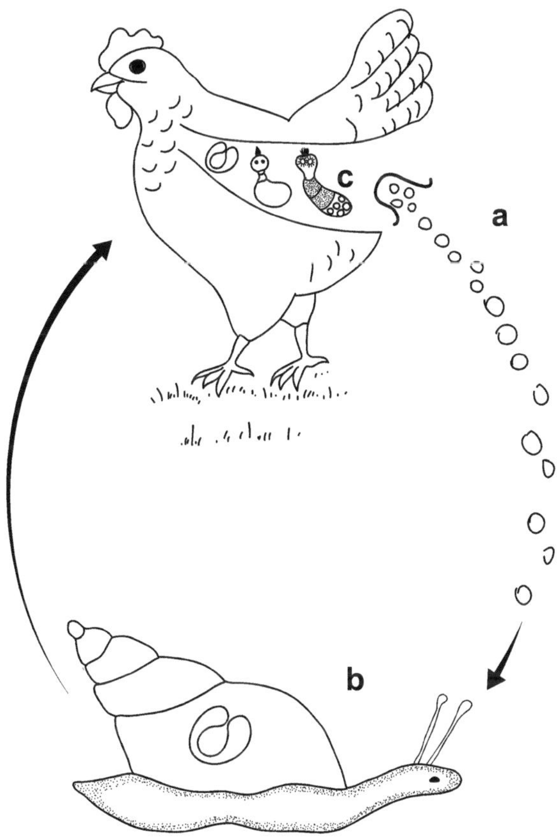

Fig. 5.2 Life cycle of *Davainea proglottina*. (**a**) Egg, (**b**) cysticercoid (**c**) Adult

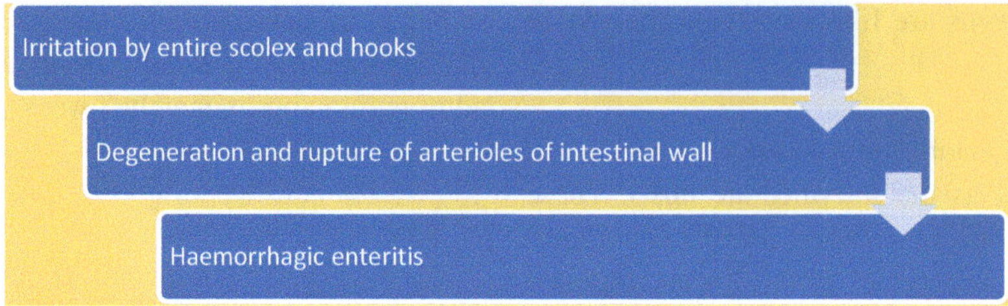

Fig. 5.3 Flowchart of pathogenesis caused by *Davainea proglottina*

Pathogenesis and Pathology

Pathogenesis

Pathogenesis has been presented in the flowchart.

Pathology

Anatomical pathology: The tapeworms have armed rostellum and suckers. The rostellum is armed with hammer-shaped hooks. Using the scolex along with these hooks, they cause irritation to the wall of the intestine. This results in denudation or sloughing of epithelial cells of the intestinal wall. This activity causes rupture of small arterioles and venules of wall of the intestine which result in hemorrhage. Hemarrhagic areas on the wall of the small intestine are found.

Hematological pathology: Hematological data displays anemia. There is low erythrocytic count, decreased Hb% and PCV%, and increased WBC count.

Chemical pathology: Hypoproteinemia may be found in long-standing case.

Immunopathology: In any cestode infection, inflammation of the gut wall is obvious. It is presumed that pro-inflammatory cytokines are produced by immune cells which cause inflammation. IL1 beta and IL6 are two important cytokines responsible for causing this inflammation. Activated macrophages produce different pro-inflammatory cytokines.

Symptoms

Primary Symptoms

- General diarrhea
- Hemorrhagic diarrhea
- Occasional detection of gravid segment in feces

Secondary Symptoms

- Debility
- Loss of body weight
- Loss of egg production
- Weakness
- Anemia

Diagnosis

Symptomatological Diagnosis

No specific clinical signs of diagnostic value are exhibited. However, hemorrhagic diarrhea may be suggestive of *Davainea proglottina* infection.

Parasitological Diagnosis

Examination of fecal sample may reveal the presence of eggs of the parasite. In case of light infection and eggs are not available in the smear, sedimentation technique comprising centrifugation floatation could be done. But detection of eggs cannot confirm the parasitic infection because egg morphology is not specific. Detection of gravid segment is a confirmatory diagnosis, but the gravid segment remains in the feces unnoticed. A few representative birds suspected to be infected can be autopsied for direct examination for presence of adult parasite. As the adult parasite remains embedded, mucosal scraping should be examined.

Immunological Diagnosis

Fragmented research work was done on immunodiagnostic works. But no immunodiagnostic test has yet been commercialized.

Molecular Diagnosis

It has been a recent trend of molecular diagnosis using specific molecular markers. Different ribosomal DNA, ribosomal RNA, and mitochondrial DNA could be used as molecular markers. But it is problematic as reference nucleotide sequence is less. Recent studies show molecular identification of *Davainea* spp. using ITS2 as molecular marker. But it is a costly and time-consuming method.

Treatment

Niclosamide	Oral	175 mg/kg

Control

General Control

The snails serve as the intermediate hosts. Control of snails should be done.

Immunological Control

No vaccine has yet been commercialized.

Molecular Control

No gene vaccine is available.

Biological Control

No biological control is available.

Genus: *Raillietina*

Species: *R. cesticillus*, *R. echinobothrida*, and *R. tetragona*

Prevalence

The worms are worldwide in distribution.

Predilection Site

Common site: The small intestine is the common site.

Erratic site: There seem no reports of erratic site.

Disease Caused

Nodular tapeworm disease is caused by *Raillietina echinobothrida*.

Morphology

Gross Morphology

Size: *R. cesticillus* is 4–15 cm long. *R. echinobothrida* is 8.5–22 cm long. *R. tetragona* is up to about 25 cm long.

Shape: They are simple dorsoventrally flattened and segmented worms. The gravid segments are rounded or spherical giving somewhat a beaded apperance.

Color: The color is white or translucent as that of other tapeworms.

Microscopic Morphology

Adult

Scolex: The scolex possesses one rostellum. *R. cesticillus* has a broad rostellum armed with 400–500 hooks. The suckers are round in shape and unarmed. *R. echinobothrida* has a small rostellum armed with double crown of 200–250-min hooks. The suckers are round in shape and armed with 8–10 rows of hooks. *R. tetragona* has a small rostellum armed with double crown of about 100 hooks. The suckers are oval and armed with 8–10 rows of hooks.

Segments: Segments are barrel shaped. Unilateral or irregularly alternating genital pores are found in mature segments of different species.

Testes: The testes are located posterior to and either side of the ovary. The segment of *R. cesticillus*, *R. echinobothrida*, and *R. tetragona* have 18–30, 20–45, and 20–30 testes, respectively.

Ovary: The ovary is lobed and located in the median field of the segment.

Vitellaria: The vitelline gland is post-ovarian.

Egg capsule: Egg capsule of *R. cesticillus*, *R. echinobothrida*, and *R. tetragona* contain 1, 8–12, and 6–12 eggs, respectively.

Genital pores: Genital pores are unilateral in *R. tetragona*, irregularly alternating in *R. cesticillus*, unilateral or irregularly alternating in *R. echinobothrida*.

Larva

Oncospheres: These are roughly round in shape and ciliated. Six hooks are present in the embryo (hexacanth embryo). The internal structure is almost same as oncospheres of other tapeworms. The whole body of oncosphere is covered with five envelopes. Germ cells are present in oncosphere. Nerve cells are also present.

Cysticercoid: The cysticercoids are pear shaped/heart shaped having a scolex. They measure 0.327–0.487 × 0.267–0.337 mm. On the way of development, different morphological stages are found. In the initial stage, elongated form is found. Later on, two other stages, i.e., bipartite and tripartite forms, are found. In mature stage, head, sucker, and rostellum are clearly visible.

Special Note

The birds are affected by different species of *Raillietina* species. But human beings are also affected by *Raillietina* sp.

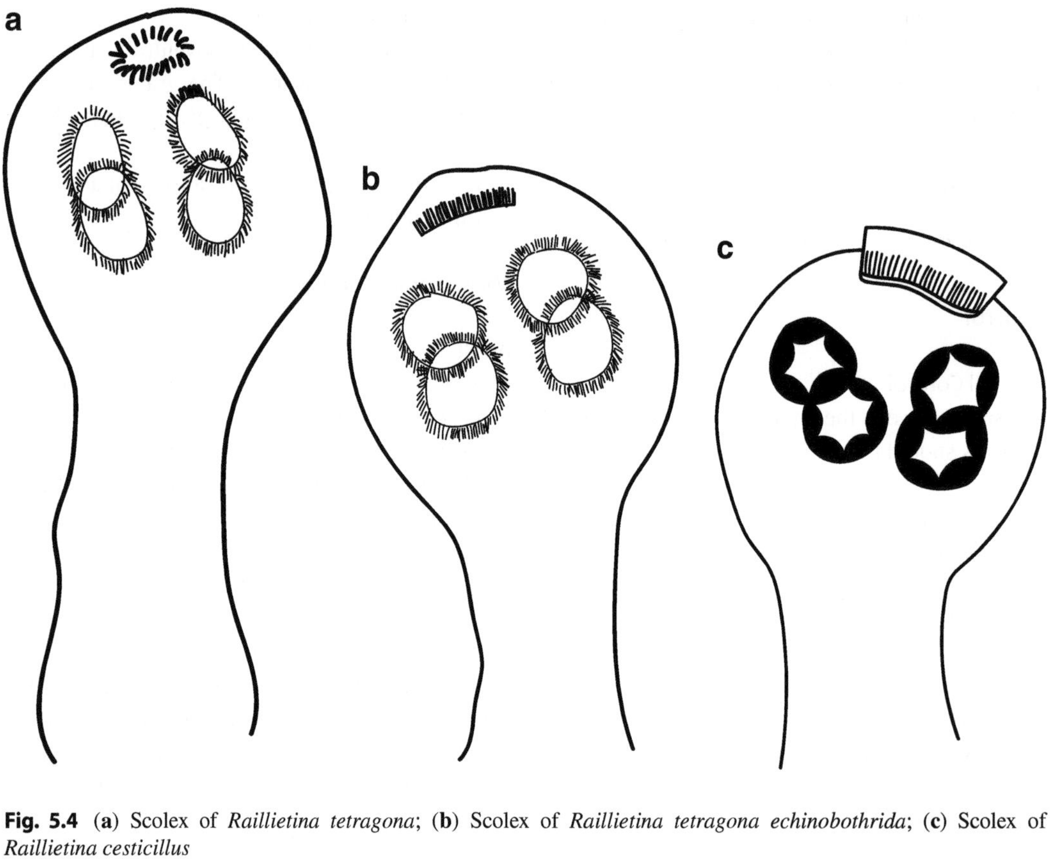

Fig. 5.4 (**a**) Scolex of *Raillietina tetragona*; (**b**) Scolex of *Raillietina tetragona echinobothrida*; (**c**) Scolex of *Raillietina cesticillus*

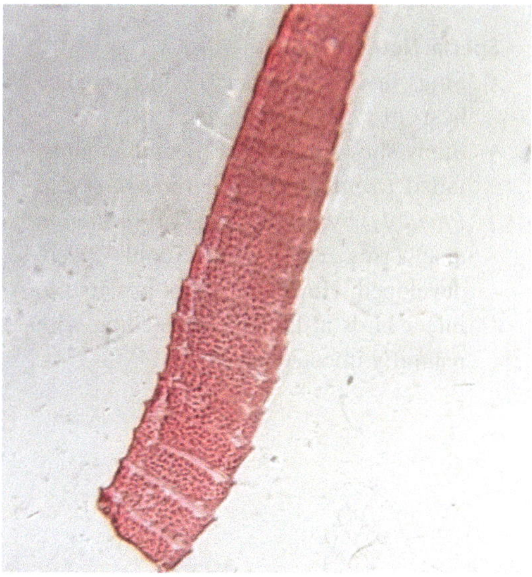

Fig. 5.5 Segment of *Raillietina* sp. (Gross)

Life Cycle

The parasites show heteroxenous life cycle which require one intermediate host for completion of their life cycle.

Stages of Life Cycle Formed

In definitive host	In intermediate host
• Adult • Egg	• Oncosphere • Metacestode

Some basic information of life cycle
• The definitive hosts are birds • The intermediate hosts are beetles, ants, and house flies • The infective stage of intermediate host is metacestode (cysticercoid) • The mode of infection of definitive host is ingestion • The mode of infection of intermediate host is ingestion

Events of life cycle: a glimpse
• Voiding of eggs with feces of definitive host • Ingestion of eggs by the intermediate hosts (beetle, ant, house fly, etc.) • Hatching of eggs in the intermediate hosts • Release of oncospheres after hatching • Oncospheres develop to cysticercoids in intermediate hosts • Ingestion of intermediate hosts by definitive hosts • Development up to adult in definitive hosts

Detailed Description

Embryonation

The worms are hermaphrodite and they self-fertilize. Embryonation of eggs occurs in the worm itself. After embryonation, segments become gravid. These gravid segments are the source of eggs.

Detachment of Gravid Segment

As is found in other tapeworms, there occurs gradual loss of integrity of junction of gravid segments. Then, the gravid segment is detached in a series.

Discharge of Gravid Segment

The gravid segments are discharged in the late afternoon. They have capability of locomotion. From the gravid segment, the eggs are released. Irrespective of any stimulus, they can move and come up to the surface of the feces and become available to different intermediate hosts.

Development in Intermediate Host

Different arthropods act as intermediate hosts. The eggs are food materials of the arthropods. The intermediate hosts ingest these eggs. Hatching of eggs occurs in the gut of the intermediate hosts (beetle in case of *R. cesticillus*, ant and house fly in case of *R. echinobothrida*, and *R. tetragona*). The beetles serving as intermediate hosts are *Amara*, *Calathus*, *Stenocellus*, *Discoiderus*, *Aphodius*, *Anisotarsus*, *Harpalus*, etc. The ants serving as intermediate hosts are *Tetramorium*, *Monomorium*, etc. The house flies (*Musca domestica*) serve as intermediate hosts. The hexacanth oncospheres are released which migrate in the body cavity of the intermediate hosts. Later, the oncospheres transform to infective stage (cysticercoid). These are also called as metacestode stage of the parasite.

Infection to Definitive Host

The definitive hosts become infected when they ingest the intermediate hosts harboring cysticercoids.

Development in Definitive Hosts

After being ingested by the definitive hosts, the body of the intermediate host is digested, and the cysticercoid becomes free in the gut. The scolex of cysticercoid gets evaginated. The head gets attached to the wall of the intestine and gradually develops to become adult.

Special Note

- Slugs and snails act as intermediate hosts of *Raillietina bonii*
- Study shows that experimental attempts failed to infect ants (*Tetramorium* and *Pheidole*) by feeding eggs of *Raillietina* tapeworms. No cysticercoids were developed. However, it was possible to infect birds artificially by feeding with naturally infected ants

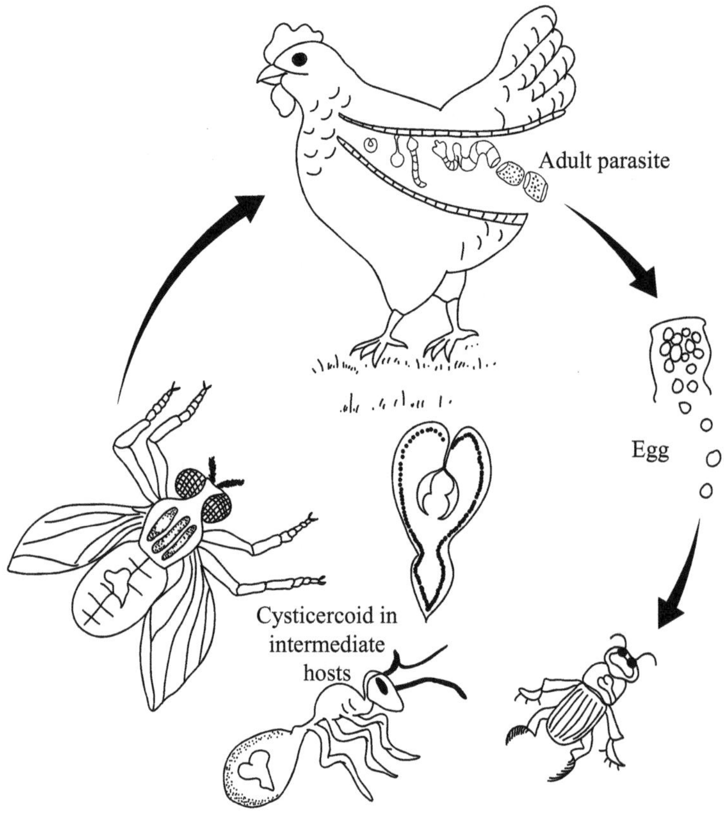

Fig. 5.6 Life cycle of *Raillietina* sp.

Pathogenesis and Pathology

Pathogenesis

The pathogenesis has been presented in flowchart.

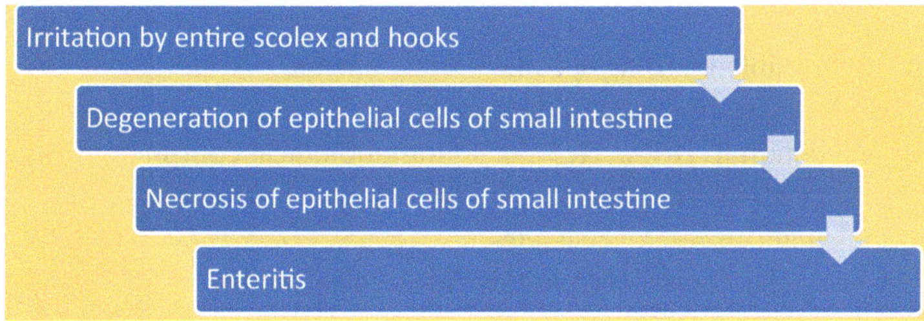

Fig. 5.7 Pathogenesis caused by *Raillietina* sp.

Pathology

Anatomical Pathology (Gross)

The tapeworms have armed rostellum and suckers. Using these structures, they damage the wall of the intestine. Constant mechanical damage results in denudation or sloughing of epithelial cells of the intestinal wall. This results in catarrhal enteritis, congestion, and infiltration of inflammatory cells.

Hematological Pathology

Usually, the tapeworms do not alter any hematological data. But in long-standing case, there is low erythrocytic count, decreased Hb% and PCV%, and increased WBC count.

Histopathology

Histopathological examination reveals epithelial desquamation, villus atrophy, hemorrhage, cellular infiltration with lymphocytes, granulocytes, and plasma cells, etc.

Chemical Pathology

The worms produce metabolic by-product which is absorbed by nervous tissues resulting in nervous disorder.

Immunopathology

Inflammation of the intestinal wall is the common pathology due to any tapeworm infection. It could be presumed that pro-inflammatory cytokines are produced which cause inflammation. IL1 and IL6 are two important cytokines responsible for causing inflammation.

Symptoms

Primary Symptoms

- General diarrhea
- Hemorrhagic diarrhea in case of *R. echinobothrida* infection
- Occasional detection of gravid segment in feces

Secondary Symptoms

- Debility
- Loss of body weight
- Loss of egg production
- Weakness
- Anemia

Diagnosis

Symptomatological Diagnosis

Clinical signs are vague. However, hemorrhagic diarrhea may be suggestive of *R. echinobothrida* infection.

Parasitological Diagnosis

Examination of fecal sample may reveal the presence of eggs of the parasite. Centrifugation floatation technique or other concentration methods can be done if eggs are not available by direct smear technique. But detection of eggs cannot confirm the parasitic infection because egg morphology is not specific. Few representative suspected birds can be autopsied for detection of tapeworms if present in the intestine.

Immunological Diagnosis

No immunodiagnostic test has yet been commercialized.

Molecular Diagnosis

Molecular diagnosis is possible by using specific molecular markers comprising ribosomal DNA, ribosomal RNA, mitochondrial DNA, etc. For confirmation, reference nucleotide sequence is required. But it is problematic as reference nucleotide sequences are less. The commonly usable markers are mainly of ribosomal DNA and RNA (ITS1, ITS2, 5.8 S rRNA, 18 S rRNA, etc.). Different markers originated from mitochondrial DNA are CO1, ND1, etc. Recent study shows molecular identification of this parasite by using ITS 2 as molecular marker.

Treatment

Name of drug	Route of administration	Dose
Niclosamide	Oral	175 mg/kg

Special Note

Different nanoparticles exhibited efficacy against different cestodes. Studies indicate more efficacy of praziquantel-loaded hydrogenated castor oil solid lipid nanoparticles. Reports also indicate efficacy of gold nanoparticles against *Raillietina* sp. Kar et al. (2014) found the efficacy of gold nanoparticles against *Raillietina* sp. Nanoparticles were produced by mixing mycelia-free culture filtrate with gold chloride. Detrimental effect on ultrastructure includes disorientation of microtriches, spines, and scales along with disfigurement and distortion of tegument. The major damages include wrinkles,lesion,rupture of surface tegument. The nanoparticles also affected physiological functions of the worms resulting in paralysis and death of the worms

Control

General Control

The affected birds should be treated properly. Routine deworming schedule should be followed. Intermediate hosts can be controlled by insecticidal agents.

Immunological Control

No vaccine has yet been commercialized.

Molecular Control

No gene vaccine is available.

Biological Control

No biological control is available.

Genus: *Cotugnia*

Species: *Cotugnia digonophora*

This tapeworm is medium in size. This tape worm is 'double pored' tape worm of poultry. The average length is 9–15 cm. The scolex is round and knob-like. The scolex bears a small rostellum which is armed with 1–2 rows of minute hooks. The suckers are unarmed. The immature segments are squarish or rectangular, whereas the gravid segments are broader than long. About 100 testes are present posterior to the ovary. The vitelline gland is present posterior to the ovary.

Class: Eucestoda—Order: Dilepididea—Family: Dipylidiidae

Family Synopsis

The members of this family are parasites of birds and mammals. They are worldwide in distribution. Rostellum is armed with one or more rows of rose thorn-shaped hooks. Suckers are unarmed. Single or double sets of reproductive organs are present. Gravid uterus filled up with egg capsule containing one or more eggs. Different fly, beetle, and flea serve as intermediate hosts. Cysticercoids are the infective stages of this tapeworm. The worms cause enteritis. Diagnosis can be done by finding eggs or gravid segments in the feces. Control mostly relies on proper treatment of the affected hosts.

History

Carl Linnaeus first described *Dipylidium caninum* in 1758 calling it *Taenia cannia*. MN Melnikov discovered that the intermediate host of *Dipylidium caninum* is *Trichodectes canis* in the year 1869.

Genus: *Dipylidium*

Species: *Dipylidium caninum*

Prevalence

The worms are worldwide in distribution.

Hosts

Domestic hosts: The domestic hosts of this tapeworm are dog, cat, etc.

Wild hosts: Foxes are also affected by this species.

Human host: The children may be affected.

Predilection Site

The predilection site is the intestine.

Common Name

- Double-pored tapeworm
- Cucumber tapeworm
- Flea tapeworm

Morphology

Gross Morphology

Size: The tapeworms are 15–17 cm long. The length may be up to 50 cm.

Shape: The mature and gravid segments are cucumber seed shaped. These segments are much longer than broad. The whole body of this tapeworm doesn't have any specific shape. The whole body is ribbon- like as other tape worms.

Color: The worms are white or creamy white in color.

Microscopical Morphology

Adult

Scolex: The scolex is pyriform shaped. They possess one conical or club-shaped rostellum which is armed with few back facing rows (2–5) of rose thorn-shaped hooks. The hooks may not be in regular arrangement. The hooks are largest in the anterior most row and smallest in the posterior most row. Suckers are armed with thorn-shaped hooks.

Segments: The mature and gravid segments are cucumber seed shaped. In each mature segment, two sets of genital organs are present.

Genital organs: There are two sets of reproductive organs.

Testes: The testes are 100–200 in number. These are distributed evenly.

Uterus: The uterus is reticulated initially. Later, the uterus forms egg capsules.

Cirrus pouch: The cirrus pouch is pyriform in shape.

Ovary and vitelline glands: The ovary is highly lobed. The ovary and vitelline glands form bunch of grapelike structure on either side of mature segment.

Vagina: The vagina opens in the genital atrium posterior to the male genital opening organ posterior to the male genital opening.

Larva

Oncospheres: The oncospheres are 0.025–0.036 mm in diameter. These are round-shaped larval stage having oncospheral membrane. This membrane is ciliated. Six hooks are present in the embryo (hexacanth embryo). As like as other cestodes, the germ cells are present in the oncospheres.

Cysticercoid: The cysticercoids are pear shaped.

Egg

The eggs are found in egg capsule. Egg capsule is round to oval in shape which contains 25–30 eggs. One egg capsule is 120–200 um in diameter. The eggs are round to oval in shape and roughly 35–60 um in diameter. The oncosphere present in the egg is 0.035–0.036 mm in diameter.

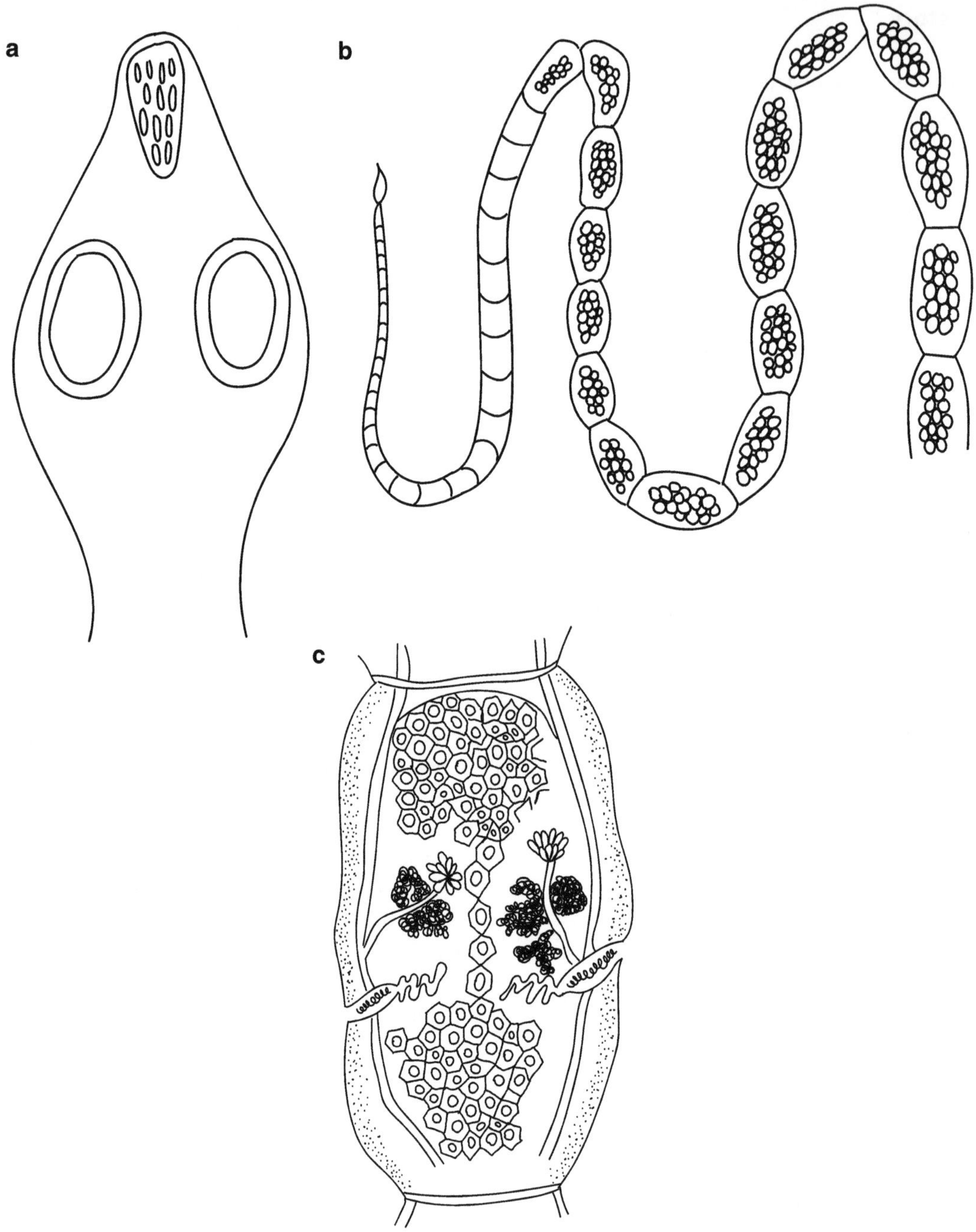

Fig. 5.8 (**a**) *Dipylidium caninum*; (**b**) Head, immature segment, gravid segment of *Dipylidium caninum*; (**c**) Details of gravid segment of *Dipylidium caninum*

Life Cycle

The parasite exhibits heteroxenous life cycle which requires intermediate host for completion of the life cycle.

Stages of Life Cycle Formed

In definitive host	In intermediate host
• Adult • Egg	• Oncosphere • Metacestode

Some basic information of life cycle
• The definitive hosts are dog, cat, etc. • The intermediate hosts are fleas • The infective stage is metacestode (cysticercoid) • The mode of infection of definitive host is ingestion • The mode of infection of intermediate host is ingestion

Events of life cycle: a glimpse
• Voiding of segments/eggs with feces of definitive host • Ingestion of eggs by the intermediate hosts (flea) • Hatching of eggs in the intermediate hosts • Release of oncospheres after hatching • Oncospheres develop to cysticercoids in intermediate hosts • Ingestion of intermediate hosts by definitive hosts • Development up to adult in the definitive hosts from cysticercoids

Detailed Description of Life Cycle

Embryonation

Embryonation of eggs occurs in the worm itself. After embryonation, segments become gravid.

Detachment of Gravid Segment

The integrity of junction of segments is lost. Then, the gravid segments are detached from the worm. The gravid segments pass through the feces. From the gravid segment, the eggs are released.

Development in Intermediate Host

Fleas (*Ctenocephalides canis*, C. *felis*, *Pulex irritans*, occasionally *Trichodectes canis*) act as the intermediate hosts. The infection of the intermediate host occurs only during its larval stage because the larval stage has chewing type of mouth parts. The adult stage has piercing type of the mouth part and for which they cannot do so. Eggs are ingested by the fleas. Hatching of eggs occurs in the gut of the intermediate hosts. The hexacanth oncospheres are released which migrate in the body cavity. The oncospheres transform to infective stages (cysticercoids). The cysticercoid is a pear shaped and somewhat solid-bodied structure.

Infection of Definitive Host

The definitive hosts get the infection by ingestion of intermediate hosts containing cysticercoid.

Development in Definitive Hosts

The body of flea is digested, and the cysticercoid gets released. The scolex of cysticercoid gets attached to the wall of the intestine and gradually develops to become adult. The patency period is 3 weeks.

Special Note

The adult stage of flea has sucking mouth parts which can suck fluid only. Actually, the larval stage of fleas is capable of ingesting the eggs of this tapeworm. The parasite grows slowly in the larvae, whereas it grows rapidly in pupa. Infective stages develop in adult stage of flea

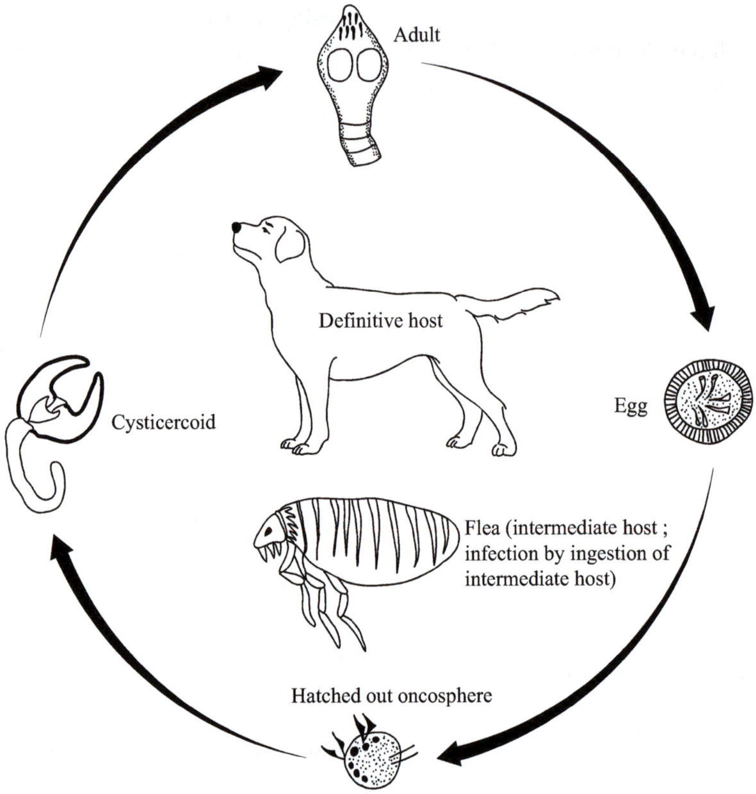

Fig. 5.9 Life cycle of *Dipylidium caninum*

Pathogenesis and Pathology

Pathogenesis

The pathogenesis has been presented in the flowchart.

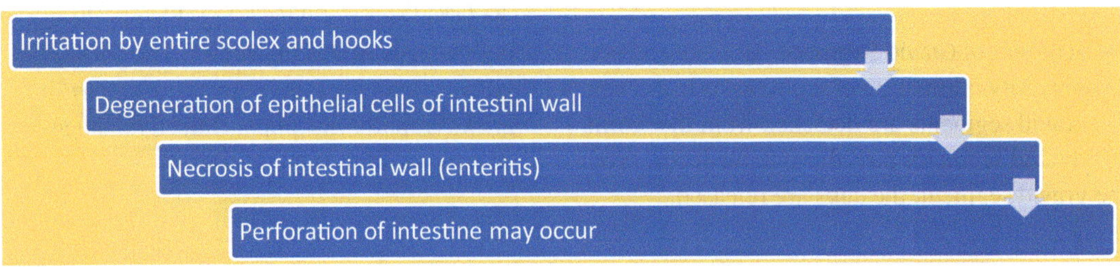

Fig. 5.10 Pathogenesis caused by *Dipylidium caninum*

Pathology

Anatomical Pathology (Gross)

Enteritis is the initial pathology found in any tapeworm infection. The tapeworms cause mechanical damage in the intestinal wall by using their armed rostellum. This results in denudation or sloughing of epithelial cells of the intestinal wall resulting in enteritis along with hepatitis. Infection by 600 worms has been described. The worms may cause perforation of the intestine and enter the peritoneum resulting in peritonitis.

Hematological Pathology

Hematological data depict anemia. The anemia is normocytic and hypochromic. Erythrocytic count is less. Hb% and PCV% are also reduced. In many cases, increased level of lymphocytes and eosinophils is found.

Histopathology

Microscopical examination reveals epithelial desquamation, villus atrophy, hemorrhage, cellular infiltration with lymphocytes, granulocytes, and plasma cells, etc.

Immunopathology

In this parasitic infection, no complicated pathology is observed except some amount of enteritis. It could be presumed that pro-inflammatory cytokines are produced which cause inflammation.

Symptoms

Primary Symptoms

- General diarrhea
- Vomiting
- Occasional detection of gravid segment in feces
- Dragging anus over the ground due to irritation caused by the gravid segment

Secondary Symptoms

- Debility
- Loss of body weight
- Weakness
- Anemia
- Nervous disorder comprising running fit, epilepsy, incoordination of movement, etc.

Diagnosis

Symptomatological Diagnosis

Clinical signs are vague and not confirmatory for diagnosis of the disease. Due to irritation caused by the passing gravid segments, the dogs drag their anus over the ground. This symptom is mostly found in case of *Dipylidium* infection.

Parasitological Diagnosis

Examination of fecal sample may reveal the presence of eggs of the parasite. If the infection load is less, centrifugation floatation technique can be done for easy detection. The eggs are found in egg capsule. Egg capsule is round to oval in shape which contains 25–30 eggs. One egg capsule is 120–200 um in diameter. The eggs are round to oval in shape and roughly 35 um in diameter. Gross examination of feces may reveal cooked rice like gravid segment. The gravid segment can be processed and stained for microscopical identification. The presence of segment on the skin around the perineum is the first indication of infection. A simple hand lens is sufficient to detect the elongated and double-pored segments. If the segment is distorted and dried, it should be placed in water to swell up. Then, by a needle, the segment is broken and placed under stereoscopic microscope, and the egg capsules containing the eggs are clearly visible.

Differential Diagnosis

Differential diagnosis of the eggs should be done because the eggs of *Dipylidium caninum* are similar to the eggs of *Hymenolepis nana*.

Immunological Diagnosis

No immunodiagnostic test has yet been commercialized.

Molecular Diagnosis

Primers are constructed for carrying out PCR. The PCR product is used for sequencing. Studies show use of 28 S rDNA as molecular markers for identification of this tapeworm

Treatment

Name of drug	Mode of administration	Dose
Praziquantel	Oral	5–7.5 mg/kg for dogs and cats
Niclosamide	Oral	100–150 mg/kg for dogs and cats

Special Note
Different nanoparticles exhibited efficacy against different cestodes. Studies indicate enhanced efficacy of praziquantel-loaded hydrogenated castor oil solid lipid nanoparticles against cestodes.

Control

General Control

The affected animals should be treated properly. Routine deworming schedule should be followed. Since fleas are intermediate hosts, they should be controlled by using ectoparasiticidal agents. Efforts should be made to destroy the larvae of fleas available on the carpet, beddings, etc. using different insecticidal agents.

Immunological Control

No vaccine has yet been commercialized.

Molecular Control

No gene vaccine is available.

Biological Control

No biological control is available.

Genus: *Choanotaenia*

Species: *Choanotaenia infundibulum*

Hosts

Domestic hosts: Chickens and other domestic birds

Wild hosts: Different wild birds are turkey, pheasants, etc.

Site

Small intestine

Morphology

Gross Morphology

Adult

Size: The worms are up to 23 cm long.

Shape: The complete worm does not have any specific shape. Like other tapeworms, they are simply flattened, segmented, and ribbonlike.

Color: They are white or creamy white in color.

Microscopic Morphology

Scolex: The scolex is roughly triangular and pointed anteriorly. It possesses one rostellum which is armed with 16–20 slender hooks. The suckers are oval in shape without hooks.

Segments: About 25 segments are found. The segments are narrow anteriorly and broader posteriorly with pointed corners. This gives saw-edge appearance of the segment.

Testes: The testes are 25–60 in number per proglottid.

Ovary: The ovary is lobed. This is located in the middle of the mature segment.

Uterus: The uterus is saclike initially and later becomes reticulated and finally forms egg capsule.

Cirrus: The cirrus bears spines and elongated.

Genital pore: The genital pores are regularly alternating.

Egg

Egg pouch contains single egg. The eggs measure 0.032–0.04 mm × 0.036–0.05 mm. They possess a delicate outer membrane. On either side of the egg pole, there have been processes which disappear after development of embryophore.

Larva

Oncospheres and cysticercoids are like that of other tapeworms. The oncospheres measure 0.032 mm × 0.022 mm.

Life Cycle

The life cycle is an indirect type which requires one intermediate host. Fly, beetle, etc. act as the intermediate hosts. The gravid segment or eggs are passed with the feces of the definitive host. In the body cavity of intermediate host, cysticercoids are formed. Different beetles, i.e., Amara, *Calathus*, *Ataenius*, *Geotrupes*, *Carabus*, etc., serve as intermediate hosts. The final hosts get the infection by ingestion of infected intermediate hosts. The tapeworms cause enteritis in the affected hosts.

Pathogenesis

See general pathogenesis of other poultry cestodes.

Treatment

Name of drugs	Mode of administration	Dose
Niclosamide	Oral	175 mg /kg bw

Class: Eucestoda—Order: Dilepididea—Family: Dilepididae

These tapeworms are usually found in birds. They are worldwide in distribution. Rostellum is armed with hook. Gravid uterus is saclike.

Genus: *Amoebotaenia*

Species: *Amoebotaenia cuneata*

Prevalence

The worms are worldwide in distribution.

Occurrence

They are less frequent compared to other poultry worms, i.e., *Davainea*, *Raillietina*, etc. Incidence is more in summer.

Morphology

Adult

Gross Morphology

Size: The tapeworms are very small and 3–4 mm long.

Shape: The whole worm is roughly triangular in shape.

Color: The worms are white or creamy white in color.

Microscopic Morphology

Scolex: The scolex is round. The rostellum is present which is armed with single row of 12–14 hooks. Prominent suckers are present.

Segments: The segments are broader than long. 13–30 segments are found.

Testes: About 12–15 testes are arranged in a row and located in the posterior region of the segment.

Ovary: The ovary is bilobed.

Vitelline glands: The vitelline gland is post-ovarian.

Genital pores: The genital pores usually alternate irregularly.

Uterus: The uterus is sac like.

Egg

They are simply round in shape. The diameter of eggs is about 35 um.

Larva

Oncospheres are round and ciliated. Cysticercoids are similar to other tapeworms.

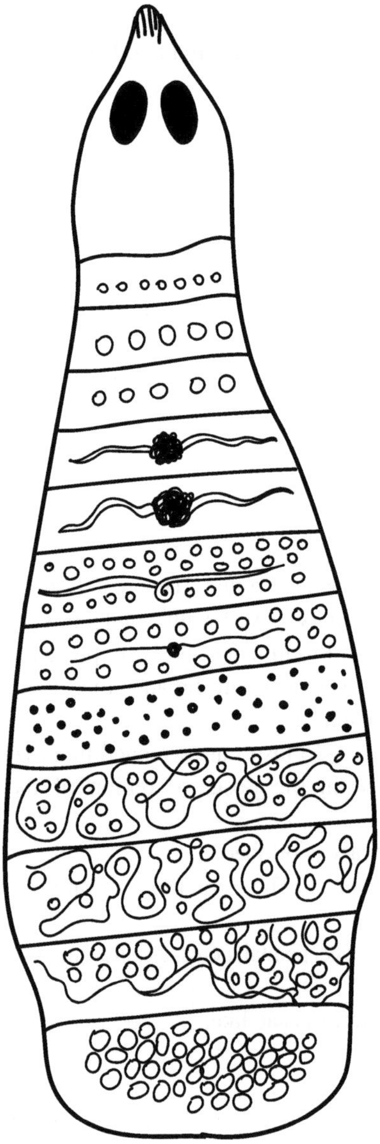

Fig. 5.11 *Amoebotaenia cuneata*

Life Cycle
The life cycle is a heteroxenous type which requires one intermediate host. The eggs or gravid segments are passed with the feces of the definitive hosts. The eggs hatch after being ingested by intermediate hosts (earthworm). Different earthworms, i.e., *Pheretima*, *Helodrilus*, *Allolobophora*, *Ocnerodrilus*, etc., serve as the intermediate hosts. The metacestodes (cysticercoids) are formed in the intermediate hosts. The final hosts get the infection by ingestion of infected earthworm containing the metacestodes.

Pathogenesis
The parasites are less pathogenic.

Class: Eucestoda—Order: Anoplocephalidea—Family: Anoplocephalidae

Family Synopsis

The members of this family are parasites of mammals occurring in the alimentary canal. They are worldwide in distribution. The tapeworms are characterized by absence of rostellum and hooks, segments broader than long, regularly or irregularly alternating genital pore when single set of reproductive organs present, the uterus not replaced and persisting as a transverse tube or network of tubes, and eggs having three layers, i.e., outermost vitelline membrane, middle albuminous coat, and inner chitinous membrane forming a pear-shaped structure with a pair of hooked projection called pyriform apparatus. Oribatidae family serves as the intermediate hosts. Cysticercoids are the infective stages of the parasites.

Genera

Anoplocephala
Paranoplocephala
Moniezia
Cittotaenia

Species: *A. magna, A. perfoliata,* and *P. mamillana*

Etymology
The word *Anoplocephala* has been originated from two Greek words, i.e., "anoplos" meaning "unarmed" and "cephale" meaning "head."

Distribution
The worms are cosmopolitan in distribution.

Common name *A. perfoliata* is called lappeted cestode. *P. mamillana* is called dwarf tape worm of horse.

Occurrence

Occurrence of the parasites is frequent.

Hosts

Domestic hosts: The horses are domestic hosts.
Wild hosts: Zebras are wild hosts.

Predilection Site

Common site: The small intestine is the common site.
Rare site: *Anoplocephala* may occur in the stomach.

Morphology

Gross Morphology

Size: *A. perfoliata* is smaller than *A. magna*, the body length ranging between 1 and 8 cm. *Anoplocephala magna* is the largest cestode of horse, the body length reaching up to 80 cm. *Paranoplocephala mamillana* is up to 5 cm.
Shape: They look like thick set tapeworm.
Color: The worms are yellowish to yellowish green in color.

Microscopic Morphology

Scolex: The scolex of *Anoplocephala perfoliata* is cubical. The scolex of *A. magna* and *Paranoplocephala mamillana* is globular. The suckers open anteriorly.
Rostellum: Rostellum is absent.
Sucker: Prominent suckers are present. Scolex of *Anoplocephala perfoliata* is small. Every sucker is attached with a lappet. The opening of suckers of *P. mamillana* is slit-like.
Segment: The segments are much wider than long.
Testes: In *Anoplocephala perfoliata*, 200 testes are present. *A. magna* possesses 400–500 testes. *P. mamillana* possesses 60–100 testes.
Ovary: The ovary is bilobed in all three species.
Vitelline gland: The vitelline gland is post-ovarian.
Uterus: Saclike uterus is transversely situated.
Genital pores: The genital pores are irregularly alternating or unilateral in *P. mammillana*. These are unilateral in *A. magna* and *A. perfoliata*.

Eggs

Eggs of *A. perfoliata* are large and irregularly spherical or triangular. The diameter is 65–80 um. The eggs have strong pyriform apparatus. The eggs of *A. magna* are same as *A. perfoliata*. The eggs of *P. mamillana* are 51 × 37 um in size and have pyriform apparatus.

Larva

It is very difficult to distinguish the cysticercoid of *Anoplocephala* from other cysticercoids. Study shows description of stagewise developmental morphology. The initial stage is a "large sphere" stage. It is spherical in shape and large in size. The next stage is "extended larval stage." Subsequent developmental stage is "segmented larva" stage. In this stage, the elongated form is divided into two parts by a constriction. The anterior part has four suckers. The posterior part forms a capsule-like structure. The hooks are placed caudally in the capsule.

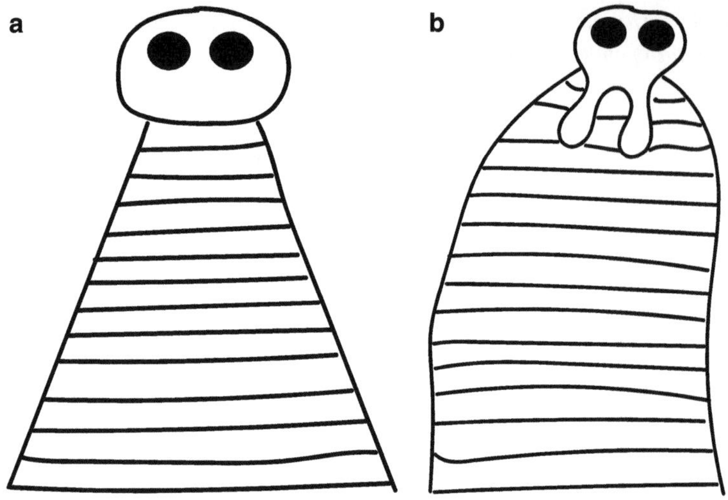

Fig. 5.12 (**a**) Anterior end of *Anoplocephala magna*; (**b**) Anterior end of *Anoplocephala perfoliata*

Life Cycle

Stages of Life Cycle Formed

In definitive host	In intermediate host
• Adult • Egg	• Oncosphere • Metacestode

Some basic information of life cycle
• The definitive hosts are equines • The intermediate hosts are mites • The infective stage is metacestode (cysticercoid) • The mode of infection of definitive host is ingestion • The mode of infection of intermediate host is ingestion

Events of life cycle: a glimpse
• Voiding of segments/eggs with feces of definitive hosts • Ingestion of eggs by the intermediate hosts (mites) • Hatching of eggs in the intermediate hosts • Release of oncospheres after hatching • Oncospheres develop to cysticercoids in the intermediate hosts • Ingestion of intermediate hosts by definitive hosts • Development up to adult in definitive hosts from cysticercoids

Detailed Description

Embryonation

In the mature segment, self-fertilization occurs. The eggs become embryonated in the segment. After embryonation, segments become gravid.

Detachment of Gravid Segment

As is found in other cestodes, the integrity of junction of gravid segments is lost, and the intersegmental junction becomes loosen due to loss of some tissues. Then, the gravid segments are separated from the worm and detached. They are passed with the feces. From the gravid segments, the eggs are released. The gravid segments can actively crawl.

Development in Intermediate Host

Oribatid mites (*Galumna, Achipteria, Scheloribates, Ceratozetes*) act as the intermediate hosts. As the eggs are the food of mite, they ingest the eggs. The eggs hatch out in the gut of the intermediate host. The hexacanth oncospheres are released which migrate in the body cavity of intermediate host. Later, the oncospheres transform to infective stage (cysticercoid). The cysticercoid is a pear-shaped somewhat solid-bodied structure containing invaginated scolex.

Infection of Definitive Host

The definitive hosts become infected by inadvertent ingestion of intermediate host containing cysticercoid.

Development in Definitive Hosts

After being ingested, the snail is digested, and the cysticercoids get released. The scolex gets

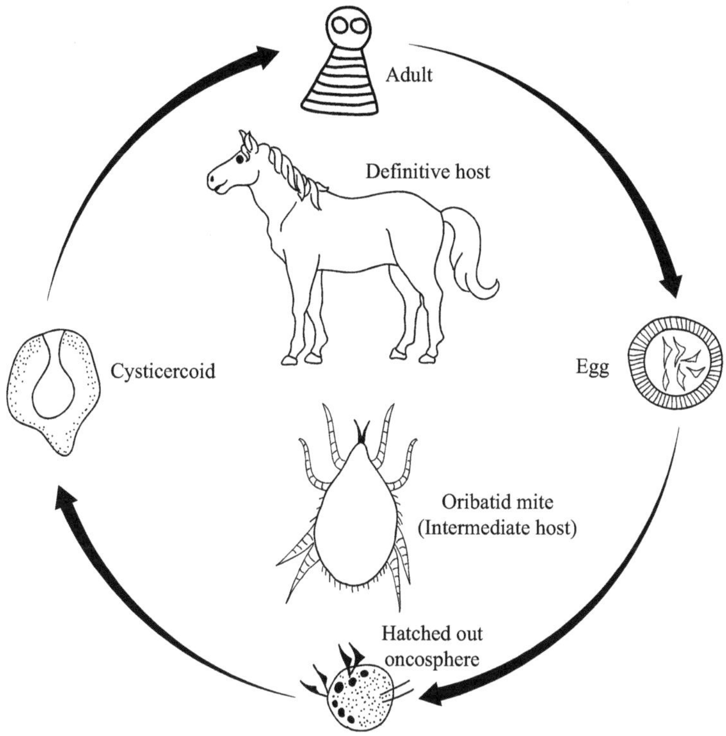

Fig. 5.13 Life cycle of *Anoplocephala* sp.

attached to the wall of the intestine and gradually develops to become adult.

Pathogenesis

A. magna may cause catarrhal and haeorrhagic enteritis. A depressed ulcer at the ileocecal region is caused by *A. perfoliata*. Deposition of large amount of granuloma tissue at this area may lead to occlusion of the orifice. Heavy infection may cause mortality of the animals. The lesions caused by *A. perfoliata* may lead to **intussusception**. These two species may cause perforation of the intestine. *P. mamillana* do not cause any pathogenesis. But heavy infection may cause digestive problem, anemia, unthriftiness, etc.

Symptoms

Primary Symptoms

- General diarrhea
- Occasional detection of gravid segment in feces

Secondary Symptoms

- Debility
- Progressive loss of body weight
- Weakness
- Anemia
- Ileocecal colic pain in *Anoplocephala perfoliata* infection

Diagnosis

Symptomatological Diagnosis

No clinical signs of diagnostic value are observed in the affected animals. Therefore, symptomatological diagnosis is not possible.

Parasitological Diagnosis

Examination of fecal sample should be done for detection of eggs. In light infection, because of less availability of eggs in the smear, centrifugation floatation technique can be done. However, confirmatory diagnosis is not possible by this because specific identifying features of eggs are lacking.

Immunological Diagnosis
No immunodiagnostic test has yet been commercialized.

Molecular Diagnosis
Diagnosis based on analysis of nucleotide sequence has been recent trend for confirmatory diagnosis. This is possible by using specific molecular markers. Different ribosomal DNA, ribosomal RNA, and mitochondrial DNA could be used as molecular markers. Studies indicate use of 18 S rRNA, 5.8 sRNA, ITS1, and ITS2 as molecular markers for identification of this tapeworm.

Treatment

Name of drug	Mode of administration	Dose
Praziquantel	Oral	50–100 mg/kg
Mebendazole	Oral	10–15 mg/kg for 3 consecutive days
Niclosamide	Oral	88–100 mg/kg

Control

General Control
The affected animals should be treated properly. Routine deworming schedule should be followed.

Immunological Control
No vaccine has yet been commercialized.

Molecular Control
No gene vaccine is available.

Biological Control
No biological control is available.

Genus: *Cittotaenia*

Species: *Cittotaenia ctenoides*

Distribution
The worms are distributed in Europe.

Site
Small intestine

Hosts
Rabbit

Species
C. ctenoides
This tapeworm has short neck.
C. denticulata
This tapeworm has no neck.
C. pectinata
This tapeworm has neck.

Genus: *Bertiella*
Species: *Bertiella studeri*
Hosts: Chimpanzee, monkey etc.
Site: Small intestine

Species: *B. mucronata*
Hosts: Monkeys
Site: Small intestine

Genus: *Moniezia*

Species: *M. expansa* and *M. benedeni*

Species
M. expansa
M. benedini

Etymology
The genus name *Moniezia* was given after a French Zoologist Frederic Moniez (1853–1918).

Distribution
The worms are cosmopolitan in distribution.

Hosts
Domestic hosts: Sheep, goat, and cattle are the domestic hosts of *Moniezia*. *M. expansa* occurs primarily in sheep, whereas *M. benedini* occurs primarily in cattle.
Wild hosts: Ibex, bison, and gazelle are some wild hosts.

Predilection Site

Common site: The small intestine is the common site.

Aberrant site: There seems no report of aberrant migration of *Moniezia*.

Morphology

Gross Morphology

Size: This tapeworm is long (6 m or more).

Shape: The worms are ribbon - like as that of other cestodes. The segments are broader than long.

Color: The color of the worm is white or creamy white.

Microscopic Morphology

Scolex: The scolex is round. The scolex does not possess any rostellum.

Suckers: Four prominent suckers are raised and these are unarmed.

Segments: The segments are broader than long. The segment of *M. expansa* is 1.5 cm wide, whereas the segments are 2.5 cm wide in case of *M. benedeni*. Two sets of reproductive organs are present in each segment. In the posterior part of mature segment, interproglottidal glands are present which are band-like in *M. benedini* and ringlike in *M. expansa*.

Reproductive organ: Each proglottid contains double set of reproductive organs.

Testes: The testes are situated in the central field. The number is 100–400 in *M. expansa* and 500 in *M. benedeni*.

Ovary and vitelline glands: The ovary and vitelline glands form ringlike structure on either side of the segment. The ovary is crescentic or fan shaped in *M. expansa*.

Cirrus pouch: Cirrus pouch is pyriform in shape.

Egg

Eggs of *Moniezia expansa* are typically triangular and eggs of *M. benedeni* are square shaped. A pyriform apparatus is present in the eggs.

Larva

Oncospheres: These are round and ciliated. Six hooks are present in the embryo (hexacanth embryo).

Cysticercoid: The cysticercoids are pear shaped.

Note

We know that *Moniezia expansa* occurs in the small intestine of sheep, goat, cattle, etc. But this species may occur in the small intestine of pig and even in human beings. There have been reports of infection of *M. expansa* in man

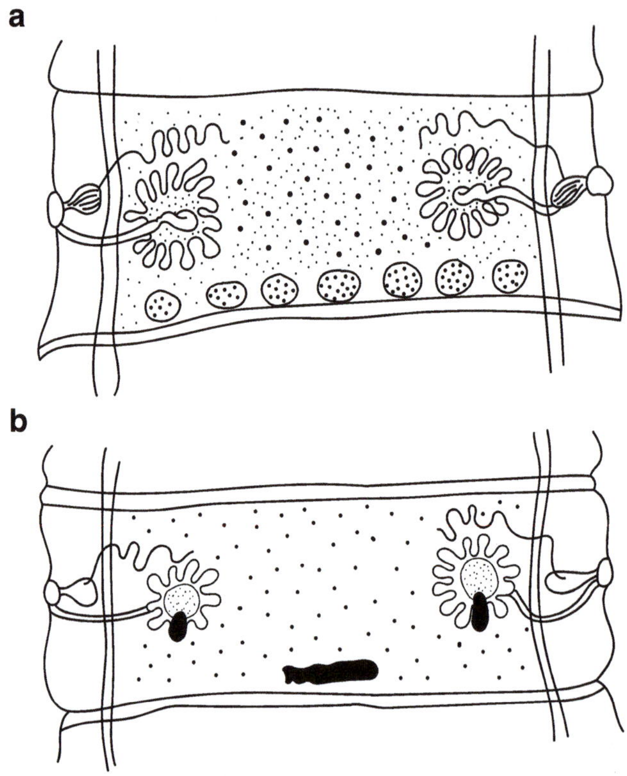

Fig. 5.14 (**a**) Mature segment of *Moniezia expansa*; (**b**) Mature segment of *Moniezia benedini*

Fig. 5.15 Segment of *Moniezia* sp. (Gross)

Life Cycle

Stages of Life Cycle Formed

In definitive host	In intermediate host
• Adult • Egg	• Oncosphere • Metacestode

Some basic information of life cycle
• The definitive hosts are sheep, goat, cattle, etc. • The intermediate hosts are mites • The infective stage is metacestode (cysticercoid) • The mode of infection of definitive host is ingestion • The mode of infection of intermediate host is ingestion

Events of life cycle: a glimpse
• Voiding of segments/eggs with feces of definitive host • Ingestion of eggs by the intermediate hosts (mite) • Hatching of eggs in the intermediate hosts • Release of oncospheres after hatching • Oncospheres develop to cysticercoids in intermediate hosts • Ingestion of intermediate hosts by definitive hosts • Development up to adult in definitive hosts from cysticercoids

Detailed Description

Embryonation

Embryonation of eggs occurs in the worm itself. After embryonation, segments become gravid.

Detachment of Gravid Segment

The integrity of junction of segments is lost. Then, the gravid segments are detached from the worm. The gravid segment passes through the feces. From the gravid segment, the eggs are released.

Development in Intermediate Host

Oribatid mites (*Gallumna, Ceratozetes, Oribatula, Peloribates*) act as the intermediate host. Eggs are ingested by the intermediate hosts. Hatching of eggs occurs in the gut of the intermediate hosts. The hexacanth oncospheres are released which migrate in the body cavity of intermediate hosts. The oncospheres transform to infective stage (cysticercoid) within about 1–4 months. The cysticercoid is a pear-shaped and somewhat solid-bodied structure.

Infection to Definitive Host

The definitive hosts get the infection by ingestion of intermediate hosts containing cysticercoid.

Development in Definitive Hosts

The body of mite is digested and the cysticercoids get released. The scolex of cysticercoid gets evaginated. After evagination, the scolex gets attached to the wall of the intestine and gradually develops to become adult. The prepatent period is approximately 6 weeks.

Pathogenesis

Symptoms

Primary Symptoms

- General diarrhea and constipation
- There may be pot-bellied condition
- Occasional detection of gravid segment in feces

Secondary Symptoms

- Debility
- Progressive loss of body weight
- Weakness
- Anemia

Diagnosis

Symptomatological Diagnosis

No specific clinical signs of diagnostic importance are observed. Therefore, confirmatory diagnosis is not possible basing on symptomatology.

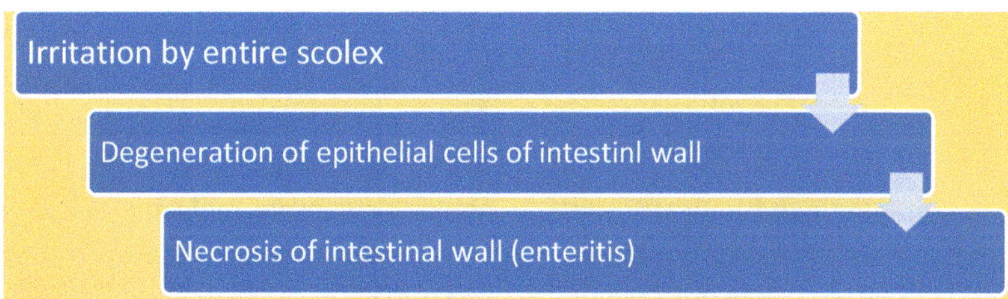

Fig. 5.16 Flow diagram of pathogenesis caused by *Anoplocephala* sp.

Parasitological Diagnosis

Examination of fecal sample may reveal the presence of eggs and segments of the parasite. If the infection load is less, centrifugation floatation technique can be done for detecting the eggs. Diagnosis basing on exhibited clinical signs is not possible. Examination of feces may reveal the presence of typical eggs. The typical shape and presence of pyriform apparatus are of diagnostic value. The eggs of *Moniezia expansa* are triangular in shape, and of *M. benedeni*, it is square shaped. The eggs of *M. benedini* are marginally bigger than *M. expansa*.

Immunological Diagnosis

Works were carried out on immunodiagnostics using whole somatic antigens by saline extraction. Immunodiagnostic works using purified antigen were also done. Abdel-Megeed et al. (2014) found serodiagnostic value of affinity-purified 164, 88, 67 kDa protein of *Moniezia expansa*. The serodiagnostic value of these antigens was checked, and this study indicated that *Moniezia expansa* affinity partially purified antigen might be successfully used in serodiagnosis for monieziosis in sheep and goat.

Molecular Diagnosis

Ribosomal DNA and RNA (ITS1, ITS2, 5.8 S rRNA, 18 S rRNA, etc.) and mitochondrial DNA (CO1, ND1, etc.) are useful molecular markers for identification of parasites. Studies indicate use of ITS2 as molecular marker for identification of *M. expansa.*

Treatment

Name of drug	Mode of administration	Dose
Fenbendazole	Oral	5 mg/kg
Albendazole	Oral	10 mg/kg
Cambendazole	Oral	20 mg/kg
Praziquantel	Oral	15 mg/kg

Special Note

Different nanoparticles exhibited efficacy against different cestodes. Studies indicate enhanced efficacy of praziquantel-loaded hydrogenated castor oil solid lipid nanoparticles

Control

General Management

Since the eggs or gravid segments are passed with the feces, the feces should be disposed properly. The affected animals should be treated properly. Routine checkup with feces examination at intervals is better to take necessary measure. Since the mites live in the soil, the following activity may reduce the population of mites:

(a) Ploughing: Ploughing can burry the mites in the depth of the soil.
(b) Reseeding.
(c) Burning of pasture.
(d) Alternation of grazing land.

Immunological Control

Many works were carried out on development of vaccine using different antigens. But till date, it seems that no vaccine has yet been commercialized.

Molecular Control

No molecular vaccine or gene vaccine is available.

Biological Control

No biological control is available in the field.

Class: Eucestoda—Order: Anoplocephalidea—Family: Thysanosomidae

Family Synopsis

The members of this family are parasites of mammals occurring in the alimentary canal and bile duct. They are worldwide in distribution. The tapeworms are characterized by presence of paruterine organs or capsule replacing the complete uterus. Psocids are considered as intermediate hosts. They cause pathology in the intestine and bile duct. Control of the parasitic infection relies on proper treatment of the affected animals.

Genus: *Thysanosoma*

Species: *Thysanosoma actinioides*

Common Name

The common name of this tapeworm is "fringed tapeworm."

Distribution

The worms are cosmopolitan in distribution.

Hosts

Domestic hosts: Domestic hosts of this tapeworm are sheep, goat, etc.

Rare host: This tapeworm seldom affects cattle.

Wild hosts: Deer, elk, etc. are wild hosts of this tapeworm.

Predilection Site

This tapeworm is found in the bile duct, pancreatic duct, and small intestine.

Morphology

Gross Morphology

Size: The worms are 15–30 cm long.

Shape: Like other tapeworms, they are simply flattened, segmented, and ribbonlike.

Color: They are white or creamy white in color.

Microscopic Morphology

Adult

Scolex: The scolex does not possess any rostellum. The scolex is roughly 1.5 mm wide. The scolex is somewhat daisy-shaped.

Segments: The segments are characteristically fringed posteriorly. That's why these are called as fringed tapeworms.

Genital organ: Double sets of genital organs are present. These are also called double-pored tapeworms.

Testes: The testes are found in the median field.

Uterus: The uterus is undulating tubelike structure which forms paruterine organs.

Cirrus: The cirrus is elongated and bears spines.

Genital pore: Two sets of genital pores are present and these worms are also double pored.

Egg

The eggs are contained in paruterine organs and the size of the eggs is 20 um × 25 um. The eggs have thick shell. One paruterine organ contains 6–12 eggs. The eggs are devoid of pyriform apparatus.

Pathogenesis

They cause diarrhea, unthriftiness, etc. Flow of bile is obstructed.

Treatment

Name of drug	Mode of administration	Dose
1. Mebendazole 2. Praziquantel	Oral	(1) Mebendazole - 10-15 mg/kg bw for 3 days (2) Praziquantel-10-15 mg/kg bw

Species: *Thysaniezia giardi*

Morphology

Gross Morphology

Size: The worms are 200 cm long.

Shape: Like other tapeworms, they are simply flattened, segmented, and ribbonlike.

Color: They are white or creamy white in color.

Microscopic Morphology

Adult

Scolex: The scolex does not possess any rostellum.

Segments: The segments are short.

Testes: The testes are located lateral to the excretory canal, whereas the testes are median to the excretory canal in case of *Thysanosoma actinioides*.

Uterus: The uterus is undulating tubelike.

Cirrus: The cirrus sac is typically bulged lateral to the segment giving an irregular appearance.

Genital organ: Single set of genital organ is usually present.

Genital pore: Genital pores alternate irregularly.

Eggs

The eggs are about 22 um. They are contained in paruterine organs. Pyriform apparatus is absent in the eggs.

Life Cycle

The life cycle is not fully understood. It is speculated that oribatid mite and bark lice might act as the intermediate hosts.

Control

General Control

Since the eggs or gravid segments passed with the feces, the feces should be disposed properly. The affected animals should be treated properly. Routine checkup with feces examination at intervals is better to take necessary measure.

Immunological Control

It seems that no vaccine is available in the field.

Molecular Control

No molecular vaccine or gene vaccine is available.

Biological Control

No biological control is available in the field.

Genus: *Avitellina*

Species: *Avitellina centripunctata*

Distribution

They are distributed in Asia, Africa, Europe, etc.

Morphology

Gross Morphology

Size: The size of the worms may reach 3 m × 3 mm long.

Shape: The posterior part is cylindrical in shape.

Color: They are white or creamy white in color.

Microscopic Morphology

Adult

Scolex: The worms have round and strong scolex with long neck. The scolex does not possess any rostellum.

Segments: Clear segmentation is absent except the posterior region.

Testes: Few numbers of testes (12–20) are located in four rows, one row lateral and another row medial to ventral longitudinal excretory tube on either side of segment.

Uterus: The uterus is transversely situated in the segment. Later, the uterus forms paruterine organs.

Genital pore: The worms have single set of reproductive organ. Genital pores alternate irregularly.

Vitelline glands: The vitelline glands are absent.

Ovary: The ovary is small and present on medial side of testes of poral group.

Egg

The eggs are 22 × 4 um in size and are devoid of pyriform apparatus. The eggs are contained in paruterine organs.

Ultrastructural Morphology

Ndom et al. (2016) carried out scanning electron microscopy. They found that the suckers had triangular openings. The scolex, suckers, and strobila were found covered with acicular and capilliform filitriches.

Life Cycle

The life cycle is not fully understood. It is suspected that book lice and bark lice might act as the intermediate hosts.

Control

General Management

Since the feces is the source of eggs or gravid segments, proper disposal of feces is an essential step. The affected animals should be treated properly. Routine feces examination at intervals is better to take necessary measure.

Immunological Control
It seems that no vaccine is available in the field.

Molecular Control
No molecular vaccine or gene vaccine is available.

Biological Control
No biological control is available in the field.

Genus: *Stilesia*

Species: *Stilesia hepatica*

Distribution
They are distributed in Asia, Africa, etc.

Site
The bile duct is affected by this worm.

Morphology

Gross Morphology
Size: The worms are up to 20–50 cm long.
Shape: Ribbon like
Color: The worms look transparent.

Microscopic Morphology

Adult
Scolex: The scolex does not possess any rostellum. The suckers are prominent.
Segments: The segments are short. Distinct segmentation is present in the posterior part.
Neck: The neck is broad.
Uterus: The uterus is dumb-bell shaped initially and later transforms into two paruterine organs.
Genital pore: The worms have single set of reproductive organs. Genital pores alternate irregularly.

Egg
The eggs are 26 um × 16–19 um in size and are devoid of pyriform apparatus. The eggs are contained in paruterine organs.

Treatment

Name of drug	Mode of administration	Dose
Praziquantel	Oral	15 mg/kg

Species: *Stilesia globipunctata*

Hosts and Site: The worms are found in the small intestine of sheep, goat, and cattle.

Morphology

Gross Morphology
Size: The worms are up to 45–60 cm long.
Shape: Ribbon like as other tape worms.
Color: They are white or creamy white in color.

Microscopic Morphology

Adult
Scolex: The scolex does not possess any rostellum.
Segments: The segments are broader than long.
Testes: Few numbers of testes (4–7) are located in rows lateral to the longitudinal excretory duct.
Ovary: The ovary is spherical in shape.
Uterus: The uterus is dumbbell shaped.
Genital pore: The worms have single set of reproductive organ. Genital pores alternate irregularly.

Egg
Each paruterine organ contains 30 eggs. The eggs are 0.058 mm × 0.027 mm in size and are devoid of pyriform apparatus.

Ultrastructural Morphology
Ndom et al. (2016) carried out scanning electron microscopy. They found that the suckers were invaginated and covered with acicular and capilliform filitriches.

Class: Eucestoda—Order: Hymenolepididea—Family: Hymenolepididae

Family Synopsis

The members of this family are parasites of birds and mammals occurring in the alimentary canal. They are worldwide in distribution. The tapeworms are characterized by the presence of armed or unarmed rostellum, single set of reproductive organs, elongated testes arranged transversely in the segment, and less number of testes commonly four and rarely more than four typically arranged in a row or in triangle, spinose cirrus. Different arthropods serve as intermediate hosts in which the cysticercoids are formed. The tapeworms cause enteritis. Diagnosis can be done by finding the eggs in the feces. Chemotherapy is the main control measures.

History

M. Bilharz discovered *Hymenolepis nana* in the year 1851.

Genus: *Hymenolepis*

Etymology

Hymenolepis is derived from two Greek words. 'Hymen' means fine membrane and 'lepis' means the scale.

Distribution

This parasite is worldwide in distribution.

Species: *Hymenolepis nana*

Common Name

The common name of this tapeworm is "dwarf tapeworm."

Hosts

Rodents, simian primates, and human beings are the hosts of this species.

Site

The seat of predilection is the small intestine.

Morphology

Gross Morphology

Segments: The tapeworm possesses 200 segments.
Size: The worms are 2.5–4 cm long.
Shape: The worms are narrow and threadlike. They do not have specific shape. Like other tapeworms, they are segmented and ribbonlike.
Color: The tapeworms are white or creamy white in color.

Microscopic Morphology

Adult

Scolex: The scolex is round in shape. The scolex bears a retractable rostellum. The rostellum is armed with a crown of hooks. The scolex possesses four suckers.
Segments: In each mature segment, single set of genital organs are present.
Testes: Number of testes is less (3–4) which are found almost in a line.
Ovary: The ovary is lobed, transversely elongated, and situated in the median field.
Uterus: The uterus is lobed and saclike.
Cirrus: The cirrus is pyriform in shape and spinosed.
Genital pore: The single set of genital pores is unilateral.

Egg

The eggs are round and oval in shape and size is 44–62 × 30–55 um.

Larva

The cysticercoid stage starts from an early cysticercoid stage. Six hooks are still found in the early cysticercoid stage. Then, post-hexacanth stage commences.

Life Cycle

Direct Life Cycle

The life cycle is direct type. Infection occurs by two modes i.e. self infection and autoinfection. In autoinfection, the eggs hatch in the gut of the host. The oncospheres penetrate the intestinal villi and they transform into cysticercoids. The cysticercoids are broken out of the intestinal villi. The scolex of cysticercoid becomes evaginated and attaches on the wall of the intestine, and gradually, they develop to adult. In self infection, egg containing developing cysticercoid is ingested.

Indirect Life Cycle

In this type of life cycle, the beetles act as the intermediate hosts. Cysticercoids are formed in the body cavity of the beetle. The final hosts get the infection by ingestion of infected beetle.

Note: Both direct and indirect life cycle are found in rodent hosts. Only direct life cycle is found in human host.

Pathogenesis

The worms are nonpathogenic or less pathogenic. They cause enteritis. Anorexia, diarrhoea, abdominal pain may occur.

Diagnosis

Symptomatological Diagnosis

Diagnosis basing on exhibited clinical signs is not possible because the exhibited clinical signs are vague.

Parasitological Diagnosis

Examination of feces may reveal eggs of the parasite.

Immunological Diagnosis

No immunodiagnostic technique is commercially available.

Molecular Diagnosis

Ribosomal DNA and RNA (ITS1, ITS2, 5.8 S rRNA, 18 S rRNA, etc.) and mitochondrial DNA (CO1, ND1, etc.) are commonly useful markers for identification. Studies indicate use of ITS1 as molecular marker for identification of *Hymenolepis.*

Treatment

Praziquantel, niclosamide, and mebendazole are some effective drugs

CSP (Chitosan Particle) Treatment

Abdel Latif et al. (2017) worked on immunoprotection effect of chitosan nanoparticles in *Hymenolepis nana*-infected mice assessing different parameters comprising worm burden, egg output, histological changes, oxidative stress markers (lipids peroxidation and reduced glutathione), and goblet cell and mucosal mast cell count. The study indicated immunoprotection effect of chitosan nanoparticle against *Hymenolepis nana.*

Control

General Management

Feces should be disposed properly because feces is the source of eggs and gravid segments. Proper deworming of the affected animals should be done. Routine checkup with feces examination at intervals is better to take necessary measure.

Immunological Control

It seems that no vaccine is available in the field.

Molecular Control

No molecular vaccine or gene vaccine is available.

Biological Control

No biological control is available in the field.

Species: *Hymenolepis diminuta*

These are rat tapeworms occurring in the small intestine. The worms may affect man also. The worm is 2–6 cm long. The rostellum is devoid of any hooks. The life cycle is indirect which requires intermediate hosts, i.e., fleas, beetles, cockroaches, etc.

Class: Eucestoda—Order: Hymenolepididea—Family: Fimbriariidae

Genus: *Fimbriaria*

The tapeworms are worldwide in distribution. They occur in the small intestine of birds. A pseudoscolex is formed by the folded anterior part of the tapeworm which is a typical morphological feature of this tape worm.

Class: Eucestoda—Order: Taeniidea—Family: Taeniidae

Family Synopsis

The members of this family are parasites of mammals occurring in the alimentary canal. They are worldwide in distribution. Rostellum is absent or present and, when present, armed with two rows of hooks. They have irregularly alternating genital pores. The ovary is situated in the posterior part of the segment. The uterus has a median longitudinal stem with lateral branch arising from either side of the longitudinal stem. The eggs are initially covered by delicate membrane which is lost and the embryophore becomes exposed. The embryophore is made of blocks and give radiated appearance. Different mammals serve as the intermediate hosts. Cysticercus, strobilocercus, coenurus, and hydatid cyst are the infective stages. The tapeworms cause marked enteritis. Diagnosis can be done by finding the eggs in the feces. The gravid segments may be found in the feces. Chemotherapy is the main control step.

History

Evidences indicate that ancient Egyptians and Greeks were aware of tapeworms in humans. Carl Linnaeus provided first description and classification of *Taenia solium*.

Species: *Taenia solium* and *T. saginata*

Distribution

The worms are cosmopolitan in distribution.

Hosts

Definitive hosts: The definitive hosts are human beings.

Intermediate hosts: The pig and cattle serve as the intermediate hosts of *T. solium* and *T. saginata*, respectively.

Predilection Site

In definitive hosts: The predilection site of adult worm in definitive hosts is the small intestine. In case of autoinfection (*T. solium*), the oncospheres migrate in different organs and form cysticerci there. The cysticerci may be formed in the brain and eye resulting in neurocysticercosis and ocular cysticercosis. The adult tapeworms are found in the small intestine.

In intermediate hosts: The cysticerci are distributed in different organs, i.e., liver, lung, diaphragm, kidney, etc. of pig (*T. solium*) and cattle (*T. saginata*).

Morphology

Gross Morphology

Size: *Taenia solium* is 3–5m long which may reach up to 8 cm. *T. saginata* is 4–8 m long which may reach up to 25 cm.

Shape: The entire worm is ribbon-like. The segments are much longer than broad.

Color: The worms are white or creamy white in color.

Microscopic Morphology

Taenia solium	*Taenia saginata*
Scolex: The scolex is round in shape. They possess one rostellum which is armed with a double crown of hooks of 26–28 in number	*Scolex*: The scolex is round in shape. Rostellum and hooks are absent
Segments: The number of segments is usually 800–900. The gravid segments are so long (1–1.2 cm)	*Segments*: The number of segments is usually 1000–2000. The gravid segments are so long (1–1.2 cm)
Testes: These are distributed throughout the segment	*Testes*: These are distributed throughout the segment
Uterus: The uterus appears as a median stem possessing 7–16 lateral branches	*Uterus*: The uterus appears as a median stem possessing 14–32 lateral branches
Cirrus: The cirrus is small in size	*Cirrus*: The cirrus is small in size
Ovary: The ovary is trilobed and located typically at the posterior part of the segment	*Ovary*: The ovary is bilobed and located typically at the posterior part of the segment
Gravid segment: The gravid segments do not leave the host spontaneously	*Gravid segment*: Gravid segments leave the host spontaneously
Eggs: One gravid segment may contain about 40,000 eggs. The eggs are round in shape	*Eggs*: The eggs are oval in shape. Gravid segment possess 80,000 eggs.

Note
Taenia saginata has a subspecies *T. saginata asiatica*. This subspecies occurs in the small intestine of man. This has a rostellum. The cysticercus is small in size and possesses primitive hooks on the rostellum

Larva

The cysticercus is small in size. A cysticercus has two components, i.e., vesicular wall and fluid. The wall has three layers. The outermost layer is an undifferentiated mantle layer. The middle layer is composed of cells like epithelial cells. The innermost layer is composed of muscle cells. The rudimentary scolex remains invaginated. The scolex possesses suckers, hooks, etc. Immediately after ingestion the scolex, few rudimentary segments are found. The mature *Cysticercus cellulosae* is 8–10 mm in diameter. About 22–32 hooks are found in it.

Egg

The embryophore is striated. The eggs of *Taenia solium* is round, whereas these are oval in case of *Taenia saginata*.

Special Note
In racemose cysticercosis, large-sized cysts are formed. They are multilobular and clustered. The scolex may be absent

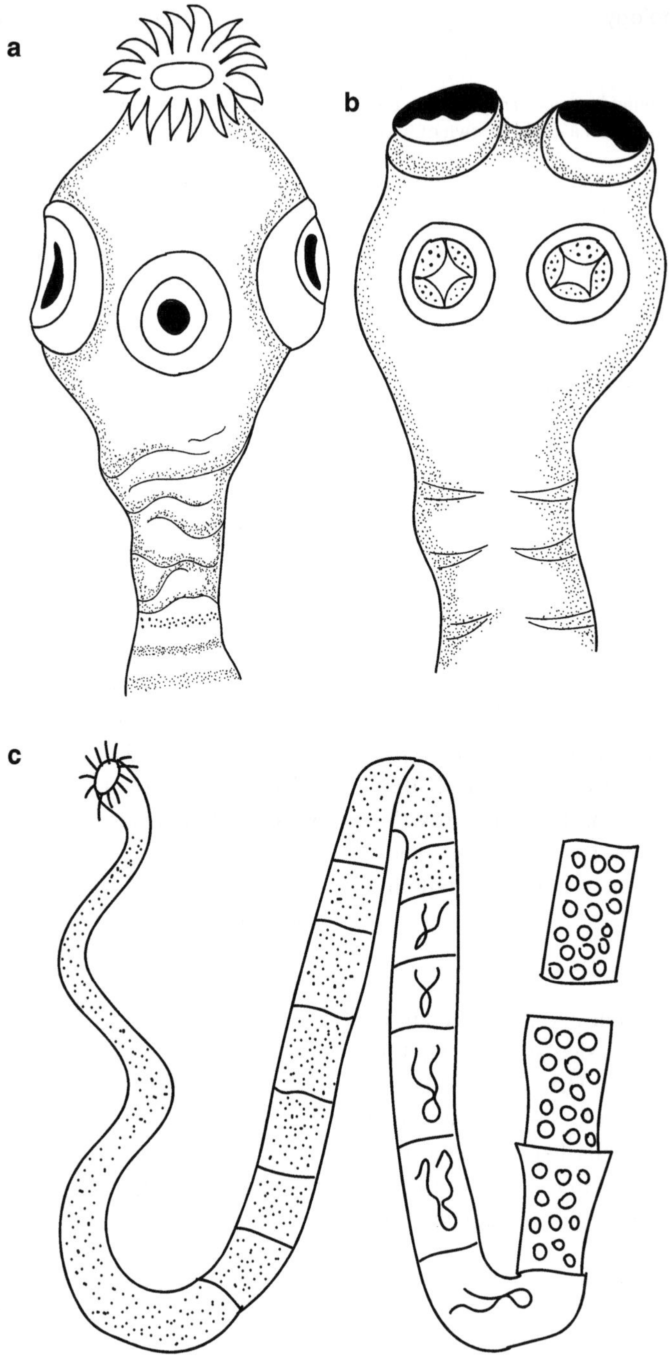

Fig. 5.17 (**a**) Head of *Taenia solium*; (**b**) Head of *Taenia saginata*(**c**) *Whole worm*

Life Cycle

The life cycle is heteroxenous which involve one intermediate host.

Stages of Life Cycle Formed

In definitive host	In intermediate host
• Adult • Egg	• Oncosphere • Metacestode

Some basic information of life cycle
• The definitive hosts are human beings • The intermediate hosts are pigs • The infective stage is metacestode (*Cysticercus cellulosae* in case of *T. solium* and *Cyst. bovis* in case of *T. saginata*) • The mode of infection of definitive host is ingestion • The mode of infection of intermediate host is ingestion

Events of life cycle: a glimpse
• Voiding of segments/eggs with feces of definitive host • Ingestion of eggs of by the intermediate hosts (pig) • Hatching of eggs in the intermediate hosts • Release of oncospheres after hatching • Oncospheres develop to cysticercus in intermediate hosts • Ingestion of muscles of intermediate hosts containing cysticercus by definitive hosts • Development up to adult in definitive hosts from cysticercus

Detailed Description

Embryonation of Eggs

The adult tapeworms remain in the small intestine. Since the worms are hermaphrodite, fertilization occurs in the same organism. Embryonation of eggs occurs in the worm itself. After embryonation, segments become gravid.

Detachment of Gravid Segment

It is the normal physiological phenomena that the integrity of junction of segments is lost due to loss of some specific tissue. Then, the gravid segments are detached from the worm. Single gravid segment or chain of gravid segments are passed by active movement. The gravid segments of *T. saginata* leave the host spontaneously.

Development in Intermediate Host

Pigs act as the intermediate host in case of *Taenia solium* and cattle serve as intermediate host of *T. saginata*. Eggs are ingested by the intermediate hosts. Hatching of eggs occurs in the intestine of the intermediate hosts. The hexacanth oncospheres are released which burrow the intestinal mucosa and by circulation they reach skeletal muscles and different internal organs, i.e., heart, lung, and diaphragm. The oncospheres transform to infective stage (*Cysticercus cellulosae* in case of *T. solium* and *C. bovis* in case of *T. saginata*). Initially small visible semitransparent gray spots of 1 mm in diameter appear within 2 weeks. The cyst gradually develops and becomes large in size which may be up to 1 cm in diameter within about 3 months when it is in the infective stage. The cysts may be viable for weeks to years. The dead cysts become calcified.

Infection of Definitive Host

The definitive hosts get the infection by ingestion of muscles of intermediate hosts containing cysticerci.

Development in Definitive Hosts

After ingestion, the muscle is digested by digestive juice of the definitive host. Thereafter, the cysticercus becomes free in the gut. It is to be noted that cysticercus is never digested. The scolex of cysticercus gets evaginated. After evagination, the scolex gets attached to the wall of the intestine. Segmentation starts at the neck region. Gradually, the length of the worm is increased. Eventually, they develop to become adult. The prepatent period is 2–3 months.

> **Note**
> The human beings may act as accidental intermediate hosts. They get the infection by ingestion of eggs along with food and drinking water. Autoinfection of *T. solium* also occurs. The gravid segment is digested and the eggs hatch. After hatching, the oncospheres are released which may migrate in the different organs including the brain. The brain is affected with numerous cysticerci. This condition is called neurocysticercosis.

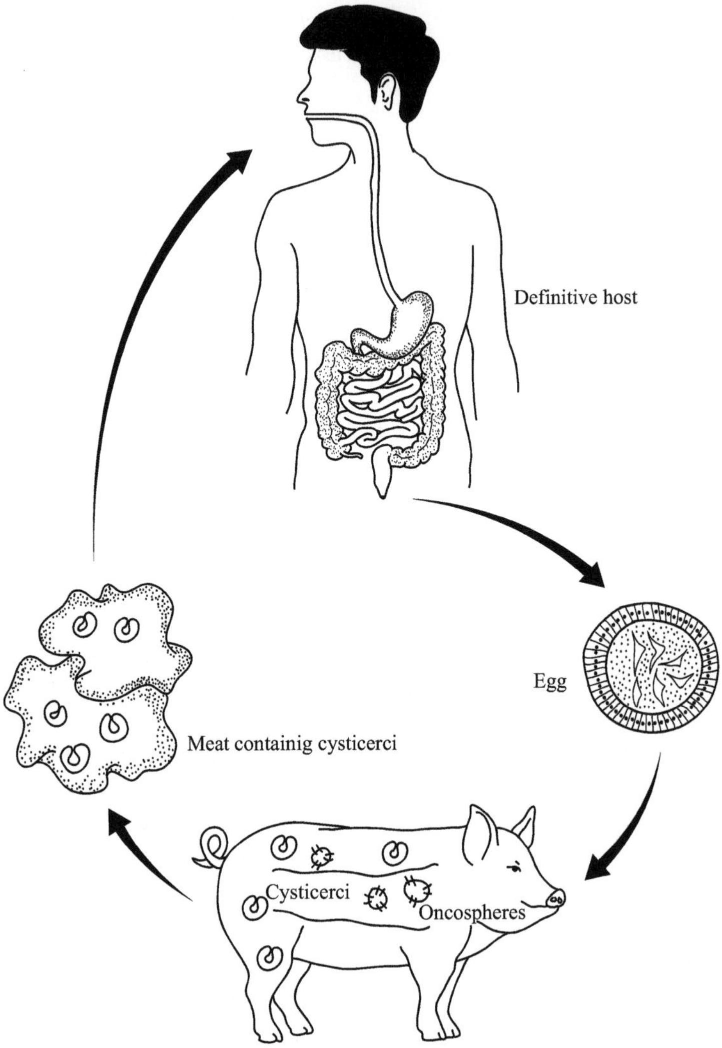

Fig. 5.18 Life cycle of *Taenia solium*

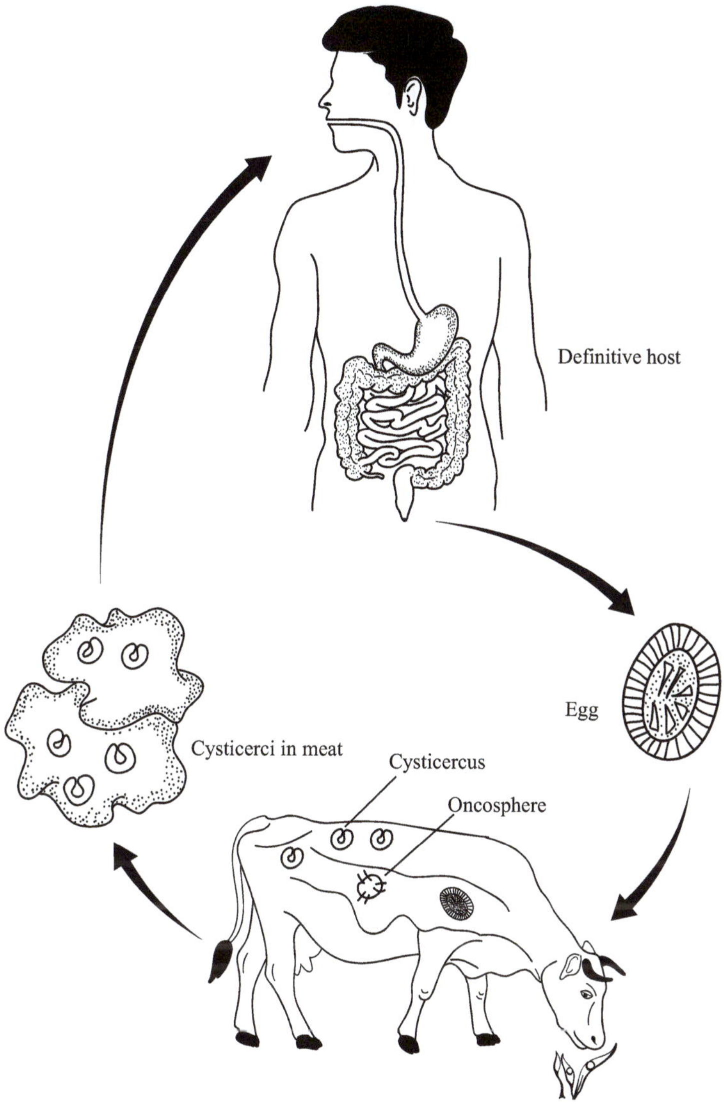

Fig. 5.19 Life cycle of *Taenia saginata*

Pathogenesis

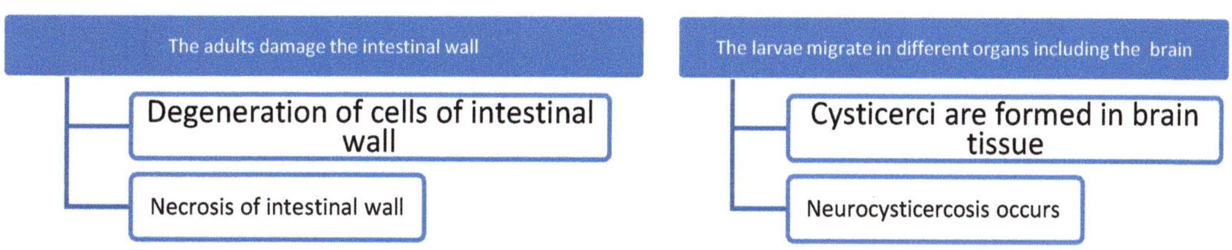

Fig. 5.20 Flowchart of pathogenesis caused by *Taenia (In normal infection and autoinfection)*

In man: The pathogenicity depends on infection load and location affected. The adult parasite causes damage to the wall of the intestinal wall resulting in enteritis. The worms may perforate the intestinal wall by the scolex and cause peritonitis. Both adult and larval stages cause pathology in man. The larvae cause inflammation in the affected parts. In prolong case, there may be calcification. The most serious complication is neurocysticercosis caused by *C. cellulosae*. The majorly affected part is the cerebrum and subarachnoid space, brain, ventricles, etc. Another one complication is ocular cysticercosis which is characterized by retinitis, iritis, conjunctivitis, choroid atrophy, etc. Cysts are formed in the affected part of different parts of eye regions.

In cattle: The cysts may occur in any organs, but they are mostly found in the heart, diaphragm, tongue, masseter muscles, intercostal muscles, etc. It is to be noted that, though many cysts may be present in those muscles, no discernible clinical features are observed.

In pigs: The infected pigs are usually asymptomatic. The pathology caused by experimental infection is characterized by myositis, myocarditis, etc.

Diagnosis

Symptomatological Diagnosis

The exhibited clinical signs do not have diagnostic importance. Confirmatory diagnosis of this tapeworm infection is not possible basing on symptomatology.

Parasitological Diagnosis

Examination of feces may reveal eggs and/or segments of the parasite. The eggs of *Taenia solium* are round and of *T. saginata* are oval. During meat inspection, the cysts are found. The cysticerci of *T. solium* is larger in size in comparison to *T. saginata*. The number of cysts *T. solium* is usually also more.

Image-based Diagnosis

Neurocysticercosis is diagnosed by CT scan which reveals the cysticerci along with their size.

Immunological Diagnosis

Works were carried out on immunodiagnostics using whole somatic antigens. Immunodiagnostic works using purified antigen were also done. Variable success was achieved. Serodiagnosis is done for diagnosis of neurocysticercosis basing on detection of antibodies against the cysticerci.

Molecular Diagnosis

Molecular diagnosis is possible by using specific molecular marker. Different ribosomal DNA, ribosomal RNA, and mitochondrial DNA could be used as molecular markers. See the note below.

> **Note**
> Serological diagnosis of human cysticercosis was done by Hubert et al. (1999) using recombinant antigens from *Taenia solium* cysticerci. In their work, NC-3 (recombinant antigen) showed 96.3% sensitivity and 91.5% specificity for serodiagnosis by ELISA.
>
> Gauci et al. (2012) did expression of recombinant antigens from the oncosphere stage of the parasitic *Taenia solium*. The recombinant proteins, i.e., TSOL16, TSOL 45-1A, and TSOL 45-1B, were produced in *E. coli* as fusion proteins with glutathione S-transferase and maltose-binding protein (MBP). The pigs were immunized twice with GST fusion antigens and boosted third times with MBP. TSOL-16 exhibited high level of protection

> **Special Note**
> Different nanoparticles exhibited efficacy against different cestodes. Studies indicate enhanced efficacy of praziquantel-loaded hydrogenated castor oil solid lipid nanoparticles

Treatment

Name of drug	Mode of administration	Dose
Niclosamide	Oral	2 gm once (in man)
Praziquantel	Oral	10 mg/kg once (in man)

Control

Do's	Don'ts
1. Affected persons should be treated properly 2. Meat should be cooked properly maintaining proper temperature and duration	1. Don't eat uncooked or undercooked meat 2. Don't eat improperly washed salad because human beings may suffer from cysticercosis due to inadvertent ingestion of eggs of the worms along with salads 3. Human night soil should be disposed properly

General Control

The following pertain to general control programs:

- Proper treatment of affected individual.
- Proper nutrition.
- Maintenance of proper hygiene.
- Proper meat inspection.
- Proper cooking of meat (65–70 °C).
- Freezing of meat in −10 °C for at least 10 days can kill the cysticerci.
- If human sludge is used in agricultural land, the cattle should not be grazed for at least 2 years.
- Public education has to be provided about hygiene maintenance and proper cooking of meat.

Immunological Control

No vaccine has yet been commercialized. *Taenia solium* is a zoonotic parasite. Pig is the intermediate host. So to break the chain, attempts were made by different researchers to control the infection. Study shows high immunoprotective potential of larval oncospheral antigen. A significant protection was conferred by three such antigens comprising Tsol 18, Tsol 45-1A, and Tsol16. Of these antigens, Tsol18 is the most protective antigen having promising immunoprotective potential. This protein was expressed as GST fusion protein in *E. coli*. GST fusion is done to isolate the Tsol18 from the whole protein mass of *E. coli*.

Molecular Control

No gene vaccine is available. Available literature shows trials of DNA vaccine alone or in combination with protein vaccine. In a work by Guo et al. (2004), significant protection was found in pigs primed with the DNA vaccine (pcDNA3-cC1) followed by two booster doses of protein (GST-cC1).

Biological Control

No biological control is available.

Species: *Taenia hydatigena*

This tapeworm occurs in the small intestine of dog. Sheep and other ruminants are the intermediate host. This is large tapeworm whose length reaches up to 5 m. The large scolex possesses rostellum. The rostellum is armed with two rows of hooks of 26 and 46 number, respectively. The uterus has 5–10 lateral branches. The eggs are oval elliptical in shape. The size is 36–39 x 31–35 um. The gravid segments pass out with the feces. The gravid segments can crawl on the hairs or surrounding the intermediate hosts. During this crawling activity, the gravid segments empty the eggs. The eggs are ingested by the intermediate hosts. Sheep, cattle, deer, and pig serve as the intermediate hosts. The eggs hatch in the gut of the intermediate hosts and oncospheres are liberated from the eggs. The oncospheres reach to the liver via blood circulation. They then migrate to the hepatic parenchyma forming different hemorrhagic tracts. They then reach to the peritoneum where the cysticerci (*Cysticercus tenuicollis*) get fully matured. The final hosts get the infection by ingestion of infected muscles. The adults may cause enteritis in dogs. The migrating oncosphere causes marked damage to the liver of the intermediate hosts. Migrating young cysticerci cause formation of hemorrhagic tracts which get fibrotic later. Calcification and caseation occur in the cysticerci. The adult tapeworms in the definitive hosts are usually asymptomatic. However, heavy infection may cause diarrhea, abdominal pain, etc. The gravid segment may cause anal pruritus at perianal region when they pass through the anus.

Species: *Taenia taeniaeformis*

This tapeworm occurs in the small intestine of cat. The length of this tapeworm is 70 cm. They have rostellum armed with hooks. The uterus has 16–18 lateral branches. The posterior segments are characteristically bell shaped. The intermediate host is rodent. Rabbits seldom act as intermediate hosts. The gravid segments are shed in the feces. The gravid segments can crawl on the hairs or surrounding the intermediate hosts. During this crawling activity, the gravid segments empty the eggs. The eggs are ingested by the intermediate hosts (rodents). The eggs hatch in the gut of the intermediate hosts. The oncosphere migrates in different tissues where they form cysts (strobilocercus) within about 9 weeks. The larval stages occur in the liver called *Cysticercus fasciolaris*. They are small cysts having an evaginated scolex attached with a segmented strobila. The final hosts get the infection by ingestion of metacestode. The matured tapeworm is formed within 6 weeks. The tapeworms fail to produce any significant pathogenesis. Diagnosis can be done by finding the eggs or the segments of the tapeworm. Control can be achieved by preventing the cats from eating rodents.

Species: *Taenia pisiformis*

This tapeworms are found in the small intestine of dog, cat, fox, etc. They are worldwide in distribution. A rostellum is present which is armed with 34–48 hooks. The gravid uterus possesses 8–14 lateral branches. The eggs are oval or elliptical in shape whose size is 43–53×43–49 um. The intermediate hosts are rabbit, hares, etc. The rabbits ingest the eggs. The eggs hatch in the gut of the intermediate hosts. The oncospheres reach to different internal organs, particularly the liver via circulation. They develop to form cysts (*Cysticercus pisiformis*) in peritoneal cavity. The final hosts become infected by ingestion of muscles of intermediate hosts. The patency period is about 6–8 weeks. This tapeworm may cause enteritis in the definitive hosts. The larval stages cause hepatitis and cirrhosis in the intermediate hosts. Diagnosis can be done by finding the eggs or segments in the feces.

Species: *Taenia multiceps*

Hosts

Definitive: Dog, fox, jackal, etc.
Intermediate: Sheep, goat, cattle, deer, pig, etc.

Site

Definitive: Small intestine
Intermediate: Brain and spinal cord

Morphology

The tapeworm may be up to 1 m long. The scolex is large having large suckers. The rostellar hooks are large in size having sinuous handle. The uterus has 14–20 lateral branches on either side of it.

Life Cycle

The adult tapeworm occurs in the small intestine of dog, fox, coyote, etc. The intermediate hosts are primarily sheep, goat, etc. The intermediate host acquires the infection by ingestion of the eggs of the tapeworm. The eggs hatch in the intestine. After hatching, the oncospheres are released from the eggs and are carried via blood to the brain and spinal cord where they form metacestodes (*Coenurus cerebralis*). The cysts develop in the brain and spinal cord if the hosts is sheep. The cysts develop in the muscles and subcutaneous tissue in goat. The fluid-filled cysts are medium sized (5 cm diameter) having few hundreds of scolices. The final hosts get the infection by ingestion of muscles containing cysts.

Pathogenesis

The cysts are formed in the brain which cause pressure atrophy. The skull becomes thin which may result in perforation. In most of the cases, chronic form of the disease status is found. If the cysts occur in the spinal cord, it may result in paralysis of hind limbs. Progressive lesions occur in the brain. Acute form of the disease occurs due to massive infection of the parasites when at a time many oncospheres arrive in the brain.

Clinical Signs

The affected intermediate hosts suffer from many nervous complications:

- Circling movement toward the affected site (due to presence of cysts in the parietal surface of one of the cerebral hemisphere)
- Holding the head against its chest (due to presence of cyst in the anterior part of the brain)
- Staggering gait and jerky movement (due to presence of cyst in the cerebellum)
- Perforation of the skull

Diagnosis

Thinning of the skull can be felt by the finger which is soft. The symptoms are guiding.

Treatment

Surgical removal of cyst.

Genus: *Echinococcus*

Etymology

The word “echinococcus” has been derived from two Greek words “echinos-” meaning “spine” and “kokkos” meaning “grain”.

Species

Echinococcus granulosus
Echinococcus oligarthrus
Echinococcus vogeli

Species: *Echinococcus granulosus*

Common Name

Dwarf dog tapeworm

Distribution

The worms are cosmopolitan in distribution.

Hosts

Definitive hosts: The dogs and different wild canids serve as definitive hosts.

Intermediate hosts: Sheep, goat, and cattle etc. act as intermediate hosts.

Site

In *definitive host*: Small intestine

In intermediate host: Larval stage is found in different organs.

Morphology

Gross Morphology

Size: The worms are small in size (2–7 mm × 0.5 mm).

Shape: Similar to other tape worms (dorsoventrally flattened and ribbon-shaped)

Color: White or creamy white in color.

Larva

The larval stage of this parasite is called hydatid cyst. The size of the cyst varies from a miniature seed to the size of a man’s head. The cyst is filled up with fluid. Within the cysts, inner cysts are produced called as the daughter cysts which may be visible from the outside. Brood capsules develop from the granular germinal membrane. Sometimes, the brood capsules are detached and float in the fluid. These detached brood capsules are called as hydatid sands.

Microscopic Morphology

Adult

Scolex: One rostellum is present which is armed with two rows of hooks (28–40 nos).

Segments: Usually, 3–4 proglottids (1 immature, 1–2 mature, 1 gravid segments) are present. The gravid segment is about half of the whole worm.

Testes: 40–60 testes are found.

Ovary: The ovary is characteristically horseshoe shaped.

Uterus: The uterus in mature segment is tubular and somewhat coiled.

Genital pore: Genital pores alternate irregularly.

Egg

40–60 eggs are found in the segment. The eggs are spherical in shape and roughly 35–40 um in diameter. It has a clear covering with a sticky layer. The embryophore is radially striated as a rule and contains six hooked oncosphere. The eggs are attached by their stickiness.

Larva

Oncospheres: These are round in shape. Six hooks are present in the embryo (hexacanth embryo).

Hydatid cyst: This is larval stage of *Echinococcus granulosus*. This is a bladder worm having two layers. The outer layer is made up of connective tissue. The inner layer is the germinal layer. From the germinal layer, the brood capsules are formed. The cyst is filled up with fluid. Within the cysts, inner cysts are produced called as the daughter cysts. Sometimes, the brood capsules are detached and float in the fluid. These detached brood capsules are called as hydatid sands. Sometimes, the cyst is ruptured and the parts of germinal layer get attached on the outer wall. From this trace of germinal layer, external cysts are produced.

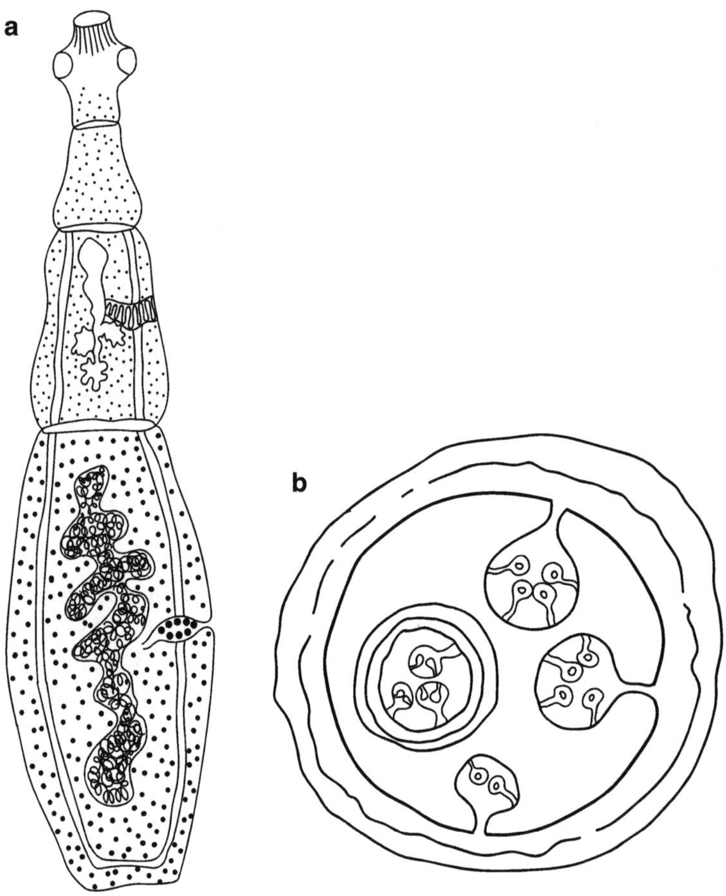

Fig. 5.21 (**a**) *Echinococcus granulosus (whole worm)*; (**b**) Hydatid cyst

Life Cycle

The parasite is heteroxenous involving one intermediate hosts for completion of life cycle.

Stages of Life Cycle Formed

In definitive host	In intermediate host
• Adult • Egg	• Oncosphere • Metacestode

Some basic information of life cycle
• The definitive hosts are dogs • The intermediate hosts are sheep, goats, cattle, etc. • The infective stage is metacestode (hydatid cysts) • The mode of infection of definitive host is ingestion • The mode of infection of intermediate host is ingestion

Events of life cycle: a glimpse
• Voiding of eggs with feces of definitive host • Ingestion of eggs by the intermediate hosts (sheep, goat, cattle) • Hatching of eggs in the intermediate hosts • Release of oncospheres after hatching • Oncospheres develop into hydatid cysts in intermediate hosts • Ingestion of muscles containing hydatid cysts by definitive hosts • Development up to adult in definitive hosts from hydatid cyst

Detailed Description

Embryonation

Embryonation of eggs occurs in the worm itself. After embryonation, segments become gravid. The gravid segment contains embryonated eggs.

Release of Eggs

One thing should be noted that in case of *Echinococcus*, the gravid segment is usually disicntegrated in the intestine and only the eggs are found in the feces, not the segments. The eggs are liberated through the genital pore. The eggs are very resistant and can survive for about 2 years.

Development in Intermediate Host

Vertebrate hosts (any ungulates, i.e., sheep, goat, cattle, etc.) act as the intermediate host. The intermediate hosts get the infection by ingestion of the eggs of the tapeworm inadvertently. Hexacanth oncospheres are released after hatching which penetrate the gut wall and reach to the liver via blood or to the lungs via lymph. These are the common sites for development of the cysts. But some oncospheres may be distributed by general circulation. The oncospheres transform to infective stage (hydatid cyst) containing viable protoscolices.

Infection of Definitive Host

The definitive hosts become infected when they ingest the meat of the intermediate host containing the cysts (hydatid cyst) or part of hydatid cyst containing viable protoscolices.

Development in Definitive Hosts

The cyst is digested and the protoscolices are released. The protoscolices develop to adult stage. The prepatent period is 40–50 days.

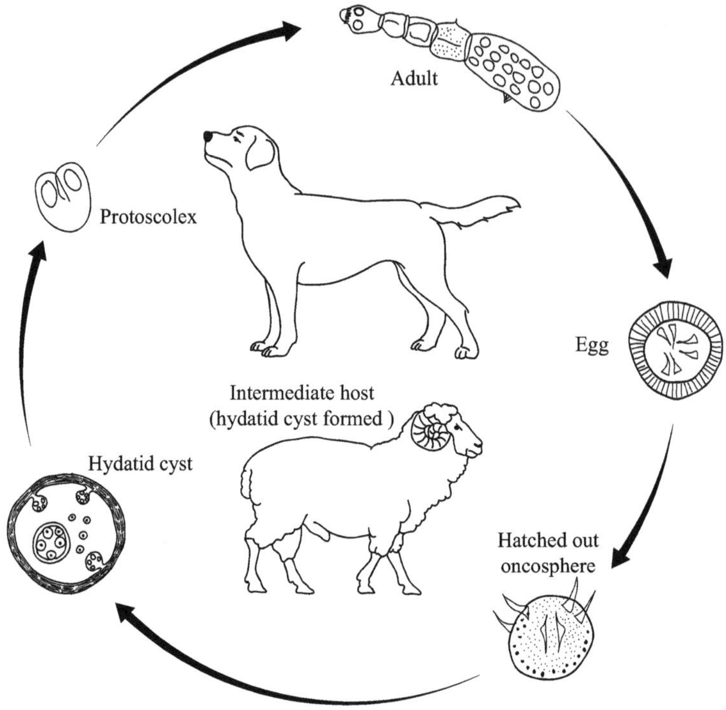

Fig. 5.22 Life cycle of *Echinococcus granulosus*

Pathogenesis

In definitive host: The adult tapeworms do not produce any marked signs in spite of infection by thousands of tapeworms.

In intermediate host: The cysts present in the lungs and liver do not cause marked pathogenesis. But cysts cause marked pathogenesis in the kidney and brain.

Clinical Symptoms

Both adult and larval stages fail to produce any symptoms. But in human beings, there occur respiratory symptoms if cyst is formed in the lungs. Large-sized cysts cause abdominal distension in man. The leakage of cysts may cause formation of new cysts in peritoneum, pleural cavity etc. The fluid of ruptured hydatid cyst is absorbed and may cause anaphylactic shock.

Diagnosis

Symptomatological Diagnosis

Diagnosis basing on exhibited clinical signs is not possible.

Parasitological Examination

Examination of feces may reveal the presence of typical eggs. In some cases, purgative is given to eliminate the tapeworms. The tapeworms can then be identified. Necropsy examination may be useful for detection of tapeworms. The intestine is opened and has to be placed in shallow water so that the worms are detached and fall in the water.

Immunodiagnosis

Antibody detection: The antibody against oncospheral antigen provides protection against *Echinococcus* worm. In chronic stage of infection, there occurs elevated level of IgG, IgM, and IgE level. The earlier immunodiagnostic tests include Casoni's intradermal test, indirect hemagglutination test, complement fixation test, latex agglutination test, etc. Later, other tests were developed which are indirect fluorescence antibody test, enzyme-linked immunosorbent assay, immunoblotting, etc. The earlier-mentioned tests are nonspecific and insensitive. The diagnostic sensitivity of the latter tests using native and recombinant antigens is more. A dipstick assay displayed more than 90% sensitivity and specificity. This test is easy to perform and very less time-consuming. ELISA is a sensitive test. Study shows development of a 3-min rapid test (dot immunogold filtration assay, DIGFA) for diagnosis of human CE and AE. The antigens for performing this test are crude and partially purified cyst fluid antigen (EgCF and AgB), *Echinococcus multilocularis* metacestode antigen (Em2), and *Echinococcus granulosus* protoscolex extract (ExP). Since variable results are found and results depend on nature of antigen and test category, rigorous standardization of these tests is urgently needed. At present, the antigens which are mostly used are hydatid cyst fluid and whole somatic antigen of adult tapeworm.

Antigen detection: Antigen detection is another one immunodiagnostic technique which is employed to detect serum antigen. The principle of antigen detection is based on detection of antigen in the sera or urine by specific antibody or monoclonal antibody using direct or indirect ELISA. The common ELISA technique done for detection of circulating antigen in cystic echinococcosis is standard double antibody-sandwich ELISA.

Special Note

Different nanoparticles exhibited efficacy against different cestodes. Studies indicate more efficacy of praziquantel-loaded hydrogenated castor oil solid lipid nanoparticles.

Norouzi et al. (2020) carried out a research work on scolicidal effects of nanoparticles against hydatid cyst protoscolices in vitro. Their findings displayed potent scolicidal effect of Ag-NPs, Fe-NPs, Cu-NPs, Si-NPs, and Zn-NPs, of which Ag-NP had the highest scolicidal effect at all concentrations. This NP at 1 mg/mL conc. exhibited 80% mortality rate after 60 min of exposure

Molecular Diagnosis

Different molecular markers are used for identification of parasite. These markers are ribosomal DNA and ribosomal RNA (ITS1, ITS2, 5.8 S rRNA, 18 S rRNA, etc) Mitochondrial DNA (CO1, ND1, etc.) are also used as molecular markers for identification of parasites.

Treatment

Name of drug	Mode of administration	Dose
a. Praziquantel b. Niclosamide	Praziquantel (Oral) Niclosamide (Oral)	a. Praziquantel-5–7.5 mg/kg orally for dog and cat b. Niclosamide-75–100 mg/kg for dog

Note

A cDNA library from mRNA of metacestode of *E. multilocularis* was constructed by Vogel et al. (1988) in *E. coli* expression vector lambda gt 11. One isolated clone (from lysogen 11/3) produced a fusion protein which was rapidly degraded and produced distinct polypeptides of Mr 31000 and Mr 33000 showing immunodiagnostic potential. The diagnostic value of 11/3 antigen was evaluated by mini-Western blot which resulted in a diagnostic sensitivity of 98% and over all specificity of 96%

Control

Do's	Don'ts
1. Affected animals should be treated properly	1. The dogs should not be allowed to eat raw and uncooked or undercooked meat
2. Routine deworming of the animals should be done	2. The dogs should not be allowed to eat offal from the slaughter house
3. Proper meat inspection should be done	3. Close contact with the dogs may cause infection to man by inadvertent ingestion of eggs of the worms
4. Dogs should be practiced to defecate at a particular place	4. Avoid the infected playground where the dogs frequently visit

General Control
The affected animals should be treated properly. They should be dewormed routinely following standard protocol. The animals should not be allowed to eat uncooked or undercooked meat. Proper meat infection should be done. The dogs should not be allowed to enter the abattoir, and strict attention should be paid so that they should not eat any meat offals. In some countries, strict legislation with provision of penalties is maintained.

Immunological Control
No vaccine has yet been available. A number of works were carried out in the past using different crude antigens for vaccination of sheep. These antigens are whole somatic extract of adult parasites, somatic extract of protoscolices, extract of tegumental antigen of protoscolices, excretory secretary antigen, etc. But no potent vaccine has yet been found to be reliable. Tegumental antigen is produced by solubilization by PBS containing N-Decanyl-N-methyl glucamine. Excretory-secretory antigen is prepared by culturing protoscolices in culture media. Attempts were also made by different workers to develop a good vaccine against this parasitic infection using different recombinant antigens. A recent experiment showed reduction of development of cysts in sheep after vaccination with a recombinant protein (*E. granulosus* myophilin, rEgmyophilin). The vaccinated animals were orally challenged with eggs of the parasite. Another study (Wang et al. 2016) displayed immunoprotective potential of a recombinant protein (rEgP29) in sheep which conferred protection up to 94.5% with rise of antibody (IgG,IgG1,IgG2,IgE) and effective cytokines (IFNgama,IL-2,IL-4). Study also shows immunoprotective potential of multiepitope multiantigenic recombinant vaccine (rEG-vac) constituting EG95, EG14-3-3, and EgEnolase (Pourseif et al. 2021). This vaccine was given to dogs (15 no) and sheep (15 no)by two subcutaneous doses 28 days apart. The vaccine was found protective both in dog and sheep. This pilot experiment indicated more immunoprotection especially in dogs but protection level in sheep was not as high as in dogs.

Molecular Control
No gene vaccine is available. However, work is in progress in this direction. Codelivery of IL12 with DNA encoding 8-kDa subunit antigen B hydatid cyst elicited Th1 response in mice which was assessed by analysis of different parameters comprising lymphocyte proliferation assay, analysis of cytokines profile, etc. (Aziz et al. 2016). IL12 acted as genetic adjuvant. This study shows efficacy against oncosphere.

Biological Control
No biological control is available.

Other Species of *Echinococcus*

Species: *Echinococcus multilocularis*

They are commonly called dwarf fox tapeworm. They are distributed in North America, Central Europe, Russia, India, China, etc. The worms are small in size (2–4 mm) possessing 3–5 proglottids. The rostellum is armed with 14–34 hooks. This species occurs in the small intestine of fox, dog, cat, etc. Microtine rodents act as the intermediate hosts. This species causes multilocular echinococcosis. This is also called as alveolar echinococcosis. The cysts are formed in the liver of the intermediate hosts. The cysts are small in size containing small amount of fluid. At first minute, small cysts are formed which are distributed in different organs, i.e., lung, liver, brain, etc. Life cycle involves two hosts comprising one definitive host and one intermediate host. The eggs pass out from the definitive hosts. The eggs are ingested by the intermediate hosts through feed, drinking water, etc. The eggs are hatched in the intestine. The oncospheres spread in different organs and form cysts. The final hosts get the infection by ingestion of infected muscles of intermediate hosts and adult worm is formed within about 5 weeks. The human beings get infection by ingestion of eggs of *E. multilocularis* through eating contaminated food and drinking water. The multilocular or alveolar cysts with many compartments are formed in the liver. The cysts are filled up with gelatinous material. The growth of the cysts has special characteristics. They invade the surrounding tissue and also metastasize in distant places comprising brain, lungs, etc. Clinical signs in man depend on load of infection, location of cysts, etc. Usually, the parasite is asymptomatic in man. The growth appears as slow-growing carcinoma. The definitive hosts should be treated with praziquantel. The human beings get the infection through food. Therefore, hand hygiene and food hygiene are very important.

Species: *Echinococcus oligarthrus*

This is small tapeworm (2.5–3.0 mm) possessing only three proglottids. The scolex has rostellum which is armed with 26–40 hooks. The uterus is saclike. Life cycle involves one intermediate host. Different wild felids, i.e., jaguar, and cougar, are the definitive hosts. Dog, cat, wolve may also act as definitive hosts. Rodents act as the intermediate hosts. Polycystic hydatid cysts are formed in rodents.

Species: *Echinococcus vogeli*

Echinococcus vogeli is similar to *E. oligarthrus*. Domestic and wild dogs serve as the definitive hosts and rodents serve as intermediate hosts. They have rostellum armed with hooks. Gravid proglottid is almost cylindrical in shape and has half of the body length. Like *E. oligarthrus*, multiple cysts are formed in different organs, i.e., lungs, liver, etc. This is called as polycystic echinococcosis. The human beings may also be infected. Multicompartmental hydatid cysts (polycystic) are produced in human beings.

Class: Eucestoda—Order: Mesocestoididea—Family: Mesocestoididae

The members of this family are parasites of mammals and birds. They occur in the small intestine of definitive hosts. The tapeworms are characterized by absence of rostellum, midventral opening of genital pore. They are heteroxenous parasites and require two intermediate hosts. The first intermediate hosts are mite and second intermediate hosts are birds, reptiles, etc. Tetrathyridium is the infective stage. The tapeworms cause enteritis. Diagnosis can be done by finding the eggs in the feces.

Genus: *Mesocestoides*

Etymology

This word has been derived from two Greek words, i.e., "mesos" meaning "middle" and "cestos" meaning "belt".

Distribution

The worms are distributed in Asia, Africa, Europe, North America, Japan, etc.

Host

Definitive hosts: There have been wide definitive host ranges comprising canids, felids, procyonids, mustelids, and opossums.

Intermediate hosts: The first intermediate hosts are mites, ants, etc. The second intermediate hosts are birds, snakes, reptiles, etc.

Species *Mesocestoides lineatus*

Morphology

Gross Morphology

Adult

Size: The size of tapeworm is 30–250 cm × 3 mm.

Shape: The shape is like other tape worms (dorsoventrally flattened and ribbon like).

Microscopic Morphology

Adult

Scolex: The scolex is anteriorly depressed. The scolex does not possess any rostellum. The suckers are characteristically oval and elongated.

Segments: The posterior segments are characteristically barrel shaped. The proglottids in the middle part are square in shape.

Testes: The testes are large and 50–60 in number located along the longitudinal excretory ducts.

Genital pore: The genital pore opens midventrally. Single set of reproductive organs is present.

Segments: Many segments are found. The segments of the mid body are square-shaped and barrel-shaped in the posterior part.

Egg: The eggs are oval in shape having two membranes. They measure 0.04–0.06 mm × 0.034–0.043 mm.

Larva

Oncospheres: These are round in shape. Six hooks are present in the embryo (hexacanth embryo).

Cysticercoid and Tetrathyridium: The cysticercoid is solid and elongated. The shape of tetrathyridium is irregular. They are several mm in length. Suckers and calcareous particles are also seen. Suckerless tetrathyridium is also found. The tetrathyridium can undergo asexual multiplication by larval budding or longitudinal fission.

Life Cycle

Stages of Life Cycle Formed

In definitive host	In 1st intermediate host	In 2nd intermediate host
• Adult • Egg	• Oncosphere and cysticercoid	• Tetrathyridium

Some basic information of life cycle
• The definitive hosts are dog, cat, etc. • The 1st intermediate hosts are mite and 2nd intermediate hosts are birds, reptiles etc. • The infective stage is metacestode • The mode of infection of definitive host is ingestion • The mode of infection of intermediate host is ingestion

Events of life cycle: a glimpse
• Voiding of segments/eggs with feces of definitive host • Ingestion of eggs by the first intermediate hosts (mite) • Hatching of eggs in the first intermediate hosts • Release of oncospheres after hatching • Transformation of oncospheres to cysticercoids in first intermediate hosts • Ingestion of first intermediate hosts by second intermediate hosts (birds, reptiles, etc.) • Development of tetrathyridium from cysticercoids • Development up to adult in definitive hosts from tetrathyridium

Detailed Description

In case of *Mesocestoides* spp., two intermediate hosts are required comprising invertebrate first intermediate host (mite) and vertebrate second intermediate hosts (birds, reptiles, amphibians). The gravid segments or eggs are passed with the feces of definitive host. The eggs are ingested by the first intermediate hosts. The eggs are hatched in the gut of mite and cysticercoids are formed. The mites harboring the cysticercoids are ingested by the second intermediate hosts. In the second intermediate hosts, one more developmental stage is produced. This stage is called tetrathyridium. It has been found that tetrathyridium can asexually multiply. The final hosts get the infection by ingestion of infected second intermediate hosts.

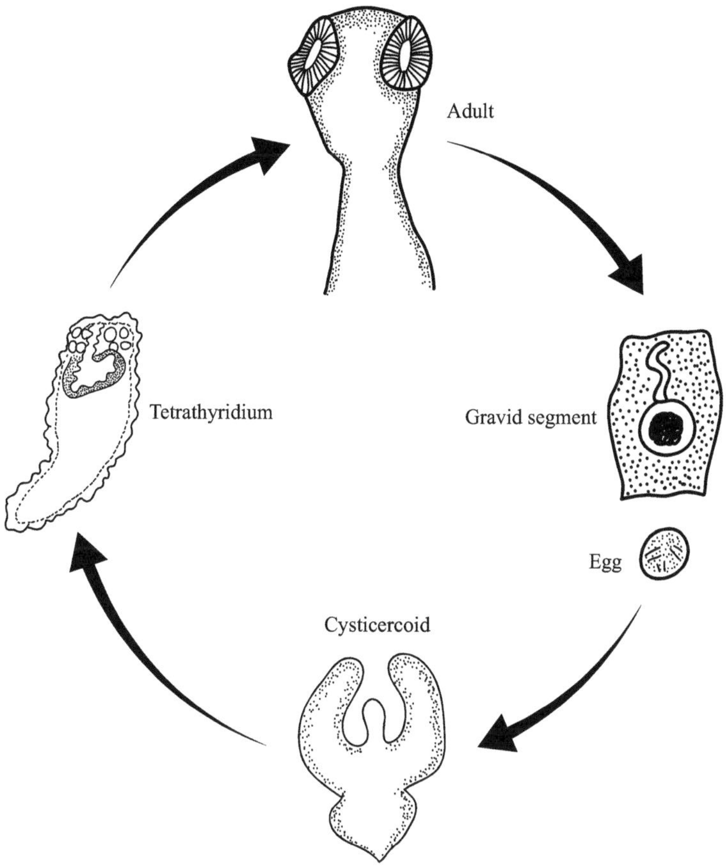

Fig. 5.23 Life cycle of *Mesocestoides lineatus*

Pathogenesis

The worms are less pathogenic. They cause enteritis which result in diarrhea.

Diagnosis

Symptomatological Diagnosis

Symptomatological diagnosis is not possible because the animals exhibit no specific clinical signs which can provide any diagnostic clue.

Parasitological Diagnosis

Gross examination may reveal the presence of gravid segment. Detection of eggs is possible by microscopical examination of fecal sample. If eggs are not found by direct smear examination, centrifugation floatation technique can be adopted.

Immunological Diagnosis

Studies indicate the use of crude somatic antigen, specific antigen. However, more and more immunodiagnostic works are required for development of reliable immunodiagnostic test in the field.

Molecular Diagnosis

Different PCRs are done for molecular characterization of *Mesocestoides* sp. Different molecular markers could be used for identification of parasite. Studies indicate use of 28 S rDNA, ND1, CO1, etc. as molecular markers for identification of *Mesocestoides* sp.

Treatment

Name of drug	Mode of administration	Dose
Praziquantel	Oral	5 mg/kg

Class: Cotyloda—Order: Diphyllidea—Family: Diphyllobothriidae

Family Synopsis

The members of this family are parasites of mammals occurring in the alimentary canal. They are worldwide in distribution. The tapeworms are characterized by small scolex, spoon shaped or spatula shaped head, midventral genital pore opening individually by way of common cirro-vaginal pore, rosette-shaped uterus, and bilobed ovary. These are heteroxenous parasites and require two intermediate hosts. Procercoid is formed in the first intermediate host and plerocercoid is formed in the second intermediate host. Plerocercoid stage is the infective stage. *Diphyllobothrium latum* is the important tapeworm of this family which causes pernicious anemia in human beings. Diagnosis can be done by finding the eggs or segments in the feces.

History

Carl Linnaeus first described *Diphyllobothrium latum* in the year 1592 and named it *Taenia lata*.

Genus: *Diphyllobothrium*

Etymology

The word *Diphyllobothrium* has been derived from three Greek words, i.e., "Di" meaning two, "phyllon" meaning "leaf", and "bothrion" meaning groove.

Species: *Diphyllobothrium latum*

Distribution

This tapeworm is worldwide in distribution.

Common Name

- Broad tapeworm
- Fish tapeworm

Morphology

Gross Morphology

Size: The tapeworms are about 2–8 m. long. This may reach up to 18 metres.

Shape: The worms are ribbon-like as that of other cestodes. The segments are much broader.

Color: When fresh, the worm is grayish yellow with dark mark in the central line due to presence of uterus and eggs.

Microscopic Morphology

Adult

Scolex: The scolex is spatula or spoon shaped. No rostellum and hooks are present. Suckers are absent. A deep groove-like structure is present called bothria. The rostellum is absent. The suckers are also absent.

Segments: Thousands of segments are present. Gravid segment is not detached. Eggs come out from the segments. Later, a small chain of empty proglottids are detached from the worms.

Testes: The testes are situated in the lateral field.

Ovary: The ovary is bilobed and situated in the posterior part of the segment.

Uterus: The uterus is rosette shaped.

Egg

The eggs are like eggs of trematodes. The eggs are oval, operculated, and golden brown colored. The size of egg is 60 um × 42 um.

Larva

Coracidium: These are round and ciliated. Six hooks are present in the embryo (hexacanth embryo).

Procercoid: The procercoid is globular in shape bearing a taillike structure called cercomer. This is solid bodied structure. Six hooks remain in inactive form. This is solid bodied. They are about 0.5 mm long.

Plerocercoid: This is an elongated larval stage with a developed scolex. The plerocercoid is about 6 mm long.

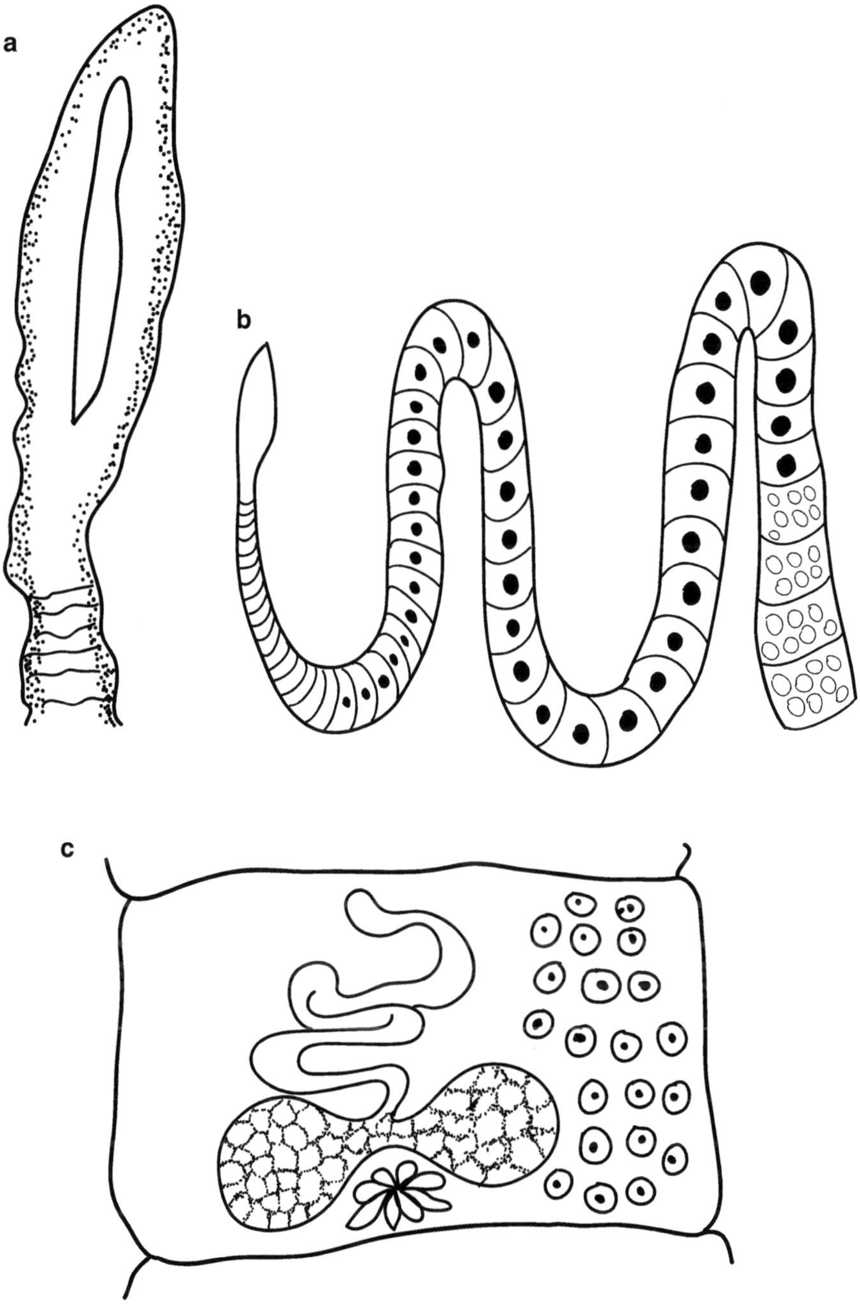

Fig. 5.24 *Diphyllobothrium latum*: (**a**) head, (**b**) whole worm, and (**c**) mature segment (detailed structure)

Life Cycle

Stages of Life cycle formed

In definitive host	In first Intermediate host	In second Intermediate host
• Adult • Egg	• Procercoid	• Plerocercoid

Some basic information of life cycle
• The definitive hosts are dog, cat, etc. • The 1st intermediate hosts are cyclops and 2nd intermediate hosts are fish. • The infective stage is metacestode (plerocercoid) • The mode of infection of definitive host is ingestion • The mode of infection of intermediate host is ingestion

Events of life cycle: a glimpse
• Voiding of eggs with feces of definitive host • Hatching of eggs in water • Release of coracidium after hatching • Ingestion of coracidium by the first intermediate hosts (cyclops) • Formation of procercoid in cyclops • Ingestion of infected cyclops by second intermediate hosts (fish) • Formation of plerocercoid in second intermediate hosts • Ingestion of infected second intermediate hosts by definitive hosts • Development up to adult in definitive hosts from plerocercoids

Detailed Description

Fertilization

Fertilization is internal which occurs in the parasite itself.

Egg Passing

After fertilization, the segments become gravid. The segments are disintegrated, and the eggs are liberated. In case of *Diphyllobothrium*, the chain of empty segments is passed out with the feces.

Embryonation

One thing should be noted that unembryonated eggs are passed with the feces and embryonation occurs in the environment.

Hatching

Hatching of eggs occurs in the water. Temperature and water are environmental stimuli for hatching of eggs. Hatching of eggs occurs in the water. The ciliated coracidia are liberated from the eggs.

Development in First Intermediate Hosts

The coracidia can swim in the water. This swimming activity attract the **zooplanktonic arthropods (Diaptomidae, Cyclopidae)**. In the gut of the cyclops, the coracidia lose their ciliated embryophores. This stage transforms to another larval stage which is called as procercoid. The procercoid is globular in shape and bears a tail-like structure called **cercomer**. Six hooks remain in inactive form.

Development in Second Intermediate Host

Fish act as the second intermediate host. The fish ingest the cyclops. The procercoid becomes free and transforms into another one larval stage called plerocercoid. The plerocercoid is elongated (5 mm) and equipped with a scolex.

Development in Definitive Host

The final hosts get the infection by ingestion of the fish harboring plerocercoids. The fish harboring the plerocercoids are ingested by second intermediate hosts. Adult tapeworm is formed in the definitive hosts.

Note

- After ingestion by the copepod, the cilia of coracidia are lost. This stage penetrates the gut and reaches the hemocoel
- This stage becomes elongated and procercoid stage is formed
- Pike, salmonids, etc. may act as paratenic hosts of the parasites. These fish ingest other fish harboring plerocercoid and they act as paratenic host

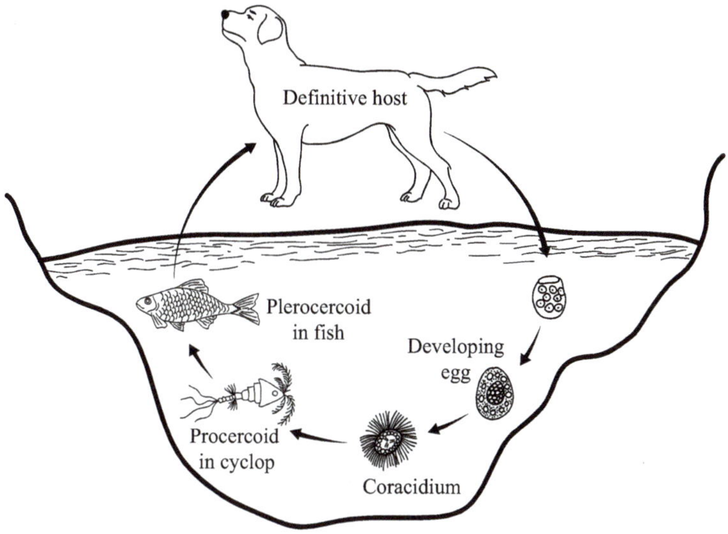

Fig. 5.25 Life cycle of *Diphyllobothrium latum*

Pathogenesis and Pathology

Pathogenesis

The pathogenesis has been presented in the flowcharts.

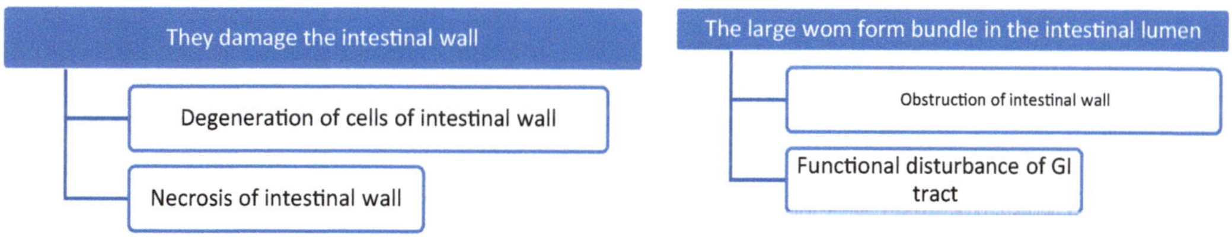

Fig. 5.26 Flowchart of pathogenesis caused by *Diphyllobothrium latum*

Pathology

Anatomical pathology: By using the scolex, the tapeworms cause irritation to the wall of the intestine. This results in denudation or sloughing of epithelial cells resulting in enteritis. Since the size of the worm is large, it may cause obstruction of the lumen of the intestine leading to functional disturbances of the lumen.

Hematological pathology: Hematological data displays anemia. There is low erythrocytic count, decreased Hb% and PCV%, and increased WBC count.

Chemical pathology: Hypoproteinemia may occur. Vitamin B12 of host is utilized by the parasite resulting in deficiency of this vitamin leading to pernicious anemia.

Immunopathology: In any cestode infection, inflammation of the gut wall is obvious. It is presumed that pro-inflammatory cytokines are produced by immune cells which cause inflammation. IL1 beta and IL6 are two important cytokines responsible for causing this inflammation. Activated macrophages produce different pro-inflammatory cytokines.

Diagnosis

Symptomatological Diagnosis

The exhibited clinical signs are vague and cannot provide confirmatory diagnosis. Chain of empty proglottids may be found in the feces. In case of *Diphyllobothrium*, single gravid proglottid is not found in the feces.

Parasitological Diagnosis

Examination of fecal sample may reveal the presence of eggs of the parasite. The eggs are identifiable. They are oval and operculated looking like trematode eggs.

Immunological Diagnosis

Fragmented research work was done on immunodiagnostic works. But no immunodiagnostic test has yet been commercialized.

Molecular Diagnosis

Molecular diagnosis is possible using specific molecular marker. Different ribosomal DNA, ribosomal RNA, and mitochondrial DNA could be used as molecular markers. But it is problematic as reference nucleotide sequence is less.

Control

Do's	Don'ts
1. Affected animals should be treated properly 2. Routine deworming of the animals should be done 3. Dogs should be practiced to defecate at a particular place	1. The dogs should not be allowed to eat raw and uncooked or undercooked fish

General Control

The affected animals should be treated properly. Routine deworming schedule should be followed. Since fish are second intermediate hosts, they should not be allowed to eat raw or uncooked fish. Dog feces should not be disseminated. The dogs should be trained to defecate at a particular place.

Immunological Control

No vaccine has yet been commercialized.

Molecular Control

No gene vaccine is available.

Treatment

Name of drug	Mode of administration	Dose
Praziquantel	Oral	5 mg/kg

Special Part

Immuno-cestodology

General Account of Immunity Against Cestodes

Both cell-mediated and humoral immune responses are found in cestode infection. Different immune activities are found in cestode-infected hosts. These are innate immunity, ADCC, CMI, etc. Concomitant immunity has been observed in some cestode infections. Concomitant immunity is the immunity which is produced by existing adult parasite in the host, and this immunity resists any new larval infection to the hosts. It has been found that infection by *Hymenolepis nana*, *H. diminuta*, and *H. microstoma* elicit concomitant immunity. High dose of infection causes expulsion of the parasites due to concomitant immunity. In humans also, the same thing is found in *Taenia* spp. infection. After expulsion of the worms, certain immunity to secondary infection occurs. In fact, the adult tapeworms are weakly immunogenic. Immature stages can produce strong immune response up to certain period. Thereafter, immune response is slowed down, and cyst formation ensues. The antigens leaked out of the cysts elicit strong immune response. This may result in massive anaphylactic reaction and may cause even death of affected hosts. The antibody or effector cells cannot destroy or penetrate the cysts. That's why the bladder worms or cysts can remain in the host for long period. Cyst formation occurs as a result of host-parasite coevolution. The larval parasites remain within the cysts and they escape the direct attack of host immune response. At the same time, hosts remain protected from the attack by the parasites. Tissue dwelling and migratory parasites elicit different types of immune response. There occurs ADCC, and in this reaction, eosinophil, neutrophil, macrophage, and platelets act as effector cells. IgE, IgA, and IgG are the antibodies which perform ADCC. After ADCC is completed, the eosinophils release eosinophilic binding proteins, eosinophilic cationic proteins, eosinophil-derived neurotoxins, reactive nitrogen intermediates, etc. to kill the parasites. Neutrophils release different cytotoxic substances. Other effector cells also release their toxic substances. Metacestodes of *Echinococcus* elicit immune response by stimulating effector cells for secretion of IL8 and IL10. It has been found that anti-inflammatory cytokines, IL10 and TGF beta, maintain tolerance and partially inhibit cytotoxic mechanism. Different immunological studies of immune response in chronic helminth infection indicate that parasites modulate host immune response in their favor.

Immunodiagnosis of Cestode Diseases

Immunodiagnosis is not specific for cestode infection. Since decades, a plethora of antigens were tried to develop immunodiagnostic tests using crude and defined antigens. But this research journey was not encouraging because of the constraint of cross-reactions and in many cases, the similarity between host protein and parasitic protein was reported.

Different Antigens of Cestodes for Immunodiagnosis

Whole Somatic Antigen

(a) **Extract of Adult Parasite**

The adult tapeworms are homogenized in PBS. Antiprotease (PMSF) is used so that proteases present in tapeworm cannot destroy the protein. Homogenized material is centrifuged at 10,000–20,000 g for 30 min. The supernatant is ultracentrifuged at 100,000 g for 30 min. The supernatant is taken out and stored at −20 °C.

(b) **Extract of Larva**

Extraction of larvae is done in similar way as mentioned before.

Excretory-Secretory Antigen (ES Antigen)

(a) **ES Antigen from Adult**

Adult tapeworms are collected from animals after autopsy. The worms are kept in culture medium (RPMI-1640) at room temperature. The culture is changed at 8 h intervals. The viability of the tapeworms is checked by observing the movement of the worms and microscopical examination. After culture of tapeworms, the material is centrifuged at 90,000 g. The supernatant is taken out and preserved after providing antiprotease (PMSF) and antibiotic (streptomycin).

(b) **ES Antigen from Larva**

ES antigen can be prepared from the larval stages of cestode. For this, the metacestodes are aseptically collected from the infected animals. These cysts are washed properly with PBS and placed in the culture medium, RPMI-1640 with antibiotics. After culture, the culture media are taken out. This is centrifuged at 90,000 g and the supernatant is collected and dialyzed. In this way, the ES antigen is concentrated. For example, *Cysticercus cellulosae* are collected from pig. These metacestodes are washed and kept in RPMI-1640. The viability is checked by observing the evagination of scolex. The cultured liquid is taken and centrifuged as described earlier. The supernatant part contains ES antigen. This is concentrated by dialysis.

Recombinant Antigen

Different works were carried out using different recombinant antigens for immunodiagnosis. **The use of recombinant antigens for immunodiagnosis of cestode infection has been a trend.**

Study on Use of Recombinant Antigen for Immunodiagnosis

Vogel et al. (1988) constructed cDNA library from mRNA of metacestode of *Echinococcus multilocularis* in *E. coli* expression vector lambda gt 11. One isolated clone (from lysogen 11/3) produced a fusion protein which was rapidly degraded and produced distinct polypeptide of Mr 31000 and Mr 33000 showing immunodiagnostic potential. The diagnostic value of **11/3 antigen** was evaluated by mini-Western blot which resulted in diagnostic sensitivity of 98% and over all specificity of 96%. Serological diagnosis of human cysticercosis was done by Hubert et al. (1999) using recombinant antigens from *Taenia solium* cysticerci. They found that antigen **NC-3** exhibited a sensitivity of 96.3% and a specificity of 91.5% for serodiagnosis of *Taenia solium*. However, NC-9 showed only 33.3% sensitivity. Study shows coproantigen assay for diagnosis of *Taenia solium*. Abdel-Megeed et al. (2014) carried out work on immunodiagnostic potential of affinity purified proteins of adult *Moniezia expansa*. Affinity chromatography was done to partially purify the protein. Antibody was raised against those proteins and antibodies were purified by dialysis. These dialyzed antibodies were conjugated to CnBr-activated Sepharose 4B beads (CnBr-Sepharose 4B for preparation of immunoaffinity column). Column-bound proteins were eluted. Three bands were found corresponding to **164, 88, and 67 kDa**. The serodiagnostic value of these antigens was checked, and this study indicated that *M. expansa* affinity partially purified antigen might be successfully used in serodiagnosis for monieziosis in sheep. Study indicates use of excretory-secretory antigens for serodiagnosis of *Anoplocephala*. Excretory and secretory antigens have also been tried for serodiagnosis of other tapeworm infections. Lightbody et al. (2016) carried out **saliva-based ELISA** for diagnosis of tapeworm infection in horses. The saliva tapeworm test displayed 83% sensitivity and 85% specificity. A series of works were carried out using hydatid cyst fluid/protoscolex/oncosphere for diagnosis with considerable success

Immunoprotection Against Cestodes

For immunoprotection, different types of antigens are used as vaccine candidates.

Vaccination Using Different Antigens

(a) **Whole Somatic Antigen**

In the earlier decades, vaccination works were carried out in a very simple way. Somatic antigen was prepared by homogenization of whole worm or a part of worm. After centrifugation, the supernatant part was used for vaccination trial. Some works were also carried out using the antigen derived from cyst fluid. It is to be noted that all these proteins are complex. These protein moieties contain many proteins. Multiple proteins are less potent as compared to specific protein. The efficacy of vaccination using the complex protein was not up to the satisfactory level. But these works have significance as these were the pioneer works in this direction.

(b) **Excretory-Secretory Antigen**

Works were also carried out using excretory-secretory antigens. The adult parasites are placed in culture media (RPMI-1640) with antibiotics. It is to be noted that the parasites should be cultured strictly in aseptic condition. The cultured material is centrifuged. After centrifugation, the supernatant material is used for vaccination trial. The metacestodes are collected from affected animals aseptically. These are cultured in culture media following standard protocol. The cultured material is centrifuged and supernatant is taken. The supernatant material is dialyzed to make it concentrated.

(c) **Defined Antigen**

Defined antigens are specific proteins which are used as vaccine candidates. The specific proteins can be isolated in native form, or this can be produced by recombinant DNA technology. Isolation of specific native proteins can be done by using different column chromatographic techniques. These are **gel filtration chromatography, ion exchange chromatography, immuno-**

affinity chromatography, etc. Before going to isolate any specific protein, it is prerequisite to identify or detect the specific immunodominant protein. For this, **Western blot** technique is performed. For this purpose, the crude protein is fractionated by SDS-PAGE technique. These fractionated proteins are transferred to nitrocellulose membrane. The immunodominant protein is detected in the blot by performing Western blot using hyperimmune sera as primary antibody. Defined antigen can be produced by recombinant DNA technology. The gene encoding a specific protein is attached with plasmid. This recombinant plasmid is inserted in the *E. coli* or yeast cells for expression of the protein.

Study on Use of Recombinant Antigen for Immunoprotection

Gauci et al. (2008) worked on immunoprotection potential of two recombinant proteins. **Tm16** and **Tm18** were expressed in *E.coli* as fusion protein with GST. This recombinant protein was used with the adjuvant, Quil A. Immunoprotection of sheep was achieved against this protein (oncosphere protein) *of Taenia multiceps*. Wang et al. (2016) carried out one work on immunoprotection studies of a recombinant antigen of *Echinococcus granulosus* (**rEgp29**). The sheep were immunized with this protein. After completion of vaccination, challenge infection was given with eggs of *E. granulosus*. Immunoprotection was achieved. up to 94.5%. Significantly increased level of IgG, IgG1, IgG2, IgE, IL2, IL4, and IFN gamma were also found. Gauci et al. (2012) carried out one work on expression and immunoprotection studies of three recombinant antigens comprising (**TSOL16, TSOL45-1A, TSOL45-1B**) from oncosphere stages of *Taenia solium*. The recombinant proteins were expressed as fusion protein with glutathione S-transferase (GST) and maltose-binding protein (MBP). The pigs were immunized twice with GST fusion antigens and boosted third times with MBP. TSOL-16 recombinant antigen exhibited high level of protection, and this protein could be an adjunct to current porcine vaccination against *T. solium* cysticercus in pigs. Zhu et al. (2016) carried out one work on immunoprotection studies of a recombinant antigen of *E. granulosus* (**rEgmyofilin**). The sheep were inoculated subcutaneously with this protein. After completion of vaccination, challenge infection was given with eggs of *E. granulosus*. The number of cysts was reduced in the vaccinated animals. Zhou et al. (2018) carried out work on immune response of recombinant proteins of *Taenia solium* (pMG36e-SP-TSOL18/*Lactococcus lactis* and pMG36e-TSOL18/*Lactobacillus lactis*. The mice were orally immunized with this vaccine candidate. This study showed cellular, humoral, and mucosal immunity elicited by this antigen of *Taenia solium*.

> Note
> **Name of some specific cestode antigens for immunoprotection**
>
> - **TSOL 16 (*Taenia solium*)**
> - **TSOL 45 A (*T. solium*)**
> - **TSOL 45 B (*T. solium*)**
> - **Tm 16 (*T. multiceps*)**
> - **Tm 18 (*T. multiceps*)**
> - **pMG36e-SP-TSOL18/*Lactococcus lactis*) (*T. solium*)**
> - **rEg29 (*E. granulosus*)**

Molecular Cestodology

Molecular Taxonomy

The field of parasitology has been progressing toward the search for the genomic structure of different species of many parasites. The nucleotide sequences of many cestodes have been derived. Different ribosomal DNA and RNA and mitochondrial DNA (mtDNA) are the molecular markers which are used for development of molecular taxonomy. The primers of these markers are commercially available or can be constructed.

Some Study on Genomic Characterization of Cestodes

Various works were carried out on molecular phylogeny. As mentioned earlier, molecular characterization of different cestodes is also done. In these works, different molecular markers are taken for studies. The molecular markers routinely used are ribosomal DNA, ribosomal DNA, mitochondrial DNA, etc. Work was carried out by Labuschange et al. (2018) on molecular characterization of *Dipylidium caninum* of canine and feline origin using 28 S rDNA as molecular marker by RFLP and hydrolysis genotyping probe-based assay. Their study showed that this species of two hosts had separate genotypes.

Some Molecular Study of Biology of Cestodes

It has been a recent trend to describe biology or life cycle of a parasite in molecular way. In life cycle of a parasite, different developmental stages are formed. These developmental stages have some specific biological activities. Therefore, it is obvious that different developmental stages would have different biochemical profiles. It is presumable that protein profile of different developmental stages will be differed somewhat in a particular period of life cycle. Therefore, gene expression in a particular period of development may be variable. Consequently, mRNA profile would be different somewhat at a particular period of biological development.

Transcriptome analysis is very useful technique in molecular biology. Transcriptome is entire set of RNA transcripts found in a cell for a specific time (developmental stages or physiological conditions). By transcriptome analysis, differential gene expression is determined. The following figure indicates different steps for transcriptome analysis:

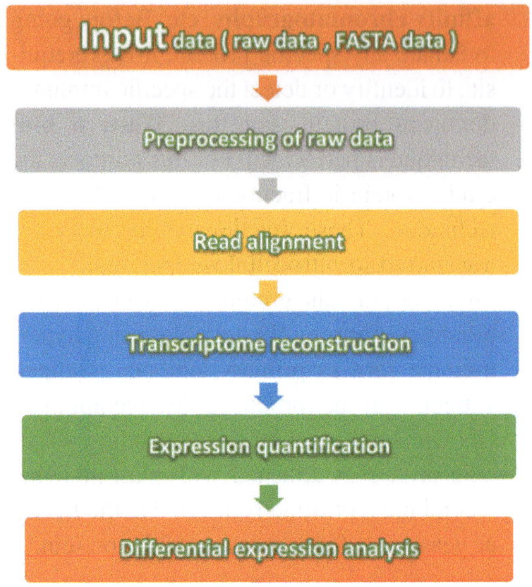

Fig. 5.27 Flowchart of differential expression

Zhang (2019) compared the transcriptome data of adult and larval stages of *Taenia pisiformis*. **Transcriptomic data analysis** revealed a large number of DEG (differentially expressed gene). The DEGs enriched in the adult stages were mainly associated with nutrient acquisition, egg development, AP axis formation. The characterization of the comparative transcriptomic data provides hints for understanding the biological and physiological mechanism of development and reproduction in *T. pisiformis*. Basika et al. (2019) compared the transcriptome data of strobilated worm and larval stages (tetrathyridium) of *Mesocestoides corti* (model parasite) by **RNA-seq. based analysis**. Transcriptomic data analysis revealed the transcripts rich in larval stages related to asexual reproduction, budding, and migration in intermediate hosts which reflects larval developmental process. Transcripts rich in strobilated worm were associated with development, cell growth, and morphogenesis. Dezaki et al. (2016) reported differential gene expression of **HoxB7** and **notch gene** in different developmental stages of *Echinococcus granulosus* sensu stricto. The findings demonstrated that HoxB7and notch gene were differentially expressed in different developmental stages of the parasite.

Molecular Pathology

Background of General Pathology

The larval stages cause massive damage to the vital organs during migration. The oncospheres of different tapeworms (*Taenia*, *Echinococcus*, etc.) migrate in internal organs. *Taenia* oncospheres may migrate to the central nervous system where they cause **neurocysticercosis**. Oncospheres of *Taenia multiceps* migrate to the brain of intermediate hosts (sheep, goat, etc.) where they cause formation of cysts (coenurus). All these cause different nervous disorders. The adult tapeworms perform erosive activity and cause local inflammation of the affected site. They use their whole scolex, suckers, and hooks and cause damage to the local tissue. The metabolic by-product of the tapeworms is absorbed by the affected tissues. The absorbed by-product damages the nervous tissue resulting in different nervous disorders. Epileptic fit, running fit, and incoordination of movement are some neurological signs exhibited by dogs infected with *Dipylidium caninum*. Sometimes, parasites die, and the degraded substance of the body of the parasite is absorbed and results in anaphylactic reaction. Leakage of tapeworm fluid and cysts also cause anaphylactic reaction.

Molecular Pathology

Different immunological reactions occur due to invading larval stages of the parasites. It has been found that when the oncospheres penetrate the tissue, the initial immunological features exhibit elevated mRNA level of different pro-inflammatory cytokines. The pro-inflammatory cytokines are IL1 beta, IL6, TNF alpha, etc. Pericystic mRNA of Thl cytokines has been found at elevated level. Gradually, the Thl response is counteracted by Th2 response later. The parasites ultimately settle and immunoregulation occurs which increases the survival of the parasite there. Treg cells are T regulatory cells which produce IL10. IL10 and TGF beta cytokines downregulate the Thl cytokines and also reduce excessive activity of Th2 cytokines. Tissue-invading immature larval cestodes enter into a resting phase after a course of migration, and they settle in a location. At these places, cyst formation and granuloma formation occur. The metacestode antigen stimulates release of some cytokines. In case of tissue-invading larval parasite of cestode, ADCC occurs which is one important immunological phenomenon. In this phenomenon, three components are required, i.e., antibody (IgG, IgE, IgM), effector cells (eosinophil, macrophage, NK cells, etc.), and target antigen. Antibody already charged with antigen binds to the effector cells through the receptors present in the effector cells. Effector cells release different toxic materials. For example, the eosinophils release different toxic substances. In this context, it is to be mentioned that eosinophils contain cysteinyl leukotrienes, arachidonic acid-based inflammatory mediators, and different cationic proteins, i.e., EBP (eosinophilic basic protein) and ECP (eosinophilic cationic protein), which are presumptively released due to ADCC. These substances cause tissue damage resulting in pathology. Complement reaction is another one immunological phenomenon which also causes tissue damage resulting in different pathological features. **These are a group of proteins which cause series of cascade reactions occurring in three ways, i.e., classical, alternative, and lectin**. The key component of complement is C1q which binds to respective binding site of antibody. Then, complement activation starts. The complement fragments are produced during cascade reactions which attract different leukocytes to the affected area.

Tissue-invading immature larval cestodes enter into a resting phase after a course of migration, and they settle in a location where cyst formation and granuloma formation occur. Bladder worms or cysts are the examples. The initial inflammatory reactions stop before cyst formation in the intermediate hosts. As a consequence, there occurs reduction of pro-inflammatory cytokines. A subset of T cells (T regulatory cells =Treg cells) produce IL10. IL10 and TGF beta cause immunoregulation. Immunoregulation is required for survival of the parasite and also for cyst formation. The cyst is formed by granulomatous tissue and fibrous tissue. IL4 has role in deposition of fibrous tissue. It has been found that metacestode antigens enhance the secretion of IL8 and IL10 but cause reduction of IL 1 and IL18. IL1 and IL18 downregulate Th2 response. There are evidences of complement-mediated lysis of tissue-invading parasites. In this context, it is felt to describe general events occurring in complement pathway as follows:

Complement is a set of proteins which carry out cascade reactions in different pathways, i.e., classical pathway, alternative pathway, and lectin pathway. In classical pathway, the antigen-antibody complex binds to C1q. This binding results in activation of a serine protease C1r. Activated C1r activates another serine protease, C1s. This C1s activates C4 and cleaves into C4a and C4b. This enzyme also cleaves C2 into C2a and C2b. Larger fragments form C4bC2a (C3 convertase). C3 convertase causes hydrolysis of C3 resulting in formation of C3a and C3b. The C3b part binds to C3 convertase and forms C5 convertase. The enzyme, C5 convertase, splits C5 into C5a and C5b. The C5b binds with C6, C7, C8, and C9 and forms membrane attack complex (MAC). In alternative pathway, C3 is automatically hydrolyzed into C3a and C3b. In **lectin pathway, lectin (mannose binding lectin=MBL) binds to the carbohydrate of the parasite cuticle and activates MASPS which forms C3 convertase**. All three pathways of complement ultimately form membrane attack complex (MAC). This MAC contains granzyme and perforin. The perforin causes pore formation, and through this pore, the granzyme enters. In this way, complement can damage or kill the parasites. Not only that, on the way of complement activation, anaphylatoxin is produced which also affects the parasites. C5b attracts the migration of different immune cells to the parasite-affected areas. All these immune activities directed toward killing or damaging parasite also damage the host tissues and cause different types of pathological features.

Note
Tissue-invading immature larval cestodes enter into a resting phase after a course of migration, and they settle in a location where cyst formation and granuloma formation occur. The initial inflammatory reactions stop before cyst formation in the intermediate hosts. As a consequence, there occurs reduction of pro-inflammatory cytokines. A subset of T cells (T regulatory cells = Treg cells) produce IL10. IL10 and TGF beta cause immunoregulation. Immunoregulation is required for survival of the parasite and also for cyst formation. The cyst is formed by granulomatous tissue and fibrous tissue. IL4 has role in deposition of fibrous tissue

Cestode Zoonoses

Diphyllobothriasis

Causative Agents
The important species is *Diphyllobothrium latum*.

Hosts Affected
The parasites affect the small intestine of dog, cat, fox, man, etc.

Mode of Infection
The life cycle includes two intermediate hosts. The first intermediate hosts are cyclops and the second intermediate hosts are fish. The eggs pass out through the feces and hatch in the environment. After hatching, the ciliated coracidium comes out. The ciliated coracidium is ingested by the cyclops in which procercoids are formed. The fish ingest the infected cyclops in which plerocercoids are formed. The plerocercoids are the infective stages of the parasite, and these develop to adult tapeworms after being ingested by the definitive hosts. Final hosts get infected with the parasites on ingestion of infected fish.

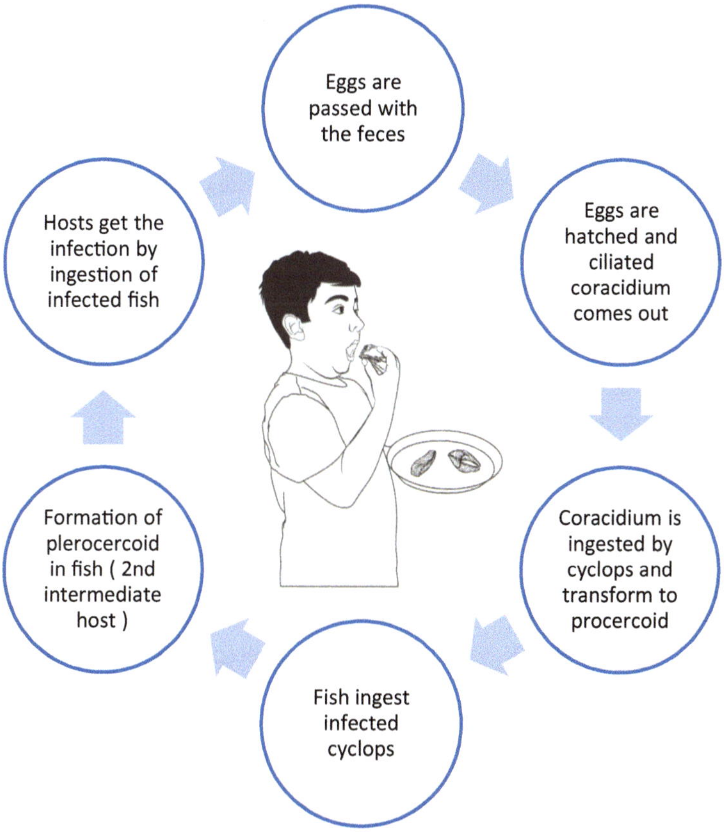

Fig. 5.28 Life cycle of *Diphyllobothrium latum*

Pathogenesis

Some people show no pathogenic effect. Depletion of nutrient is caused due to competition between host and worm resulting in deficiency of vitamin B12 causing pernicious anemia. Mechanical obstruction of the lumen and absorption of by-product of worm by the host are other reasons for causing pathology. Unusual location of occurrence of this parasite is the gallbladder. Vomiting with several meters of long worm may occur. The clinical symptoms include diarrhea, nausea, abdominal discomfort, etc. Intermittent diarrhoea and constipation occur in animals.

Diagnosis

Symptomatological Diagnosis

The exhibited clinical signs are not so specific for confirmation of this parasitic infection.

Parasitological Diagnosis

Feces should be examined properly which may reveal the presence of the eggs of the parasites. Examination of fecal sample of the affected hosts may reveal the presence of egg capsules or the eggs. The eggs are characteristic of trematode eggs which are oval and have operculum.

Immunological Diagnosis

Several works have been carried out using crude and purified antigens. But no immunodiagnostic technique has been commercialized so far for field application.

Molecular Diagnosis

Molecular diagnosis is possible by using different molecular markers. Different ribosomal DNA, ribosomal RNA, and mitochondrial DNA could be tried as molecular markers. **Copro-**

PCR could be tried using specific molecular markers.

Control

General Control

(a) Proper treatment should be done using specific drug available so far. The affected persons should be treated with anti-cestodal drug.
(b) Uncooked, undercooked, and unsmoked fish should be avoided. Proper cooking of fish and deep freezing for a duration of 24–48 h in −20 °C can kill the plerocercoids. In Jewish population, **gefilte fish (minced)** is spread on bread while serving. This minced fish should be properly frozen.
(c) Public education regarding danger of eating uncooked and undercooked fish should be done.

Immunological Control

Different crude and defined antigens have been tried but no vaccine has yet been commercialized.

Molecular Control

So far, no naked DNA vaccine or gene vaccine is available.

Dipylidiasis

Causative Agent

The main causative agent is *Dipylidium caninum* occurring in the small intestine of man and dog.

Mode of Infection

Eggs of the parasites are ingested by flea (*Ctenocephalides canis*, *C.felis*). Cysticercoids are formed in the body of the flea. Man get the infection by accidental ingestion of infected second intermediate hosts (flea).

Pathological Features

Minor intestinal disturbances including indigestion and loss of appetite occur due to this tapeworm infection. Proglottids migrating from the anus cause **anal pruritus**. The worms may cause mechanical obstruction of food passage causing functional disturbances. In man, abdominal discomfort, diarrhea, nausea are frequently observed. Occasionally macrocytic hypochromic anemia occurs as a result of depletion of Vit B12 by the parasites. In long standing case, neurological symptoms manifest.

Diagnosis

Diagnosis is based on examination of fecal sample and detection of eggs or egg capsule. Eggs, if required, can be recovered from egg capsule. Detection of gravid segment in feces is also helpful for diagnosis.

Control

Ectoparasiticidal agents should be applied to kill flea as these ectoparasites act as the intermediate hosts.

Spirometrosis

Causative Agents

Spirometra erinacei, *S. fali*, etc. have zoonotic importance.

Mode of Infection

Two intermediate hosts are required for completion of the life cycle of the parasite. The first intermediate hosts are cyclops in which the procercoids develop. Subsequent larval stage is plerocercoid or spargana which develops in the second intermediate hosts, i.e., snakes, tadpoles, etc. Human beings get the infection through different routes, i.e., accidental ingestion of infected crustacea, ingestion of flesh of second intermediate hosts, and rural practicing of wound dressing with flesh of snake.

Pathological Features

It has been found that the plerocercoids are not markedly pathogenic. Abscess may be formed. The degenerated parasite may cause inflammation, erythema, urticaria, and edema in human beings. If the parasites occur in the eye region, there may be edema, conjunctivitis, etc.

Diagnosis

Examination of the lesion may reveal the presence of larvae of the parasite.

Control

The flesh of the second intermediate hosts should be properly cooked before eating. Food and drinking water should not be contaminated with infected cyclops.

Hymenolepiasis

Hosts Affected

H. nana affects rodents and man. *H. diminuta* commonly occurs in rodents. But man may also be infected.

Mode of Infection

Human beings get the infection by ingestion of eggs. Autoinfection may occur in man. The eggs are hatched in the lumen and the oncospheres come out. They affect the wall of the intestine where cysticercoids are formed. **Cysticercoids break out the intestinal villi** and develop in the lumen of the intestine of man. The rodents get the infection by ingestion of infected beetle. Both direct and indirect life cycles occur in rodents.

Pathological Features

In man, this parasite does not cause severe pathogenesis even with large numbers. However, infection may result in different clinical signs comprising allergic reaction, headache, periodic diarrhea, dizziness, pruritus of the nose and anal region, and abdominal discomfort. Tumor formation by *H. nana* has been reported. Internal autoinfection has also been reported which causes different complications. *H. diminuta* fails to produce pathogenesis in human beings.

Diagnosis

Examination of fecal sample for detection of eggs can be done for diagnosis.

Taeniasis

Causative Agents

The important zoonotic species are *T. solium*, *T. saginata*, *T. taeniaeformis*, *T. multiceps*, and *T. crassiceps*.

Mode of Infection

The metacestodes (bladder worms) occur in the body of intermediate hosts. The final hosts get the infection on ingestion of metacestodes present in the muscles of intermediate hosts.

Taenia solium Infection in Man

The adult parasites do not produce marked pathogenesis. Enteritis in definitive hosts occurs. Other symptoms are dullness, flatulence, abdominal discomfort etc. However, larval stage of *Taenia solium* causes neurocysticercosis in human beings. Autoinfection of *Taenia solium* in man occurs due to constipation and some other reasons.

Taenia saginata Infection in Man

Though the worm is large in size, it does not cause severe pathology. Intestinal obstruction by this worm causes functional disturbances. Segments of *T. saginata* actively crawl out usually at evening time which may cause anal pruritus. Dullness, hunger, pain, loss of body weight, and abdominal discomfort are the clinical signs. The stool should be examined properly for presence of segment of tapeworms. The segment should be pressed in between two slides and lateral uterine branches should be counted. In *T. saginata*, usually 18 lateral branches on each side of uterine stem may be found. In many cases, distinctive features of identification are absent due to loss of eggs from the gravid segment.

Taenia taeniaeformis Infection in Man

This is usually a parasite of cat. But scanty reports of human infection with presence of strobilocerci in the liver are available.

Taenia crassiceps infection in man

These are common parasites of fox. The intermediate hosts are rodents in which cysticerci are formed. But infection in human beings with cysticerci was reported.

Taenia multiceps Infection in Man

The definitive hosts are dogs. In sheep and goat, coenuri are formed in the brain. However, human infection with coenuri was reported in the brain, subcutaneous tissues, eye, etc. The cysticerci cause different damages in the nervous system. The clinical features include epilepsy, chronic meningitis, encephalitis, and different psychological disorders. Vision may be impaired due to presence of cysts in eye.

Diagnosis of adult and cyst of different *Taenia* spp

Parasitological Diagnosis

(a) Typical *Taenia* eggs can be detected by examination of fecal sample.
(b) **Radiodiagnosis** can be done for detection of cysticercosis.
(c) **CT scan** can be done for detection of cysticercosis. Hypodense area (low signal intensity) displays the vesicle and hypersensitive area (high signal intensity) displays the scolex. The image of calcified cyst by CT is appreciable.
(d) **MRI**. Higher-contrast resolution is found by MRI technique. In MRI, lesion is more clearly visible. Extraparenchymal cysts are detectable by MRI.
(e) **Biopsy examination**
Biopsy material is taken from the eye, muscle, subcutaneous tissue, etc. for detection of cysticercosis.
(f) **FNAC (Fine-needle aspiration cytology)**
Fluid is taken and examined for detection of cysticercus.

Immunodiagnosis

(a) ELISA is done for detection of antibody in serum and CSF. For performing the test, crude antigen derived from cyst and vesicular fluid is used. **Quick ELISA kit** is commercially available. T24 H recombinant antigen is used for diagnosis of neurocysticercosis having 96% sensitivity and 99% specificity.
(b) **Western blot** is done for using lentil-lectin purified seven glycoproteins. Seven bands are formed in this Western blot technique.
(c) Antigen detection is done by using monoclonal *T. solium* antibody.

Control

(a) Treatment of the infected persons should be done.
(b) Proper meat inspection should be done.
(c) Mass education about the harm of these parasites and control of the parasites should be done.
(d) Proper sewage treatment and disposal are some important control techniques.

Coenurosis

Causative Agents

The causative agent is *Coenurus cerebralis*.

Transmission

Human beings get the infection accidentally. They get the infection by ingestion of eggs of *Taenia multiceps*.

Pathological Features

Coenuri appearing in human beings are mainly of *Taenia multiceps*. But other species should also be taken into account. Several human cases of coenurosis have been recorded. These cysts occur in the subcutaneous tissues, muscle tissues, eye, central nervous system, etc. Most pathological features are due to pressure effect exerted by cysts. But allergic reaction also occurs which result in pyrexia, recurrent rash, night sweats, etc. Subcutaneous or intramuscular cysts are revealed by swelling in the affected area. Cerebral coenurosis is more severe involving different parts of the brain. There occurs pressure necrosis. Headache and vomiting are other clinical signs.

Diagnosis

CT scan and ultrasonography may reveal the presence of cysts.

Echinococcosis

Causative Agents

Some important species are *Echinococcus granulosus*, *E. multilocularis*, and *E. vogeli*. Human beings may act as the accidental intermediate hosts of the parasites.

Mode of Infection

Human beings may act as the intermediate hosts, and they get the infection by ingestion of eggs. Metacestodes (hydatid cysts) are formed in the human body.

Pathological Features

Adult is less harmful to dog. But marked pathology occurs in the intermediate hosts when vital organs, i.e., brain, heart, etc., are affected by the larval stages of the parasites (hydatid cysts). Hydatid cysts may occur in different organs of human beings, i.e., subcostal muscles, abdomen, thoracic muscles, lung, liver, heart, kidney, brain, etc. Rupture of the cysts may cause **anaphylactic reaction** resulting in severe problems. Man may act as intermediate host of *Echinococcus multilocularis* and they get infection through contaminated food and drinking water. Alveolar echinococcosis or alveococcosis occur due to this infection. This is invasive in nature and cause serious condition like malignant tumor with metastasis. The larval stages of *E. vogeli* cause polycystic forms of hydatidosis.

Diagnosis

Symptomatological Diagnosis

The exhibited clinical signs are not so specific for confirmation of this parasitic infection.

Parasitological Diagnosis

(a) Feces should be examined properly which may reveal the presence of the eggs of the parasite. This is done for diagnosis in definitive hosts.
(b) **Image-based diagnosis** by X-ray analysis can reveal the presence of the parasitic cysts. This X-ray displays hepatomegaly or calcified cysts.
(c) **CT** scan differentiates hydatid cyst from other cysts.
(d) **MRI** has higher-contrast resolution. MRI has less efficiency to detect the calcified cyst.
(e) Direct microscopy or acid-fast stain of hydatid sand can be done.
(f) Cyst fluid is centrifuged and one drop is placed between two slides. Grating of sound is felt.
(g) Histopathology of the removed cyst is done.

Immunological Diagnosis

Many works have been carried out using crude and purified antigens.

Two-step Diagnosis

(a) Initially, ELISA is done and then **DIGFA (dot immunogold filtration assay)** is done for further confirmation.
(b) Commercially available ELISA kit basing on the use of a recombinant antigen can be used. It has more than 90% sensitivity and specificity.
(c) Western blot

h-HCF-IB (human hydatid cyst fluid immunoblotting) is a commercially available assay technique.

Molecular Diagnosis

Molecular diagnosis is possible by using different molecular markers. Different ribosomal DNA, ribosomal RNA, and mitochondrial DNA could be tried as molecular markers. Studies indicate the use of ribosomal DNA (ITS1, ITS2, etc.), mitochondrial DNA, i.e., Cox 1 (cytochrome oxidase 1), and Nad 1 (nicotinamide adenine dinucleotide subunit 1) for characterization/identification of this parasite. PCR is done using mitochondrial DNA.

Control

General Control: Proper treatment should be done using specific drug available so far.

Immunological control: Different crude and defined antigens have been tried but no vaccine has yet been commercialized. A series of works were carried out using crude somatic protein of whole parasitic extract, ES protein (excretory and secretory protein) of metacestodes, etc.

Molecular control: So far, no naked DNA vaccine or gene vaccine is available.

Mesocestoidiasis

The important zoonotic species are *Mesocestoides variabilis* and *M. lineatus*.

Mode of Infection

Two intermediate hosts are required for completion of life cycle of this parasite. The first intermediate hosts are arthropods, i.e., beetles, oribatid mites, etc. Small vertebrates, i.e., snakes, frogs, lizards, rats, mice, and birds, act as the second intermediate hosts. The definitive hosts get the infection by ingestion of second intermediate hosts harboring the infective stage of the parasites (**tetrathyridia**). Human beings may get infection by ingestion of second intermediate hosts.

Pathological Features

This tapeworm appears to be less pathogenic. However, enteritis may be caused which may result in loss of appetite, diarrhea, etc. Anemia, poor appetite, and abdominal pain are common clinical features.

Diagnosis

Diagnosis can be done by examination of fecal sample for detection of eggs.

In Vitro Culture of Cestodes

Background of Cestode Culture

It is really very difficult to culture cestode in vitro unless the medium is not properly aseptic. However, if aseptic condition is maintained, cestodes can be cultured up to certain extent. For example, plerocercoid larvae can be cultivated for a period of 10 months at room temperature. In the earlier period, the physiology of cestodes in all aspects was not known clearly. For that reason, culture of cestodes was very difficult earlier. In the recent time, culture techniques of cestodes have been improved as compared to previous situation due to updated knowledge. The osmotic pressure of the surrounding of tapeworm is higher than the osmotic pressure of the internal body of the cestodes. This may be the reason that the tapeworms can absorb food material through their tegument. The outer cuticle contains the scleroprotein. The outer cuticle is permeable to water, electrolytes, some amino acids, etc.

On ingestion of the eggs of tapeworms, the embryophore has to be exposed in acid followed by alkali; otherwise, the oncosphere will not be free. This occurs in the intestine but not in the stomach. If the cysts are ingested by the definitive hosts, only the bile can stimulate the scolex to get out from invagination state. To reach the duodenum, the scolex has to pass through the stomach. The scolex remains invaginated in such a way that the gastric juice cannot affect the scolex. On reaching the duodenum, the bladder is digested, but the scolex remains undigested. The tapeworms cannot tolerate all pH range. By experiment, it has been found that they can tolerate the pH range of 4–11.

As has been stated earlier, it is difficult to culture cestodes in vitro, which is due to lack of maintenance of aseptic condition. Once aseptic condition is maintained perfectly, some cestodes can be grown up to certain extent. For example, culture of plerocercoids is possible up to 10 months in strict aseptic condition. Raising temperature up to 40 °C results in formation of adult sexually matured strobila. It has been found that cestode can utilize glucose and the excess glucose is stored in the form of glycogen. Cestodes can accept oxygen if offered. But anaerobic respiration has also been detected. No report is available on fat requirement of cestodes. It has been reported that vitamin G is essential for cestode development.

Some Examples of Previous Attempts for Culture of Cestodes

Name of cestode	Medium used	Stage maintained
Proteocephalus percae	Seawater	Adult
Schistocephalus solidus	Tap water	Plerocercoid
Moniezia expansa	Albumen + beef extract	Adult
Diphyllobothrium latum	Ringer-Locke	Plerocercoid
Taenia taeniaeformis	Saline + chick extract + horse serum	Cysticercus
Taenia serialis	Saline + horse serum	Metacestode
Echinococcus granulosus	Hydatid fluid + serum	Metacestode

Culture Methods of Some Cestodes

Echinococcus granulosus

Smyth and Davies (1974) reviewed in vitro culture of *Echinococcus granulosus*. Culture of *Echinococcus granulosus* from protoscolices to adult stage is possible. For this purpose, protoscolices are collected aseptically from brood capsules. These are treated with pepsin and washed repeatedly in saline. Evagination of scolex occurs in the solution of sodium taurocholate. The protoscolices so collected can be aseptically cultured in a diphasic medium comprising approximately 80% parker 858 with additional glucose, K^+ and Yeast extract, 20% fetal calf serum over a solid base of bovine or calf serum coagulated at 76 °C. Static bottle culture or continuous circulating system with a gas phase of 5% CO2 + 10% O2 in N2 are used.

Hymenolepis diminuta

Bertzen (1961) carried out in vitro culture of *Hymenolepis diminuta* using Tyrode solution. The infected beetle harboring cysticercoids were used in this experiment. The cysticercoids were dissected out and pushed in the Parker's Tyrode solution. After two changes of Tyrode solution, the cysticercoids were placed in the incubator at 37 °C. They were excysted at pH 8–9. The sterile condition was maintained with constant change of Tyrode solution. The worms grow from cysticercoids to adult with matured segments containing eggs. In a study, *Hymenolepis nana* was grown successfully in vitro from cysticercoid stage up to the sexually matured tapeworm harboring eggs. The culture was done for growing *H. nana* in vitro by a method of recirculating continuous flow using a controlled medium-gas-phase pumping system. This arrangement continuously recirculates the medium over the worms.

Ligula intestinalis

Study indicates culture of *Ligula intestinalis* is possible using peptone broth at 40 °C. The *Ligula* attained its growth from plerocercoids to sexually matured tapeworm. However, the length of maturation as compared to the growth in birds is twice. Another notable thing is that the eggs of cultured tapeworms were found infertile

Taenia spp.

For culture of *Taenia* spp., RPMI medium 1640 is useful without addition of serum. Budding of cysts of *Taenia crassiceps* was studied.

Emerging and Reemerging Cestodes

Echinococcus Infection

Cystic echinococcosis occurs with high incidences in many parts of the world and in some areas, for example, Bulgaria, Kazakhstan, People's republic of China; this is re-emerging in humans (Reviewed, Eckert et al. 2000).

Taenia Infection

Taenia solium occurs in the small intestine of human beings. The intermediate host of this tapeworm is pig. The larval stages of this tapeworm are formed in different organs of pig. The definitive hosts of this tapeworms are human beings which get the infection by ingestion of raw, undercooked meat. This pork tapeworm, *T. solium*, is also spreading in the places where there were no occurrences of taeniasis earlier. The increasing level of popularity of pork consumption and influx of international workers from large cross-border migration causes new endemic foci in nonendemic areas.

Cysticercus Infection

Cysticercus cellulosae is the larval stage of *Taenia solium*. This cysticercus formation may occur in the brain of human beings. It is endemic in Central and South America, Mexico, parts of Africa, and Asia. It is more common in India particularly in Northern India. Neurocysticercosis (NCC) is one of the serious complications of this infection. Unfortunately, the cases of neurocysticercosis are being increased due to migration of persons from endemic countries to nonendemic countries. Cysticercosis was long absent in Southern Europe and other developed countries, but cysticercosis was re-emerged due to imigration, commerce, etc. (Symeonidou 2018).

Cestodes of Wild and Zoo Animals

Parasites	Wild hosts	Site
Cittotaenia ctenoides	Rabbit	Small intestine
Bertiella studeri	Rhesus monkeys, baboon, gibbons, Japanese macaque, chimpanzee, etc.	Small intestine
B. mucronata	Monkey rhesus macaques,gray langurs	
Stilesia hepatica	Wild ruminants (sheep, goat, cattle, deer, antelopes, gazelles)	Bile duct
Thysanosoma actinioides	Deer, antelope, elk, mountain goats	Bile duct, pancreatic duct
Atriotaenia procyonis	Mustelids, procyonids	Bile duct
Raillietina cesticillus	Turkey, guinea fowl, peafowl	Small intestine
R. georgiensis	Wild turkey	Small intestine
Houttuynia struthionis	Ostrich	Small intestine
Choanotaenia infundibulum	Turkey, game birds, pheasant, quail, patridge	Small intestine
Dipylidium caninum	Indian fox, hyenas, jackal	Small intestine
Hymenolepis nana	Rodents	Small intestine
Taenia hydatigena	Wolves, wild carnivores	Small intestine
T. pisiformis	Fox (definitive host) Wild carnivores, lagomorphs, rabbits, hares (intermediate hosts)	Small intestine
T. multiceps	Fox, coyotes (definitive hosts)	Small intestine
Echinococcus multilocularis	Fox, wolves, coyotes (definitive hosts); Mice, rodents (intermediate host)	Small intestine
E. oligarthrus	Jaguar, puma, jaguarundi (definitive hosts); Rodents (intermediate hosts)	
E. vogeli	Bush dogs (definitive hosts); Rodents (intermediate hosts)	
Diphyllobothrium latum	Polar bear, wolves, fox, otter	Small intestine

Clinical Cestodology

(For Clinical Cestodology, the readers are required to read different aspects i.e. pathogenesis, clinical signs, diagnosis, treatment, control of different tape worms described in this chapter. Here some relevant information have been given below)

Specimens Required for diagnosis of cestode in fection

Feces

The feces should be collected afresh for gross examination and microscopical examination. The gravid segments are also found in the fresh but they crawl to the distant place if the feces is not used afresh. Usually 100–200 g feces should be collected so that all type of examinations can be done, i.e., gross examination, direct microscopical examination, concentration techniques, etc. For collection of feces from large animal, the hand after wearing a plastic glove is inserted to take out some amount of feces per rectum. See also the description in the chapter Trematodology regarding preservation of feces and other aspects.

Blood

Examination of blood is not required for diagnosis of any cestode infection. But blood is collected to study the hematology if required in cestode infection. See also the chapter of Trematodology for collection technique. For performing any immunodiagnostic test of cestode infection, blood should be collected in test tubes without any anticoagulant. The serum is prepared in this way and used for performing immunological tests.

Urine

Urine examination does not have role for diagnosis of any cestode infection.

Lacrimal Secretion

Examination of lacrimal secretion does not have any role for diagnosis of any cestode infection.

Nasal Discharge

Examination of nasal discharge does not have any role for diagnosis of any cestode infection.

Examination of Biopsy Material

Examination of biopsy material is very important in some cestode infection. Cysticercosis, sparganosis, etc. can be diagnosed by examination of biopsy material.

Egg Character of Cestodes

Cattle

Parasite	Eggs/parts of worms
Moniezia expansa	Eggs of *M. expansa* are triangular
M. benedini	Eggs of *M. benedeni* are square

Horse

Parasite	Egg/parts of worms
Anoplocephala perfoliata, A. magna, P. mamillana	The eggs are thick walled with one or more flattened sides. These are three layered eggshell. Pyriform apparatus is present in the hexacanth embryo

Dog and Cats

Parasite	Eggs/parts of worms
Dipylidium caninum	Thousands of cucumber seed-shaped gravid segments are present. When the gravid segments are dried up, they appear as uncooked rice. Eggs remain in egg capsule, each containing about 30 eggs.
Taenia pisiformis, T. hydatigena, T. ovis	The eggshells are striated (embryophore) containing six hooks
Taenia multiceps	The eggs are 29 × 37 um in diameter
T. serialis	The eggs are elliptical in shape and the size is 31–34 um × 29–30 um
Echinococcus granulosus, E. multilocularis	The eggs of *Echinococcus* are ovoid (32–36 um × 25–30 um. Single oncosphere with six hooks is found
Mesocestoides	The mature segments are often separated along the longitudinal axis for a short distance. The common name of the tapeworm is "zipper tapeworm." Zipped and unzipped proglottids appear in the feces of pet

References

Abdou AH (1956) Observations on the life cycle of Davainea proglottina in Britain. J Helmintho 30(4):189–202. https://doi.org/10.1017/S0022149x00033162

Abdel-Megeed KN, Hassan SE, El-Ezz NMTA, Farag TK (2014) Diagnosis of monieziasis using adult *Moniezia expansa* affinity partially purified antigen. Glob Vet 13(5):814–819. https://doi.org/10.5829/idosi.gv.2014.13.05.86221

Abdel-Latif M, EI -Shahawi G, Aboelhadid SM, Abdel-Tawab H (2017) Immunoprotection effect of chitosan particles on Hymenolepis nana-infected mice. Scand J Immunol 86(2):83–90. https://doi.org/10.1111/sji.12568. PMID:28513991

Aziz H, Kazemi B, Bandehpour A, Khamesipour A et al (2016) Modulation of the immune response to DNA vaccine encoding gene of 8 kDa subunit of Echinococcus gran ulosus antigen B using murine interleukin 12 plasmid in balb/c mice. Iran J Parasitol 11(4):1480–1488

Basika T, Paludo GP, Araujo FM, Salim AC et al (2019) Transcriptomic profile of two developmental stages of the cestode parasite *Mesocestoides corti*. Mol Biochem Parasitol 229:35–46. https://doi.org/10.1016/j.molbiopara.2019.02.006. PMID: 30797776

Bertzen AK (1961) The in vitro cultivation of tape worms .I.Growth of Hymenolepisdiminuta (Cestoda:Cyclophyllidea). J Parasitol 47(3):351-355. https://doi.org/10.2307/327534

Dezaki ES, Yaghoobi MM, Taheri E, Almani PG (2016) Differential expression of Hox and Notch genes in larval and adult stages of *Echinococcus granulosus*. Korean J Parasitol 54(5):653–658. https://doi.org/10.3347/kjp.2016.54.5.653. PMID: 27853123

Eckert J, Conraths FJ, Tackmann K (2000) Echinococcosis: an emerging or re-emerging zoonosis? Int J Parasitol 30ae:13(1283–1294)

Gauci C, Vural G, Oncel T, Varcasia A et al (2008) Vaccination with recombinant oncosphere antigens reduces the susceptibility of sheep to infection with *Taenia multiceps*. Int J Parasitol 38(8–9):1041–1050. https://doi.org/10.1016/j.ijpara.2007.11.006. PMID: 18160069

Gauci CG, Jayashi CM, Gonzalez AE, Lakenby J et al (2012) Protection of pigs against *Taenia solium* cysticercosis by immunization with novel recombinant antigens. Vaccine 30(26):3824–3828. https://doi.org/10.1016/j.vaccine.2012.04.019. PMID: 22521850

Guo YJ, Sun SH, Zhang Y, Chen ZH, Wang KY, Huang L, Zhang S, Zhang HY, Wang QM, Wu D, Zhu WJ (2004) Protection of pigs against *Taenia solium* cysticercosis using recombinant antigen or in combination with DNA vaccine. Vaccine 22(29–30):3841–3847. https://doi.org/10.1016/j.vaccine.2004.07.012. PMID: 15364430

Hubert K, Andriantsimahavandy A, Michault A, Frosch M, Muhlschlegel FA (1999) Serological diagnosis of human cysticercosis by use of recombinant antigens from *Taenia solium* cysticerci. Clin Diagn Lab Immunol 6(4):479–482. PMID: 10391846

Kar PK, Murmu S, Saha S, Tandon V et al (2014) Anthelmintic efficacy of gold nanoparticle derived from phytopathogenic fungus, *Nigrospora oriza*. PLoS One 9(1):e84693. https://doi.org/10.1371/journal.pone.0084693

Labuschange M, Beugnet F, Rehbein S, Guillot J et al (2018) Analysis of *Dipylidium caninum* tapeworms from dogs and cats, or their respective fleas-Part 1. Molecular characterization of *Dipylidium caninum*: genetic analysis supporting two distinct species adapted to dogs and cats. Parasite 25:30. https://doi.org/10.1051/parasite/2018028. PMID: 29806592

Lightbody KL, David PJ, Austin CJ (2016) Validation of a novel saliva- based ELISA test for diagnosing tapeworm burden in horses. Vet Clin Pathol 45(2):335–346. https://doi.org/10.1111/vcp.12364. PMID: 27218436

Ndom M, Diop G, Quilichini Y, Yanagida T et al (2016) Prevalence and scanning electron microscopic identification of *Anocephalid cestodes* among small ruminants in Senegal. J Parasitol Res 2016:3937292. https://doi.org/10.1155/2016/3917292

Norouzi RR, Ateei A, Hazazy M, Noreddin A, Elzowalaty ME (2020) Scolicidal effects of nanoparticles against hydatid cyst protoscolices in vitro . Int J Nanomedicines 15:1095–1100

Poursif MM, Moghaddam G, Nematollahi A, Khordadmehr M et al (2021) Vaccination with rEG vaccine elicits immunoprotection against different stages of Echinococcus granulosus life cycle: A pilot study. Acta Trop 218:105883. https://doi.org/10.1016/Jactatropica.2021.105883. PMID: 33676937

Smyth JD, Davies Z (1974) In vitro culture of the strobilar stage of Echinococcus granulosus (sheep strain): a review of basic problems and results. Int J Parasitol 4(6):631–644. https://doi.org/10.1016/0020-7519(74)90028-9.PMID:4609935

Symeonidou I, Arsenopoulos K, Tzilves D, Soba B et al (2018) Human taeniasis/cysticercosis: a potentially emerging parasitic disease in Europe. Ann Gastroenterol 31(4):406–412. https://doi.org/10.20524/aog.2018.0260. PMID: 29991885

Vogel M, Gotstein B, Muller N, Seebeck T (1988) Production of a recombinant antigen of *Echinococcus multilocularis* with high immunodiagnostic sensitivity and specificity. Mol Biochem Parasitol 31(2):117–125. https://doi.org/10.1016/0166-6851(88)90162-4. PMID: 3185617

Wang H, Li Z, Gao F, Zhao J et al (2016) Immunoprotection of recombinant Eg.p29 against *Echinococcus granulosus* in sheep. Vet Res Commun 40(2):73–79. https://doi.org/10.1007/s11259-016-9656-7. PMID: 27094043

Zhang S (2019) Comparative transcriptomic analysis of the larval and adult stages of Taenia pisiformis. Genes 10(7):507. https://doi.org/10.3390/genes/0070507

Zhang W, Che F, Tian S, Shu J, Zhang X (2015) Molecular identification of *Diphyllobothrium nihonkaiense* from 3 human cases in Heilongjiang province with a brief literature review in China. Korean J Parasitol 53(6):683–688. https://doi.org/10.3347/kjp.2015.53.6.683. PMID: 26797434

Zhou BY, Sun JC, Li X, Zhang Y et al (2018) Analysis of immune response in mice orally immunized with recombinant pMG36e-SP-TSOL 18/*Lactococcus lactis* and pMG36e-TSOL18/*Lactococcus lactis* vaccines of *Taenia solium*. J Immunol Res 2018:9262631. https://doi.org/10.1155/2018/9262631. PMID: 30581878

Zhu M, Gao F, Li Z, Wang X et al (2016) Immunoprotection of recombinant Eg.myophilin against *Echinococcus granulosus* infection in sheep. Exp Ther Med 12(3):1585–1590. https://doi.org/10.3892/etm.2016.3501. PMID: 27602078

6 Veterinary Nematology

Nematodes are free-living or parasitic animals having a cylindrical body. They are not segmented as cestodes. They are not hermaphrodite. They occur in different organic systems of a wide variety of animals. Both direct and indirect life cycle are found. In this chapter, description of discovery, distribution, hosts, site, morphology, biology, pathogenesis, diagnosis, treatment, and control have been provided.

General Description of Nematode

Size

Nematodes are widely variable in size. The length of nematodes may vary from few centimeters to more than 1 m. The female *Dracunculus medinensis* is approximately 1 m long. Nematode of whale is 30 ft long.

Shape

Nematodes are cylindrical in shape. Both ends are tapering. The body is bilaterally symmetrical except the head. Radial symmetry is found in the head region. There have exceptions of shape of roundworms. *Tetrameres* females appear globular in shape after copulation.

Body Covering

The body covering is the barrier between the internal structure of the nematode and environment. The cuticle is tough but flexible which gives the normal shape of a nematode. The cuticle is acellular structure and is formed by extracellular matrix of collagen. The collagens are typically cross-linked. There are different genes which express collagen proteins. The body covering has three layers. These are the cuticle, hypodermis, and muscle layer. The hypodermis secretes the cuticle. The cuticle not only forms the outer covering but also forms different specific structures, i.e., epaulette, teeth, cordons, cervical papillae, cervical alae, caudal alae, precloacal papillae, postcloacal papillae, etc. The cuticle is syncytial. The followings are formed by cuticle:

Epaulette: These are typical prominences found around the buccal rings of some nematodes. These are found in the members of Setariidae family.

Cordons: These are specific ornamental structures found in some nematodes. Cordons may be recurrent or nonrecurrent. These may anastomose each other or may proceed separately and do not anastomose.

Buccal capsule: In some nematodes, buccal capsule is found. The depth, shape, and thickness of wall of buccal capsule are important for taxonomy.

S. C. Mandal, *Textbook of Veterinary Parasitology*, https://doi.org/10.1007/978-981-99-2924-5_6

Lips: Some nematodes are devoid of lips and some are equipped with lips. If lips are present, the number may be two, three, six, etc. In some nematodes, two lips are present each being trilobed. In some nematodes, small lip is present in between the lips called as interlabia.

Teeth: Teeth may be present or absent in nematodes. Teeth are important morphological feature used in taxonomy. The location of the teeth is variable. These may be subventral, ventral, centrolateral, dorsal, etc. In some nematodes, the teeth are simply located at the base of the buccal capsule. The teeth may be bifid, cusped, triangular, ear-shaped, etc.

Lancet: These are teeth-like structures found in some nematodes.

Alae: Cuticular expansion of nematode is called as alae. The name of alae is as per the location. These are cervical alae (head region), caudal alae (tail region), lateral alae (lateral region), etc. These morphological structures have also taxonomic importance.

Papillae: Papillae are also the cuticular structures having taxonomic importance. In some nematodes, a pair of cervical papillae are present. In some nematodes, papillae are present anterior or posterior to cloaca called precloacal and postcloacal papillae. The shape of papillae is also variable. This may be round, ovoid, conical, etc.

Collar: In some nematodes, mouth collar is present. The shape of it is like collar. This has taxonomic importance for identification of some nematodes.

Scales, spines: These are also cuticular structures found in some nematodes. These structures also have taxonomic importance.

Bursa: One of the most important morphological features of male nematodes is the presence of bursa. Basing on this morphological character, nematodes are broadly grouped as bursate nematode (nematodes with bursa) and nonbursate nematode (nematodes without bursa).

Digestive System

In nematodes, the digestive system is tubular. In the anterior end, the mouth or buccal capsule is present. The mouth is a simple opening without any lips; or the mouth may be surrounded by two lips or three lips. The next part of the digestive system is the esophagus. In some nematodes, the mouth opens into the buccal capsule. There may be teeth or lancets in the buccal capsule. In some nematodes, the buccal capsule bears leaflike structures called the corona radiata. Next to the mouth is the esophagus. The esophagi are of different types:

Rhabditiform: Anterior club-shaped part is connected with pear-shaped posterior bulb by narrow neck.

Bulb shaped: The esophagus has posterior bulb (either single bulb or double bulb).

Ventriculus or muscular-glandular: This type of esophagus has anterior muscular part and posterior glandular part.

Club-shaped or strongyliform (strongylid): This type of esophagus has posterior club-shaped end.

Filriform: This type of esophagus is tubular or cylindrical.

Trichuroid or stichosome: A row of secretary cells or glands are present outside the esophagus.

Nervous System

The nervous system includes the ganglia, nerve fibers, and sensory papillae. The ganglia connected with fiber form a ring on the esophagus. Different sensory papillae are found in nematodes.

General Pattern of Life Cycle of Nematode

- The eggs are voided with the feces. The eggs develop in the environment. Infective larval stage is formed in the egg. The hosts get the infection by ingestion of eggs containing the infective larvae. Development up to adult occurs in the hosts.
- The eggs are voided with the feces. The eggs hatch in the environment. Infective larval stage is formed in the environment. The hosts get the infection by ingestion of infective larvae. Development up to adult occurs in the hosts.
- The eggs are voided with the feces. The eggs are ingested by intermediate hosts. The eggs hatch in the intermediate host. The infective

larvae are formed in the intermediate host. The final hosts get the infection by ingestion of intermediate hosts. Development up to adult occurs in the final hosts.

- The eggs are voided with the feces. The eggs are hatched in the environment. The larvae are ingested by the intermediate hosts and develop into infective stage. The final hosts get the infection by ingestion of intermediate hosts. Development up to adult occurs in the final hosts.
- The adult parasite lives in the lymphatic or circulatory system. The worms are larviparous. The larvae become available in the blood. The mosquitoes or bloodsucking flies take blood meal and get the infection of larvae. The larvae develop up to infective stage in these intermediate hosts. The healthy hosts get the infection when the infected mosquitoes take blood meal from the hosts.

Classification of Nematodes of Veterinary Importance

Phylum: Nemathelminthes
Class: Nematoda

Order: Ascaridida
Superfamily: Ascaridoidea
Family: Ascarididae
Family: Anisakidae
Superfamily: Oxyuroidea
Family: Oxyuridae
Family: Kathlanidae

Superfamily: Subuluroidea
Family: Heterakidae
Family: Subuluridae
Order: Rhabditida
Superfamily: Rhabditoidea
Family: Rhabditidae
Family: Strongyloididae
Order: Strongylida

Superfamily: Strongyloidea
Family: Strongylidae
Family: Trichonematidae
Family: Amidostomidae
Family: Stephanuridae
Family: Syngamidae

Superfamily: Ancylostomatoidea
Family: Ancylostomatidae

Superfamily: Trichostrongyloidea
Family: Trichostrongylidae
Family: Dictyocaulidae
Family: Ollulanidae

Superfamily: Metastrongyloidea
Family: Metastrongylidae
Family: Protostrongylidae
Family: Filaroididae
Family: Skrjabingylidae.
Family: Crenosomatidae

Order: **Spirurida**
Superfamily: Spiruroidea
Family: Spiruridae
Family: Thelaziidae
Family: Acuariidae
Family: Tetrameridae

Superfamily: Physalopteroidea
Family: Physalopteridae
Family: Gnathostomatidae

Superfamily: Filaroidea
Family: Setariidae
Family: Filariidae
Family: Onchocercidae

Superfamily: Dracunculoidea
Family: Dracunculidae

Order: Enoplida
Superfamily: Trichuroidea
Family: Trichinellidae
Family: Trichuriidae
Family: Capillaridae

Superfamily: Dioctophymatoidea
Family: Dioctophymatidae
Family: Soboliphymidae
Superfamily: Mermithoidea
Family: Mermithidae
Family: Tetradonematidae

Order: Ascaridida—Superfamily: Ascaridoidea—Family: Ascarididae

Family Synopsis

The members of this family occur in alimentary canal of mammals. They are worldwide in distribution. These are common large roundworms. They have three distinct lips. One lip is dorsal in position and another two are subventral in position. At the base region, there may be interlabia in between the lips. Both the pharynx and buccal capsule are absent. The esophagus is usually without bulb and is club-shaped in appearance. The male tail is without any bursa. Well developed caudal alae is also absent in the male tail. Life cycle is direct without involving any intermediate hosts. They cause variable pathogenesis. Diagnosis is mostly based on examination of feces for detection of typical ascarid eggs. Control is done by chemotherapy, manure disposal, etc.

History

The roundworm, *Ascaris lumbricoides*, is known to the people since antiquity. Linnaeus named the worm *Ascaris lumbricoides*. Henry Ranson detected eggs of *Ascaris* in the year 1855.

Genus: *Ascaris*

Species: *A. suum*

Introduction

These are important worms occurring in the small intestine of pigs. Sheep, cattle, man are rare hosts. In pigs, the immature worms cause damage to the liver resulting in massive fibrosis called milk spot. They also cause pneumonia.

Common Name

These are large intestinal roundworms of swine.

Distribution

The worms are cosmopolitan in distribution.

Morphology

Gross

Shape: The worms are cylindrical.

Size: The male worm is 15–25 cm long and the female may reach up to 41 cm.

Color: The worms are white in color.

Microscopic

Adult

They have three lips. The esophageal bulb is absent. Pair of spicules are stout and about 2 mm long. Large number of precloacal papillae are present.

Egg

The eggs are oval. Eggshell is thick and mammillated. The size of the eggs is 40–70 × 35–50 um. The eggs are thick and multilayered. The eggs are brownish yellow in color. The thick albuminous layer shows projection.

Note: There is much controversy whether *A. suum* and *A. lumbricoides* are same or different species. The origin of *Ascaris* spp. is not well clarified. Leles et al. (2012) concluded that only a single interbreeding population of *Ascaris* exists, and the genotypic and phenotypic differences are only manifested at population level, and they recommended synonymizing these two species, with the name *Ascaris lumbricoides* Linnaeus 1758 taking priority over *Ascaris suum* Goeze 1782.

Life Cycle

Some Basic Information of Life Cycle

Type of parasite	Homoexenous parasite
Type of life cycle	Direct life cycle
Type of definitive host	Vertebrate host
Location of fertilization	In the parasite
Location of embryonation	In the environment
Name of infective stage	Eggs containing second larval stage
Mode of infection	Ingestion
Alternative mode of infection (hyperinfection, autoinfection, etc.)	No

Stages of Life Cycle Formed

In host	In environment
• Third stage larva • Fourth stage larva • Adult • Egg	• Egg containing first stage larva • Egg containing second stage larva

Events of life cycle: a glimpse
• Voiding of eggs with feces
• Development up to L2 within the eggs in the environment
• Infection of hosts by ingestion of infective eggs
• Hatching of eggs in the gut
• Penetration of gut wall by L2
• Transportation of L2 to the liver occurs by hepatoportal circulation. Then they are transported to the heart. From the heart, they are further transported to the lungs
• Formation of L3 in the lungs after one molting
• Ascending migration of larvae to the trachea by the larvae
• L3 coughed up
• Development up to adult in the intestine

Detailed Description

Laying Eggs

Adult male and female remain in their predilection site. The eggs are passed with the feces. Many number of eggs are laid by the female parasites which may range from 100,000 to 200,000 per day. Both fertilized and unfertilized eggs may be laid by the female, but understandably unfertilized eggs cannot set up any infection in the hosts. The eggs are too resistant and remain viable even in the extreme environmental temperature. They are viable in the environment for about 4 h.

Embryonation

Embryonation of eggs occurs in the environment.

Larval Development in Environment

The eggs pass out with the feces. Larvae develop within the eggs. First larval stage transforms to second larval stage. The eggs containing second larval stage are infective stages of the parasite. Usually 4 weeks are required for development of infective eggs.

Infection of the Host

The animals get the infection by ingestion of infective eggs.

Development in the Host

The larvated eggs hatch in the gut. The freed larvae migrate through the wall of the intestine and gain entry into the circulation. Through the hepatoportal circulation, the larvae reach to the liver. From the liver, the larvae are transported to the heart. Through pulmonary circulation, the larvae reach the lungs. In the lungs, one molting occurs here and third stage larvae are formed. These larvae break out of the alveolar capillary. They gradually pass through the alveolar duct to the bronchiole and ascend to the upper respiratory tract and eventually reach the trachea and pharynx. The larvae are then coughed up and reach the intestine where they got sexual maturity. The prepatent period is 7–9 weeks.

> **Note**
>
> In different experiments, it has been proven that larvae of *T. canis* can migrate in different tissues of invertebrates, i.e., arthropods, snails, etc.

Pathogenesis

The migrating larvae cause hemorrhagic lesions in the liver. In the chronic stage, deposition of fibrous tissue occurs in the liver resulting in formation of "milk spot" or "white spot" condition. The larvae damage the lung parenchymatous tissue. The alveolar epithelium gets desquamated which results in emphysema and pneumonia (*Ascaris* pneumonia). Massive cellular infiltration occurs at the affected area. The adults cause enteritis. In rare case, the worms may reach the bile duct causing obstructive jaundice. Heavy infections may occlude the lumen of the intestine resulting in functional disturbance of the intestine.

Pathology

Anatomical Pathology

The anatomical pathology includes hemorrhagic lesions in intestinal wall, deposition of fibrous tissue in the liver, and damage of the lung.

Hematopathology

Hematological study displays low Hb%, RBC, and PCV. High eosinophilia is observed.

Chemical Pathology

Low serum iron level has been reported.

Immunopathology

Eosinophilia is common incidence of parasitic infection. Two cytokines, i.e., IL3 and IL5, play a role for production and activation of eosinophils.

Treatment

Name of drug	Mode of administration	Dose
Piperazine compounds	Oral	200–300 mg/kg bw
Fenbendazole	Oral	5 mg/kg bw
Cambendazole	Oral	20 mg/kg bw

Control

All sows should be screened and treated accordingly. They should be thoroughly washed and scrabbed to remove any adhered eggs. The feces should be disposed properly. Infected pen should be sterilized by use of hot caustic soda.

Genus: *Toxocara*

Species: *Toxocara canis*

Introduction

These are important worms occurring in the small intestine of dog, cat, cattle, buffalo, etc.

Hosts: dog, fox

Site: small intestine

Distribution: The worms are worldwide in distribution.

Etymology

The generic name *Toxocara* has been derived from the Greek word "*toxon*" meaning "bow or quiver" and the Latin word "*caro*" meaning "flesh".

Morphology

Adult

Gross

Shape: The worms are cylindrical in shape.

Size: The male and female worms are up to 12 cm and up to 18 cm, respectively.

Color: Color is white or creamy white.

Microscopic

The head has strong cervical alae which give the appearance of "arrow." That's why they are called as arrow headed worm. The mouth is equipped with three lips. Buccal capsule is absent. Interlabia are absent. The esophagus does not have any esophageal bulb. The male tail has caudal alae. A pair of spicules are present which are alate and subequal. They are non-bursate worms.

Egg

The eggs are subglobular in shape and dark brown in color. The size is 90 um × 75 um. The egg shells are pitted or mamillated.

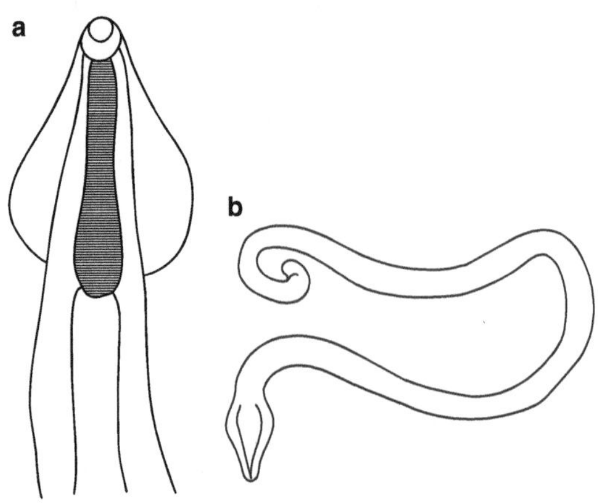

Fig. 6.1 (**a**) Head of *Toxocara canis*. (**b**) Whole worm (*Toxocara canis*)

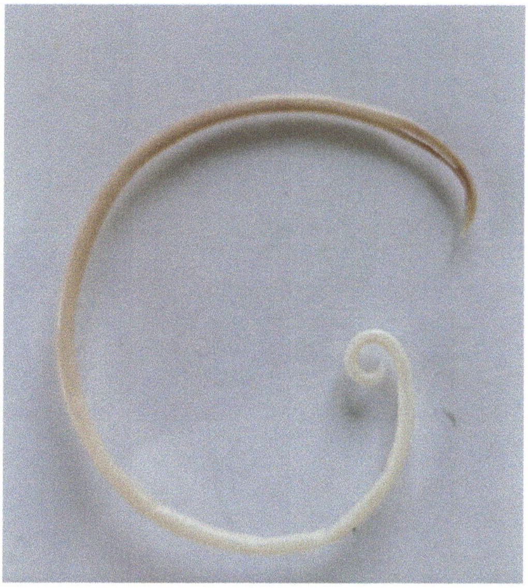

Fig. 6.2 *Toxocara canis (Gross)*

Note

The rodents act as the paratenic hosts of *Toxocara* sp. But chicken may also act as paratenic hosts. Chicken gets the infection by eating the infected arthropods. The arthropods come in contact with embryonated eggs. Human beings get the infection by ingestion of raw or undercooked chicken. There have reports on recovery of *Toxocara* larvae from earthworms.

Life Cycle

Some Basic Information of Life Cycle

Type of parasite	Homoexenous parasite
Type of life cycle	Direct life cycle
Type of definitive host	Vertebrate host
Location of fertilization	In the parasite
Location of embryonation	In the environment
Infective stage	Second larva in eggs
Mode of infection	Ingestion
Alternative mode of infection (hyperinfection, autoinfection, etc.)	No

Stages of Life Cycle Formed

In the host	In the environment
• Third stage larva • Fourth stage larva • Adult • Egg	• Egg containing first stage larva • Egg containing second stage larva (infective stage)

Events of life cycle: a glimpse
• Voiding of eggs with feces
• Development up to L2 in the eggs in the environment
• Infection of hosts by ingestion of infective eggs
• Hatching of eggs in the gut
• Penetration of the gut wall by L2
• Transportation of L2 to the liver occurs by hepatoportal circulation. Then the larvae are transported to the heart. From the heart, the larvae are further transported to the lungs
• Formation of L3 in the lungs after one molting
• Ascending migration of larvae to the trachea occurs
• L3 are coughed up
• Development up to adult occurs in the intestine
• Somatic migration of larvae occurs if the animal is adult one. The placenta may be infected by the larvae and prenatal infection occurs in this way. Newborn animals may get infection through milk or colostrum which is called transcolostral infection. Rodents may act as transport hosts

Detailed Description of Life Cycle

If the dogs are less than 3 months of age, tracheal migration occurs. The eggs are passed with the feces. Larvae develop within the eggs. First larval stage transforms into second larval stage. The eggs containing second larval stage is infective stage of the parasite. Hatching of eggs occurs in the gut. The larvae penetrate through the wall of the gut and gain entry into the circulation. Through the hepatoportal circulation, the larvae reach the liver. Then the larvae are transported to the heart. From the heart, the larvae are transported to the lungs. In the lungs, one molting occurs and third larvae are formed here. These larvae gradually ascend to the upper respiratory tract and eventually reach the trachea. The larvae are then coughed up completing "tracheal route of migration" and reach the intestine and achieve sexual maturity after two more moltings.

If the dog is adult, somatic migration of larvae occurs after hatching of eggs. The hatched-out larvae migrate in different somatic tissues (liver,

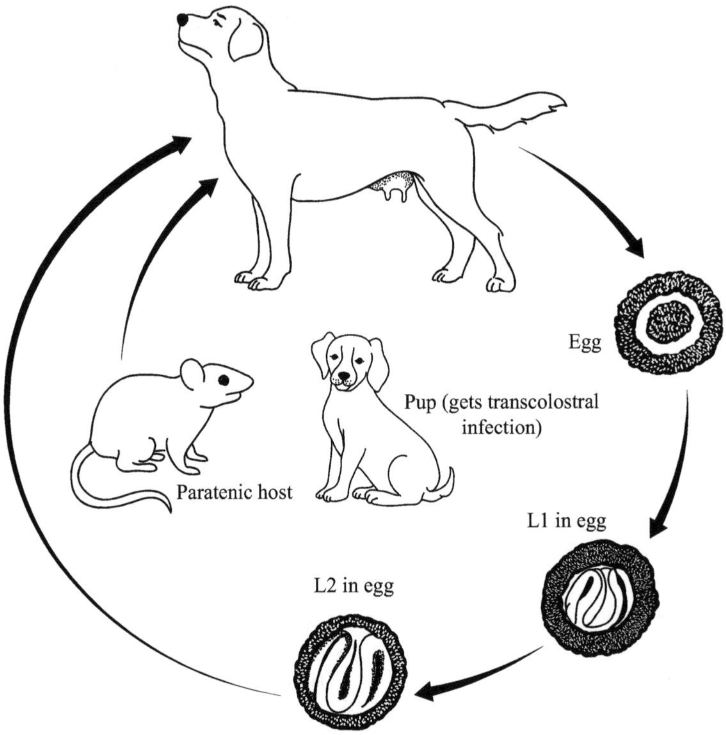

Fig. 6.3 Life cycle of *Toxocara canis*

lungs, kidneys) and remain in dormant status. In case, the bitch is pregnant, after 42 days of pregnancy, the encapsulated larvae are mobilized and become active. The larvae may reach the placenta, and the larvae enter the fetus which is called as transuterine infection or congenital infection. As a result of transuterine infection, the larvae reach the liver of the fetus. In the liver, the second stage larvae transform into third stage after second molting. After parturition, the larvae reach the lungs, and the third larvae molt to become fourth stage larvae after one molting in the lungs. From the lungs, the larvae ascend to the upper part of the respiratory system and later become coughed up and reach to the stomach. They moult to L5 stage by the end of 2nd week of age.

After birth, the young pups may get the infection through colostrum during first 3 weeks of lactation. This type of infection is called **transcolostral** or **lactogenic** infection.

The hosts may get the infection by ingestion of the paratenic hosts. Rodents act as the paratenic hosts. In the **paratenic hosts**, the larvae remain encapsulated.

Pathogenesis

By Larvae

The migrating larvae cause hemorrhagic lesions in the liver. In the chronic stage, deposition of fibrous tissue occurs in the liver. The larvae damage the lung parenchymatous tissue. Alveolar epithelium gets desquamated which results in occurrence of pneumonia. Massive cellular infiltration occurs. Visceral larva migrans is another pathological feature which is found in human beings (children) due to migration of larvae of *Toxocara canis* in different organs. However, larvae of *Toxocara cati* and *Capillaria hepatica* are also responsible for VLM (visceral larva migrans). This is characterized by chronic granulomatous lesion with massive infiltration of eosinophils. In addition to liver, lungs, brain, the eyes may also be affected resulting in ocular larva migrans (OLM).

By Adult

The adult parasites cause enteritis. A large number of worms sometimes cause occlusion of the lumen of the intestine. The worms may perforate the intestinal wall and escape the intestine and enter the peritoneum resulting in peritonitis. They may reach the bile duct causing obstructive jaundice.

Pathology

Anatomical Pathology (Gross)

The anatomical pathology includes hemorrhagic lesions in the liver, deposition of fibrous tissue in the liver, damage of lung parenchymatous tissue, desquamation of alveolar epithelium, and pneumonia.

Hematopathology

Hematopathological study depicts low TEC, Hb%, and PCV. Leukocytosis is detected. Increased level of neutrophil and eosinophil is found.

Histopathology

Histopathology of the liver displays destruction of parenchymatous tissues, hemorrhage, and infiltration of inflammatory cells comprising neutrophils, eosinophils, etc. Fibrosis is predominant in chronic phase of the stage.

Chemical Pathology

Hypoproteinemia and hypoalbuminemia are observed.

Symptoms

In less infection, no apparent clinical signs are found. The affected animals suffer from diarrhea, vomiting, malaise, etc. The affected dogs exhibit different respiratory symptoms. The clinical signs include diarrhea, potbellied condition, anemia, tucked up abdomen, etc. Sometimes, the worms may migrate to the stomach, and the intact worm gets vomited out of the host. Heavy infection causes extensive damage to the lungs and exhibits different clinical signs comprising coughing, high respiratory rate, nasal discharge, etc.

Diagnosis

Symptomatological Diagnosis

Clinical signs of diagnostic value are not found. However, few symptoms of newly born pups provide clues for diagnosis of the case. The affected animals suffer from respiratory disorder, persistent vomiting, pendulous abdomen, soiling of the whole body with vomitus, etc. which are suggestive of toxocarosis.

Parasitological Diagnosis

Examination of fecal sample may reveal the presence of eggs of the parasite. If the eggs are not found by direct smear technique, concentration method comprising centrifugal floatation should be done. The eggs are typical. These are round, suboval with pitted wall. In case of *Toxocara* infection, eggs are easily detected by direct smear technique because a large number of eggs are produced by the parasites.

Immunological Diagnosis

Many works have been carried out using crude and purified antigens. All sorts of immunodiagnosis for toxocarosis are in progress. The antigens which are used for this purpose are whole somatic antigen of adult parasites, eggs, etc. Excretory-secretory protein of *Toxocara* is more satisfactory for immunodiagnosis. Studies indicate use of defined antigen with considerable success. These are TcES-57 (57 kDa protein of ES product of *T. canis*), rTES 26 recombinant *T. canis*, 26 kDa excretory-secretory protein, and rTES 30 and rTES 120 of *T. canis*.

Molecular Diagnosis

It has been a trend to diagnose parasitic infection using specific molecular markers. Different ribosomal DNA, ribosomal RNA, and mitochondrial DNA could be used as molecular markers. Studies indicate the use of ribosomal DNA

(ITS2) and mitochondrial DNA (Cox 1) for identification of this parasite.

Treatment

Name of drug	Mode of administration	Dose
Piperazine adipate	Oral	100 mg/kg is effective for pups to kill the adults; 200 mg/kg is effective for immature
Fenbendazole	Oral	100 mg/kg as single dose for dogs and cats
Mebendazole	Oral	10 mg/kg twice daily for two consecutive days
Pyrantel pamoate	Oral	5 mg/kg

Control

General Management

The affected animals should be treated properly. Since the feces is the source of eggs of the parasite, the feces should be disposed properly. The floor of animal house should be cleaned. The pups should be dewormed within 2 weeks of infection. The pregnant animals should be treated with safe anthelmintic.

Immunological Control

Different crude and defined antigens have been used in different trials, but no vaccine has yet been commercialized. The antigens which are used for this purpose are whole somatic antigen of adult extract and egg extract.

Molecular Control

No naked DNA vaccine or gene vaccine is available.

Biological Control

No biological control has been introduced for field application. Some mites are predators of nematode eggs.

Treatment

Piperazine -50–140 mg /kg bw (In dog)

Fenbendazole - 50 mg /kg (In dogs and cats for 3 consecutive days)

Ivermectin - 0.3 mg /kg bw s/c on 0.30 and 60 th day of gestation in bitches (prevention of prenatal infection)

Species: *Toxocara cati*

Hosts: Cat

Site: These worms are found in the small intestine of cat.

Common name: They are also called as "arrow headed worm."

Distribution: The worms are worldwide in distribution.

Morphology

The length may reach up to 3–6 cm (male) and 4–10 cm (female). The cervical alae of *Toxocara cati* are short but wide and striated. The cervical alae is much broad posteriorly and the posterior margin is almost right angle to the body. The eggs are subglobular and pitted. The size is 65–75 um.

Life Cycle

The life cycle is direct type. The eggs containing second stage larvae are the infective stage of the parasite. The animals get the infection by ingestion of infective eggs. In the gut, the eggs hatch and the second stage larvae are liberated. The larvae penetrate the gut and reach the lung via the liver. They ascend to the upper respiratory tract. They are coughed up and enter the gut and develop to adult stage after 3 moultings. Rodents may serve as paratenic host as for *T. canis*. Prenatal infection has not been reported. But transmammary infection is found. No migration of larvae occurs neither in case of transmammary infection or in case infection through ingestion of parataenic host. The prepatent period is 8 weeks.

Clinical Signs

The clinical signs include inappetence, vomiting, intermittent diarrhea, and pendulous abdomen.

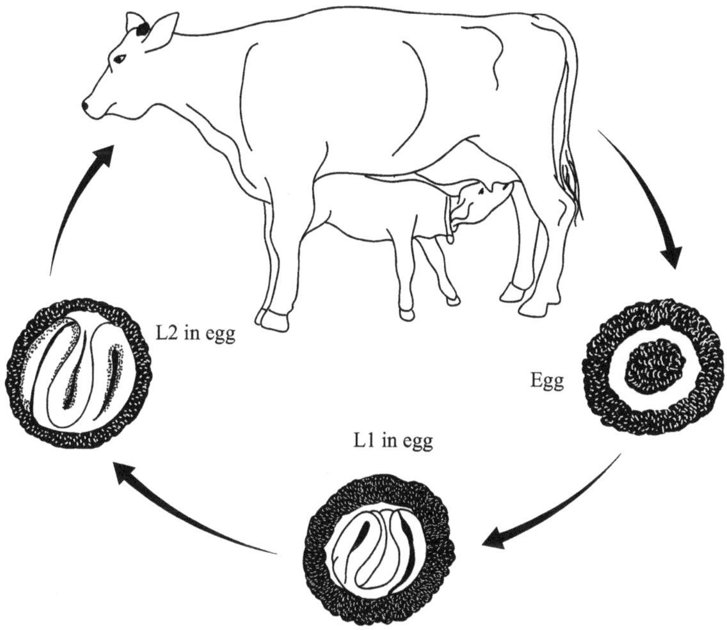

Fig. 6.4 Life cycle of *Toxocara vitulorum*

Species: *Toxocara vitulorum*

Hosts

Common host: cattle, buffalo
Rare hosts: sheep, goat
Site: small intestine

Morphology

The male is 25 cm and female is 30 cm

The cuticle of the worm is not thick as other species. The cuticle is translucent and soft. The ends of body are not tapering as that of *Toxocara canis*. The worms possess three lips. The esophagus has posterior glandular ventriculus. A variable number of precloacal papillae are present, and about five pairs of postcloacal papillae are observed. The male tail forms a small spikelike appendage. The eggs are subglobular with pitted shell. The egg size is 75–95 × 60–74 um.

Life Cycle

The life cycle is direct type. Prenatal infection is doubtful or rare. Transmammary infection is common. Ingestion of larvated eggs by the calves more than 6 months' old results in less patency. The larvae may migrate in different tissues in bitch and remain dormant until the late pregnancy. The larvae resume their activity in late pregnancy which results in transmammary infection.

Pathogenesis

In general, the worms do not produce any marked pathogenicity. But a large number of worms may cause enteritis. The infected calf shows clinical symptoms within two weeks of birth comprising diarrhea, steatorrhaea, mud colored foul smelling feces, colic etc. The worms in large number occlude the lumen of the intestine and cause functional disturbance. The worms may even cause perforation of the intestinal wall and enter the peritoneum resulting in peritonitis.

Clinical Signs

The clinical signs include diarrhea, steatorrhea, and foul-smelling feces. The severely affected animals may die.

Diagnosis

Diagnosis can be done by finding the characteristic eggs with thick and pitted shell.

Treatment

Piperazine -200–300mg/kg bw (In cattle, buffalo)

Genus: *Parascaris*

Species: *Parascaris equorum*

Hosts: horse, donkey, ass, pony, zebra
Site: small intestine
Common name: Large intestinal round worm of equine

Morphology

Gross
Color: white
Consistency: stout and rigid
Size: The male is 15–25 cm and the female is 40 cm long.
Shape: The worms are cylindrical, typical of round worm. The ends of the body is not tapering as that of *Toxocara canis*.

Microscopic

The worms possess three large lips. The mouth opening is simple. The lips are separated by interlabia. The male tail possess small lateral alae. The spicules are stout, long and equal. The eggs are subglobular, thick, and pitted. The color is brownish.

Life Cycle

- The life cycle is direct type. The eggs are voided with the feces.
- The eggs reach to infective stage in the environment in about 10–14 days. The eggs containing second larval stage are the infective stage.
- The hosts are infected by ingestion of the infective eggs.
- Hatching of the eggs occurs in the gut and the liberated larvae penetrate the wall of the intestine.
- The larvae reach the liver by circulation.
- From the liver, they reach the lungs.
- From the lungs, the larvae ascend to the upper respiratory tract and reach the trachea via the bronchi.
- The larvae are coughed up and enter the gut.
- In the intestine, they develop to adult. The prepatent period is 10 weeks.

Pathogenesis

Experimental infection revealed respiratory symptoms associated with coughing nasal discharge due to migration through the respiratory system. In general, the worms do not produce any marked pathogenicity. But a large number of worms may cause enteritis. The migratory larvae cause damage to the respiratory system causing hemorrhage. There may be pneumonia in turn.

Clinical Signs

The clinical signs include diarrhea, lassitude, harsh coat, unthriftiness, coughing, etc. There may be fever.

Diagnosis

Diagnosis can be done by finding the characteristic eggs with thick and pitted shell. Helminthic purgation may result in adult intact worm in the feces.

Genus: *Toxascaris*

Species: *Toxascaris leonina*

The male and female are 7 cm and 10 cm long, respectively. The large sized cervical alae gives an elliptical shape of head which is an important identifying feature. Three large lips are present. The worms occur in the small intestine of dog, cat, etc. The life cycle differs from *T. canis* in that no migration of larvae occurs in different organs. The hosts get the infection by ingestion of infective eggs. The eggs are hatched in the intestine. They enter the mucosa where they grow up to fourth stage and then come to the lumen and develop to adult. The eggs of *Toxascaris leonina* may be ingested by rodents, and the hatched-out 2nd stage larvae migrate in different organs of rodents and develop to L3 stage. The dogs, cats, etc. may be infected by ingestion of the rodents carrying L3 larvae. The affected animals suffer diarrhea, pot belly condition etc.

Order: Ascaridida—Superfamily: Ascaridoidea—Family: Anisakidae

Genus: *Anisakis*

Hosts: marine animals

Human beings get infection (herring worm disease) by ingestion of fish harboring larvae of the parasite.

Genus: *Porrocaecum*

Species: *P. crassum*

Hosts: domestic and wild ducks

Site: intestine

These are reddish white colored worms.

Genus: *Contracaecum*

Species: *Contracaecum rudolphii*

Host: Pelican, heron, cormorant etc.

Site: proventriculus

Morphology

They are medium sized, white colored nematodes. Three lips are present. Three bifid interlabia are found. There have fine annulations posterior to the head. Both precloacal and postcloacal papillae are present. The male tail end is conical and sub-pointed.

Species: *Contracaecum spiculigerum*

Hosts: ducks, geese

Site: proventriculus

Morphology

Anterior caecum is present. Esophageal appendix is present.

Biology

The life cycle is indirect type. Two intermediate hosts are required for completion of life cycle. The copepods are first intermediate hosts in which second larval stages are formed. Fish act as second intermediate hosts in which third larval stages are formed. The birds acquire infection on ingestion of infected fish.

Pathogenesis

Diarrhea, anemia

Order: Ascaridida—Superfamily: Oxyuroidea—Family: Oxyuridae

Family Synopsis

The members of this family occur in the large intestine of different mammals. The worms are characterized by the presence of inconspicuous three lips, well-developed esophageal bulb, presence of cloacal papillae surrounding the cloaca, one spicule found in male, and vulva found in the anterior end. The life cycle is direct type. They cause anal pruritus. Examination of perineal region leads to the detection of eggs.

History

Schrank first described *Oxyuris equi* in the year 1788 as *Trichocephalus equi*. Rudolphi named the worm *Oxyuris curvula* in the year 1803.

Genus: *Oxyuris*

Introduction

These are medium sized worms occurring in the large intestine of horse.

Species: *Oxyuris equi*

Hosts: horse, donkey

Site: cecum, colon, rectum

Common name: Horse pin worm

Morphology

Adult

Gross

Color: The color is grayish white opaque.

Shape: The shape of the worms is whiplike.

Microscopic

Anteriorly they are thick and posteriorly thin giving a reverse appearance of *Trichuris*. That's why they are called as false *Trichuris*. The mature female has long narrow tail which is three times of the rest of the body. A single spicule is present which is pin-shaped. The male tail bears large and small papillae. They are non-bursate worms.

Egg

The eggs are elongated and flattened at one side. The eggs possess one operculum at one end. The size is 0.074–0.099 × 0.038–0.045 mm.

Life Cycle

Some Basic Information of Life Cycle

Type of parasite	Homoexenous parasite
Type of life cycle	Direct life cycle
Type of definitive host	Vertebrate host
Location of fertilization	In the parasite
Location of embryonation	In the environment
Name of infective stage	Eggs containing third larval stage
Mode of infection	Ingestion
Alternative mode of infection (hyperinfection, autoinfection, etc.)	No

Detailed Description of Life Cycle

Egg Laying

Adult male and female remain in their seat of predilection. The fertilized females migrate to the anal region of the hosts and lay eggs in clumps creating yellowish white gelatinous streaks at the perianal region. This is a typical egg laying behavior of *Oxyuris equi*. The eggs are sticky. The eggs get attached on the hairs of the animals.

Embryonation

Embryonation of eggs in the environment.

Larval Development in Environment

Larvae develop within the eggs. The egg containing third larval stage is infective stage of the parasite.

Infection of the Host

The animals get the infection by ingestion of infective eggs along with feed or drinking water. The animals may lick the perianal region and ingest the eggs.

Development in the Host

The eggs hatch in the small intestine and the larvae are released there. The larvae then move to the large intestine where they enter in the mucosal crypts.

Pathogenesis

The gravid females cause anal pruritus. To alleviate the irritation, the animals rub their tail which results in rat-tailed appearance. The adults do not attach to the intestinal wall but feed on the content of the intestine. The parasites may cause diarrhea, anemia, etc.

Clinical Signs

The adult parasites in the large intestine usually do not cause any marked symptoms. The eggs laid in the perianal region cause intense irritation. To alleviate the irritation, the animals rub their perianal region which causes broken hairs and patches. Scaling is formed in the rump and tail head.

Diagnosis

Symptomatological Diagnosis

The exhibited clinical signs are not specific for confirmatory diagnosis.

Parasitological Diagnosis

Yellow, greenish, orange patch at the anal region is indicative of the presence of eggs in these patched areas. Examination of fecal sample is of no value since we know that the gravid females lay eggs at anal region which get attached to the anal region and fall down to the ground after maturity. A "scotch tape" technique is useful to collect the attached eggs. In this technique, a parafilm tape is attached on the suspected area with sticky area of the tape down, and then it is taken out and placed under the microscope for detection of eggs.

Immunological Diagnosis

No vaccine against this parasite is available.

Molecular Diagnosis

Molecular diagnosis is possible by using specific molecular marker.

Treatment

Name of drug	Mode of administration	Dose
Mebendazole	Oral	5–10 mg/kg
Cambendazole	Oral	20 mg/kg

Genus: *Enterobius*

Introduction

These are medium sized worms occurring in the large intestine of man. These are called pinworm of man.

Species: *Enterobius vermicularis*

Hosts: man
Site: large intestine
Common name: Human pin worm or seat worm

Morphology

The worms are creamy or creamy white in color. The morphology is similar to *Oxyuris equi*. The female tail is long and pointed. They have single spicules looking like a pin. The worms are called pinworm of man.

Life Cycle

- The gravid females migrate to the perianal and perineal region where they lay eggs.
- Eggs reach to the infective stage outside the host.
- The hosts get the infection by ingestion of the infective stages.

Pathogenesis

The gravid females cause irritation at the anal region resulting in anal pruritus.

Genus: *Passalurus*

Introduction

These are medium sized worms occurring in the large intestine of rabbits.

Species: *Passalurus ambiguus*

Hosts: rabbits
Site: cecum and colon

Morphology

The worms have triangular mouth. Papillae are present in the lips. The esophageal bulb is equipped with triradiate valvular apparatus. Posterior end of male bears caudal papillae.

Life Cycle

Life cycle is direct. The eggs hatch in the intestine and larvae are released. The larvae develop into adult in the gut.

Pathogenesis

The affected animals are asymptomatic even after having a large number of parasites (1000 or more). Inflammation of mucosa and dystrophic changes are observed during necropsy. Threadlike mucosal streaks are found in the feces.

Genus: *Skrjabinema*

Species: *Skrjabinema ovis*

Hosts: sheep, goat
Site: cecum
Common name: pinworm
Distribution: worldwide

Morphology
The male worm is 3 mm and the female is 6–7 mm long. They possess three lips. Three intermediate lips are also present. The esophagus has a large esophageal bulb. The male tail bears cuticular caudal expansion supported by two pairs of processes. The male has single spicule.

Life Cycle
The adult female lays eggs at the perineal region. The hosts are infected by ingestion of infective eggs.

Pathogenesis
No significant pathogenesis is found.

Genus: *Aspiculuris*

Species: *Aspiculuris tetraptera*
Common name: pinworms of rodent
Site: large intestine
Family: Kathlaniidae
Genus: *Probstmayria*

Order: Ascaridida—Superfamily: Subuluroidea—Family: Heterakidae

Family Synopsis

The worms mostly occur in the alimentary canal of birds. They are characterized by the presence of three lips around the mouth, a small buccal cavity, and pharynx and presence of rim-like preanal sucker in male tail end. The life cycle is direct. In some cases, the earthworms act as transport hosts. Diagnosis can be done by finding the eggs in the feces.

Genus: *Ascaridia*

Introduction
They are very common worms of birds. They do not produce marked pathogenesis. But a large number of worms cause enteritis and occlusion of lumen of the intestine.

Species: *Ascaridia galli*
Site: small intestine
Hosts: poultry birds, chicken, fowl, guinea fowl, duck, goose
Distribution: The worms are prevalent in most parts of the world.

Morphology

Gross
Shape: cylindrical in shape
Size: The male is 50–76 mm long. The female is 72–116 mm long. This is by far largest nematode of poultry.
Color: They are white in color.

Microscopic

Adult
Head: The head is simple. Three large lips are present.
Esophagus: The esophagus is club-shaped. It does not have posterior bulb.
Alae, ***papillae,*** etc.: Prominent caudal alae are present in male tail. Male tail bearing 10 pairs of anal papillae are characteristic features. A precloacal sucker with rim-like structure is present.
Spicule: A pair of spicules are present which are subequal.

Egg
The size of egg is 45–55 × 75–90 um. The eggs are oval in shape. The shell is thick, albuminous, and smooth. They are resistant to desiccation.

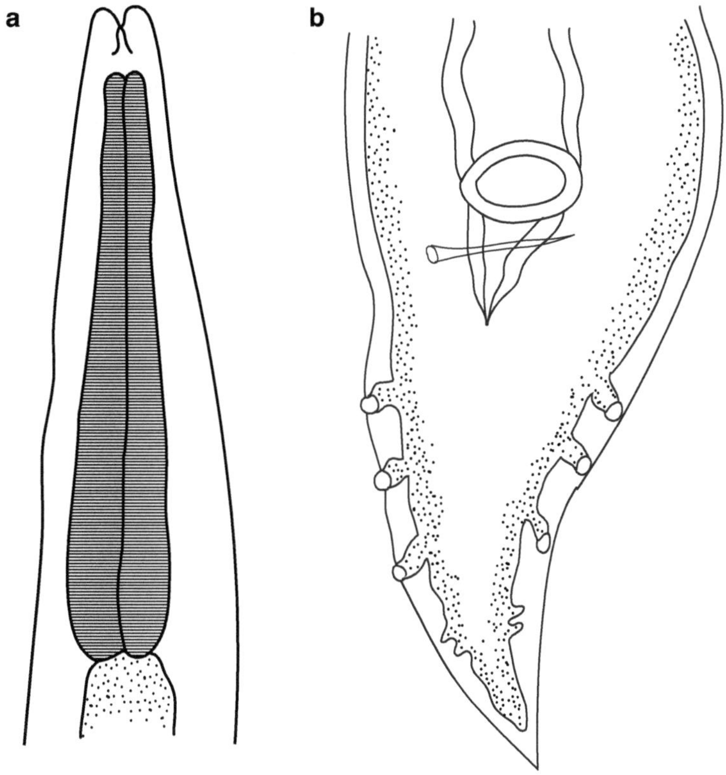

Fig. 6.5 (**a**) Head of *Ascaridia galli*. (**b**) Hind end of male *Ascaridia galli*

Life Cycle

Some Basic Information

Type of parasite	Homoexenous parasite
Type of life cycle	Direct life cycle
Type of definitive host	Vertebrate host
Location of fertilization	In the parasite
Location of embryonation	In the environment
Name of infective stage	Eggs containing second larval stage
Mode of infection	Ingestion
Alternative mode of infection (hyperinfection, autoinfection, etc.*)*	No

Stages of Development

In the host	In the environment
• Third stage larvae • Fourth stage larvae • Adult	• Eggs • First stage larvae in egg • Second stage larvae in egg • Third stage larvae

Detailed Description of Life Cycle

Egg Passing

They are homoexenous parasites. They do not require any intermediate host for completion of life cycle. The adult parasites live in their predilection site where the female parasites lay eggs. The eggs are voided with the feces.

Embryonation

Embryonation of eggs occurs in the environment.

Development up to Infective Stage

Development up to second larval stage occurs within the eggs. The egg containing second larval stage is the infective stage of the parasite. The infective stage is formed in 3 weeks or more.

Development in the Host

In the host, the eggs hatch, and the second stage larvae are released which develop up to adult in the intestine. The second molting occurs after about 7 days and third molting after about 2 weeks of infection. The prepatent period is 8 weeks.

Pathogenesis

The adult birds do not suffer. The young birds suffer somewhat. The parasites cause damage to the intestinal wall during developmental stge. The pathological features include inflammation, hemorrhage, and fibrosis on the wall of the intestine. Heavy infection causes extensive damage of the glandular epithelium. The affected intestinal wall becomes flabby as a result of loss of muscle tone. *Ascaridia galli*, if it occurs in large number, causes occlusion of the lumen of the small intestine. In this situation, the birds may die. There may be aberrant migration of this parasite in the peritoneum resulting in peritonitis.

Symptoms

The birds show intermittent diarrhea and anorexia. Loss of body weight and loss of egg production unthriftiness are some associated clinical signs.

Diagnosis

Symptomatological Diagnosis

No confirmative clinical signs are exhibited by the affected animals.

Parasitological Diagnosis

Examination of fecal sample may reveal the presence of eggs of the parasite. If the infection load is less, centrifugal floatation technique should be done. The morphological features have been described before.

Immunological Diagnosis

No immunological diagnosis has yet been commercialized.

Molecular Diagnosis

Molecular diagnosis can be possible using specific molecular markers. Different ribosomal DNA, ribosomal RNA, and mitochondrial DNA could be used as molecular markers. Studies indicate the use of mitochondrial DNA and CO1 (cytochrome oxidase 1) for molecular identification.

Treatment

Name of drugs	Mode of administration	Dose
Piperazine adipate	Oral	300–440 mg/kg in the feed

Control

General Management

The affected birds should be treated properly. The litter of the birds should be kept dried. Litter should be changed intermittently.

Immunological Control

Different crude and defined antigens have been used for trial of vaccination, but no vaccine has yet been commercialized.

Molecular Control

No naked DNA vaccine or gene vaccine is available.

Biological Control

No biological control is available in the field.

Other Species

Species: *Ascaridia numidae*

This is found in the small intestine of guinea fowl. One papilla is present on the posterior part of the preanal sucker. Ten pairs of caudal papillae are also present. The vulva is present at the midline.

Species: *Ascaridia dissimilis*

Hosts: domestic and wild turkey
Site: small intestine

Morphology

The worms are smaller than *Ascaridia galli*. The length is 3–7 cm. The shape of the worm is like a drumstick. Papillae are present at the site of cloaca. First pair of ventral caudal papillae are present opposite to the precloacal sucker. Fourth pair of ventral papillae are present posterior to cloaca.

Pathogenesis

The parasites are nonpathogenic.

Diagnosis

During necropsy, adult parasites may be identified.

Species: *Ascaridia compar*

Hosts: grouse, partridge, pheasant

Genus: *Heterakis*

These worms are usually found in birds. They are small to medium sized worms. They have three lips, a short buccal cavity, and a short esophagus. A large caudal alae are present in male. The esophagus has strong posterior bulb equipped with valvular apparatus. *Heterakis gallinarum* occurs in the cecum of turkey and causes enteritis.

Species: *Heterakis gallinarum*

Hosts: chicken, pigeon, turkey, duck, goose, fowl, guinea fowl, quail, partridge

Site

Common site: ceca
Rare site: small and large intestine

Morphology

Adult

Gross

Color: They are whitish in color.
Shape: cylindrical
Size: The male is 7–13 mm and the female is 10–15 mm long.

Microscopic

Strong esophageal bulb is present. Prominent caudal alae are present in male tail. There are 12 pairs of caudal papillae. A large sized precloacal sucker is present in the male. A pair of spicules are present which are unequal in size. The left spicule has broad alae.

Egg

The eggs are oval having smooth and thick shell. The size is 65–80 × 35–46 um.

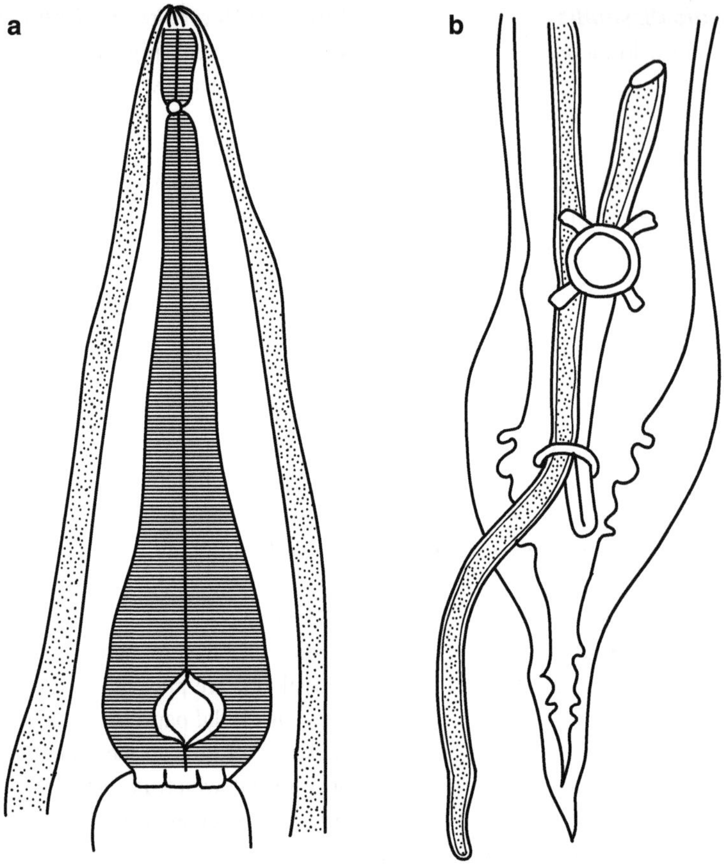

Fig. 6.6 (**a**) Head of *Heterakis gallinarum*. (**b**) Hind end of male *Heterakis gallinarum*

Life Cycle

Some Basic Information of Life Cycle

Type of parasite	Homoexenous parasite
Type of life cycle	Direct life cycle
Type of definitive host	Vertebrate host
Location of fertilization	In the parasite
Location of embryonation	In the environment
Name of infective stage	Eggs containing second larval stage
Mode of infection	Ingestion
Alternative mode of infection (hyperinfection, autoinfection, etc.)	No

Stages of Life Cycle Formed

In the host	In the environment
• Third stage larva • Fourth stage larva • Adult • Egg	• Egg containing first stage larva • Egg containing second stage larva

Detailed of Life Cycle

Egg Passing

The parasite is homoexenous, and they do not require any intermediate hosts for completion of life cycle. The adult parasites live in their predilection site. The females lay unembryonated eggs there. The eggs are voided with the feces.

Embryonation

Embryonation of eggs occurs in the environment.

Development up to Infective Stage

Development up to second larval stage occurs within the eggs. The eggs containing second larval stage is the infective stage of the parasite which is formed in about 2 weeks. Earthworm may serve as transport host in which the second stage larvae remain in the earthworm.

Infection of the Host

The hosts get the infection by ingestion of infective eggs.

Development in the Host

In the gut of the host, the eggs hatch, and the second stage larvae are liberated which develop up to adult after three moltings in the gut. The prepatent period is about 4 weeks.

Pathogenesis

Occurrence of this parasite in the birds is very common and regarded as nonpathogenic. This parasite is not pathogenic, but a large number of parasites may cause thickening of the cecal wall. They cause diarrhea. The notable importance of this roundworm is that the eggs of these worms carry one important pathogenic protozoan parasite, *Histomonas meleagridis*, causative agent black head of turkey or enterohepatitis.

Symptoms

The parasite does not produce any clinical sign in the affected birds.

Diagnosis

The diagnosis can be done by finding the eggs in the feces and identification of the adult worm during necropsy.

Control

- Maintenance of hygiene.
- Turkeys should be segregated from other poultry.

Other Species

Species: *Heterakis dispar*

Hosts: goose and duck

The worms are larger as compared to *H. gallinarum*. The male is 11–18 mm and the female is 16–23 mm. The life cycle is similar to *Ascaridia* spp. The eggs reach the infective stage in the environment in 2 weeks. The hosts get the infection by ingestion of the infective eggs. Earthworm may act as the transport hosts. The prepatent period is about 4 weeks.

Species: *Heterakis brevispiculum*

Hosts: chicken and guinea fowl
Site: caecum

Morphology

The spicules are barbed at the tip. The spicules are equal in size.

Species: *Heterakis isolonche*

Hosts: pheasant, quail
Site: cecum

Order: Ascaridida—Superfamily: Subuluroidea—Family: Subuluridae

Genus: *Subulura*

Introduction

These are cecal worms of birds.

Species: *Subulura brumpti*

These worms occur in the cecum of birds. They are cylindrical worms. The anterior and posterior ends are curved. These are called **pinworms of fowl**. The base of the buccal capsule is equipped with three teeth. The lateral alae extend up to the level of the esophagus. The oesophagus has a distinct bulb. It has a small swelling followed by a deep constriction and then a prominent bulb. The beetles act as the intermediate hosts. The worms cause pathogenesis in the cecum comprising hemorrhagic enteritis. The infected birds look dull. The birds do not feed and drink properly.

Order: Rhabditida—Superfamily: Rhabditoidea—Family: Strongyloididae

Family Synopsis

The worms of this family occur in the alimentary canal of different mammals. They are cosmopolitan in distribution. The worms are characterized by the presence of three lips in the parasitic generation and two lips in the free living generation, esophagus long and cylindrical in parasitic generation and having vulvulated bulb in free living generation. They exhibit alternation of free-living and parasitic life cycle. The worms are parthenogenetic. Parasitic males are not found. The worms cause anorexia, diarrhea, and loss of production. Diagnosis can be done by the presence of eggs or larvae in the feces.

Genus: *Strongyloides*

Introduction

The worms are found in the small intestine of different types of animals, i.e., cattle, sheep, goat, horse, etc. Both homogonic and heterogonic type of life cycles are found. The parasites are parthenogenetic.

Species	Hosts	Site
S. stercoralis	Man, different primates, dog, fox, etc.	Small intestine
S. papillosus	Sheep, goat, cattle	Small intestine
S. westeri	Horse, pig, etc.	Small intestine
S. ransomi	Pig	Small intestine
S. cati	Cat	Small intestine

Species: *Strongyloides papillosus*

Common name: The common name is intestinal threadworm of ruminants.

Distribution: The worms are distributed worldwide.

Morphology

Gross

Shape: The worms are hairlike slender. They look like thin white thread.

Size: The worms are less than 1.0 cm long

Adult

These worms are found in the small intestine particularly in the duodenum and jejunum of sheep, goat, cattle, etc. The esophagus is long enough which may be one-third of the whole length. The esophagus of the parasitic female is cylindrical without posterior bulb. Uterus and intestine give a twisted thread like appearance due to interwining of the uterus by the intestine. A pair of spicules are present which are unequal in size. The intestine occupies the posterior two-thirds of the body. The intertwining of the uterus and intestine results in twisted threadlike appearance. The anus opens midventrally. Parasitic males are broader and shorter.

Egg

The fresh eggs are thin walled having larvae within the eggs. Both ends of the eggs are blunt. In herbivorous animals, larvated eggs pass out with the feces whereas hatched out L1 in case of other animals.

Larva

The filariform larvae have long esophagus up to half of the body length. The tail of larva is notched.

Life Cycle

Some Basic Information of Life Cycle

Type of parasite	Homoexenous parasite
Type of life cycle	Direct life cycle
Type of definitive host	Vertebrate host
Location of fertilization	Fertilization occurs in the parasite in case of free living life cycle; Fertilization does not occur in parasitic life cyle.
Location of embryonation	In the parasite
Name of infective stage	Third larval stage (L3)
Mode of infection	Ingestion, skin penetration, autoinfection, hyperinfection

Description

The parthenogenetic females lay eggs. The eggs are oval and transparent containing a larva which is ready to hatch. The rhabditiform larvae immediately come out from the eggs. They undergo homogonic or heterogonic type of development. In homogonic type of development, the rhabditiform larvae transform to infective filariform larvae. These infective larvae penetrate the skin of the host and the host thus gets infection. In heterogonic type of life cycle, the rhabditiform larvae transform to free-living sexually matured male and female in the environment. Copulation occurs between the male and female parasites. The gravid females lay eggs. After hatching, rhabditiform larvae come out. The rhabditiform larvae transform to infective filariform larvae. These filariform larvae infect the hosts through skin penetration. They enter skin capillary or venule and are carried to the lungs where they break out into alveoli. They migrate from bronchiole to bronchi to trachea and then descend to oesophagus. They then reach to the intestine. Further development up to adult occurs there.

Autoinfection and Hyperinfection

However, **autoinfection** or **hyperinfection** also occurs. In the lumen of the bowel, the rhabditiform larvae may change to filariform larvae. These larvae may penetrate into the perianal skin.

Pathogenesis

By Larvae

Local dermatitis and papule formation may occur due to penetration by the infective larvae. The larvae cause erythematous lesion in the skin through which causative organisms of foot-rot enter. Experimental work revealed damage of the lungs associated with hemorrhage due to migration of larval stages in the lung.

By Adult

The adult parasites occur in the duodenum and proximal jejunum. Heavy infection results in erosion of epithelial layer, inflammation, edema, etc. This causes functional disturbances of the intestine, i.e., digestion, absorption, etc.

By Autoinfection

In case of autoinfection, the pathological features are of higher magnitude. Severe enteritis, severe pneumonitis, and severe general allergy are found in case of autoinfection.

Pathology

Anatomical Pathology

Anatomical pathology includes dermatitis, enteritis, and papule formation. The adult worms cause denudation or erosion of intestinal epithelium. The adult worms cause formation of tunnels in the intestine in which they remain resulting in villous atrophy and hypertrophy of the epithelium.

Hematopathology

Low Hb% with microcytosis, eosinophilia, etc. have been reported.

Chemical Pathology

Low iron level, normal ferritin level, hypoalbuminemia, hyponatremia, and high ESR have been reported.

Secondary Pathology

There have been reports of mixed infection of *Strongyloides* spp. and ascarid worms. Both these worms may cause pathology simultaneously.

Symptoms

The clinical signs include intermittent diarrhea, anorexia, anemia, weight loss, lassitude, etc.

Note

1. Female parasites produce eggs parthenogenetically in the host, whereas in the environment, female and male adults reproduce sexually.
2. There have been reports of autoinfection of *Strongyloides* sp. Autoinfection occurs when the dogs are infected with human strain of *S. stercoralis*. But after few generations, it reverts to normal infection.
3. In case of autoinfection, eggs hatch in the gut and L1 are liberated. L1 transform to L2 and then to L3 after two consecutive moltings. This L3 is the infective stage.

Diagnosis

Direct fecal smear examination may reveal typical eggs of *Strongyloides*. For further confirmation, fecal culture can be done. Different molecular markers, i.e., ribosomal DNA and RNA, mitochondrial DNA, etc., are used as molecular markers. Studies indicate use of cytochrome c oxidase (cox 1) as molecular marker for genomic characterization of this parasite.

Treatment

Name of drug	Mode of administration	Dose
Thiabendazole	Oral	50–75 mg/kg is highly effective for dogs; 75 mg/kg is highly effective for sheep; 50 mg/kg for pigs mixed in feed
Cambendazole	Oral	20 mg/kg mg for horses
Fenbendazole	Oral	50 mg/kg for horses
Albendazole	Oral	5–7.5 mg/kg for cattle, buffalo, sheep, goat, horse, pig

Order: Strongylida—Superfamily: Strongyloidea—Family: Strongylidae

Family Synopsis

The roundworms of this family commonly occur in the alimentary tract of equines. They are worldwide in distribution. Mouth opening is usually surrounded by leaf crown of two types, i.e., external leaf crown and internal leaf crown. They are called palisade worms as the leaf crown looks like fence. The teeth may be found in the depth of buccal capsule. Copulatory bursa is well developed having a pair of large lateral lobes and a dorsal lobe. Bursa is supported by bursal rays. Two equal spicules are found. The life cycle is direct. The worms cause marked anemia by sucking blood.

Genus: *Strongylus*

Introduction

These are large strongyles of horses, mules, donkeys, and other equids. The parasites inhabit the large intestine. The arteries are markedly affected by *Strongylus vulgaris*. *Strongylus equinus* causes formation of nodules.

Distribution

The worms are worldwide in distribution.

Species: *Strongylus equinus*

Morphology

Gross

Shape: The worms are cylindrical in shape.

Size: The male is 26–35 mm and the female is 38–47 mm long.

Color: They are red to dark red in color.

Microscopic

Adult

Head: The head is simple which is not marked off from the rest of the body.

Buccal capsule/buccal cavity: A buccal capsule is oval in outline and is well developed.

Leaf crown: Both external and internal leaf crown are present.

Teeth: A pair of subventral teeth and a large dorsal tooth with bifid tips are present at the base of buccal capsule.

Dorsal gutter: Dorsal wall of the buccal capsule possess a median thickening called dorsal gutter. A distinct dorsal gutter is present.

Spicule: A pair of spicules are present which are simple and slender.

Bursa: They are bursate worm. The male worms possess well-developed bursa.

Egg

The eggs (75–92 × 41–54 um) are oval and thin shelled. The eggs are laid in segmenting stage.

Species: *Strongylus vulgaris*

Morphology

Gross

Shape: The worms are cylindrical in shape.
Size: The male worm is 14–16 mm long. The size of female worm is 20–24 mm.

Color: The worms are red to dark red in color.

Microscopic

Head: The head is simple and not broad as *Strongylus edentatus*.

Buccal capsule/buccal cavity: A deep and large sized buccal capsule is present.

Leaf crown: Both external and internal leaf crown are present.

Teeth: Two ear-shaped dorsal teeth are present at the base of buccal capsule.

Dorsal gutter: A distinct dorsal gutter is present.

Spicule: A pair of simple spicules are present.

Bursa: They are bursate worms. The male worms possess bursa.

Eggs: The eggs are oval and thin shelled and the size is 80–90 um × 47 um.

Species: *Strongylus edentatus*

Morphology

Gross

Shape: The worms are cylindrical in shape.

Size: The size of the male worm is 23–28 mm × 2 mm. The size of female worm is 33–44 mm × 2 mm.

Color: They are red to dark red in color.

Microscopic

Head: The head is simple.
Buccal capsule/buccal cavity: A buccal capsule is present which is typically cup-shaped.
Leaf crown: Both external and internal leaf crown are present.
Teeth: Teeth are absent.
Dorsal gutter: A distinct dorsal gutter is present.
Spicule: A pair of spicules are present which are simple and slender.
Bursa: They are bursate worm. The male worm possesses bursa.

Egg

The eggs are oval in shape having 24–26 embryonic masses. Outer covering is smooth and thin. The embryonic mass occupies almost the whole egg. The size is approximately 90 um × 50 um.

Larva of *Strongylus* spp.

The tail sheath of larvae of all three species is long and filamentous. Intestinal cells are 28–32 in number and these are rectangular and well defined in case of *Strongylus vulgaris*. The intestinal cells of larvae of *Strongylus edentatus* are elongated and not well defined. The larvae of *Strongylus equinus* have long filamentous sheath, and the intestinal cells are rectangular and not well defined.

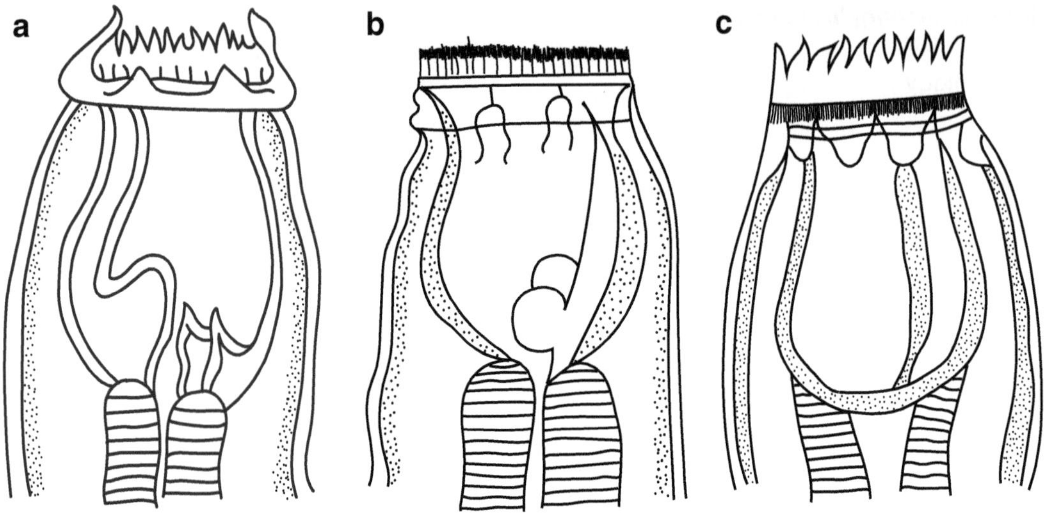

Fig. 6.7 (**a**) Head of *Strongylus equinus*. (**b**) Head of *Strongylus vulgaris*. (**c**) Head of *Strongylus edentatus*

Life Cycle

These nematodes are homoexenous, and they do not require any intermediate host for completion of life cycle. The adult parasites live in their predilection site where they lay eggs. The eggs are voided with the feces. In the presence of suitable environmental condition, the eggs are hatched. The first stage larvae come out. The first stage larvae transform to second stage larvae after one molting. The second stage larvae further transform into third stage larvae after one more molting. The third stage larvae retain the sheath of second stage larvae which provides additional protection to the infective third stage larvae. The infective larvae vertically migrate along the grass blades or other vegetations. The hosts get the infection when they inadvertently ingest the infective third stage larvae. The rest part of life cycle is as follows:

Strongylus vulgaris

- After exsheathment, the third stage larvae penetrate the wall of the intestine.
- Third molting occurs in the intestinal wall, and formation of fourth stage larvae occurs after one molting.
- Fourth stage larvae enter in small intestinal arteries.
- Those larvae enter in large arteries.
- They follow cranial mesenteric artery.
- They reach to the submucosa of the cecum and colon and 5th stage larvae are formed.
- The young adult then come to the lumen of the cecum and colon and get matured to adult stage.

Strongylus edentatus

- The exsheathed larvae enter the wall of the intestine.
- The larvae follow hepatoportal circulation and reach the liver where they molt to 4 th stage larvae.
- The larvae reach to the hepatic ligament.
- Through the hepatic ligament, the larvae reach the parietal peritoneal region of right abdominal flank. In this vicinity, they form hemorrhagic nodules.
- From the right abdominal flank, the larvae reach the cecum and colon and form haemorrhagic nodules.
- Sexual maturity occurs there.

Strongylus equinus

- The exsheathed larvae enter in the wall of the large intestine.
- Formation of nodules occurs in the large intestine.

- Molting occurs in the nodules and fourth stage larvae are formed.
- Larvae reach the liver via peritoneal cavity.
- The larvae then reach the pancreas via liver.
- From the pancreas, the larvae reach the peritoneal cavity.
- From the peritoneal cavity, the larvae reach the lumen of the cecum and colon and sexual maturity occurs there.

Pathogenesis

Strongylus equinus

By Adult

The adult parasites cause enteritis. They ingest plug of mucosa in their buccal capsule. The site of attachment may get ulcerated. The adult parasites suck blood which results in anemia.

By Larvae

The migrating larvae of *S. equinus* produce hemorrhagic tracts in liver parenchyma and also cause deposition of fibrous tissue on the hepatic capsule. Omental adhesion may occur as a result of larval migration. The larvae cause formation of nodules in the subserosa of cecum and colony.

Strongylus edentatus

By Adult

The adult parasites cause enteritis. Other pathological features caused by the adults are same as *S. equinus*.

By Larvae

Early larval stages cause hemorrhagic nodule formation in the hepatic ligament and right abdominal flank. The late larval stages cause hemorrhagic nodules in the subperitoneal tissues. It is to be noted that nodule formation at all these mentioned areas rarely causes to display any clinical signs.

Strongylus vulgaris

By Adult

This is the most pathogenic one. The adult worms cause damage to the intestinal mucosa during feeding. The worms have large and strong buccal capsule and ingest plug of mucosa moving on the surface of the intestinal wall. During this feeding activity, the blood vessels are ruptured and considerable hemorrhage occurs on the wall. The lesion caused by the worms may later turn into ulcer. However, ulcers are normally healed up leaving scar area on the intestinal wall.

By Larvae

The larvae of these parasites cause lesions in the artery. They affect the submucosal arteriole and then they reach to the cranial mesenteric artery. Tortuous tract is formed there. There occurs thrombus formation. The most serious condition is verminous aneurism in the anterior mesenteric artery as a result of loss of elasticity of the artery.

Anatomical pathology (Gross): Different structural alternations occur due to this parasitic infection.

Hematopathology: Hematopathology displays a normochromic normocytic anemia. There occurs reduction of TEC, Hb%, PCV%, etc. Eosinophil count is increased.

Histopathology: In *Strongylus vulgaris* infection, histopathology reveals marked thickening of intimal part along with infiltration of plasma cells, lymphocytes, neutrophils, mast cells, etc. Affected arterial wall displays fibrosis with disruption of intimal wall. Histopathology also shows thrombi formation.

Clinical Signs

The animals exhibit the clinical signs of diarrhea, anorexia, anemia, fever, weight loss, etc.

Diagnosis

Symptomatological Diagnosis

The clinical signs exhibited by the affected animals are vague and not so specific for confirmatory diagnosis.

Parasitological Diagnosis

Examination of fecal sample may reveal the presence of eggs of the parasite. If the infection load is less, concentration method, i.e., centrifugal floatation technique, should be done. But detection of eggs cannot confirm the parasite because egg morphology is not specific. For specific identification, coproculture can be done and identification of larvae can lead to specific identification.

Immunological Diagnosis

Works were carried out by different researchers using crude and purified antigens, i.e., whole somatic antigen, surface antigen, recombinant antigen for performing different immunodiagnostic tests and variable success has been achieved. But no immunodiagnostic test has yet been commercialized. Study shows diagnostic potential of a recombinant antigen and rsvSXP for diagnosis of *Strongylus vulgaris*.

Molecular Diagnosis

Molecular diagnosis is possible using specific molecular markers. Different ribosomal DNA, ribosomal RNA, and mitochondrial DNA are used as molecular markers for diagnosis of different parasitic identification. Studies indicate the use of mitochondrial DNA (CO1) for molecular identification of this parasite.

Treatment

Name of drug	Mode of administration	Dose
Fenbendazole	Oral	7.5 mg/kg (effective against adult); 60 mg/kg is effective against migrating larvae of *S. vulgaris*
Mebendazole	Oral	10 mg/kg (effective against adult worms)
Cambendazole	Oral	20 mg/kg (effective against adult
Ivermectin	S/C	0.2 mg/kg

Control

General Management

The affected animals should be treated properly. The feces should be disposed properly. The floor of animal house should be cleaned so that it remains free from any contamination of parasitic egg.

Immunological Control

Different crude and defined antigens have been used in different trials, but no vaccine has yet been commercialized. Study shows protection against *Strongylus vulgaris* using different types of antigens, i.e., somatic extract of adults and larvae and radiation-attenuated third larvae. But no vaccine has been commercialized. Klei et al. (1989) reported protection of ponies at the age of 8–10 months challenged with 5 × 1000 L3 ivermectin treated yearlings. The animals did not show any typical signs of acute verminous arteritis, and at the time of necropsy, 89% reduction of arterial larval burden was found.

Molecular Control

No naked DNA vaccine or gene vaccine is available.

Biological Control

No biological control is available.

Genus: *Triodontophorus*

Distribution

The worms are worldwide in distribution.

Hosts: horse and other equids

Site: large intestine

Species: *Triodontophorus serratus*

Site: large intestine

Hosts: horse and other equids

The head of the worms appear like inverted saucer. The buccal capsule is cup shaped. A buccal capsule is present.Three large esophageal teeth are present which project in to the the buccal capsule.

Species: *Triodontophorus brevicauda*

Site: large intestine

Hosts: Horse and other equids

Salient Morphology

The worms possess subglobular buccal capsule. Three oesophageal teeth are present; each has three elevations which project into the buccal cavity.

Genus: *Craterostomum*

Distribution
Craterostomum acuticaudatum is worldwide in distribution. *C. tenuicauda* is found in India.

Species: *C. acuticaudatum*
Hosts: horse and other equids

Salient Morphology
The buccal capsule looks globular but the top of the buccal capsule appears as flames of gas burner. Three small triangular oesophageal teeth are present but do not project into the buccal cavity.

Genus

Oesophagodontus

Species
Oesophagodontus robustus

Morphology
Buccal capsule is funnel-shaped. Oesophageal tunnel bears three lancet-like teeth. These teeth do not project into the buccal capsule.

	Triodontophorus	*Craterostomum*	*Oesophagodontus*
Hosts	Horse and other equids	Horse and other equids	Horse and other equids
Site	Large intestine	Large intestine	Large intestine
Buccal capsule	Thick walled subglobular buccal capsule is found	The buccal capsule is globular, but the top of the buccal capsule appears as flames of gas burner	Goblet-shaped buccal capsule is found
Teeth of buccal capsule	Three pairs of teeth are present at the base of the buccal capsule	There have no teeth in the buccal capsule	Three toothlike folds are present at the base of the buccal capsule. There is shallow constriction between the anterior end and rest of the body

Order: Strongylida—Superfamily: Strongyloidea—Family: Trichonematidae

Family Synopsis

The roundworms of this family commonly occur in the alimentary tract of a wide variety of animals. They are worldwide in distribution. Mouth opening is usually surrounded by leaf crown of two types, i.e., external leaf crown and internal leaf crown. Buccal capsule is short cylindrical or annular. Teeth are absent. Dorsal gutter is also absent. The spicules have barbed teeth. The parasites of equine hosts have been described below.

	Cyathostomum	*Cylicocyclus*	*Cylicodontophorus*	*Cylicostephanus*
Hosts	Horse and other equids	Horse and other equids	Horse and other equids	Horse and other equids
Site and morphology	Cecum and colon	Cecum and colon	Cecum and colon	Cecum and colon
	They have both external and internal leaf crowns. The elements of external leaf crown are larger and broader. The number is less than internal leaf crown elements.	Small sized rod shaped internal leaf crown elements are present. The elements of external leaf crown are larger and broader and less in number.	Long and broad internal leaf crown elements are present. The elements of internal leaf crown are less numerous than in external leaf crown.	Rod shaped thin internal leaf crown elements are present. The elements of external leaf crown are longer, broader and less numerous.

Genus: *Cyathostomum*

Distribution
The worms are worldwide in distribution.

Species: *Cyathostomum catinatum*
Hosts: horse, mule, donkey
Site: cecum and colon
Prevalence: They are worldwide in distribution. The worms are markedly prevalent in Europe.

Morphology
The mouth collar is prominently marked off from the rest of the body. The buccal capsule is wider than long.

Genus: *Cylicodontophorus*

Introduction
This is horse strongyle worm affects the large intestine. The parasites cause pathological features in the wall of the large intestine. The species are *C. bicoronatus*, *C. mettami* etc.

Distribution
The worms are worldwide in distribution.

Genus: *Cylicocyclus*

Morphology
The worms are small in size. That's why these are called small red worms. They are less than 1.5 cm. The buccal capsule is short. The spicules are barbed at the anterior tip. The oesophageal funnel of *C. elongatus* is large which may be almost same size of buccal capsule. The esophagus is markedly elongated.

Life Cycle
Type of life cycle: direct

The following events are observed in life cycle of *Cylicocyclus* worms.

- The adult parasites are found in the large intestine. Copulation between the male and female occurs there. The eggs are passed with the feces.
- Further development of eggs and larvae is dependent on environmental temperature.
- Eggs hatch and first stage larvae are liberated in the environment.
- The first stage larvae transform to infective third larvae after two moltings. The third stage larva is the infective stage of this roundworm.

Diagnosis
Examination of feces for identification of *Cylicocyclus* always remains impossible. Mitchell et al. (2016) reported immunodiagnostic potential of a recombinant protein Cy-Gala-1 for the diagnosis of *Cylicocyclus ashworthi*.

Pathogenesis
The third stage larvae penetrate the mucosa of the intestine and transform to fourth stage larvae after one molting. The parasites cause colitis. Morbidity and mortality are observed.

Genus: *Cylicostephanus*

Distribution
C. calicatus are worldwide in distribution.

Morphology
The worms are small (4–10 mm) in size. The internal leaf crown is formed by thin and rodlike elements. 8–18 triangular elements are present in external leaf crown.

Other Species
Other species are *C. barbatus, C. goldi, C. longibarbatus etc.*

Pathogenesis
The parasites cause enteritis.

Diagnosis
Examination of fecal sample can be done for detection of eggs, but confirmatory diagnosis is not possible because the eggs do not have any specific identifying features. Mitchell et al. (2016) reported immunodiagnostic potential of a recombinant protein Cy-Gala for diagnosis of *Cylicostephanus goldi* and *C. longibursatus*.

Genus: Gyalocephalus

Species: *Gyalocephalus capitatus*

Hosts: Equines. Site : Large intestine

Morphology

The head is distinctly separated off the rest of the body by a well marked neck. The buccal capsule is thick walled and short. Three teeth are projected in the buccal capsule from the large esophageal funnel which is a characteristic morphological feature.

Genus: *Chabertia*

Introduction

These are strongyle worms of sheep, goat, etc. affecting the large intestine (cecum and colon). The parasites are accidental bloodsuckers.

Distribution

The worms are worldwide in distribution. But occurrence is more in temperate countries.

Species: *Chabertia ovina*

Hosts: cattle, sheep, goat

Site: cecum and colon

Distribution: The worms are cosmopolitan in distribution.

Morphology

The worms possess one cup-shaped buccal capsule. The head is slightly bent. The buccal capsule is surrounded by a leaf crown equipped with double rows of small cuticular elements. A cephalic vesicle and a cervical groove are present. Bursa is well developed.

Life Cycle

Type of life cycle: direct

The following events are observed in life cycle of *Chabertia* worms.

- The adult parasites are found in the large intestine. Copulation between male and female parasites occurs there. The eggs are passed with the feces.
- Further development of eggs and larvae is dependent on environmental temperature.
- Eggs hatch and first stage larvae come out.
- The first stage larvae transform to infective third larvae after two moltings.
- The hosts get the infection by ingestion of infective larvae inadvertently with feed or drinking water.
- The larvae enter in the wall of small intestine and transform to fourth stage.
- The 4th larvae undergo moulting in the lumen of caecum and 5 th stage larvae are formed. They reach to the colon.

Pathogenesis

Usually less number of worms affect the intestine. But sometimes more parasites may affect the hosts. They firmly attach to the intestinal wall. The worms take out plug of mucosa into their buccal capsule. They are not intentional blood feeder. But they can ingest blood when the blood vessels are ruptured. The affected wall becomes thickened. Hemorrhage occurs at the site of attachment. Protein is lost through the bite opening.

Symptoms

The affected animals exhibit symptoms of hemorrhagic diarrhea.

Diagnosis

Examination of fecal sample reveals eggs of the parasite. In case of massive diarrhea, intact worm may be expelled with the feces. During necropsy, lesion is mainly visible. But the worms may be found attached on the wall of the intestine.

Name of drug	Mode of administration	Dose
Fenbendazole	Oral	5–7.5 mg/kg for cattle, sheep, goat
Mebendazole	Oral	15 mg/kg for cattle, sheep, goat
Cambendazole	Oral	20 mg/kg
Ivermectin	S/C	0.2 mg/kg for cattle

Genus: *Bourgelatia*

Species: *Bourgelatia diducta*

Hosts: The parasites occur in cecum and colon of pig

Morphology

The worms are roughly 1 cm long. The buccal capsule is cylindrical. They have both external and internal leaf crowns. Characteristic J shaped type II ovijector is found in female. Type II dorsal ray with two rami on each side of the median fissure is found in male. A pair of alate spicules are equal in size.

Pathogenesis

Marked enteritis is found.

Genus: *Oesophagostomum*

Species: *Oesophagostomum columbianum*

Introduction

These are very important parasites occurring in the large intestine of sheep, goat, cattle, etc. They cause formation of nodules in the large intestine. The condition is called nodular gut or knotty gut condition. The worms are important in livestock economy.

Distribution

Though *O. columbianum* is worldwide in distribution, they are commonly available in tropical and subtropical countries.

Morphology

Gross

Shape: The worms are cylindrical in shape. The anterior end is often bent to form hooklike structure looking like a walking stick.

Size: The length of the male worm is 12–16 mm and the female worm is 14–18 mm.

Color: The worms are red to dark red in color.

Microscopic

Adult

Head: Anteriorly they have mouth collar. The shape of this mouth collar is like a truncate cone. Behind the mouth collar, a cervical groove is present after certain distance. Before cervical groove, the cephalic vesicle is formed by cuticular inflation. Anterior to the cervical groove, the cephalic vesicle is present. Posterior to the cervical groove, cervical alae are present which are pierced by a pair of cervical papillae at their anterior extremity. Both external and internal leaf crown are present. External leaf crown contains 20–24 elements.

Bursa: The male has bursa having bursal rays.

Spicule: A pair of spicules are present which are equal in size and have pointed end. The spicules are alate.

Tail end: The male has bursa and the female tail is long and tapering.

Egg

The eggs are oval and thin shelled. They have 11–27 cells. The eggs have darkly granulated embryonic mass with cleave stage and are similar to eggs of other GI nematodes. The eggs cannot be differentiated from hookworms. The size of the egg is 65–88 um × 4–54 um.

Larva

The body length of L3 is 0.79 mm. The length of the esophagus is 0.16 mm. Sixteen triangular cells are present in the intestine. Tail sheath extends 0.125–0.16 mm beyond the tail end of the larvae. The third stage infective larvae have sheath forming a long whiplike appendage, while the tail of the larva is short and pointed. The third stage larvae have 20–21 gut cells.

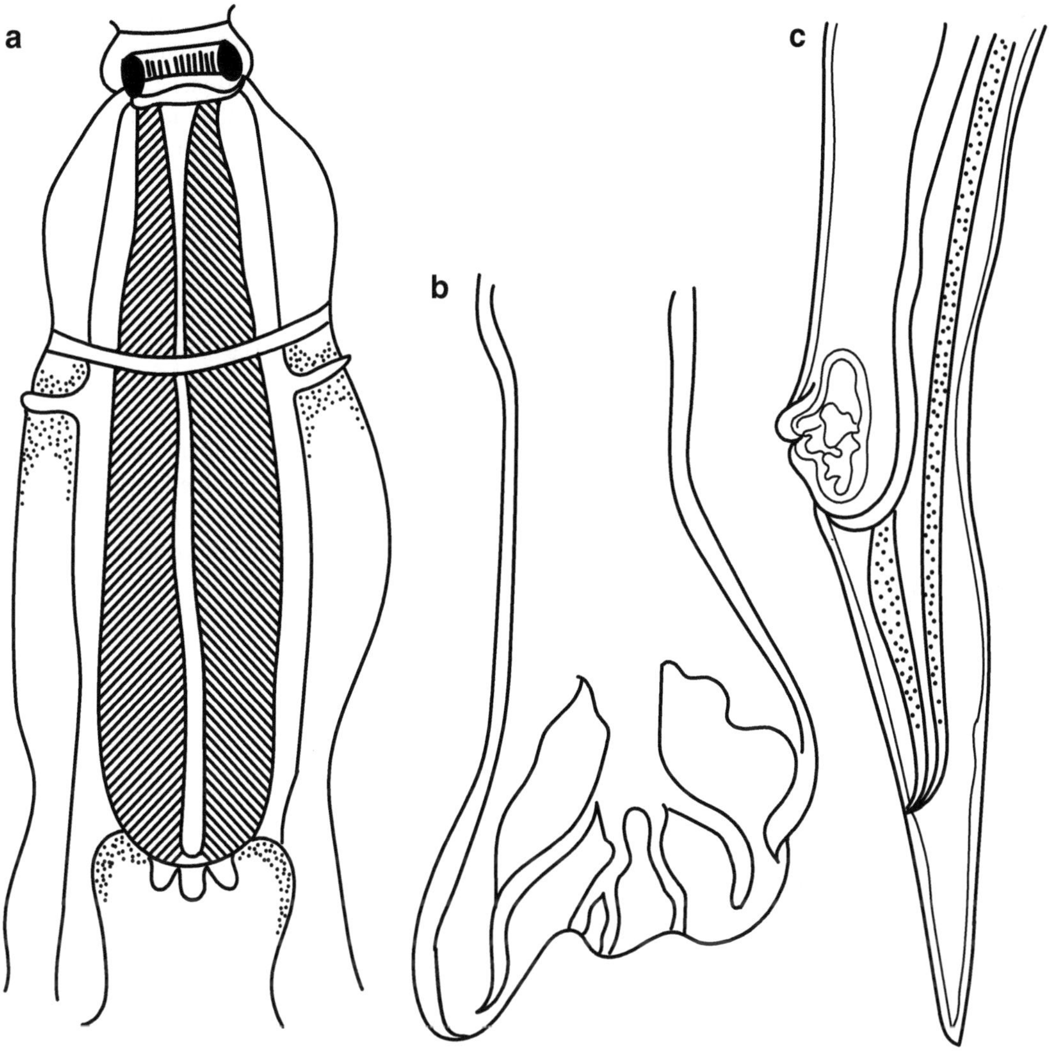

Fig. 6.8 *Oesophagostomum columbianum*. (**a**) Head (**b**) bursa (**c**) tail of female

Life Cycle

They are homoexenous parasites and do not involve any intermediate hosts for completion of life cycle. The following are the events occurring in the life cycle:

Basic information	• It is a direct type life cycle. • The hosts are sheep, goat, camel, etc. • The infective stage is third stage larvae (L3). • Infection of the hosts occurs by ingestion.

Stages of Life cycle formed

In the host	In the environment
• Fourth stage larvae • Adult and egg	• First stage larvae • Second stage larvae • Third stage larvae

Events of life cycle: a glimpse	• Voiding of eggs with the feces of hosts • Hatching of eggs in environment • Transformation of L1 to L3 after two moltings • Infection of hosts by ingestion of L3 • Development to adult in the respective location

Detailed Description of Life Cycle

Egg Passing

The adult parasites live in their predilection site where they lay eggs. The eggs are voided with the feces.

Embryonation

The eggs laid by the parasite commence embryonation immediately. Embryonation of eggs occurs in the environment itself. The eggs are passed with the feces.

Hatching and Development up to Infective Stage

In the presence of suitable environmental condition, the eggs hatch. The first stage larvae are liberated after hatching. First stage larvae transform to second stage larvae after one molting. The second stage larvae transform to third stage larvae after one more molting. The third stage larvae retain the sheath of second stage larvae which provide additional protection to the infective third stage larvae. Formation of third stage larvae requires about 7 days.

Infection and Maturity

The hosts get the infection when they inadvertently ingest the infective third stage larvae usually during grazing. The third stage larvae penetrate the mucosa of wall of the intestine where they form cysts within which the larvae transform to fourth stage after one molting at 4 days after infection. Then the larvae reach the lumen of the intestine after 5–7 days. Then the juveniles migrate to the colon where they grow to adult after fourth molting.

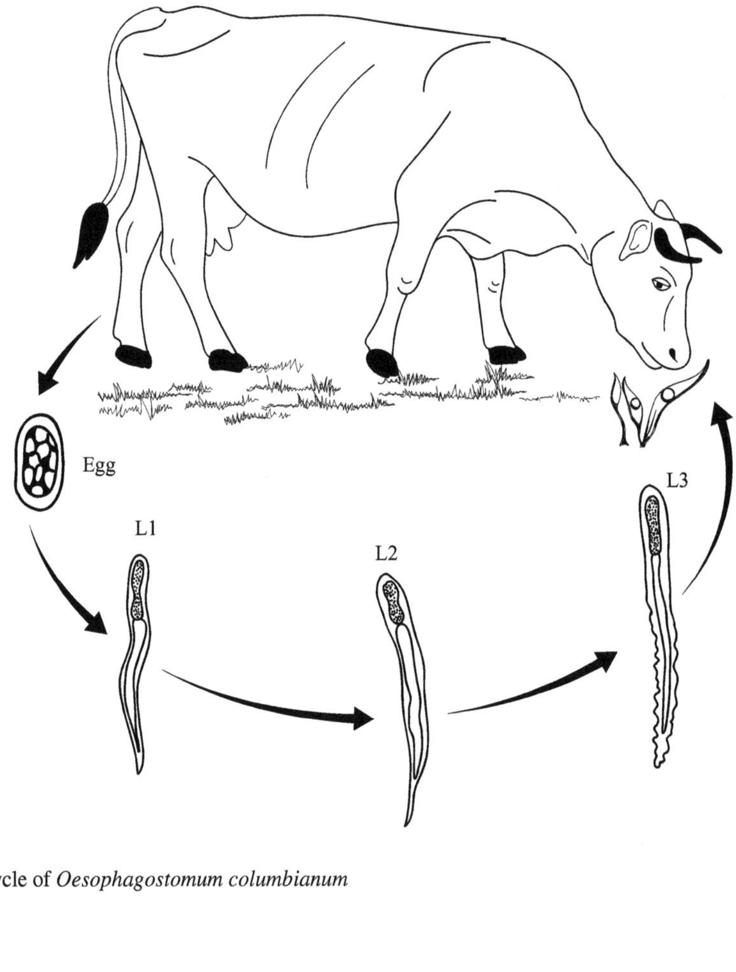

Fig. 6.9 Life cycle of *Oesophagostomum columbianum*

Pathogenesis

In the initial phase, hyperemia occurs when the larvae penetrate the wall of the intestinal mucosa within 12 hours of infection. The wandering larvae cause marked damage by destroying the cells and producing hemorrhage at the blood vessels of the damaged tissue. Hemorrhagic lesions occur within 2 days in the wall of the intestine. The juveniles become encysted in the mucosa in about 4 days. Thereafter, deposition of white granulation tissue occurs and cyst formation commences. The larvae remain within the cysts. Suppurative material is found in the cysts. Most of the larvae reenter the lumen of the intestine. But some larvae remain in the deeper tissues and cause formation of nodules. Some investigators described that the intensity of the nodules depends on the number of exposure of infection and not on the number of larvae entering the tissues. The size of the nodules is like a pea. The nodules may be found singly or in confluent form. A cheesy and caseous mass is found inside the nodules, and the juveniles and adults may not be found in the nodules.

Clinical Signs

The affected animals exhibit clinical signs comprising dark greenish diarrhea. Severe diarrhea results in dehydration, emaciation, weight loss, etc. In chronic case, the animals become emaciated and anemic. The animals suffer from intermittent diarrhea.

Diagnosis

Symptomatological Diagnosis

Clinical signs do not have any diagnostic value for confirmation. The worms cause formation of nodules in the wall of the large intestine. Due to this, absorption is hampered and results in greenish diarrhea. This symptom may provide clues to proceed for taking further diagnostic step.

Parasitological Diagnosis

Examination of fecal sample may reveal the presence of eggs of the parasite. Sometimes, fourth stage larvae may be detected in the feces. Centrifugal floatation technique can be done for better availability of the eggs. For specific diagnosis, fecal culture can be done for collection and identification of larvae. Autopsy of a selected animal may reveal characteristic nodules and the worms. This is the most reliable diagnosis.

Immunological Diagnosis

Many works have been carried out using crude and purified antigens for performing immunodiagnosis. Works on use of coproantigen have also been done with considerable success. Different types of ELISA have been done. But by far no immunodiagnostic technique has yet been commercialized for specific diagnosis of *Oesophagostomum* sp.

Molecular Diagnosis

Molecular diagnosis is possible using specific molecular markers. Different ribosomal DNA, ribosomal RNA, and mitochondrial DNA could be used as molecular markers. Studies indicate the use of mitochondrial DNA (cytochrome oxidase 1) and ribosomal DNA (ITS2) as molecular markers for identification of this parasite.

Treatment

Name of drug	Mode of administration	Dose
Fenbendazole	Oral	5–7.5 mg/kg b/w
Albendazole	Oral	7.5 mg/kg b/w
Mebendazole	Oral	15 mg/kg b/w
Cambendazole	Oral	20 mg/kg b/w
Oxfenbendazole	Oral	5 mg/kg b/w
Ivermectin	S/C	0.2 mg/kg b/w

Control

General Management

The affected animals should be treated properly. Since the feces is the source of eggs of parasites, the feces should be disposed properly. The floor of animal house should be cleaned properly to keep the floor free from any contamination of eggs. Rotational grazing should be practiced.

Immunological Control

In different trials, crude and defined antigens have been used. But no vaccine has been developed.

Molecular Control

No naked DNA vaccine or gene vaccine is available.

Biological Control

No biological control is available.

Species: *Oesophagostomum radiatum*

Hosts: cattle, zebu, water buffalo

Site: colon

Morphology

Gross

Shape: slender

Color: whitish

Size: Male is 14–17 mm long and the female is 16–22 mm.

Microscopic

The worms possess rounded mouth collar. The cephalic vesicle is large which is characteristically constricted behind the middle by a groove. Internal leaf crown is present consisting of 38–40 triangular denticles. The external leaf crown is absent. Lateral alae are well developed.

Fig. 6.10 *Oesophagostomum radiatum*

Life Cycle

Type of life cycle: direct

The following events are observed in life cycle of this worm:

- The adult parasites are found in the colon part of the intestine where copulation occurs. The eggs are passed with the feces.
- Further development of eggs occurs in the environment dependent on environmental temperature.
- Eggs hatch and first stage larvae come out.
- The first stage larvae transform into infective third stage larvae after two moltings.
- The hosts get the infection by ingestion of infective larvae inadvertently with food.
- Exsheathment of larvae occurs in the small intestine.
- The larvae get attached to wall of the small intestine and one molting occurs there. The fourth stage larvae return to the lumen. Some of the larvae fail to return to the lumen and form nodules there.
- Then the larvae migrate to the large intestine and get matured there.

Pathogenesis

The worms are markedly pathogenic. They cause formation of nodules both in the wall of the small intestine and large intestine. Young animals suffer more compared to the adult one. Absorption is completely hampered. Initially they suffer from intermittent diarrhea which turns to constant purging resulting in severe emaciation and prostration. The affected animals may die.

Symptoms

The affected animals exhibit symptoms of fetid diarrhea.

Diagnosis

Examination of fecal sample reveals eggs of the parasite.

Control

Immunoprotection

Gasbarre and Canals (1989) studied on immunoprotection against *Oesophagostomum radiatum*. A soluble fraction was obtained from disrupted adult worms, and this fraction was used to vaccinate calves. The vaccination protocol included two immunizations. The first was administered intramuscularly with complete Freund's adjuvant. The second was given intraperitoneally with antigen and alum. This immunization reduced the number of worms developing from a subsequent challenge infection by 85% and also reduced clinical signs associated with infection with adult worms.

Oesophagostomum multifoliatum

Hosts: sheep, goat

Site: large intestine

Species: *Oesophagostomum venulosum*

Hosts: sheep, goat, deer, etc.

Site: colon

Common name: The common name is large bowl worm.

Morphology

The male worm is 11–16 mm and the female is 13–24 mm. The worms possess both external and internal leaf crown. The external leaf crown possesses 18 elements. The internal leaf crown contains 36 elements. One inflated cuticular vesicle is present at the anterior part of esophageal region which terminates into the cervical groove. The cervical groove is present laterally and ventrally. The cervical papillae are present below the level of the esophagus. Lateral cervical alae are absent.

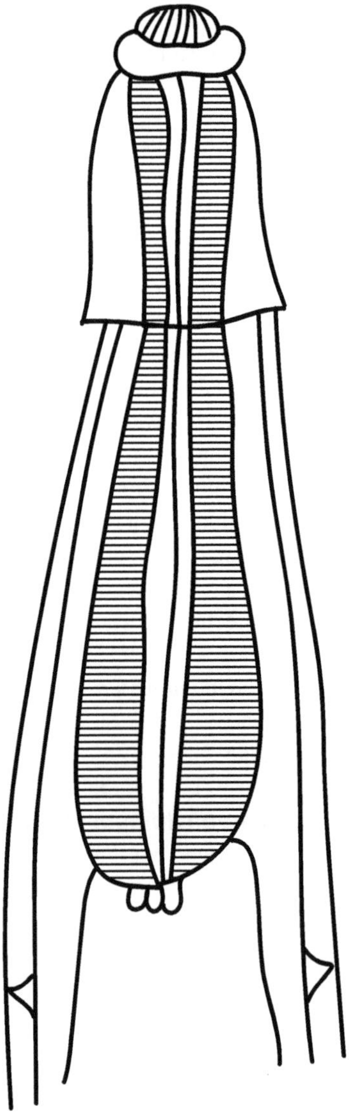

Fig. 6.11 *Oesophagostomum venulosum*

Life Cycle

Type of life cycle: direct

The following events are observed in the life cycle of this worm:

- Development up to infective third stage larvae is similar to other species of *Oesophagostomum*.
- The hosts get the infection usually during grazing by ingestion of infective third stage larvae. **There is limited evidence of skin penetration**.
- Exsheathment of larvae occurs in the small intestine.
- The larvae get attached to wall of the small intestine and lie coiled up within 1 day in the muscularis mucosa.
- The juvenile then gets encysted there within about 4 days. At this place, one molting occurs and fourth stage larvae are formed.
- The fourth stage larvae return to the lumen of the small intestine.
- Then the fourth larvae migrate to the large intestine and get matured there.

Pathogenesis

The worms are nonpathogenic. The worms may produce nodules occasionally. If produced, the nodules are small in size.

Symptoms

The affected animals are apparently asymptomatic. Rarely nodule formation occurs.

Diagnosis

Diagnosis can be done by grazing history, clinical signs, signs, etc. But all these cannot provide confirmative diagnosis. Examination of fecal sample reveals the eggs of the parasites.

Species: *Oesophagostomum asperum*

Hosts: sheep, goat

Site: cecum and colon

Morphology

The worms possess both internal and external leaf crown. They possess cone shaped mouth collar and inflated cephalic vesicle.

Species: *Oesophagostomum dentatum*

Hosts: pig

Site: large intestine

Distribution: They are worldwide in distribution.

Morphology

The male worms are 8–10 mm long and female worms are 11–14 mm. They are white in color. The worms possess both external and internal leaf crown. Cervical papillae are present at the posterior end of the esophagus. Cephalic vesicle is well developed. Lateral alae are absent. The esophagus is club-shaped.

Life Cycle

Type of life cycle: direct

The following events are found in the life cycle of this worm:

- Development up to infective third stage larvae is similar to other species of *Oesophagostomum*. The infective stage is produced within 3–6 days. The third stage larvae are produced in three days.
- The pigs get the infection by ingestion of infective larvae inadvertently with food material.
- Exsheathment of larvae occurs in the small intestine.
- The larvae get attached to wall of the intestine and form cyst within which they perform one molting. After molting, the third stage larvae transform into fourth stage larvae. Third moulting takes place after about 4th day of infection.
- The fourth stage larvae return back to the lumen of the intestine and travel to the large intestine. Fourth ecdysis usually occurs approximately between the 20th and 30th day after infection.

Pathogenesis

The worms cause nodule formation which is less compared to other species. They cause thickening of the large intestine along with catarrhal enteritis.

Diagnosis

Examination of fecal sample reveals eggs of the parasite.

Species: *Oesophagostomum brevicaudum*

Hosts: pig

Site: large intestine

Morphology

The lengths of male and female worms are 6.2–8.75 mm and 6.4–11.4 mm, respectively. Buccal capsule is large. The worms possess both external and internal leaf crown. They have well developed cephalic vesicle. Cervical papillae are located at the posterior part of oesophagus.

Pathogenesis

Pin head shaped nodules are produced by the third and fourth stage larvae. However, some nodules may be large in size. There may be inflammatory zone surrounding the nodules. General hyperplasia and edema may be seen. The nodules interfere with the normal intestinal function.

Symptoms

Blood stained hemorrhagic diarrhea occurs. Other clinical signs are inappetence, weight loss, etc.

Oesophagostomum granatensis

This species occurs in the large intestine of pig. The worms are prevalent in Spain.

Oesophagostomum quadrispinulatum

Hosts: This species occurs in large and small intestine of pig

Oesophagostomum bifurcum

Hosts: primates

Site: large and small intestine

Oesophagostomum aculeatum

Hosts: primates

Site: large Intestine

Order: Strongylida—Superfamily: Strongyloidea—Family: Amidostomidae

Family Synopsis

The roundworms of this family commonly occur in gizzard, proventriculus, crop of duck, and geese. The worms are characterized by absence of leaf crown and short, shallow, thick walled buccal capsule. Life cycle is direct. The worms are highly pathogenic to the young birds, but the adults act as the carrier. Diagnosis is best done by autopsy of the ducks and detection of worms in the respective locations. Different benzimidazoles are effective drugs.

Genus: *Amidostomum*

Introduction

These are important parasites occurring in the gizzard of duck, geese, etc. They cause marked loss of production.

Distribution

It was first reported in Germany. Since then, the worms were reported in Europe, South America, etc.

Species: *Amidostomum anseris*

Hosts: domestic and wild ducks and geese

Site: gizzard (horny linings), proventriculus, oesophagus

Distribution: The worms are worldwide in distribution

Morphology

The male worm is 10–17 mm and the female is 15–25 mm. These are red colored thin worms. Three pointed teeth are present in the base of the short and stout buccal capsule. The middle tooth is much prominent. The spicules are of equal length having two branches at the posterior region.

Life Cycle

Type of life cycle: direct

The following events are observed in life cycle of this worm:

- The adult parasites are found in the gizzard part of the gut where copulation occurs. The eggs are passed with the feces.
- Eggs hatch and first stage larvae come out.
- The first stage larvae transform into infective third stage larvae after two consecutive moltings.
- The hosts get the infection per os.
- There have reports of development up to L3 within eggs and L3 are liberated after hatching of the eggs.
- Further development occurs in the gizzard.

Pathology

Anatomical Pathology

Young birds suffer more. The worms actively penetrate the wall of gizzard and proventriculus. The irritation leads to localized inflammation. The area becomes necrosed with hemorrhage.

Hematological Pathology

Total RBC count is reduced. There is increased eosinophil and heterophile count.

Chemical Pathology

There is reduction in total serum protein, cholesterol, and triglyceride level.

Symptoms

The symptoms include weakness, emaciation, anemia, loss of body weight, loss of production, etc.

Diagnosis

Examination of autopsied birds reveals the pathological features along with the presence of worms in gizzard.

Treatment

Name of drug	Mode of administration	Dose
Cambendazole	Oral	60–80 mg/kg
Tetramisole	Oral	30–40 mg/kg

Genus: *Epomediostomum*

Introduction

These are important parasites occurring in the gizzard of duck, geese, etc. They cause marked loss of production.

Species: *Epomidiostomum uncinatum*

Hosts: domestic and wild ducks and geese

Site: gizzard and proventriculus

Distribution: The worms are distributed in Asia, Africa, Europe, North America, etc. The worms are also found in other parts of the world.

Morphology

The worms are filiform. The teeth are absent in the buccal capsule. The head end possesses four flap-like epaulettes. The spicules are of equal length, and the end of which bears three branches. The female tail is conical in shape and ends in small rounded tip.

Life Cycle

Type of life cycle: direct

The following events are observed in the life cycle of this worm:

- The adult parasites are found in the gizzard part of the gut. The females lay eggs which are passed with the feces.
- Development up to L3 occurs within eggs.
- Then the eggs hatch and L3 are liberated.
- Infection occurs by ingestion of eggs.
- Rest of the development occurs in the gizzard whereby two more ecdyses occur for formation of adult worms.

Pathogenesis

Young birds suffer more. The worms actively penetrate the wall of the gizzard and proventriculus. The irritation leads to localized inflammation. The area becomes necrosed with hemorrhage.

Symptoms

The symptoms include weakness, emaciation, anemia, loss of body weight, loss of production, etc.

Diagnosis

Examination of autopsied birds reveals the pathological features along with the presence of worms in gizzard.

Species: *Epomidiostomum skrjabini*

This worm occurs in gizzard of domestic and wild goose. The head bears four spines on cuticular prominences. The mouth also possesses four spines. The bursa has three lobes. The spicules are of equal size and end in three sharp-tipped branches.

Order: Strongylida—Superfamily: Strongyloidea—Family: Syngamidae

Family Synopsis

These worms are found in the respiratory system of mammals and birds. The buccal capsule is cup-shaped. They do not have any leaf crown. The members of *Syngamus* are found in the trachea of different types of birds, i.e., fowl, guinea fowl, turkey, pigeon, etc. The members of *Cyathostoma* are found in the trachea and bronchi of duck, goose, etc. The *Mammomonogamus* spp. occur in the larynx of cattle, buffalo, etc. The animals suffer from marked respiratory problem.

Genus: *Syngamus*

Species: *Syngamus trachea*

Introduction

These are important parasites occurring in the trachea of different types of birds. They cause marked loss of production. These are called gape worm, forked worm, Y-shaped worm

Distribution

Syngamus trachea is distributed worldwide.

Morphology

Gross

Shape: The male and female worms remain permanently copulated forming Y-shaped appearance.

Size: The male worm is 2–6 mm long and the female worm is 5–20 mm long.

Color: They are red colored and called red worm.

Microscopic

Adult

Head: Leaf crown is absent. They have a cup-shaped buccal capsule having six to ten teeth at the base.

Leaf crown: The leaf crown is absent.

Bursa: The bursa is slightly bent having short and stout rays. Bursal rays are usually tridigitate.

Spicule: A pair of spicules are present which are equal in size.

Tail end: The male tail has bursa and the female tail end is conical and pointed.

Egg

The eggs are roughly barrel shaped having opercular plugs on either ends. The size is 78–100 um × 43–50 um. The eggshell is smooth. Eight to 16 blastomeres are found in the eggs.

Larva

Development up to third stage occurs in the eggs.

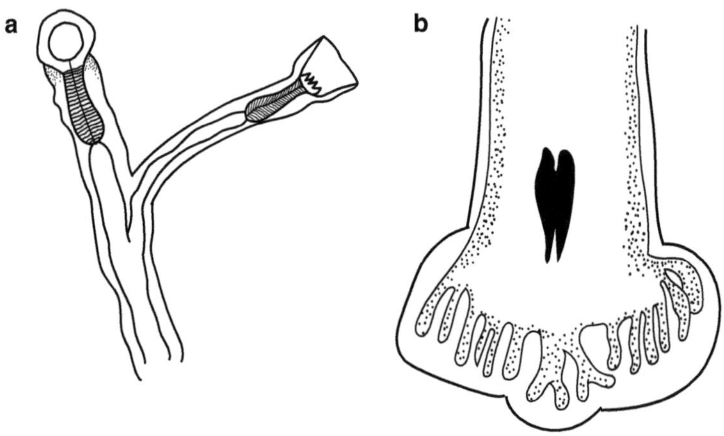

Fig. 6.12 *Syngamus trachea*. (**a**) Head end (**b**) bursa

Life Cycle

Basic information of life cycle
• It is a direct type life cycle • The hosts are different types of birds • The infective stage is egg containing L3 or hatched-out L3 • Infection of the hosts occurs by ingestion of infective stage (L3), or egg containing infective stage (L3) or transport hosts (earth worm, snail etc.)

Stages of Found

In the host	In the environment
• Third stage larvae or eggs containing third stage larvae • Fourth stage larvae • Adult	• Egg and egg containing L3 • First stage larvae • Second stage larvae • Third stage larvae

Events of life cycle: a glimpse
• Voiding of eggs with the feces of hosts • Larval development (up to L3) within the eggs in environment • Eggs containing L3 are infective stages. The eggs may hatch and larvae are liberated from the eggs. Earthworm, snails, slugs, and arthropods may serve as transport hosts and ingest the eggs. The larvae hatch out and remain encysted in the transport hosts • Infection to the hosts occurs by ingestion of eggs, infective larvae, transport hosts, etc. • L3 reach the lungs through blood circulation • Development to adult occurs in the respective location

Detailed Description

They are homoexenous parasites, and they do not require any intermediate host for completion of the life cycle. The following are the events occurring in the life cycle:

Egg Passing

The seat of predilection of the parasites is the trachea. The female lays the eggs in their predilection site. The eggs are coughed up and voided through the feces.

Embryonation

Embryonation of eggs occurs in the environment.

Development up to Infective Stage

Development up to third stage occurs in the eggs. The egg containing the third larva is infective stage. The eggs may be hatched and the larvae are liberated. Earthworm, slugs, snails, etc. may act as the transport host which harbors the infective larvae for long time. The larvae may remain viable for as long as 3 years.

Infection of Hosts

The hosts get the infection when they inadvertently ingest the infective eggs containing third stage larvae or hatched out larvae or ingestion of transport host harboring infective larvae.

Development in the Host

The hatching of eggs occurs in the gut. The hatched-out larvae penetrate the wall of the gut and enter into the circulation and reach the lung. Two moultings occur in the lungs within 5 days of infection. Then they reach to the trachea and become adult. The prepatent period is 16–20 days.

Pathogenesis

Syngamus trachea occurs in the trachea. The parasites cause irritation there. Inflammation and other pathological features occur. There may be formation of nodules there. A large amount of mucus is secreted resulting in occlusion of the lumen of the trachea. Migratory worms cause emphysema, edema, and pneumonia in lungs in case of heavy infection. In turkeys, the worms may induce formation of nodules.

Symptoms

The tracheal passage is occluded by mucus material. The birds try to remove it by tossing their head. They move with the head tilted above and the legs widen apart. This type of movement is called gaping movement. There may be pneumonia which results in dyspnea. Asphyxia and respiratory distress are common clinical signs. Severe phase of the disease may cause mortality of the birds. In less severe cases, weakness, anemia, and emaciation are some clinical features.

Diagnosis

Symptomatological Diagnosis

The affected birds perform gaping movement. The birds walk with the legs wide apart and the head directed forward. This characteristic movement is indicative of this parasitic infection.

Parasitological Diagnosis

Though the parasite remains in the trachea, the eggs come out through the feces because the eggs are coughed up. Examination of fecal sample may reveal the presence of eggs of the parasite. The eggs are large and bipolar. If the infection load is less, concentration technique, i.e., centrifugal floatation technique, should be done. At necropsy, red worms are readily visible attached to the tracheal mucosa.

Immunological Diagnosis

No immunodiagnostic test is available. Complement fixation test was described earlier.

Molecular Diagnosis

Molecular diagnosis is possible by using specific molecular marker. Different ribosomal DNA, ribosomal RNA, and mitochondrial DNA could be used as molecular markers. Studies indicate the use of ITS1, ITS2, and CO1 as molecular markers for characterization/identification of *Syngamus trachea*.

Treatment
Albendazole, fenbendazole, cambendazole, oxfenbendazole, etc. are some effective drugs.

Control

General Management

The affected birds should be treated properly. Prophylactic treatment can be initiated when there is chance of outbreak. The feces should be disposed properly. The litter should be replaced to avoid infection to other birds because the life cycle of the parasite is direct. We know that earthworm, snails, and insects act as the transport hosts. Therefore, these biological transmitting agents should be controlled properly.

Immunological Control

No vaccine has yet been developed. Earlier study failed to immunize birds against syngamosis using irradiated larvae of *Syngamus trachea.*

Molecular Control

No naked DNA vaccine or gene vaccine is available.

Biological Control

No biological control technology is available.

Treatment
No satisfactory treatment is available

Genus: *Cyathostoma*

Introduction
These are roundworms occurring in the trachea and bronchi of ducks and geese. They cause marked loss of production.

Species: *Cyathostoma bronchialis*

Distribution
Cyathostoma bronchialis occurs in North America and Europe.

Hosts: duck and geese

Site: trachea and bronchi

Morphology
The male worms are 4–5.8 mm and the females are 16–31 mm. The buccal capsule is cup-shaped and has six to seven teeth at its base. The bursa is well developed. The worms are not permanently copulated as is found in *Syngamus trachea.*

Life Cycle
Type of life cycle: direct

The following events are observed in the life cycle of this worm:

- The adult parasites are found in the trachea and bronchi part of respiratory system.
- The eggs are coughed up and pass with the feces.
- Development up to L3 stage occurs within the eggs itself. The eggs may hatch.
- Infection occurs by ingestion of larvae or eggs.
- Eggs are hatched and the larvae come out. The larvae reach to the peritoneum and then to bronchi and trachea.

Pathogenesis
Young birds suffer more. The affected birds display different respiratory symptoms which may even lead to death of the birds due to asphyxiation.

Diagnosis
Examination of autopsied birds reveals the pathological features along with the presence of worms in the gizzard.

Genus: *Mammomonogamus*

Species: *Mammomonogamus laryngeus*
Hosts: cattle, buffalo, sheep, goat

Site: The worms are found in the trachea, larynx, and bronchi of cattle, sheep, goat, etc.

Distribution
Mammomonogamus laryngeus are found in Asia, South Africa, South America, Asia, etc.

Morphology

The worms are reddish-colored. They have cup-shaped buccal capsule. At the depth of the buccal capsule, there are eight to ten teeth. The worms are permanently copulated and have Y-shaped appearance as that of *Syngamus* trachea. The small male worm is firmly attached with the vulva of the female by its bursa. The female has pointed tailed end. The eggs are ellipsoidal and the size is 75–85 × 42–45 um. The worms do not cause severe pathology. The infection is asymptomatic, but the youngs may suffer from respiratory distress comprising bronchitis, aspiration pneumonia, etc.

Order: Strongylida—Superfamily: Strongyloidea—Family: Stephanuridae

Family Synopsis

These are nematodes of the kidney, perirenal fat, etc. of pigs. They possess cup-shaped buccal cavity having teeth. The life cycle is direct. Earthworms may act as transport hosts.

Genus: *Stephanurus*

Introduction

These are important roundworms occurring in the kidney, perirenal fat, and ureter of pig. They cause marked loss of production.

Species: *Stephanurus dentatus*

Distribution

The worms are distributed in tropical and subtropical countries.

Hosts

Common hosts: pig and wild boar

Rare hosts: cattle

Morphology

Gross

Shape: The worms are cylindrical having a curved anterior end.

Size: The male worm is 20–40 mm long and the female worm is 25–54 mm long.

Color: The worms are mottled in appearance due to having translucent cuticle. The internal organs may be visible partially.

Microscopic

Adult

Head: They have buccal capsule containing six cusped teeth at the base. The rim of the buccal capsule bears small leaf crown. Six external cuticular thickenings (epaulettes) are present.

Bursa: The male has small bursa which is supported by stout short bursal rays.

Spicule: A pair of spicules are present which have striated alae. The tips of the spicules are typically swollen.

Tail end: The male tail has bursa and the female tail is constricted in the middle.

Larva

The larvae are of strongyle type. The L3 are 0.607 mm × 0.028 mm in size. They have strongyliform oesophagus. Research study revealed sexual differentiation in the fourth stage larvae. Spicular primordia are found in male. The fourth larvae display the stoma, esophagus, excretory gland cells, etc.

Egg

The eggs are similar to other strongyle worms. The shell of the egg is smooth. The size is 90–120 um × 43–70 um. A number of blastomeres are found in the egg which are 4–16 in number.

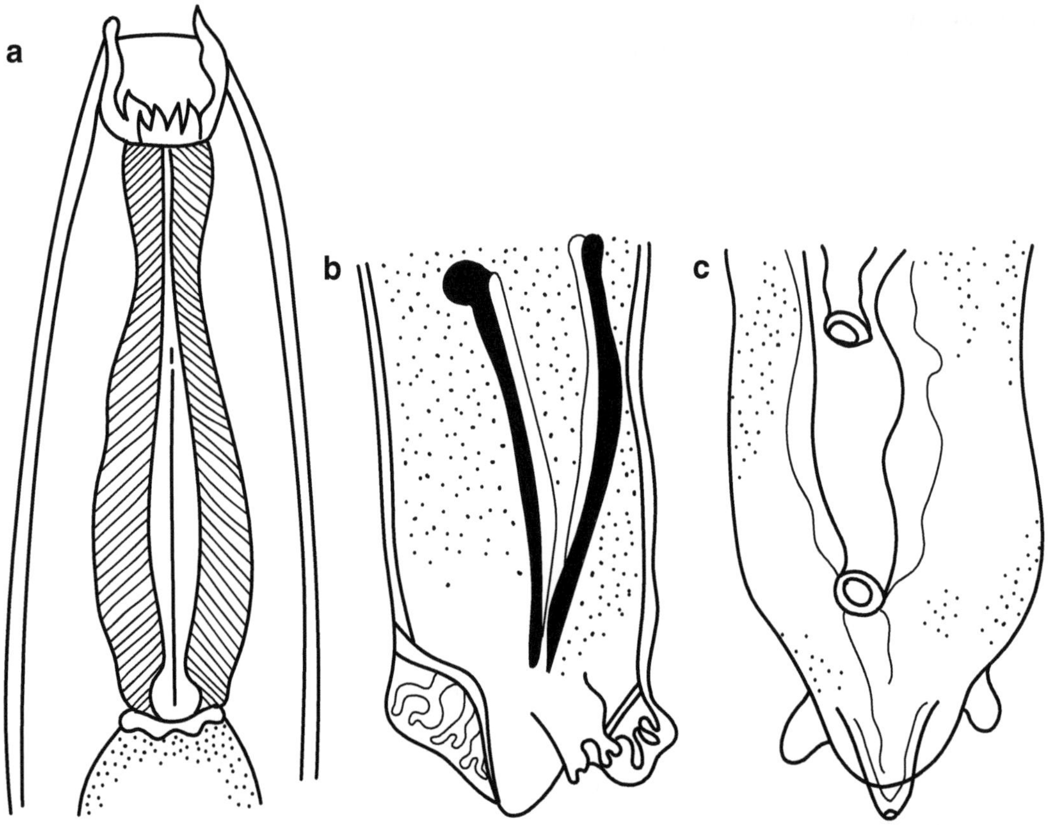

Fig. 6.13 *Stephanurus dentatus*. (**a**) head end (**b**) hind end of male (**c**) hind end of female

Life Cycle

Basic information life cycle
• It is a direct type life cycle • The hosts are pigs • The infective stage is third stage larvae • Infection of the hosts occurs by ingestion of infective stage of the parasite or by skin penetration

Stages of life cycle formed

In the host	In the environment
• Fourth stage larvae • Adult • Eggs	• First stage larvae • Second stage larvae • Third stage larvae

Events of life cycle: a glimpse
• Voiding of eggs with the urine of hosts
• Hatching of eggs in environment
• L1 transform to L3 after two moltings in the environment
• Infection of host by ingestion of L3 or through transcutaneous route (penetration of skin by larvae)
• Molting either in the abdominal muscles or stomach according to mode of entry
• Development to adult in the respective location

Detailed Description

Egg Passing

The parasite is homoexenous type, and it does not require any intermediate host for completion of life cycle. The adult parasites remain within the cyst. The cyst is connected with the urinary passage by connecting channels, and the eggs are voided with the urine.

Embryonation
Embryonation of eggs occurs in the environment.

Development up to Infective Stage
In the presence of suitable environmental condition, the eggs hatch. The first stage larvae come out. The first stage larvae transform to third stage larvae after passing two moltings. The third stage larvae are infective stage of the parasite.

Infection of the Hosts
The hosts get the infection mainly by two ways, i.e., ingestion and percutaneous. The third stage larvae molt to fourth stage in abdominal muscles if they follow percutaneous route. But if infection occurs by ingestion, molting occurs in the stomach and they become fourth stage after one molting. The fourth stage larvae reach the liver from which they go to the peritoneum. From the peritoneum, they go to the ureter and ascend to the kidney.

Special Note

- Earthworm may act as transport host for third stage larvae. The hosts may get the infection by ingestion of infected earthworm
- Prenatal infection is found in the life cycle of *Stephanurus dentatus*. Batte et al. (1966) reported prenatal infection with the swine kidney worm (*Stephanurus dentatus*) and associated lesion

Pathogenesis

By Larvae
The main pathogenesis is caused by the larval migration. Percutaneous infection by larvae causes nodule formation and enlargement of lymph nodes at the affected site. Erratic migration results in the presence of larvae in different organs, i.e., lung, spleen, spinal cord, etc., whereby they remain encapsulated. The late fourth stage larvae are more dangerous having a functional buccal capsule. They cause severe liver damage by rupturing the liver tissues resulting cirrhosis, ascites, etc.

By Adult
It is notable matter that the adult parasites are mostly nonpathogenic. They form cysts and remain simply within the cyst filled up with greenish pus.

Pathology

Structural Pathology (Gross)
Stephanurus dentatus is very important parasite. These parasites occur in the kidney where they cause cyst formation. The cysts are moderately hard and are found in the perirenal fat of the ureter. Portal phlebitis with thrombosis occurs.

Histopathology
Microscopical examination reveals fibrosis, granulomatous inflammation, and perivascular infiltration with eosinophil.

Symptoms
The symptoms include emaciation, weight loss, and ascites due to liver damage.

Diagnosis

Symptomatological Diagnosis
No specific symptoms are found which can provide diagnostic clue.

Parasitological Diagnosis
Examination of urine sample may reveal the presence of eggs of the parasite. Concentration technique should be done if the eggs are very scanty. Centrifugal floatation technique can be done for better availability of eggs during microscopical examination.

Immunological Diagnosis

Studies indicate use of different antigens for serodiagnosis purpose, but no reliable test has yet been commercialized. Literatures show fragmented research works on immunodiagnosis. Earlier work shows failure of development of immunodiagnosis by immunodiffusion and immunoelectrophoresis using crude somatic antigen of *Syngamus trachea.*

Molecular Diagnosis

Specific molecular marker comprising ribosomal DNA, ribosomal RNA, and mitochondrial DNA could be used for genomic characterization or identification of parasites. Genus specific primer can be designed for specific diagnosis of this parasite.

Treatment

Modern benzimidazole, levamisole, and ivermectin are some effective drugs.

Control

General Management

The affected animals should be treated properly. The urine should be disposed properly. The floor of animal house should be cleaned regularly. The larvae are more susceptible to desiccation. Therefore, the floor and feeding areas should be kept dried so that the larvae may be died. One interesting control step is gilt only technique. This technique consists of using only the gilts for breeding purposes. The gilts are reared on dry areas and they are marketed as soon as they are weaned.

Immunological Control

Different antigens have been tried for development of vaccine. But so far, no vaccine is available against this parasite.

Molecular Control

No naked DNA vaccine or gene vaccine is available.

Biological Control

So far, no biological control technique is available in the field.

Order: Strongylida—Superfamily: Ancylostomatoidea—Family: Ancylostomatidae

Family Synopsis

These are the worms of different mammals, i.e., cattle, sheep, goat, pig, etc. They are worldwide in distribution. The head of this worm is bent dorsad giving a hooklike appearance. That's why these worms are called hookworms. Leaf crown is absent. Well-developed subglobular buccal capsule is present. Life cycle is direct. Ventral margin of buccal capsule bears teeth or chitinous cutting plates. The hosts get the infection mainly through percutaneous route. Oral, prenatal, lactogenic are other routes of infection. These worms are very important. They are hematophagous worms and suck blood consistently from the blood vessels present in the wall of the intestine. They cause loss of morbidity and mortality. The gold standard diagnosis is by examination of feces for detection of eggs.

Genus: *Ancylostoma*

Species: *Ancylostoma caninum*

Introduction

These are important roundworms occurring in the small intestine of different animals (dog, fox etc.) and man. They are hematophagous parasites. They are voracious bloodsuckers and suck blood from the blood vessels of the intestine of the hosts. The parasites have zoonotic importance also.

Distribution

The worms are worldwide in distribution. They are mostly prevalent in tropical and subtropical countries.

Morphology

Gross

Shape: The worms are cylindrical in shape. The head is bent dorsad giving a characteristic hook shape and earned the name "**Hook worms**".

Size: The male worm is 9–14 mm long and the female worm is 15–20 mm long.

Color: The worms are red or gray in color in accordance with the stage of digestion of blood in the intestine.

Microscopic

Adult

Head: They have a well-developed buccal capsule which is subglobular in shape. At the ventral margin of the buccal capsule, three teeth are present on either side. At the depth of buccal capsule, a pair of dorsal teeth are present. They also possess a pair of centrolateral teeth in the depth of the buccal capsule.

Dorsal gutter: A dorsal gutter is present

Esophagus: The esophagus is club-shaped.

Spicule: A pair of spicules are present which are unequal in size.

Hind end: The male has a well-developed bursa at the posterior end. Female hind end is tapering.

Egg

Eggs are oval or elliptical. The eggshell is thin. They contain 8 blastomeres while passing through the feces. The size of eggs of *A. caninum* is 55–75 um × 34.4–44.7 um.

Larva

The rhabditiform hookworm larvae have long buccal cavity which is short in rhabditiform larvae of *Strongyloides* sp. The infective larvae are 630–664 um long. The larvae are sheathed.

Special Information

Xie et al. (2017) reported a new *Ancylostoma ailuropodae* n. sp. as a fourth species of hookworms described from Ursidae and 15th species assigned to the genus *Ancylostoma*. Morphologically the hookworm specimens differed from other congeneric species primarily based on the structure of the buccal capsule in males and females characterized by two pairs of dorsolateral teeth and two pairs of ventrolateral teeth. Male differed in shape and structure of copulatory bursa.

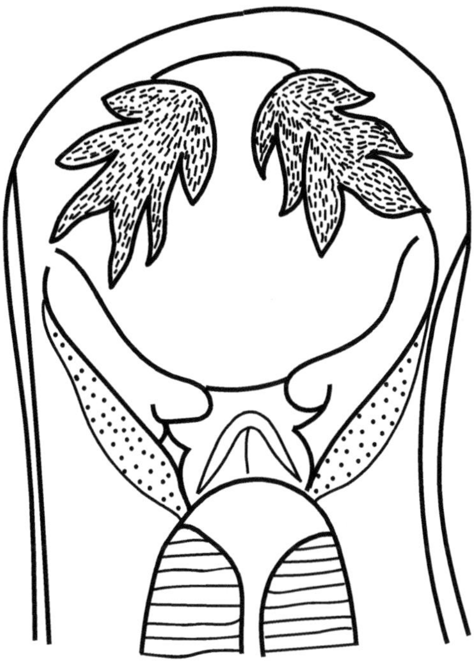

Fig. 6.14 Head end of *Ancylostoma caninum*

Life Cycle

Basic information of life cycle
• It is a direct type life cycle • The hosts are dogs • The infective stage is third stage larvae • Infection of the hosts occurs by ingestion of infective stage of the parasite, penetration of host's skin by infective larvae, prenatal infection, lactogenic infection

Stages of life cycle formed

In the host	In the environment
• Fourth stage larvae • Adult and egg	• First stage larvae • Second stage larvae • Third stage larvae

Events of life cycle: a glimpse
• Voiding of eggs with the feces of hosts
• Hatching of eggs in the environment
• L1 transform to L3 after two moltings in the environment
• Infection of host through percutaneous route or oral route
• Larvae reach the lung alveoli and molt to become L4
• L4 are coughed up
• Development to adult in the intestine
• Somatic migration may occur. Placenta may be infected. Prenatal and lactogenic infection may occur

Detailed Description

Egg Passing

The parasite is homoexenous and do not require any intermediate host. The adult parasites live in their predilection site where they lay eggs. The eggs are voided through the feces.

Embryonation

Embryonation of eggs occurs in the environment.

Hatching and Development up to Infective Stage

Hatching of eggs occurs in the environment. In the presence of suitable environmental condition, the eggs hatch. The first stage larvae come out. The first stage larvae transform to second stage larvae after one molting. The second stage larvae transform to third stage larvae after one more molting. Formation of third stage larvae takes about minimum 5 days.

Development in Hosts

The puppies become infected when the infective larvae enter through percutaneous route. The larvae enter the blood circulation and reach the lungs. From the lungs, the larvae follow the tracheal migration. The larvae burrow the lung alveoli. Later, they leave the alveoli and transform into fourth stage larvae after molting. The larvae are coughed up and reach to the intestine and develop into adult in the small intestine.

Infection of the hosts may take place through ingestion of infective third stage larvae; the larvae penetrate the buccal mucosa and follow the pulmonary migration as described above. Or these larvae directly enter the intestine where they develop into adult.

In case of bitch, some third stage larvae from the lungs migrate in different skeletal muscles and remain in dormant stage. They resume their activity when the animals become pregnant. If the animal is pregnant, placental infection occurs resulting in prenatal infection.

The neonatal pups may get lactogenic infection also. The newly born animals suck colostrum and get infection.

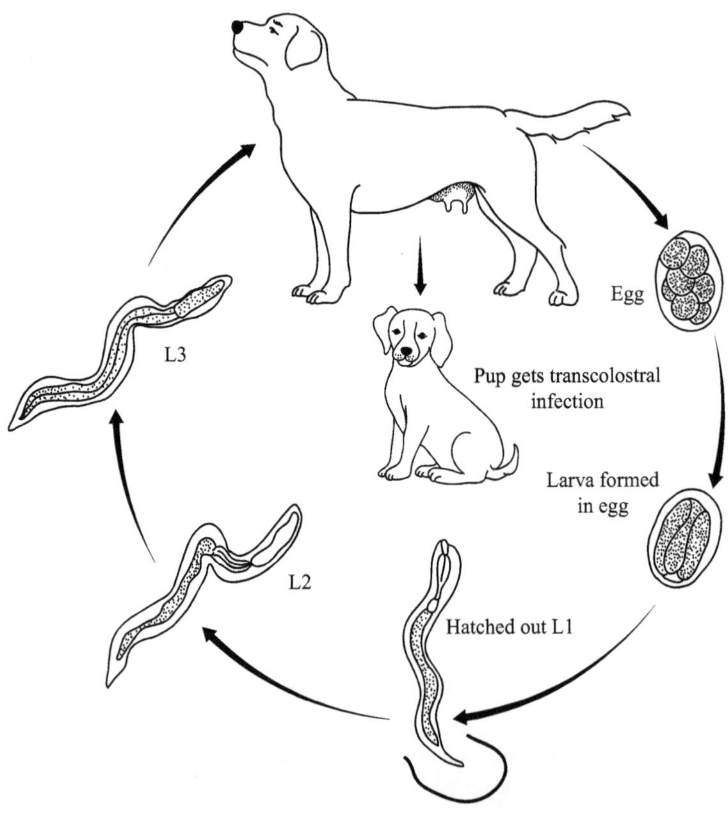

Fig. 6.15 Life cycle of *Ancylostoma caninum*

Pathogenesis and Pathology

Pathogenesis

Hookworms are bloodsuckers. They suck blood voraciously from the hosts. Severe anemia occurs in the affected animals. Anemia leads to hypoproteinemia resulting in edema formation.

By Larvae

Both larvae and adult parasites cause damage to the hosts. The migrating larvae cause erythema in the skin. Inflammation occurs in the skin. The condition may be more pronounced if the animals are sensitized with previous infection. In such situation, edema occurs. Necrosis and exudation occur at the migratory area. If many juveniles migrate in the skin, marked pathology occurs comprising of massive inflammation with different inflammatory cells, i.e., macrophages, eosinophils, lymphocytes, etc. There may be secondary bacterial infection which aggravates the condition. The migratory juveniles cause severe damage to the lungs. They break into the alveoli and cause petechial hemorrhage. There may be lobar pneumonia.

By Adults

The adult hookworms are bloodsuckers. They use their strong buccal capsule and take out the mucosal plugs. The wall of the intestine is ragged, or due to frequent change of their holds, the wall is completely torn away. They suck blood voraciously from the hosts. Severe anemia occurs in the affected animals. A single parasite can suck 0.36–0.8 ml of blood per day. Anemia leads to hypoproteinemia resulting in edema formation. To compensate the blood loss, the hematopoietic system is stimulated and hyperplasia of the bone marrow occurs. Secondary bacterial infection on the affected wall is more aggravated.

Pathology

Structural Pathology

The worms cause lesion on the wall of the stomach. Bite marks are produced by the bloodsucking activity.

Hematological Pathology

Hematological pathology reveals low PCV, low total erythrocyte count, and low Hb%. The anemia is macrocytic hypochromic type.

Chemical Pathology

There occur hypoproteinemia, hypoalbuminemia, etc. Loss of serum iron has also been recorded.

Symptoms

The common clinical signs include dermatitis, pneumonia, anemia, paleness of the mucus membrane, diarrhea, progressive emaciation, weakness, cardiac failure, death, etc. Anemia is very severe in the suckling pups. Depletion of iron reserve is common. There may be bloody diarrhea, respiratory syndromes, dehydration, loss of skin condition, etc.

Diagnosis

Symptomatological Diagnosis

The clinical signs displayed by the affected animals are not specific for confirmatory diagnosis. Paleness of the eye mucus membrane may be indicative of this parasitic infection. Bottle jaw condition is another symptom.

Parasitological Diagnosis

Examination of fecal sample may reveal the presence of eggs of the parasite. The eggs are rectangular having eight clear embryonic cells in fresh feces. If no eggs are found by direct smear examination, one should proceed for concentration technique. By examination of fecal sample, diagnosis up to the genus level is not possible. It could be possible only when the feces is cultured and larval morphology is studied critically. It is very difficult to get any egg in the feces of suckling pups because they suffer severely due to the larval stages, and when the most prominent symptoms are exhibited, the eggs are not found. This matter should be taken into account by the examiner.

Immunological Diagnosis

Many works have been carried out using crude and purified antigens. But so far, no immunodiagnostic technique has been commercialized for field application.

Molecular Diagnosis

Molecular diagnosis is the current perspective. Different ribosomal DNA, ribosomal RNA, and mitochondrial DNA are useful molecular markers. Studies indicate the use of ribosomal DNA (ITS1, ITS2) and mitochondrial DNA (cytochrome oxidase subunit 1) for characterization/identification of this parasite.

Treatment

Name of drug	Mode of administration	Dose
Fenbendazole	Oral	50 mg/kg for three consecutive days for dog and cat
Ivermectin	Oral	0.5 mg/kg for dog; 0.024 mg/kg for cat

Control

General Management

The affected animals should be treated properly. The feces should be disposed properly since the life cycle is direct type and feces are the source of eggs. Deworming schedule should be followed perfectly. The floor of the kennel should be cleaned and kept dried.

Immunological Control

Different crude and defined antigens have been used in different research trials, but no vaccine has yet been commercialized. A series of works were carried out using attenuated live vaccine comprising irradiated third stage larvae, crude somatic protein of whole parasitic extract, ES protein (excretory and secretory protein), etc. Different purified proteins, i.e., cysteine protease and glutathione s-transferase, were also used with different levels of success.

Molecular Control

Works were carried out on gene immunization. But so far, no naked DNA vaccine or gene vaccine is available.

Biological Control

No biological control has entered for field application.

Other spp. of Hookworms

Species: *Ancylostoma braziliense*

This species occurs in dog, cat, and fox. Wild canid animals and man may also be affected. They are distributed in the tropical and subtropical regions. The male and female are 7.5 mm and 9–10 mm long, respectively. The seat of predilection of the worm is the small intestine. They are smaller than *Ancylostoma caninum*. Buccal capsule has narrow opening. The buccal capsule contains one pair of large and one pair of small rudimentary teeth. Human beings are not natural hosts and the parasites do not get maturity in human hosts. The larvae of this worm cause creeping eruption. Life cycle is mostly similar to *A. caninum*. The hosts are infected through transcutaneous and oral route. Transmammary infection has not been reported. The prepatent period is about 2 weeks. **Rodents may serve as paratenic host**. See cutaneous larva migrans. Its pathological significance is less which is only enteritis. The parasite is not bloodsucker. But plasma drainage through the bite opening results in hypoproteinemia.

Species: *Ancylostoma ceylanicum*

The worms are mostly prevalent in Asia. The seat of predilection of the worm is the small intestine. The hosts are dog, cat, and tiger. Previously it was considered as *A. braziliense*. But this species differs by morphological features of larger inner pair of the ventral teeth and differently directed ventral ray of bursa. The cuticular striations are more wider than *A. braziliense*. The parasites do not cause marked pathogenesis and subclinical form of the disease prevails. Anemia and diarrhea are common clinical features.

Species: *Ancylostoma tubaeforme*

This is a hookworm of cat. The seat of predilection of the worms is the small intestine. The species is similar to *A. caninum* and previously this species was considered as *A. caninum*. They are slightly smaller than *A. caninum*. The male and female are 10 mm and 12–15 mm, respectively. They have deep buccal capsule. But the teeth of ventral margin are slightly larger than *A. caninum*. The bursa is well developed. The spicules are larger than *A. caninum*. Eggs are similar to *A. caninum*.

Ancylostoma duodenale
The host is man. This also affects wild carnivores.

Ancylostoma paraduodenale
The hosts are lion and other carnivores.

Genus: *Bunostomum*

Introduction

These are important roundworms occurring in the small intestine of sheep and goats. They are hematophagous parasites.

Species: *Bunostomum phlebotomum*

This worm occurs in the small intestine of cattle and buffalo. They are large hookworms reaching 1–3 cm long. The color is grayish white. The anterior end is bent dorsally giving a characteristic hooked appearance. The worms are apparently similar to *B. trigonocephalum*. But this species differs from *B. trigonocephalum* by having short dorsal cone and the presence of two pairs of subventral lancets in the buccal capsule. The bursa is well developed having two lateral and a dorsal lobe. The male worm has slender and long spicule. The eggs are elliptical (size 97 × 50 um) having four to eight blastomeres. The larvae are small having filamentous tail. They possess 16 gut cells. The larvae (L3) are 0.525–0.575 mm long having 0.14–0.15 mm tail of sheath and 0.06–0.08 mm extension of sheath beyond tail of juvenile. Life cycle is direct. The infective stage is third larval stage as that of other hookworms. The mode of infection is ingestion or percutaneous route. The larvae reach the lung and get molted to become fourth stage. These fourth stage larvae reach the gut where they develop into adult. The prepatent period is 6 weeks in case of transcutaneous mode and 7–10 weeks in case of oral mode of infection. The adult worms are potent bloodsuckers which causes anemia hypoproteinemia in affected animals.

Species: *Bunostomum trigonocephalum*

This worm occurs in the small intestine of sheep and goat. They are stout worm of 1–3 cm length. The color is grayish white. They have large funnel-shaped buccal capsules opening anterodorsally. Dorsal teeth are absent. They have a pair of chitinous plates on ventral margin and a pair subventral lancets near the base of the buccal capsule. The bursa is well developed having two lateral and one asymmetrical dorsal lobe. The spicules are slender and short. The hosts get the infection through larval (third) skin penetration or ingestion of third stage larvae. In case of transcutaneous infection, the larvae reach the lungs whereby they get molted to fourth stage larvae. They then reach the gut and develop into adult. In case of oral infection, the larvae directly develop into adult without migration. The adult worms are bloodsuckers and cause anemia and hypoproteinemia. The clinical symptoms include paleness of the eyes, anemia, weight loss, roughness of the body coat, hair loss, and bottle jaw condition. The affected animals exhibit diarrhea with dark feces. The affected animals may die. Literatures show characterization of different crude antigens. Further work is essential for development of immunodiagnosis and immunoprotection using that antigen.

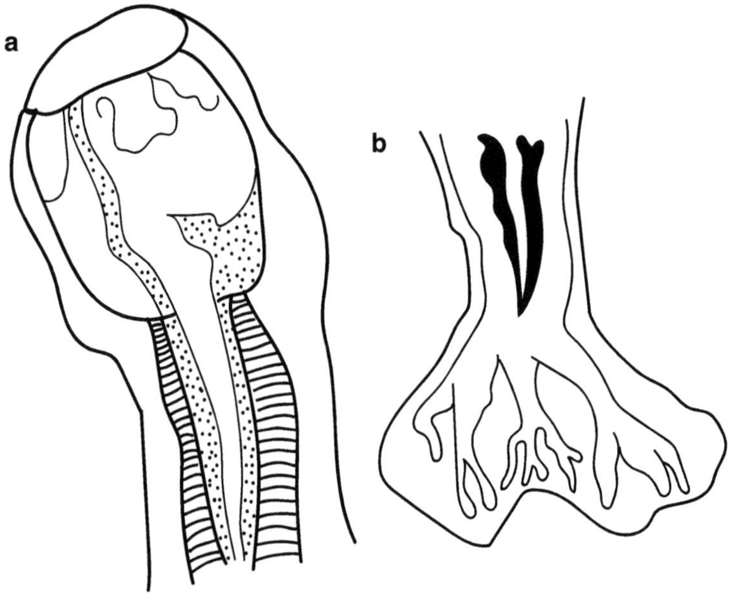

Fig. 6.16 *Bunostomum trigonocephalum* (**a**) head end (**b**) hind end

Genus: *Agriostomum*

Species: *Agriostomum vryburgi*

Hosts: zebu, ox
Site: small intestine

Prevalence

The parasites occur in India, Sumatra, etc.

Morphology

The male worm is about 9–11 mm and the female worm is 13–16 mm long. Buccal capsule is shallow which is followed by a large esophageal funnel. The esophagus is club-shaped. Two subventral lancets are present in the esophageal funnel. Four pairs of large teeth are present in the oral margin of buccal capsule. A rudimentary leaf crown is present on oral margin.

Life Cycle

Probably the life cycle is direct.

Pathogenesis

Probably anemia occurs due to bloodsucking activity of the parasites.

Genus: *Necator*

Species: *Necator americanus*

Hosts: The common hosts are human beings. The worms are reported in animals (dog, pig).
Site: small intestine

Genus: *Uncinaria*

Species: *Uncinaria stenocephala*

Hosts: dog, cat, fox
Site: small intestine

Morphology

The male is 5–8.5 mm long and female worm is 7–12 mm long. The buccal capsule is funnel-shaped. Two large chitinous plates are present at the ventral border of buccal capsule. A pair of subventral teeth are present in the base of the buccal capsule. The dorsal teeth are absent. The bursa is well developed and equipped with two lateral lobes and a small dorsal lobe.

Life Cycle

The usual mode of infection is ingestion of third stage larvae. However, infection through transcutaneous route may also occur. No transmammary and transuterine infection have been proved by experiment.

Pathogenesis

The pathogenesis is similar to other hookworms. They are not strong bloodsuckers. They cause anemia and hypoalbuminemia. Pedal dermatitis occurs in the sensitized animals.

Diagnosis

Postigo et al. (2003) demonstrated four larval antigens, i.e., 20, 25, 30, and 38 kDa larval proteins, as immuno-diagnostic by immunoblotting and immunoblotting inhibition using sera of naturally infected animals.

Genus: *Gaigeria*

Introduction

These are important roundworms occurring in the small intestine of sheep and goat. They are hematophagous parasites.

Species: *Gaigeria pachyscelis*

Hosts: sheep, and goat
Site: duodenum

Morphology

Gross

Shape: The worms are cylindrical in shape. The anterior end is bent dorsally.

Size: The male worm is 11–18 mm long and the female worm is 15–24 mm long.

Color: The worms are somewhat brown in color.

Microscopic

Adult

Head: They have well-developed buccal capsule which is equipped with large dorsal cone. Buccal capsule is well developed.

Cutting plates and teeth: A pair of cusped subventral lancets are present. Dorsal tooth is absent.

Esophagus: The esophagus is slender.

Bursa: The bursa has two small lateral lobes and a dorsal lobe.

Spicules: The spicules are stout. Teeth-like structures are present at the outer edges.

Hind end: The male hind end has a bursa. The female tail ends in blunt tip.

Eggs

The eggs are brownish in color. The shape is characteristic having one long side flattened and these have bluntly rounded ends. The size of egg is 0.105–0.129 × 0.05–0.066 mm. When deposited, the egg contains 16–32 cells.

Larva

The first juvenile has rhabditiform oesophagus. The oesophagus of second juvenile is filariform without any vulvular apparatus. The third juvenile is 0.58–0.67 mm long and has filariform oesophagus with characteristic swelling in the middle. The esophagus is filariform which has a small swelling in the middle.

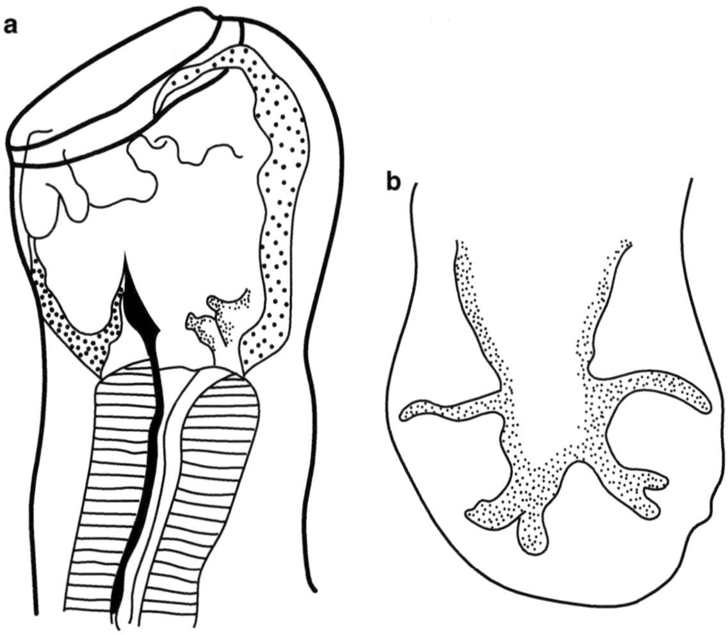

Fig. 6.17 *Gaigeria pachyscelis*. (**a**) head (**b**) bursa

Life Cycle

Type of life cycle: direct

The following are the events:

- The eggs are passed with the feces.
- The eggs hatch in the environment. The larvae develop into infective third stage after two moltings. The infective juveniles are negatively geotropic and positively phototropic and migrate vertically.
- The larvae penetrate the skin. Thus, the hosts get the infection by skin penetration. Experimental infection by ingestion of infective larvae has not been established.
- The larvae reach the lungs by circulation. One molting occurs there and fourth stage larvae are produced after 13–14 days of infection. The fourth stage larvae display the presence of buccal capsule with a pair of subventral lancets.
- The larvae ascend from the lower respiratory tract to the upper respiratory tract. They are coughed up and reach the gut.
- Then the larvae develop into adult.

Pathogenesis

By Larvae

The larvae feed on blood in the lung areas during migration.

By Adult

The worms are bloodsuckers. The affected animals suffer from anemia. The anemia is normocytic and normochromic. The animals may even die. There occur hemoglobinemia, oligocythemia, fall in PCV, etc.

Diagnosis

Examination of fecal sample reveals eggs of the parasites.

Genus: *Globocephalus*

Species: *Globocephalus longemucronatus*

Introduction

These are important roundworms occurring in the small intestine of pigs.

Hosts: pig

Site: intestine
Distribution: North America, South America, etc.

Morphology

Gross

Shape: The worms are cylindrical in shape.
Size: They are small sized worms. The male worm is 5.7 mm long and the female worm is 6–8 mm long.

Microscopic

Cuticle: They have thick striated cuticle.
Buccal capsule: The buccal capsule is globular.
Esophagus: The esophagus is club-shaped.
Teeth: The bottom of the buccal capsule bears two triangular teeth having pointed ends.
Bursa: The bursa is well developed.
Spicules: The spicules are slender.

Life Cycle

Type of life cycle: direct. The life cycle is not known completely.

- The eggs are passed with the feces.
- The eggs hatch in the environment. The larvae are susceptible to strong sunlight, low temperature, dryness, etc. The larvae develop into infective third stage larvae after 2 moltings.

Pathogenesis

The worms are bloodsuckers. They change their place while sucking blood and leave bleeding marks.

Genus: *Bathmostomum*

Species: *Bathmostomum sangeri*

Hosts: Indian elephants
Site: cecum

Morphology

Setasuban (1976) described the morphology of *Bathmostomum sangeri*. The worms are small. The buccal capsule is well developed and funnel-shaped. The inner wall of the buccal capsule is with number of raised ridges. A pair of subventral teeth is found. The spicules are wing shaped. Approximately the male is 12.5–14.5 mm long and the female is 14.98–17.68 mm long. Gubernaculum is absent and telamon is pyriform. Two pairs of papillae are present on either side of the cloaca.

Order: Strongylida—Superfamily: Trichostrongyloidea—Family: Trichostrongylidae

Family Synopsis

This is a large family with many genera. The roundworms are worldwide in distribution occurring in different mammals and avians. Buccal capsule is either absent or much reduced. The leaf crown is absent. The male has large bursa with two lateral lobes and a distinct dorsal lobe. The lobes are supported by prominent and strong bursal rays. The life cycle is direct type involving no intermediate hosts. Many members of this family are pathogenic to the hosts. Diagnosis is done by finding the eggs in the feces. The control of the roundworms is based on chemotherapy, grazing management, etc.

Genus: *Trichostrongylus*

Introduction

The worms occur in the stomach and intestine of different animals. They cause incessant loss of livestock economy. They are small, slender worms with white or reddish brown color. Buccal capsule is absent. The excretory pore is situated in a ventral notch near the anterior extremity.

Species: *Trichostrongylus axei*

Hosts and site: It occurs in the abomasum of sheep, goat, cattle, and deer.

Morphology

These are small and slender worms. The color is pale reddish brown. The head is small without having any buccal cavity. The leaf crown and teeth are absent. A distinct excretory notch is present on the esophageal region. The bursa is well developed having large lateral lobes but the dorsal lobe is not well marked. The stout spicules are ridged and pigmented brown. The eggs have tapered ends on either side. The eggs are oval in shape.

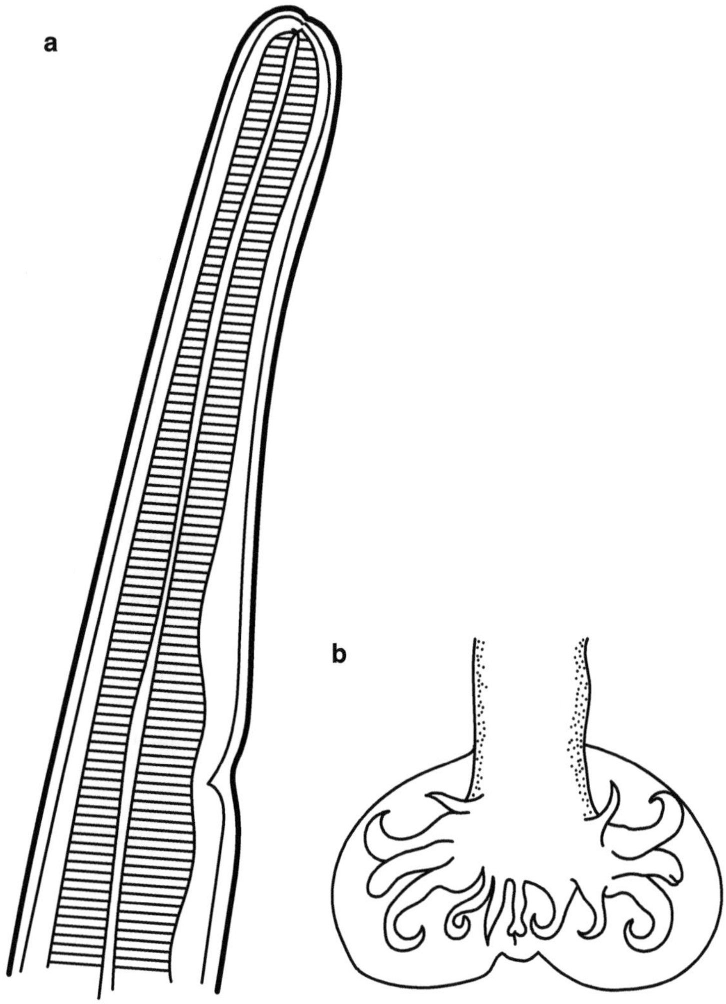

Fig. 6.18 *Trichostrongylus* sp. (**a**) head (**b**) bursa

Life Cycle

Some Basic Information of Life Cycle

Type of parasite	Homoexenous parasite
Type of life cycle	Direct life cycle
Type of definitive host	Vertebrate host
Location of fertilization	In the parasite
Location of embryonation	In the environment
Name of infective stage	Third larval stage (L3)
Mode of infection	Ingestion
Alternative mode of infection (hyperinfection, autoinfection, etc.)	No

Stages of life cycle formed

In the host	In the environment
• Fourth stage larvae • Adult and egg	• First stage larvae • Second stage larvae • Third stage larvae

Events of life cycle: a glimpse
• The eggs are voided with the feces of hosts
• Hatching of eggs occurs in the environment
• L1 transforms to L3 after two moltings
• Infection of host occurs by ingestion of L3
• L3 transforms to adult after two moltings

Description in Detail

The parasite is homoexenous and does not require any intermediate hosts for completion of their life cycle.

Embryonation

Embryonation of eggs occurs in the environment.

Passing Eggs

The females lay eggs in their predilection sites. The eggs are voided with the feces.

Hatching and Development of Infective Stage

In the presence of suitable environmental condition, the eggs hatch. Rhabditiform first stage larvae come out. The first stage larvae transform to second stage larvae after one molting. The second stage larvae transform to third stage larvae after one more molting. Development of infective third stage larvae takes place in about 7–10 days. The third stage larvae retain the sheath of second stage larvae which provide additional protection to the infective third stage larvae. These third stage larvae are filariform which are negatively geotropic and positively phototropic. The infective larvae crawl on the vegetation.

Infection of the Hosts

The hosts become infected when they inadvertently ingest the infective third stage larvae. Exsheathment of larvae occurs following ingestion, and exsheathed larvae penetrate the mucosa of the abomasum. The third larvae develop into adult worm after two moltings in their respective site. The prepatent period is 2–3 weeks.

Pathology

The worms cause damage to the glandular area. Hyperemia of mucosa, catarrhal inflammation, necrosis, and ulceration are some pathological features. The parasites remain embedded on the mucosa forming "ringworm lesions" characterized by raised plaques with sharp demarcation. The affected animals suffer from protein losing gastropathy. The animals suffer from severe diarrhea which may even cause death of the animals.

Diagnosis

History-Based Diagnosis

Seasonal occurrence and previous records might be supportive for diagnosis. Obviously, this does not provide any confirmative diagnosis.

Postmortem Diagnosis

Detection of lesions (flattened red thickened areas on the intestinal wall) and recovery of worms during postmortem can provide confirmative diagnosis if the worm is microscopically examined.

Symptomatological Diagnosis

The affected animals do not exhibit clinical signs, and confirmatory diagnosis is not possible by clinical signs.

Parasitological Diagnosis

Examination of fecal sample may reveal the presence of eggs of the parasite. If no eggs are found in direct smear examination, concentration technique comprising centrifugal floatation should be done. By examination of fecal sample, specific identification of the parasite is not possible up to genus level. It could be possible only when the feces is cultured and larval morphology is studied.

Immunological Diagnosis

Many works have been carried out using crude and purified antigens. So far no immunodiagnostic technique has been commercialized for field application.

Molecular Diagnosis

Molecular diagnosis of parasites has been a trend using specific molecular markers. Different ribosomal DNA, ribosomal RNA, and mitochondrial DNA could be used as molecular markers.

Treatment

Name of drug	Mode of administration	Dose
Fenbendazole	Oral	5–7.5 mg/kg for cattle, buffalo, sheep, goat
Albendazole	Oral	5–7.5 mg/kg for cattle, buffalo, sheep, goat
Mebendazole	Oral	10–15 mg/kg for cattle, buffalo, sheep, goat
Levamisole	Oral	7.5 mg/kg for cattle, buffalo, sheep, goat
Tetramisole	Oral	15 mg/kg for cattle, buffalo, sheep, goat
Ivermectin	S/C	0.2 mg/kg for cattle, buffalo

Control

General Management

The affected animals should be treated properly. The feces should be disposed properly. The floor of animal house should be cleaned properly. The animals should be provided with underground water.

Immunological Control

Different crude and defined antigens have been used in different trials, but no vaccine has yet been commercialized.

Molecular Control

No naked DNA vaccine or gene vaccine is available.

Biological Control

No biological control is commercialized for field application.

Other Species: *Trichostrongylus capricola*

Hosts: sheep, goat

Site: small intestine

Morphology

The spicules are proximally thicker. The distal ends of the spicules are not triangular as that of *T. colubriformis*.

T. colubriformis

Hosts: sheep, goat, cattle, etc.

Site: small intestine and sometimes abomasum

Morphology

The spicules are equal in size. The spicules have small boatlike structure with an outgrowth which caps the root proximally. The gubernaculum is present laterally.

T. vitrinus

Hosts: sheep, goat, deer; occasionally rabbit, camel, and even man

Site: small intestine

The spicules are short and straight and have sharp tapering distal end. Gubernaculum is simple.

T. probolorus

The two spicules are approximately of the same length. Two triangular projections are present on the ventral side. The shiny gubernaculum is dark brown in color.

T. longispicularis

The spicules are thicker proximally compared to distal end. The gubernaculum is light brown in color and normal in structure.

T. falculatus

Hosts: sheep, goat, antelope
Site: small intestine

T. rugatus

Hosts: sheep, goat
Site: small intestine

T. retortaeformis

Hosts: rabbit, hare, goat
Site: small intestine and rarely abomasum

T. tenuis

Hosts: duck, goose
Site: small intestine and cecum

Genus: *Graphidium*

Introduction

These are roundworms of rabbits. The worms occur in the stomach and intestine.

Species: *Graphidium strigosum*

Hosts: rabbit and hare
Site: stomach and small intestine
Distribution: The worms are prevalent in Europe.

Morphology

The average length of male and female parasites is 1.2 cm and 1.6 cm, respectively. The worms are red in color. Many longitudinal lines are present along with transverse striations. The bursa has two large lateral lobes and one small dorsal lobe. The male possesses two thin spicules.

Life Cycle

Very little information is available on the life cycle of this parasite. The eggs are passed with the feces. The eggs hatch in the environment in the presence of suitable environmental condition. The larvae reach the infective stage after two moltings. The hosts become infected by ingestion of infective larvae along with food and drinking water. The larvae are found in the gastric glands but the adults are found in the mucosa.

Pathogenesis

The parasites affect the stomach, small intestine, cecum, etc. Catarrhal inflammation and fibrosis are some pathological features.

Genus: *Obeliscoides*

Introduction

These are worms of rabbits. The worms occur in the stomach.

Species: *Obeliscoides cuniculi*

Hosts: rabbit
Site: stomach

Morphology

The worms are brownish red in color. The body is striated with longitudinal ridges. Cervical papillae are present. The spicules are bifurcated at distal part and end in hooked process.

Life Cycle

The life cycle is direct. The hosts get the infection by ingestion of infective larvae. The larvae develop into adult in the stomach.

Pathogenesis

The parasite is apparently harmless. But heavy infection may cause harm to the hosts.

Diagnosis

Diagnosis can be done by finding the eggs in the feces. The adults can be identified during necropsy.

Genus: *Ostertagia*

Common name: Brown stomach worm

Morphology

Gross

Shape: cylindrical
Size: medium sized worm (about 5–10 mm)
Color: They are brown in color. They are called brown stomach worm.

Microscopic

Adult

Cuticle: The cuticle of the anterior extremity of the worms has transverse striations. The remaining part of the body is devoid of any striation but has longitudinal ridges.

Head: The cuticle of the head may have a vesicle. They have small buccal cavity.

Bursa: The bursa is small having lateral and dorsal lobe.

Spicule: A pair of spicules are present which are unequal in size and brown in color. *O. ostertagi* has spicules with three stubby hooked process. *O. circumcincta* has spicule with a large knobbed tip along with an acute process. *O. trifurcata* has spicule with stout knobbed tip and a pair of spurs behind the middle.

Eggs

The eggs are simple strongyle type.

Larva

The third stage juvenile of *O. ostertagi* has club-shaped esophagus and has bulged appearance at the junction with the intestine. The tail of the sheath differs from other two species in that it gradually ends in blunt tip and it is longer and more blunter than other two species.

Life Cycle

- The eggs are passed with feces.
- Hatching of eggs occurs in the environment.
- Formation of infective third stage larvae after two moltings in 2 weeks.
- Infection of the hosts by ingestion of infective larvae.
- Exsheathment of larvae occurs in the stomach.
- The infective juveniles penetrate the abomasal mucosa found after 6 h of infection.
- They are found in the gastric glands after 12 h of infection.
- Ecdysis is completed 24 hours later.
- The development of early fourth stage may be arrested for about 6 months called **hypobiosis** under certain circumstances.
- The L5 emerge out from the glands and mature sexually about 18 days of infection.

Pathogenesis

The juveniles cause serious damage to the glandular cells. They cause nodule formation. The nodules contain cysts within which the fourth stage larvae are found. Hemorrhages are found on the wall of the abomasum. Zymogen cells are denuded. Due to this condition, functional disturbance of the abomasum occurs. The parasites are bloodsuckers. They cause emaciation and anemia. The persisting pathological features are loss of glandular epithelial cells.

One important pathological feature has to be noted here. Heavy infection causes drastic pH change. The pH is increased from 2.0 to 7.0. Pepsinogen is not activated in this high pH and pepsin is not formed. This results in elevated level of pepsinogen concentration in the blood. Leakage of endogenous protein in the lumen of the GI tract results in hypoproteinemia. High load of infection results in hyperplasia of mucosa giving 'morocco leather' like appearance.

Clinical Signs

Type 1: In northern hemisphere, Type 1 ostertagiosis occurs from mid July onward due to intensive grazing during 1st grazing season. During this time, L3 is abundantly available in the pasture and they get matured in short period. In this type of ostertagiosis, morbidity is high but mortality is very less. Profuse diarrhea is common. Thirst is increased. Loss of body weight is common.

Type 2: Type 2 ostertagiosis occurs usually in late winter or spring. This phase of the disease occurs due to maturation of the larvae which were ingested in the autumn and underwent arrested development in early fourth stage. Profuse diarrhea occurs as that of Type 1 ostertagiosis. During Type 2 ostertagiosis, severe hypoproteinemia occurs. Edema is found in the intermandibular region. High level of mortality is found.

Diagnosis

Clinical signs: Clinical signs include diarrhea, inappetence, and loss of body weight along with grazing history.

Seasonal History

The occurrence of the disease in the seasons of a particular year is indicative.

Grazing History

Grazing history in the autumn season is a supportive diagnosis.

Plasma-Pepsinogen Level

Estimation of plasma pepsinogen level can be done in case of the young animals up to 2 years of age. The normal level is 1.0 IU. It becomes excess of 3.0 IU in parasitized animals.

Postmortem Diagnosis

Adult worms are found attached on wall of the abomasal surface which may reach the number of more than 40,000.

Genus: *Cooperia*

Species: *Cooperia oncophora*

Hosts: The worms occur in small intestine and rarely in abomasum of cattle, sheep, camel, horse

Morphology

Gross

Shape: The worms are cylindrical in shape. The anterior end is coiled.

Size: They are medium sized. The male is 5.5–9 mm long and the female is 6–8 mm long.

Color: Color is pink or reddish when fresh.

Microscopic

Adult

The anterior extremity has cephalic swelling or cephalic vesicle. In the esophageal region of the worm, there have transverse striations. Longitudinal ridges are also present in the rest of the body. A pair of spicules are present which are brown in color and have winglike expansion and are unequal in size. They are bursate worms having two large lateral lobes and a dorsal lobe. The eggs are comparatively large. The female tail is sharply pointed.

Egg

The eggs are segmented when laid. The size of the egg is 75–95 um × 36–45 um. Sides of the eggs look almost parallel. The eggs contain 16–32 embryonic cells.

Larva

The third juveniles contain 16 cells. According to some workers, the infective larvae possess two conspicuous oval bodies on either sides of the lumen of the esophagus.

Life Cycle

- Un embryonated eggs are passed with feces and undergo embryonationin the environment.
- Hatching of eggs occurs in the environment.
- The infective third stage larvae are formed after two moltings.
- The hosts get infection by ingestion of infective larvae.
- The infective juveniles penetrate and feed the intestinal wall.
- They enter the crypts in between the villi. Third molting occurs there and they transform to fourth stage after about 4 days of infection.
- The fourth juveniles come to the lumen of the intestine. Fifth stages are formed after 9 days of infection.

Pathogenesis

C. oncophora is less pathogenic in calves. The pink color of the worms indicate that they are blood-feeders. Repeated infections cause formation of nodules. Immunological reaction results in inflammation on the wall of the intestine.

Diagnosis

Diagnosis can be done by finding the eggs in the feces. For specific identification, fecal culture should be done.

Species: *Cooperia punctata*

They occur in the abomasum, small intestine of cattle, sheep and deer. The morphology is similar to *C. oncophora*. The males are 4.7–5.9 mm long and the females are 5.7–7.5 mm long. The life cycle is similar to *C. oncophora*. They are more pathogenic than *C. oncophora* as the adults penetrate the epithelial layer of the intestine resulting in villous atrophy leading to functional disturbances of the intestinal wall.

Species: *Cooperia pectinata*

They occur in the small intestine of cattle, sheep and deer. The morphology is similar to *C. oncophora*. The males are 7 mm long and the females 7.5–9 mm long. The life cycle is similar to *C. oncophora*. Pathogenicity is similar to *C. punctata*

Species: *C. curticei*

The worms occur in abomasum and small intestine of sheep, goat, cattle.

Genus: *Haemonchus*

Species: *Haemonchus contortus*

Introduction

These are important worms occurring in the stomach of sheep, goat, cattle, etc. They are called stomach worm, wire worm, barber's pole worm.

Morphology

Gross

Shape: The worms are cylindrical in shape.

Size: The males are 10–20 mm long and the females are 18–30 mm long.

Color: Color is pink or gray in accordance with state of digestion of the blood in the GI tract.

Microscopic

Adult

Head: The head is simple. A small buccal cavity is present. A pair of cervical papillae are present.

Teeth/lancet: A tooth or lancet is present in the buccal cavity.

Specific feature: These worms are called barber's pole worm. The white ovary winds around the red intestine giving the appearance of barber's pole.

Spicules: A pair of spicules are present which are barbed posteriorly.

Bursa: They are bursate worms having a pair of large lateral lobes and small dorsal lobe. A Y-shaped dorsal ray is present in the dorsal lobe.

Egg

Eggs of *Haemonchus* are oval in shape. The size is 70–85 × 41–48 um. The color of the egg is yellowish. Sixteen to 32 blastomeres are present in the eggs.

Larva

The first stage larvae are 0.34–0.35 mm long. The intestine contains one row of eight roughly triangular cells. The tail of sheath of third juvenile is 0.119–0.165 mm long. The sheath extends 0.065–0.084 mm beyond the tail of the larvae. The larvae possess an oval buccal cavity. The esophagus is filariform. The intestine has 16 rectangular cells.

Note

The striped poles are found outside the barber's shop. The white colored ovary coils around the red blood-filled intestine giving striped appearance. The similarity with barber's pole led to coin the name barber's pole worm.

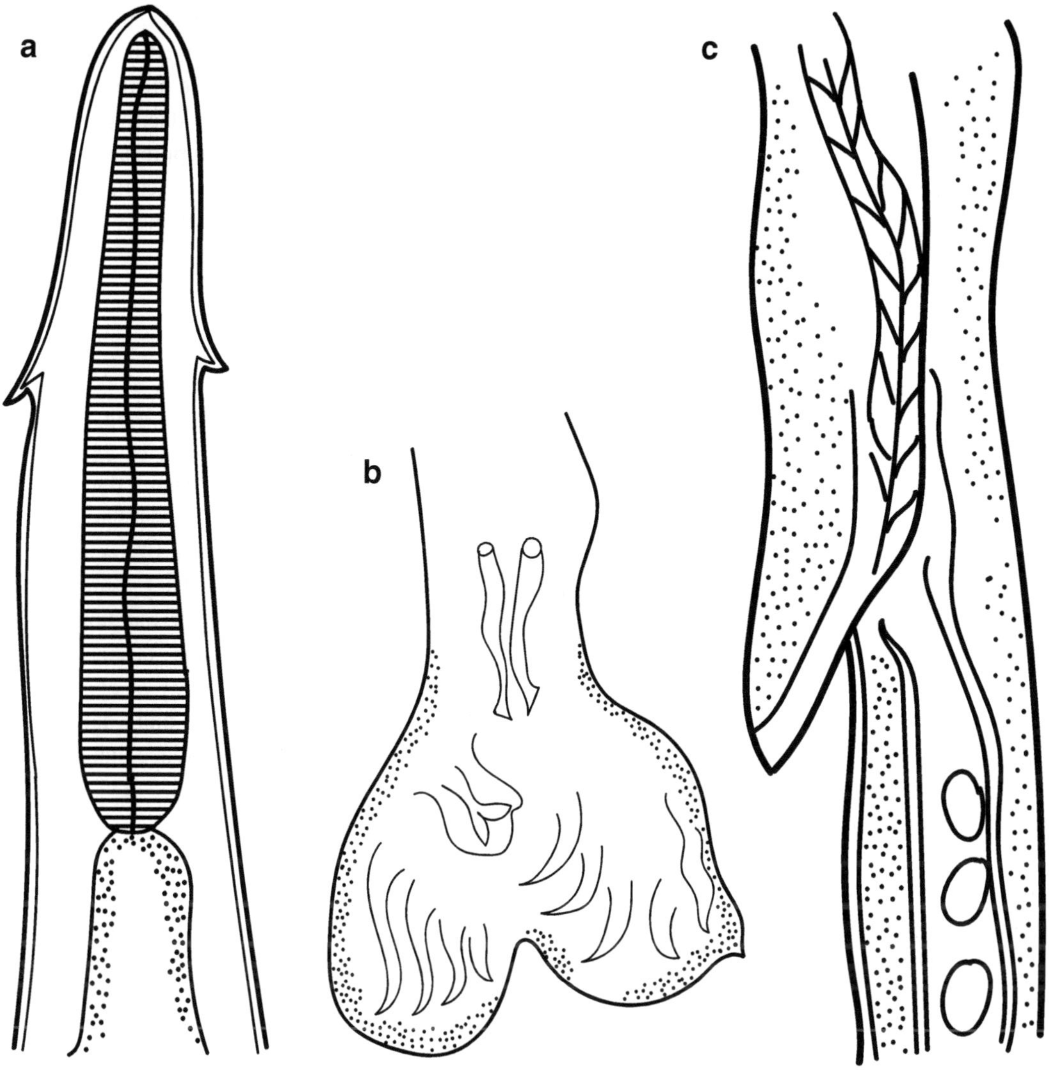

Fig. 6.19 *Haemonchus contortus*. (**a**) anterior end (**b**) bursa (**c**) vulval flap

Life Cycle

Some Basic Information of Life Cycle

Type of parasite	Homoexenous parasite
Type of life cycle	Direct life cycle
Type of definitive host	Vertebrate host
Location of fertilization	In the parasite
Location of embryonation	In the environment
Name of infective stage	Third larval stage (L3)
Mode of infection	Ingestion
Alternative mode of infection (hyperinfection, autoinfection, etc.)	No

Stages of life cycle formed

In the host	In the environment
• Fourth stage larvae • Adult and egg	• First stage larvae • Second stage larvae • Third stage larvae

Events of life cycle: a glimpse
• Voiding of eggs with the feces of hosts • Hatching of eggs in environment • Transformation of first stage larvae to third stage larvae after two moltings • Infection of hosts by ingestion of infective L3 • Development up to adult stage in the gut of the hosts

Detailed Description

The parasite is homoexenous and does not require any intermediate host.

Egg Passing

The adult parasites live in their predilection site. The female parasites lay eggs in the predilection site. The female parasite may lay about 10,000 eggs per day. The eggs are voided with the feces.

Embryonation

Embryonation of eggs occurs in the environment.

Hatching and Development of Infective Stage

The eggs are passed in the feces. The eggs are subjected to the exposure of different environmental factors. Humidity and temperature are very important for further development. It has been found that the eggs hatch within 4–6 days at temperature ranging between 24 and 29 °C. As per reports, 80% humidity is optimal for development. In the presence of suitable environmental condition, the eggs hatch. Rhabditiform first stage larvae come out. The first stage larvae transform to second stage larvae after one molting. The second stage larvae transform to third stage larvae after one more molting. About 5 days are required for formation of third stage larvae. The third stage larvae retain the sheath of second stage larvae which provide additional protection to the infective third stage larvae. These third stage larvae are filariform which are negatively geotropic and positively phototropic. They crawl on the vegetations. In the winter season, the development of larval stage is arrested. This is called hypobiosis. In the spring, the larvae resume their development. At a time, many eggs are shed in the pasture, and the existing larvae also start development which causes maximum larval precipitation in the pasture. This condition is called "spring rise." Periparturient rise attributes to the immunosuppression of animals due to parturition of the animals. The immunosuppressed animals pick up more infection which is termed as periparturient rise which is contributory factor of spring rise.

> **Note**
>
> Rogers and Brooke (1977) carried out a work on hatching mechanism of eggs of *Haemonchus contortus*. They detected lipase in the collected fluid of hatching eggs. This enzyme hydrolyzed 2-naphthyl laurate and L-leucinamide. They found the role of these substances for hatching of eggs. They also found the efficacy of exsheathing fluid for hatching of eggs. In that experiment, lipase and leucine aminopeptidase did not induce hatching eggs of *H. contortus*.

Infection of Hosts

The hosts get the infection when they inadvertently ingest the infective third stage larvae. The larvae pass through the three chambers of the stomach and reach the fourth chamber. Exsheathment of larvae occurs there. The exsheathed larvae then get attached to the wall of the stomach. The third stage larvae develop into adult worm after two moltings in the close apposition of the gastric glands. The prepatent period is 2–3 weeks in sheep and 4 weeks in cattle.

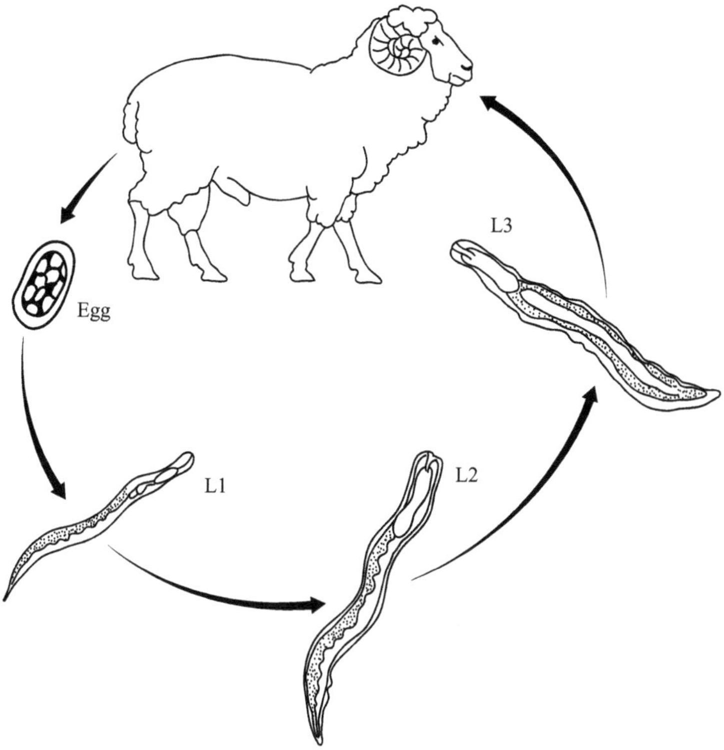

Fig. 6.20 Life cycle of *Haemonchus contortus*

> **Note**
>
> Emery et al. (2016) reviewed *Haemonchus contortus* with an interesting aspect which is highlighted below:
>
> *Haemonchus contortus* has significantly changed its distribution in Australia over the past 70 years.
>
> Inhibition of *H. contortus* L4s over winter has largely been abolished by more effective anthelmintics.
>
> Inhibition of *H. contortus* L4s over winter has resulted in the advent of yearly rather than seasonal haemonchosis.
>
> Customized on-farm control and monitoring programs should arise from rapid advances in *H. contortus* genomics.

Self-Cure

The predisposing factor of self-cure is an immunological reaction mediated by a hypersensitivity reaction. It occurs as rapid expulsion of existing adult worms due to ingestion of challenge infection of larvae. In the pre-sensitized animals which have already got an infection, if subsequent infection occurs, Fc part of IgE (already antigen-charged) binds to different effector cells, i.e., mast cell, basophil, etc. The effector cells release different vasoactive amines, i.e., histamine. This histamine increases the peristaltic movement. Simultaneously goblet cells increase in number and produce large amount of mucus. This causes expulsion of the worms. This phenomenon is called self-cure phenomena.

Spring Rise

The arrested larvae resume their activity when the spring season commences, and this season is suitable for survival of the larvae. During this period, parturition of cows causes lowering of immunity and results in maturity of the larvae. The animals pass a large number of eggs. This double action increases the number of larvae in the pasture. This is called spring rise.

Pathogenesis

Haemonchus contortus is a bloodsucking parasite. The seat of predilection of this parasite is the stomach. In the stomach, the buccal cavity is attached to the blood vessels. They suck blood continuously. A single worm may cause loss of 0.05 ml per day by sucking and seepage. An anticoagulant substance is secreted by these parasites which inhibit clotting of blood. The parasites cause lesions in the wall of the stomach. As the parasites frequently change their sites, they can cause many bite marks. In acute stage of infection, dramatic fall of PCV value is observed. Thereafter, gradual compensation of erythropoiesis occurs but in lesser extent which cannot meet up the loss of blood. Continuous drainage of protein and iron makes the bone marrow exhausted, and it completely fails to replenish the loss of blood and results in death of the animals. Sudden death of sheep occurs due to heavy infection called hyperacute haemonchosis.

Pathology

Anatomical Pathology (Gross)

Lesions in the stomach wall and bite marks occur.

Histopathology

Microscopical observation reveals infiltration of mononuclear cells, i.e., lymphocytes, plasma cells, etc. in the gastric glands along with necrosis. Periglandular hyperemia and hemorrhage are also found. Hyperplasia of the mucous glands and proliferation of connective tissue are also observed.

Hematopathology

Hematopathological study displays decreased TEC, Hb%, and PCV. Increased count of circulating eosinophil and decreased count of neutrophil are also observed.

Diagnosis

Symptomatological Diagnosis

No specific diagnostic symptoms are found. However, marked anemic condition of many animals at the season of *Haemonchus* infection reflected by pale mucus membrane of the eye is indicative of *Haemonchus* infection. During the season of *Haemonchus* infection, the feces color is also changed which appears black.

Parasitological Diagnosis

Examination of fecal sample may reveal the presence of eggs of the parasite. If direct smear technique fails, centrifugal floatation technique should be done. Coproculture technique can be done for collection of specific larvae. The isolated larvae should be perfectly identified.

Immunological Diagnosis

Many works have been carried out using crude and purified antigens. All sorts of immunodiagnosis for haemonchosis are in progress. The antigens which are used for this purpose are whole somatic antigen of adult extract, whole somatic antigen of larval extract, and excretory-secretory antigen of both adult and larva. Different purified proteins were also immunologically characterized. These are cysteine protease, cathepsin-like protein, etc.

Molecular Diagnosis

Different ribosomal DNA, ribosomal RNA, and mitochondrial DNA could be used as molecular markers. Studies indicate the use of ITS1, ITS2, etc. as molecular markers for identification of this parasite.

Treatment

Name of drug	Mode of administration	Dose
Fenbendazole	Oral	5–7.5 mg/kg for cattle, buffalo, sheep, goat
Albendazole	Oral	5–7.5 mg/kg for cattle, buffalo, sheep, goat
Mebendazole	Oral	10–15 mg/kg for cattle, buffalo, sheep, goat
Levamisole	Oral	7.5 mg/kg for cattle, buffalo, sheep, goat
Tetramisole	Oral	15 mg/kg for cattle, buffalo, sheep, goat
Ivermectin	S/C	0.2 mg/kg for cattle, buffalo

Control

General Management

The affected animals should be treated properly. Proper disposal of feces is essential because feces is the source of eggs of this parasite.

Immunological Control

Different crude and defined antigens have been used in different trials. But no vaccine has yet been commercialized. The antigens which are used for this purpose are whole somatic antigen of adult extract, whole somatic antigen of larval extract, and excretory-secretory antigen of both adult and larva. Different purified proteins have also been used for this purpose, and variable success was achieved. Some of the research trials conferred satisfactory results. These antigens are H11, P46, P52, P100, H-gal-GP, glutathione s-transferase, etc. The vaccine, Barbervax is commercially available.

Molecular Control

No naked DNA vaccine or gene vaccine is available.

Biological Control

No biological control has been introduced for field application. Works were carried out on biological control of *H. contortus* using *Duddingtonia flagrans, Arthrobortrys oligospora, Poconia spp. etc.*

Note

There has been a quest for search of new drug for treatment of hookworms. An interesting work was carried out by Bouchery et al. (2018), and they showed that the hookworms might be treated in another way. Like many other bloodsucking parasites, hookworms also produce dark brown pigment, hemozoin (detoxified form of heme). The detoxification pathway is necessary for survival of the worms, otherwise, they may be damaged by the toxin heme. Any substance blocking this detoxification pathway can be useful to reduce the growth of or kill the parasite. The present workers were successful in this field using quinoline. Research works in this direction can be done to develop new drug against haemonchosis.

See the Chapter "Trends in control of livestock and poultry parasites" for more information.

Treatment

Nanomedicine

Tomar and Preet (2017) found in vitro anthelmintic activity of silver nanoparticle (AgNPs) synthesized by using the aqueous extract of *Azadirachta indica*. Assessment of efficacy was based on egg hatch assay, adult motility inhibition assay, etc. They found potent anthelmintic efficacy of that nanoparticle.

Haemonchus placei

This species occurs in the abomasum of cattle, sheep, goat, etc. This species is similar to *H. contortus*. This species differs from other species in having longer spicules. Valval flap is rounded or knob-like.

Haemonchus similis

The worms are distributed in North America, Europe, etc. This species occurs in the abomasum of cattle, sheep, deer, etc. They are reddish in color. This is smaller than *H. contortus*. It has longer terminal process of dorsal ray which is the differentiating morphological feature from *H. contortus*. Each branch of dorsal ray is bidigitate. The pathogenesis is more or less similar to *H. contortus*.

Haemonchus longistipes

This species occurs in the abomasum of camel. The esophagus is longer than in *H. contortus*.

Genus: *Nematodirus*

These are comparatively long worms having inflated cuticle at the anterior end. The bursa have long lateral lobes. The spicules are fused at the anterior tip. The species are *N. filicollis, N. pathiger* occurring in the sheep, goat, and cattle. *N. battus* occurs in sheep, deer etc. The inner surface of bursa has oval or round cuticular bosses.

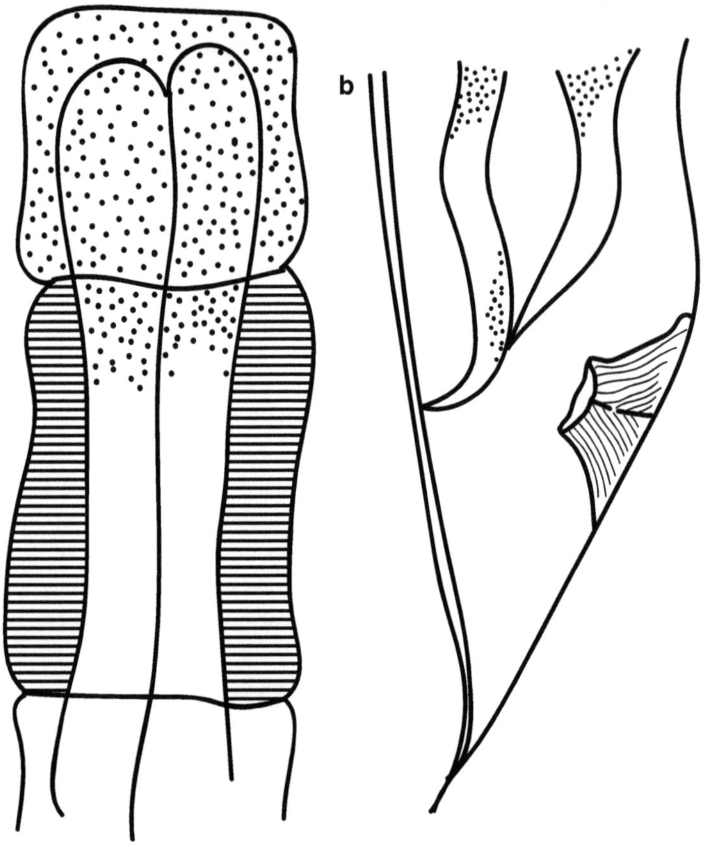

Fig. 6.21 *Nematodirus battus* (**a**) head end (**b**) female tail end

Genus: *Mecistocirrus*

Species: *Mecistocirrus digitatus*

Hosts: sheep, goat, cattle, zebu, buffalo, etc.
Site: abomasum, duodenum, intestine

Morphology

The male and female are 16–30 mm and 19–40 mm long, respectively. These are slender worms. Transverse striations are present on the cuticle at the head region. Many longitudinal ridges are also present in the cuticle. A lancet is present in the buccal capsule as that of *H. contortus*. The spicules are long. Most of the morphological features are same as that of *Haemonchus*. This species also have the ovary wound around the intestine. The spicules of *Mecistocirrus* are different from *Haemonchus*. They are long and narrow and fused at most part of them. The characteristic differentiating feature is that the female lacks any vulval flap.

Life Cycle

Some Basic Information

Type of parasite	Homoexenous parasite
Type of life cycle	Direct life cycle
Type of definitive host	Vertebrate host
Location of fertilization	In the parasite
Location of embryonation	In the environment
Name of infective stage	Third larval stage (L3)
Mode of infection	Ingestion
Alternative mode of infection (hyperinfection, autoinfection, etc.)	No

Detailed Description
Type of life cycle: direct

The following events are observed in life cycle of this worm:

- Voiding of eggs with the feces
- Hatching of eggs in the environment
- Transformation of first stage larvae to infective third stage larvae after two consecutive moltings
- Infection of the hosts by ingestion of infected third stage larvae
- Development up to adult in the gut of the hosts

Pathogenesis
The parasites cause marked hemorrhage at the affected part. Degeneration, necrosis, edema, desquamation, etc. are some pathological features which occur at the predilection site. Like *Haemonchus*, they also suck blood and cause marked anemia.

Diagnosis
Examination of fecal sample may reveal the eggs.

Genus: *Marshallagia*

Species: *Marshallagia marshalli*
Hosts: sheep, goat
Site: abomasum and duodenum

Morphology
The length of male and female is 10–13 mm and 12–20 mm, respectively. Cervical papillae are present. Thirty to 35 longitudinal cuticular ridges are present. Transverse striations are absent. The spicules end in three processes. A copulatory bursa is present. The dorsal ray is noticeably long and bifurcated near the distal extremity.

Life Cycle
Type of life cycle: indirect

The following events are found in life cycle of this worm:

- The eggs are passed with the feces.
- Development up to second stage larvae occurs in the eggs.
- The eggs hatch in the environment and second stage larvae come out.
- The second stage larvae transform into infective third stage after one molting.
- The hosts get the infection by ingestion of the infective larvae.
- Following ingestion, exsheathment occurs. The larvae penetrate the abomasal mucosa wherein they form grayish white nodules. In the nodules, the development up to L5 occurs.
- Then the L5 stages come out from the nodules.

Pathogenesis
Serum pepsinogen content is increased. Hyperplasia of mucosal cells, loss of parietal cells, inflammation of affected area with predominant infiltrated lymphocytes, and replacement of functional parietal cells by undifferentiated cells have been described.

Genus: *Hyostrongylus*

Species: *Hyostrongylus rubidus*
Hosts: pig
Site: stomach
Common name: red stomach worm

Morphology
Fresh specimen is reddish in color. Both longitudinal and transverse striations are present on the body cuticle. The worm has cephalic vesicle like *Ostertagia* sp. Well-developed bursa has small dorsal lobe. 40–45 longitudinal striations are also present.

Life Cycle
Eggs are passed with the feces. The eggs are hatched in the environment. Larvae reach to infective stage in the environment. The hosts get the infection by ingestion of infective larvae. The larvae penetrate the gastric glands and they develop to adult there after two moltings. Life cycle is similar to *Ostertagia* sp.

Pathogenesis and Clinical Signs
The parasites penetrate the gastric glands and suck blood. Like *Ostertagia*, the parietal cells are replaced by the developing undifferentiated cells

resulting in formation of nodules. They cause gastritis which may turn to ulceration later. Light infection fails to produce marked infection. But heavy infection causes different clinical signs comprising diarrhea, inappetence, vomiting, anemia, etc.

Diagnosis

Examination of fecal sample may reveal the eggs. But confirmatory diagnosis can be done only by autopsy.

Genus: *Ornithostrongylus*

Species: *Ornithostrongylus quadriradiatus*

Hosts: pigeons

Site: crop, proventriculus, small intestine

Distribution: The worms are distributed in Europe, Australia, North America, South Africa, etc.

Morphology

The worms are up to 2.5 cm long. Fresh specimen is red in color. The head cuticle is inflated. The body cuticle has longitudinal striations. The spicules end in three processes.

Life Cycle

Type of life cycle: direct

The following events are found in the life cycle of this worm:

- The eggs are passed with the feces.
- Hatching of eggs occurs in the environment.
- The first stage larvae transform into infective third stage after two consecutive moltings in the environment.
- The hosts become infected by ingestion of the infective third stage larvae.

Pathogenesis

The parasites are bloodsuckers. Degeneration of cells of affected area results in necrosis. Hemorrhagic enteritis may occur. The affected area may get ulcerated.

Diagnosis

Examination of fecal sample may reveal the eggs.

Genus: *Libyostrongylus*

Species: *Libyostrongylus douglassii*

Hosts: ostrich

Site: Proventriculus, gizzard

Morphology

The male worms are 4–6 mm and the female worms are 5–6 mm. Fresh specimen is yellowish red in color. The bursa is large in size. The spicules end in a large and a small spine.

Life Cycle

Type of life cycle: direct

The following events are found in the life cycle of this worm:

- The eggs are passed with the feces.
- Hatching of eggs occur in the environment.
- The first stage larvae transform into infective third stage after two moltings. The development of larvae depends on environmental temperature ranging from 10 °C to 37 °C.
- The hosts become infected by ingestion of the infective third stage larvae.
- The larvae enter the proventricular glands and grow up to adult there.

Pathogenesis

The worms cause severe irritation at the wall of proventriculus. They cause destruction of proventricular glands. They suck blood there. The affected birds may be died. The clinical signs include anemia, emaciation, stunted growth etc.

Treatment

Different benzimidazole compounds are effective.

Libyostrongylus dentatus

Hosts: ostrich

Site: proventriculus, gizzard

Morphology

The male worms are 6–8 mm and the female worms are 10–12 mm. A dorsal esophageal tooth is a characteristic morphological feature. The bursa is large in size.

Order: Strongylida—Superfamily: Trichostrongyloidea—Family: Ollulanidae

Family Synopsis

The members of this family occur in the stomach of dog, cat, etc. The worms are characterized by the presence of two, three, or more caudal processes. They are ovoviviparous or viviparous. The life cycle is direct type involving no intermediate hosts. They are apparently not harmful but may cause enteritis sometimes. Diagnosis is done by finding the larvae in the feces. The benzimidazole compounds are found effective against this infection.

Genus: *Ollulanus*

Species: *Ollulanus tricuspis*

Hosts: cat, fox
Site: stomach

Morphology

The head of this worm is spirally coiled. The buccal cavity is small. The male has well-developed bursa. The spicules split into two for a certain distance. The female tail ends in three or more short cusps.

Life Cycle

Type of life cycle: direct

The following events are found in life cycle of this worm:

- The females are larviparous. The eggs hatch within the uterus only. The larvae develop into third stage in the uterus.
- The third stage larvae are passed with the vomitus.
- The healthy cats get fresh infection by ingestion of vomitus containing third stage larvae or contaminated food and drinking water.
- The third stage larvae further develop into adult in the stomach itself.

Pathogenesis

The worms cause gastritis which may turn to chronic gastritis. Initially inflammation with different infiltrative cells occurs. There occurs erosion of mucosal layer. Fibrosis in mucosal layer occurs later. There may be carcinogenesis in long-standing cases.

Diagnosis

Gross examination reveals the adult worms in the vomitus. Microscopical examination of vomitus material reveals larvae. Nothing is found by fecal examination because if any larvae and adults pass to the intestine, they die and are destroyed. Artificial induction of emesis is done by xylazine and vomitus material is examined. Stomach irrigation and washing are done for detection of parasite or parasitic stages. Baermann technique can be done for concentration of live larvae and adults.

Order: Strongylida—Superfamily: Trichostrongyloidea—Family: Dictyocaulidae

Family Synopsis

The members of the family are the lungworms of cattle, sheep, goat, horse, etc. These worms are important as they cause loss of morbidity and mortality of the affected animals. Bursa of the worms has long bursal rays; the rays may be fused at the distal end in some members. The worms have short and stout spicule. The life cycle is direct type. The worms cause marked damage to the respiratory system of the hosts. Diagnosis is done by examination of feces for detection of first stage larvae in the fresh feces. Control is done by chemotherapy and grazing management.

Species: *Dictyocaulus filaria*

Introduction

These are important roundworms occurring in the lungs of sheep, goat, and cattle. They are called lungworms. The worms cause marked pathology resulting in significant loss in livestock economy.

Common name: Thread lung worm of sheep

Hosts: sheep, goat etc.

Site: bronchi and lungs

Distribution: The worms are cosmopolitan in distribution.

Morphology

Gross

Shape: These worms are cylindrical in shape and threadlike in appearance.

Size: The male is 3–8 cm long and female is 5–10 cm long.

Color: Color is milky white with the intestine visible from outside as a dark band.

Microscopic

Adult

Buccal capsule: They have small buccal capsule having four lips.

Esophagus: The esophagus is club-shaped.

Bursa: The bursa is short. The ventral rays are fused proximally. Fused bursal rays appear as fingers.

Spicule: The spicules are dark brown and boot shaped.

Egg

The eggs are oval or ellipsoidal. The size is 0.112–0.138 × 0.052–0.094 mm. The eggs are larvated when laid. The eggs do not have any specific identifying character.

Larva

First juvenile is 0.55–0.585 mm long. A characteristic knob-like button is present at the anterior end. Second juvenile is almost same in size as first juvenile. Knob -like button is absent in second juvenile. Third juvenile has two sheaths. Mouth opening and oesophagus with bulb are found. The intestine contains 8 cells. The larvae of *Dictyocaulus filaria* have brown food granules in the intestinal cells.

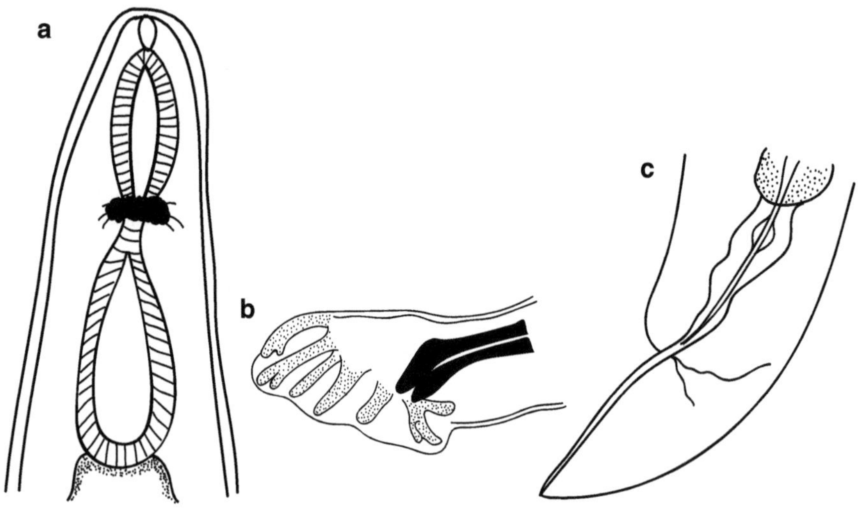

Fig. 6.22 *Dictyocaulus filaria*. (**a**) head end (**b**) bursa (**c**) female tail end

Knowledge Hunting Box

Johnson et al. (2003) studied to see the cross infectivity of *Dictyocaulus* sp. between cattle and red deer. They demonstrated that both cattle and red deer could be infected with either *D. viviparus* or *D. eckerti*. However, *D. eckerti* larvae that originated from deer established more successfully in deer, and *D. viviparus* larvae that originated from cattle established more successfully in cattle

Life Cycle

Basic information of life cycle
• It is a direct type life cycle • The hosts are sheep, goat etc • The infective stage is third stage larvae • Infection of the hosts occurs by ingestion of infective stage of the parasite

Stages of Life Cycle Formed

In host	In environment
• Third stage larva • Fourth stage larva • Adult • Egg • First stage larva (while passing through GI tract)	• Second stage larva • Third stage larva

Events of life cycle: a glimpse
• Hatching of eggs while passing through through GI tract
• L1 transform into L3 after two moltings
• Infection of hosts by ingestion of L3
• L3 reach the lymphatic vessel and mesenteric lymph glands
• Formation of L4 after molting in the mesenteric glands
• Larvae reach to the lungs. Then they develop to adult in the bronci and bronchioles

Detailed Description

Embryonation

Embryonation of eggs occurs in the worm itself.

Egg Laying

The adult parasites reside in the bronchi where the gravid females lay embryonated eggs. The eggs are coughed up and enter the GI tract.

Egg Hatching

The eggs hatch while passing through the GI tract. The eggs are hatched in the bronchioles also.

Larval Development in Environment

The first stage larvae transform into third stage larvae after two moltings within 1 week. The infective larvae are slow moving and very much vulnerable to desiccation. The larvae remain nearer to defecation but may be spread by bursting of sporangium of pilobolus to expel the spores. The third stage infective larvae exhibit vertical and horizontal migration. The juveniles have least chance to reach the herbage unless the larvae are spread in this way or some other means.

Infection of Hosts

The hosts get the infection by ingestion of infective stage larvae along with grass while grazing. The beddings and hays may be contaminated by the larvae, from where the hosts may get infection.

Development in Host

After ingestion, the larvae penetrate the wall of the gut and reach the lymphatic vessel and then to mesenteric lymph gland after about 4 days of infection. They then reach to the thoracic duct from where they are transported to the lungs. In the lungs, they migrate into the capillaries and destroy the air ways.They then ascend the upper respiratory passage and eventually reach to the bronchi where they get molted to become adult at about 4 weeks of infection.

Pathogenesis

By Larvae

After exsheathment, the juveniles penetrate the intestinal wall and cause damage, thereby forming hemorrhagic spots. The mesenteric lymph nodes are also affected by the larvae causing congestion and edema.

By Larvae and Adult

Both larvae and adult parasites cause severe pathology in the lungs. They rupture the lung alveoli. There occurs production of copious exudate and mucus material which block the bronchioles resulting in alveolar emphysema. Consolidation and atrophy of some regions of the lungs occur. Hypertrophy of bronchial tissue and hyperplasia of the peribronchial lymphatics are observed. Pneumonia is also associated with pathological feature. Adhesion of the alveoli with the body wall and diaphragm is found.

Symptoms

- Coughing
- Dyspnea
- Tachypnea
- Nasal discharge
- Unthriftiness

Diagnosis

Symptomatological Diagnosis

The clinical signs are not specific for confirmatory diagnosis. The affected animals produce typical "husk and hoose" sound which is indicative of *Dictyocaulus* infection.

Parasitological Diagnosis

Examination of fecal sample may reveal the presence of first stage of larvae. We can perform Baermann technique for better isolation of larvae. Nasal discharge and bronchoalveolar lavage could be examined for identification of eggs and larva of the parasite.

Immunological Diagnosis

Different immunodiagnostic tests have been carried out using crude and purified antigens. ELISA test has been found better. But no immunodiagnostic technique has yet been commercialized for confirmatory diagnosis.

Molecular Diagnosis

Different molecular markers are used for identification of parasites. These markers are mainly of ribosomal DNA and RNA. Mitochondrial DNA are also used as molecular markers for identification of parasites. Studies indicate use of cytochrome c oxidase (cox 1) and ITS2 as molecular markers for genomic characterization.

Treatment

Name of drug	Mode of administration	Dose
Fenbendazole	Oral	5–7.5 mg/kg for cattle, buffalo, sheep, goat
Albendazole	Oral	7.5 mg/kg for cattle, sheep, goat
Levamisole	Oral	7.5 mg/kg for cattle, sheep, goat
Tetramisole	Oral or parenteral	15 mg/kg for cattle, sheep, goat
Ivermectin	S/C, I/M	0.2 mg/kg for cattle, buffalo, sheep, goat

Control

General Management

The affected animals should be treated properly. Since the parasites perform direct life cycle, feces should be disposed properly. The floor of the animal house should be cleaned properly. There is chance of getting infection from the pasture; so the animals should be grazed in clean and uncontaminated pasture. Pasture management comprising pasture resting, rotational grazing, etc. should be done.

Immunological Control

Different crude and defined antigens have been used in research trial for development of vaccine. Earlier attempts comprised of radiation-attenuated third stage larvae as vaccine candidate. Two vaccines were developed. **Difil** was the vaccine against *D. filaria* and was found efficacious after performing different immunological tests. A similar vaccine was prepared named as Dictol against *D. viviparus*. These vaccines are not available in the market due to different constraints. Later, efforts were made to produce vaccines using purified antigens. **Recombinant paramyosin** was found to reduce considerable worm load.

Molecular Control

No naked DNA vaccine or gene vaccine is available.

Biological Control

No biological control technology has been implemented for field use.

Species: *Dictyocaulus viviparus*

Introduction

These are important roundworms occurring in the lungs of cattle. They are called lungworms. The worms cause marked pathology resulting in significant loss in livestock economy.

Common name: bovine lungworm
Disease caused: husk-hoose disease
Hosts: Catte, buffalo, deer, etc.
Site: bronchi and lungs
Distribution: The worms are cosmopolitan in distribution.

Morphology

Gross

Shape: These worms are cylindrical in shape and threadlike in appearance. The body is characteristically enveloped in a teguminal sheath.
Size: The male is 4–5.5 cm long and female is 6–8 cm long.
Color: Color is white with the intestine visible from outside as a dark band.

Microscopic

Adult

Buccal capsule: They have small buccal capsule having four lips.
Esophagus: The esophagus is club-shaped.
Bursa: The bursa is short. The ventral rays are fused proximally. Fused bursal rays appear as fingers.
Spicule: The spicules are dark brown and boot shaped.

Egg

The eggs are oval and larvated when laid. The females are ovoviviparous.

Larva

The larvae of *Dictyocaulus viviparus* have brown food granules in the intestinal cells. The tail is blunt. Cuticular knob is absent in L1. The larvae (L1) are 300–360 um in length. They have brown food granules in the intestinal cells. The tail is straight. The 3rd stage larvae are similar to *D. filaria*.

Life Cycle

Basic information of life cycle
• It is a direct type life cycle • The hosts are cattle, buffalo, etc. • The infective stage is third stage larvae • Infection of the hosts occurs by ingestion of infective stage of the parasite

Stages of Life Cycle Formed

In host	In environment
• Fourth stage larvae • Adult • Eggs • First stage larva (while passing through GI tract)	• Second stage larvae • Third stage larvae

Events of life cycle: a glimpse
• Hatching of eggs while passing through the gut
• L1 transform into L3 after two moltings
• Infection of hosts by ingestion of L3
• L3 reach the lymphatic vessel and transform to L4
• Then the larvae reach the lungs
• Formation of L5 after molting in the bronchi, bronchioles

Detailed Description

The life cycle is essentially similar to *D. filaria*.

Embryonation

Embryonation of eggs occurs in the worm itself.

Egg Laying
The females lay their eggs in their predilection site. The eggs pass in the alimentary canal.

Egg Hatching
The eggs hatch while passing through the GI tract. Even the eggs are hatched in the bronchiole.

Larval Development in Environment
The first stage larvae transform into second stage larvae after one molting. The second stage larvae molt to become third stage larvae. The infective larvae are slow moving and very much vulnerable to desiccation. The larvae remain nearer to defecation but may be spread by bursting of sporangium of pilobolus to expel the spores. The juveniles have least chance to reach to the herbage unless the larvae are spread in this way or some other means.

Infection of Hosts
The hosts get the infection by ingestion of infective stage larvae along with grass while grazing. The beddings and hays may be contaminated by the larvae, from where the hosts may get infection. The drinking water may be another source of infection.

Development in Host
Exsheathed larvae penetrate the wall of the gut and reach the lymphatic vessel. From the lymphatic vessel, they reach to the mesenteric lymph glands where one molting occurs. They are transported to the lungs via thoracic duct. In the lungs, they get molted to become fifth stage after one molting. Then they migrate to the bronchi.

Pathogenesis

By Larvae
After exsheathment, the juveniles penetrate the intestinal wall and cause damage, thereby. There may be gastroenteritis and diffused peritonitis.

By Larvae and Adult
The pathogenesis is described in three phases i.e. pre-patent phase, patent phase, and post-patent phase. In the pre-patent phase, damage of lung and bronchi result in alveolitis and bronchilitis. Immature worms are found in the alveolar airways. In the patent phase, hundreds or thousands of adult worms are found in the bulk of white mucus in the bronchial lumen. Detectable dark red collapsed areas are found in the bronchi. Post- patent phase is called recovery phase. The adult worms are expelled. Bronchopulmonary system become normal. Some heavily infected animals may be died.

Symptoms

- Coughing (Characteristic spasmodic sound is heard called "**husk**" or "**hoose**".)
- Dyspnea
- Tachypnoea
- Nasal discharge
- Unthriftiness

Diagnosis

History-Based Diagnosis
History-based diagnosis comprising the season, grazing history, previous records of lungworm infection, etc. is extensively helpful for diagnosis of the disease.

Symptomatological Diagnosis
The clinical signs are not specific for confirmatory diagnosis. The affected animals produce typical "husk and hoose" sound which is indicative of *Dictyocaulus* infection.

Parasitological Diagnosis
Examination of fecal sample may reveal the presence of first stage of larvae. We can perform Baermann technique for better isolation of larvae. Nasal discharge and bronchoalveolar lavage could be examined for identification of eggs and larva of the parasite.

Immunological Diagnosis

Different immunodiagnostic tests have been carried out using crude and purified antigens. ELISA test has been found better. But no immunodiagnostic technique has yet been commercialized for confirmatory diagnosis.

Molecular Diagnosis

Different molecular markers are used for identification of parasites. These markers are mainly of ribosomal DNA and RNA. Mitochondrial DNA are also used as molecular markers for identification of parasites. Studies indicate use of cytochrome c oxidase (cox 1) and ITS2 as molecular markers for genomic characterization.

Treatment

See the treatment of *D. filaria*

Control

General Management

The affected animals should be treated properly. Since the parasites perform direct life cycle, feces should be disposed properly. The floor of the animal house should be cleaned properly. There is chance of getting infection from the pasture; so the animals should be grazed in clean and uncontaminated pasture. Pasture management comprising pasture resting, rotational grazing, etc. should be done.

Immunological Control

Different crude and defined antigens have been used in research trial for development of vaccine. Earlier attempts comprised of radiation-attenuated third stage larvae as vaccine candidate. Two vaccines were developed. **Difil** was the vaccine against *D. filaria* and was found efficacious after performing different immunological tests. A similar vaccine was prepared named as **Dictol** against *D. viviparus*. These vaccines are not available in the market due to different constraints. Later, efforts were made to produce vaccines using purified antigens. **Recombinant paramyosin** was found to reduce considerable worm load.

Molecular Control

No naked DNA vaccine or gene vaccine is available.

Biological Control

No biological control technology has been implemented for field use.

D. arnfieldi

This species occurs in horse and donkeys. Pathogenicity is less compared to other spp. Old animals suffer from persistent coughing, nasal discharge, etc.

Order: Strongylida—Superfamily: Metastrongyloidea—Family: Metastrongylidae

Family Synopsis

The members of the family are the lungworms of pig. But sheep, deer etc. are also affected. These worms are important as they cause marked loss of morbidity of the affected animals. Bursa is reduced or absent. Bursal rays are fused. The worms have long and filiform spicules. The life cycle is indirect type. The worms cause marked damage to the respiratory system of the hosts. Diagnosis is done by examination of feces for detection of first stage larvae in the fresh feces. Control is done by chemotherapy and grazing management.

Genus: *Metastrongylus*

Introduction

These are important roundworms occurring in the bronchi of pigs. They are called lungworms of pigs. The worms cause significant pathology resulting in significant loss in the pig industry.

Species

Metastrongylus elongatus
M. salmi
M. pudendotectus

Species: *Metastrongylus elongatus*

Distribution: The worms are worldwide in distribution.

Hosts

Natural hosts: The common host is pig. The worms have also been recorded in sheep, deer etc.

Accidental hosts: man

Site: bronchi and bronchiole

Morphology

The worms are white in color. They possess six small lips. The bursa is small. The anterolateral ray is large having a swollen tip. The mediolateral and posterolateral rays are fused. The female tail has typical swelling. Each spicule ends in hooked structure. The egg wall has corrugated appearance.

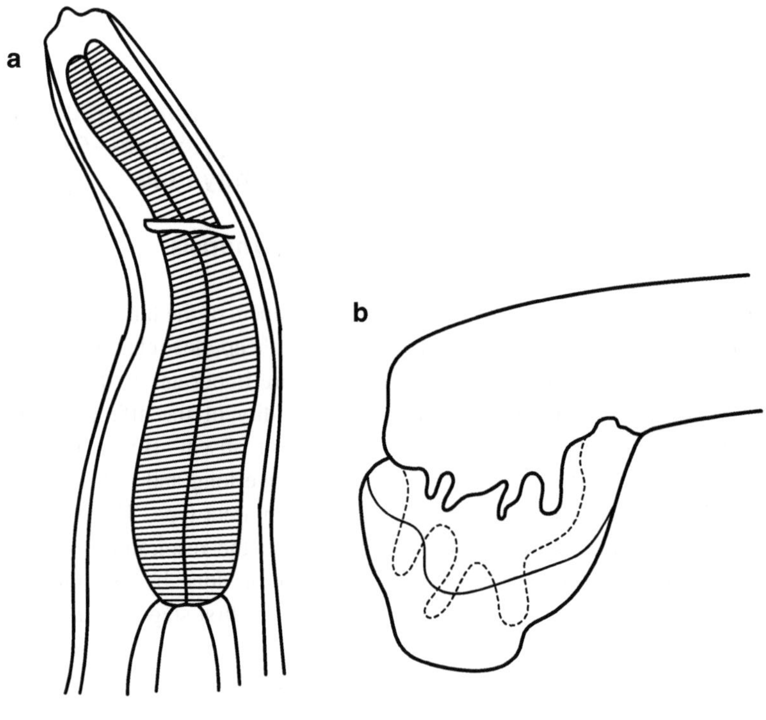

Fig. 6.23 *Metastrongylus* sp. (**a**) head end (**b**) bursa

Life Cycle

Basic Information on Life Cycle

Type of parasite	Heteroxenous parasite
Type of life cycle	Indirect life cycle
Type of definitive host	Vertebrate host
Type of intermediate hosts	Invertebrate hosts
Location of fertilization	In the parasite
Location of embryonation	Embryonation of eggs start in the host and is completed in the environment
Name of infective stage	Third larval stage (L3)
Mode of infection	Ingestion of the intermediate host harboring the infective larvae
Alternative mode of infection (hyperinfection, autoinfection, etc.)	No

Stages of Life Cycle Formed

In definitive host	In environment
• Fourth stage larva • Adult • Egg • First stage larva (after hatching in feces)	• Second stage larva • Third stage larva

The following events are observed in the life cycle of this worm:

- The embryonated eggs are coughed up and are passed with the feces. The larvated eggs hatch in the feces. The larvae can remain in feces and moist soil for long period.
- The larvae are ingested by earthworms (*Helodrilus*, *Lumbricus*). The eggs may be ingested by the earthworm and hatch in it's gut.
- Two ecdyses pass on in the earthworm and infective third stage larvae are produced. The larvae may remain viable in the earthworms for 1 year or more.
- The final hosts become infected by ingestion of the infected earthworms. Infection may also occur by ingestion of accidentally liberated larvae from earthworm.
- The larvae penetrate the gut and reach the mesenteric lymphatic glands where one molting occurs and third stage larvae transform into fourth stage larvae. The larvae then reach the lungs and grow to adult there.
- The adult females lay larvated eggs in the pharynx.
- The eggs are coughed up there.

Pathogenesis

The abnormal migratory route by the larvae may produce gray to white scar in the liver resulting in milk spot lesions there. When the larvae migrate in the lungs, they break the lung alveoli. The parasites mature there and further migrate to the bronchiole. The pathological features include alveolitis, bronchiolitis, etc. Atelectasis, alveolar emphysema, and bronchopneumonia are other features. It is reported that this roundworm harbors swine influenza virus. Individual effect, either by parasite or by virus, is less severe. But joint effect by these two causes severe pathology in the affected hosts. The parasites cause lobar pneumonia. They affect both the bronchioles and lungs. Desquamation of epithelial cells occurs. Localized inflammation with infiltration of cells takes place. All these result in formation of exudation. Associated purulent staphylococcus infection has been reported.

Symptoms

Clinical signs comprise of chronic and sometimes paroxysmal coughing. Dyspnea and abdominal type of respiration are noted. The exhibited symptoms and pathology are similar to the pathology caused by *Ascaris* worms.

Diagnosis

Examination of fecal sample may reveal the embryonated eggs. Postmortem examination of affected pig is useful for confirmatory diagnosis.

Control

- The pigs eagerly search for earthworms by digging the soil using their noses. Ringing the nose may be useful for controlling the worms.
- The pigs should not be allowed in the soil where earthworms are present.
- The affected pigs should be treated properly.
- The pigs should be farrowed and raised in clean area.
- Deworming prior to furrowing and then moving the animals in clean pasture can be effective for controlling the worms.

Species: *Metastrongylus salmi*

Hosts: pigs
Site: bronchi

M. pudendotectus

This species occurs in the bronchi of pig. The bursa is much larger compared to *M. apri*. Different earthworms serve as the intermediate host.

Order: Strongylida—Superfamily: Metastrongyloidea—Family: Protostrongylidae

Family Synopsis

The members of the family are the lungworms of various species of mammals. Bursa of the worms may be very small or absent. Spicules usually have membranous expansion. The life cycle is indirect involving snails as intermediate hosts. The worms cause marked damage to the respiratory system of the hosts.

Genus: *Protostrongylus*

Species: *Protostrongylus rufescens*

Common name: red lungworm
Hosts: sheep and goat
Site: lung parenchyma, bronchioles
Distribution: The worms are distributed in North America, Europe, Africa, etc.

Morphology

Adult

Shape: The worms are hairlike.
Color: The worms are reddish brown colored.
Size: The males are 2–4 cm long and the females are 2.5–5 cm long.
Bursa: Bursa is short.
Spicules: The two spicules are short and have membranous expansions.

Egg

The eggs are unsegmented when laid. The size is 75–120 × 45–82 um.

Larva

The larvae (L1) have well-developed rhabditiform esophagus. The tip of the larvae has wavy outline. The tail is elongated without any spine.

Life Cycle

Basic Information of Life Cycle

Type of parasite	Heteroxenous parasite (intermediate host required)
Type of life cycle	Indirect life cycle
Type of definitive host	Vertebrate host
Type of intermediate hosts	Invertebrate hosts
Name of infective stage	Third larval stage (L3)
Mode of infection	Ingestion of the intermediate host harboring the infective larvae
Alternative mode of infection (hyperinfection, autoinfection, etc.)	No

Stages of Life Cycle Formed

In definitive host	In environment
• Fourth stage larva • Adult • Egg • First stage larva	• Second stage larva • Third stage larva

Type of Life Cycle: Indirect

The following events are observed in the life cycle of this worm:

- The adults live in the bronchi and unembryonated eggs are laid by the females. There, the eggs get embryonated and hatch. After hatching, the larvae escape from the eggs.
- The larvae are coughed up and enter the GI tract.
- Then the larvae are passed with the feces.
- The larvae enter into the snails by penetration of the soft tissues of the snail. *Theba*, *Helicella*, and *Zebrina* spp. act as the intermediate hosts.
- Two moltings of the larvae take place in the snail and first stage larvae transform into infective third stage larvae.
- The final hosts become infected by ingestion of the snails harboring the larvae.
- After the snail is being digested, the larvae become free and they penetrate the intestine of the final host.
- They enter the mesenteric lymph node wherein the third stage larvae undergo one molting and transform into fourth stage larvae.
- From the mesenteric lymph nodes, they reach the lung alveoli and then the bronchi.
- There have been records of transplacental infection and the larvae are found in the liver and lungs of fetus.

Pathogenesis

The parasites cause damage to the lung alveoli. Localized inflammation is found in the bronchi. Both the alveolar and bronchiolar epithelium get desquamated. Small foci of lobar pneumonia occur which are yellowish gray in color. Nodules are formed on the surface of the lungs.

Diagnosis

Examination of fecal sample may reveal the larvae. The larvae should be differentiated from larvae of *Dictyocaulus* spp. by observing the absence of protoplasmic knob. The L1 has wavy outline at the tip.

Genus: *Cystocaulus*

Species: *C. nigrescens*

Hosts: sheep and goat
Site: lungs

Pathology

Subpleural nodules are formed in the lungs.

Genus: *Muellerius*

Introduction

These are important roundworms occurring in the lungs of sheep and goat. The worms cause marked pathology. The disease causes significant loss in sheep and goat industry.

Species: *Muellerius capillaris*

Common name: hair lungworm
Hosts: sheep, goat, etc.
Site: lungs

Morphology

Adult

The posterior end of male is spirally coiled. The male has no bursa. The spicules are curved. Proximal half part of the spicules have striated alae, and the distal unequal arms are serrated which terminates into sharp end. Vulva opens near the anus and has a typical swelling at the posterior border. The esophagus of the first stage larva has swelling in the middle and distal extremity.

Egg

The unsegmented eggs are laid.

Larva

The larvae have undulating tip equipped with dorsal spine. They have S-shaped tail. First stage of larva has two characteristic in the middle and distal extremity of the esophagus.

Life Cycle

Basic Information of Life Cycle

Type of parasite	Heteroxenous parasite
Type of life cycle	Indirect life cycle
Type of definitive host	Vertebrate host
Type of intermediate hosts	Invertebrate hosts
Name of infective stage	Third larval stage (L3)
Mode of infection	Ingestion of the intermediate host harboring the infective larvae
Alternative mode of infection if any (hyperinfection, autoinfection, etc.)	No

Stages of Life cycle Formed

In definitive host	In environment
• Fourth stage larva • Adult • Egg • First stage larva	• Second stage larva • Third stage larva

Intermediate hosts: *Helix*, *Helicella*, *Limax*, *Agriolimax*

Type of life cycle: *indirect*

The following events are observed in life cycle of this worm:

- The adult parasites are found in the lungs and copulation occurs there.
- The eggs are laid by the females in their predilection site.
- Embryonation and hatching of eggs occur in the lung parenchyma.
- The first stage larvae are coughed up.
- First stage larvae are passed with the feces.
- The larvae penetrate the exposed tissues of the snails (*Helix*, *Limax*, *Agriolimax*, *Arion*, etc.). The larvae may be swallowed.
- Two ecdyses pass on in the snail and infective third stage larvae are produced in the snails.
- The final hosts become infected by ingestion of the infected snail.

Pathogenesis

The parasites affect the lung tissues and cause pathological damage there. There occurs infiltration of inflammatory cells which results in necrosis. The parasites may also lead to nodule formation in the connective tissue of the lungs. Many nodules may be formed, and closely associated nodules give grayish patches in the lungs. Other clinicopathological features are pulmonary emphysema, perivascular sclerosis, etc. The worms cause formation of focal lesion in the lungs. Initially, small, soft, and hemorrhagic nodules are formed. Later the nodules become hard and changed to gray in color. In goat, the worms do not cause focal lesions but cause interstitial lesions

Symptoms

The animals are generally asymptomatic. Occasionally clinical symptoms are exhibited with coughing and dyspnea.

Genus: *Neostrongylus*

Species: *Neostrongylus linearis*

Hosts: sheep and goats

Site: lung

Morphology: The worms are slender. The males are 5–8 mm long and the females are 13–15 mm. The larvae possess one dorsal and two lateral spines.

Genus: *Bicaulus*

Species: *Bicaulus saggitatus*

Hosts: deer

Site: lung

Morphology: They are slender worms. The length is 1.4–3.4 cm.

Life cycle: The slugs and snails serve as the intermediate host in which third infective larvae are formed. The final hosts get the infection by ingestion of infected intermediate hosts. The larvae reach the lungs via the lymph or blood.

Pathogenesis

The common clinical features are pulmonary edema, emphysema, etc. Secondary bacterial infection may aggravate the condition.

Diagnosis

Diagnosis can be done by finding the larvae having posteriorly directed spine.

Genus: *Elaphostrongylus*

Species: *Elaphostrongylus cervi*

Hosts: different deers, i.e., red deer, fallow deer, watipi, sambar deer, etc.

Site: The parasites affect the intermuscular connective tissue of the thorax, back, rib cages, forelimbs, etc. The central nervous system and spinal cords are also affected.

Morphology

The worms are thin and threadlike. They are approximately 5–6 cm long. The first stage larvae possess dorsal spine on the tail.

Life Cycle

The eggs are laid by the females in the blood vessels and carried to the lungs. The eggs hatch in the lungs. The larvae migrate in the lungs and eventually are coughed up and are passed with the feces. The snails and slugs serve as the intermediate hosts in which infective third stage larvae are formed. The final hosts get the infection by ingestion of infected snails and slugs.

Pathogenesis

The animals display two main symptoms, i.e., hind limb paralysis due to CNS involvement and pneumonia due to larval migration. Spinal demyelination and meningoencephalitis occur due to CNS involvement.

Diagnosis

Diagnosis can be done by finding the larvae in the feces. The first stage larvae possess one dorsal spine on their tail region.

Species: *Elaphostrongylus rangiferi*

Hosts: different deers

Genus: *Parelaphostrongylus*

Species: *Parelaphostrongylus tenuis*

Hosts: White tailed deer are the common hosts. The parasite also affect elk, caribou, moose, etc.

Site: cranial venous sinuses and subdural spaces

Life Cycle

Type of life cycle: indirect

The following events are observed in the life cycle of this worm:

- The eggs are laid in the meninges. The eggs enter the circulation and reach the lungs. In the lungs, the eggs hatch.
- The larvae then reach the bronchi and trachea and are coughed up. They are passed with the feces.
- Subsequent development of the larvae occur in the intermediate hosts (snails). The larvae penetrate the exposed tissue of the snails and enter into the snails (*Arion*, *Cionella*). The larvae may be ingested by the snails also.
- The larvae undergo two ecdyses and transform into infective third stage larvae.
- The final hosts become infected by ingestion of the infected snail.
- The larvae then reach the spinal cord and further development takes place in the gray matter. Adults reach the cranium subsequently.

Pathogenesis

The parasites do not cause any disease in natural hosts (white tailed deer). But the parasites cause fatal neurological disease in the abnormal hosts causing "moose disease."

Diagnosis

Examination of fecal sample may reveal the larvae.

Species: *Parelaphostrongylus odocoilei*

Hosts: mule deer, black tailed deer, etc.

Site: connective tissue lining the lymphatics and blood vessels

Species: *Parelaphostrongylus andersoni*

Hosts: white tailed deer

Site: musculature

Order: Strongylida—Superfamily: Metastrongyloidea—Family: Filaroididae

Species: *Aelurostrongylus abstrusus*

Common name: cat lungworm
Hosts: cats
Site: lungs
Distribution: The worms are available in most parts of the world.

They are small sized (up to 1 cm) worms. The bursa is so short that the lobes are indistinct. The larvae have kinked tail with a spine. The worms are very often non-pathogenic. But a large number of parasites cause clinical signs.

Life Cycle

Events of Life Cycle

- The adults live in bronchioles and alveolar ducts whereby they lay their eggs. The eggs hatch there and the larvae are coughed up and enter the GI tract.
- The first stage larvae are passed with the feces of the host.
- Different snails act as the intermediate hosts, i.e., *Helix*, *Agriolimax*, etc. In the snail, the L1 undergo two moltings there and L3 stages are produced. Birds and rodents may serve as paratenic hosts which ingest the infected snails harboring the third stage larvae.
- The final hosts usually get the infection by ingestion of the paratenic host and less frequently by ingestion of infected intermediate hosts.

Pathogenesis

This is feline lungworm causing chronic respiratory disease. The pathology is characterized by formation of subpleural hard and grayish nodules of about 1 cm diameter. The secondary bacterial infection results in severe dyspnea. Multifocal, amber, and subpleural nodules are formed elsewhere in the lungs. Alveolitis, alveolar fibrosis, and fibromuscular hyperplasia are other pathological features.

Symptoms

Coughing and severe dyspnea are important symptoms.

Diagnosis

Examination of feces may reveal the presence of L1 stage of the parasite. L1 has to be differentiated from other lungworm larvae. This L1 bears a subterminal spine on the S-shaped tail. During necropsy, the adult worms can be recovered by squeezing the pieces of the lung. Then the worms can be identified. Radiography can be done which does not confirm but may be a supportive diagnostic step.

Genus: *Perostrongylus*

P. pridhami

This roundworm is found in the lungs of wild mink

Genus: *Filaroides*

Introduction

These are important roundworms occurring in the bronchi and trachea of dogs. These are called lungworms of dogs.

Species: *Filaroides osleri*

Morphology

The parasites occur in the bronchi and trachea of dogs. They may also affect the lungs. The worms are small in size. The hind end of male has rounded tail bearing few papillae. The spicules are somewhat unequal. The larvae do not have any caudal spine.

Life Cycle

The worms are ovoviviparous and hatch in the trachea. The larvae are coughed up and pass with the feces. Direct transmission occurs through infected feces, vomitus, cough material, etc. These materials are contaminated with infective larvae. The larvae are passed from the affected hosts through the feces and saliva. No intermediate host is required for completion of life cycle. The hosts get the infection by ingestion of infective 1st stage larvae. The bitch may transmit infection when she licks or cleans its pups. On ingestion, the larvae are molted to L2 in the small intestine. Then the L2 follows the lymphatic-vascular route and reach the lungs. They develop up to L5 in the bronchi and lungs.

Pathogenesis

The worms form pinkish-gray granular wartlike nodules in the bronchi and trachea. The adult worms remain within the nodules.

Clinical Signs

The disease runs in a chronic course. Dry rasping persistent cough and other respiratory anomalies are common clinical features. The affected animals may die. The household resting dogs show less respiratory distress.

Diagnosis

Examination of sputum and feces may reveal the presence of the larvae. Bronchoscopy can also be done. Bronchoscopy reveals the nodules. The tracheal mucus can be collected during bronchoscopy. The mucus is examined properly for detection of larvae having S-shaped tail.

Name of drug	Mode of administration	Dose
Thiabendazole	Oral	64–140 mg/kg/day can be given in two divided doses for consecutive 23 days

Genus: *Angiostrongylus*

These are roundworms of the lungs of rat. Only *A. vasorum* is found in the pulmonary artery of dog, cat, etc.

Etymology

The generic name *Angiostrongylus* has been derived from two Greek words: *angio* and *strongulos*. *Angio* means "blood vessel" and *strongylos* (Greek) means "round."

Species: *Angiostrongylus vasorum*

Hosts: dog, fox
Site: pulmonary artery and right ventricle of the heart
Common name: French heartworm

Morphology

The worms are small in size (up to 2.5 cm). The bursa is short in size. The white ovary of the female is found coiled surrounding the red intestine. The first stage larvae have wavy tail and cephalic button. A notch-like structure is present subterminally.

Life Cycle

Type of life cycle: indirect

The following events are found in the life cycle of this worm:

- The females are ovoviviparous. The eggs are laid in the lung capillaries wherein they hatch. After hatching, the larvae are liberated which pass to the trachea from which they are coughed up and pass with the feces.
- The larvae develop in the snail host (*Arion* spp.).
- Two moltings occur in the snail and infective third stage larvae are produced in 17 days.
- The final hosts become infected by ingestion of the infected snail.
- The larvae penetrate the intestine of the definitive hosts and enter the mesenteric lymph node.
- Two moltings occur in the mesenteric lymph nodes.
- The parasite reaches the pulmonary artery via the liver and right ventricle. The prepatent period is 7 weeks.

Pathogenesis

The parasites cause nodule formation in the lungs. Other clinicopathological features are pulmonary emphysema, perivascular sclerosis, etc. The worms and eggs block the blood vessels. The affected animals show cardiorespiratory symptoms comprising tachypnea, dyspnea, etc.

Symptoms

The resting dogs usually do not exhibit any symptoms if infection load is less. But heavy infection results in tachycardia, tachypnea, and bloody sputum. Profuse productive cough with occasional bloody sputum is noticed. The condition takes up a chronic course and the animals display inappetence, weakness, and ascites. The affected animals may die.

Diagnosis

Examination of fecal sample or sputum may reveal the larvae. The L1 stage has wavy tail with subterminal notch. A small cephalic button is also present.

A. ondatrae

These are roundworms occurring in the muskrat.

Angiostrongylus cantonensis

This is lungworm of rat. It may also infect human beings and causes eosinophilic meningoencephalitis.

Angiostrongylus costaricensis

The worms affect rats. The seat of predilection of the parasite is subserosal artery of the cecum and cranial mesenteric artery. The worms have zoonotic importance and cause eosinophilic granuloma in the intestine of man.

Genus: *Parafilaroides*

Species: *P. decorus*

Hosts: sea lion
Site: lungs

Order: Strongylida—Superfamily: Metastrongyloidea—Family: Skrjabingylidae

Genus: *Skrjabingylus*

They occur in the mustelids (badger, otter, ferrets, minks, etc.). *Skrjabingylus nasicola* occur in the nasal sinuses of mink. The snails act as the intermediate hosts and the rodents act as transport hosts. The first stage larvae are passed with the feces. The infective third stage larvae are formed in the intermediate hosts (gastropods). The transport hosts or paratenic hosts are frogs, snakes, fish, etc. The final hosts get the infection by ingestion of intermediate hosts or paratenic hosts.

Genus: *Metathelazia*

These are primarily found in the lungs of carnivores.

M. felis

Hosts: cat
Site: lungs

M. californica

Hosts: cat
Site: lungs

Order: Strongylida—Superfamily: Metastrongyloidea—Family: Crenosomatidae

Genus: *Crenosoma*

Species: *Crenosoma vulpis*

Hosts: fox; dogs are also affected

Site

Common site: bronchi
Rare site: trachea
Common name: fox lungworm

Morphology

The cuticle of anterior end of both male and female bears backwardly directed overlapping circular folds bearing small spines. This morphological feature is identifying feature.

Life Cycle

Type of life cycle: indirect

The following events are found in the life cycle of this worm:

- The worms are ovoviviparous.
- The larvae are passed with the feces.
- The snails (*Helix*, *Agriolimax*, *Arion*, *Cepea*, *Arianta*, sp.) act as the intermediate hosts. The larvae penetrate the exposed tissues of the snail.
- Infective third stage larvae are formed in the snail after two moltings.
- The final hosts become infected by ingestion of the infected snail.
- The larvae penetrate the intestinal wall and enter the mesenteric lymph node.
- From the mesenteric lymph node, they follow hepatic circulation and reach the lungs and develop to become adult.

Pathogenesis

The parasites cause bronchitis, tracheitis, etc. The affected animals exhibit different respiratory symptoms comprising coughing and sneezing. Tachypnea is also observed.

Diagnosis

Examination of fecal sample may reveal the larvae. Baerman technique can be done for easy collection of the larvae. Floatation technique is also helpful for easy availability. The larvae have straight tail.

Treatment

Name of drug	Mode of administration	Dose
Levamisole	Oral	8 mg/kg

Genus: *Troglostrongylus*

Species: *T. subcrenatus*

Introduction

These are important roundworms occurring in the lungs of cats.

Morphology

The cuticle of the worms is inflated with folds. The esophagus of the worm is club-shaped. The bursa is undivided. The spicules are equal in size.

Hosts: domestic and wild cat
Site: lungs

Life Cycle

- The eggs are laid in the lungs wherein they hatch.
- The larvae are passed with the feces.
- Different land snails, i.e., *Limax*, *Helicella*, *Retinella*, *Theba*, etc., are intermediate hosts of this parasite.
- The final hosts become infected by ingestion of the infected snail.
- Mice may act as the paratenic hosts. The mice ingest the infected snail and harbor the parasite.

Species: *Troglostrongylus brevior*

Hosts: cat
Site: lung

Order: Spirurida—Superfamily: Spiruroidea-Family: Thelaziidae

Family Synopsis

The worms of this family occur in the eye, conjunctival sac, and nictitating membrane of mammals and birds. They are cosmopolitan in distribution. The worms are characterized by the absence of pseudolabia, prominent mouth capsule, and pre- and postanal papillae at hind end. Life cycle is indirect. Different flies serve as the intermediate hosts. The third larval stage is the infective stage of the parasite. The worms cause marked pathogenesis in the eye. Diagnosis is done by naked eye detection of worm in the eye.

History

Railliet and Henry first documented *Thelazia callipaeda* in the year 1910.

Genus: *Thelazia*

Introduction

These worms are found in the lacrimal duct, conjunctival sac, and gut of mammals and birds. Precloacal and postcloacal papillae are present in the hind end of male. There may be striation in the body.

Etymology

The generic name *Thelazia* has been derived from renowned scientist Alphonse Thelaz. He first described the species, *T. rhodesii* in the year 1890.

Morphology

The worms are cylindrical in shape. Color is milky white. They have truncated cuticle. A small buccal capsule is present. A pair of spicules are present which are unequal in size. They are non-bursate worms. The body may be transversely striated.

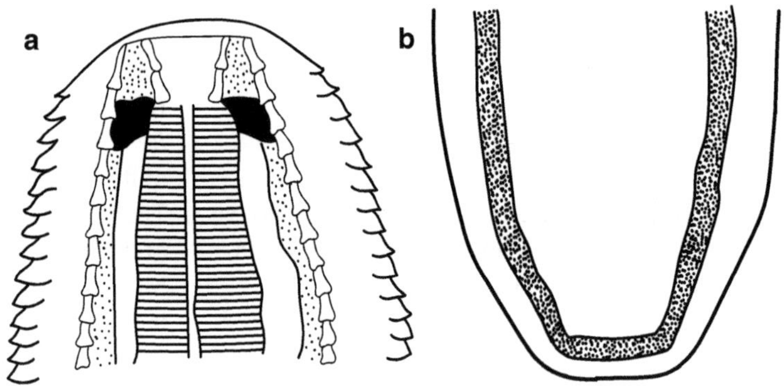

Fig. 6.24 *Thelazia* sp. (**a**) Head end. (**b**) Hind end of female

Life Cycle

Some Basic Information of Life Cycle

Type of parasite	Heteroxenous parasite
Type of life cycle	Indirect life cycle
Type of definitive host	Vertebrate host
Type of intermediate hosts	Invertebrate host
Location of fertilization	In the parasite
Location of embryonation	In the uterus of female
Name of infective stage	Third larval stage (L3)
Mode of infection	Deposition of infective larvae by the infected fly in the eye of the hosts
Alternative mode of infection (hyperinfection, autoinfection, etc.)	

Stages of Life Cycle Formed

In host	In intermediate hosts
• Fourth stage larva • Adult • Egg • First stage larva	• Second stage larva • Third stage larva

Description

The adult worms reside in the nictitating membrane of the eye. The worms are ovoviviparous and the larvae are laid in the eye. The first stage larvae are taken up by different flies during feeding. Different *Musca* flies act as the intermediate hosts. Being ingested, the larvae reach the ovarian follicles. In the ovary, the larvae transform into third stage after two consecutive two moltings. The infective stage of the parasite is formed within a month or so in the summer months. The larvae then migrate to the mouth of the flies. The flies then infect the eye with the larvae when they visit the eyes of the hosts for food.

Pathogenesis

Thelazia spp. are called eye worms which cause different pathological features in eye comprising conjunctivitis, keratitis, corneal opacity, etc. Moderate infections fail to produce marked pathogenesis. The striated cuticle of the worms causes lesions in the eyes which result in continuous lachrymation. Heavy infection may lead to corneal ulcer. Corneal ulcer may be recovered automatically. The parasites may enter the lacrimal duct and cause lacrimal secretion. Secondary bacterial infection may aggravate the condition. Thus, a white cheesy material is collected around the area.

Diagnosis

Naked eye observation reveals the presence of worms. The movement of worms may be detected.

Treatment

Name of drug	Mode of administration	Dose
Levamisole	Oral	5 mg/kg for cattle, buffalo
Fenbendazole	Oral	10 mg/kg for horse for 5 days
Ivermectin	S/C, I/M	0.2 mg /kg cattle, buffalo, etc.

Control

Since the flies act as intermediate hosts, control of flies is important. The affected animals should be treated properly.

Species	Hosts	Site
T. rhodesii	Cattle, buffalo, sheep, goat	Eye
T. callipaeda	Dog	Nictating membrane
T. skrjabini	Cattle	Eye
T. gulosa	Cattle	Eye
T. alfortensis	Cattle	Eye

Genus: *Oxyspirura*

The important species under this genus is *Oxyspirura mansoni*. The common name of this worm is poultry eye worm. The worms occur in the nictating membrane of eye. The pharynx is hourglass-shaped. They are transmitted by cockroach. The bird are infected by ingestion of infected cockroach.

Genus: *Spirocerca*

Species: *Spirocerca lupi*

Introduction

These are very important worms occurring in the stomach, esophagus, and aorta of dog, fox, wolf, etc. They cause formation of tumor in the esophagus, aorta, etc. There may be occurrence of malignancy. They are called '**esophageal tumor worm of dog**'.

Morphology

Adult

The male is about 30–55 mm and the female is 55–80 mm. The lips are trilobed. A short pharynx is present. The worms are spirally coiled and has lateral alae. A pair of spicules are present which are unequal in size. The male tail has lateral alae. Anterior to the cloaca, four paired and one unpaired median precloacal papillae are present. Postcloacal papillae are two pairs. At the tip of the tail, a group of minute papillae are present. They are non-bursate worms.

Egg

The eggs are small capsule shaped. These are elongated. The sides are parallel. The eggs are larvated when laid. The size is 20–37 um × 11–18 um.

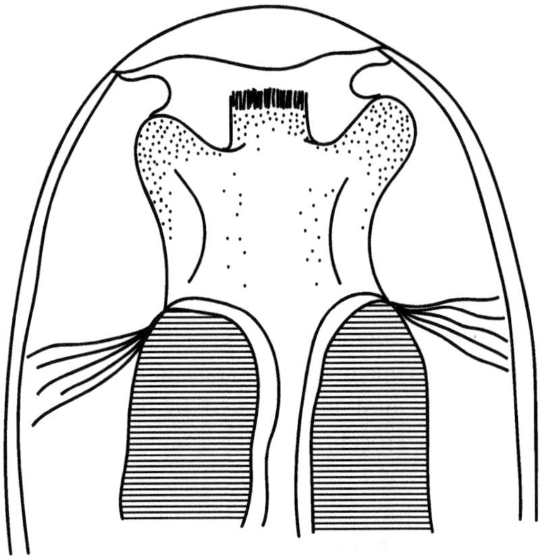

Fig. 6.25 Head end of *Spirocerca lupi*

Some Information of Life Cycle

Type of parasite	Heteroxenous parasite
Type of life cycle	Indirect life cycle
Type of definitive host	Vertebrate
Type of intermediate hosts	Invertebrate
Location of fertilization	In the parasite
Location of embryonation	In the parasite
Name of infective stage	Third larval stage (L3)
Mode of infection	Ingestion of beetle carrying the infective larvae
Alternative mode of infection (hyperinfection, autoinfection, etc.*)*	No

Detailed Description of Life Cycle

Egg Passing

These are heteroxenous parasites which involve one intermediate host. The adult parasites live in their predilection site (esophagus). Females lay eggs in the predilection site. The eggs are voided with the feces.

Embryonation

Embryonation of eggs occurs in the parasite itself.

Development up to Infective Stage

The eggs are ingested by beetle (intermediate host). In the beetle, the eggs hatch. Coprophagous beetles which serve as intermediate hosts are *Scarabeus*, *Atenchus*, *Gymnopleurus*, *Akis*, etc. After hatching, the larvae (L1) are liberated. Larvae develop to L3 after two moltings in the beetle. Some organisms serve as paratenic hosts. These are rodents, birds, reptiles, etc.

Infection of the Definitive Host

The hosts get the infection by inadvertent ingestion of infected beetle.

Development in the Definitive Host

In the host, L3 come out after digestion of beetle by the action of stomach acid and other biochemical factors. The larvae then penetrate the gastric glands. Then they migrate to the gastric and gastroepiploic artery, coeliac artery, and arrive to the thoracic aorta within 7–10 days via the coeliac artery. In the aorta, the larvae remain here for a few months and pass on a molting there and become fourth stage larvae. Then the larvae reach the lower part of the thoracic duct and develop to become L5 after one more molting. This stage then reaches the esophageal mucosa and matures there.

Pathogenesis

The larvae cause marked damage to the wall of then thoracic aorta. The aortic wall may be ruptured. Stenosis and aneurysm are common pathological features. There occurs formation of aortic nodules. The adult worms affect the esophagus. The esophageal wall up to the muscularis layer may be affected. Initially inflammation prevails. But serious complication occurs due to malignancy. Esophageal tumor is formed. Sometimes, the esophagus may be ruptured, and it results into discharge of esophageal content in the thoracic cavity. Hypertrophic pulmonary osteoarthropathy and spondylosis also occur. They may cause osteosarcoma and fibrosarcoma on the wall of the esophagus.

Note

Rajas et al. (2018) reported a new species of *Spirocerca* (*Spirocerca vulpis* sp. nov.) of fox. They characterized spirurid worms collected from red foxes and are compared with *S. lupi* from domestic dogs by morphometric and phylogenetic analysis. Nematodes of red fox differed from *S. lupi* by the presence of six triangular teeth-like buccal capsule structures which are absent in *Spirocerca lupi*

Clinical Signs

The interesting aspect is that despite high grade pathogenicity, the affected animals apparently do not exhibit clinical signs. The animals face problem of swallowing. In some cases, the vomit with expelling the worms with the vomitus. The animal may die due to rupture of the aorta.

Diagnosis

Symptomatological Diagnosis

The exhibited clinical signs are not so specific for confirmatory diagnosis.

Parasitological Diagnosis

Examination of fecal sample may reveal the presence of egg. Vomitus should also be examined for the presence of whole worm and/or eggs.

Endoscopy

Diagnosis can be done by endoscopy for detection of tumor or granuloma in the affected area.

Radiography

Radiological diagnosis reveals tumors on the esophageal wall, aorta, etc.

Immunological Diagnosis

No immunodiagnostic test is available for field application.

Molecular Diagnosis

Molecular diagnosis is possible by using specific molecular markers. Different ribosomal DNA, ribosomal RNA, and mitochondrial DNA could be used as molecular markers. Studies indicate the use of mitochondrial DNA and CO1 (cytochrome oxidase 1) for molecular identification.

Treatment

Ivermectin 0.6 mg/Kg S/C two doses at 2 weeks interval is effective.

Control

General Management

The affected animals should be treated properly. The dogs should be prevented from eating beetle.

Immunological Control

No vaccine is available.

Molecular Control

No naked DNA vaccine or gene vaccine is available.

Biological Control

No biological control is available for field application.

Genus: *Ascarops*

Introduction

These are roundworms occurring in the stomach of pigs. The worms are characterized by triple and quadruple thickenings of the pharynx.

Species: *Ascarops strongylina*

Hosts: pig, wild boar

Site: stomach

Distribution: The worms are distributed worldwide.

Morphology

The worms are cylindrical in shape. The male is 10–15 mm and the female is 16–22 mm long. A cervical ala is present which is located at the left side of the worm. Triple and quadruple thickenings are present in the pharynx. A pair of spicules are present which are unequal in size. They are non-bursate worms.

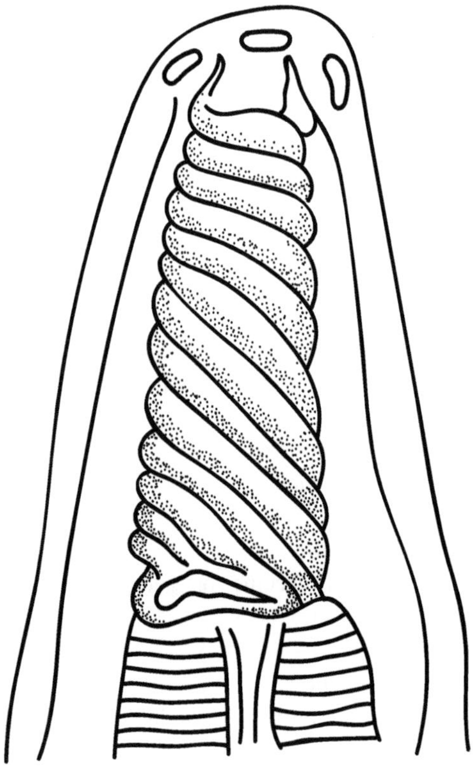

Fig. 6.26 Head end of *Ascarops strongylina*

Life Cycle

Embryonation

Embryonation of eggs occurs in the uterus of female.

Egg Passing

The worms are heteroxenous which requires one intermediate host for completion of life cycle. The adult parasites live in their predilection site (stomach). The females lay eggs in their predilection site. The eggs are voided through the feces.

Development up to Infective Stage

The eggs are ingested by beetle (intermediate host). In the gut of beetle, the eggs hatch. After hatching, the larvae (L1) come out. Larvae develop to L3 after two moltings in the beetle.

Infection of the Definitive Host

The hosts get the infection by inadvertent ingestion of infected beetle.

Development in the Host

In the host, L3 come out after digestion of the intermediate host by the action of stomach acid and other biochemical factors. L3 develop into L5 after two moltings.

Pathogenesis

The parasites are nonpathogenic. But the worms may cause catarrhal gastritis.

Diagnosis

Diagnosis can be done by presence of eggs in the feces.

Ascarops dentata

Hosts: pig
Site: stomach

Morphology

Yadav and Tandon (1992) studied the morphology of this parasite. The worms have two pseudolips, each being equipped with three lobes. Of the outer two lobes, each bears dome shaped submedian cephalic papilla. The body cuticle is transversely striated. The middle lobe bears a buccal tooth.

Physocephalus

Introduction

These are roundworms of the stomach of pigs.

Species: *Physocephalus sexalatus*

Site: stomach

Host

Common hosts: pig
Occasional hosts: rabbit, hare
Distribution: The worms are commonly available in many parts of the world.

Morphology

The male parasite is 10–12 mm and the female is 13–22 mm long. A cuticular inflation is present in the pharynx region. After the inflated area, three cervical alae are present on either side. The teeth are absent. Single spiral thickenings are present on the pharynx which break up into several ring structures. A pair of spicules are present which are unequal in size. Four pairs of precloacal and postcloacal papillae are present.

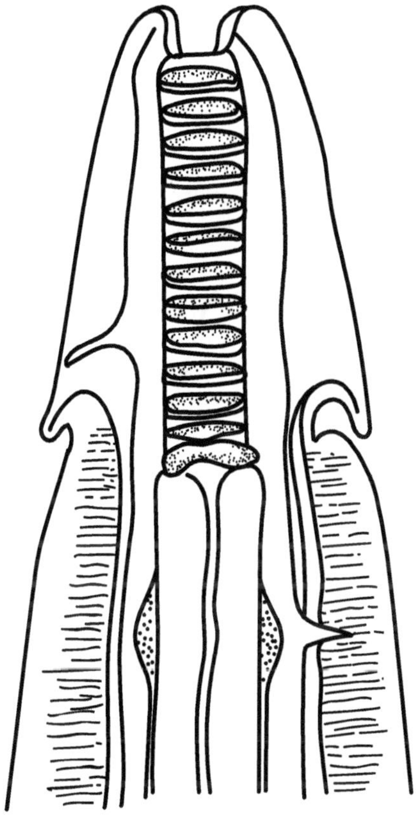

Fig. 6.27 Head end of *Physocephalus sexalatus*

Life Cycle

Some Information of Life Cycle

Type of parasite	Heteroxenous parasite
Type of life cycle	Direct life cycle
Type of definitive host	Vertebrate
Type of intermediate hosts	Invertebrate
Infective stage	Third larval stage (L3)
Mode of infection	Ingestion
Alternative mode of infection (hyperinfection, autoinfection, etc.*)*	No

Description

Egg Passing

The parasite is heteroxenous which involves one intermediate host. The adult parasites live in their predilection site (stomach) where they lay eggs. The eggs are voided in the feces.

Development up to Infective Stage

The eggs are ingested by intermediate hosts (different beetle genera, i.e., *Gymnopleurus*, *Geotrupes*, *Onthophagus*). In the beetle, the eggs are hatched. After hatching, the larvae (L1) come out. Larvae develop to L3 after two moltings in the beetle.

Infection of the Host

The hosts get the infection by inadvertent ingestion of infected beetle.

Development in the Host

In the host, L3 come out after digestion of beetle by the action of stomach acid and other biochemical factors. L3 develop into L5 after two moltings. The prepatent period is 6 weeks.

Pathology

The parasites cause gastritis.

Treatment of *Ascarops* and *Physocephalus*

Different benzimidazole compounds can be tried.

Genus: *Simondsia*

Species: *Simondsia paradoxa*

Hosts: pig
Site: stomach
Distribution: The worms are usually distributed in tropical and subtropical countries.

Salient Morphology

The male is 12–15 mm and the females are up to 20 mm long. Females possess large lateral cervical alae. The gravid female is globular posteriorly having the eggs. They form small cysts in the stomach wall and remain there. Male parasite has spirally coiled tail.

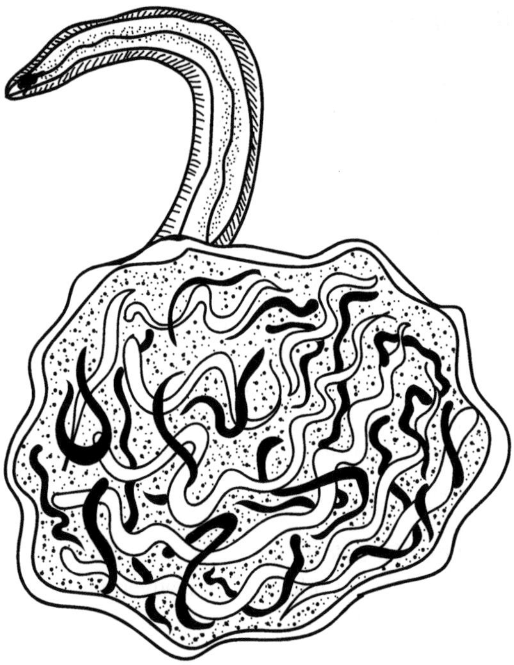

Fig. 6.28 *Simondsia paradoxa* (female)

Life Cycle

- Life cycle is indirect.
- The eggs are passed with feces.
- The eggs are ingested by the beetles.
- In the beetles, the infective stage is formed.
- The final hosts get the infection by ingestion of the infected beetle.
- The characteristic location of the parasites is that the male worm remains on the surface of the mucosa whereas the female worm lives in the cyst keeping their anterior end protruding.

Pathogenesis

They cause gastritis and nodule formation in gastric mucosa.

Diagnosis

Diagnosis can be done by finding the elongated eggs in the feces.

Genus: *Gongylonema*

Introduction

These are very common roundworms occurring in different parts of the GI tract of different animals. The worms have cuticular thickenings on the body. They are commonly called 'gullet' worm or stich worm.

Species: *Gongylonema pulchrum*

Hosts: sheep, goat, cattle, buffalo, horse, donkey, etc.
Site: oral epithelium, esophagus and subcutaneous tissue

Morphology

Round or oval cuticular thickenings are present at the anterior end called cuticular bosses or scutes. Cervical alae are present. Anterior end is round. The lips are absent. The pharynx is short.

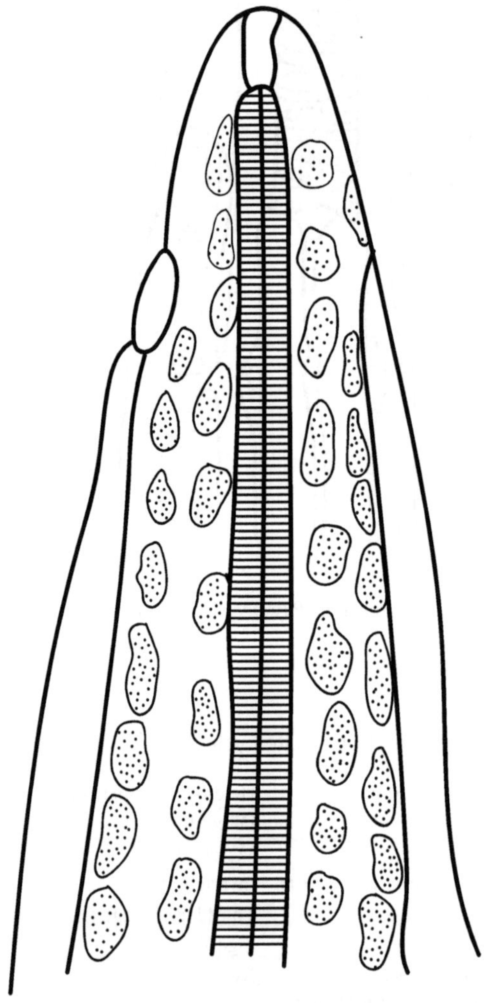

Fig. 6.29 Head end of *Gongylonema pulchrum*

> **Note**
>
> The common name of this parasite is **gullet** worm. Another common name of this parasite is **stich** worm.

Life Cycle

Some Important Information of Life Cycle

Type of parasite	Heteroxenous parasite
Type of life cycle	Indirecti life cycle
Type of definitive host	Vertebrate host
Type of intermediate host	Invertebrate host
Location of fertilization	In the parasite
Location of embryonation	In the parasite
Name of infective stage	Third larval stage (L3)
Mode of infection	Ingestion of intermediate host harboring the infective larvae
Alternative mode of infection (hyperinfection, autoinfection, etc.*)*	No

Life Cycle in Detail

Embryonation

Embryonation of eggs occurs in the parasite itself.

Egg Passing

Eggs pass out with the feces.

Development in Intermediate Host

The eggs are voided through the feces. The voided eggs are ingested by different intermediate hosts (beetle, cockroaches). Development up to third larval stage occurs in the intermediate hosts. The third stage larvae produced in the intermediate hosts are infective stage of the parasite.

Infection to the Final Hosts

The final hosts get the infection by ingestion of the intermediate hosts harboring the infective larvae of the parasite.

Pathogenesis

The parasites cause enteritis. They remain on esophageal mucosa in a zig-zag fashion looking like 'tailor's stiches'.

Diagnosis

Symptomatological Diagnosis

No specific clinical signs are exhibited by the affected animals which can confirm diagnosis of this parasitic infection.

Parasitological Diagnosis

Examination of fecal sample may reveal the presence of eggs of the parasite.

Immunological Diagnosis

No specific immunodiagnostic test has been commercialized.

Molecular Diagnosis

Molecular diagnosis is possible using specific molecular marker.

Treatment

Different benzimidazole compounds can be tried.

Control

The affected animal should be treated properly.

Species: *Gongylonema verrucosum*

Hosts: sheep, goat, cattle, zebra, deer, etc.

Site: Rumen, reticulum, omasum

Distribution: The worms are distributed in India, the USA, and South Africa.

Salient Morphology

The fresh worms are reddish in color. The males are about 3.5 cm and the females are 7.0–9.5 cm. Cuticular bosses are found in the left side of the anterior end of the parasite. The cervical alae are festoon shaped. The spicules are unequal in size of which the left is longer.

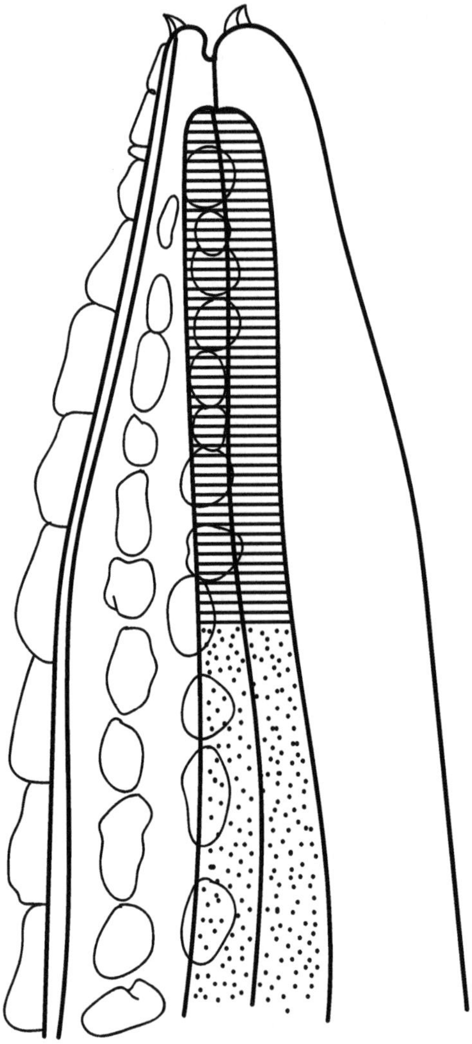

Fig. 6.30 Head end of *Gongylonema verrucosum*

Life Cycle

- Life cycle is indirect.
- The eggs are passed with feces.
- The eggs are ingested by the intermediate hosts (beetles and cockroaches).
- In the beetles, the infective stages are formed within about 4 weeks.
- The final hosts get the infection by ingestion of the infected intermediate hosts.
- The adults are formed in their respective location.

Pathogenesis and Clinical Signs

The worms are apparently nonpathogenic. But the parasites bury in the layer of the stomach epithelium and cause damage.

Species: *Gongylonema ingluvicola*

Site: crop
Hosts: birds

Life Cycle

Type of life cycle: indirect

Different Events of Life Cycle

- Embryonated eggs are voided with the feces.
- Eggs are ingested by the intermediate hosts (cockroaches, beetles).
- Two moltings occur in the intermediate hosts, and third infective stage is formed in the body of the intermediate hosts.
- The final hosts become infected by ingestion of infected intermediate hosts harboring the infective larvae.
- The adult parasites are developed in the gut of the final hosts. The prepatent period is 8 weeks.

Pathogenesis

Catarrhal gastritis is caused which turns into chronic hypertrophic gastritis. The gastric glands display wartlike circumscribed marked thickenings. Cornification of epithelium of crop occur. The worms remain embeded in the mucosa and submucosa of crop in a 'zipper' fashion.

Order: Spirurida—Superfamily: Spiruroidea-Family: Spiruridae

Family: Spiruridae (two lips each being trilobed, esophagus having anterior muscular and posterior glandular part, spicules unequal, hind end of male usually coiled bearing lateral alae and papillae)

Genus: *Habronema*

Introduction

These are stomach worms of horse.

Etymology

The generic name "Habronema" has been derived from two Greek words i.e., '*habros*' meaning 'fine' and '*nema*' meaning 'filament'.

Morphology

The worms are cylindrical in shape. The dorsal and ventral teeth are present in the pharynx in case of *Habronema majus*. The head of *H. majus* is not constricted off as is found in *Draschia megastoma*. *Habronema muscae* has cylindrical pharynx. A pair of spicules are present which are unequal in size. They are non-bursate worms.

Genus: *Draschia*

Introduction

These are stomach worms of horse. The worms cause tumors in the stomach. The important species is *Draschia megastoma*.

Etymology

The generic name has probably been derived from Richard v. Drasche.

Morphology

The head is typically constricted off. The pharynx is funnel-shaped. A pair of spicules are present which are unequal in size.

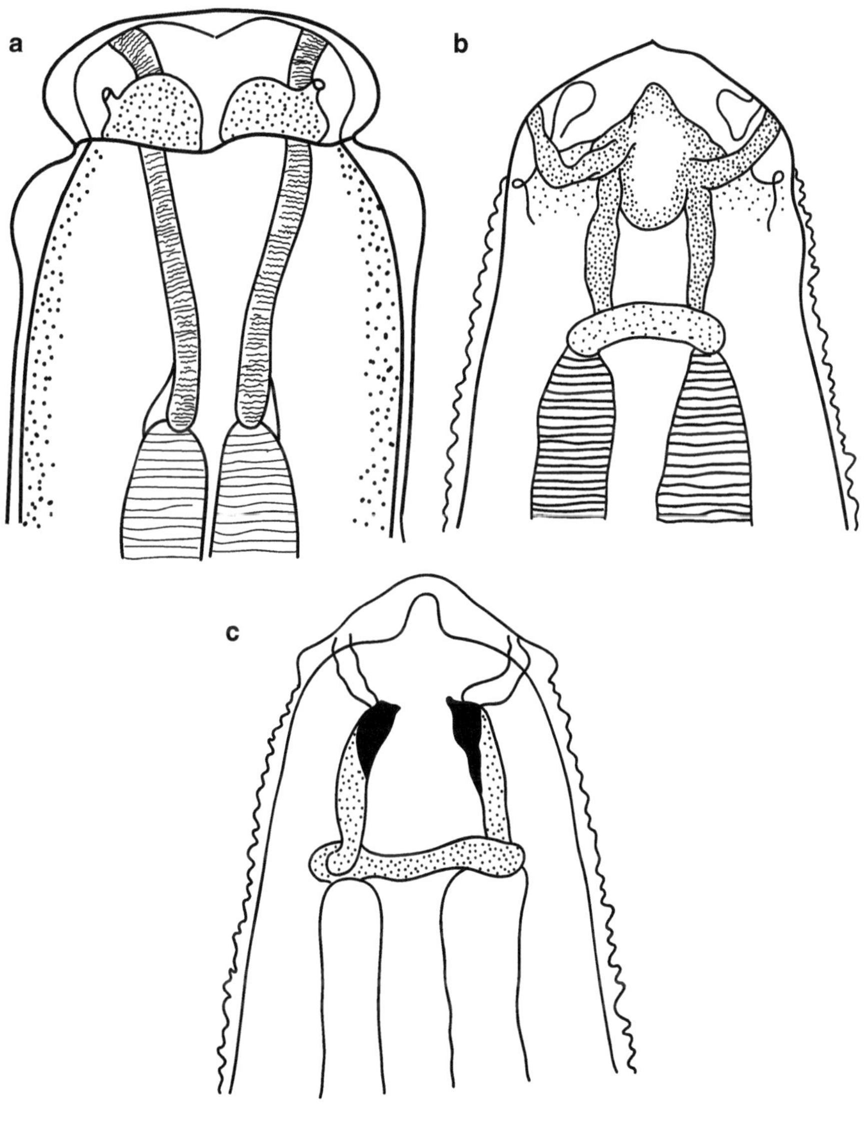

Fig. 6.31 (**a**) Head end of *Draschia megastoma*. (**b**) Head end of *Habronema muscae*. (**c**) Head end of *Habronema majus*

Life Cycle

Some General Information of Life Cycle

Type of parasite	Heteroxenous parasite
Type of life cycle	Indirect life cycle
Type of definitive host	Vertebrate host
Type of intermediate host	Invertebrate intermediate host
Location of fertilization	In the parasite
Location of embryonation	In the host while the eggs pass with the feces
Name of infective stage	Third larval stage (L3)
Mode of infection	Ingestion
Alternative mode of infection (hyperinfection, autoinfection, etc.)	No

Stages of life cycle formed

In the host	In the intermediate hosts
• Fourth stage larvae • Adult • Egg and first larave	• Second stage larvae • Third stage larvae

Description

The larvae or eggs are voided with the feces. They are ingested by the intermediate hosts. Different flies (*Musca domestica, Stomoxys calcitrans*) serve as the intermediate hosts of this parasite. *Musca domestica* serves as intermediate hosts of *H. muscae* and *D. megastoma*. *Stomoxys calcitrans* serves as intermediate hosts of *H. majus*. The final hosts get the infection by ingestion of infected fly. Or the fly may deposit eggs at the nostril area of the final hosts. The larvae develop into adult in the stomach.

Pathogenesis

The larvae of *Habronema* and *Draschia* cause cutaneous habronemosis, summer sore, granular dermatitis, or bursati. In the existing wound, the flies deposit the infective larvae. Legs, withers, prepuce, etc. are common sites because these areas are commonly injured. Since this skin disease occurs in summer due to high population of flies, the condition is called **summer sore**. The lesion is characterized by uneven surface with hard granulation. D. *megastoma* is the most pathogenic one which causes tumor-like abscesses and the adults remain within the caseous and pus material of the abscesses. The affected area may be perforated. Gastritis is caused by *H. muscae*.

Diagnosis

Examination of fecal sample may reveal the presence of eggs or larvae.

Control

The affected animals should be treated properly. Since the flies act as the intermediate hosts, they should be controlled by using insecticidal agents.

Treatment of *Habronema* and *Draschia*

Name of drugs	Mode of administration	Doses
Piperazine	Oral	200–300 mg/kg
Fenbendazole	Oral	10 mg/kg
Ivermectin	I/M	0.2 mg/kg

Genus: *Cyrnea*

Long esophagus and well-developed caudal alae are important morphological features. The members occur in the proventriculus and gizzard of birds. *C. piliata* and *C. colini* occur in the proventriculus and gizzard of birds.

Genus: *Spirura*

Species

S. rytipleurites

Hosts: cat and fox

Site: stomach

Intermediate host: Cockroach and beetle are intermediate hosts.

Genus: *Hartertia*

Species: *H. gallinarum*

Hosts: birds

Site: small intestine

Morphology

Two lateral lips are present. Each lip is trilobed. Pre- and postcloacal papillae are present in the hind end of male. The male has caudal alae.

Intermediate hosts: Termites are intermediate hosts of this parasite.

Genus: *Streptopharagus*

Species: *S. armatus*

Hosts: monkey and apes
Site: stomach

Salient Morphology

The cuticle bears longitudinal ridges.

Order: Spirurida—Superfamily: Spiruroidea-Family: Acuariidae

Family Synopsis

The worms of this family occur in the gizzard, proventriculus, and intestine of different birds. They are characterized by ornamental structures called cordons. The cordons may be recurrent or nonrecurrent. These either may anastomose or do not anastomose. The life cycle is indirect type. Different insects serve as the intermediate hosts in which infective stage develops.

Genus: *Cheilospirura (Acuaria)*

Species: *Cheilospirura (Acuaria) hamulosa*

Hosts: different types of birds, i.e., chicken, fowl, guinea fowl, turkey, etc.
Site: gizzard
Distribution: The worms are worldwide in distribution.

Morphology

This worm is small in size. The male is up to 15 mm and the female is up to 30 mm long. They have specific cuticular structures looking like ornaments. These are called cordons. The cordons appear as cuticular ridges extending far back. The male tail has double contoured caudal alae. The male parasites have four pairs of precloacal and six pairs of postcloacal papillae. The eggs are embryonated while passing with the feces.

Life Cycle

Embryonation

Embryonation of eggs occurs in the uterus of the female itself.

Egg Passing

The worms are heteroxenous which involves one intermediate host for completion of the life cycle of the parasite. The adult parasites live in their predilection site (stomach) where the female parasite lays eggs. The eggs are voided with the feces.

Development up to Infective Stage

The eggs are ingested by grasshoppers (*Oxya, Melanoplus*). Beetle also acts as intermediate hosts. In the gut of the grasshoppers, the eggs hatch. After hatching, the larvae (L1) come out. Larvae develop to L3 after two moltings in the beetle. The time required for formation of infective third stage larvae is 3 weeks.

Infection to the Host

The hosts get the infection by inadvertent ingestion of infected beetle.

Development in the Host

In the host, the body of the intermediate host is digested and L3 are liberated. L3 develop into L5 after two moltings. The prepatent period is about 3 weeks.

Pathogenesis and Clinical Signs

The worms are less pathogenic. But heavy infection causes damage to the gut. The worms penetrate in the gizzard wall and forms nodules which are orange-colored. Anemia, emaciation, and weakness are some associated signs.

Genus: *Dispharynx*

Introduction

These are roundworms occurring in the wall of the proventriculus, esophagus, and intestine of birds. The worms are morphologically characterized by the presence of cordons.

Species: *Dispharynx spiralis*

Morphology

This worm is small. The length of this worm is approximately 1 cm. They have four wavy specific cuticular structures called cordons which have a recurrent sinuous course and do not anastomose. The right spicule is shorter than the left spicule. The right spicule is boat-shaped.

Fig. 6.32 Head end of *Dispharynx spiralis*

Site: The worms occur in the proventriculus, stomach, and intestine.

Hosts: different types of birds, i.e., chicken, fowl, guinea fowl, turkey, etc.

Distribution: The worms are distributed in Asia, Africa, and America.

Life Cycle

The life cycle is indirect type. Different isopods act as intermediate hosts of the parasite. The eggs are voided along with feces. The voided eggs are ingested by the intermediate hosts. Eggs are hatched in the gut of intermediate hosts. Development up to third larval stage occurs in the intermediate hosts. This third larval stage is the infective stage of the parasite. The final hosts are infected by ingestion of infected intermediate hosts. The parasites cause enteritis.

Pathogenesis and Clinical Signs

The worms cause formation of tumor like bodies. They cause inflammation, deep ulceration etc. The affected birds may expel mucus material.

Genus: *Echinuria*

Introduction

These are roundworms occurring in the different parts of the GI tract of ducks, geese, etc. The worms are characterized by nonrecurrent cordons which anastomose in pairs. The worms possess both precloacal and postcloacal papillae.

Species: *Echinuria uncinata*

Hosts: duck, goose

Site: proventriculus, esophagus, gizzards, small intestine

Morphology

Their four nonrecurrent wavy cordons anastomose in pairs. Four longitudinal rows of spines are present on the cuticle. Four pairs of precloacal and four pairs of postclocal papillae are present. The right spicule is smaller than the left spicule.

Life Cycle

The life cycle is indirect. *Daphnia*, *Gammarus*, etc. act as the intermediate hosts. The infective stage is formed in the intermediate hosts. The final hosts get the infection by ingestion of intermediate hosts harboring the infective stage of the parasite.

Pathogenesis

The parasites cause inflammation and nodule formation at the predilection site. These parasites may cause formation of caseous nodules which can cause obstruction if the nodules are large in size.

Order: Spirurida—Superfamily: Spiruroidea-Family: Tetrameridae

Family Synopsis

They are mostly found in the alimentary tract of different birds. The worms are morphologically similar to Acuariidae. But they do not possess any cordon. Exceptionally, they have marked sexual dimorphism. Male and female look dissimilar. The male is a slender bodied parasite but the female is globular in shape. The life cycle is indirect. Cockroaches, grasshopper, etc. act as the intermediate hosts in which infective third stage is formed.

Genus: *Tetrameres*

Species: *Tetrameres mohtedai*

The worms commonly occur in the proventriculus. In case of female, the anterior and posterior ends are filiform but the body is globular and pink in color. Life cycle is indirect. Grasshoppers and cockroaches act as the intermediate hosts. The eggs are passed with the feces. The eggs are ingested by the intermediate hosts and are hatched in the intermediate hosts. The hatched-out larvae reach the infective third stage after two moltings. The definitive hosts get the infection by ingestion of infected intermediate hosts harboring infective larvae. The worms affect the proventriculus. Nodules are formed in the proventriculus and damage the glands of wall of the proventriculus. Greenish diarrhea and anemia are the clinicopathological features.

Species: *Tetrameres fissispina*

Hosts: duck, fowl, turkey

Morphology

Four rows of cuticular spines are present in the male parasites. The spicules are unequal in size. Two inconspicuous lateral alae are present in male parasites. The female is blood red in color and globular in shape. Four longitudinal furrows are present. The body is divided into three parts. Middle is globular shaped. Anterior and posterior parts are slender. *Water crustacea (Daphnia, Gammarus* etc.) Cockroach and grasshopper are the intermediate hosts of the parasite.

T. crami

Hosts: domestic and wild ducks

T. pattersoni

Hosts: quail

Intermediate hosts: grasshopper and cockroach

T. americana

Hosts: fowl, turkey

Intermediate hosts: Amphipods (*Gammarus, Hyalella*)

Order: Spirurida—Superfamily: Physalopteroidea—Family: Physalopteridae

Family Synopsis

They are mostly found in the stomach of cat, wild felids, etc. The worms are morphologically characterized by the presence of collar-like projection at the head region; the lips are equipped with teeth. The members of this family do not have any pharynx. The life cycle is indirect. Orthoptera and coleoptera members serve as intermediate hosts in which third infective larval stage is formed. They cause marked damage to the stomach. Diagnosis can be done by finding spirurid eggs in the feces.

Morphology

The members of this family do not have any pharynx. The head region bears a collar-like structure.

Genus: *Physaloptera*

Etymology

The generic name *Physaloptera* has been derived from two Greek words *physalis* (bubble) and *pteron* (wing).

Species

P. rara

Species: *Physaloptera praeputialis*

The worms are found in the stomach of cat. In both sexes, the cuticle is extended beyond the posterior end. A typical morphological structure, cuticular collar, is present at the anterior extremity. The lips are simple having teeth on the medial surface. The pharynx is absent. In the fertilized female, brown cement material is present on the vulva. A pair of spicules are present which are unequal in size. They are non-bursate worms. Life cycle of the parasite is indirect type involving cricket, cockroach, and beetle as intermediate hosts. The larvated eggs pass out. Development up to third infective stage of larvae occurs in the intermediate hosts. The final hosts get the infection of this parasite by ingestion of infected intermediate hosts. The parasites bite the gastric mucosa by their teeth. The lesion caused by the parasites tends to end in ulcer formation. The parasites leave their places and cause hemorrhagic spots. The digested blood may come out with feces making the feces dark in color.

Order: Spirurida—Superfamily: Physalopteroidea—Family: Gnathostomatidae

Family Synopsis

They are mostly found in the stomach of different mammals. The worms are morphologically characterized by the presence of trilobed large lips; teeth-like ridges are present on the cuticle. The life cycle is indirect and involves two intermediate hosts. The first intermediate host is cyclop and the second intermediate host is fish. The worms have pathogenic significance as they cause formation of gastric tumors.

Genus: *Gnathostoma*

This is a common roundworm occurring in the stomach of different animals. The worms are characterized by large head bulbs and ballonets.

Species: *Gnathostoma spinigerum*

Hosts: dog, cat, pole cat, mink, man
Site: stomach

Etymology

The generic name *Gnathostoma* has been derived from two Greek words *gnathos* (jaw) and *stoma* (mouth).

Morphology

Adult

Gross

Shape: The worms are thick bodied with cylindrical structure.
Size: The male is 1–2.5 cm and the female is 0.9–3 cm long.
Color: The anterior side is reddish and posterior side is grayish.

Microscopic

The worms have large head bulb containing four submedian cavities which are called ballonets. Typical cuticular hooks (6–11 rows) are present on the head bulb. Below the head bulb, cuticular spines are present approximately two-thirds of the body. A pair of spicules are present which are unequal in size. The right spicule is smaller than the left spicule.

Egg and Larvae

The eggs are oval in shape having one cap-like structure at one pole. The eggshell is greenish and ornamented. The hatched out juveniles are 0.223-0.275 mm long. The anterior end is provided with bulb. Spines are present on the bulb.

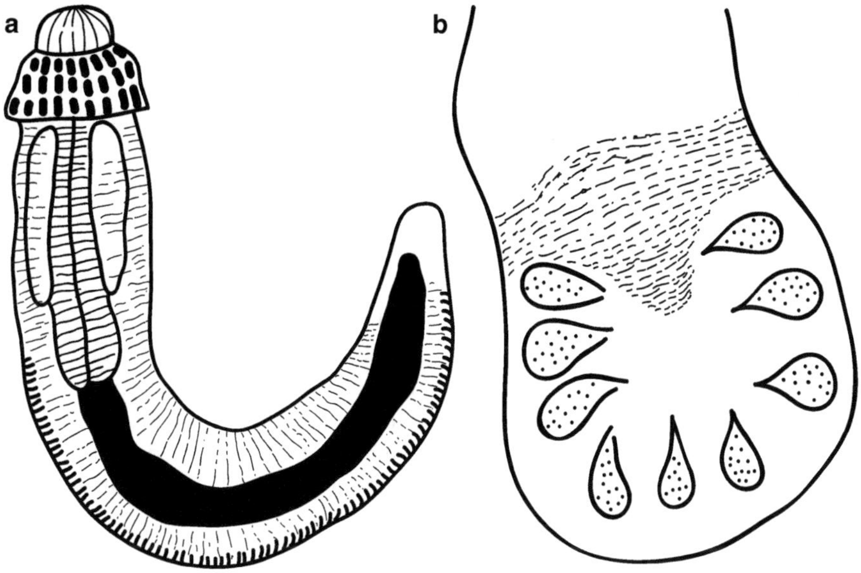

Fig. 6.33 *Gnathostoma spinigerum*. (**a**) Whole body. (**b**) Hind end

Life Cycle

Some Important Information of Life Cycle

Type of parasite	Heteroxenous parasite
Type of life cycle	Indirect life cycle
Type of definitive host	Vertebrate
Type of intermediate hosts	Invertebrate
Location of fertilization	In the parasite
Location of embryonation	In the environment
Name of infective stage	Third larval stage (L3)
Mode of infection	Ingestion of intermediate host carrying infective stage
Alternative mode of infection (hyperinfection, autoinfection, etc.)	No

Description of Life Cycle

Egg Passing

The parasite is heteroxenous which requires one intermediate host for completion of the life cycle. The adult parasites live in their predilection site (stomach) where the female parasites lay eggs. Tumors are formed in the wall of the stomach. The tumors are connected with the lumen of the stomach. The eggs are passed through the connecting duct. Eventually the eggs are passed with the feces.

Development up to Infective Stage

The eggs are passed with the feces. The eggs hatch in contact with water. After hatching, the larvae (L1) come out. Larvae are ingested by cyclops and develop'into early L2 after one molting in the first intermediate hosts. Fish act as the second intermediate hosts. The first intermediate hosts are ingested by the second intermediate hosts. In the body of the second intermediate hosts, the matured third stage larvae are produced.

Infection to the Host

The final hosts get the infection by ingestion of infected fish.

Development in the Host

In the host, the body of the fish is digested and L3 become free. L3 develop into L5 after two moltings.

Pathogenesis

The L3 stages migrate in different internal organs and cause formation of lesions. The worms cause formation of tumor in the stomach. The parasites remain in the stomach. As is found in other cases, the tumor is filled with serosanguinous fluid. The tumors are connected with the lumen of the stomach by small connecting ducts. They may be connected with the peritoneum in similar fashion and cause peritonitis. The worms may cause perforation in the stomach and cause peritonitis.

Diagnosis

Symptomatological diagnosis: No specific clinical signs are observed so that the parasitic infection can be diagnosed with confirmation.

Parasitological diagnosis: Examination of fecal sample may not always reveal any egg.

Immunological diagnosis: Some works were carried out using crude and purified antigens.

Control

The animals should not be allowed to eat uncooked or undercooked fish.

G. hispidum

Hosts: pig
Site: stomach

Salient Morphology

The whole body is covered with spines. This is a characteristic morphological feature.

Pathogenesis

Young parasites migrate in the liver where they cause hepatitis.

G. doloresi

Hosts: pigs

G. nipponicum

Hosts: weasel

Order: Spirurida—Superfamily: Filaroidea—Family: Filariidae

Family Synopsis

The members of this family contain very important filarial worms. The worms are devoid of lips, buccal cavity, and pharynx. Peribuccal chitinous ring on head is also absent. They are ovoviviparous worms. Several genera of worms are found. The genera under the family are *Dirofilaria*, *Wuchereria*, *Parafilaria*, *Ornithofilaria*, *Suifilaria*, *Elaeophora*, *Brugia*, and *Mansonella*. Of these genera, *Dirofilaria* and *Stephanofilaria* are important genera. The worms exhibit indirect life cycle which involves mostly different mosquitoes as intermediate host. Infective microfilariae develop in mosquitoes and are transmitted to the hosts when they suck blood from the hosts. Diagnosis is done by finding the microfilariae in the blood. Control of filarial worms relies on chemotherapy, control of mosquitoes, etc.

Genus: *Dirofilaria*

Introduction

This is a common filarial roundworm occurring in the heart of dog, man, etc. Some species under this genus are *Dirofilaria immitis*, *D. repens*, *D. tenuis*, *D. conjunctivae* etc. They cause marked pathology. The hosts od *D. immitis* are man, dog, etc. occurring in heart. *D. repens* is a parasite of dog occurring in the subcutaneous tissue. However, infection in man has also been reported. *D. tenuis* is a parasite of racoon occurring in subcutaneous tissue forming nodules there. Occurrence of this parasitic infection has also been reported in man. *D. conjunctivae* occurs in conjunctival tissue where they may form hazel nut sized nodules.

Etymology

The generic name of *Dirofilaria* has been derived from two Latin words, i.e., *dirus* (fearful) and *filum* (thread).

Species: *Dirofilaria immitis*

Hosts

Common hosts: dog, cat, fox, wolf

Occasional hosts: horse, man

Site

Cardiovascular system comprising the right ventricle, right atrium, pulmonary artery, posterior vena cava, etc. is affected.

Morphology

Adult

Gross

Size: The male worms are roughly 12–16 cm long and females are 25–30 cm long.
Shape: The shape of the worms is cylindrical.
Color: The color is white-gray.

Microscopic

The worms are slender. Lips, buccal capsule and pharynx are absent. The esophagus has anterior muscular and posterior glandular portion. A pair of spicules are present which are unequal in size. In the posterior part of the male worm, characteristic papillae, i.e., ovoid papillae, finger-like papillae, minute conical papillae, are present. The male tail is spirally coiled. The male tail bears lateral alae. The spicules are unequal in size. The left spicule is pointed and longer than the right spicule which ends bluntly.

Microfilaria

The microfilariae are unsheathed. The length is 0.218–0.329 mm. The microfilariae show the periodicity. Some workers found maximum microfilariae at 6 PM in the peripheral blood.

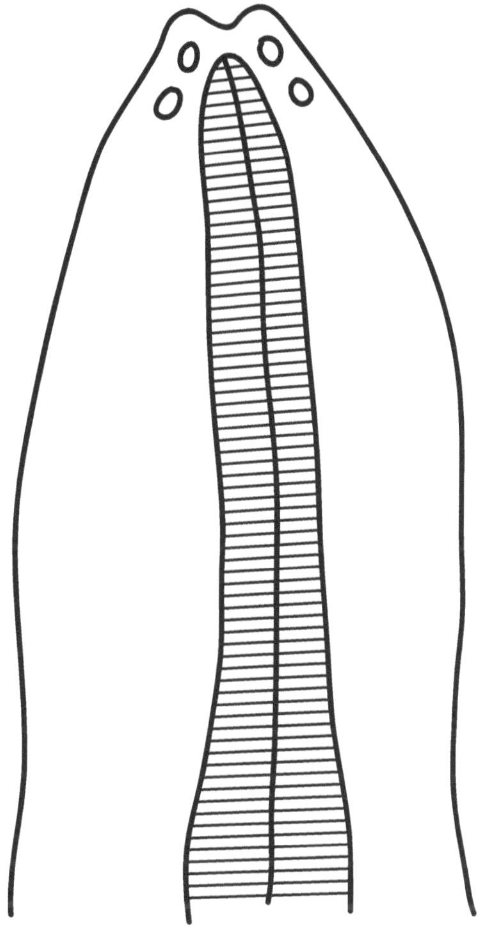

Fig. 6.34 Head end of *Dirofilaria immitis*

Life Cycle

Some Important Information of Life Cycle

Type of parasite	Heteroxenous parasite
Type of life cycle	Indirect life cycle
Type of definitive host	Vertebrate
Type of intermediate hosts	Invertebrate
Location of fertilization	In the parasite
Location of embryonation	In the parasite
Name of infective stage	Third larval stage (L3)
Mode of infection	Inoculation
Alternative mode of infection (hyperinfection, autoinfection, etc.*)*	No

Stages of Development

In the host	In the intermediate hosts
• Fourth stage larvae • Adult, egg and first stage larvae	• Second stage larvae • Third stage larvae

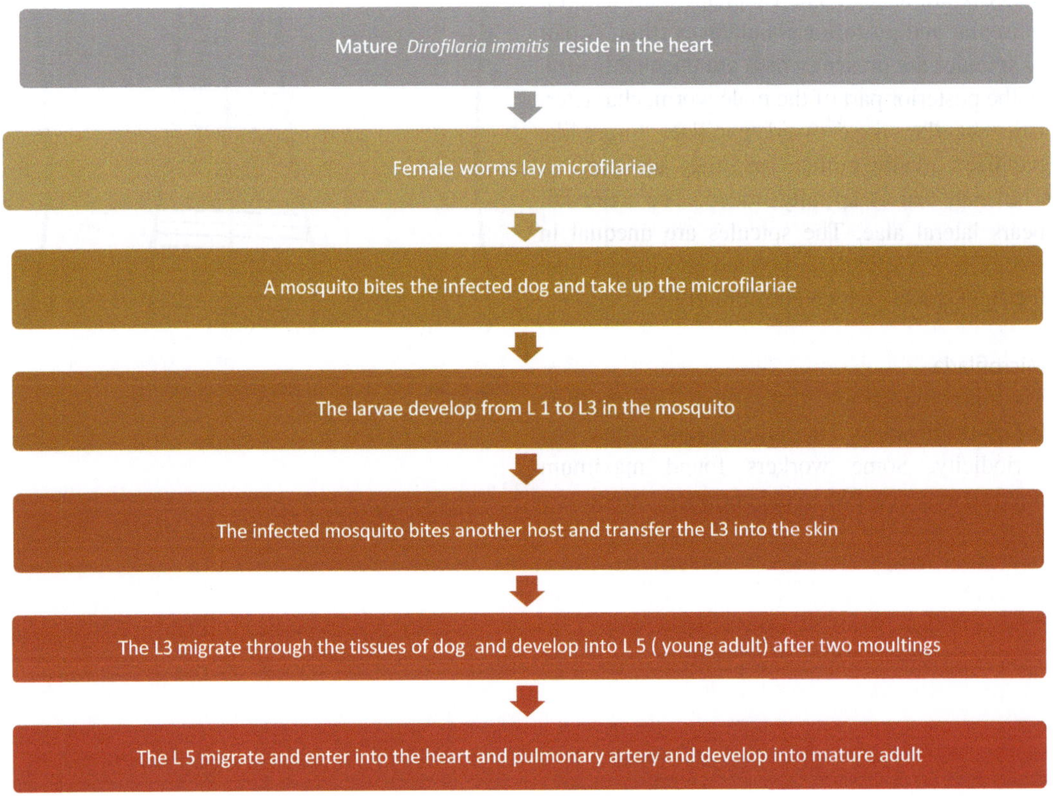

Fig. 6.35 Life cycle of *Dirofilaria immitis*

Description of Life Cycle

Embryonation
The adult parasites remain in their predilection site. Embryonation of eggs occurs in the parasite itself.

Hatching
Hatching of eggs occur in the uterus of the female. The female worms lay their larvae in their predilection site. The microfilariae are available in the blood.

Development in Mosquitoes
The mosquitoes (*Anopheles*, *Culex*, *Aedes*) serve as intermediate hosts. They pick up the infection when they suck blood from infected hosts. The larvae become available in the gut of the mosquitoes within 24 h of infection. During the next 24 h, the larvae penetrate the gut and reach the Malpighian tubules of the intermediate hosts. Then the larvae reach the distal part of the gut. The larvae at first develop to a sausage-shaped larvae and then transform to elongated sausage-shaped stage. Study shows the 1st molting at about 10th day and second molting at about 13th day of infection. About 15th day of infection, the larvae migrate to the head region and are ready to give infection to the definitive hosts. The development up to third infective larval stage of the parasite occurs in the mosquitoes.

Infection of the Definitive Hosts
The final hosts get the infection when the mosquitoes take blood meal from the hosts. The larvae then develop into adult in the definitive hosts.

Pathogenesis
Pulmonary hypertension, hypertrophy of the right ventricle, congestive heart failure, hemoptysis, etc. are some common clinicopathological features occurring in dog. Hemoglobinuria and jaundice are other clinicopathological features. Lodgement of adult and larval parasites in the pulmonary artery results in pulmonary hypertension. The adult and larval stages of parasite are found in between the right atrium and the right ventricle. Congestive heart failure occurs due to blockage of the valvular passage. Passive congestion also occurs. The dead worms may cause pulmonary embolism. One more very important pathological feature is vena cava syndrome which is due to lodgment of masses of worms in the posterior vena cava. Pulmonary hypertension may cause right ventricular hypertrophy leading to congestive heart failure. This results in occurrence of edema and ascites. This condition results in hemoptysis, hemoglobinuria, bilirubinemia, icterus, anemia, etc.

Symptoms
The clinical signs include the following:

- Shallow coughing with hemoptysis
- Hemoglobinuria
- Jaundice
- Pulmonary hypertension
- Weakness
- Listlessness
- Dyspnea
- Tachypnea
- Ascites
- Hemoglobinuria

Diagnosis

General Diagnosis
General diagnosis is not possible basing on symptomatology because the clinical signs do not have diagnostic importance.

Parasitological Diagnosis

- Direct examination of wet smear of blood may reveal the movement of the microfilariae.
- Knott's method is done in case of low parasitemia. 1 ml of blood from infected animal is taken at first. The blood is mixed with 9 ml of 2% formalin. Centrifugation of the mixed material is done. The sediment is stained by methylene blue and microfilariae are detected.

Immunological Diagnosis
Some works have been carried out using crude and purified antigens. All sorts of immunodiagnosis are in progress. The antigens which are

used for this purpose are whole somatic antigen and excretory-secretory antigens of adult parasite. Different purified proteins were also immunologically characterized. IgE and IgG specific ELISA and EITB have been described. Solgi et al. (2018) worked on immunodiagnostic potential of a new **recombinant *protein Dirofilaria immitis*** **(rDgK)**. The recombinant protein was expressed in *Escherichia coli*, and the target protein was isolated by Ni2+ Hitrap column. The protein was tested by indirect ELISA using dog sera.

Molecular Diagnosis

Molecular diagnosis is possible by using specific molecular marker. Different ribosomal DNA, ribosomal RNA, and mitochondrial DNA could be used as molecular markers. Studies indicate the use of ITS1, ITS2, etc. as molecular markers for identification of this parasite. This similar technique could be used for identification of microfilariae. Molecular identification of larval stage of *Dirofilaria immitis* using panfilarial specific primer in different mosquitoes, i.e., *Anopheles*, *Culex*, *Aedes*, has been reported.

Treatment

Name of drugs	Route of administration	Dose
Levamisole	Oral	10 mg/kg for 2–3 weeks
Ivermectin	Oral	6 μg/kg bw at monthly interval

Other drugs are ivermectin, melarsomine, doxycycline, etc.

Control

Control of mosquitoes should be done by using different insecticidal agents.

Species: *Dirofilaria repens*

Hosts: dog, man

Site: subcutaneous tissue

Morphology

The cuticle of the head may be thick and granular in appearance. The esophagus has two distinct regions i.e. anterior muscular and posterior glandular.

Intermediate hosts: *Aedes*, *Culex*, and *Mansonia*

Pathogenesis: The parasites form nodules in the subcutaneous tissue of man and dog.

Life Cycle

Type of life cycle: indirect

- The adult parasites remain in their predilection site. Embryonation of eggs occurs in the parasite itself.
- The parasites are larviparous and lay their larvae.
- The larvae become available in the circulation.
- The mosquitoes (*Culex, Aedes, Mansonia*) act as the intermediate hosts. They pick up the infection when they suck blood from infected hosts. Development up to third infective larval stage occurs in the mosquitoes.
- The final hosts get the infection when the mosquitoes take blood meal from the hosts. The larvae then develop into adult in the definitive hosts in their respective predilection site.

Pathogenesis

The affected dogs remain symptomless. However, the parasites may cause dermal swelling, nodule formation, etc. The parasites occur under the skin forming coiled knots. Human beings may act as unnatural hosts who may get the infection by mosquito bites. The larvae do not develop into adult stage in most of the cases.

D. conjunctivae

These worms cause formation of tumor in the eyes of man. Hazelnut-sized nodules are formed in the head and eyelids of man.

D. roemeri

Hosts: kangaroo and wallabies

Site: subcutaneous and intramuscular connective tissue

Intermediate hosts: Tabanid fly

D. tenuis

Hosts: racoons

Site: subcutaneous tissue

Genus: *Parafilaria*

Species: *Parafilaria bovicola*

Hosts: cattle and buffalo
Site: skin

Morphology

The females are 4–5 cm long and males are 2–3 cm. Anterior end is equipped with 13 prominent transverse cuticular ridges with irregular interruptions. A small number of round tubercles are present at the anterior tip. Rest of the body is striated transversely.

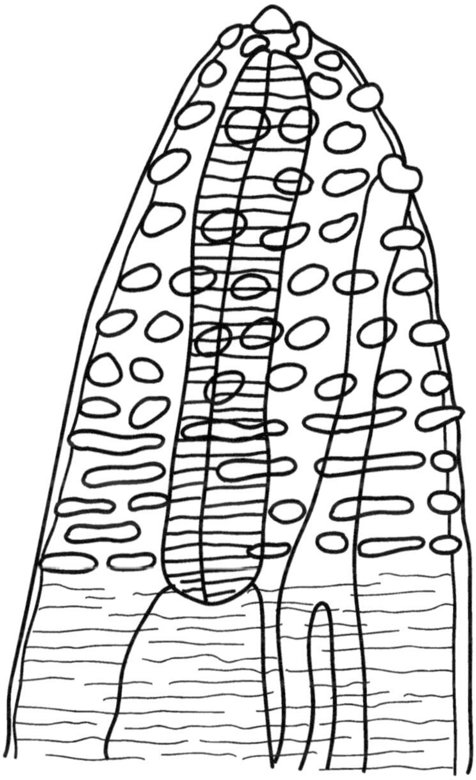

Fig. 6.36 Head end of *Parafilaria multipapillosa*

Life Cycle

Type of life cycle: indirect

The following events occur during the life cycle:

- The parasites lay their eggs at their predilection site.
- The larvae become available in the wounds very soon after deposition of eggs.
- Probably house fly act as intermediate host.

Pathogenesis

The parasites cause formation of subcutaneous nodules. These nodules are typical. These appear suddenly in summer and break open later resulting in bleeding. In the next year, the same pathological features occur. The pathogenesis is briefly mentioned below:

- The worms live in the subcutaneous tissue.
- Immunological reaction leads to local inflammation.
- Local inflammation results in dermal swelling.
- Later nodules are developed.
- Nodules get ruptured and produce different bleeding points.
- The dermal lesions attract the *Musca* flies. This aggravates the condition due to feeding activity at the wounds.

Diagnosis

Microscopic examination of smear for oozing material may reveal egg or microfilaria

Treatment

Name of drugs	Route of administration	Dose
Nitroxinil	S/C	20 mg/kg for cattle
Ivermectin	S/C	0.2 mg/kg for cattle and horse

Genus

Suifilaria

Species: *Suifilaria suis*

Hosts: pig
Site: subcutaneous and intermuscular connective tissue

Morphology

The worms are slender. The hind end of female is equipped with small tubercles. The male tail is spirally coiled.

Pathogenesis

The worms produce characteristic small whitish nodules in the connective tissues. They are non-pathogenic. But secondary bacterial infection may occur when the vesicles burst, and it turns into formation of abscess.

Species: *Loa loa*

Disease name: calabar swelling, fugitive swelling
Hosts: man, monkey, baboon
Site: subcutaneous tissue

Life Cycle

Some Important Information of Life Cycle

Type of parasite	Heteroxenous parasite
Type of life cycle	Indirect life cycle
Type of definitive host	Vertebrate
Type of intermediate host	Invertebrate
Location of fertilization	In the parasite
Location of embryonation	In the parasite
Name of infective stage	Third larval stage (L3)
Mode of infection	Inoculation
Alternative mode of infection (hyperinfection, autoinfection, etc.*)*	No

Description of Life Cycle
Type of life cycle: indirect

The following events occur during the life cycle:

- The parasites are larviparous and lay their larvae at their predilection site.
- The larvae enter into the blood circulation.
- The microfilariae from the hosts are taken up by the bloodsucking flies (*Chrysops silacea*, *C. dimidiata*) during taking blood meal.
- They transform into third infective larvae after two moltings in the intermediate hosts.
- The healthy animals become infected when these infected flies take blood meal.

Pathogenesis

Angioedema, eosinophilia, pruritus, etc. are the pathological features. The microfilariae are available in the blood, urine, spinal fluid, etc.

Genus: *Ornithofilaria*

O. fallisensis
Hosts: anatid birds

Transmission of the parasites occurs by *Simulium* spp.

Genus: *Elaeophora*

Species: *Elaeophora schneideri*

Hosts: deer, sheep
Site: common carotid, internal maxillary artery

Morphology

The oesophagus is markedly long. The male tail is coiled. The spicules are unequal.

Life Cycle

Life cycle is indirect. Horseflies act as the intermediate hosts of the parasite. They take up microfilariae found in capillaries of the forehead and face.

Pathogenesis

In the normal hosts, less or no clinical features are found. The parasites cause sorehead condition in sheep. Filarial dermatitis is a common pathological feature. In the skin of sheep, there occurs granulomatous inflammation infiltrated with eosinophil, lymphocytes, neutrophils, giant cells, etc. There also occurs crust formation. Hemorrhagic lesions occur due to dislodgement of crusts. Dermatitis on the head region is called 'sore head'.

Diagnosis

Diagnosis can be done by demonstration of microfilariae in the skin. Pieces of skin is macerated in warm saline. This macerated material is examined after 2 h for detection of microfilariae.

Species: *Elaeophora poeli*

Hosts: cattle, zebu, buffalo, water buffalo
Site: aorta

Salient Morphology

Lips are absent. The esophagus is markedly long. The uterus is typically branched and have four branches. The parasites cause nodule formation in the aorta within which the males remain.

Genus: *Wuchereria*

Species: *Wuchereria bancrofti*

Common Name

This is called Bancroft's worm.

Distribution

The worms are distributed in South America, Central Africa, West Indies, Japan, India, etc.

Site: They occur in the lymph nodes and lymphatic vessels.

Hosts: The worms are recorded in man and monkey.

Morphology

They are threadlike creamy white colored worms. The female tail end is markedly pointed whereas the male tail end is curved ventrally. The microfilariae of these worms are highly active. Several morphological features of stained microfilariae are found which are a structureless hyaline sheath, and cuticula is lined by subcuticular cells. From head to the tail, there are a number of somatic cells appearing as granules in the central axis (a rudimentary excretory system represented by anterior V spot, anus, and cloaca represented by posterior V spot), an oblique space representing the nerve ring. Different genital cells are present which are called G cells (G1, G2, G3, G4, etc.). The microfilariae show the periodicity. They show nocturnal periodicity between 10 PM and 4 AM. They appear in the peripheral blood during this period; however, they remain in the cardiorespiratory system, i.e., heart, lung, etc., during daytime. Periodicity may vary from place to place. The oriental microfilariae show periodicity whereas periodicity is absent in pacific region.

Life Cycle

Anopheles, *Culex*, and *Aedes* act as the intermediate hosts of this parasite. Mosquitoes take up microfilariae during blood meal. The microfilariae penetrate the gut wall and reach the thoracic muscles. The larvae grow and become thick sausage shaped within 2 days. These are the first stage larvae possessing digestive tract. Within about 11 days, third stage larvae are produced after two more moltings. These third stage larvae are the infective stages. The infected mosquitoes transmit the microfilariae in the skin when they suck blood from the hosts. The larvae gain entry into the lymphatic channels and are transported to different locations, i.e., scrotal, inguinal, and abdominal regions. The juveniles get matured to adult there.

Pathogenesis

The adult and the larvae cause mechanical obstruction of lymphatic channel resulting in lymphangitis. Absorption of the toxic material as a result of disintegration of dead worms or by absorption of the metabolic by-products are also responsible for pathological features. Hypertrophy of the lymph node results in elephantiasis. Occult filariasis is characterized by marked eosinophilia and different pulmonary symptoms like coughing, sneezing, etc. Classical filariasis is characterized by lymphangiovarix (dilatation of lymphatics). This lymphangiovarix may rupture and cause several pathological complications, i.e., lymphocele, lymphuria, chylothorax, chylous diarrhea, etc. Fever, hydrocele, and elephantiasis are the major clinical signs in this filarial infection. Anaphylactic reaction may occur due to dead adult worms. The larvae are not able to cause any marked pathogenesis. However, the metabolic by-products of the larvae may cause allergic reaction in the skin.

Diagnosis

Examination of blood smear may reveal microfilariae in the peripheral blood. Smears can be prepared with the exudates from the lymph varix, hydrocele fluid, etc. Different immunodiagnostic tests comprising ELISA, CFT, FAT, etc. can be done.

Genus: *Brugia*

Hosts

Many species are found in the genus *Brugia*. The most important zoonotic parasitic species of this genus is *Brugia malayi*. *B. timori* also occurs in man. Two other important species are *Brugia pahangi* occurring in the lymphatic system of cat, dog, monkeys, etc. and *B. patei* occurring in the lymph nodes of dog, cat, etc.

Life Cycle

Anopheles, *Mansonia*, and *Aedes* serve as intermediate hosts of this parasite. The mosquitoes take up microfilariae from the infected hosts during blood meal. The microfilariae develop in the mosquitoes and reach the infective stage. The infected mosquitoes transmit the microfilariae to the new hosts during blood meal from the same.

Pathogenesis

The adults and larvae cause blockage of the lymphatic system resulting in lymphangitis, lymphadenoma, and elephantiasis. The skin of the affected area becomes fibrotic. Recurrent fever is another symptom. Hydrocoele and chyluria also occur in this case. *Brugia timori* causes acute lymphangitis in man. Other spp. such as *Brugia malayi* cause leg elephantiasis and scrotal elephantiasis. Adenolymphangitis is another clinicopathological feature.

Diagnosis

Enlargement of the inguinal lymph node provides diagnostic clue. Examination of the blood smear may reveal the presence of microfilariae. For better demonstration, Knott's method is done. Many workers are doing research on use of recombinant antigens for diagnosis of brugiosis. BmR1 is one such antigen. Immunochromatographic rapid dipstick test is useful. Molecular identification using different ribosomal and mitochondrial DNA is possible.

Treatment

Treatment not only includes the filaricidal drugs, but attempt has been made to kill the endosymbiont (*Wolbachia* spp.). Rifampicin and chloramphenicol have been found to affect the development of the parasite during in vitro study.

Control

The breeding foci should be exterminated. For this, different herbicidal agents can be used to clear up the plants from the breeding zones. Spraying of insecticidal agents can be done for control of mosquitoes.

Order: Spirurida—Superfamily: Filaroidea—Family: Setariidae

Family Synopsis

They are mostly found in the peritoneal cavity of different mammals. They are morphologically characterized by head equipped with peribuccal ring bearing cuticular epaulette-like structure. The life cycle of the worms is indirect and involves different insects as intermediate hosts. They cause variable pathological features. Diagnosis is done by finding the microfilariae in the blood. Control of the parasites can be done by chemotherapy and control of bloodsucking insects which are the intermediate hosts.

Species and Hosts

Species	Hosts
S. labiatopapillosa	cattle, buffalo
S. digitata	cattle, buffalo, zebu
S. equina	Horse
S. cervi	Deer

Genus: *Setaria*

General Morphology

Color is milky white. The salient features are characterized by presence of peribuccal cuticular ring. At the head region of the parasite, cuticular prominences (epaulette) are present. A pair of spicules are present which are unequal in size. The male tail bears four pairs of precloacal and usually four pairs of postcloacal papillae. The female tail bears minute spines and conical projections.

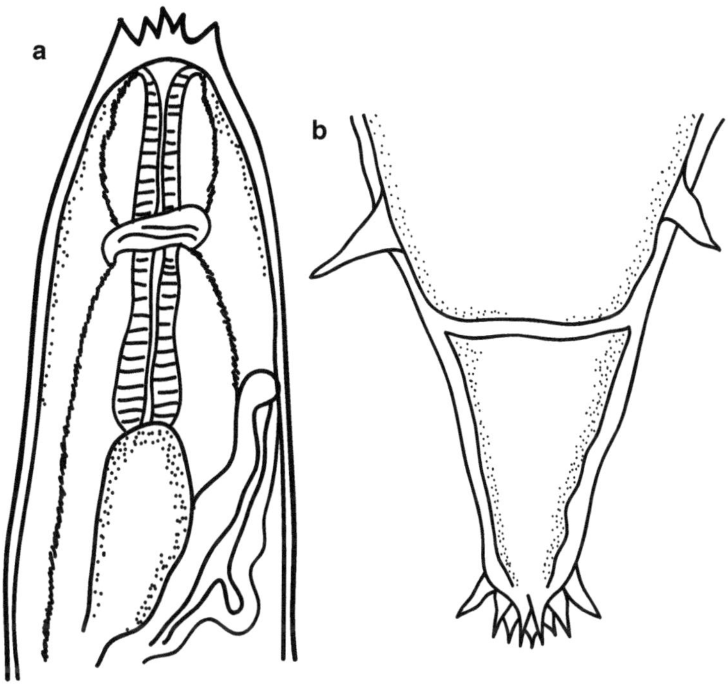

Fig. 6.37 *Setaria labiatopapillosa*. (**a**) Anterior end of female. (**b**) Hind end of female

Etymology

The generic name *Setaria* has been derived from Latin word *seta* meaning bristle.

> **Note**
> Kausar and Khan (2018) reported comparative efficacy of diethylcarbamazine, nitazoxanide, nanocomposite of nitazoxanide, and silver nanoparticle on dehydrogenase of TCA cycle in *Setaria cervi* in vitro. Inhibition of TCA cycle was observed both in microfilariae and adult stage. NTZ + AgNPs was found most effective synergistic combination against TCA cycle enzymes which blocked the isocitrate and malate dehydrogenase almost completely and succinate dehydrogenase to a large extent both in adult and larva.

General Life Cycle

Some Important Information of Life Cycle

Type of parasite	Heteroxenous parasite
Type of life cycle	Indirect life cycle
Type of definitive host	Vertebrate
Type of intermediate host	Invertebrate
Location of fertilization	In the parasite
Location of embryonation	In the parasite
Name of infective stage	Third larval stage (L3)
Mode of infection	Inoculation
Alternative mode of infection (hyperinfection, autoinfection, etc.*)*	No

Description of Life Cycle

Life cycle of most of the spp. is unknown. But the general events of life cycle are as follows:

Life cycle: indirect

The following events occur:

- The adult parasites remain in their predilection site. Embryonation of eggs occurs in the parasites itself.
- Hatching of eggs occurs in the uterus of the female. The female worms lay their larvae in their predilection site. The microfilariae are available in the blood.
- The mosquitoes (*Anopheles*, *Culex*, *Aedes*) serve as the intermediate hosts. They got the infection when they suck blood from infected hosts. Development up to third infective larval stage occurs in the mosquitoes.
- The final hosts acquire the infection when the mosquitoes take blood meal from the hosts. The larvae undergo development up to adult in the definitive hosts.

General Pathogenesis

By Adult

The adults are apparently nonpathogenic. But fibrinous peritonitis may be caused in lesser extent. There may be erratic migration of adult parasites in different organs, i.e., liver, kidney, urinary bladder, eye, etc. Adult *S. digitata* may migrate to the urinary bladder and causes granulomatous lesions there. The adults may migrate to the eye causing severe damage to the eye resulting in blindness.

By Larvae

The larval stages of *S. digitata* and *S. labiatopapillosa* migrate erratically to the central nervous system of abnormal hosts such as sheep, goat, horse, etc. The damage to the nervous system causes incoordination of movement, paralysis, severe weakness, and death. The larvae of *S. digitata* cause enzootic cerebrospinal nematodosis in sheep, and it occurs in the summer and rainy season. The affected animals exhibit acute encephalomyelomalacia resulting in locomotor disturbance in the hind legs (lumber paralysis). The affected animals may even die. The disease is called as KUMRI.

General Treatment

Name of drugs	Route of administration	Dose
Fenbendazole	Oral	5 mg/kg for treatment of enzootic cerebrospinal nematodosis of sheep, goat and horse
Ivermectin	S/C	0.2 mg/kg

Setaria equina

Hosts: equines
Site: peritoneal cavity

Salient Morphology

The male is 8 cm and the female is 13 cm long. The worms possess peribuccal prominences. The female tail is pointed. Microfilariae are sheathed. The life cycle is indirect. The infective larvae develop into mosquitoes, i.e., *Anopheles*, *Culex*, *Aedes*. The final hosts get the infection when the mosquitoes suck blood. The peritoneum is the main predilection site whereby they do not produce any pathogenesis. However, occasionally the larvae migrate to the brain and eye causing encephalomyelitis and blindness, respectively. Examination of blood smear may reveal the presence of microfilariae.

Species

S. digitata

Hosts: cattle, buffalo
Site: peritoneal cavity

Salient Morphology

The worms are prevalent in Asian countries. The male is 4–5 cm and the female is 6–8 cm long. The female tail ends in a button-like structure. The life cycle is indirect. The infective larvae develop into mosquitoes, i.e., *Anopheles*, *Culex*, *Aedes*. The final hosts get the infection when the mosquitoes suck blood. In the natural hosts, they do not apparently produce any harm. But the larvae cause cerebrospinal nematodosis in unnatural host. The associated pathological features are focal encephalomyelomalacia, microcavitation, and hemorrhage in the brain of aberrant host.

Genus: *Dipetalonema*

Etymology

The generic name dipetalonema has been derived from new Latin words di (prefix meaning two), petalon (Greek word meaning leaf), nema (Greek word meaning thread).

Morphology

The body is filiform. The mouth is simple without any lip. But the mouth is surrounded by flat cuticular shield. The esophagus is usually divided into one anterior short muscular part and posterior long glandular part. Caudal alae are absent. Preanal and postanal papillae are present. Spicules are unequal.

Life Cycle

Some Important Information of Life Cycle

Type of parasite	Heteroxenous parasite
Type of life cycle	Indirect life cycle
Type of definitive host	Vertebrate
Type of intermediate host	Invertebrate
Location of fertilization	In the parasite
Location of embryonation	In the parasite
Name of infective stage	Third larval stage (L3)
Mode of infection	Inoculation
Alternative mode of infection (hyperinfection, autoinfection, etc.*)*	No

Description of Life Cycle

Different insects serve as the intermediate hosts. *Ctenocephalides canis*, *C. felis*, and *Pulex irritans* etc. are the intermediate hosts of *D. reconditum*. Different *Culicoides* flies (*C. austeni*, *C. grahami*, etc.) are the intermediate hosts of *D. perstans*. The flies take up the microfilariae from the definitive hosts during blood meal. The microfilariae reach the infective stage in the flies. The infective stages are transmitted to the new hosts when they take blood meal from the same.

Pathogenesis

Dipetalonema streptocerca causes lymphadenopathy, elephantiasis, etc. In most of the cases, these parasites do not produce any specific symptoms. *D. perstans* causes cerebral filariasis in man.

Diagnosis

Examination of blood smear may reveal the presence of microfilariae. Knott's method is very useful. Microfilariae can also be identified by examination of skin scraping. Differential identification from other microfilariae should be done.

Control

Control of the intermediate hosts can be done using different insecticidal agents. *Culicoides* flies are the intermediate hosts. So man should not come in contact with these flies.

Species

D. reconditum

Hosts: dogs

Site: connective tissue, lung, liver and kidney

Salient Morphology

The adults are small and slender. The spicules are unequal. Microfilaria with button hook tail is characteristic morphological feature.

Intermediate hosts: Fleas (*Ctenocephalides canis*, *C. felis*) and ticks (*Rhipicephalus sanguineus*) are the intermediate hosts.

D. dracunculoides

Hosts: dogs

Site: peritoneal membrane

Morphology

The male is 2.4–3 cm and the female is 3.2–6 cm. Microfilariae are similar to *D. immitis*; they only differ in having a short blunt tail. The microfilariae are unsheathed. The life cycle is indirect. The spicules are unequal. The ticks and fleas serve as the intermediate hosts. The worms are regarded as nonpathogenic.

D. perstans
Hosts: man
Hosts: peritoneal and pleural cavities

Morphology
The microfilariae are unsheathed and do not show any periodicity.

Intermediate hosts: Different *Culicoides* flies are the intermediate hosts of this parasite.

D. evansi
Hosts: camels
Site: heart, spermatic artery, and pulmonary artery
Intermediate hosts: *Aedes spp.* are reported as intermediate hosts.

Morphology
The male is 8–11 cm and the female is 14.5–18.5 cm. The microfilariae are sheathed. The life cycle is indirect. The infective larvae develop into the mosquitoes. *Aedes* is considered to be the intermediate hosts of this filarial species. The final hosts get the infection when the mosquitoes suck blood. They cause orchitis in the spermatic vessel. They also cause different cardiac pathology. Other pathological features include thrombosis, arteritis, and phlebitis. Diagnosis can be done by detection of microfilariae in the blood. Knott's method is very useful technique.

D. streptocerca
Hosts: man
Site: skin
Intermediate hosts: *Culicoides austeni* is the intermediate host.

Pathogenesis
The parasites cause elephantiasis, edema formation, etc. in human beings.

Genus: *Stephanofilaria*

Morphology
A pair of spicules are present which are unequal in size. They are non-bursate worms.

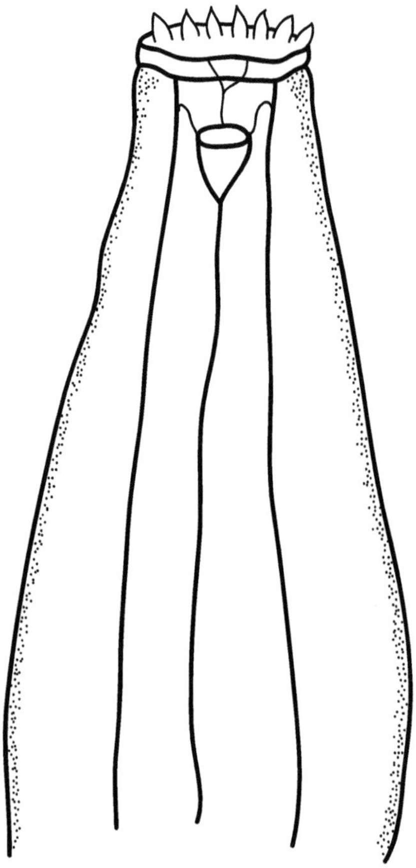

Fig. 6.38 Head end of *Stephanofilaria assamensis*

Name of species	Hosts	Site	Pathology
Stephanofilaria assamensis	Cattle	Hump	Hump sore
S. kaeli	Cattle	Leg	Leg sore
S. stilesi	Cattle	Abdomen	Lesion on the abdominal area
S. dedosi	Cattle	Skin	Sore
S. zaheeri	Buffalo	Ear	Ear sore
S. okinawaensis	Cattle	Muzzle and teat	Lesion on muzzle and teeth

Hump Sore

The lesions mainly occur in the hump region. The parasites cause continuous irritation. The animals rub the affected part which aggravates the condition. The lesions become ulcerative due to repeated rubbing. At first, small papules are formed. The papules coalesce to form lesion in large area which is covered by crust material. Hemorrhagic ulceration, hyperkeratosis, hair loss are some clinical features. The lesions become quiescent in the winter but it becomes active during rainy season.

Ear Sore

The infection starts from the meatus. Gradually the lesions increases in size. Chronic form of lesions is characterized by papule, scab, ulceration, encrustation, etc.

Treatment of Hump Sore and Ear Sore

Name of drugs	Route of administration	Dose
Oxfendazole	Oral	5 mg/kg
Levamisole	Oral	7 mg/kg
Ivermectin	S/C	0.2 mg/kg

Order: Spirurida—Superfamily: Filaroidea—Family: Onchocercidae

Family Synopsis

The family contains very important filarial worms. They are morphologically characterized by presence of striated cuticle with transverse thickenings and genital opening located anteriorly. Male tail is spirally coiled and female tail is bluntly conical. The worms exhibit indirect life cycle which involves mostly different bloodsucking flies as intermediate hosts. Infective microfilariae develop in the intermediate hosts and are transmitted to the definitive hosts when they suck blood from the hosts. Diagnosis is done by finding the microfilariae in the blood. Control of filarial worms mostly relies on chemotherapy and control of intermediate hosts.

Genus: *Onchocerca*

Morphology

They are filiform. The mouth does not possess any lip. The esophagus is short and is indistinctly divided into two portions. The cuticle has transverse striations and thickenings. The male tail end is spirally coiled. The female tail end is bluntly conical.

Some Information of Life Cycle

Type of parasite	Heteroxenous parasite
Type of life cycle	Indirect life cycle
Type of definitive host	Vertebrate
Type of intermediate host	Invertebrate
Location of fertilization	In the parasite
Location of embryonation	In the parasite
Name of infective stage	Third larval stage (L3)
Mode of infection	Inoculation
Alternative mode of infection (hyperinfection, autoinfection, etc.*)*	No

Description

- The worms are larviparous and lay larvae in the vicinity.
- The larvae are taken up by the intermediate hosts (*Simulium* spp., *Culicoides* spp.).
- The microfilariae reach the thoracic muscles.
- The microfilariae transform to sausage shaped form. They later become elongated and thinner and develop to infective stage.
- The infective stages reach the head region of the flies and lodge in proboscis.
- The files transmit the infection to the hosts when they suck blood.

Species: *Onchocerca volvulus*

Hosts: These are filarial worms occurring in man.
Site: subcutaneous tissues

Distribution: The worms are distributed in Central America, Africa, etc.

Pathogenesis

The pathology is caused both by the microfilariae and the adult parasites. The dead microfilariae cause hypersensitive reaction in the skin which leads to swelling of the skin area. Subcutaneous nodules are formed in the skin. This species causes ocular onchocerciasis called river blindness. This parasite also causes mazzotti reaction. This is characterized by dermatitis following treatment with diethylcarbamazine.

O. gibsoni

Hosts: cattle, zebu, buffalo

Site: subcutaneous tissue of brisket region and hind legs

Morphology

The male tail has small alae. On either side of the tail of male parasite, six to nine papillae are present. Microfilariae are unsheathed.

Pathogenesis: The parasites cause nodule formation in brisket region. Nodules are called worm nests containing coiled up worm.

O. armillata

Hosts: cattle, buffalo, sheep, goat, etc.

Site: aorta

Morphology

Microfilariae are unsheathed.

Intermediate hosts: not known

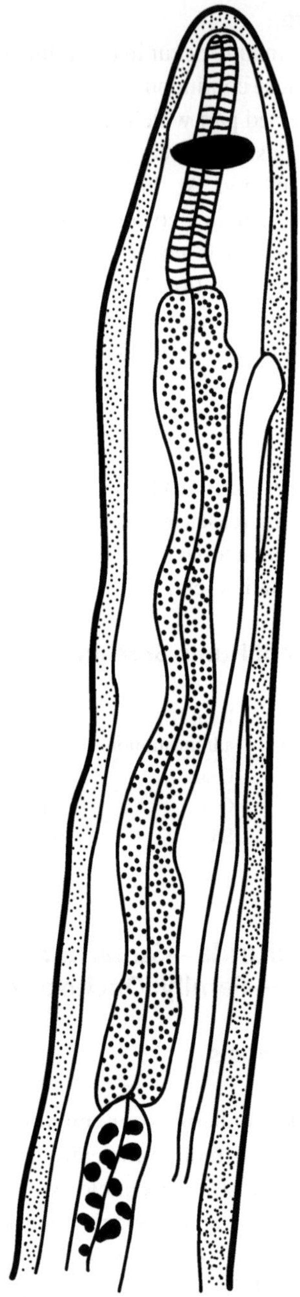

Fig. 6.39 *Onchocerca sp.* (head end of female)

O. dukei

Hosts: cattle
Site: subcutaneous and perimuscular tissue
Pathogenesis: Nodule formation occurs in the thorax, abdomen, diaphragm, etc.

O. gutturosa

Hosts: cattle, buffalo
Site: ligamentum nuchae, scapular cartilage, hip, stipple, shoulder region
Intermediate hosts: *Simulium ornatum*
Pathogenesis: Nodule formation occurs in the affected area.

O. cervicalis

Hosts: Horse and mule
Site: Ligamentum nuchae
Intermediate host: *Culicoides nebeculosus*

Order: Spirurida—Superfamily: Dracunculoidea—Family: Dracunculidae

Family Synopsis

The members of this family are the parasites of mammals and birds occurring mostly in the subcutaneous tissue. The important species is *Dracunculus medinensis* which is called dragon worm. The worms of this family are morphologically characterized by large sized female, atrophied intestine, vulva, and vagina. The worms exhibit indirect life cycle which involves cyclops as intermediate hosts. Infective microfilariae develop in the intermediate hosts and are transmitted to the hosts when the definitive hosts drink water contaminated with cyclops.

History

The Guinea worm was known to the people since antiquity. Russian naturalist Alexei Fedchenko detected larvae of Guinea worm in cyclops in the year 1870.

Genus: *Dracunculus*

Etymology

The generic name *Dracunculus* is deminutive of the Latin word "draco" meaning "dragon" or "serpent."

Species: *Dracunculus medinensis*

Common name: dragon worm, Guinea worm, serpent worm

Host

Common hosts: man
Occasional host: dog, cattle, horse
Site: Subcutaneous tisue

Morphology

Color is white or creamy white. Female parasite is too large as compared to male parasite, and it has tapering hooklike posterior extremity. The male worm is 1.2–4.0 cm and female worm is 100–400 cm long. A pair of spicules are present which are subequal. The larvae are 500–750 um long having broad anterior end and slender tapering posterior end.

Life Cycle

Some Important Information of Life Cycle

Type of parasite	Heteroxenous parasite
Type of life cycle	Indirect life cycle
Type of definitive host	Vertebrate
Type of intermediate host	Invertebrate
Location of fertilization	In the parasite
Location of embryonation	In the environment
Name of infective stage	Third larval stage (L3)
Mode of infection	Ingestion of intermediate host carrying infective stage
Alternative mode of infection (hyperinfection, autoinfection, etc.)	No

Stages of life cycle formed

In the hosts	In the intermediate host
• Fourth stage larvae, Adult, Egg • First stage larvae	• Second stage larvae • Third stage larvae

Description of Life Cycle

Embryonation: Eggs get embryonated within the parasite.

Hatching: Hatching of eggs occurs in the uterus of the parasite.

Development in intermediate hosts: Adult female parasites get matured in the deep connective tissue. Then the female parasites migrates to the peripheral subcutaneous tissue after about 9 months of infection. They cause swelling and blister formation. The blister bursts and lesion is produced. When these lesions come in contact with water, the gravid uterus of the worm gets prolapsed and the larvae enter water. The discharge of larvae may continue for several weeks whenever the lesion comes in contact with water. These larvae (first stage) are ingested by the cyclops. Infective third stage larvae are produced in the cyclops.

Infection of definitive hosts: The definitive hosts get the infection by ingestion of infected cyclops.

Development in final host: The body of cyclops gets digested and the larvae become free in the stomach. The liberated larvae penetrate through the gut wall and reach the retroperitoneal connective tissue. In this vicinity, they get sexually matured. Fertilization occurs there. After fertilization, the male parasites die. The gravid females migrate to the skin of the hosts. The prepatent period is about 1 year.

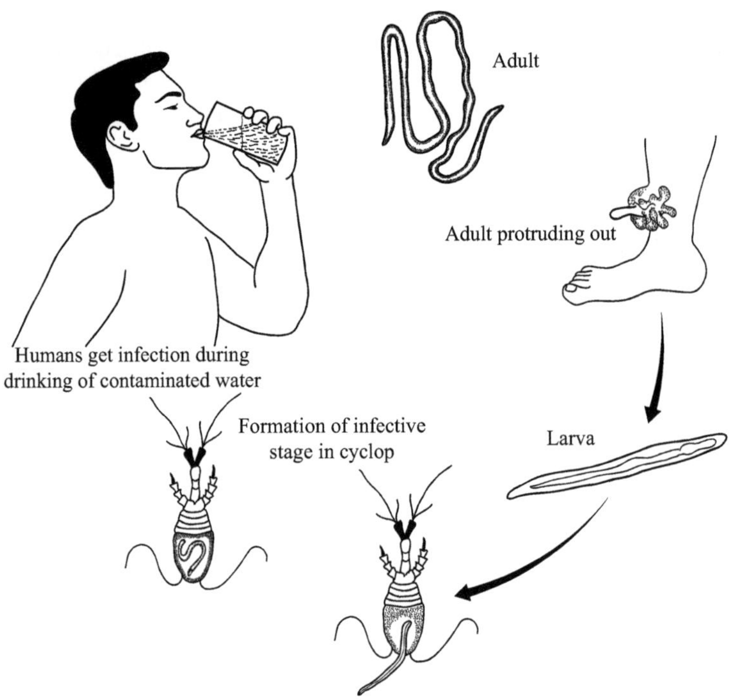

Fig. 6.40 Life cycle of *Dracunculus medinensis*

Pathogenesis
As such, no symptoms are found initially. But clinical symptoms are exhibited when female worms come to the skin. The pathology starts with swelling of the skin which results in lesion. The toxic material secreted by the parasites causes allergic reaction which results in blister formation in different areas like the head, neck, arm, buttock, limbs, etc. The lesion eventually leads to ulcer formation. In man, erythema, urticaria, and pruritus are common clinical features. Other symptoms are dyspnea, vomiting, diarrhea, etc. If the worm is dead, severe symptoms are found due to anaphylactic reaction.

Diagnosis
Observation of emerging parasite from the lesions of the legs of the affected individuals is of diagnostic value. Identification of the larvae of the parasites can be done by microscopical examination. However, radiography can reveal the calcified stage of the worm. Extraction of the worm and identification of the same are possible.

Treatment
Manually the emerging parasite can be removed by a specific technique using a stick to roll up the worm gradually. Every day some length of worm is rolled up until the whole worm is removed out. Surgical intervention can also be done for removal of the worm.

Control
Care should be taken to drink uncontaminated pure water. Underground water should be used for drinking.

Order: Enoplida—Superfamily: Dioctophymatoidea—Family: Dioctophymatidae

Family Synopsis

The members of this family are mostly the parasites of different mammals. *Dioctophyma renale* occurs in the kidney of dog and called kidney worm. The worms of this family are morphologically characterized by presence of cup-shaped bursa without rays. The worms exhibit indirect life cycle which involves mostly the oligochaetes as intermediate hosts. Infective stage develops in the intermediate hosts and is transmitted to the hosts through oral route.

History
Dioctophyma renale was first described by Johann Goeze in a dog kidney in the year 1782.

Genus: *Dioctophyma*

Species: *Dioctophyma renale*
Hosts

Common hosts: dog, fox etc.
Occasional hosts: horse, pig, cattle, man

Common name: kidney worm of dog (largest nematode of dog)

Morphology
The worms are blood red in color. This is the largest known nematode occurring in the kidney of dog. They have anterior tapered end and truncated posterior end. Male worms possess bursa which is oval in shape and does not have any bursal ray. Only one spicule is found which is seta like. Brownish eggs are barrel-shaped.

Life Cycle

Some Important Information of Life Cycle

Type of parasite	Heteroxenous parasite
Type of life cycle	Indirect life cycle
Type of definitive host	Vertebrate
Type of intermediate host	Invertebrate
Location of fertilization	In the parasite
Location of embryonation	In the environment
Name of infective stage	Third larval stage (L3)
Mode of infection	Ingestion
Alternative mode of infection (hyperinfection, autoinfection, etc.)	No

Description of Life Cycle
Life cycle is indirect.

Egg Passing
Eggs are passed with the urine.

Embryonation
Embryonation of eggs occurs in the environment when these come in contact with water.

Infection of Intermediate Hosts
Earthworms act as the intermediate hosts. The intermediate hosts ingest the embryonated eggs.

Development in Intermediate Hosts
The infective stage of the parasites develops in the earthworms, i.e., oligochaete annelid (*Lumbriculus variegatus*). The eggs hatch in the intermediate hosts. After hatching, the larvae (first stage larvae) are liberated. The larvae develop into third stage larvae after two moltings in the earthworm. The third stage larvae are infective stage.

Development in Hosts
The final hosts get the infection by ingestion of earthworm. The larvae develop into adult after two moltings. The eggs are expelled out of the hosts through urine.

Pathogenesis
The penetrating parasite secretes proteolytic and lipolytic enzymes which cause coagulative necrosis of the renal parenchyma. Destruction of parenchyma of the pelvis, blockage of the ureter, uraemia, calcification, etc. are some pathological features. Destruction of the kidney tissue reaches to such an extent that only the capsule is left filled up with hemorrhagic fluid.

Diagnosis
Since the seat of predilection of the parasite is the kidney, examination of urine may reveal the presence of eggs of the parasite.

Treatment
The worms can be removed surgically. Ivermectin is effective but the dead worm may cause damage to the tissue.

Order: Enoplida—Superfamily: Trichuroidea—Family: Trichuridae

Family Synopsis

The members of this family are the parasites of different mammals occurring mostly in the alimentary canal. The worms of this family are morphologically characterized by slender anterior part and thick posterior part called whipworm. One spicule is present which is enclosed in spicule sheath. The worms exhibit direct life cycle.

Family: Trichuridae

History
The credit of first recorded discovery goes to GB Morgani in the year 1740.

Genus: *Trichuris*

The worms are whiplike. The posterior end is much thicker, and anterior end is slender and threadlike giving a whiplike appearance. Only one spicule is present. The spicule is covered by spicule sheath which may or may not be equipped with cuticular spines. The eggs are barrel-shaped.

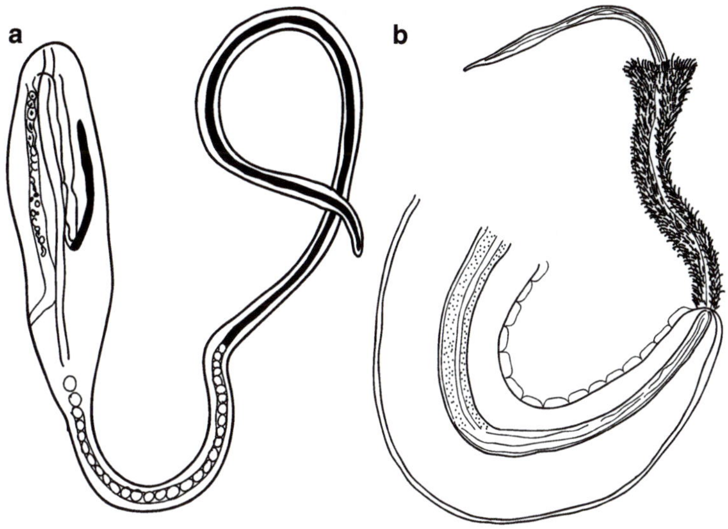

Fig. 6.41 *Trichuris* sp. (**a**) Whole body. (**b**) Hind end of male

Fig. 6.42 *Trichuris* worms (Gross)

Trichuris ovis	*Trichuris vulpis*	*Trichuris discolor*	*Trichuris suis*	*Trichuris trichiura*
Hosts: sheep, goat, cattle, camel	Hosts: dog, cat, etc.	Hosts: cattle, buffalo, sheep, goat	Hosts: pig	Hosts: man, primate
Site: cecum	Site: cecum and other parts of the large intestine	Site: cecum and colon	Site: cecum and colon	Site: cecum
Spicule sheath: The spicule sheath has an oblong swelling and is spinose	Spicule sheath: The spicule sheath is tubular and spinose proximally	Spicule sheath: The spicule sheath has distal inflation and covered by spines	Spicule sheath: The spicule sheath is variably shaped	Spicule sheath may be cylindrical or expanded and wholly or partially spinose

General Life Cycle

Some Information of Life Cycle

Type of parasite	Homoexenous parasite
Type of life cycle	Direct life cycle
Type of host	Vertebrate
Location of fertilization	In the parasite
Location of embryonation	In the environment
Name of infective stage	Eggs containing first larval stage
Mode of infection	Ingestion
Alternative mode of infection (hyperinfection, autoinfection, etc.*)*	No

Description of Life Cycle

The parasite is homoxenous and does not require any intermediate hosts for completion of the life cycle.

Egg passing: Unembryonated eggs are passed with the feces.

Embryonation: Embryonation of eggs occurs in the environment. The first stage larvae are developed in the eggs. The eggs containing first stage larvae are infective stage of the parasite which develops within 1 or 2 months.

Infection: The hosts get the infection by ingestion of infective eggs.

Hatching of eggs: In the gut, the eggs hatch and the larvae come out.

Development in the hosts: The larvae penetrate the cecal mucosal glands wherein they pass four moltings. Later, the young adults emerge out from the mucosal glands and remain attached on the mucosa with the anterior end embedded in the mucosal layer.

Pathogenesis

The parasites cause enteritis. Apparently the infections are asymptomatic. But a large number of parasites precipitate diphtheritic inflammation in the caecal mucosa.

Diagnosis

Examination of fecal sample reveals the presence of typical eggs. The *Trichuris* eggs are barrel shaped.

Treatment

Name of drugs	Route of administration	Dose
Fenbendazole	Oral	5–20 mg/kg for sheep, cattle; 3–25 mg/kg for pig
Oxfenbendazole	Oral	5 mg/kg for sheep, cattle
Mebendazole	Oral	100 mg/kg twice a day for 5 days for dog

Species: *Trichuris ovis*

Hosts: sheep, goat, cattle

Site: cecum

Morphology

Spicule sheath has typical oblong swelling. Fully evaginated spicule is found.

Species: *Trichuris suis*

Hosts: pig

Site: cecum and colon

Spicule has blunt tip. Spicule sheath has variable shape.

Trichuris trichiura

Morphology

The worms are found in the colon, cecum, and appendix of human beings. The worms look like a whip. The anterior part is threadlike which remains embedded into the mucosa, but the posterior portion is thick and remains free outwardly. The slender part of the worm possesses the esophagus, but the posterior part contains the digestive tract and sexual organs. Eggs are bile stained giving the brownish color.

Life Cycle

The worms show direct type of life cycle. No intermediate host is required for completion of the life cycle. Eggs are passed with the stool of infected man. The eggs subsequently develop in the moist earth to the infective stage. Human beings get the infection by ingestion of the embryonated infective eggs. Hatching of the eggs occurs in the GI tract of the host. The larvae are liberated from the eggs and get localized in the cecum.

Type of life cycle: direct

The following events occur during the life cycle:

- Male and female parasites live in the large intestine. Thousands of unembryonated eggs are laid by the female in a day.
- Embryonation starts in the environment and larvae develop within the eggs. Development depends on environmental temperature. 26–30 °C is optimal for development. Once the infective larvae are formed in the eggs, the eggs remain viable in infective stage for about 2 years.
- Hosts become infected by ingestion of infective eggs.
- Hatching of eggs occurs in the gut and first stage larvae are released.
- The larvae then attach to the intestinal wall and pass on four moltings to become adult.

Pathogenesis

The worms are not harmful. However, they may show the symptoms of acute appendicitis. Diarrhea and abdominal discomfort are other symptoms.

Diagnosis

Diagnosis can be done by examination of fecal sample. High level of eosinophilia occurs. However, this is not a confirmative diagnosis.

Treatment

Mebendazole, thiabendazole, pyrantel, and albendazole are the drugs of choice.

Species: *Trichuris discolor*

Hosts: ox, zebra, buffalo, sheep, goat

Site: cecum, colon

Species: *Trichuris campanula*

Hosts: cat

Site: cecum, colon

Order: Enoplida—Superfamily: Trichuroidea—Family: Trichinellidae

Genus: *Trichinella*

Species: *Trichinella spiralis*

Hosts: man, pig, rat

Site: small intestine

Morphology

The worms are small in size. The males are 1.5 mm and the females are about 4 mm long. The male has two bursa-like appearance on either side of the cloacal opening at the posterior end. These are called **pseudobursa**. The anus is present terminally. The vulva is located at the middle of the esophageal region.

Life Cycle

The female worms invade the intestinal mucosa to lay their larvae. The larvae gain entry in the lymphatic vessels, and then via blood circulation, they get disseminated in different muscles. In the skeletal muscle cells, the larvae become encapsulated. Vascular endothelial growth factors cause transformation of affected cells to nurse cells. Humans get the infection by ingestion of infected pork. Transmission of the parasite may occur in between the pigs. The pigs may sniff of the tail muscle and thus become infected.

Pathogenesis

When adults invade the intestinal mucosa during laying of larvae, it causes inflammation as a result of immunological reaction. When the larvae spread in different organs, they cause pathology there. The larvae may reach the brain where they cause encephalitis. Massive larval invasion causes myositis, myocarditis, periorbital edema, etc. The affected individual suffers from diarrhea, muscle pain, etc.

Diagnosis

Humans get the infection by ingestion of infected pig muscle. Larvae are released in the duodenum and jejunum and get matured to adult stage within a week. The fecal sample may be examined for detection of any larvae or even the adult worm in rare case. On fourth week of infection and onward, muscle biopsy can be done for detection of larvae. It is better to take biopsy material from tendinous insertions of the deltoid and gastrocnemius muscle. Skin test can also be done by intradermal injection of 0.1 ml of 1:10000 dilution of Bachman's antigen. Bachman's antigen is prepared from *Trichinella* larvae from infected rabbit's muscle. An immediate reaction with visible erythematous patch develops within 15–20 min. X-ray can also be done for detection of calcified cysts. Examination of pressed preparation of suspected muscle (1gm) in between two slides can be done. Trichinoscope can be used. Muscle digestion using pepsin, Hcl etc. can be done for examination of the sediment.

Treatment

Name of drugs	Route of administration	Dose
Fenbendazole	Oral	5–7.5 mg/kg
Levamisole	Oral	7.5 mg/kg

Control

1. Educational campaign regarding danger of trichinosis and safe cooking of meat.
2. Heating pork above 55 °C and freezing at −15 °C for a period of 20 days kill the parasite present in the muscle.
3. The pigs should not be allowed to eat raw garbage.

Order: Enoplida—Superfamily: Trichuroidea—Family: Capillariidae

Family: Capillariidae (almost equal diameter throughout the entire length)

Genus: *Capillaria*

This is a genus under the family Capillariidae. They are similar to *Trichuris* but posterior part is not thicker as *Trichuris*. The eggs are barrel-shaped. Life cycle is either direct or indirect. The different species are described below.

Species: *Capillaria aerophila*

This species is found in different parts of the respiratory system (trachea, bronchi, etc.) of dog, cat, fox, and other related animals. The male has one sheathed spicule armed with spines. Life cycle is direct. The infective larvae develop within the eggs. The hosts get the infection by ingestion of the infective eggs. Low level of infection of this parasite fails to produce significant pathogenesis. However, severe infection causes different respiratory anomalies comprising rhinitis, trachitis, bronchitis, etc.

Species: *Capillaria anatis*

This species is found in the ceca of different types of birds. Severe infection causes hemorrhagic enteritis resulting in hemorrhagic diarrhea.

Species: *Capillaria annulata*

This species occur in the esophagus and crop of different types of birds. There is a characteristic cuticular swelling behind the head of the parasite. The life cycle is indirect. The earthworms act as the intermediate hosts of the parasites. The hosts get the infection by ingestion of the earthworms harboring the infective larvae.

Species: *Capillaria bilobata*

This species occurs in the abomasum of zebu.

Species: *Capillaria caudinflata*

They occur in the duodenum and ileum of different types of birds. Life cycle is indirect. Earthworms act as intermediate host.

Species: *Capillaria contorta*

They occur in the mouth, esophagus, and crop of different types of birds. Mild infection fails to produce marked pathogenesis. But heavy infection results in marked thickenings of the wall of the predilection site.

Species: *Capillaria erinacea*

They occur in the intestine of pig, hedgehog, dog, cat, etc.

Species: *Capillaria obsignata*

They occur in the small intestine of different types of birds. The esophagus is remarkably large which is more than half of the body of the worm. The life cycle is direct. The infective larvae develop in the eggs. The hosts become infected by ingestion of the infective eggs.

Species: *Capillaria plica*

This species occurs in the urinary bladder and kidney of dog, cat, fox, etc. Apparently, the worms are nonpathogenic but occasionally they may cause cystitis.

Species: *Capillaria putorii*

This species occurs in the small intestine of cat.

Special Part of Nematodes

Immuno-Nematology

Details of Immunity

In case of nematode infection, the immunity is mainly associated with Th2 type of response. In most of the cases, no single substance, cell, or molecule can kill or restrict the nematode infection. Different immune actions jointly kill or eliminate the parasites. There are different immunological phenomena which trigger role for killing and/or elimination of nematode parasites. These are hypersensitivity reaction, complement cascade reaction, ADCC, nitric oxide activity, etc.

Important Points

- The whole GI tract is covered with intact mucosal layer. The mucosal layer provides a primary barrier against the entry of the nematode in the GI wall. This GI layer is formed by mucin secreted by the goblet cells.
- The GI nematode causes injury to the epithelial cells of the GI tract. The injured epithelial cells produce two cytokines, namely, IL25 and IL33. These cytokines activate the ILC2s to produce different cytokines, i.e., IL4, IL5, IL9, IL13, etc.
- The epithelial cells secrete galectins (Gal-15) which increase the mucus viscosity. This results in the impairment of nematode movement.
- C3a and C5a are produced in the pathway of ADCC. These recruit the eosinophils at the site of infection. The eosinophils play great role in elimination of nematode infection.
- Macrophage and dendritic cells perform phagocytosis. These cells also perform antigen presentation. They have toll-like receptors which recognize pathogen associated molecular pattern (PAMP) of invading parasites.
- The activated macrophages are of two types, i.e., M1 (killer macrophage) and M2 (repair macrophage). The killer macrophages are activated by IFN γ and produce different important cytokines, i.e., tumor necrosis factor (TNF), IL1, IL6, IL8, IL12, etc. The M2 is activated by IL4 and IL13. These macrophages induce tissue repair by the action of IL10.
- There are different types of T lymphocytes. These are CD4, CD8, and γ δ subsets. The CD4 T lymphocytes express MHC-II antigens and recognize extracellular antigens. CD4 T lymphocytes have three subpopulations, i.e., Th1, Th2, Th17. IL12 is produced by dendritic cells and macrophages. This IL12 stimulates Th1 lymphocytes. These stimulated lymphocytes produce IFN γ, IL2, and TNF β. The Th2 cells release IL3, IL4, IL5, IL9, IL10, and IL13.
- MMC, IgA, IgE, and eosinophils result in distortion of survival of gastrointestinal nematode parasites.

Role of Eosinophil

Eosinophils have great role for eliminating nematode infection. Different studies proved increased level of eosinophils in nematode infection. The activated eosinophils release IL4, IL5, IL6, IL10, and IL13. After degranulation, the eosinophils also release different granular proteins, i.e., major basic protein, eosinophil cationic protein, eosinophil peroxidase, eosinophil derived neurotoxin, etc., which damage the parasites.

Role of Mucosal Mast Cells

The mucosal mast cells have significant role for nematode and other parasitic infection. A type-1 hypersensitivity reaction occurs. Mast cells have FC epsilon R1 and are used for IgE binding. When this antigen-captured IgE binds with mast cells, a type-1 hypersensitivity occurs as a consequence. The mast cells degranulate and release histamine, serotonin, prostaglandin, leukotrienes, bradykinin, 5-hydroxy tryptamine, etc. They also release IL4, IL5, IL6, etc.

Immunity to Adult Nematodes

Different adult nematodes are eliminated by a specific immunological phenomenon called "self-cure" phenomena. Self-cure phenomena are found in case of *Haemonchus* and some other nematode infection. A strong type-2 response, goblet cell hyperplasia, and mast cell degranulation trigger self-cure phenomena. For occurrence of self-cure phenomena, a primary infection with this nematode is essential which sensitizes the animals. When subsequent infection occurs, hypersensitivity-1 reaction ensues. In hypersensitivity-1 reaction, different events occur. The antibody molecules (IgE) bind to the antigen molecules. In this state, different effector cells (basophil, mast cells, etc.) bind to Fc part of antibody molecule through their Fc receptors. Then degranulation of basophils and mast cells occurs. **After degranulation of these cells, histamine, leukotriene, prostaglandin, etc. are released**. The high level of histamine increases the peristaltic movement causing contraction of the intestinal muscles. The histamine also increases the permeability of the GI tract resulting in efflux of a large amount of fluid in the lumen. Hyperplasia of goblet cells occurs. This hyperplastic goblet cells secrete large amount of mucus material. All these cause dislodgement and expulsion of nematodes.

When the nematode parasites damage the gut epithelial cells, these damaged epithelial cells release IL25 and IL33. IL25 cytokine stimulates type-2 innate lymphoid cells (ILC2s). These stimulated ILC2s release different type-2 cytokines. These cytokines augment differentiation of Th2 cells. IL33 stimulates a set of cells, i.e., ILC2 cell, mast cells, and basophils for production of different cytokines, i.e., IL4, IL13, and IL5.

Immunity to Tissue Migratory Parasite

Tissue migratory parasites are mainly controlled by ADCC. In ADCC, different cells, i.e., eosinophil, neutrophil, macrophages, and platelets, act as effector cells. IgE and IgG antibodies already charged with antigen bind to aforesaid effector cells through the receptors. When binding of the effector cells with the immunoglobulin is completed, toxic substances are released from the effector cells which damage or kill the parasites. In case of nematode infection, eosinophil plays great role. Eosinophilic toxic products, i.e., eosinophilic basic protein, eosinophilic cationic protein, eosinophil-derived neurotoxin, etc., are released which affect the parasites. Migratory parasites may also be killed by NO (nitric oxide) released by macrophage.

Granuloma/Tumor Formation

In some nematodes, the common immunological phenomena fail to eliminate or kill the nematodes. And in this case, the parasites are walled of with formation of granuloma, tumor, nodule, etc. In case of formation of nodule and tumor, generation of fibrous tissue is obvious. *Oesophagostomum* spp. cause formation of nodules in the wall of the large intestine. *Gnathostoma spinigerum* and *Draschia megastoma* cause formation of tumor in the stomach. There are many nematode infections where fibrosis occurs. Fibrosis is caused by deposition of fibroblasts and formation of extracellular collagen matrix. **IL13 has role in production of fibroblasts**. There are a number of nematodes which cause deposition of fibrous tissue. Therefore, presumably the aforesaid immune reactions are related to deposition of fibrous tissue. **AAM (alternatively activated macrophages)** are produced.

These alternatively activated macrophages produce molecules which are very much toxic to the parasites and cause fibrosis and tissue repair.

Immunoregulation

In acute stage of inflammation, immunopathology continues by the inflammatory substances. But this inflammatory process subsides after certain period because immune response is controlled or regulated by a subset of T cells. These are T regulatory cells (Treg cells). IL10 is released by Treg cells which control the acute phase of inflammation because IL10 is an anti-inflammatory cytokine. That's why, IL10 is called regulatory cytokine and Treg cells are called regulatory immune cells. After initial immune reaction, a modified Th2 response prevails. Due to this modified Th2 immune response, the nematodes can enter into chronic stage. TGF is another cytokine which also regulates immune response. This immunomodulatory Th2 or Treg response causes survival of parasites by down-regulating the host's inflammatory and host's immunopathological reactions.

Note: Steps of Self-Cure Phenomena

- Self-cure is mediated by hypersensitivity I reaction.
- Antibody molecule binds to antigen molecule by Fab part.
- Effector cells (mast cells, basophil) bind to Fc part of antibody through their receptors.
- Degranulation of mast cells and basophil occurs.
- Histamine, prostaglandin, and leukotriene are released after degranulation of mast cells.
- Goblet cell hyperplasia occurs due to hypersensitivity I reaction which results in production of a large amount of mucus.
- Histamine increases peristaltic movement.
- End result is expulsion of adult worms and larvae.

Immunodiagnosis

Many immunodiagnostic tests are done for diagnosis of different nematode infection. Of these immunodiagnostic tests, ELISA tests are found better as compared to other tests. An array of research works are being carried out using different antigens. Antigens which are used for immunodiagnosis are crude antigens and defined antigens. The crude antigens comprise of whole somatic antigen of eggs, larvae, and adult parasites. Excretory-secretory products were also used. But unfortunately, specific diagnosis of nematode parasites is a constraint due to problem of cross-reaction. For diagnosis of any parasite, antigenic specificity is primary requisite. However, recent trend has been pinpointed to the use of defined antigens either in native form or recombinant form. The antigens used for different diagnostic works are as follows:

Whole Somatic Antigen

In the earlier decades, works on immunodiagnosis for any parasitic infection were carried out using simple proteins comprising somatic antigen, excretory-secretory antigen, etc.

(a) **Somatic antigen**
   1. **Somatic antigen from whole adult parasite**
      The whole parasite is homogenized and sonicated in PBS solution. This homogenized material is centrifuged and the supernatant is used as antigen.
   2. **Somatic antigen from larvae/eggs**
      This antigen is also prepared in similar way. The larvae/eggs are homogenized and sonicated in PBS solution. This homogenized material is centrifuged and the supernatant is used as antigen.
   3. **Somatic antigen from part of parasite**
      Antigen can be prepared from any part of the body of nematode. The gut of nematode can be used as antigen.

(b) **Excretory-secretory antigen**
   Works were also carried out using excretory-secretory antigens for immunodiagnosis. The adult/larval stages of parasites are placed in culture media. These

stages are cultured in culture media following standard protocol. The cultured material is centrifuged and supernatant is taken as antigen. The supernatant material is dialyzed to make it concentrated.

(c) **Defined antigens**

Defined antigens are specific proteins which are used for immunodiagnosis. The specific proteins can be isolated in native form, or it can be produced by recombinant DNA technology. Isolation of specific native protein can be done by using different column chromatographic techniques. These are gel-filtration chromatography, ion-exchange chromatography, immuno-affinity chromatography, etc. Before proceeding to isolate any specific protein, it is prerequisite to identify or detect the specific protein. For this, Western blot technique is performed. The crude protein is fractionated by SDS-PAGE technique. These fractionated proteins are transferred to nitrocellulose membrane. The immunodominant protein is detected in the nitrocellulose membrane by performing Western blot using hyperimmune sera as primary antibody.

Defined antigen can be produced by recombinant DNA technology. The gene encoding a specific protein is attached with plasmid. This recombinant plasmid is inserted in the *E. coli* or yeast cells for expression of the protein.

Antigens used for immunodiagnostic studies
Crude antigen
(a) Soluble extract of whole parasite
(b) Soluble extract of larvae of parasite
(c) Soluble extract of eggs of parasite
(d) Excretory-secretory antigen of adult parasites
(e) Excretory-secretory antigen of larvae of parasite
(f) Molting fluid of parasite
Defined antigens
• *Toxocara cati* recombinant antigen (TES-120)
• *Strongylus vulgaris* recombinant antigen (rSvSXP)
• *Strongyloides stercoralis recombinant antigen* (SsNIE-1)

Different immunodiagnostic tests performed
• Double immunodiffusion test
• Counterimmunoelectrophoresis
• Indirect hemagglutination test
• Complement fixation test
• Indirect fluorescence antibody test
• ELISA
• Indirect ELISA
• Dot ELISA
• FAST-ELISA

Some Studies of Immunodiagnosis

Recombinant based analysis for immunodiagnosis of strongyloidiasis was done. A **recombinant protein of *S. stercoralis* SsNIE-1** without the luciferase tag was used by Rascoe et al. (2015) in a luciferase immunoprecipitation system (LIPS) to detect IgG4 specific antibodies. As per the report of the study, the sensitivity and specificity of SsNIE-1 IgG4 ELISA were 95% and 93%, respectively. Zahabiun et al. (2015) cloned *T. cati* recombinant **TES -120 (rTES-120)** and expressed it. This recombinant protein was used for diagnosis of human toxocariasis. They compared diagnostic efficacy of this antigen with TES-120 of *T. canis*. The diagnostic sensitivity and specificity of *T. cati*/rTES-120 was 70% and 100%, respectively, and for *T. canis*, it was 57.4% and 94.4%, respectively. Anderson et al. (2013) used recombinant protein **rSvSXP** *Strongylus vulgaris* as a diagnostic antigen. The work displayed the evaluation of diagnostic accuracy of ELISA with sensitivity of 73.3% and specificity of 81.0%. They concluded that IgG(T) antibodies to recombinant SvSXP had potential for prepatent diagnosis of migratory larval stages of *S. vulgaris* which depicted moderate to good diagnostic accuracy.

Immunoprotection

Different antigens are used for immunoprotection studies. These are crude antigen, purified antigen, recombinant antigen, etc. Different antigens used for immunoprotection studies are as follows:

Crude antigens

Different types of crude antigens are used for immunoprotection studies against nematode infections. Different types of antigens have been described below:

a. Somatic Antigen

1. **Somatic antigen from whole adult parasite**
   The whole parasite is homogenized and sonicated in PBS solution. This homogenized material is centrifuged and the supernatant is used as antigen.
2. **Somatic antigen from larvae/eggs**
   The antigen is prepared in similar way. The larvae/eggs are homogenized and sonicated in PBS solution. This homogenized material is centrifuged and the supernatant is used as antigen.
3. **Somatic antigen from part of parasite**
   Antigen can be prepared from any body parts of nematode. The gut of nematode has been used as antigen.

b. Excretory-secretory antigen

Excretory-secretory antigens are used for immunoprotection studies. The adult/larval parasites are placed in culture media. These are cultured in culture media following standard protocol. The cultured material is centrifuged and supernatant is taken. The supernatant material is dialyzed to make it concentrated.

Defined Antigens

Defined antigens are specific proteins which may be used for immunoprotection. The specific proteins can be isolated in native form. Or specific proteins can be produced by recombinant DNA technology. Isolation of specific native protein can be done by using different column chromatographic techniques. These are gel-filtration chromatography, ion-exchange chromatography, immuno-affinity chromatography, etc. Detailed description for preparation of recombinant antigen is same as mentioned earlier.

Antigens used for immunoprotection studies
Crude somatic antigen
• Soluble extract of whole parasite
• Soluble extract of larvae of parasite
• Soluble extract of eggs of parasite
Excretory-secretory antigen of adult parasites
• Excretory-secretory antigen of adult and larvae of parasite
• Molting fluid of parasite
Defined antigens
• Recombinant antigen of *Ascaris suum* (rAs-16)
• *Haemonchus contortus* defined antigen (H11)
• *Haemonchus contortus* defined antigen (H110D)
• *Haemonchus contortus* defined antigen (H-gal-GP, recent hidden antigen)
• *Haemonchus contortus* recombinant somatic protein 26/23 (rHcp 26/23)
• *Haemonchus contortus* recombinant protein 14-3-3 isoform 2 (rHcftt-2)
• *Haemonchus contortus* recombinant antigen (rHc23)
• *Ancylostoma secreted* protein 1 and 2 (ASP1 and ASP2)

Some Immunoprotection Studies

Tsuji et al. (2001) performed intranasal immunization of **As 16 (recombinant *Ascaris Suum* 16 coupled with cholera toxin)** which induced protective immunity in mice. Studies showed immunoprotection values of recombinant As16 against *Ascaris suum*. Wei et al. (2017) carried out a study on immunoprotection potential of a recombinant protein of *Ascaris suum*. The recombinant antigen **rAs 16** was expressed in yeast cells, and the same was formulated with ISA 720. The formulation displayed significant reduction (36.7%) of larvae. Stunted development of larvae was also recorded. Mice were challenged, and the immune response achieved was due to Th2 skewed immune response. Newton (1995) reviewed the progress of vaccination against *Haemonchus contortus*. He mentioned that promising results had been found using different hidden gut antigens, i.e., **H11, H110D, H-gal-GP (recent hidden antigen)**. Kandil et al. (2017) carried out an immunoprotection study using a recombinant protein of *Haemonchus contortus*. Sheep were immunized with this protein at day 0 and day 14. Challenge infection was given with infective larvae on day 42. **Recombinant somatic protein 26/23 (rHcp 26/23)** conferred protection in sheep with 92.2% egg count reduction and 59.9% worm burden reduction. Another one recombinant protein GTP is **ADP-ribosylation factor-1 (Hc ARF1)** that modulates the cell mediated immune response (Gadahi et al. 2017). These ADP-ribosylation factors (ARFs) are members of the Ras-related small GTPase family which is involved in vesicular trafficking regulations. The gene of *H. contortus* small GTPase ADP-ribosylation 1 gene (HcARF1) was cloned, and recombinant proteins of Hc ARF1 (rHcARF1) was expressed in *Escherichia coli*. In vitro study using PBMC revealed immunomodulation. This protein modulated cytokine production profile by increasing IL4, IL10, and IL17. At the same time, IFN γ production was decreased significantly. The study further displayed significantly increased NO production. All these indicated immunomodulatory effect of CMI by the aforesaid recombinant protein. Bu et al. (2020) carried out work on immunization of goats using one **recombinant protein 14-3-3 isoform 2(rHcftt-2)** which induced moderate protection against *Haemonchus contortus* challenge. They evaluated the immunoprotection efficacy of a recombinant Hcftt-2 against *Haemonchus contortus* in goats. The animals were immunized twice using this antigen along with Freund's adjuvant. The recombinant antigen rHcftt displayed 26.46% reduction of mean eggs per gram and 32.33% reduction of worm burden.

Molecular Nematology

Molecular Characterization of Nematode

In the earlier period, taxonomy was based on external and internal morphological features, biochemical characterization, immunological characterization, etc. But molecular phylogeny has taken a new dimension for characterization of different organisms. Classification of organisms is being reoriented basing on molecular profiles. For this, different marker genes are taken for molecular characterization of different nematode parasites. For the purpose of genomic characterization of nematodes, molecular markers frequently used are **ITS1, ITS2, ND1, CO1, 5S, 5.8S, 18S**, etc. Nucleotide sequences of these marker genes are derived. The nucleotide sequence sequence of the marker gene of a particular nematode is compared with other nematodes of the same species or similar species available in NCBI database. The similarity or any difference of nucleotide sequence of the marker genes is analyzed. Interspecific variation, intraspecific variation, host related strain variation, and strain variation due to geographic locations are vividly studied for molecular characterization of different nematodes. The relevant output of those studies is taken for development of molecular taxonomy.

Some Studies of Molecular Characterization

Sadaow et al. (2018) carried out a study on molecular characterization of *Ascaris lumbricoides* and *Ascaris suum*. Nuclear ribosomal DNA (ITS1 and ITS2 regions) and mitochondrial DNA (cox1 gene) of *Ascaris* spp. were amplified and sequenced. Nucleotide sequence of ITS1 region revealed a probable hybrid genotype from two human ascariasis cases from Chiang Mai Province, northern Thailand. But nucleotide sequences of ITS2 of two species were identical and did not differ. Phylogenetic trees and haplotype analysis of cox1 sequences revealed 3 clusters with 99 haplotypes. The first internal transcribed spacer (ITS 1) and 18S genes of ribosomal DNA of *Toxocara vitulorum* of Egypt were amplified by Sultan et al. (2018). The amplified product was sequenced. The so derived nucleotide sequence was compared with nucleotide sequences deposited in databases and also used to construct the phylogenetic trees. *T. vitulorum* isolated from cattle in Egypt was found identical to those recorded in other countries. The constructed phylogenetic trees showed a close relationship among different species of *Toxocara*, including the zoonotic species. Molecular and phylogenetic characterization of *Dirofilaria immitis* was done by Liu et al.

(2014) based on ND1 and 16 SrDNA complete gene sequence. The result depicted more mutation of 16 SrDNA compared to ND1. (Shi et al. (2018) conducted one study to amplify complete mitochondrial genome sequence of *Ancylostoma ceylanicum* from dog. And the sequence of this species was compared with *A. tubaeforme*, *A. duodenale*, and *A. caninum*. The phylogenetic tree displayed *A. ceylanicum* on a branch which was external to other three *Ancylostoma* species. In the same study, a close phylogenetic relationship between *A. caninum* and *A. tubaeforme* than *A. duodenale* was found.

Some Molecular Study of Biology

It has been a trend to describe biology of a parasite in molecular way. In life cycle of a parasite, different developmental stages are observed. Biological activities of different developmental stages are variably different. Gene expression profile differs in different developmental stages. Consequently mRNA profile also differs in different period of biological development. Huang et al. (2014) carried out one study in which the **O7E12** gene was identified from the subtractive DNA library for the infective larvae of *Ascaris suum*. The gene was amplified and characterized. Bioinformatic analysis revealed differential expression in third stage larvae which was much higher as compared to other stages. From BLASTN and BLASTP searches in the gene bank, the said gene showed 99% identity with *A. suum* cre-nlp-2 protein. Joachim et al. (1998) performed single dimensional SDS-PAGE and found stage specific protein profile of *Oesophagostomum dentatum*. The stages of development comprised of preparasitic, histotrophic, and intestinal stages. Wang et al. (2010) performed transcriptome analysis of *Ancylostoma caninum* by **high throughput technique**. The study revealed broad diversity of transcriptomes from larval to adult stages. Developmental expression profile displayed major transition in expression from larva to adult stage.

Molecular Pathology

At first, general pathogenesis caused by important nematodes has been described here. Then we will understand the basis of molecular pathology. Pathogenesis is caused both by adult and larval stages of nematodes. It starts from general inflammation which may further proceed to state of nodule and tumor formation. Even, there may be formation of malignant tumors in some cases. Granuloma formation is also observed. In case of ascarid nematodes, the migrating larvae cause hemorrhagic lesions in the liver. In the chronic stage, deposition of fibrous tissue occurs resulting in formation of **milk spot** condition in the liver. The larvae damage the lung parenchymatous tissue. Alveolar epithelium gets desquamated which results in occurrence of pneumonia. Massive cellular infiltration occurs there. Larvae of *Toxocara canis* also cause destruction of lung alveoli during migration. Pneumonia may be caused. The liver is also damaged by the larvae as is found in *Ascaris* infection. **Visceral larva migrans** is another pathological feature which is found as a consequence of larval migration of *Toxocara* in man. Larvae of *Toxocara cati* and *Capillaria hepatica* are also responsible for VLM (visceral larva migrans). *Heterakis gallinarum* occurs in the cecum of turkey. *Ascaridia galli* occurs in the small intestine of birds. Both these parasites cause enteritis. *Strongylus equinus* causes nodule formation in the large intestine. More severe pathology is caused by *Strongylus vulgaris*. **Thrombosis of submucosal artery, verminous pneumonia, verminous aneurism**, etc. are some pathological features associated with *Strongylus vulgaris* infection. Different *Oesophagostomum* spp. cause nodule formation in the wall of the large intestine of animals. Different cells are collected around the parasite, and eventually the mass is encapsulated by fibrous tissue resulting in formation of nodules. The condition is called **knotty gut, nodulated gut, or pimply gut**. *Syngamus trachea* occurs in the trachea of different types of birds. The parasites cause irritation there. Inflammation and other pathological features occur. There may be formation of nodules there. A large amount of mucus is secreted which results in occlusion of the lumen. *Stephanurus dentatus* is a very important parasite. This parasite occurs in the kidney where they cause cyst formation. *Thelazia* spp. are called eye worms which cause different pathological features in the eye comprising conjunctivitis, keratitis, corneal opacity, etc. Severe pathology is caused by *Spirocerca lupi*. They may cause **osteosarcoma** and **fibrosarcoma** on the wall of the esophagus. Malignant tumors may develop on the wall of the esophagus. Aneurysm and hypertrophic pulmonary osteoarthropathy are some other associated pathological features. *Habronema* larvae cause cutaneous habronemosis which is also called **summer sore, granular dermatitis, or bursati**. *Draschia megastoma* causes formation of tumor in the stomach. Tumor formation also occurs in the stomach of dogs by *Gnathostoma spinigerum*.

Different types of pathology caused by different nematodes have been described in the earlier paragraph. We will now understand how these pathological features occur by immunological way. Different immune cells including lymphocytes play role in inflammation in the GI tract. TNFα is an important pro-inflammatory cytokine. IL1 is an important cytokine to cause inflammation in the gut. IL1 has two members, i.e., IL1α and IL1β. Pro-inflammatory effect is also found by the action of IL6. It starts action when the same is bound to its receptor (sIL6 R). The binding of **IL6 and sIL6R results in production of a specific protein, gp 130. This protein has role in initiation of signal transduction. This protein initiates the trans-signaling process via JAK/STAT pathway**. This trans-signaling activity causes pro-inflammatory effect.

TNF α, IL-1β, and IL-6 cause expression of adhesion molecules, procoagulant factors, etc. Different cytokines and other protein molecules are responsible for initiation and progression of different pathological features. The genes encoding those proteins are transcribed in different events of pathology.

Nanotherapy

Development of Nanodrug

Very limited works have been carried out on application of nanotechnology in the field of veterinary parasitology, but very recently a speedy research work has been directed toward nanoparasitology with sufficient fund. The following points are important facts:

1. Enhancement of the efficacy of different drugs with the help of nanoparticles has been proven.
2. Use of different nanovehicles has been proven to reduce toxicity of antiparasitic drugs.
3. Use of nanoparticles and nanovehicles to increase the bioavailability of the drug for long period. Thus, duration of drug activity can be increased.
4. Development of **smart antiparasitic drug** can be possible.
5. The active gold coated glass particles commonly known as nanoshells which can open a new window to target the parasite without affecting/reducing host tissue damage. Ample works were carried out, and many present works are already in progress to envisage the activity of nanoshells. The nanoshells are being used as biosensors in the study of host parasite association.

Anthelmintic efficacy of biologically synthesized AgNPs against *H. contortus* was reported. AgNPs produced by *Ziziphus jujuba* leaf extract displayed anthelmintic activity against *H. contortus*. The synthesis of nanoparticles using microbes such as actinomycetes, bacteria, and fungi is undoubtedly an easy and less time-consuming method for getting novel nanomolecules. It is needless to mention that fungi can be grown easily in the laboratory as well as on a large scale as mycelia can withstand extreme conditions of the environment. Different fungi are used for production of different nanoparticles. These fungi are *Fusarium* sp., *Aspergillus* sp., *Penicillium* sp., etc. The use of different nanoparticles using different fungi is different.

Nematodal Zoonosis

Ascariasis

Causative Agents

The important species under the genus *Ascaris* having zoonotic significance is *Ascaris suum*. This is large roundworm of pig occurring in the small intestine. Sporadic zoonotic *Ascaris suum* infection has been reported. Cross infection of

Ascaris suum and *A. lumbricoides* occurs in between human and pig.

Mode of Infection

The eggs are expelled out with the feces of the infected hosts. Development up to second larval stage occurs in the egg itself. The pigs get the infection by ingestion of the infective eggs containing second stage larvae. Humans also get the infection by ingestion of infective eggs containing the second stage larvae.

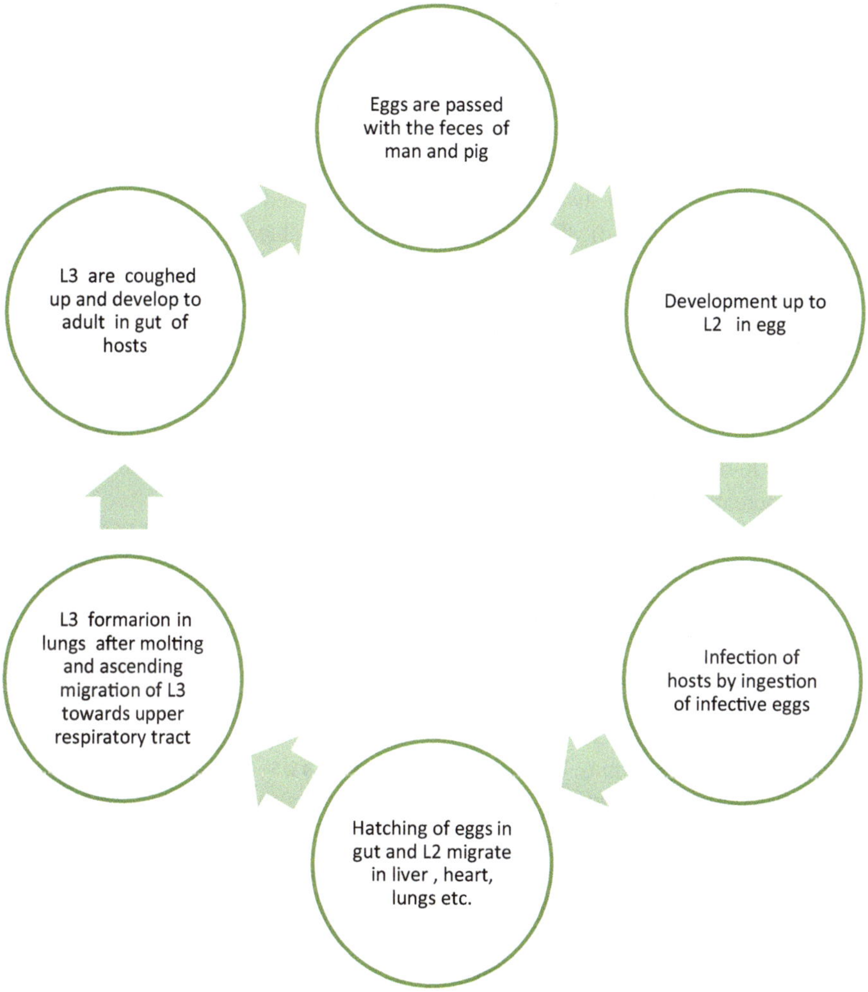

Fig. 6.43 Life cycle of *Ascaris suum*

Pathological Features

In Pigs

In the pig, the larval parasites cause little damage to the intestinal mucosa resulting in enteritis. But they cause marked damage in the lungs. During migration through the lungs, they break out the lung capillaries and cause mild hemorrhage. There occurs infiltration of inflammatory cells. Congestion also occurs. All these result in *Ascaris* pneumonia. Deposition of fibrous tissue in liver cause milk spot or white spot condition.

In Humans

The symptoms in humans include fever, respiratory troubles comprising asthmatic breathing, spasmodic coughing, diarrhea, abdominal discomfort, vomiting, colic pain, etc.

Diagnosis

Symptomatological Diagnosis

The exhibited clinical signs observed in humans are not specific for confirmation of this parasitic infection. However, nonproductive cough and fever nonresponsive to antibiotic therapy might provide some indication of this parasitic infection.

Parasitological Diagnosis

(a) Examination of fecal sample may reveal the eggs of the parasite. Initially direct smear technique is done. At least two to three smears should be examined. If direct smear examination reveals negative finding, then concentration method should be applied using sedimentation technique. Floatation technique is of no value as the unembryonated eggs do not float.
(b) Image-based diagnosis can be done by barium meal X-ray of the GI tract. Cholangiopancreatography can be done for detection of extraintestinal migration.
(c) Gastric aspiration can be done, and the aspirated material can be examined for detection of any larva.
(d) Intestinal biopsy material can be examined after hematoxylin-eosin stain. This test may reveal the eggs of the parasites.
(e) Sometimes, adult parasites may be found in the sputum or vomitus material.

Supportive Diagnosis

Some additional findings may be supportive:

- Detection of **Charcot-Leyden crystals** in sputum and stool
- Examination of blood for DLC and detection of eosinophilia

Immunological Diagnosis

Many works have been carried out using crude and purified antigens for performing different immunodiagnostic tests, i.e., complement fixation test, ELISA, etc. But no immunodiagnostic technique has been commercialized so far for field application. ELISA test is sensitive but it lacks specificity.

Molecular Diagnosis

Molecular diagnosis is possible by using different molecular markers. Different ribosomal DNA, ribosomal RNA, and mitochondrial DNA could be used as molecular markers. **Copro-PCR** is reliable for diagnosis in which ITS1 and CO1 are used as molecular markers.

Control

General control: Proper treatment should be done using specific drug available so far. Personal hygiene should be maintained. The children should specially be taken care of; they should not eat any soil, dirt, etc.

Immunological control: Different crude and defined antigens have been tried, but no vaccine has yet been commercialized.

Molecular control: Works were also carried out on gene immunization. But so far no naked DNA vaccine or gene vaccine is available.

Toxocariasis

Hosts Affected

The important zoonotic species is *Toxocara canis* occurring in dogs. Humans get the infection accidentally by ingestion of infective eggs of the parasite.

Mode of Infection

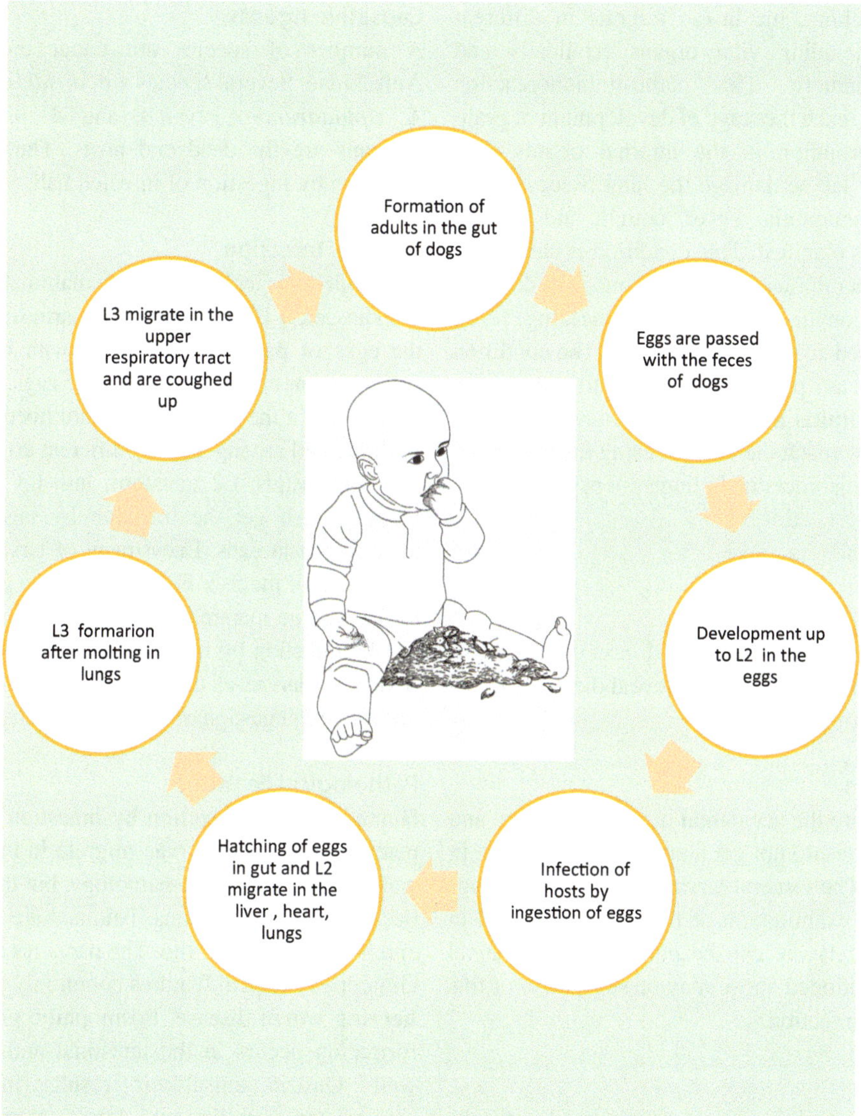

Fig. 6.44 Life cycle of *Toxocara canis*

Note

1. **If the dog is pregnant, the pups may get the infection congenitally. The offsprings get the infection through colostrum/milk also.**
2. **Dogs may get the infection through ingestion of infected paratenic hosts(rodents).**
3. **Humans get the infection by ingestion of infective eggs.**

Pathological Features

In Dogs

In dogs, the adult parasites cause enteritis. The larval stages of the parasite migrate in the hepatic tissue and lung tissue and thereby cause maximum damage. The lungs are massively damaged as the larvae break out the alveolar wall resulting in pneumonia.

In Humans

Humans get the infection by ingestion of eggs. After hatching, the larvae migrate in different organs including vital organs erratically and indiscriminately. The patho-immunoreaction may even reach the stage of development of granuloma formation in the internal organs. The migratory larvae damage the lung tissues resulting in pneumonia. Fever, cough, and allergic symptoms manifest. This condition is called visceral larva migrans. The symptoms include inappetence, convulsion, coughing, wheezing, fever, etc. Marked eosinophilia is found. The condition may be as prolong as 2 years or more. **Chorioretinitis and blindness** may occur. The larvae may reach the brain causing different neurological disorders including epilepsy.

Diagnosis

In Dogs

In case of diagnosis of infection in dogs, examination of fecal sample may reveal the presence of typical *Toxocara* eggs.

In Humans

Humans are the accidental or aberrant hosts, and the parasites do not get matured to adult stage in humans. The visceral larva migrans can be diagnosed by examination of the biopsy material in which the larvae can be detected. The clinical signs mentioned above may be suggestive of this parasitic infection.

Control

The feces of dogs should be disposed properly. Children should not be allowed to play in the infected ground. Personal hygiene should be maintained. They should not be allowed to defecate elsewhere. Dogs should be allowed to defecate in a particular place, and the feces should be disposed properly.

Anisakiasis

Causative Agents

A number of species are under the family Anisakidae. Several species are *Anisakis simplex*, *A. ziphidarum*, *A. typica*, and *A. physeteris*. Humans are the dead end hosts. They get the infection by ingestion of infected fish.

Mode of Infection

The adult parasites affect marine mammals (whale, dolphins, etc.). From the affected marine mammals, the eggs of parasites are passed with the feces. Development up to L2 occurs in the eggs. Then the eggs hatch in the water. The L2 are liberated after hatching and are ingested by different crustaceans, i.e., euphausiids. L2 transform into L3 after one molting. Fish get the infection by ingestion of infected crustaceans. Encystment of larvae occurs in the skin of the fish. **Fish serve as the paratenic hosts**. Marine mammals are the final hosts which get the infection by ingestion of infected fish harboring the larvae of the parasite. Humans get the infection by ingestion of infected sea fish.

Pathological Features

Humans get the infection by ingestion of marinated raw fish. The larvae migrate in the gastric wall and thereby cause pathology, but they never become adult in humans. Humans are the dead end hosts of this parasite. The parasites cause the clinicopathological features commonly named as **herring worm** disease. Eosinophilic granuloma formation occurs in the intestinal and stomach wall. Gastric anisakiasis results in gastric pain,nausea, vomiting, etc. Intestinal anisakiasis causes abdominal pain, nausea, vomiting, etc.

Diagnosis

Examination of biopsy material may reveal presence of parasitic stage.

Control

Fish should be preserved at −20 °C for 60 h. Salting of the material has added effect. Completely cooked fish should be eaten.

Strongyloidiasis

Causative Agent

The important zoonotic parasite is *Strongyloides stercoralis*. This parasite occurs in man, dog, etc. They are found in the small intestine particularly in the duodenum and jejunum of humans. *S. fuelleborni* has also zoonotic importance.

Transmission

The eggs are oval and transparent containing a larva which is ready to hatch. The rhabditiform larvae are immediately liberated from the eggs after hatching. The rhabditiform larvae are passed out with feces. They undergo homogonic or heterogonic type of development. In homogonic type of development, the rhabditiform larvae transform into infective filariform larvae. These infective larvae penetrate the skin of the hosts and the hosts thus get infection. In heterogonic type of life cycle, the rhabditiform larvae transform to free-living sexually matured male and female in the environment. Copulation occurs between the male and female parasite. The gravid females lay eggs and the eggs hatch to release rhabditiform larvae. The rhabditiform larvae transform into infective filariform larvae in the environment. These filariform larvae infect the hosts by skin penetration. However, there occurs **autoinfection** or **hyperinfection** also. In the lumen of the bowel, the rhabditiform larvae may change to filariform larvae and penetrate intestinal wall (autoinfection). These larvae may penetrate into the perianal skin (hyperinfection). Man acquire infection while walking bare foot. Children may get infection through eating soil. The larvae penetrate the buccal mucosa.

Pathology

In Animals

The worms affect dogs; particularly the puppies are commonly affected in summer seasons. Marked enteritis occurs which results in moderate to severe diarrhea. Diarrhea may be blood stained, and the affected animals may die due to severe dehydration. Purulent conjunctivitis and soft cough are also absorbed.

In Humans

Enteritis, occasional pneumonitis, and general allergy are the common clinical features occurring in humans. Local dermatitis may occur due to penetration by the infective larvae. However, in the autoinfection, the pathological features are of higher magnitude. Severe enteritis, marked pneumonitis, and massive general allergy are found in the case of autoinfection. Severe general allergy may lead to fatal condition of the patient.

Diagnosis

Symptomatological Diagnosis

The exhibited clinical signs are not so specific for confirmation of this parasitic infection.

Parasitological Diagnosis

- Diagnosis of this infection can be done by examination of the stool for detection of the rhabditiform larvae. Rhabditiform larvae may be seen in the sputum in case of pulmonary infection. **L1 has double bulbed esophagus which has diagnostic value. The L3 has notched tail**. L1 is similar to hookworm larvae.
- Enterotest (string test) is done by performing duodenal aspiration. Larvae or larvated eggs may be found in the duodenal aspirate.
- High number of rhabditiform larvae is found during biopsy examination.
- **Harada-Mori's stool culture** is done for collection and identification of larvae.

Immunodiagnosis

- ELISA can be done using crude antigen. ELISA cannot differentiate early and late infection.
- **Luciferase Immunoprecipitation System (LIPS) Assay** has been developed which is done using the 31 kDa recombinant protein of *S. stercoralis* and/or *S. stercoralis* immunoreactive antigen (SsIR).

Molecular Diagnosis

Molecular diagnosis is possible by using different molecular markers. Different ribosomal DNA, ribosomal RNA, and mitochondrial DNA could be used as molecular markers. Studies indicate the use of ribosomal RNA, i.e., 18S rRNA, 28S rRNA, and CO1 (cytochrome oxidase 1), as molecular markers for characterization/identification of this parasite.

Control

General Control

Proper treatment should be done using specific drug available so far.

Immunological Control

Different crude and defined antigens have been used for research trial, but no vaccine has yet been commercialized.

Oesophagostomiasis

Causative Agent

Oesophagostomum bifurcum, *O. stephanostomum*, etc. occur in the intestine of primates and humans.

Mode of Infection

The infected hosts pass the eggs with the feces. The eggs are hatched in the environment. Development up to L3 occurs in the environment. The animals get the infection by ingestion of infective larvae with the food or drinking water. Humans get the infection accidentally.

Pathological Features

Diarrhea, dysentery, and nodule formation on the wall of the intestine occur.

Hookworm Infection

Causative Agents

The zoonotic hookworms are *Ancylostoma caninum*, *A. braziliense*, *A. duodenale*, *Necator americanus*, and *Bunostomum phlebotomum*.

Mode of Infection

The eggs are voided along with feces. The eggs hatch and rhabditiform larvae are produced. The first stage larvae transform into second stage larvae after first molting. The second stage larvae further transform into third stage larvae after second molting. The third stage larvae are the infective stage. The filariform larvae leave their sheath and the larvae then penetrate the skin. The hosts get the infection in this way. The larvae then penetrate the subcutaneous lymphatics. Then they enter into the small venules and gain entry into the venous circulation. They reach the right heart from where the larvae reach the pulmonary capillaries and enter into the alveolar space. Then they follow the route of bronchiole, bronchi, trachea, and larynx and reach the epiglottis. Then the larvae are coughed up. The larvae transform into fourth stage after third molting during this migration or in the esophagus. Oral infection, prenatal infection, and transcolostral infection are other route of infection.

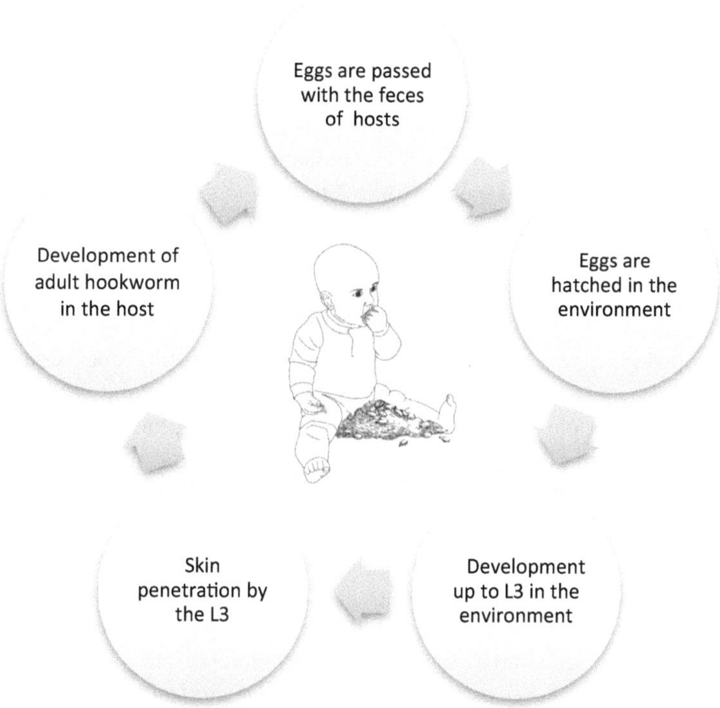

Fig. 6.45 General life cycle of hookworms

Note: Humans (children) get the infection when they inadvertently ingest larvae or the larvae may penetrate the skin due to contact with the contaminated soil or other material.

Pathological Features

In Animals

Ancylostoma caninum causes significant pathology in dogs. The larval stages enter the lungs and may cause pneumonia and coughing. The adult worms are voracious bloodsuckers and make the process enable by secreting **AcAPs (*Ancylostoma caninum* anticoagulant proteins)**, and as a result of this, the blood fails to clot. Edema is caused as a result of hypoproteinemia. Anemia and hemorrhagic enteritis occur in animals by the adult parasites.

In Humans

Hookworms are important zoonotic parasites and may cause maternal and child morbidity. Ground itch in humans is caused due to larval migration through the skin. Larvae of hookworms, i.e., *A. caninum*, *A. duodenale*, *Necator americanus*, *Bunostomum phlebotomum*, and *Uncinaria stenocephala*, cause cutaneous migrans (**creeping eruption)** in humans. The larvae migrate in tortuous tunnel between the stratum germinativum and stratum corneum and thereby cause serpiginous vesicular lesions. Looss in 1898 first described *Ancylostoma duodenale*. They occur in the duodenum and jejunum of man, dog, and cat. This is also called old hookworm. The larvae cause "ground itch" dermatitis during skin penetration. However, this situation lasts only for about 15 days. The adult worms are firmly attached to the mucosa particularly of the jejunal mucosa. The worms are bloodsuckers. The worms secrete anticoagulating agents. They voraciously suck blood and cause anemia. **Anemia occurs due to loss of blood and iron deficiency. Microcytic hypochromic, dimorphic anemia occur due to adult *A. duodenale*.**

Hematopoietic vitamin deficiency is another clinical feature.

Diagnosis

Symptomatological Diagnosis

The exhibited clinical signs are not so specific for confirmation of this parasitic infection. However, symptoms like progressive anemia and melena may indicate this parasitic infection.

Parasitological Diagnosis

- Examination of fecal sample may reveal the eggs of the parasite. Initially direct smear technique is done. At least two to three smears should be examined. If direct smear examination reveals negative finding, concentration method should be applied later.
- Eggs of *Ancylostoma duodenale* and *Strongyloides stercoralis* are similar. Eggs of *Ancylostoma duodenale* and *Necator americanus* are also similar. That's why, for specific identification, Harada-Mori's fecal culture technique could be done.
- Fecal sample can be diagnosed by Kato-Katz technique. This is a very simple technique. Some amount of feces is pressed in between sieves to remove the coarse materials. The sieved material is then transferred in the template of glass slide. The transferred material is then covered by cellophane soaked in glycerol with malachite green or glycerol with methylene blue. Then the material is examined under the microscope.

Immunological Diagnosis

Works have been carried out using crude and purified antigens. Different immunological tests have been described. Of these tests, ELISA test is better. Somatic antigen and excretory-secretory antigen are used for this test.

Molecular Diagnosis

Molecular diagnosis is possible by using different molecular markers. Different ribosomal DNA, ribosomal RNA, and mitochondrial DNA could be used as molecular markers. Studies indicate that ribosomal DNA (ITS1, ITS2) and mitochondrial DNA (CO1) could be used for specific identification of the parasite. The molecular technique can differentiate *Ancylostoma duodenale* and *Necator americanus*. Copro-PCR can be tried using the aforesaid markers.

Control

General control: Proper treatment should be done using specific drugs available so far. Personal hygiene should be maintained. The children should specially be taken care of, so that they should not play in contaminated ground.

Immunological control: Different crude and defined antigens have been used in different research trial, but no vaccine has yet been commercialized.

Molecular control: Works were also carried out on gene immunization. But so far, no naked DNA vaccine or gene vaccine is available.

Angiostrongyliasis

Causative Agents

Angiostrongylus costaricensis and *A. cantonensis* are two causative agents responsible for occurrence of zoonosis. *A. costaricensis* is responsible for causing abdominal angiostrongyliasis, and *A. cantonensis* causes eosinophilic meningitis in man. Rats and rodents are nonhuman hosts.

Mode of Infection

The larvae of the parasites are voided with the feces. The larvae develop in the snails, i.e., *Arion* spp., *Bradybaena* spp. In the intermediate hosts, larvae attain infective stage. The definitive hosts get infected by ingestion of the infected snail. **Humans get the infection through eating raw or undercooked snails or the vegetables on which the small snails remain attached**.

Pathological Features

A. costaricensis causes formation of eosinophilic granulomatous lesion in the intestine of humans. Abdominal angiostrongyliasis caused by *A. costaricensis* results in fever, abdominal pain, etc. There may be occurrence of granuloma in the intestine which may even cause partial or complete obstruction of intestinal passage. Anorexia, diarrhea, and vomiting are common clinical features frequently detected in the affected persons. *A. cantonensis* causes eosinophilic meningoencephalitis and meningitis in humans. Humans exhibit neurological signs.

Diagnosis

- Histopathological examination of tissue biopsy material reveals predominant eosinophilic infiltration. Lung lesions may be detected by radiography which may be indicative along with other diagnostic steps.
- Examination of CSF may reveal the presence of the parasitic stage.
- CT scan and MRI of the brain can be done for diagnosis of eosinophilic meningoencephalitis.
- Antibody detection by ELISA or dot ELISA using specific antigen has been reported.

Control

Control steps should be taken toward eradication of rats and mollusks which can potentially reduce the problem. The persons should be made aware about the danger of eating raw and undercooked terrestrial mollusks. Eating raw vegetables and vegetable juices should be avoided.

Dirofilariasis

Causative Agents

Some important species are *Dirofilaria immitis, D. repens, D. tenuis, D. conjunctivae etc.* The hosts of *D. immitis* are humans, dogs, etc. occurring in the heart. *D. repens* is a parasite of dog occurring in the subcutaneous tissue. However, infection in human has also been reported. *D. tenuis* is a parasite of raccoon occurring in the subcutaneous tissue. Occurrence of this parasitic infection has also been reported in humans. *D. conjunctivae* occurs in conjunctival tissues where they may form tumor.

Mode of Infection

Life cycle of *D. immitis* has been described here. Mosquitoes act as the intermediate hosts of the parasites which pick up the microfilariae (larval stage of parasite) when they take blood meal from the infected hosts. L3 infective stages are formed in mosquitoes. The infective mosquitoes transmit the larvae to the final hosts during blood meal from new host. In humans, incomplete development occurs and the parasites do not develop into adult.

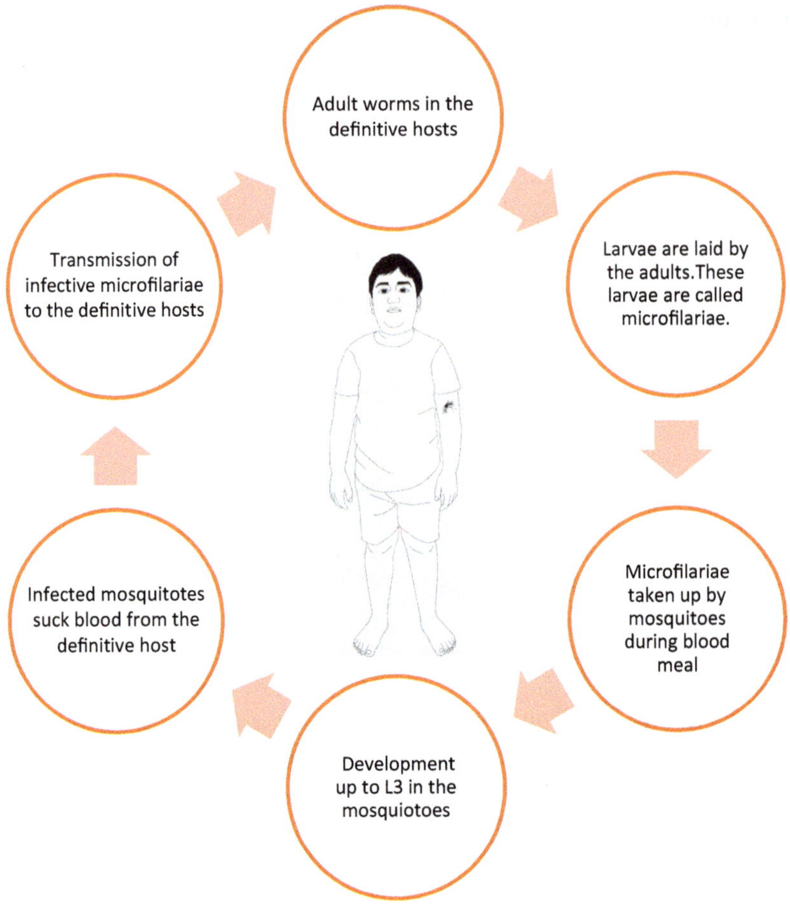

Fig. 6.46 Life cycle of *Dirofilaria immitis*

Pathogenesis

In Dog

Pulmonary hypertension, hypertrophy of the right ventricle, congestive heart failure, hemoptysis, etc. are some common clinicopathological features occurring in dog due to *D. immitis* infection. *D. repens* may cause formation of nodules in the subcutaneous tissues in dog.

In Humans

Adult D. tenuis are found in subcutaneous tissue of humans. *D. repens* may cause formation of nodules in the subcutaneous tissues of humans. Nodules may also be formed in the conjunctiva of the eyes of humans due to infection with *D. conjunctivae. D. immitis* are found in the lung and pleural cavity of humans. Dying *D. immitis* may cause formation of granuloma in the pulmonary artery which appears as coin lesion. Humans suffer from fever, cough, etc. In humans, parasites remain in immature stage. They reside in the pulmonary artery, lungs, etc. In the lungs, nodule and granuloma may be formed.

Diagnosis

Symptomatological diagnosis: The exhibited clinical signs are not so specific for confirmation of this parasitic infection. However, nonproductive cough and fever nonresponsive to antibiotic therapy might provide some indications of this parasitic infection.

Parasitological diagnosis: Identification of larvae of *Dirofilaria* sp. in blood smear should be done. Knott's method is very useful for easy detection of parasitic larvae.

Immunological diagnosis: Some works have been carried out using crude and purified antigens. ELISA test has been described for diagnosis of this parasitic infection. ELISA test is done for detection of circulating antigen of canine *Dirofilaria immitis*. The infection should be of more than 5 months old to display any result. The result is consistently found if the infection is 9 months old. There is possibility of false negative result due to antigen-antibody complex which traps the antigen. Heat treatment and acid treatment are recommended to make the antigen free from antigen-antibody complex. False positive result is also found due to residual antigen in the blood. ELISA antigen test for *Dirofilaria immitis* infection is commercially available. DiroCHEK is commercially available ELISA test kit. This is a canine heart worm antigen test kit.

Biopsy examination: In humans, pulmonary infection with *Dirofilaria immitis* is infrequently found. Major pathological feature is characterized by formation of granuloma in the lung tissue. Examination of biopsy material of the lung tissue is done for diagnosis of granuloma due to *Dirofilaria immitis* infection. ELISA test is also done for diagnosis of human infection with *Dirofilaria immitis*.

Molecular diagnosis: Molecular diagnosis is possible by using different molecular markers. Different ribosomal DNA, ribosomal RNA, and mitochondrial DNA could be used as molecular markers for identification of microfilariae.

Control

General control: Care should be taken for protection from mosquito bites. Since mosquitoes are the intermediate hosts, they should be controlled by using insecticidal agents.

Immunological control: Different crude and defined antigens have been tried, but no vaccine has yet been commercialized. Crude somatic protein of whole parasitic extract and excretory and secretory (ES) protein have been used in different trials of immunoprotection.

Molecular control: Works were carried out on gene immunization. But, so far, no naked DNA vaccine or gene vaccine is available.

Dipetalonemiasis

Hosts Affected

The important zoonotic parasites are *Dipetalonema streptocerca* and *D. perstans* occurring in humans and different anthropoid animals. *D. streptocerca* is found in subcutaneous tissue of chimpanzee and man *D. perstans* occurs in human and apes.

Mode of Infection

The *Culicoides* flies (*C. austeni*, *C. grahami*, etc.) are the intermediate hosts of this parasite. The flies take up the microfilariae from the definitive hosts during blood meal. The microfilariae reach the infective stage in flies. The infective stages are transmitted to the new hosts when they take blood meal from the hosts.

Pathological Features

Dipetalonema streptocerca causes lymphadenopathy, elephantiasis, etc. in humans. In most of the cases, these parasites do not produce any specific symptom. Other clinical features are allergic reaction, pruritis etc. *D. perstans* cause swelling of hand and face, eosinophilia, etc. in humans.

Diagnosis

Examination of blood smear using Romanowsky stains may reveal the presence of microfilariae. **Knott's method** is very useful. Examination of wet smear of blood is also done for detection of microfilariae. Microfilariae can also be identified by examination of skin scraping. Differential identification from other microfilariae should be done.

Control

Control of the intermediate hosts can be done using different insecticidal agents. *Culicoides* flies are the intermediate hosts. Humans should protect themselves from bites of *Culicoides* flies.

Wuchereriasis

The worms are called Bancroft's worm. The worms are distributed in South America, Central Africa, West Indies, Japan, India, etc. These are the filarial worms of humans occurring in the lymph nodes and lymphatic vessels. The worms are also recorded in monkeys. The microfilariae show the periodicity. They show nocturnal periodicity between 10 PM and 4 AM. They appear in the peripheral blood during this period. During daytime, microfilariae are found in the cardiorespiratory system (heart, lung, etc.). Periodicity may vary from place to place.

Transmission

Anopheles, *Culex*, and *Aedes* act as the intermediate hosts of this parasite. Mosquitoes take up microfilariae during blood meal. The microfilariae cast off their sheath. Then they penetrate the gut wall of the mosquitoes and subsequently migrate to the thoracic muscles. The larvae grow and become thick and sausage shaped within 2 days. These are the first stage larvae possessing a digestive tract. Within around 11 days, third stage larvae are produced after two more moltings. These third stage larvae are the infective stages. The infective larvae are deposited in the skin during blood meal by the mosquitoes. The larvae then reach the lymphatic channels and are distributed in different places (scrotal, inguinal, abdominal regions). Adult worms are formed there. Copulation of the worms occurs and the gravid female lays the larvae.

Pathology

Adult worms cause marked pathogenesis. Anaphylactic reaction may occur due to dead adult worms. The larvae fail to produce any marked pathogenesis. However, the metabolic by-product of the larvae may cause allergic reaction in the skin. Lymphangitis is a main pathognomonic feature occurring in this infection. The reason for lymphangitis is mechanical damage by the adult and the larvae, absorption of the toxic material as a result of disintegration of dead worms, or absorption of the metabolic by-products released by the worms. There occurs hypertrophy of the lymph nodes resulting in elephantiasis. Occult filariasis is characterized by marked eosinophilia and different pulmonary symptoms like coughing, sneezing, etc. Typical filariasis is characterized by lymphangiovarix (dilatation of lymphatics). This lymphangiovarix may rupture and cause several pathological complications, i.e., **lymphocele, lymphuria, chylothorax, chylous diarrhea, etc. Fever, hydrocele, and elephantiasis** are the major clinical signs in this filarial infection.

Diagnosis

Diagnosis can be done in different ways. Detection of microfilariae can be done by examination of blood smears from the peripheral blood. Smears can be prepared with the exudate from the lymph varix, hydrocele fluid, etc. Different immunodiagnostic tests can be performed. These are ELISA, CFT, FAT, etc.

Brugiasis

Causative Agents

The important zoonotic parasite is *Brugia malayi*.

Hosts Affected

The most important zoonotic species of this genus is *Brugia malayi*. *B. malayi* can infect humans, monkeys, cats, etc.

Mode of Infection

Different species of the genera *Anopheles*, *Mansonia*, and *Aedes* are responsible for transmission of this parasitic pathogens. The mosquitoes get the infection of microfilariae from the infected hosts during blood meal. The microfilariae develop in the mosquitoes and develop into infective stage. The infected mosquitoes transmit the microfilariae to the new hosts during blood meal.

Pathological Features

Lymphangitis, lymphadenoma, and elephantiasis are common clinicopathological features which occur as a result of limb swelling due to blockage of lymphatic system. The skin of the affected area becomes fibrotic. Recurrent fever is another symptom. Hydrocoele and chyluria also occur in this case.

Diagnosis

Enlargement of the inguinal lymph node provides some diagnostic clue. Examination of the blood smear may reveal the presence of microfilariae. Knott's method is very much useful. Molecular diagnosis, i.e., PCR based assays, can be done for specific identification of the parasite. Different investigators worked immunodiagnosis using recombinant antigen for diagnosis of brugiasis. BmR1 and BmSXP are antigens which were used by Rahman et al. (2007) for detection of brugian filariasis. 99–100% specificity was found in that study. Immunochromatographic rapid dipstick test is useful. Immunochromatographic test **(Brugia Rapid)** is available commercially. By this test, parasite specific IgG4 antibody can be detected. Detection of eosinophilia and leukocytosis is supportive diagnostic feature.

Control

The breeding foci of mosquitoes should be exterminated. For this, different herbicidal agents can be used to clear up the plants from the breeding zones. Spraying of insecticidal agents can be done for control of mosquitoes.

Thelaziasis

Causative Agents

Thelazia callipaeda, *T. californiensis*, etc. are zoonotic species of *Thelazia*.

Mode of Transmission

The larvae of this parasite are taken up by *Musca convexifrons*, *M. autumnalis*, *M. larvipara*, and *Stomoxys calcitrans* during feeding of eye exudate. In these flies, infective stages of the parasite are formed. Humans get the infection when the flies deposit larvae in the eyes.

Clinical Features

Irritation, lacrimation, photophobia, corneal opacity, and corneal ulcer are some clinical features.

Gongylonemiasis

Hosts Affected

Gongylonema pulchrum is a common species affecting the esophagus of cattle, zebu, buffalo, sheep, goat, etc. This parasite may infect humans also.

Mode of Infection

Coprophagous beetles act as the intermediate hosts of the parasite. The hosts get the infection by ingestion of coprophagous beetle harboring the infective stage of the parasite. *Humans are the accidental hosts. They get the infection by accidental ingestion of infected coprophagous beetle.

Pathological Features

In Animals

In animals, hypertrophy and cornification of the epithelium are commonly observed. The parasites cause enteritis resulting in diarrhea.

In Humans

In humans, the larvae affect the lips, gums, tonsils, etc. where they cause irritation and inflammation. Pharyngitis and stomatitis are other symptoms.

Diagnosis

For Animals

Examination of fecal sample may reveal the presence of parasitic eggs. Differential diagnosis should be done from the other spp., i.e., *Ostertagia* spp., *Oesophagostomum* spp., *Cooperia* spp. For differential diagnosis, larvae culture is done for characterization of the larvae by morphometry.

For Humans

Detection of the parasites moving in the tissues of the buccal cavity is important visual approach. Microscopical identification of the recovered worm from the patient's mouth or tissue should be done.

Control

Control of the intermediate hosts is an important step which can be done by using insecticidal agents, proper disposal of feces, etc.

Dracunculiasis

Causative Agents

The important causative agent is *Dracunculus medinensis*.

Mode of Infection

The hosts get the infection by ingestion of infected cyclops. The cyclops are the intermediate hosts harboring infective larvae (L3).

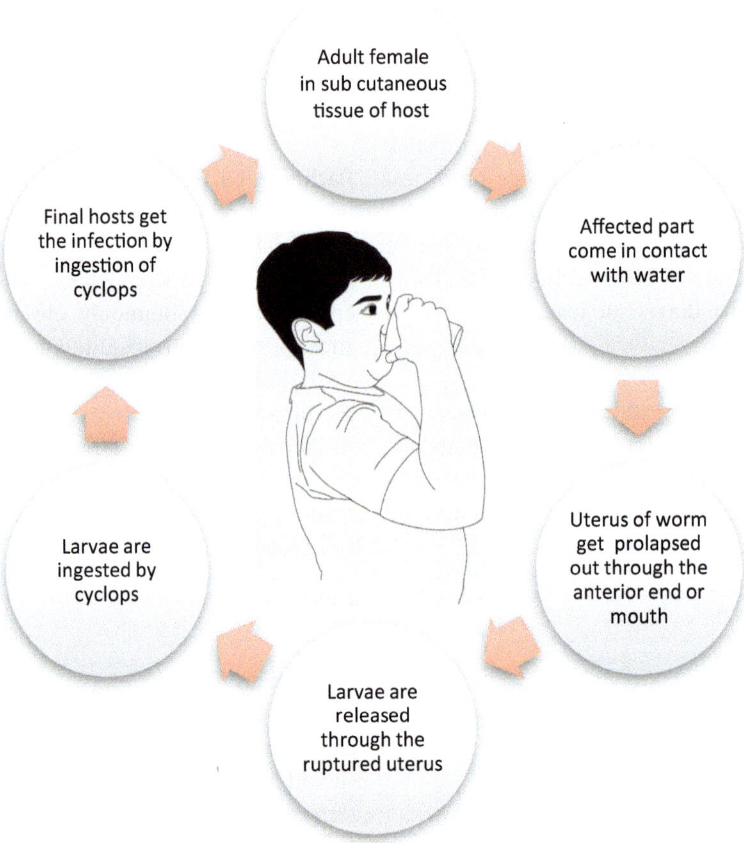

Fig. 6.47 Life cycle of *Dracunculus medinensis*

Pathological Features

In humans

No symptoms are found initially. But clinical symptoms are exhibited when female worms come to the skin. The pathology starts with swelling of skin which results in lesion. The lesion eventually leads to ulcer formation. Erythema, urticaria, and pruritus are common clinical features. Other symptoms are dyspnea, vomiting, diarrhea, etc. If the worm is dead, severe symptoms are found due to anaphylactic reaction.

In dogs

Subcutaneous inflammatory tract is formed leading to non-healing crater-like ulcers.

Diagnosis

Observation of emerging parasite from the lesions of the legs of the affected individuals is of diagnostic value. Identification of larvae of the parasites can be done by microscopical examination. Radiography can reveal the calcified stage of the worm. Extraction of the worm and identification of the same are possible.

Control

Care should be taken to drink uncontaminated pure water. Underground water should be used for drinking. Manually the emerging parasite can be removed by a specific technique using a stick to roll up the worm gradually. Every day, some length of worm is rolled up until the whole worm is removed out. Surgical intervention can also be done for removal of the worm from the subcutaneous tissue.

Trichinelliasis

Causative Agent

Trichinella spiralis is a zoonotic parasite occurring in humans and pigs.

Mode of Infection

Humans get the infection by ingestion of infected pig muscle. Larvae are released in the duodenum and jejunum and get matured to adult stage within a week. Females invade the intestinal wall of pig and lay larvae there. The larvae are carried to different skeletal muscles by intestinal lymphatics and mesenteric venous circulation. The larvae remain encysted in the skeletal muscles. Humans get the infection on ingestion of undercooked pork containing encysted larvae of the parasite.

Pathological Features

In humans

Humans get the infection on ingestion of undercooked pork. Pathology depends on the organs affected. Eosinophilia, myositis, etc. occur. The clinical signs include muscular pain, headache, fever, weakness, etc.

In pigs

The larval stages may cause lesions in the intestine. The parasites cause formation of muscle cysts within which they remain.

Diagnosis

Parasitological Diagnosis

- The fecal sample should be examined for detection of any larvae or even the adult worm in rare case.
- Muscle biopsy can be done for detection of larvae. It is better to take biopsy material from tendinous insertions of the **deltoid** and **gastrocnemius** muscle. The fresh muscle sample is compressed in between two slides, and the same is placed under low power microscope for detection of any larvae.

Immunological Diagnosis

- ELISA test is done to detect parasite specific antibody using excretory-secretory antigen derived from muscle larvae. But there is evidence of cross-reactivity with other parasites.
- A modified double sandwich ELISA is available to detect larval antigen in stool.
- Skin test can also be done by intradermal injection of 0.1 ml of 1:10000 dilution of **Bachman's antigen**. Bachman's antigen is prepared from *Trichinella* larvae from infected rabbit's muscle. An immediate reaction with visible erythematous patch develops within 15–20 min.

Animal Inoculation Test

Laboratory rats are infected with suspected muscle tissue of patients. Larvae in the diaphragm tissue of rat can be demonstrated.

Supportive Diagnostic Test

Serum creatinine phosphokinase (muscle enzyme) is increased due to this parasitic infection. Marked eosinophilia and leukocytosis are found.

Image Based Diagnosis

Calcified muscle cyst can be detected by X-ray.

Control

1. Educational campaign should be done regarding danger of trichinosis and safe cooking of meat.
2. Heating pork above 55 °C and freezing at −15 °C for a period of 20 days kill the parasite present in the muscle.
3. The pigs should not be allowed to eat raw garbage.
4. Carcass of pig should be disposed properly.
5. Proper meat infection should be done.

Trichuriasis

Causative Agents

Trichuris vulpis and *T. trichiura* are zoonotic parasites. The worms are found in the colon and cecum.

Mode of Infection

The worms show direct type of life cycle. No intermediate host is required for completion of the life cycle. Eggs are passed out with the stool of infected human. The eggs subsequently develop in the moist earth to the infective stage. Humans get the infection by ingestion of the infective eggs. Hatching of the eggs occurs in the GI tract of the host. The larvae are liberated from the eggs and get localized in cecum and develop to adult.

Pathogenesis

In Animals

The parasites cause enteritis which leads to diarrhea.

In Humans

The parasites cause enteritis. Diarrhea, dysentery, abdominal tenesmus, rectal prolapse, etc. are some clinical signs found in *Trichuris vulpis* infection.

Diagnosis

Diagnosis can be done by examination of fecal sample. The eggs are typically barrel shaped and bile stained. In case of *Trichuris* infection, it is observed that egg output is high. Therefore, direct smear examination is sufficient for detection of eggs. However, concentration methods can be done if the eggs are not found in direct smear technique. High level of eosinophilia occurs.

Gnathostomiasis

Causative Agents

Gnathostoma spinigerum occurs in the stomach of dog. *Gnathostoma spinigerum* also occurs in humans. Other *Gnathostoma* spp., i.e., *G. doloresi*, *G. hispidum*, occur in pigs. These two species may also occur in humans.

Mode of Infection

In life cycle of *G. spinigerum*, two intermediate hosts are required. First intermediate hosts are cyclops and second intermediate hosts are fish, snakes, birds, etc. The eggs are passed out with the feces. They hatch in the water. The first stage larvae are ingested by the cyclops. The infected cyclops are ingested by the fish in which infective larvae are formed. The final hosts get the infection by ingestion of the infected fish. In case of *G. hispidum*, one intermediate host is required in its life cycle. Development up to second larval stage occurs in the eggs. The cyclops ingest the eggs and infective stage of the parasite is formed in the cyclops. The final hosts get the infection by ingestion of infected cyclops. In humans, the larvae do not develop into adults.

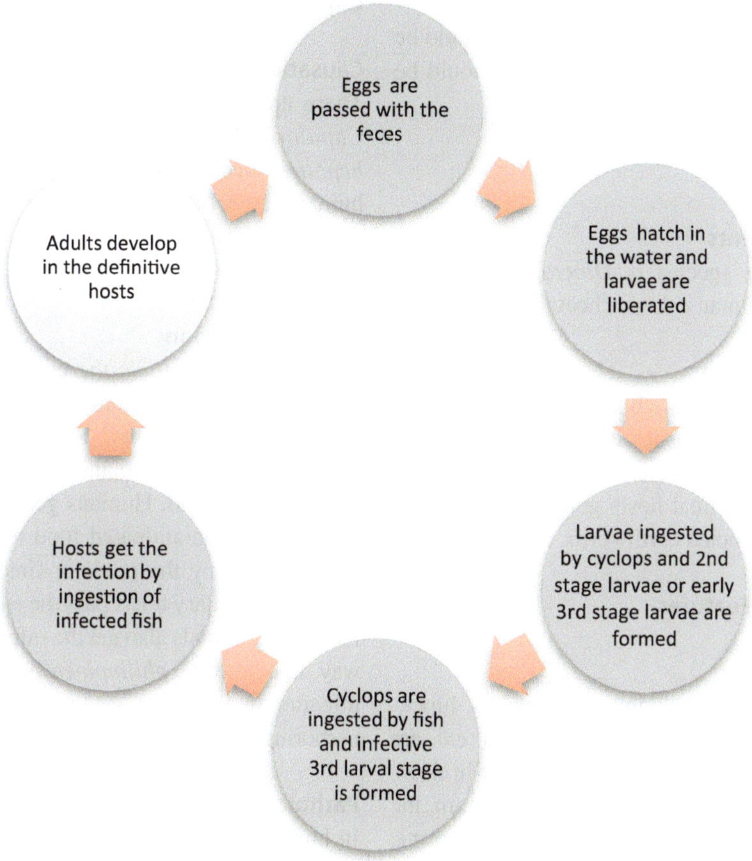

Fig. 6.48 Life cycle of *Gnathostoma spinigerum*

Pathological Features

In Animals

Gnathostoma spinigerum affects the wall of the stomach of dogs. Tumors are formed in the stomach. *G. hispidum* causes gastritis of affected animals. *G. nipponicum* causes formation of granuloma in the esophagus in the weasels.

In Humans

Cutaneous gnathostomiasis is common manifestation of human infection. Migrating larvae may reach to many internal organs and cause different damages there. These may be occular, auricular, gastrointestinal, genitourinary, pulmonary, and CNS manifestation.

Diagnosis

In Animals

Diagnosis is very much tedious as eggs may not be found during coproscopical examination because they remain in the cavity of the tumors formed by the parasites.

In Humans

Characteristic lesions on the subcutaneous tissue may be suspected for occurrence of this parasitic infection. Cutaneous swelling along with dietary history may be supportive diagnostic approach. Biopsy material can be taken from subcutaneous and ocular tissue and examined for detection of the larvae.

Control
Proper treatment of the definitive hosts should be done. Eating raw or undercooked fish should be avoided.

Dioctophymiasis

Causative Agents
The important species is *Dioctophyma renale* occurring in humans, dogs, cheetah, pigs, etc.

Mode of Infection
The eggs are expelled with the urine. The infective stage of the parasites develops in the earthworms, i.e., oligochaete annelid (*Lumbriculus variegatus*). The final hosts get the infection by ingestion of infected earthworm.

Pathological Features

In pigs
Destruction of parenchyma of the pelvis, blockage of the ureter, uraemia, calcification, etc. are some pathological features which occur in dogs. Due to anatomical location of the duodenum, the right kidney is more often affected by worms. The whole kidney may be lost leaving the capsule containing the worm and hemorrhagic exudate.

In humans
The right kidney of humans is severely damaged. Loin pain, haematuria, pyuria are commonly found. Renal carcinoma has been reported.

Diagnosis
Since the seat of predilection of the parasite is the kidney, examination of urine may reveal the presence of eggs of the parasite. History of eating uncooked, undercooked fish, frog (paratenic host) coupled with radiological observation of calcification reveals *Dioctophyma* infection.

Control
The worm can be removed surgically. Ivermectin is effective but the dead worm may cause damage to the tissue.

Capillariasis

Causative Agents
There are two important zoonotic species, i.e., *Capillaria hepatica* and *C. philippinensis*. *C. hepatica* occurs in dogs, monkeys, rodents, humans, etc. Humans are not the natural hosts of *C. philippinensis*, but they get the infection from the infected fish.

Mode of Infection
In case of *C. hepatica*, the eggs are laid by the female parasites in the liver. These eggs cannot embryonate in the liver. The liver gets decomposed after death of the animal. Embryonation of eggs occurs in soil. Humans get the infection by ingestion of contaminated food or water. Or the liver is eaten by the predator (intercalary hosts) and the intercalary hosts pass the eggs through the feces. The animals also get the infection in similar way. In case of *C. philippinensis*, humans get the infection by ingestion of raw or undercooked fish harboring the infective larvae of the parasites.

Pathological Features
In humans, hepatitis and liver dysfunction occur in case of hepatic capillariasis caused by *Capillaria hepatica*. In case of intestinal capillariasis caused by *C. philippinensis*, enteritis occurs. The symptoms include abdominal pain, vomiting, distension of the abdomen, etc.

Diagnosis
In case of hepatic capillariasis, detection of parasite can be possible by examination of liver biopsy or at necropsy. In case of intestinal capillariasis, examination of fecal sample may reveal the presence of eggs, larvae, adult parasites, etc.

Control
Eating of raw and undercooked fish should be discouraged.

Culture of Nematodes

Culture of Larvae of Nematodes
Larvae of roundworms can be cultured by simple fecal culture technique (Harada-Mori's tech-

nique). Researchers tried to carry out one new fecal culture technique and compared it with the conventional fecal culture technique. In this culture technique, the culture media is more defined (agar plate fortified with Earle's medium and yeast extract). Study shows more efficiency of "on-agar" cultures as compared to the fecal culture. Larvae of different roundworms are cultured in different specific medium, i.e., Dulbecco's Modified Eagle Medium, MEM Eagle, NCTC-135, RPMI-1640, M-199, etc.

Culture of Adult Roundworms

Adult roundworms can be cultured in RPMI-1640. The worms are collected directly from the organs. The collected worms are washed with PBS. Then the worms are transferred in specific culture medium.

Different culture media used for culture of nematode larvae are presented in the following table (several information are from Ahmed 2014).

Name of parasite	Name of culture	Culture condition/ suppliment	Development of parasite
Different hookworms	Harada-Mori	Condition 25–28 °C	From egg to first stage larva
Strongyloides spp.	Harada-Mori	Condition 25–28 °C	From egg to first stage larva
Haemonchus sp., *Trichostrongylus* sp., *Nematodirus* sp., etc.	Harada-Mori	Condition 25–28 °C	From egg to first stage larva
Trichinella spiralis	Caco-2 cell line. Caco-2 cell line is continuous line of heterogenous human epithelial colorectal adenocarcinoma cells		From first stage larva to adult
Dirofilaria immitis	1. NCTC-135 2. F12(K) (This is Kaighn's modification of Ham's F12 medium) 3. CMRL1066 4. Dulbecco's Modified Eagle Medium	1. Different amino acids, nucleotides, salts, etc. 2. Glutamine, sodium bicarbonate, etc. 3. Amino acids, enzymes, vitamin, salts, etc. 4. Fourfold amino acids, vitamin, D-glucose, phenol red sodium salt, etc.	Third stage to fourth stage larva
Wuchereria bancrofti	1. NCTC-135 2. Iscove's Modified Dulbecco's Medium	10% human serum, antibiotic, antifungal, etc.	From infective larva to fourth stage larva
Brugia malayi	Monkey kidney/mosquito cell line	Preferred temperature for monkey kidney cell culture 37 °C and for mosquito cell line 26-28 °C	From infective larva to young adult
Angiostrongylus costericensis	Ham's F12	8% CO2	From egg to first stage larva

Emerging and Reemerging Parasitic Disease

Baylisascaris Infection

Baylisascaris is a raccoon roundworm. But this worm has recently been recognized as a human pathogen. The normal hosts of this parasite are racoons. But this is emerging as an important helminthic zoonosis, and the young children are mainly affected (Sorvillo et al. 2002). Raccoons have increasingly become peridomestic animals as they are living in close proximity to humans (Sorvillo et al. 2002). In humans, the parasites cannot reach up to sexual maturity. The immature stages migrate in different organs causing visceral migrans and neural larval migrans.

Anisakis Infection

Anisakiasis is an emerging zoonosis caused by fish nematode, *Anisakis*. The highest number of reports of anisakiasis is from Spain. Humans get the infection by consumption of infected raw or unsmoked fish. Transmission of this parasite to humans occurs in this way. The incidences of infection are being increased in the USA. Monte Carlo simulations estimated *postmortem* migration of *Anisakis* from the viscera to flesh increases the disease burden by >1000% while an education campaign to freeze anchovy before consumption may reduce cases by 80% (Bao et al. 2017).

Dirofilaria Infection

Dirofilaria immitis is an important filarial worm of dogs called heartworm. *D. immitis* and *D. repens* are emerging in some places where these were not endemic. *Dirofilaria immitis* and *Dirofilaria repens* are two important species causing human dirofilariasis. Hungary was not considered to be a heartworm (*Dirofilaria immitis*) endemic country until 2007, when the first autochthonous canine infection was described (Tolnai et al. 2014). In this study, they reported additional autochthonous heartworm infections in 2 dogs (*Canis familiaris*), 20 red foxes (*Vulpes vulpes*), and 2 golden jackals (*Canis aureus*) coming from 8 counties (Tolnai et al. 2014). Their study comprising morphometry and molecular characterization by DNA sequencing explored Hungary as endemic for *Dirofilaria immitis* infection and obviously recognized dirofilariasis as an emerging disease.

Ancylostoma Infection

Ancylostoma caninum is an important hookworm which mainly affects the small intestine of dogs. In the USA, this hookworm is well controlled by application of approved common drugs. But in the recent years, recurrent/persistent infection by this parasite has been increased dramatically which led that the reason might be due to drug resistance (Jimenez Castro et al. 2019). *Ancylostoma ceylanicum* is a hookworm of dog and cat in Asia. But this parasite was found emerging in Australia (Smout et al. 2017). They found this parasitic infection with high proportion in soil sample, fecal samples, etc.

Trichinella Infection

Cases of trichinelliasis is increasing gradually in Europe. This has been considered as an emerging and reemerging disease in Europe during the past decades. Not only in Europe, the disease is reemerging in many other parts of the world.

Gnathostoma Infection

Gnathostomiasis is a foodborne zoonotic disease. This nematode is transmitted by ingestion of raw or undercooked fish harboring larvae of this parasite. It is noticed that infection of gnathostomiasis in humans is increasing in countries where there were no records of gnathostomiasis earlier. Southeast Asia and Central and South America are endemic countries of this disease. But it seems that geographical boundaries of endemicity are increasing gradually.

Nematodes of Wild Animals

Parasites	Wild hosts	Site
Parascaris equorum	Zebra, donkey, horse	Small intestine
Toxascaris leonina	Fox, wild felids	Small intestine
Lagochilascaris *Baylisascaris*	Wild felids, didelphoids, rodents Do	Small intestine
Toxocara canis *T. cati*	Fox, wolve, coyote Wild felids	Small intestine Small intestine
Contracaecum spiculigerum	Wild waterfowl, swan, duck, cormorant, heron, pelican	Stomach, small intestine
Porrocaecum crassum	Wild duck, heron, egret, cormorant	Small intestine
Passalurus ambiguus	Rabbits, hares, lagomorphs	Cecum, colon

Parasites	Wild hosts	Site
Skrjabinema ovis	Antelopes, European mouflon	Intestine
Skrjabinema tarandi	Caribok and reindeer	Intestine
Leiperenia leiperi	Hippopotamus, African elephant	Intestine
Subulura brumpti	Fowl, guinea fowl, wild birds	Ceca
Pseudoaspidodera paronis *P. uncinata*	Peafowl Guineapig	Ceca Large intestine
Ascaridia dissimilis *A. compar* *A. columbae*	Wild turkeys Ptarmigan, pheasant, partridge, grouse Wild pigeons	Small intestine
Strongyloides papillosus *S. westeri* *S. stercoralis* *S. avium* *S. fuelleborni* *S. procyonis* *S. ratti* *S. venezuelensis*	Wild ruminants, deer, antelope Zebra, donkey Primate, (Chimpanzee, monkey), fox Turkeys, wild birds Macaque, chimpanzees Raccoons Rats, brown rats Rats, brown rats	Small intestine
Craterostomum acuticaudatum *C. tenuicauda*	Ass, mule, mule, zebra, donkey Pony	Large intestine
Poteriostomum imperidentatum *P. ratzii*	Wild equids Wild equids	Cecum, colon Cecum, colon
Codiostomum struthionis	Ostrich	Large intestine
Decrusia additicta	Indian elephants	Large intestine
Bourgelatia diducta	Wild boar	Cecum and colon
Quinolia travancra	African elephants	Cecum
Murshidia murshida	Indian elephants	Cecum
Choniangium epistomum	Indian elephants	Cecum
Khalilia pileata *K. buta*	Indian elephants African elephant	Cecum Large intestine
Chabertia ovina	Roe deer, antelope	Cecum
Ternidens deminatus	Chimpanzees, gorilla, macaque	Large intestine
Oesophagostomum columbianum	Camel, wild antelopes	Large intestine
O. bifurcum	Primates	Large intestine
O. venulosum	Deer	Large intestine
O. radiatum	Water buffalo	Large intestine
O. aculeatum	Macaques	Large intestine
Ancylostoma caninum	Foxes, coyotes, wolves	Small intestine
O. bifurcatum	Primates	Large intestine
Amidostomum anseris	Wild geese, duck	Gizzard
Dictyocaulus arnfieldi	Horse, mule, donkey, zebra	bronchi, lungs
Dirofilaria immitis	Foxes, coyotes, wolve	Heart, pulmonary artery
Gnathostoma spinigerumm	Foxes, tigers, leopards, wolves, wild cats	Stomach
Epomidiostomum uncinatum	Wild ducks	Gizzards
Epomidiostomum skrjabini	Wild geese	Oesophagus, proventriculus

***Toxocara* spp.**

Alexander et al. (2018) identified *Toxocara tanuki* in Korean racoon dog by means of morphology and molecular technique. Morphological study displayed small lateral alae on the cranial end of the body of adults of both sexes. SEM revealed 18 pairs of proximal precloacal, a preclocal median, a postcloacal median papillae. Five pairs of postcloacal papillae are found in posterior portion of male. Proximal papillae were not found in lateral end of female. Acharjyo (2004) in his review article enlisted three species of *Toxocara*, i.e., *T. canis* in jackal, lion, leopard, and tiger; *Toxocara cati* in lion, tiger, leopard, clouded leopard, fishing cat, leopard cat, jungle cat, golden cat, civet, etc.; and *T. leonina* in lion, tiger, leopard, snow leopard, fox, fishing cat, and leopard cat. The disease caused by *Toxocara* may cause ill health leading to fatality in young ones particularly in the felids. *T. mystax* from tiger, leopard, jungle cat, leopard cat and T. transfuga from Himalayan black bear, sloth bear, red panda etc. was reported.

Strongyloides spp.

Strongyloides procyonis Little 1966 was detected about 45 years ago in raccoons (*Procyon lotor*) from Southern Louisiana, USA, and was demonstrated experimentally to cause creeping eruption and a short-lived intestinal infection in a healthy volunteer (Sato et al. 2006). In a study of survey on *S. procyonis* infection in feral raccoon in Japan, Sato et al. (2006) found female parasites in 66 animals (28.3%) of 233 racoons collected between May 2004 and January 2005. Molecular nucleotide sequence of the small and large subunit ribosomal RNA genes (SSU/LSUrDNA) of *S. procyonis* closely resembled those of zoonotic *Strongyloides stercoralis*. The sequences of internal transcribed spacer (ITS 1) and 28 rDNA of these two species depicted clear difference.

Oesophagostomum spp.

Oesophagostomum is an important roundworm occurring in a wide range of animals, i.e., cattle, sheep, goat, pig, etc. They not only affect the domestic animals but also affect wild animals. *Oesophagostomum aculeatum* and *O. bifurcatum* have been found in old world monkeys. *O. stephanostomum* occurs in chimpanzees and gorillas. Species of Oesophagostomum in nilgai at Maharajabag Zoo was reported. *Oesophagostomum* have been recorded in chimpanzee, red colobus, and olive baboons of Uganda by Ghi et al. (2014). They also performed molecular characterization of *Oesophagostomum* of nonhuman primates in Uganda. Terio et al. (2018) reported *Oesophagostomum* sp. in nonhuman primates (chimpanzees, baboons, colobus, cercopithecid monkeys). They found moderate to numerous nodules in some chimpanzees. Nodules were also found in baboon but were absent in cercopithecid monkeys. Nodules occur in chimpanzees. There is report of diagnosis of primate *Oesophagostomum* by fecal examination from blue monkey, black and white colobus, gray-cheeked mangabey, olive baboon, red tailed guenon, chimpanzee, etc.

Strongylus spp.

Different *Strongylus* spp. occur in horse, ass, mule, donkey, etc. *Strongylus vulgaris* occurs in Indian rhinoceros. *S. vulgaris* also occurs in wild ass and zebra.

Haemonchus spp.

Haemonchus spp. are very important nematode parasites causing significant economic loss in livestock industry. There are some reports of *Haemonchus* infection in wild animals. This has been reported in white tailed deer. The animals exhibit different symptoms comprising anemia, weakness, weight loss, etc. Accumulation of fluid occurs in the submandibular region. Diagnosis can be done by examination of fecal sample. Fecal culture can be useful for identification of the parasite up to genus level. DNA based techniques can be done for identification of this parasite. Albendazole, fenbendazole, cambendazole, and oxfenbendazole are some effective drugs.

Dictyocaulus spp.

Dictyocaulus spp. are important lungworms of cattle, sheep, goats, etc. Different spp. of *Dictyocaulus* have been recorded in wild animals also. Looking into the importance of ecosystem in the environment, conservation of animals is very important. To maintain the health of the wild animals, the parasites concerned to health of wild animals should be studied very well. *Dictyocaulus africana*, *D. filaria*, and *D. viviparous* are found in the bronchi and trachea of antelopes. *D. capreolus* from roe deer and *D. viviparous* from European bison have also been reported. Pyziel et al. (2017) described *Dictyocaulus cervi* n. sp. recovered from red deer which is distinguished from *D. eckerti* on the basis of absence of cervical papillae and occurrence of a single ring of four symmetrical submedian cephalic papillae. Some other morphological parameters were also different, i.e., length of female tail, morphometry of female reproductive system, and measurement of gubernaculum in male. Fenbendazole, thiabendazole, tetramisole, etc. are some effective drugs.

Hookworm Infection

It has been found that zoo animals harbor hookworm infection. A hookworm species, *Haemonchus pernicious*, was reported in the intestine of lion, tiger, and leopard from Nandankanan zoo. The worms were recovered during PM examination. Hookworm infections occur in the small intestine of different wild animals, i.e., wild dog, wolf, jackal, fox, jungle cat, leopard cat, clouded leopard, hyena, lion, tiger, civet cat, honey badger, etc. The worms recovered from the small intestine of those animals were *Ancylostoma caninum*, *A. ceylanicum*, *A. braziliense*, *A. duodenale*, *A. paraduodenale*, *Arthocephalus gambiense*, *A. herpestis*, *Uncinaria stenocephalus*, and *U. fidelis*. Reviewed Acharjyo, (2004). Another one hookworm, *Arthocephalus longespiculum*, was recovered from a hog badger (Reviewed, Acharjyo, 2004). The worms caused fibromatous growth in the pancreas. Many works were carried out on hookworm infection in wild animals in Eastern Asia and North America. Animals infected were red fox, coyote, raccoons, and northern fur seals. In Asia, *Ancylostoma kusimaense* and *Arthrostoma miyazakiense* are common GI nematodes of raccoon dogs. Here *Uncinaria stenocephala* is found in red foxes. *Uncinaria fidelis* and *U. maya* are most commonly found in leopard and Iriomote cats. *Arthrostoma humanensis* infects the bile duct of leopard cats. *Ancylostoma kusimaense* and *Arthrostoma miyazakiense* and *Uncinaria rauschi* are most commonly found in black bears of Canada. *U. yukonensis* is more commonly found in brown bears of North America. *A. malayanum* has been reported from brown bears in Japan. Interestingly, human hookworm, *A. duodenale*, has been found in spotted hyenas in Ethiopia. *Ancylostoma ceylanicum* has been reported from Malay civet. *Agriostomum* sp. has been reported from blue wildebeest. *Globocephalus urosubulatus* has been described in wild boars of Central America, Brazil, Korea, etc.

Anemia is rare in wild animals due to hookworm infection. Retarded growth and hemorrhagic enteritis are common consequences of *A. caninum* infection in coyotes. Mortality has been commonly recorded in canine, feline, etc. which is as a result of chronic anemia, tissue damage, retarded growth, secondary bacterial infection, etc.

Syngamus spp.

Syngamus spp. are important roundworms which affect the trachea of birds. The affected birds exhibit different respiratory disorders. There are reports of gapeworm (*Syngamus trachea*) infection in wild birds, i.e., carrion crow, jackdaw, magpie starling, house sparrow, and purple sandpiper. Respiratory distress has been reported in gulls by this parasitic infection.

Filaroidosis spp.

Filaroides osleri is an important lungworm of cat. But there are reports on *Filaroides* infection in wild cats. Earlier reports reveals *Filaroides osleri* infection in the lungs of a leopard cat in Nandankanan zoo. The worms caused formation of nodule and pneumonia in the lung.

Dirofilaria spp.

Dirofilaria spp. are important worms which commonly occur in the circulatory system of dogs. These worms have also been reported in wild animals. The worm is found in the lumen of the right atrium of the heart and the connective tissue of the lung of captive Humboldt penguin (*Spheniscus humboldti*). *Dirofilaria immitis* occurs in wild animals, i.e., fox, tiger, wolf, dingo, red fox, gray fox, Bengal fox, pale fox, raccoon, etc. This worm has also been reported from lion, leopard, golden cats, wild dog, jackal, etc.

Troglostrongylus sp.

The important species is *Troglostrongylus brevior*. This parasite occurs in the trachea and bronchioles of wild cats. The parasites have folded cuticle. A well-developed copulatory bursa is present in the posterior end. Dorsal, externodorsal, lateral, and ventral rays are found. The dorsal rays are elongated and have four apical papillae. The lateral ray is further divided into three branches, i.e., anterolateral, mediolateral, and posterolateral.

Spirocerca sp.

Spirocerca lupi is an important roundworm found in dog. However, its occurrence in wild animals has also been reported. This species has been reported in fox. In fox, *S. lupi* causes gastric lesions and nodule formation in the stomach. *S. vulpes* sp. nov. is a recently proposed species recorded from fox (Rojas et al. 2018). This species differs from *S. lupi* in having eight teeth-like structures in the buccal capsule which are absent in *S. lupi*. The distance of the vulva in the anterior extremity of this new species is larger compared to *S. lupi*. The molecular characterization of this species also displayed separate identity. *Spirocerca lupi* infection in jungle cat, bear cat, lion, and jackal was reported.

Gnathostoma spp.

Gnathostoma spp. are roundworms usually found in the stomach of dogs. The worms have head bulbs. Four submedian cavities are present in the head bulb. *G. spinigerum* have been reported in wild animals too, i.e., wild dogs, foxes, tigers, jackals, etc. *G. spinigerum* causes gastric tumor in tiger. *G. hispidum* occurs in wild pigs, wild dogs, and wild cats. *G. doloresi* occurs in wild boars. *G. nipponicum* occurs in tigers and leopard.

Dioctophyma sp.

Dioctophyma renale occurs in different wild animals, i.e., fox, coyote, mink, ferrets, otters, etc.

Trichinella Infection

Trichinella are roundworms of man and pig. *Trichinella* spp. are widely distributed in all continents except Antarctica; but the full spectrum of *Trichinella* species and their hosts in sub-Saharan African countries has not been fully documented (Mukuratirwa et al. 2019). Mukuratirwa et al. (2019) conducted their study to determine the prevalence of *Trichinella* in wildlife from the Greater Kruger National Park (GKNP) and adjacent areas located in the Limpopo and Mpumalanga provinces of South Africa. In their study based on molecular characterization by multiplex PCR amplification of two important molecular markers (ITS1 and ITS2); the isolates from spotted hyaena, lion, leopard, small spotted genet, and Nile monitor lizard was conformed to *Trichinella zimbabwensis*. Earlier reports reveals *Trichinella spiralis* infection in civet cat occurring in the diaphragm, skeletal muscle, etc.

Clinical Nematology: A Glance

(For Clinical Nematology, the readers are required to study different aspects i.e. pathogenesis, clinical signs, diagnosis, treatment and control of different nematodes described in this chapter. Here, some other information has been given).

Nematode Infection

Dogs/Cats

Parasite	Egg character/clinical features
Spirocerca lupi	Unique paper clip shaped egg (30–37 × 11–25 um). Eggs are larvated when laid. Floatation technique is done taking vomitus or feces. Radiography or endoscopy are done to reveal nodule or granuloma in the esophagus or stomach
Ollulanus tricuspis	Worm identification could be done from vomitus. Floatation technique is done using Sheather's flotation solutions. Eggs are larvated and third stage when released from the hosts
Physaloptera praeputialis	The observed clinical signs are anorexia, vomition, and dark tarry stool. Worms may be found in vomitus. The worms are confused with *Ascaris* worms. Eggs may be taken out and the eggs should be examined. The eggs are 30–34 × 49–59 um in size. The eggs are larvated when laid

Parasite	Egg character/clinical features
Toxocara canis/*T. cati*, *T. leonina*	The adult worms may be found in vomitus. The vomitus may have "spaghetti" in it. The eggs have deeply pigmented center with pitted outer shell. The size of eggs of *T. canis* are 75 × 90 um. The eggs of *Toxocara cati* is smaller having 65–75 um diameter. *Toxascaris leonina* eggs have smooth outer shell and a hyaline or ground glass central position. The size of the eggs is 75–85 × 60–75 um.
Ancylostoma spp. (*A. caninum* in dogs, *A. tubaeforme* in cats, *A. braziliense* in dogs and cats, *Uncinaria stenocephala* in dogs and cats)	The eggs are oval or ellipsoidal (8–16 cells in morula). Fresh feces should be examined. Fecal flotation could be done
Trichuris vulpis are canine whipworms. *T. campanula* and *T. serrata* are feline whipworms	The eggs are trichuroid or trichinelloid. Typically barrel-shaped eggs having polar plugs on either sides
Dirofilaria immitis	Modified Knott's method is done for identification of microfilariae. Commercially available filter technique is also done. But sometimes false positive may occur
Aelurostrongylus abstrusus These are feline lungworms	Larvae come out through feces. Examination of feces reveal the larvae. The larvae are 360 um long. The tail has S shaped bend. Fecal flotation and Baerman technique are also done. Larvae can be recovered by tracheal wash
Filaroides osleri/*F. hirthi*	The larvae are 232–266 um in length. The tail has S shaped appendage
Dioctophyma renale	Eggs are recovered from urine by centrifugation. These are barrel shaped, bipolar, and yellow brown in color. The shells exhibit pitted appearance
Capillaria plica, *C. feliscati* These are found in the urinary tract of dogs and cats	Eggs are found in urine which are 63–68 × 24–27 um in size. They have flattened, bipolar end plugs. Outer surface is rough in appearance
Thelazia californiensis	The larvae can be recovered from conjunctival sac and lacrimal duct. Lacrimal secretion may reveal first stage larvae. *D. immitis* can also be recovered from anterior chamber of the eye

Different Ruminants and Equines

Parasite	Egg character/clinical features
Gongylonema pulchrum	Fecal examination may reveal larvated eggs. Fecal culture is done for larval identification
Bunostomum spp., *Chabertia* spp., *Cooperia* spp., *Haemonchus* spp., *Oesophagostomum* spp., *Ostertagia* spp., *Trichostrongylus* spp.	All these worms produce thin shelled eggs
Elaeophora schneideri These worms occur in the common carotid artery of sheep	Microfilariae are detected in the skin. A small piece of skin is macerated in saline. The fluid is examined after 2 h. Microfilariae are rarely found
Dictyocaulus viviparous is found in cattle. *D. filaria* is found in sheep and goats	Examination of feces may reveal larvae in feces Larvae of *Dictyocaulus filaria* have brown granules in intestinal cells. They have blunt tail with anterior cuticular knob. Larvae of *Dictyocaulus viviparous* have brown food granules in intestinal cells. They have straight tail. They have no cuticular knob.
Muellerius capillaris These are called hair lungworm of sheep, goats, etc.	The tail of larvae (230–300 um) have undulating tip and a dorsal spine
Stephanofilaria spp.	Initially red papules develop which later coalesce to form large pruritic areas of alopecia with thick moist crusts
Thelazia rhodesii and *T. gulosa*	The worms are recovered from conjunctivital sac. Examination of lacrimal secretion reveals eggs or first stage larvae
Parascaris equorum	Eggs are round to oval and deeply pigmented
Strongylus spp.	Fecal culture is required for larva identification. Flotation is done

Parasite	Egg character/clinical features
Oxyuris equi	The eggs may be larvated while passing. The eggs have smooth and thick shell with a distinct plug at one pole. Cellophane tape impression is taken for identification of eggs. Pruritus occurs in the anal region
Habronema	The larvae cause wounds on the skin. Larvae are aberrant or in off course condition known as cutaneous habronemiasis

Pig

Parasite	Egg character/clinical signs
Strongyloides ransomi	Fecal flotation is done for identification of eggs
Oesophagostomum dentatum	Fecal flotation is done for identification of eggs
Trichuris suis	The eggs are trichuroid or trichinelloid. They have thick brown barrel shaped shell
Metastrongylus elongatus	Eggs are recovered by fecal flotation using flotation medium with specific gravity greater than 1.25

Phylum: Acanthocephala

The members under this phylum are called **thorny headed worms**. The worms possess one proboscis. The proboscis is a hollow organ armed with longitudinal rows of recurved hooks.

Life Cycle

- Life cycle is indirect type. Beetle, crustacea, and cockroach serve as the intermediate hosts.
- The females lay fully embryonated eggs containing hooked larvae (acanthor).
- The eggs containing acanthor are ingested by the intermediate hosts.
- In the intermediate hosts, the eggs are hatched. After hatching, the acanthors are liberated. The acanthors are further transformed into acanthella. The acanthella are matured to infective cystacanth.
- The final hosts get the infection by ingestion of infected intermediate hosts.

Important Species

Macracanthorhynchus hirudinaceus	*Fillicollis anatis*	*Polymorphus boschadis*
It comes under the family Oligacanthorhynchidae (order: Archiacanthocephala)	Family: Polymorphidae (order: Palaeacanthocephala)	Family: Polymorphidae (order: Palaeacanthocephala)
Pig	Duck, fowl, geese	Duck, fowl, geese
Small intestine	Small intestine	Small intestine
Intermediate hosts: dung beetles (Scarabaeidae)	Intermediate hosts: different isopods	Intermediate host: crustacean (*Gammarus*)
The eggs are dark brown and pitted	The eggs are simple	The eggs are spindle shaped

References

Acharjyo LN (2004) Helminthiasis in captive wild carnivores and its control in India. Rev Zoos' Print J 19:1540–1543

Ahmed NH (2014) Cultivation of parasites. Trop Parasitol 4(2):80–89. https://doi.org/10.4103/2229-5070.138534. PMID: 25250227

Alexander U, Lim CW, Kim B, Hong EJ, KimHC, Park BK (2018) Morphological and molecular characterization of Toxocara tanuki (Nematoda: Acaridae) from Korean Racoon dog, Nyctereutes procyonoides koreensis. Korean J Parasitol 56(6):567–575. https://doi.org/10.3347/kjp.2018.56.6.568. PMID:30630277; PMCIDPMC6327194

Anderson UV, Howe DK, Dangoudoubiyam TN (2013) SvSXP: a *Strongylus vulgaris* antigen with potential for prepatent diagnosis. Parasit Vectors 6(1):84. https://doi.org/10.1186/1756-3305-6-84

Batte EG, Moncol DJ, Barber CW (1966) Prenatal infection with the swine kidney worm (*Stephanurus dentatus*) and associated lesions. J Am Vet Med Assoc 149(6):758–765. PMID: 6008248

Bao M, Pierce GJ, Pascual S, Mattiucci MGS (2017) Assessing the risk of an emerging zoonosis of worldwide concern: anisakiasis. Sci Rep 7:43699. https://doi.org/10.1038/srep43699. PMID: 28287609

Bouchery T, Filbey K, Shepherd A, Chandler J, Patel D, Schmidt A, Camberis M, Peignier A, Smith AAT, Johnston K, Painter G, Pearson M, Giacomin P, Loukas A, Bottazzi ME, Hotez P, LeGros G (2018) A novel blood-feeding detoxification pathway in *Nippostrongylus brasiliensis* L3 reveals a potential checkpoint for arresting hookworm development. PLoS Pathol 14(3):e1006931. https://doi.org/10.1371/journal.ppat.1006931. PMID: 29566094; PMCID: PMC5864084

Bu Y, Jia C, Tian X, Aimulajiang K et al (2020) Immunization of goats with recombinant protein 14-3-3 isoform 2(rHcftt-2) induced moderate protection against *Haemonchus contortus* challenge. Pathogens 9(1):46. https://doi.org/10.3390/pathogens9010046. PMID: 31935869

Emery DL, Hunt PW, Le Jambre LF (2016) Haemonchus contortus: The then and now ,and where to from here? Int J Parasitol 46(12):755–769. https://doi.org/10.1016/j-ijpara.2016.07.001. PMID: 27620133

Gadahi JA, Eshan M, Wang S, Zhang Z et al (2017) Recombinant protein of *Haemonchus contortus* small GTP ase ADP- ribosylation factor 1(HcARF1) modulate the mediated immune response in vitro. Oncotarget 8(68):112211–112221. https://doi.org/10.18632/oncotarget.22662. PMID: 29

Gasbarre LC, Canals A (1989) Induction of protective immunity in calves immunized with adult *Oesophagostomum radiatum* somatic antigens. Vet Parasitol 34(3):223–238. https://doi.org/10.1016/0304-4017(89)90053-8. PMID: 2617827

Ghi RR, Champan CA, Omeja PA, Davies TJ, Goldberg TL (2014) Nodule worm infection in humans and wild primates in Uganda: Cryptic Species in a newly identified region of human transmission. PLOS Negl Trop Dis 8(1):e2641. https://doi.org/10.1371/Journal.pntd.0002641

Huang C, You J, Nai F (2014) Identification and characterization of a differentially expressed gene (07E12) in the infective larvae of the parasitic nematode *Ascaris suum*. Iran J Parasitol 9(2):209–217. PMID: 25848387

James JL (1997) Emerging parasitic infections. FEMS Immunol Med Microbiol 18(4):313–317. https://doi.org/10.1111/j.1574-695X.1997.tb01061.X

Jimenez Castro PD, Howel SB, Shaefer JJ et al. (2019) Multiple drug resistance in the canine hook worm Ancylostoma caninum: an emerging threat? Parasites Vec 12:576. https://doi.org/10.1186/s13071-019-3828-6

Joachim A, Ruttkowski B, Daugschies A (1998) Changes in antigen and glycoprotein patterns during the development of *Oesophagostomum dentatum*. Int J Parasitol 28(12):1853–1860. https://doi.org/10.1016/s0020-7519(98)00163-5. PMID: 9925263

Johnson M, Mackintosh CG, Labes RE, Taylor MJ, Wharton DA (2003) *Dictyocaulus* species: cross infection between cattle and red deer. N Z Vet J 51(2):93–98. https://doi.org/10.1080/00480169.2003.36346. PMID: 16032306

Kandil OM, Abdelrahman KA, Shalaby HA, Hendawy SHM (2017) Evaluation of a crude larval protein and recombinant somatic protein 26/23 (rHcp26/23) immunization against *Haemonchus contortus* in sheep. Vet World 10(7):758–763. https://doi.org/10.14202/vetworld.2017.758-763. PMID: 28831218

Kausar S, Khan W (2018) Comparative efficacy of diethylcarbamazine, nitazoxanide and nanocomposite of nitazoxanide and silver nanoparticles on the dehydrogenases of TCA cycle in *Setaria cervi*, in vitro. Iran J Parasitol 13(3):399–405. PMID: 30483331; PMCID: PMC6243168

Klei TR, French DD, Chapman MR, McClure JR, Dennis VA, Taylor HW, Hutchinson GW (1989) Protection of yearling ponies against *Strongylus vulgaris* by foalhood vaccination. Equine Vet J Suppl (7):2–7. https://doi.org/10.1111/j.2042-3306.1989.tb05645.x. PMID: 9118099

Leles D, Gardner SL, Reinhard K et al. (2012) Are Ascaris lumbricoides and Ascaris suum a single species? Parasit Vec. 5,42. https://doi.org/10.1186/1756-3305-5-42

Liu T, Liang Y, Zhong X, Wang N, Hu D et al (2014) Intraspecific genetic variation and phylogenetic analysis of *Dirofilaria immitis* samples from Western China using complete ND1 and 16 SrDNA gene sequences. Am J Trop Med Hyg 90(6):1176–1179. https://doi.org/10.4269/ajtmh.13-0579. PMID: 24639299

Mitchell MC, Tzelos T, Handel IG, Macwillium HEG (2016) Development of a recombinant protein-based ELISA for diagnosis of larval cyathostomin infection. Parasitology 143(8):1055–1066. https://doi.org/10.1017/S0031182016000627

Mukuratirwa S, LaGrange LJ, Malatji MP, Reininghaus B, Lamb J (2019) Prevalence and molecular identification of Trichinella species isolated from wild life originating from Limpo and Mpumalanga provinces of South Africa. J Helminthol 93(1):50–56. https://doi.org/10.1017/S0022149x17001079. PMID:29168444

Newton SE (1995) Progress on vaccination against *Haemonchus contortus*. Int J Parasitol 25(11):1281–1289. https://doi.org/10.1016/0020-7519(95)00065-a. PMID: 8635880

Postigo I, Martínez J, Cardona G, Fernández-Pérez I, Guisantes JA (2003) Uncinaria stenocephala: antigenic characterization of larvae and adults worms using sera from naturally infected dogs. Exp Parasitol 103(3–4):171–173. https://doi.org/10.1016/s0014-4894(03)00089-4. PMID: 12880595

Pyziel AM, Laskowski Z, Hoglund J (2017) Interrelationship of *Dictyocaulus* spp. in wild ruminants with morphological description of Dictyocaulus cervi n.sp. (nematode:Trichostrongyloidea)from red deer, Cervus elaphus. J Parasitol 103(5):506–518. https://doi.org/10.1645/16-75. Epub 2017. PMID 28585897

Rahman RA, Hwen-Yee C, Noordin R (2007) Pan LF-ELISA using BmR1 and Bm SXP recombinant antigens for detection of lymphatic filariasis. Filaria J 6:10. https://doi.org/10.1186/1475-2883-6-10

Rascoe LN, Price C, Shin SH, McAuliffe I et al (2015) Development of Ss-NIE-1 recombinant antigen based assays for immunodiagnosis of Strongyloidiasis. PLoS Negl Trop Dis 9(4):e0003694. https://doi.org/10.1371/journal.pntd.0003694. PMID: 25860665

Rogers WP, Brooks F (1977) The mechanism of hatching of eggs of Haemonchus contortus. Int J Parasitol 7(1):61–65. https://doi.org/10.1016/0020-7519(77)90026-1

Rojas A, Sanchis-Monsonis G, Alic A, Hodzie A, Otranto D (2018) Spirocerca vulpis sp.nov (Spiruridae:Spirocercidae):description of a new nematode species of the red fox, Vulpes vulpes(Carnivora:Canidae). Parasitol 145(14):1917–1928. https://doi.org/10.1017/S0031182018000707. PMID:29781422

Sadaow L, Sanpool O, Phosuk I, Rodpai R et al (2018) Molecular identification of *Ascaris lumbricoides* and *Ascaris suum* recovered from humans and pigs in Thiland, Lao PDR, and Myanmar. Parasitol Res 117(8):2427–2436. https://doi.org/10.1007/s00436-018-5931-6. PMID: 29860571

Sato H, Suzuki K, Osanai A, Kamiya H, Furuoko H (2006) Identification and characterization of the threadworm, *Strongyloides procyonis*, from feral racoon (*Procyon lotor*) in Japan. J Parasitol 92(1):63–68. https://doi.org/10.16451/GE-623R.I. PMID:16629317

Setasuban P (1976) Light microscopy and scanning electron microscopy of Bathmostomum sangeri Cobbold, 1879, of elephants. Southeast Asian J Trop Med Public Hlth 7(3):390–4. PMID:1035986

Shi X, Wang M, Abdullahi AY et al (2018) Comparative analysis of *Ancylostoma ceylanicum* mitochondrial genome with other Ancylostoma species. Infect Genet Evol 62:40–45. https://doi.org/10.1016/j.meegid.2018.04.012

Smout FA, Skerratt LF, Butler JRA, Johnson CN et al (2017) The hookworm *Ancylostoma ceylanicum*: an emerging public health risk in Australian tropical rainforests and indigenous community. One Health 3:66–69. https://doi.org/10.1016/j.onehlt.2017.04.002. PMID: 28616506

Solgi R, Sadjjadi SM, Mohebali M, Zarei Z et al (2018) Development of new recombinant DgK antigen for diagnosis of *Dirofilaria immitis* infections in dogs using ELISA technique and it's comparison to molecular methods. Iran Biomed J 22(4): 283–289. https://doi.org/10.22034/ibj.22.4.283. PMID: 29031244

Sorvillo F, Ash LR, Berlin OG, Morse SA (2002) *Baylisascaris procyonis*: an emerging helminthic zoonosis. Emerg Infect Dis 8(4):355–359. https://doi.org/10.3201/eid0804.010273. PMID: 11971766

Sultan K, Omar M, Desouky AY, El-Sheify MA (2018) Molecular and phylogenetic study on *Toxocara vitulorum* from cattle in the mid-Delta of Egypt. J Parasit Dis 39(3):584–587. https://doi.org/10.1007/s12639-013-0373-7. PMID: 26345077

Terio KA, Lonsdorf EV, Kinsel MJ, Raphael Y, LIpende I et al (2018) Oesophagostomiasis in non-human primates of Gomba National Park, Tanzania. Am J Primatol 80(1):10.1002/ajp.22572. https://doi.org/10.1002/ajp.22572. PMCID:PMC5161720/NIHMSID:NIHMS810771/PMID:27309976

Tolnai Z, Szell Z, Sproch A, Szeredi L, Sreter T (2014) Dirofilaria immitis: an emerging parasite in dogs, red foxes and golden jackals in Hungary. Vet Parasitol 203(3–4):339–342. https://doi.org/10.1016/j.vetpar.2014.04.004. PMID:24810374

Tomar RS, Preet S (2017) Evaluation of anthelmintic activity of biologically synthesized silver nanoparticles against the gastrointestinal nematode, Haemonchus contortus. J Helminthol 91(4):454–461. https://doi.org/10.1017/S0022149X16000444. PMID 27374616

Tsuji N, Suzuki K, Kasuga-Aoki H, Matsumoto Y et al (2001) Intranasal immunization with recombinant *Ascaris suum* 14-kilodalton antigen coupled with cholera toxin B subunit induces protective immunity to *a. suum* infection in mice. Infect Immun 69(12):7285–7292. https://doi.org/10.1128/IAI.69.12.7285-7292.2001. PMID: 11705899

Wang Z, Abubucker S, Martin J, Wilson et al (2010) Characterizing *Ancylostoma caninum* transcriptome and exploring nematode parasitic adaptation. BMC Genomics 11:307. https://doi.org/10.1186/1471-2164-11-307. PMID: 20470405

Wei J, Versteeg L, Liu Z, Liu Z et al (2017) Yeast-expressed recombinant as 16 protects mice against *Ascaris suum* infection through induction of a Th2 skewed immune response. PLoS Negl Trop Dis 11(7):e0005769. https://doi.org/10.1371/journal.pntd.0005769. PMID: 28708895

Xie Y, Hoberg EP, Yang Z, Urban JF Jr, Yang G (2017) *Ancylostoma ailuropodae* n. sp. (Nematoda: Ancylostomatidae), a new hookworm parasite isolated from wild giant pandas in Southwest China. Parasit Vectors 10(1):277. https://doi.org/10.1186/s13071-017-2209-2. PMID: 28576124; PMCID: PMC5457663

Yadav AK, Tandon V (1992) Morphology of Ascarops dentata (Nematoda:Spirocercidae) male by scanning electron microscopy. Jpn J Parasitol 41(2):118–121

Zahabiun F, Sadjjadi SM, Yunus MH, Rahamatullah A et al (2015) Production of *Toxocara cati* TES -120 recombinant antigen and comparison with its *T. Canis* homolog for serodiagnosis of toxocariasis. Am J Trop Med Hyg 93(2):319–325. https://doi.org/10.4269/ajtmh.15-0190

Veterinary Entomology and Acarology

7

Arthropods are the largest group of animals (80% of all animal spp.) occurring in the water, soil, air, etc. The main characteristics of arthropods are hard chitinous exoskeletons, segmented body, and jointed legs. The word Arthropoda has been derived from two Greek words: *arthros* meaning "joint" and *podos* meaning "foot." The outer covering of arthropods is called exoskeleton formed by chitin. The exoskeleton is composed of different chitinous plates which are called sclerites. The dorsal sclerite is called tergum; ventral sclerite, sternum; and lateral sclerite, pleuron. The old cuticle is cast off periodically and a new cuticle is formed. This biological phenomena of arthropod is called ecdysis. The body of arthropod is metamerically segmented.

Origin and Evolution of Arthropod

The arthropods share many structural and functional features with annelids; both are segmented. In both groups, a tubular dorsal heart is found. The fossilized arthropod was of **Cambrian** period, millions of years ago.

Seasonal and Regional Distribution of Arthropod

Distribution and occurrence of arthropods depend on season. A season constitutes many abiotic factors, i.e., temperature, humidity, etc. These abiotic factors directly relate to prevalence of arthropods. Temperature is one of the most important abiotic factors. Tolerance of arthropod toward temperature is important for occurrences of arthropods in a particular region. The temperature determines the **microclimate** which influences the evolution of arthropods. The temperature variation causes alternation of other environmental factors which leads to cause adaptation of any organism for survival. Humidity is also an important factor which affects the biology of many organisms. Therefore, season has role on population of arthropods. For example, a climate with 85% or higher humidity is ideal for ticks. At this humidity, a tick can absorb moisture from the air to make them hydrated. Ticks die at low humidity below 80%. Therefore, ticks require a warm temperature in order to quest. Less than 44 °F temperature makes it difficult for the tick to move around and find a host.

Population dynamics of insects and acarines in relation to biotic and abiotic factors are important aspects of study of any arthropod. A concept

S. C. Mandal, *Textbook of Veterinary Parasitology*, https://doi.org/10.1007/978-981-99-2924-5_7

is frequently used in insect population biology to determine how pest population increases as a result of environmental factors. The intrinsic rate of increase is the rate at which a population increases in size in the absence of density dependent forces regulating the population. The body systems and reproduction have been described below:

Alimentary System

The alimentary canal is divided into three regions comprising the foregut (stomodeum), midgut (mesenteron), and hindgut (proctodeum). The foregut includes the buccal cavity, pharynx, esophagus, and proventriculus and gizzard. In most of the cases, the esophagus is dilated posteriorly and forms a special structure called crop. The proventriculus acts as a valve preventing regurgitation. A number of excretory tubules are found at the junction of mid- and hindgut called Malpighian tubules. The Malpighian tubules perform excretory function.

Nutrition is holozoic type. The arthropods take up their food material through their mouth. The digestion and absorption of nutrients occur in the digestive system. Egestion occurs through the anus. The food is taken up by the mouth which gradually reaches to the crop via the esophagus. The muscular proventriculus serves as a valve which prevents regurgitation. The proventriculus may be equipped with teeth for grinding and disintegration of food material. The midgut secretes a number of enzymes which help for digestion. The hindgut includes the ileum, colon, and a dilated rectum. The rectum possess several papillae which serve for reabsorption of water from the feces.

Excretory System

Excretion is performed by different organs, i.e., Malpighian tubules, nephridia, and coxal glands. Malpighian glands are lined with the alimentary canal in ring form surrounding it. Coxal glands are found in the legs. The arachnids have both Malpighian tubules and coxal glands. Crustaceans have a pair of nephridia which perform excretion. These nephridia open either at the base of second maxilla or in the second antennae.

Circulatory System

The circulation is open type. The body cavity is called hemocoel. Hemolymph is found instead in the blood. All the organs bathe in the hemolymph. A small tubelike structure is found called the heart. This is primitive tubular heart. The heart is enclosed in a compartment of hemocoel called pericardium. In the heart, small openings are found called ostia. The hemolymph enters the heart through the ostia. When the heart contracts, the ostia remains closed.

Respiratory System

Respiratory organs of different arthropods are the book lung, tracheal system, gill, body surface, etc. In insects, tracheal system comprising a fine network of tracheal tubes is found. The minute tracheal tubes penetrate the cells. Air enters the body through some openings called spiracles. Two pairs of spiracles are found in the thorax region, and eight pairs of spiracles are found in the abdominal regions. The gills are found in the aquatic arthropods. Lung books are found in spiders and gill books are found in crabs.

Nervous System

The arthropods have complex nervous system which is more developed as compared to other invertebrate organisms. This consists of a central mass (brain) present on the stomodeum, nerve trunk, or commissure surrounding the esophagus. Ganglions are present in different segments, each having a single ganglion. A ventral nerve trunk is present comprising nerve cords supplying segmental ganglions. The nervous system further includes different sense organs (e.g., sensory setae).

Reproduction

The arthropods are unisexual organisms. However, hermaphrodism and parthenogenesis are also found. The life cycles are of three types: hemimetabolous, holometabolous, and ametabolous. Complete metamorphosis occurs in holometabolous life cycle (e.g., Diptera, Siphonaptera, etc.). In this life cycle, different stages, i.e., egg, larva, pupa, and adult, are found. Three or more larval stages are found. In hemimetabolous life cycle, incomplete metamorphosis occurs (Hemiptera, Mellophaga). The developmental stages are egg, nymph, and adult. In ametabolous life cycle, metamorphosis is absent.

The phylum has three subphyla of veterinary and medical importance.

Subphylum: Mandibulata

The members of this subphylum have mandibles. Insects, crustacea, centipedes, millipedes, etc. are in this subphylum.

Subphylum: Chelicerata

The members of this subphylum possess chelicerae. Ticks, mites, king crabs, etc. are in this subphylum.

Subphylum: Pentastomida

Tongue worms are in this subphylum.

Subphylum: Mandibulata

Class: Insecta

The body of this arthropod is generally divided into three parts, i.e., head, thorax, abdomen.

Head: The head of insect has taxonomic importance. The shape and size of it are largely variable. The head may be round, oval, triangular, flattened, etc.

Eyes: The eyes of insect are very characteristic. These are generally composed of several units called ommatidium. The eyes are holoptic in male and dichoptic in female. Therefore, the eyes of arthropod provide clues for sexual identification of arthropods.

Proboscis: This is a flexible and tubular structure. In many members proboscis is present. This may be long and short in size. The proboscis is used for piercing and sucking.

Antenna: A pair of antenna are present on either side of the head. These are sensory organs used for touching, smelling, hearing, etc. The nerve is present in the antenna which generates action potential when it touches the surrounding molecules. The antenna are segmented. The number of segments is used for taxonomic identification. The antenna is divided into three basic parts, i.e., scape (base), pedicel (stem), and the last segment (flagellum). The flagellum may have several units called flagellomeres. The antennae may possess bristlelike structures called aristae.

Arista: This bristlelike structures are found on the third segment of antenna. This acts as hydro- and thermoreceptors of some arthropods. This is found in the members of suborder Brachycera of order Diptera. The arista may be segmented. The segments are called aristameres. The arista may be plumose having long and many hairs over it or may have less hairs (pilose).

Thorax: The thorax is the second body segment of Insecta which is also called mesosoma. The thorax may be divided into three parts, i.e., prothorax, mesothorax, metathorax. In some members, thoracic segments are fused together. In some cases, only pro- and mesothorax are fused together. In dragonflies, the meso- and metathorax parts are fused together. In some members, the first abdominal segment is fused with metathorax which is called propodeum found in some hymenopteran members. In some larval stages, the head and thorax are fused together which is called cephalothorax. The shape of the thorax has taxonomic importance. In some dipteran members, the thorax is humped. In some members, the thorax is covered with hairs.

Wings: Wings arise from thoracic segments. If two pairs of wings are present, front pair arises from the mesothorax and the hind pair arises from the metathorax. The forewings of cockroaches are called tegmina. These are parchment-like. The hind wings are thin and membraneous having veination. Individual species have their characteristic wing veination. The pattern of wing veins is important for taxonomy. In Diptera, the hind pair of wings are reduced to form halters or balancers.

Legs: The legs are also very important body parts used for locomotion. The legs also have taxonomic importance. There may be presence of hairs, bristles, and scales on the legs. Three pairs of legs are usually found. The length and size of the legs are variable. The legs are segmented. These segments are the coxa, trochanter, femur, tibia, pretarsus, tarsus, and pulvilli.

Different Types of Mouthparts of Insecta

Chewing Type

This is a basic type of mouthparts found in some insects (e.g., Mallophaga). This type of mouthparts consist of large mandible and a pair of separate maxillae I and a fused maxillae II. This type of mouthparts is used for biting and chewing purpose.

Sponging or Lapping Type

This type of mouthparts is found in nonbiting dipteran members (*Musca* spp.). In this type of mouthparts, mandibles and maxillae are nonfunctional. The proboscis has three parts i.e., rostrum, haustellum, and flabellum. The labellum is sponge like having minute channels called pseudo-trachae.

Cutting-Sponging Type (Lapping)

This type of mouthparts consists of sharp bladelike mandible and long stylet form maxillae; both these are used for cutting the host's skin. Blood is taken up by lapping method using a spongelike labium. This type of mouthparts is found in *Tabanus* spp.

Piercing-Sucking

In this type of mouthparts, two different channels are formed. One large channel is used as food canal, and the other is used to provide saliva containing anticoagulant. The elongated labium act as sheath and encloses the mandibles and maxillae. The mandibles and maxillae are modified to needle like stylet.

Digestive System

The digestive system is usually divided into three parts comprising the stomodeum (foregut), mesenteron (midgut), and proctodeum (hindgut). Generally, the gut of Diptera includes the mouth, pharynx, esophagus, crop, proventriculus, gizzard, intestine, and anus.

Circulatory System

The circulation is open which possesses the heart and aorta. The body cavity is called hemocoel. Hemolymph is found instead in the blood. The blood of arthropod does not contain any hemoglobin. Thats why the color of blood is not red as in higher animals. The blood of arthropod is copper-based. A small tubelike structure is found called the heart. In the heart, small openings are found called ostia. The hemolymph enters the heart through the ostia. When the heart ripples, the ostia remains closed. The circulatory fluid enter the head through aorta and then it reaches to haemocoel and bathes all the organs. In arthropod, the heart does not contract like in higher animals. Rather, the heart ripples which causes forward movement of the blood.

Respiratory System

Respiration is performed by different respiratory structures found in arthropods. These are book lung, tracheal system, gill, body surface, etc. In insects particularly, tracheal system is found which is a fine network of tracheal tubes. The minute tracheal tubes penetrate the tissues and even the cells. The terminal endings of tracheal tubes are called tracheoles. Air enters the body through some openings called spiracles. Two pairs of spiracles are found in the thorax region, and eight pairs of spiracles are found in the abdominal regions.

Nervous System

They have complex nervous system which is more developed as compared to other invertebrate organisms. This consists of a central mass (brain) present on the stomodeum, nerve trunk, or commissure surrounding the esophagus. Ganglions are present in different segments. A ventral nerve trunk is present comprising nerve cords supplying segmental ganglions.

Reproductive System

The male reproductive system consists of two testes, vas deferens, vesicula seminalis, common median duct, ejaculatory duct, and copulatory organ. The female reproductive system consists of a pair of ovary consisting ovarian tubules, oviduct, accessory gland, common duct bearing recepticulum seminis, genital atrium, gonopore, ovipositor.

Life Cycle

The male and female organisms are different. The life cycles are of two types: hemimetabolous, holometabolous. Complete metamorphosis occurs in holometabolous life cycle. In this life cycle, different stages, i.e., egg, larva, pupa, and adult, are found. In hemimetabolous life cycle, incomplete metamorphosis occurs. The developmental stages are egg, nymph, and adult. In ametabolous, metamorphosis is absent.

Classification of Class Insecta

Class: Insecta

The class Insecta has two subclasses, i.e., Apterygota and Pterygota.

The subclass Pterygota has two divisions, i.e., Exopterygota and Endopterygota.

Division: Exopterygota

Order: Mallophaga (biting lice)
Order: Siphunculata (sucking lice)
Order: Hemiptera (bugs)
Order: Orthoptera (grasshopper, cockroaches)
Order: Odonata (dragonfly)
Division: Endopterygota
Order: Diptera (true fly)
Order: Coleoptera (beetle)
Order: Hymenoptera (bees, wasps)
Order: Siphonoptera (fleas)

Order: Mallophaga

General Introduction

These are also called chewing lice because the mouthparts are adopted for biting or chewing. They affect mainly birds. But some spp. affect mammals, i.e., dogs, cats, cattle, horses, sheep, goats, etc.

Morphology

The head is broad anteriorly, not tapering as that of sucking lice. They have a pair of small eyes. The pair of antennae are short having three to five segments. The front pair of legs are short which are adapted to take the food material and deliver it to the mouth. Meso and metathorax are mostly fused which is called pterothorax. The tarsi have one or two segments. Each tarsus is equipped with one or two claws.

Feeding Behavior

They feed on feathers, hairs, epidermal cells, etc. Biting lice feed on growing feathers, feather fiber, scab, blood, etc. The lice of birds can digest keratin material. Lice of mammals also eat the same food as that of birds, i.e., epidermal scales, wool, wax, skin scurf, etc.

Harm Caused

As such they do not produce much pathogenesis. But high load of infection leads to dermatitis. They may cause intense pruritus which results in distraction of the animals from feeding.

Disease Transmission

They are not effective transmitters of disease producing pathogens. But some pathogens, i.e., rickettsiae are transmitted by the biting lice.

Life Cycle

The biting lice lay eggs on the hosts as the lice are permanent ectoparasite. The eggs are glued on the feathers or hairs by a sticky material secreted by the lice. The eggs hatch within a few days. But hatching may take more time depending on different factors. After hatching, the nymph stages come out which are quite similar to adult except the small size. They pass three nymph stages. The last nymph stage develops into adult. The adults are brown to dark tan in color.

Importance

Lice directly harm the hosts by causing destruction of the epidermal layer. The exudate comes out and this exudate is dried up. The biting lice feed on the epidermal cells and debris material mixed with dried fluid. Lice affect the root of the feathers or hairs. Hairs fall down. Intense pruritus occurs due to allergic reaction. The birds scratch that affected areas.

Suborder

Amblycera
Ischnocera
Rhychophthirina

Suborder: Amblycera—Family: Menoponidae

Family Synopsis

The lice of this family are ectoparasites of birds. There are two important genera of this family, i.e., *Menopon* and *Menacanthus*. Life cycle is direct type. The developmental stages are egg, nymph, and adult. Control of these lice can be done by application of ectoparasiticidal agents.

Genus: *Menopon*

Species

Menopon gallinae

These are called “shaft louse” of poultry. They are characteristically found on the shaft of feathers of birds. They are yellowish to pale yellow in color. In the segment of the thorax and abdomen, single row of bristles is present. They mostly affect the head, underside of wings, vent, etc. The lice are much active and cause harm to the epidermal layer. There are no reports regarding transmission of any pathogen by this lice to their hosts. But their mechanical damage is important. The birds loose body weight and health condition if a large number of lice affect and pass without treatment.

Genus: *Menacanthus*

Species

Menacanthus stramineus

These are called “body louse” of poultry. They are dark yellow in color. In each segment of the abdomen, double rows of bristles are found. These lice affect the skin of domestic birds, ducks, goose, etc. The lice suck blood from the birds. They affect the head, underside of wings, vent, breast, thighs, etc. where dense feathers are absent. *Menacanthus stramineus* is slightly larger than *Menopon gallinae* and more hairy than the latter. Like *Menopon gallinae*, these lice also cause marked harm to the hosts.

Genus: *Trinoton*

Species

Trinoton anseris
These lice occur in ducks, geese, etc.

Suborder: Amblycera—Family: Gyropidae

Genus: *Gyropus*

Species

Gyropus ovalis
Gyropus ovalis affects guinea pigs. These are oval in shape. That's why they are called "oval louse." They do not cause pathogenesis. But high infection level causes pruritus, alopecia, etc. They abrade the skin by the mouthparts and ingest fluid. They cause scab formation in the affected areas but more predominantly in the ear.

Suborder: Amblycera—Family: Boophidae

Genus: *Heterodoxus*

Genus: *Boophia*

Species

Heterodoxus longitarsus
These are biting lice of kangaroos. They have strong mandibles. The antennae are placed in deep grooves. These lice eat the dermal cells. But blood has been detected in the gut of these lice. They do not cause much pathogenesis. But a large number of lice cause some pathology of the skin. The animal losses its condition.

Suborder: Ischnocera—Family: Philopteridae

Genus: *Lipeurus*

Species

Lipeurus caponis
These are biting lice occurring in birds. The common name of the lice is "wing lice" or "depluming lice." They are gray in color and move slowly. The lice are very common compared to other spp. They affect the feathers of wings. These are slender and elongated lice.

Cuclotogaster (Lipeurus) heterographus

These are found on the skin of birds. They affect the head and neck of birds. Since they are almost invariably found on the head and neck region of birds, they are called "head louse" of poultry. They feed on base of the feathers. Notably these lice do not suck blood.

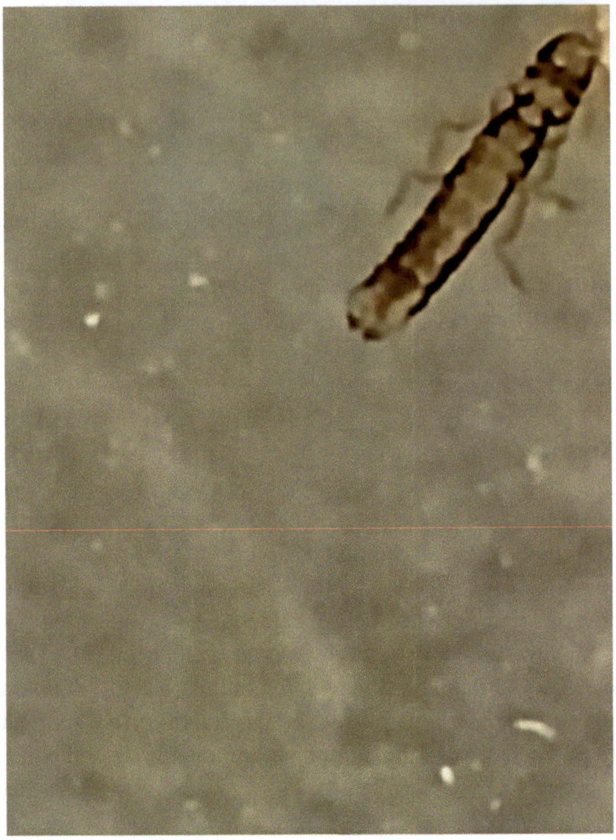

Fig. 7.1 *Lipeurus* sp.

Genus: *Goniodes*

Species

Goniodes gigas
These lice occur on the body of chicken. These are large sized lice. Characteristic triangular areas are found on either sides of the abdomen.

Genus: *Goniocotes*

Species

Goniocotes gallinae
These lice are smaller than *G. gigas*. They are called "fluff louse" as they occur in the fluff region of base of feathers of birds.

Genus: *Columbicola*

Species

Columbicola columbae
These are common lice occurring in the pigeons. The shape is typically elongated and the lice are easily identifiable.

Suborder: Ischnocera—Family: Trichodectidae

Family Synopsis

The lice of this family are ectoparasites of cattle, sheep, goat, dog etc. The important genera are *Damalinia*, *Trichodectes*, and *Felicola*. The antennae are trisegmented and the tarsi have single claw. The developmental stages are egg, nymph, and adult. Control of these lice can be done by application of ectoparasiticidal agents.

Genus: *Trichodectes*

These lice are short and broad. The important species is *Trichodectes canis*. The color is yellowish. They are important lice as they transmit very important tapeworm, *Dipylidium caninum*.

Genus: *Damalinia*

There are four important species of this genus, i.e., *D. bovis*, *D. caprae*, *D. ovis*, and *D. equi*, occurring in cattle, goat, sheep, and horse, respectively. *D. bovis* are called "red louse" of cattle. Dark band-like structure is found in each abdominal segment.

Genus: *Felicola*

Species

Felicola subrostratus
These lice occur in cats. They are distinguishable among all Mallophagian lice in having pointed head which is somewhat similar to sucking louse.

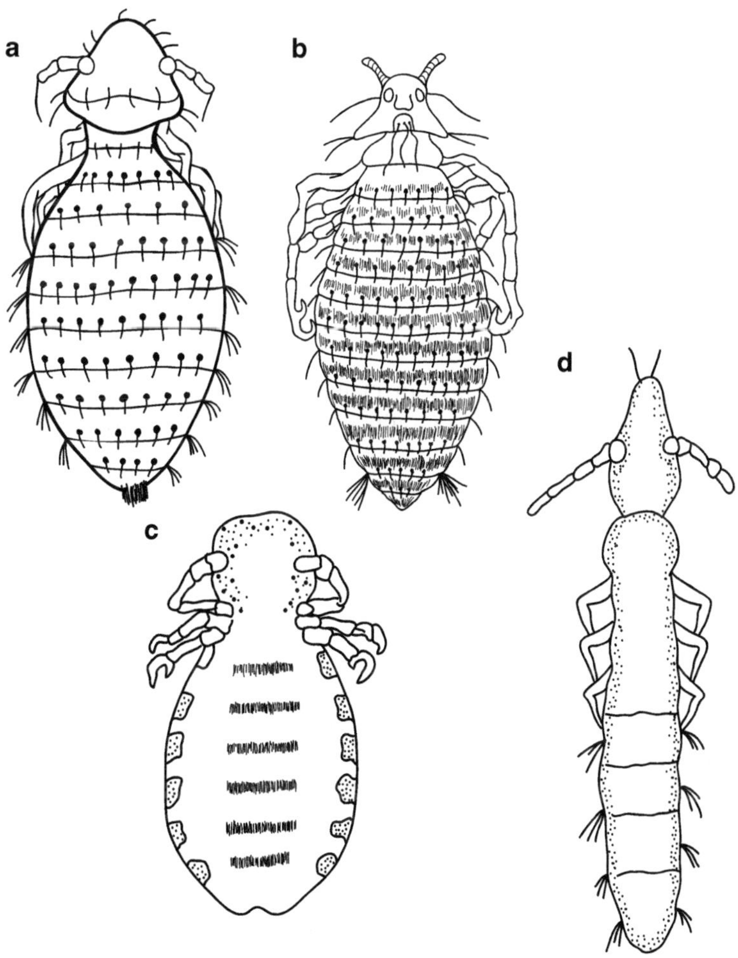

Fig. 7.2 Live (**a**) *Menopon gallinae* (**b**) *Menacanthus straminеus* (**c**) *Damalinia bovis* (**d**) *Columbicola columbae*

Suborder: Rhynchophthirina—Family: Haematomyzidae

Genus: *Haematomyzus*

Species

Haematomyzus elephantis
This louse is found in elephants.

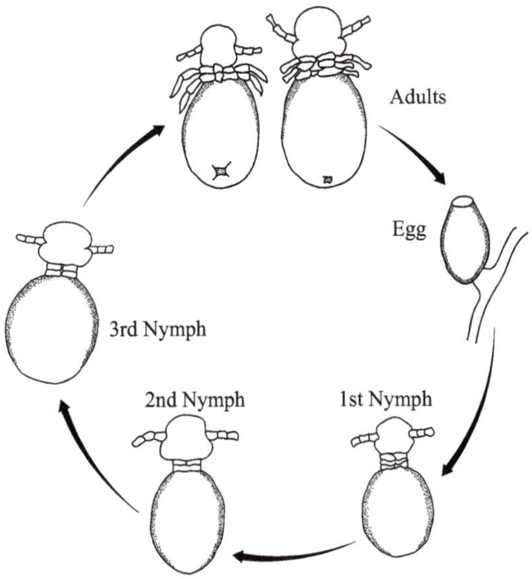

Fig. 7.3 *Haematomyzus* sp.

Order: Siphunculata

These are bloodsucking lice of mammals and humans. They are dorsoventrally flattened as that of biting lice. The head is tapering anteriorly forming a conical or pointed structure and adopted for sucking blood or tissue fluid. The head is not broad as that of biting lice. Three segments of the thorax are fused together. The abdomen is clearly segmented and very often bears paratergal plates on either side of each abdominal segment. Paratergal plates are thickened chitinous dark brown or black colored platelike structures. This prominent morphological feature is helpful for identification of lice of this category. Eyes may be absent or present (e.g., human lice). Antennae have three to five segments. The first pair of legs are small and the third pair of legs are large. Each tarsus is equipped with a claw. Tibial pad is found in the legs of some members.

Order: Siphunculata—Family: Linognathidae

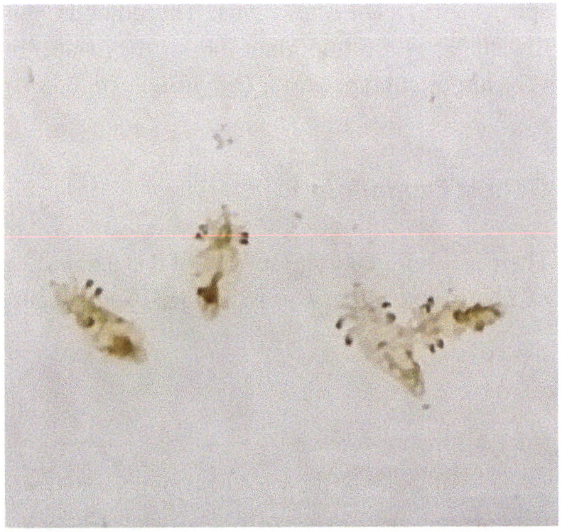

Fig. 7.4 *Linognathus* sp.

Genus: *Linognathus*

Species

Linognathus stenopsis
These lice are found on the body of goats.

Linognathus setosus
These lice are found in dogs, foxes, etc.

Linognathus pedalis
These are called "foot louse" of sheep as they affect the feet of the animals (sheep).

Linognathus ovillus
These lice are called "body louse" or "blue louse" of sheep. They mostly affect the face and they are also called face louse.

Linognathus vituli

These are sucking lice of cattle. The head is elongated and tapering. That's why these lice are called "long nosed cattle lice." First pair of legs are short but last two pairs are notably large compared to the first pair of legs. They are slow moving compared to biting lice. The lice are important as they are seriously concerned to livestock economy. Since these are sucking lice, apart from pruritus (only symptom of biting lice infection), they cause naked anemia. The animals lose body weight. Scratching and irritation distract the animals from normal behavior. The brown colored batches of lice are found attached on the neck and groin region. The affected animals exhibit poor immunity and succumb to other infection. The affected area caused by lice becomes prone to fungal infection. A huge number of eggs (nits) are found attached on the shaft of the hairs.

Order: Siphunculata—Family: Haematopinidae

Family Synopsis

The lice of this family are ectoparasites of mammals. The important genus of this family is *Haematopinus*. The lice are morphologically characterized by presence of head prolongation (temporal angle), absence of eyes, five segmented antennae, broad thorax, presence of paratergal plates, and presence of tibial pad in claws. Life cycle is direct type. The developmental stages are egg, nymph, and adult. Control of these lice can be done by application of ectoparasiticidal agents.

Genus: *Haematopinus*

Species

Haematopinus eurysternus

These are very important sucking lice. The lice are very much common in cattle, buffalo, etc. They have comparatively short head bearing sucking type of mouthparts. The abdomen is larger in proportion to other lice. They have large and strong legs with powerful claws. Brown to dark brown prominent paratergal plates are present in the abdominal segments. The lice are important as they are seriously concerned with livestock economy. They cause anemia as a result of long-standing infection and bloodsucking activity. Feed intake of animals is decreased. The animals lose body weight. The brown colored batches of lice are visible attached on the neck and groin region. Long-standing infection results in poor immunity, and they succumb to other infections. The affected area caused by lice invites secondary infection (fungal infection). A huge number of eggs (nits) are found attached on the shaft of the hairs.

Fig. 7.5 *Haematopinus* sp.

Haematopinus suis

This is common species found in India. This louse affects pigs. They are large in size and gray in color.

Haematopinus asini

These lice occur on the body of equines.

Order: Siphunculata—Family: Pediculidae

Genus: *Pediculus*

Species

P. humanus humanus
P.h. capitis

P. humanus humanus (body louse) and *P. human capitis* (head louse) are two important lice of humans. These lice are very important as they transmit *Borrelia recurrentis*, causative agent of relapsing fever, and *Rickettsia prowazekii*, the causative agent of endemic typhus.

Genus: *Pthirus pubis*

This louse is called "human pubic louse" or "crab louse."

Harm Caused by Lice

Both biting and sucking lice cause damage to the hosts in different ways:

Annoyance

Both biting and sucking lice cause constant irritation on the affected skin which results in great annoyance to the animals. The affected animals take less food and water due to disturbance caused by the lice. This ultimately affects body weight and milk and egg production.

Intake of Blood and Tissue Fluid

Anemia may be caused if a large number of sucking lice infest the animals. Lice with high population occurs during the winter season. The animals become weak and debilitated due to severe loss of blood and tissue fluid.

Self-infliction

To alleviate the irritation, the animals scratch the affected parts. This may lead to cause wound of the affected area. Self-infliction causes secondary bacterial infection. This causes lesion with formation of exudate. The wound may attract myiasis-causing flies.

Ball Formation

During scratching, the animals ingest hairs and feathers which cause formation of balls. The balls obstruct the lumen of the food passage. This causes functional disturbance of food canal.

Disease Transmission

Reports of disease transmission by lice is very limited. *Haematopinus suis* transmits African swine fever, and *Eperythrozoon suis* transmits virus of swine pox. *Trichodectes canis* transmits *Dipylidium caninum*. Some pathogens of humans are transmitted by some lice. *Pediculus humanus* transmits *Rickettsia prowazekii* and *Borrelia recurrentis* in humans.

Control of Lice

General Control

- Good nutrition should be provided to the animals for maintenance of good health. Healthy animals are immunocompetent and less prone to any kind of infection.
- The infected animals should be treated properly (See end part of this chapter).
- The uninfected stock should be kept separately so that they cannot get any lice infection from the infected stock.

Immunological Control

No immunological control technique or vaccine is available so far.

Molecular Control

No molecular vaccine or gene vaccine is available in the field.

Order: Hemiptera

The members of this order are called bugs. There are two important families under this order, i.e., Cimicidae and Reduviidae. Two pairs of wings are present. First pair of wings are hard (hemelytra) The second pair (hind) of wings are membraneous.

Order: Hemiptera—Family: Cimicidae

Family Synopsis

The important genus of this family is *Cimex*. They are called bed bugs. They suck blood from humans and animals. The life cycle is direct type. The developmental stages are egg, nymph, and adult. They are called bed bugs.

Morphology

They are oval and flattened. The body is divided into three parts, i.e., head, thorax, and abdomen. The head bears a pair of antennae. The antennae has four joints. The mouth parts are modified for piercing and sucking. The prothorax is deeply notched in which the head is inserted. The wings are vestigial. The body is covered by hairs and bristles. A pair of ventral thoracic stink glands are present which secrete some substance emitting a characteristic odor.

Behavior

The bugs live on blood of their hosts. They suck blood from the hosts and hide themselves in their nests. They digest the blood for few days to a week. Few bugs can live for years without food. Bugs live mostly near human dwelling place. They are attracted to the hosts by different cues, i.e., temperature, smell, etc.

Biology

They perform hemimetabolous type of life cycle. The stages are egg, nymph, and adult. The fertilization of eggs occurs by a specific activity. This is called traumatic fertilization. In this way, males insert their sperm in the abdomen into the spermalege. The sperms are then carried to the eggs through paragenital organs. Then the sperms fertilize the eggs.

Order: Hemiptera—Family: Reduviidae

Family Synopsis

The members of this family are called cone nosed bugs, kissing bugs or assassin bugs. They are morphologically characterized by presence of conspicuous eyes and four segmented antennae. The important genera are *Triatoma*, *Rhodnius*, and *Panstrongylus*. The life cycle is direct having the stages of egg, nymph, and adult. They are serious blood feeder and transmit important pathogen, *Trypanosoma cruzi*.

Morphology

The bugs are morphologically different. They are not oval as cimicids. The head is elongated and typically cone shaped, thus called "cone nosed bugs" with narrow neck. The antenna has four joints. The proboscis is curved. They are larger than Cimicidae. They have well-developed wings. The abdomen of Reduviid members is less flattened compared to Cimicidae.

Biology

The bugs exhibit hemimetabolous type of life cycle. The stages are egg, nymph, and adult. Several nymphal stages are found before becoming adult. Fertilization is specific which is called traumatized fertilization as described earlier.

Importance

The *Triatoma* bugs are important arthropods. These are called "kissing bugs" or "assassin bugs." They suck blood of humans from the lips. During bloodsucking, they defecate there which contaminates the pierced area. The trypomastigote stages of *Trypanosoma cruzi* are passed with the feces. In this way, humans get the infection of Chagas disease.

Order: Orthoptera

The important members are cockroaches, grasshoppers, etc.

Morphology

The body is elongated. The antenna is filiform and has segments. They have large eyes. Two pair of wings are present. The forewings are hardened at the base which are called tegmina. The hind wings are membraneous.

Importance

Grasshopper

They transmit *Cheilospirura hamulosa* and *Tetrameres americana.*

Cockroach

They have veterinary importance. They are the intermediate hosts of *Gongylonema* spp., *Oxyspirura mansoni* etc.

Order: Odonata

These are large sized flies having long legs and antennae. The nymph (naiad) stages act as the intermediate hosts of an important fluke, *Prosthogonimus pellucidus.*

Division: Endopterygota—Order: Siphonaptera

Introduction

The members of this order are commonly called flea. In the veterinary field, the fleas are very important. They affect different animals including birds. They cause direct and indirect damage to the hosts. They transmit different important pathogens to the hosts.

Morphology

Adult

Shape: The body of the fleas is roughly oval in shape.

Size: Approximately they are 1–4 mm long. The size is widely variable.

Color: The color is dark reddish brown, dark brown, etc.

Body: The body is laterally compressed. The body is divided into three parts, i.e., head, thorax, and abdomen. Spines or bristles are present on the body. The body is segmented.

Head: The head is comparatively small in size. Biting type of mouthparts is found. Genal combs may be found at the mouth region.

Thorax: The thorax is distinct. The spines may be present in the thorax. These are called ctenidium or combs. The combs present in the pronotum are called pronotal combs.

Legs: Three pairs of legs are present. The legs are unequal in size. The first pair of legs is the smallest and last pair is the largest.

Wings: The wings are absent.

Abdomen: The abdomen is segmented. In the nineth abdominal segment, a sensilium or pygidium is present covered with sensory setae.

Egg

The eggs are oval and elongated. These are about 0.5 mm long. The glistening eggs are pearly white in color having rounded ends.

Larva

The larvae are apodous and look like maggot. No appendage is found. The abdomen is clearly segmented. They are elongated maggot-like yellowish white in color. The larva has three thoracic and ten abdominal segments. The mouth of a larva contains a pair of mandibles, maxillae, and a labium.

Pupa

The larval stage proceeds to become pupal stage. A tough cocoon is formed, within which the larva remains hidden in the cocoon. The pupa is sticky. Dusts, debris material get adhered to the pupa.

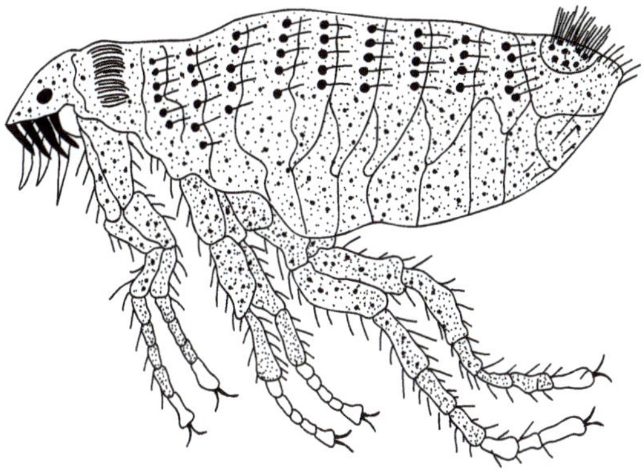

Fig. 7.6 *Ctenocephalides* sp.

Fig. 7.7 *Ctenocephalides* sp. (Gross)

Biology

Eggs, larva, pupa, and adult are four stages found in the life cycle of flea. Hence, the life cycle is holometabolous type.

Egg Laying

The adult flea must take blood meal for its reproductive performance. Without blood meal, egg laying does not occur. The female flea lays about 3–18 eggs in a single oviposition. The fleas lay their eggs in dust, dirt, debris etc. The eggs may be deposited on the hosts, and the eggs immediately drop off the ground.

Egg Development

The development of eggs depends on temperature, humidity, etc.

Egg Hatching

After a few days (2–16 days) of egg laying, hatching of eggs occurs. After hatching, the larvae are liberated. The apodous larvae are active. They take different food materials comprising of dry blood, exudate, feces, etc. Two ecdyses pass on and the larvae are transformed into opaque white matured larvae.

Larval Development

After hatching, the larvae come out. The larvae are apodous and look like maggot. The larva does not have any appendages. The abdomen is segmented. The larvae take different food material, i.e., dry blood, dry exudate, and other proteinaceous material. Moderate teperature and high humidity are required for development of larvae.

Late Larval Stage

The larvae become quiescent and leave taking any food. The larval stage lasts for about 7–10 days.

Pupa Formation

The larval stages last for a few days to a few weeks. 21–29 °C is suitable for growth. But larval stages may be prolonged if the environmental factors are not suitable. The larval stage proceeds to become pupal stage. A tough cocoon is formed, within which the larva remains hidden in the cocoon. The glycogen stored during the larval stage is utilized for pupal stage. The cocoon is loosely spun and the imago is easily visible from outside the cocoon. The pupal stage may last for 10 days to about a year depending on environmental factors.

Adult Release

Adults come out from the pupa later. The adults can sense nearby hosts by recognizing different stimuli. The pre-emergent adult waits for suitable trigger factors for emergence. These are sound vibration, heat, (of host), CO_2 etc. The emerged adult must find a host soon for its survival; otherwise, it may die due to starvation.

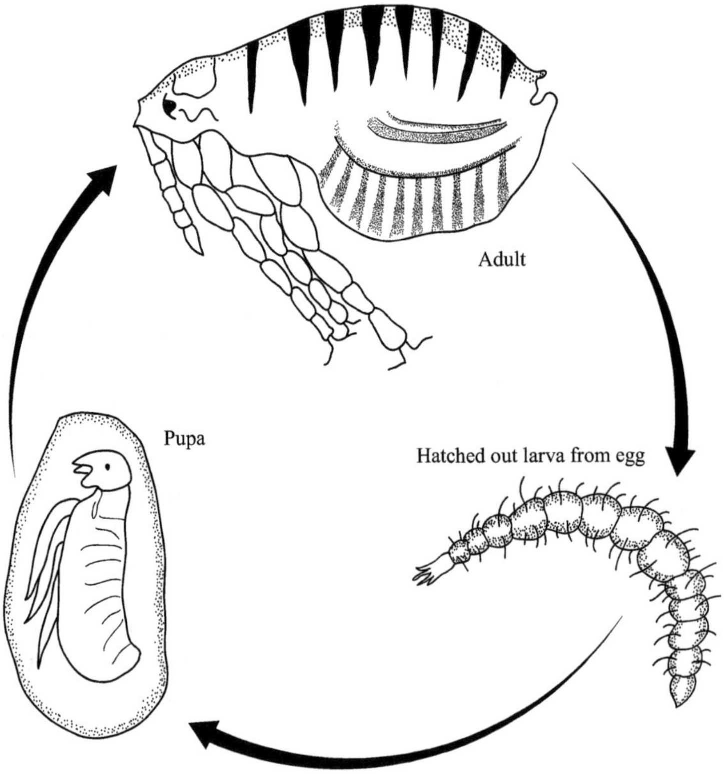

Fig. 7.8 Life cycle of *Ctenocephalides* sp.

Importance

Direct Damage

Irritation

Fleas have much importance as they cause both direct and indirect damage. Directly they cause dermatitis. They suck blood, lymph, etc. The debilitated animals suffering from a chronic disease are affected mostly by the fleas. The fleas cause allergic dermatitis mediated by hypersensitivity I reaction. Flea saliva acts as hapten which binds to collagen and acts as immunogen. It causes allergic reaction. Hypersensitivity reaction results in production of histamine which causes irritation in the skin. *Echidnophaga gallinacea* are called sticktight flea which attaches on the comb, wattles, and eye regions. High load of infection may cause death of the birds.

Self-infliction

To alleviate the irritation, the affected animals scratch or bite the coat and lose the condition.

Indirect Damage

Ctenocephalides canis, *C. felis*, and *Pulex irritans* act as the intermediate hosts of *Dipylidium caninum*. *Xenopsylla cheopis* transmits *Yersinia pestis*, a causative agent of bubonic plague. *Spilopsyllus cuniculi* transmits myxomatosis in rabbit. The filarial worm, *Dipetalonema reconditum*, is transmitted by *C. canis*, *C. felis*, *P. irritans*.

Genus: *Xeopsylla*

Xenopsylla cheopis

These are called oriental rat flea. The primary hosts are rodents. They transmit bubonic plague. The fleas are predominantly found in tropical and subtropical region because this climate is suitable for their survival. They may occasionally be found in the temperate climate. The fleas are brown to dark brown in color. This color is suitable for their camouflaging on the host's body. The adults have no pronotal and genal combs. Approximately they are 2–4 mm long. The fleas have sexual dimorphism. The male and female fleas are distinguished by presence of different reproductive system. Gravid females lay eggs. In suitable environmental factors, the eggs hatch and the larvae come out. They pass three larval stages to finally become a stage of pupa. The larval stage may be as prolong as 200 days. From the pupa, the adult flea comes out.

Genus: *Ceratophyllus*

Species

C. gallinae

This species is called "hen flea" or "European chicken flea." The fleas are brown colored. Typically they have bristles at the femur of hind legs. The fleas pass four stages during their life cycles, i.e., eggs, larva, pupa, and adult. The fleas attack the hosts by jumping. The jumping activity is found in less sunlight.

Genus: *Echidnophaga*

The important species is *E. gallinacea*. The common name of this flea is "sticktight flea." They affect chicken, dog and even man. These fleas unlike other genera members remain attached to a particular location and cause inflammation and ulcer of that area of body. The fleas are dark brown in color. Genal and pronotal combs are absent.

Some Common Names

Sticktight flea	*Echidnophaga gallinacea*
Chigger, chigoe flea	*Tunga penetrans*

Genus: *Pulex*

The important species is *Pulex irritans*.

Genus: *Tunga*

The important species is *Tunga penetrans*. These are called Jigger, Chigoe.

General Control of Flea

Application of Ectoparasiticidal Agent

Different compounds, i.e., malathion, carbaryl, are effective ectoparasiticidal agents which can kill the fleas. Dusts and solution form of this agent are used. Solution form is used mainly for treatment of *Echidnophaga gallinacea*. The affected heads of the birds are dipped in the medicated solutions.

Management of Poultry Litter

The poultry litter is removed or burned if infection is detected. Insecticide can be sprayed routinely in the poultry house.

Management of Floor of Animal House

The cracks and crevices should be sealed.

Use of Insect Growth Regulator

Different insect growth regulators (e.g., methoprene) can be used. These are nontoxic to the vertebrates.

Use of Flea Collar

The flea collars could be used for control of flea infection in pets.

Control of Rats

In the plague-prone areas, the rats and rodents should be controlled using rodenticides or by applying other rodent control techniques.

Order: Diptera

Different Characters of Diptera

- The body is divided into three parts, i.e., head, thorax, abdomen.
- The thorax is divided into prothorax, mesothorax, and metathorax.
- The head is round, oval, or suboval.
- The head possesses a pair of eyes, a pair of antennae, and a proboscis.
- The mouthparts are adopted for sucking. The mandibles are usually absent.
- The members possess a pair of wings present on the mesothorax. The metathoracic pair of wings are reduced to form halters or balancers.
- Three pairs of legs are present. The tarsi usually have five joints.
- The metamorphosis is complete (holometabolous). The larvae are apodous. The pupae are of obtectate or coarctate type.

The order Diptera contains three suborders as following:

Brachycera

Brachyceran members are identified by presence of small arista. The head of larva is not much developed. The antenna is shorter than thorax. The antenna has three segments. The last segment is annulated.

Nematocera

Nematoceran members are identified by noting the absence of arista. The antenna is longer than head and thorax. The head of larva is much developed. The antenna has more than eight segments.

Cyclorrhapha

Cyclorrhaphan members have arista. The antenna is trisegmented.

Suborder: Brachycera—Family: Tabanidae

Family Synopsis

The members of this family are called horsefly, deerfly, breeze fly, cleg, etc. There is only one important family, Tabanidae. The flies are morphologically characterized by a small arista, trisegmented antenna, annulations in the last antenal segment, and presence of piercing and lapping type of mouthparts. The life cycle is direct type. The developmental stages are egg, larva, pupa, and adult. They are extremely important as they transmit trypanosomosis in different animals. The control of these flies is mostly based on use of insecticidal agents.

Genus: *Tabanus*

(**Species**: *T. latipes*, *T. rubidus*, *T. striatus*, *T. orientis*, *T. bicinctus*, *T. macer*)

Species

Tabanus latipes

Morphology

Adult

Shape: These are large robust flies and easily identified by the naked eyes.

Size: They are 2.5–3 cm long.

Color: The color is brown.

Head: The head is large. Minute hairs are found in the head and thorax.

Proboscis: The proboscis is short and soft. The proboscis typically hangs down.

Mandible: A pair of mandibles are bladelike.

Other mouthparts: A pair of maxillae, labrum, and labium are found.

Antenna: A pair of antennae are present. The initial two segments are small. The third segment has toothlike projection and is equipped with four annulations.

Eyes: The eyes are large in size. These are striped manner. The eyes are holoptic in male and dichoptic in female.

Wings: They possess large wings.

Abdomen: They have large segmented abdomen.

Egg

The eggs are cigar-shaped. They are light in color but turn to dark after some time of laying. The length of egg is 2 mm.

Larva
The larvae are elongated and segmented. They are whitish in color. They have three thoracic and eight abdominal segments. Fleshy tubercles are present in the segments. Eighth abdominal segment is equipped with an elongated retractile siphon tube.

Fig. 7.9 *Tabanus* sp.

Biology
Holometabolous life cycle is found. The flies pass four developmental stages during the period of life cycle, i.e., egg, larva, pupa, and adult.

Egg Laying
The eggs are laid on tree leaves, hanging waterdrops, stones, etc. The eggs are laid in batches. Each batch may contain about 300–600. The light colored eggs soon turn dark.

Egg Hatching
The eggs are hatched after about 1–2 weeks. After hatching, the larvae drop on the water or mud. They come in contact in water. Semiaquatic climate is preferred.

Larval Development
The larvae are aquatic. They are sluggish in movement. They intake decaying organic matter. They may eat small arthropods. They may even intake other tabanid larvae. Larval development may take up to 3 months. The larvae pass on several ecdyses.

Late Larval Stage
The larvae enter into a nonfeeding stage. They become quiescent. This is the preparatory stage of pupa.

Pupa Formation
The larvae take food material voraciously and the same is stored as glycogen. The stored glycogen is utilized during pupal stage. The pupa is obtectate. They are subcylindrical in shape and brown in color. The pupae are formed in the mud. Pupal stage lasts for 10–14 days.

Adult Release
The adults come out from the pupa. A dorsal slit is formed on the pupa through which the adult emerges out. Pupal development depends on the environmental temperature. The low temperature delays the pupal development.

Importance
They transmit different pathogens.

Protozoa
Tabanus flies transmit different species of protozoa. These are *Trypanosoma evansi*, *T. equinum*, etc.

Bacteria
They transmit anthrax bacteria.

Virus
The virus of equine infectious anemia is transmitted by these flies.

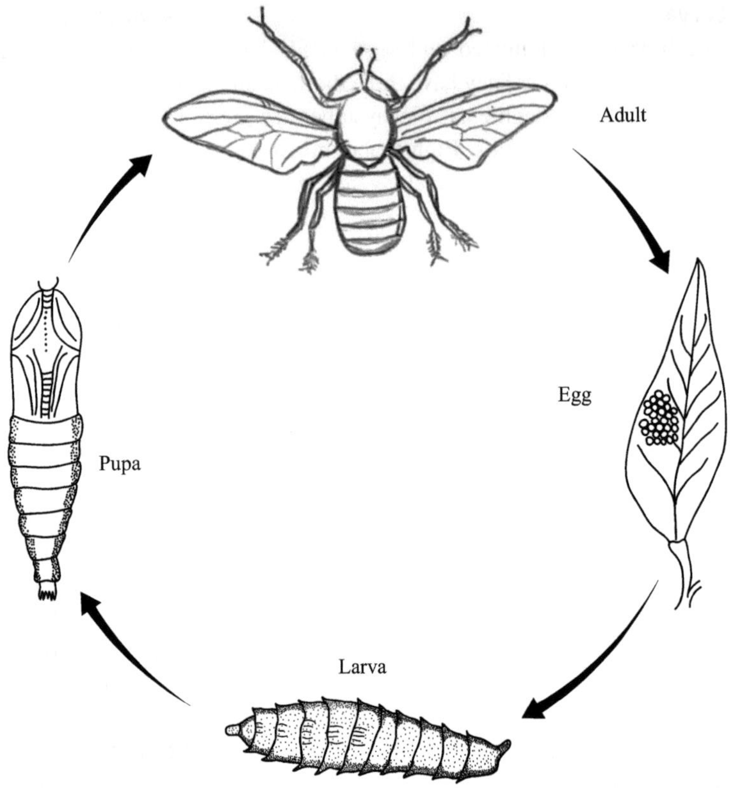

Fig. 7.10 Life cycle of *Tabanus* sp.

Habitat

The flies are very much active in hot summer. The females suck blood from the animals. The flies utilize different cues, i.e., CO_2, host smell, and visual cues. The flies mainly attack the large animals. The male flies feed on sugar, honeydew, excretion, secretion, etc. The female flies take blood either by cutting the skin or they may take blood if available on the skin. The bites are extremely painful.

Genus: *Chrysops*

Species

Chrysops discalis

These are small flies with bright large eyes. They are fairly larger than houseflies and stable flies but smaller than *Tabanus* flies. They possess clear and prominent dark band passing anterior to the posterior border of the wing. Both male and female flies are necator feeder, but females require blood meal for reproductive purpose. That's why the flies are called anautogenous flies. The flies lay hundreds of eggs in different batches on the vegetation and wet land areas where semiaquatic zone is available. After hatching, the larvae come out. The larvae are apodous and take rotten, decayed protein material. The larval stage may be as prolong as 1 year or more. The flies are abundant in the summer season. They are attracted to the bright sunlight and avoid shade areas with absence of or less light. The flies are important because they transmit important filarial worm, *Loa loa*.

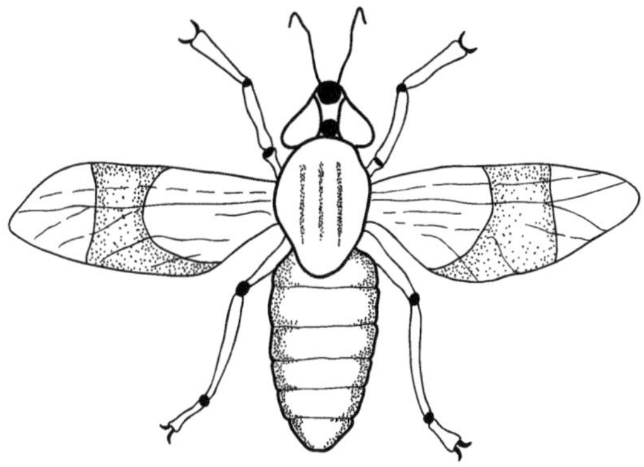

Fig. 7.11 *Chrysops* sp.

Haematopota

These are small to medium sized flies. They have bright colored patterned eyes. The most characteristic feature for identification is presence of gray mottled brown wings. The abdomen is spotted. The first antennal segment is bulbous. The third antennal segment is partly reddish yellow colored in some species. In others, the third antennal segment is entirely black. Both male and female flies are necator feeders. The females suck blood from cattle and horse. The flies are commonly called clegs. They may also attack humans. The female flies lay eggs in moist soil. They avoid damp area for egg laying. The hatched out larvae feed on other insects for getting proteinaceous material. Different stimuli attract the flies to the animals. These are smell, color, warmth, CO_2, etc.

Pangonia

This is a genus of family Tabanidae. The proboscis is longer compared to other genera members and project forward. The third antennal segment is equipped with six or seven annulations. The distal tibial end of third pair of legs bears two spurs. While at rest, the wings are divergently placed.

Suborder: Nematocera

The members of this suborder are characterized by presence of long filamentous antennae having 14–15 segments, absence of arista, maxillary palp with 4–5 segments, and absence of discal cell. The mandibles of larva bite horizontally and pupa is obtectate type. This suborder includes biting midge, sand fly, black fly, and mosquito.

Suborder: Nematocera—Family: Culicidae

Family Synopsis

The members of this family are mosquitoes. They are characterized by long filiform antennae having 14–15 segments, long proboscis of piercing and sucking type, characteristically wedge shaped thorax, and fused segments. They severely affect both humans and animals. The important genera are *Anopheles*, *Culex*, and *Aedes*. The life cycle is direct type. The developmental stages are egg, larva, pupa, and adult. Their bite is painful. But their main importance lies on transmission of very important pathogens(dengue, malaria, etc.). Control of mosquitoes is mainly based on application of insecticidal agents.

Genera

Anopheles, *Culex*, *Aedes*, *Mansonia*

Morphology

Eggs

There are some variations in shape of eggs of mosquitoes. These may be oval, elongated, or spindle shaped. Some characteristic morphological features of the eggs obviously have taxonomic importance for identification. The typical raft form of a batch of eggs is found in case of *Culex* flies. The eggs of *Anopheles* are easily identifiable, because the eggs have lateral floats.

Larva

Like other dipteran members, larvae of mosquitoes are apodous. But different characteristic features are important for identification of the larvae of different mosquitoes. Like adult, the body of larva has three distinct divisions, i.e., head, thorax, and abdomen. The head is competitively large in size having a pair of distinct eyes. There is a broad thorax without any legs. The thorax has three inconspicuous divisions. The abdomen of larva of mosquitoes is large enough having eight to nine segments. Except anopheline mosquitoes, others are equipped with siphon tube. *Anopheles* have palmate hairs present on abdominal segments which are used for clinging or floating in the water. Palmate hairs are absent in the culicine mosquitoes.

Pupa

The pupae of mosquitoes are characteristically comma shaped. Unlike larva, the body of pupa is divided into two parts, i.e., cephalothorax and abdomen. The abdomen is curled below the cephalothorax forming comma shaped appearance. A pair of air trumpet is present on the mesothorax part of the pupa.

Adult

Size: The mosquitoes are 2–10 mm long.

Body division: The adults are slender having three broad divisions as are found in other dipteran members.

Head: The head is globular in shape and they are highly mobile.

Mouthparts: The mouthparts are piercing and sucking type which include a pair of mandible, a pair of maxillae, labrum, labium, epipharynx, and hypopharynx, proboscis, palps etc.

Eyes: The head is small bearing a pair of eyes which are holoptic type in male and dichoptic type in female. The eyes have small divisions called ommatidium.

Antenna: A pair of antennae are present on the head which are long filiform. The antennae perform sensory functions.

Proboscis: A single proboscis is found. The proboscis is used for bloodsucking.

Thorax: The thorax is characteristically wedge shaped. Dorsally this is broad in appearance. The thorax is divided into three parts comprising prothorax, mesothorax, and metathorax.

Wings: A pair of wings arise from the mesothorax. The wings are long having typical scales over it. During rest, the wings remain folded flat. A pair of drumstick shaped halters are present on either side of metathorax. They are actually the reduced form of hind pairs of wings.

Legs: The legs are long enough. These are slender and soft.

Abdomen: The abdomen is segmented having ten segments. The initial segments are broad, and remaining segments are narrow compared to initial segments.

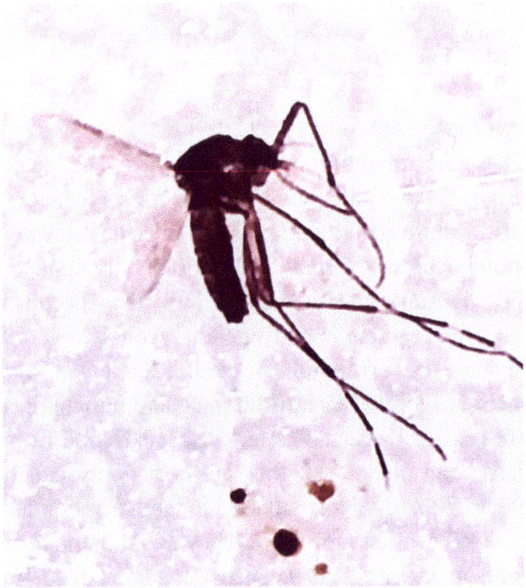

Fig. 7.12 *Culex* mosquito (female)

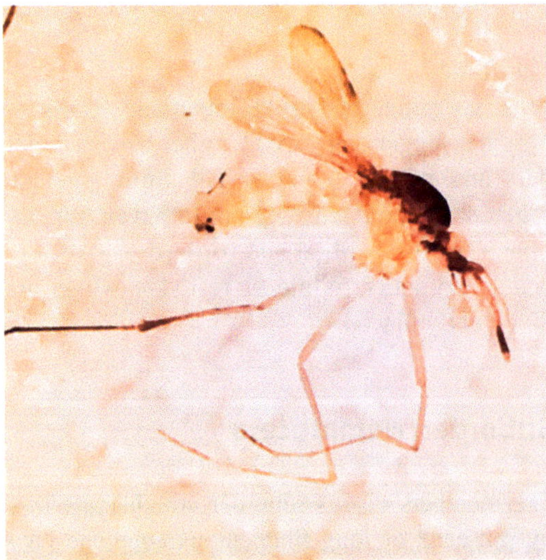

Fig. 7.13 *Anopheles* mosquito (male)

Life Cycle

Mating

Adults are ready for mating after few hours to 1 day of emergence from pupa. It is to be noted that mating occurs during flight.

Oviposition

Each female mosquito has two ovaries. Each ovary has 50–200 ovarioles. Every ovariole gives rise to only one egg. Only 4–5 ovipositions occur and each oviposition gives about 30–500 eggs.

Egg Laying (in Detail)

The female mosquitoes lay their eggs in different sites. Blood meal is essential in some species. The members of those species cannot lay eggs without blood meals. Blood is utilized for gathering nutrient for development and maturation of eggs. In some species, it is found that the female can lay eggs without blood meal. This is called autogeny. In some species, blood meal is not required for first gonotrophic cycle, but in the subsequent cycle, blood meal is required. For autogenous mosquitoes, blood meal is not required because during larval stage, they store up much nutrient in the fat bodies. And this nutrient is utilized for egg maturation. But it has been found that this autogeny depends on environmental factor and genetic factor. Mosquitoes lay their eggs in stagnant water or still water. The still water is preferred because the laid eggs can incubate undisturbed. That's why the mosquitoes lay their eggs in ponds, swamps, or any other water reservoir where the water is not disturbed. The mosquitoes lay eggs on water bodies because water is required for hatching. Albeit, the eggs may survive in dry place but water is must for hatching. The mosquitoes may also lay eggs at places where water level would rise. There are different site preferences for different mosquitoes. The *Anopheles* mosquitoes lay their eggs in clean water i.e. rain, pool, unused well. The eggs are laid singly by the *Anopheles* mosquitoes. The eggs of *Anopheles* mosquitoes have lateral floats filled with air. By using this structure, the eggs can float in the water. *Culex* mosquitoes lay their eggs in rafts.

Egg Hatching

Different environmental factors play on hatching of eggs of the mosquitoes. Water is essential for hatching of eggs. Different hatching experiment proves that presence of bacteria in the water stimulates hatching. The reason is not clearly known yet. Experiment has also proven that vibration of water enhances the hatching. Temperature is very important for hatching. The eggs hatch within 15–27 °C. Below 15 °C, hatching is reduced. However, there are reports of massive hatching of eggs at 13 °C. Reports are available on hatching of eggs even at 4 °C. Though hatching occurs in low temperature, the larvae are not viable.

Larval Development

After hatching, the larvae come out from the eggshell. The larvae cannot live without water. They eat different microorganisms present in the water. Siphon tube is present in *Culex* mosquitoes. They remain in hanging position from the water surface being suspended by the siphon tube. By the siphon tubes, they take oxygen from the air. They eat the microorganisms available at the water surface. When disturbed, the larvae quickly dip inside the water after leaving the water surface. During this time, they perform a characteristic movement forming an S shaped pathway. This characteristic movement provides the common name wrigglers or wigglers. Four larval stages pass on, and the larvae gradually increase in size and cast off their old cuticle during molting. In between the moltings, the stage is called instars. The larvae are highly active

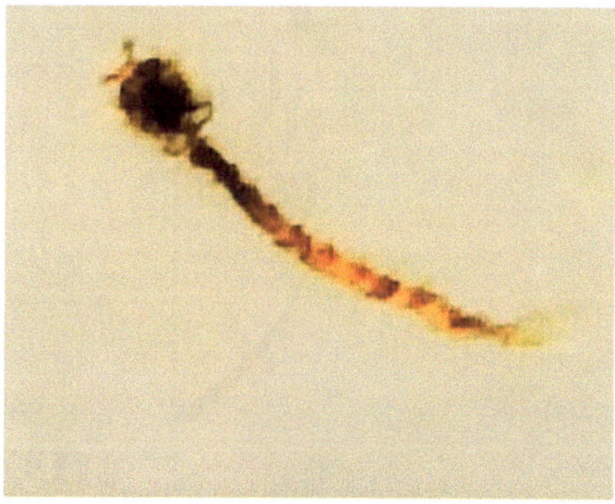

Fig. 7.14 Larva of mosquito

Pupa Formation

The last larval stage transforms into pupal stage. Unlike other insects, the pupa of mosquitoes can move. Development of pupa is temperature dependent. It may last for 1 day to 3 weeks.

Adults Release

After complete maturity, the adults come out from the pupa.

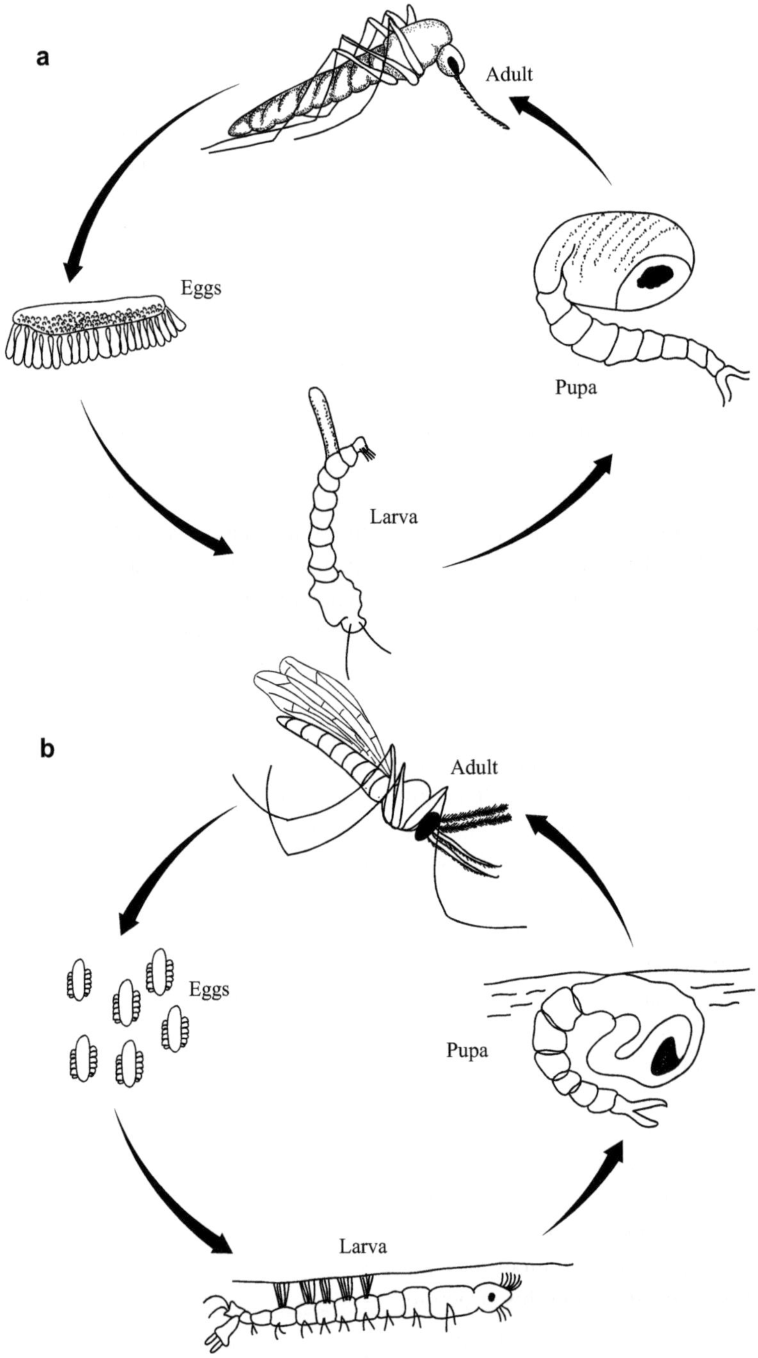

Fig. 7.15 (**a**) Life cycle of *Culex* (**b**) Life cycle of *Anopheles*

Importance

Virus Transmission

La Crosse encephalitis
Severe illness occurs in children.

West Nile Virus
This is a viral disease transmitted by mosquitoes. This disease occurs in humans. The disease is characterized by fever, rash, diarrhea, vomiting, etc.

Jamestown Canyon Virus
Humans of different age groups are affected. The incidences are very less.

Western Equine Encephalitis
This is a viral disease transmitted by mosquitoes. This disease occurs in horse.

Eastern Equine Encephalitis
This is a viral disease transmitted by mosquitoes. This disease occurs in horses and humans.

St. Louis Encephalitis
This viral disease occurs in humans.

Chikungunya
Humans are affected. The disease is rarely fatal but causes severe debilitating condition with severe joint pain. *Aedes aegypti* and *A. albopictus* are two important species attributed to transmit this disease.

Yellow Fever Virus
This is a viral disease occurring in humans. The occurrence is rare.

Dengue Fever Virus
This is a viral disease occurring in humans. The disease is characterized by fever, rash, brushing, gum bleeding, weakness, etc.

Zika Virus
The virus causes disease in humans. This disease is characterized by fever, rash, joint pain, etc. The babies suffer seriously due to congenital infection.

Protozoa

Malaria
Different *Plasmodium* spp. are transmitted by mosquitoes. These are *Plasmodium vivax*, *P. falciparum*, etc. *Anopheles quadrimaculatus* and *A. freeborni* have been found to be the primary vectors of *Plasmodium vivax*.

Helminths

Lymphatic Filariasis in Human
Humans are affected by different filarial worms, i.e., *Wuchereria bancrofti*, *Brugia malayi*, and *Brugia timori*. These filarial worms are transmitted by different species of *Anopheles*, *Culex*, and *Aedes* spp.

Filarial Worms in Animal

Dirofilaria immitis and *D. romeri* are two important spp. transmitted by different mosquitoes, i.e., *Anopheles*, *Culex*, and *Aedes*.

Control

General Management

The following should be done:

- Extermination of water from water logging areas, canals, ditches, culvert, etc.
- Removal of jungles, bushes, etc.
- Use of mosquito nets
- Use of repellants
- Use of smokes
- Less removal of animals to mosquito prone area
- Use of double doors

Use of Insecticidal Agents

Since different bloodsucking mosquitoes act as the vectors, they should be controlled using different insecticidal agents. A series of insecticidal agents are available in the market. These are eldrin, dieldrin, DDT, BHC, coumaphos, crufomate, crotoxyphos, pyrethrins, cypermethrin, dimethoate, etc. These insecticidal agents are sprayed locally.

Insect Growth Regulator
Insect growth regulator is also used for control program. Methoprene is an important insect growth regulator.

Biological Control
Different biological agents can be used for biological control of mosquitoes.

(a) *Bacillus thuringiensis var. israelensis* is one important bacteria, the spores of which contain a crystalline endotoxin capable of inducing lysis of midgut cells of larvae ultimately causing the death of the larvae. In this way, biological control of mosquito larvae is possible.
(b) *Gambusia* fish (*Gambusia affinis*) can be used to clear the mosquito larvae because the *Gambusia* fish eat the larvae of mosquitoes. Guppies can also be used for the same purpose. Other promising fish species are *Cyprinus carpio*, *Aphanius dispar*, *Catla catla*, *Tilapia* sp., etc.
(c) Larvae and nymphs of dragonflies feed on mosquito larvae. Adult dragonflies feed on adult mosquitoes.
(d) Water bugs (*Diplonychus indicus*) feed on mosquito larvae.
(e) Laboratory study shows that Asian house gecko, *Hemidactylus frenatus*, preys on mosquitoes.
(f) Different fungal spores are useful biocontrol agents of mosquitoes. The fungal spores are used both for indoor and outdoor control of mosquitoes. The product based on *Beauveria bassiana* is available commercially.

Study of Olfactory Physiology
Recently, study on olfactory physiology of vector is progressing. The output of study could be harnessed to create new trap or modify the existing trap of vector flies.

Prophylactic Management
Mosquito nets should be used.

Educational Campaign
Educational campaign can be done in different places in different times to make people aware about the possibility of controlling mosquitoes by simple managemental practices, i.e., use mosquito net, anti-mosquito coils, etc.

Endosymbionts
Different studies on use of endosymbionts of different insects is in progress. *Wolbachia* is one such bacterial endosymbionts found in most mosquitoes. The knowledge derived from study on endosymbionts can be exploited to design control regimen against mosquitoes.

Sterile Insect Technique
In this technique, a large number of flies are produced in large sized insectaries following standard protocol. The newly produced flies are sexually assorted. The male flies are subjected to sterilization. This is done by irradiation or genetic manipulation. However, chemosterilization is another way to produce sterile male. For performing this technique, one should have crucial knowledge about the biology of the mosquitoes which can be harnessed to develop the perfect protocol. A lot of questions pertain to the development of this technology. The relevant questions are as follows: When are the sterile males be released? How many sterile males are to be released in a particular area? What is the reproductive potential of the mosquitoes? Many and many such questions are there. Sterile insect technique is nonintrusive to the environment. It has no adverse effect against the nontarget organisms. Moreover, this is species specific. This technique does not have any residual effect. The modified organisms do not produce any poisonous or toxic material. In totality, this technique is environmentally friendly.

Anti-mosquito Vaccination
Different works were carried out on anti-mosquito vaccination. Several investigators found that saliva proteins could be used as vaccine which can protect the hosts against the mosquito-transmitted pathogens. It has been found that

saliva protein recruits different cells, i.e., neutrophils, eosinophils, etc., within which the pathogens grow. The anti-saliva antibody can neutralize the saliva, and recruitment of those specific cells does not occur and the pathogens cannot be able to grow. Not only the pathogens, the anti-saliva antibodies may cause reduction of fecundities of mosquitoes. These mosquitoes with reduced fecundity results in reduction of mosquito population.

Another one anti-mosquito vaccination is use of peritrophic membrane as vaccine candidate. Peritrophic membrane lines the gut of many arthropods. This membrane separates ingested food from the gut epithelial layer. This membrane performs characteristic role of creating partition of the gut lumen into different regions, and this membrane limits the macromolecules and other particles due to its permeability property. The anti-peritrophic membrane antibody perhaps affects the peritrophic membrane by increasing its permeability, thus causing hamper of food utilization.

The midgut-specific protein antigens have been targeted as vaccine candidate by several workers, because the insect-transmitted pathogens interact with vector-midgut for infection and development. That's why the molecules associated with midgut epithelium of mosquito may be potential target.

Suborder: Nematocera—Family: Ceratopogonidae

Family Synopsis

The members of this family are tiny flies. They are commonly called biting midges. The flies are morphologically characterized by piercing or sucking type of mouthparts, presence of very short piercing type of proboscis, long antennae having 13–15 segments, and hairy body without any scale, etc. The life cycle is direct type. The developmental stages are egg, larva, pupa, and adult. They are highly important as they transmit important pathogens in different animals. The control of these flies is mostly based on use of insecticidal agents.

Genus: *Culicoides*

Morphology

Color: The color is gray.

Shape: No typical shape is found.

Size: The flies are about 3 mm long. These are the smallest biting flies of the world.

Head: The head is small.

Proboscis, mouth parts, eyes etc.: The flies have small hanging proboscis. The eyes are large in size and contiguous. The antenna of the flies holds the Johnston's organ. The mouthparts are adapted for bloodsucking.

Antenna: The antenna is long and slender. The antenna has 15 segments. Like mosquitoes, the antenna is plumose in male and pilose in female.

Arista: The arista is absent.

Thorax: The thorax is typically humped over the head.

Wings: The wings are dappled and have long profuse hairs. The wings have spotted areas which are important identifying character. The pigment pattern and hairs are used for species identification.

Legs: The legs are short.

Egg

The fresh eggs are cigar, banana, or sausage shaped. Initially they are white in color. But later, the eggs turn to brown or black.

Larva

The hatched out larvae are cylindrical and maggot-like. The larvae are 2–5 mm long. The larva has small head, three thoracic and nine abdominal segments.

Pupa

The pupa is obtectate type. The whole body bears spine and tubercles. The pupae are yellow or brown in color. The pupae anchor themselves with rotten vegetations, manure, and other substrate using two spines present in the last abdominal segment. They possess two long respiratory trumpets.

Life Cycle

The flies prefer the semiaquatic areas. Both male and female are nectar feeder. But only the females suck blood from the hosts. They attack cattle, horse, etc. The dogs may also be affected. The sucked blood is useful for reproductive purposes. All four stages are found during their life cycle. The stages are egg, larva, pupa, and adult. The females lay their eggs in moist areas where it is continuously splashed with water.

These are the following events occurring in life cycle of flies.

Egg Laying

The adult females lay their eggs in moist soil in batches which may be up to seven batches. The eggs cannot survive desiccation. Males feed on nectar. The females take nectar but also suck blood from the hosts. Some members are autogenous. That means that they can produce first batch of eggs without blood. The reserved food stored during larval stage is used during the first gonotrophic cycle. But some are anautogenous.

Hatching of Eggs

Egg hatching period is widely variable depending on relative humidity, temperature, etc. It takes about 2–10 days for hatching.

Larval Development

The hatched out larvae are cylindrical and maggot-like. After hatching, the minute tiny larvae come out. The larvae eat bacteria, algae, and fungi and take nutrients from them. They pass on four larval stages. The larvae are 2–5 mm long. They cannot survive desiccation. Moisture is essential for survival of the larvae. The mangroves are water holding plants. This is a very suitable place for life of the larvae. The mud, debris of water edges, and rotten fruits are the good habitat for the organisms.

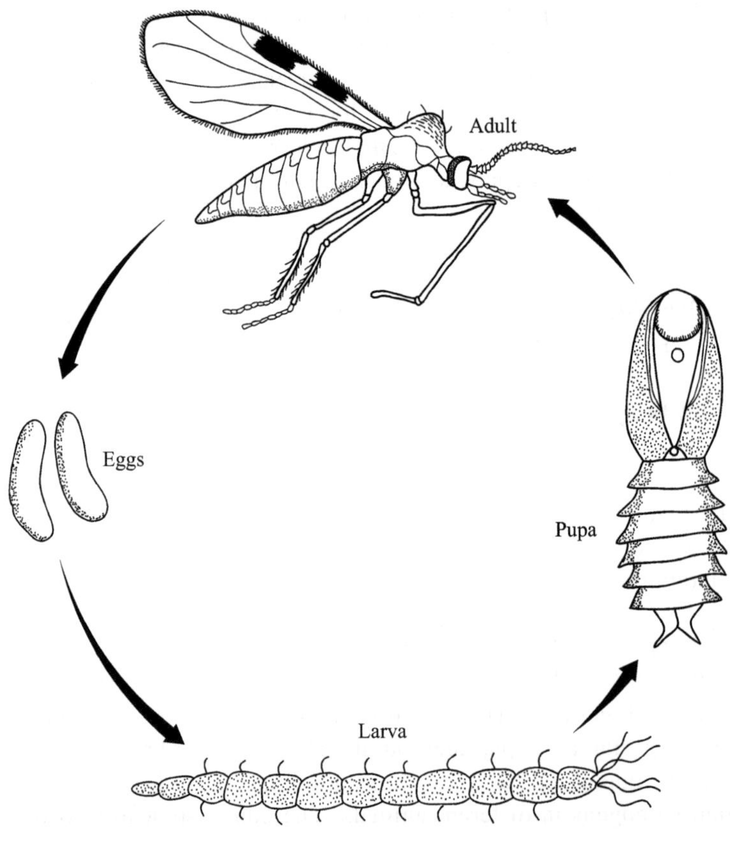

Fig. 7.16 Life cycle of *Culicoides* fly

Pupa Formation

Pupa is formed from the larval stage. Compared to other stages, the pupal stage is short which is 2–3 days. The adult flies come out from the pupa. The color of pupa is variable which may be yellow or brown in color. The pupae have broadly two parts, a cephalothorax and abdomen. The body is spiny. A pair of respiratory horns are found.

Adults Release

The adult is formed within pupa. It is interesting that the adult male flies come out from the pupa first. The female adults emerge later. The male gets ready to mate the female immediately after emergence of pupa. The time taken for development from egg to adult is more or less 1–2 months. The male mates the female in swarms. It is observed that mating occurs after the females finish their blood meal.

Habitat

The flies are not found active in the daytime. They are found in the dusk or dawn during dim light. The females suck blood from different animals. They also attack humans. The good habitat of *Culicoides* flies is highly salt marsh areas wherein large population of flies are found in this area. The members fly a short distance. Mating occurs when the flies are in swarm.

Importance

Mechanical Damage

These are crepuscular flies and are active during early morning and evening. However, cloudy weather is also preferred by these flies. During this time, they attack humans and animals. The fishermen, gardeners, and hunters are affected. Since the flies are very small in size, they are not easily recognized by humans when they bite. Only the affected person can feel that something is biting. The biting is painful. Bites of *Culicoides robertsi* cause marked allergic reaction in horse in Australia. This is called 'Queensland itch'.

Disease Transmission

Virus

Bluetongue virus, Akabane virus, African horse sickness virus, bovine ephemeral virus, etc. are transmitted by these flies.

Protozoa

Leucocytozoon caulleryi are transmitted by these flies. The flies also transmit *Haemoproteus meleagridis* in turkey and *Hepatocystis* sp. in monkey.

Helminth

The flies transmit different helminthic parasites, i.e., *Onchocerca gibsoni*, *O. cervicalis*, *Dipetalonema perstans*, *D. streptocerca*, etc.

Control

Habitat Modification

Diking and drainage of marshlands can be done to reduce the habitats used by the larvae. Control of larvae is very difficult because of extensive area which is inhabited by the larvae. Change of water channel can modify the habitat.

Insecticides

Application of insecticidal agents is not effective for controlling adults. Aircraft application of insecticidal agents can be done to cover large area for controlling larvae. Insecticides can be spread in the animal's house and its surrounding places.

Repellants

Some repellants, i.e., DEET (*N,N*-diethyl-meta-toluamide), can be applied for control of mosquitoes.

Use of Net, Screen

Proper screen should be used so that the biting midges cannot enter the houses. Smaller mesh size (less than 16-mesh) should be used because the biting midges are very small in size. Use of net is effective. The no-see-ums can easily pass through the mosquito nets. Therefore, care should be taken about the mesh size of net.

Use of Trap

Different types of traps are used for attracting the biting midges using different baits, i.e., CO_2. The flies are trapped into an insecticide-treated target.

Suborder: Nematocera—Family: Simuliidae

Family Synopsis

This family contains about 12 genera; of them *Simulium* is the most important genus. The members are called blackflies or buffalo gnats. The flies are morphologically characterized by black body, hornlike antennae and short piercing proboscis. The life cycle is direct type. The developmental stages are egg, larva, pupa, and adult. They are important as they transmit filarial worms. The control of these flies is mostly based on use of insecticidal agents.

Common name: buffalo gnats, turkey gnat, blackflies

Genus: *Simulium*

Morphology

Adult

Color: The flies are black in color. The thorax is shiny gray. Though the flies are called blackflies, the color may be yellow, tan, etc.

Size: The flies are small in size (1–4 mm).

Body: The body is divided into three parts, i.e., head, thorax, and abdomen. The whole body is covered with golden or silvery short hairs.

Head: The head is roughly suboval.

Antennae: The antennae are 11 segmented.

Mouth: They have serrated stylets along with tooth structures.

Eyes: The eyes are round and dark.

Thorax: The thorax is humped which is markedly visible at first glance. The size of hump is variable. The hump is large sized in male than female.

Abdomen: The abdomen has nine segments. The genitalia is situated in the last segment.

Wings: The wings are broad without any hairs or scales. However, some bristles are present on the wings.

Egg

The eggs are triangular or oval in shape. They are laid in mass and attached with a sticky substance.

Larva

The larvae are yellowish green in color. The larvae possess a pseudo-leg called proleg. At the end of the proleg, a circle of hooks are present. The anterior end of the larvae possesses a pair of brushlike structure.

Pupa

This is a triangular cocoon within which the imago remains. One dorsal and one ventral respiratory tubes are present.

Life Cycle

Egg Laying

The flies lay eggs underside the leaves, aquatic vegetation, etc. in the running water just below the water surface. Eggs are laid by the female in masses containing 200–500 triangular eggs. The flies insert their ovipositors in the water and glue the eggs on solid objects under the running water.

Egg Hatching

The larvae come out from the eggs after hatching. Hatching depends on environmental temperature. The duration may range from few days to 1 month.

Larval Development

The larvae remain attached on rocks, logs, etc. They eat bacteria, fungi, etc. They have no wings. They have leglike appendages called proleg. The proleg is armed with a circle of hooklets. They take food materials in aquatic zone. There several larval stages. They pass six or more larval stages. The larval period is short up to about 7 days.

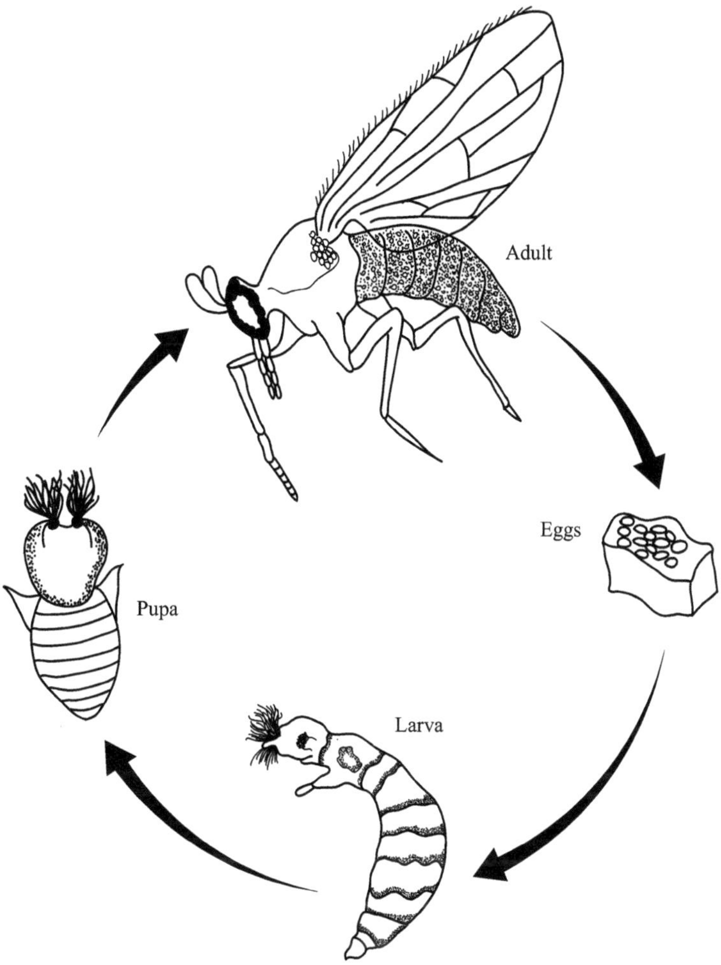

Fig. 7.17 Life cycle of *Simulium* fly

Pupa Formation

The pupal stage is formed by a silken cocoon attached to aquatic vegetation. After complete maturity, the adult comes out from the pupa. Immediately after emergence, they mate.

Adult release

After complete maturity, the adult releases from the pupa.

Habitat

The flies are found both in tropical and temperate countries. The flies are found at the area where there is running water, i.e., stream, river, etc. They prefer clean water and refuse contaminated water. It is notable that oxygenated water is essential for adult and larval development. That's why they prefer running water. One more thing should be noted that slow running water is preferred to restrain their body because they cannot restrain their body in fast running water. The flies are strong fliers. They can fly several kilometers from their dwelling place (breeding site) of running water. Both male and female flies are nectar feeders. The male has simple stylets. But females are blood feeders because blood meal is required for egg maturation. That's why they are called anautogenous. The flies cut the skin and blood is pooled there and the flies accomplish blood meal. The flies use anticoagulant for preventing the clotting of blood. They attack different types of animals. Cattle, sheep, goat, and birds are affected by these flies. The flies follow the hosts by the host's cues comprising of color, smell, heat, etc. They also attack wild animals. They attack man also.

Control

Habitat Modification

The possibility of habitat modification is very less. The female *Simulium* lays eggs in the running water stream. Attention can be paid to the running water stream. The direction of water channel could be changed so that this running water stream does not become available near the animal farms.

Insecticides

For Adults

Application of insecticidal agents for controlling adults is very difficult because the adults are strong fliers and can fly for few kilometers. That's why the program of application of insecticidal agents for specific area is not possible which is possible for controlling the mosquitoes. Different "pour on" and "spray formulation" of insecticides can be applied on the animals so that the flies do not attack the animals.

For Larvae

Insecticide application from aircraft directed to the rivers is done with regular intervals for controlling larvae. However, practical implication is very difficult and expensive. This control program was applied in different places for controlling onchocerciasis.

Repellants

Application of repellant on the clothing can be done because the *Simulium* flies have tendency to get into the clothing before biting.

Use of Screen

Use of medicated screen is useful.

Use of Trap

Different types of traps are used for attracting the flies using olfactory and visual cues. Yeast and sugar are used for production of CO_2, and this CO_2 acts as a bait to attract the flies. The flies are trapped to an insecticide-treated target. Use of "host decoy trap" is also used for controlling *Simulium* flies.

Biological Control

Use of *Bacillus thuringiensis* var *israelensis* is used for control of *Simulium* flies.

Season Specific Management

The animals should be kept indoor during peak time of *Simulium* population.

Sterile Insect Technique

Sterile insect technique can be applied for production of sterile male *Simulium*. By this technique, sterile flies are produced by irradiation or chemical. This technique does not have any residual effect. The modified organisms do not produce any poisonous or toxic material.

Clothings

Since humans are also attacked by these flies, the garment should cover the whole body. Light colored clothings are less attractive to the flies.

Importance

Direct Damage

The female flies are strong blood feeders and can visit hosts after covering a distance of several miles. In some cases, they can march up to 40 km. Their bite is very much painful. The itching may persist for a few days. A severe allergic reaction may occur. The allergic reaction may be so severe as to cause death of the animals. Different important pathogens are transmitted by these flies. These are the following:

Indirect Harm

Helminth

The flies transmit different *Onchocerca* spp. *Simulium damnosum* and *S. neavi* transmit *Onchocerca volvulus*.

Protozoa

Different *Leucocytozoon* spp. are transmitted by *Simulium* flies. *Leucocytozoon simondi* is transmitted by *Simulium rugglesi, S. venustum*. A non-pathogenic *Trypanosoma* is also transmitted by *Simulium* fly.

Virus

The *Simulium* flies transmit vesicular stomatitis in cattle, horse, etc. This disease is characterized by ulceration of oral mucosa. *S. notatum* and *S. venustum* transmit this virus.

Immunological Reaction

During biting, the flies introduce some immunogenic material to the host which elicits allergic and anaphylactic reaction. *S. arcticum, S. ornatum* cause toxemia in the hosts called simulitoxicosis.

Suborder: Nematocera—Family: Psychodidae

Family Synopsis

The members of the family are called sand flies. The flies are morphologically characterized by small size, hairy body, long antennae having 16 segments, short and piercing proboscis, and hairy wings without any scales. The life cycle is direct type. The developmental stages are egg, larva, pupa, and adult. They are extremely important as they transmit *Leishmania* spp. causative agent of visceral leishmaniasis (kala-azar) and cutaneous leishmaniasis. The control of these flies is mostly based on use of insecticidal agents.

Genus: *Phlebotomus*

Common names of the flies are **owl midge**, **sand fly**. The species under this genus are *P. argentipes, P. sergenti, P. papatasi, P. longipalpalis, P. orientalis etc.*

Identification

Color: The flies are yellow or yellowish brown in color.

Size: The flies are small in size (1.5–3 mm).

Body: The body of the phlebotomine flies is more fragile. The body is markedly elongated.

Mouthparts and antenna: There are five subfamilies of this family. Of these, the subfamily Phlebotominae has piercing type of mouthparts. The antenna has 16 segments.

Eyes: The eyes are large and black.

Thorax: The thorax is humped. The whole body has profuse hairs.

Legs: The legs are long and stilt-like.

Wings: The wings are hairy and lanceolate-shaped. During rest, the wings extend at 40° angle over the body.

Clasper: A long genital terminalis is found in male which is called claspers.

Egg: The eggs are simple. They are white in color at the time of laying, but turn to dark later.

Larva: The larvae are elongated having a cephalothorax and an abdomen. The head is typically black. The body has 12 segments. Many branched bristles are found in the head and body.

Life Cycle

Mating: Mating of flies occurs soon after the emergence. Males capture the female either at the resting site or on the hosts.

Egg Laying

After emergence, the adult flies soon mate. Blood meal is required for maturation of eggs. A typical egg laying habit is found in *Phlebotomus* flies. The eggs are laid in cracks and crevices so that the eggs are not dried up because the eggs are sensitive to drying. Eggs are laid in different batches each containing a range of 30–70 eggs.

Egg Hatching

Within 2 weeks or more, the eggs hatch and the larvae come out. However, the time of hatching depends on the environmental temperature.

Larval Development

The larvae are elongated having a cephalothorax and an abdomen. The head is typically black. The body has 12 segments. Many branched bristles are found in the head and body. Four larval instars are found during their larval period. They feed on fungi, decaying vegetations, decomposing arthro-

Fig. 7.18 Life cycle of *Phlebotomus* fly

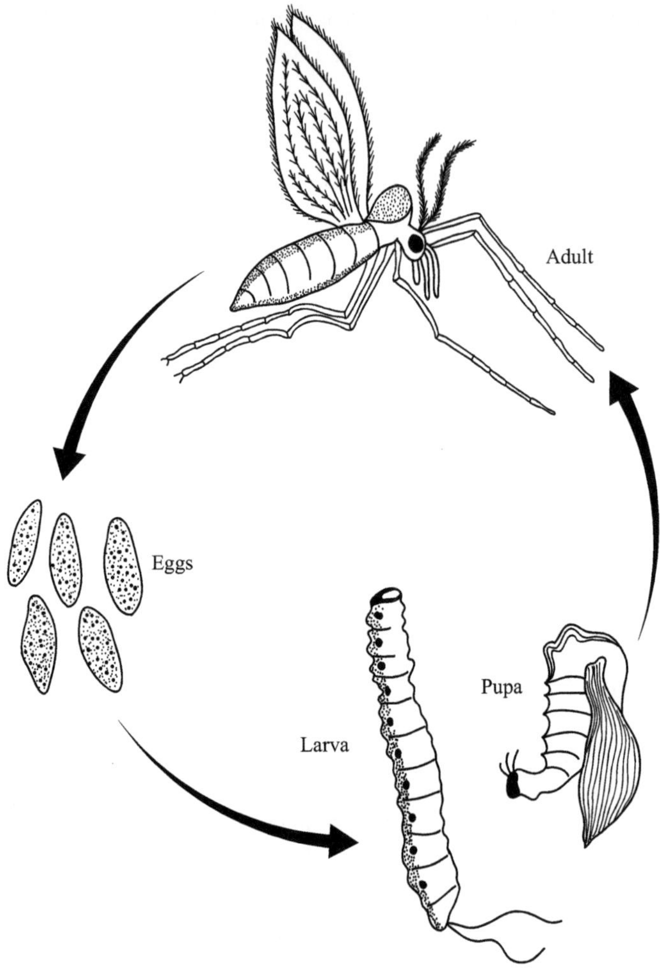

pods, animal feces, etc. The larvae may undergo diapause due to unfavorable hot and cold climate. There are four larval stages which last for about 20–30 days.

Pupa Formation
The next stage is pupa. The pupal stage lasts for 1–2 weeks.

Adult Release
After complete maturity, the adults are released from the pupa.

Habitat
The flies are found in different parts of the world, i.e., tropical, subtropical, temperate, etc. But they are most abundantly found in the tropical countries. The flies prefer humid climate. They are weak fliers and fly 100 m or more. The flies are found in rural, semi-urban, and urban places. They remain in cracks, crevices, tree holes, and rodent burrow where moisture is retained. In these areas, the females lay their eggs because the eggs are sensitive to dessication. The flies attack a wide range of animals. Humans are also affected by the flies. Both male and female take sugary material, fruit juices, etc. Female flies of some species are also blood feeders. They have piercing mouthparts. By this mouthparts, they pierce the skin of the hosts. A pool of blood comes out which is taken up by the flies. The feeding depends on temperature and air current. The force of air current is important because the flies are so small that strong airflow can displace

them from the feeding site. Dusk and night are the usual time of blood feeding. Daytime feeding is found in indoor place. The male flies capture the female flies during rest or on the host and then mate.

Seasonality and Dispersals

The population of sand flies depends on local climate and seasonal changes. The flies are abundantly found during the summer season. The population is the lowest in the winter season. In tropical countries, some species are commonly found throughout the year. Adult flies are weak fliers and can travel only a few hundred meters. Most flies fly horizontally over the ground.

Host Preference

The flies are opportunistic because they affect the hosts to which they are accessible. Female flies feed on a wide variety of animals, i.e., cattle, sheep, goat, dogs. They also affect humans and rodents.

Importance

Protozoa

Phlebotomus flies transmit different *Leishmania* spp. *Phlebotomus argentipes* transmits *Leishmania donovani*. *P. papatasi* transmits *L. major*.

Pathogen	Species of sand flies
a. Leishmania infantum	*a. P. balcanicus, P. halepensis, P. kandelakii*
b. L. donovani	*b. P. argentipes*
c. l. major	*c. P. papatasi*

Virus

P. duboscqi transmits Rift Valley fever.

Control

Habitat Modification

The possibility of habitat modification is very less. Elimination of breeding or resting sites by removal of organic waste material, wood piles, and unwanted vegetations, sealing sewage drain, and plastering holes of walls of buildings can be done for modification of habitat.

Insecticides

For Adults

Application of insecticidal agents for controlling adults can be done. Insecticides are sprayed on the resting sites of the flies. Pyrethroids and other effective insecticidal agents are applied on the vegetations and water collection drainage system.

For Larvae

Insecticide application from aircraft directed to a large breeding area can be done.

Repellants

Use of different repellants has been advised.

Use of Screen

Use of medicated screen is useful.

Use of Trap

Different types of traps are used for attracting the flies using olfactory and visual cues. The traps are kept surrounding the animal house. Yeast and sugar are used for production of CO_2 and this CO_2 acts as a bait to attract the flies. The flies are trapped to an insecticide-treated target. "Host decoy trap" is also used for controlling *Phlebotomus* flies.

Season Specific Management

The animals should be kept indoor during peak time of *Phlebotomus* population.

Sterile Insect Technique

Sterile insect technique can be applied for production of sterile male *Phlebotomus*. By this technique, sterile flies are produced by irradiation or chemical. This technique does not have any residual effect. The modified organisms do not produce any poisonous or toxic material.

Suborder: Cyclorrhapha—Series: Aschiza

They have no veterinary importance.

Series: Schizophora—Section: Acalyptratae—Family: Gasterophilidae

Family Synopsis

The important genus of this family is *Gasterophilus*. Different species of this genus cause intestinal myiasis. The flies are hairy and the mouthparts are functionless. The life cycle is direct type. The developmental stages are egg, larva, pupa, and adult. The larvae of some species of this genus occur in equines called bots.

Genus: *Gasterophilus*

The adults have nonfunctional mouthparts. Dark band passes from anterior to posterior border of the wings. The flies are dark in color. They are hairy. Different species are *G. intestinalis*, *G. haemorrhoidalis*, *G. nasalis*, *G. inermis*, *G. pecorum.*

Life Cycle

Egg Laying

Gasterophilus intestinalis lays their eggs on the hairs of the face, fetlock, forelegs, weithers, etc. *G. nasalis* lays eggs on the muzzle, intermandibular region, etc.

Egg Ingestion

The eggs are ingested due to the activity of licking and grooming. The mechanical activity and heat produced due to licking act as stimuli for hatching of eggs for some species.

Larvae Penetration

The released larvae penetrate the buccal region, tongue, etc. The larvae penetrate the buccal mucosa or tongue. They wander the tissues for several weeks. The larvae then reach the pharynx and esophagus and eventually enter the stomach. The larvae of some species preferably attach to cardiac part of the stomach, and rarely the fundus and pylorus region and gain maturity there.

Passing of Larvae

The larvae are present in the stomach being attached to the gastric or duodenal wall and remain there for about a year. The larvae get detached and pass out with the feces after gaining maturity in the spring and early summer. *Gasterophilus haemorrhoidalis* larvae may reattach the gastric wall further.

Pupa Formation

The larvae are passed out with feces. Later, pupa formation occurs.

Adults Release

The adults are liberated from the pupa after about a month or more.

Fig. 7.19 Larva of *Gasterophilus*

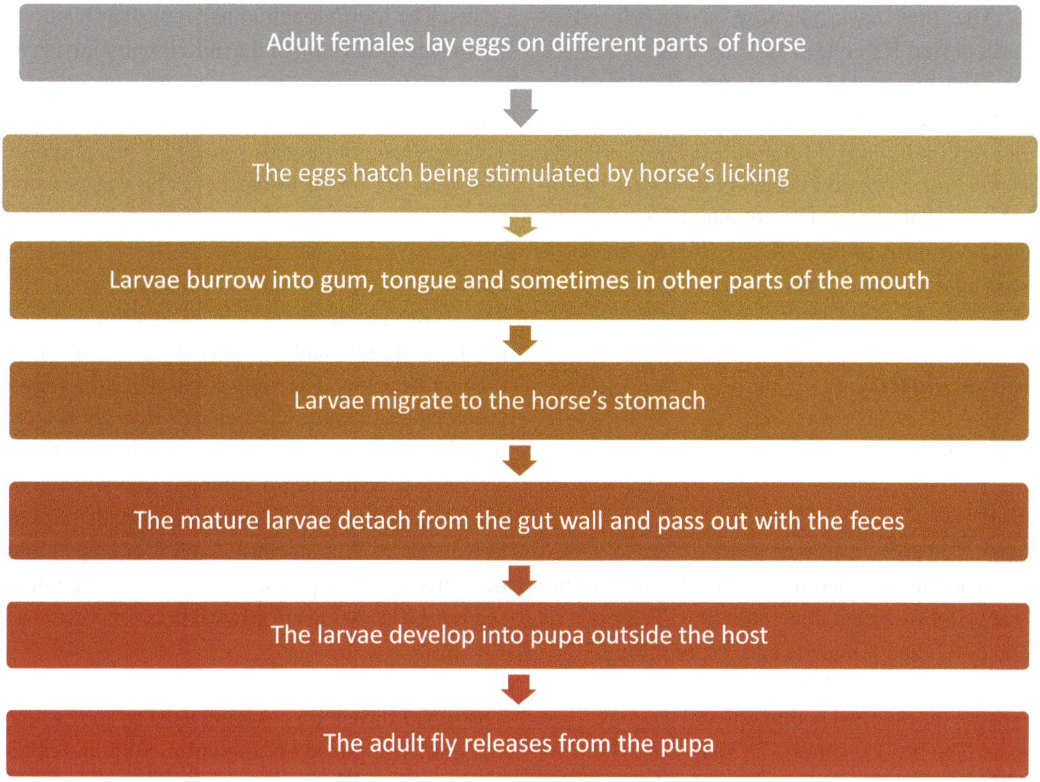

Fig. 7.20 Life cycle of *Gasterophilus intestinalis* in flow chart

Importance

Annoyance

The adults create great annoyance to the animals while laying eggs, particularly by the flies which lay eggs on the muzzle or other space of the head.

Ulceration of Tongue

The larvae penetrate the mucosa of gum, tongue, and other parts of the mouth. Stomatitis with ulceration of the tongue occurs.

Ulceration of the Stomach

The larvae initiate an inflammatory reaction. The affected area may be ulcerated later forming a typical funnel-shaped lesion. The ulcer may extend up to duodenum causing duodenal ulcer. There may be perforation of the stomach wall.

Control

- Old treatment consists of introduction of carbon disulfide by the stomach tube at a rate of 2–5 ml per Kg body weight.
- If eggs are detected on the animals, sponging with hot water mixed with insecticide is done. The warmth of the water initiates hatching of the eggs.
- Since the larvae are voided with the feces, the feces should be disposed properly.

Section: Calypterae—Family: Muscidae

Family Synopsis

This family includes both biting and nonbiting genera. The important genera of veterinary importance are *Musca*, *Stomoxys*, *Hydrotaea*, etc. The flies have plumose arista, closed or nearly closed R5 cell, distal curvature of M1 + 2 vein. The life cycle is direct type. The developmental stages are egg, larva, pupa, and adult. They are extremely important as they transmit *Leishmania* spp. causative agent of visceral leishmaniasis (kala-azar) and cutaneous leishmaniasis. The control of these flies is mostly based on use of insecticidal agents.

Genus: *Musca*

Morphology

Adult

Color: The color is gray.

Size: The flies are medium sized. The female flies are slightly larger than male flies. The male fly is 6–6.5 and female is 6.5–7.5 mm long.

Body: The body is divided into three parts, i.e., head, thorax, and abdomen.

Head: The head is suboval in shape.

Mouth: Musca flies do not suck blood. They have lapping type of mouthparts. The mouthparts are adopted for lapping or imbibing liquid food. The mouthparts are spongelike. The mouthparts contain a pair of labella on which grooves are present. The labella are attached with labium under which a number of transverse grooves are present and these transverse grooves serve as the food channels. The *Musca* flies take liquid food material.

Eyes: The head bears a pair of prominent eyes.

Antennae: A pair of antennae are present which are small in size. The arista is present on the antenna which is plumose both dorsally and ventrally.

Thorax: The thorax is easily identifiable having four dark longitudinal stripes which extend to the posterior border of the scutum.

Abdomen: The abdomen is gray or yellowish colored having a median line, and this is diffused on the fourth segment. These features are frequently used to identify *Musca* fly.

Wings: As that of other dipteran members, *Musca* flies have a pair of translucent wings. The wing's base is yellowish in color. The distal pair of wings have been reduced to halteres or balancers. The fourth longitudinal vein sharply curves and reaches the tip of the third longitudinal vein.

Egg

The eggs are elongated oval and white in color. The eggs are found in batches. The egg is 1 mm long.

Larva

The larvae are apodous pointed anteriorly with a pair of hooks internally connected with cephalopharyngeal skeleton. The cephalopharyngeal skeleton is composed of kitin, sclerotin etc. They are posteriorly flat bearing stigmal plates. A pair of spiracles are present at the last abdominal segment (metapneustic) in first larval stage, whereas the second larval stage is amphipneustic and is equipped with the spiracles on the prothorax and on one abdominal segment. The larvae are whitish in color.

Pupa

The pupae have hard covering used by the last larval skin. The pupae are brown in color and about 8 mm long. They are characteristically barrel shaped. The two spinelike spiracular processes are found in between the fourth and fifth segment.

Fig. 7.21 *Musca* fly

Life Cycle

Mating

A polygynous type of mating system is found. A male can mate many females but the females require only one male for mating. The characteristic feature is that the female participates in copulation only once and the remaining sperms are kept stored for later.

They are holometabolous having four developmental stages during their life cycle. The stages are egg, larva, pupa, and adult. A warm climate is the most suitable for biological development. *Musca* flies display high reproductive potential compared to other flies. In a season, they can perform several generations of flies to enhance the population. The duration of life cycle is also very less.

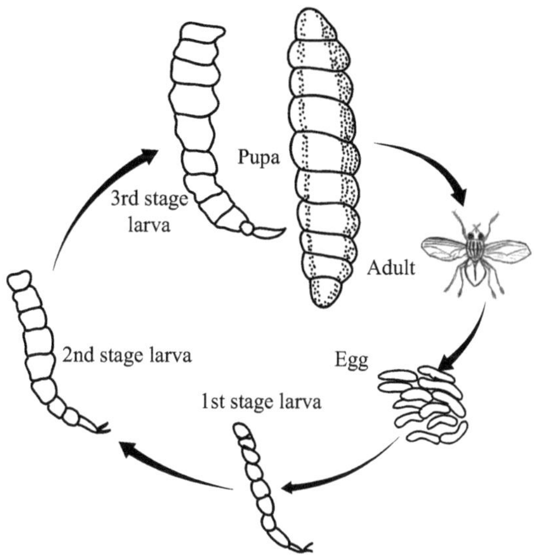

Fig. 7.22 Life cycle of *Musca* fly

Egg Laying

The eggs are laid in batches having 100 eggs in a batch. The eggs are laid in decaying vegetation, carrion, animal feces, etc.

Hatching

The duration of hatching may be prolonged in low temperature, which hatches within 12 h or so.

Larval Development

After hatching, the larvae come out. In fact, the female flies lay their eggs in nutrient-rich area so that the larvae can get sufficient nutrient. These are decaying vegetations or other food substances contaminated with feces and animal excreta. The larvae pass on four larval stages after three ecdyses.

Pupa Formation

The fully matured larvae reach the dry place for pupation. The pupae are reddish brown in color.

Adult Emergence

Thee adult flies come out from the pupa. The cycle takes 7–10 days (from egg to adult).

Habitat

They are found both in tropical and temperate countries. But they are most abundantly found in warm climate. Extreme heat and cold is unsuitable for them. The population is high in household areas where food, sugar, and waste materials are easily available. They are found in the market, vegetable shop, fruit shop, sweet shop, etc.

Pathogens Transmitted by *Musca* Flies

Virus	Polio virus
Bacteria	Cholera, typhoid, *Salmonella*, *Shigella*, *Escherichia coli*, bacillary dysentery causing bacteria, tuberculosis bacteria, anthrax causing bacteria
Helminthes	• **Roundworms** *Habronema muscae*, *Draschia megastoma* • **Tapeworms** *Raillietina echinobothrida*, *R. tetragona*, *Choanotaenia infundibulum*

Control

Cultural Control

Cultural control includes modification of habitats. These are as follows:

(a) Excreta, stall litter, residual feed, and fodder should be properly disposed. The feces should be disposed in a particular pit.
(b) Any sugary material or other food material spread smell which attracts the flies. The food material should be kept covered in proper place so that the food is not contaminated with the pathogens transported by the flies.
(c) The waste material should be disposed in the garbage can. The garbage can should be washed routinely.

Household Control

(a) The windows should be screened properly to prevent the entry of the flies in the room.
(b) The floor of the room should be cleaned properly.
(c) Proper sanitation and aeration are essential both for human and animal house.
(d) Automated closing of door is useful to reduce flies in house.

Use of Traps

Different light traps and baited traps are very much useful. The light traps should be kept in dark areas where other light source is not available. Traps baited with sugary material are very much useful.

Chemical Control

(a) *Insect growth regulator*: Different insect growth regulators are used at the areas where the insect population is high. The insect growth regulator inhibits the growth of immature stages of insects.
(b) *Insecticides*: Different insecticidal agents are sprayed at the fly-populated areas. Residual spray of insecticides in the animal house should be done. Application of insecticidal agents can be done in the form of "pour on" and "spot-on" preparation.
(c) *Ear tags:* Ear tags impregnated with insecticidal agents can be used. But this is mostly useful in case of horn fly.

Biological Control

Different ants and wasps are biocontrol agents which are effective up to certain level. In some countries, the immature stages of insects are commercially available. Several mite species are potent predators of eggs and larvae of houseflies (egg and larva stages). A mite species, *Macrocheles muscaedomesticae*, can eat eggs of ten houseflies per day. Some predatory flies (*Hydrotaea*) are used for indoor control of houseflies. These are commercially available in some northern European countries and used widely for control of houseflies. Different dung beetles of the family Scarabaeidae eat eggs and larvae of *Musca* spp.

Genus: *Stomoxys*

Morphology

Adult

Color: These flies are brownish gray in color.

Size: They are medium sized flies, slightly smaller than *Musca* fly.

Mouth: They look like *Musca* flies. But some features are different. The mouthparts are sucking type and look bayonet like by which they pierce the skin, whereas the *Musca* flies have sponging and lapping type of mouthparts. Mouthparts contain bulbous labium with small oval labella, hairy swordlike labrum epipharynx, and fine needlelike hypopharynx.

Proboscis: A proboscis is present which is stout and projected anteriorly.

Antennae: A pair of antennae are present. The arista is plumose only dorsally.

Thorax: The thorax is gray in color and has four dark longitudinal stripes. The lateral two stripes do not reach the end of the scutum.

Abdomen: The abdomen of these flies have seven distinct spots arranged in checkerboard pattern, whereas these spots are absent in *Musca* flies.

Wings: The wing veination is characteristic. The third and fourth longitudinal veins are open which is not as that of *Musca* flies.

Egg

The eggs are sausage shaped and white in color measuring about 1 mm in length. The eggs are typical in shape having one side straight with a longitudinal groove and one side is smooth and curved. The eggs are dirty -white to yellow in color.

Larva

Larvae of different instars vary in size and color. The first instar larvae are small and translucent. The larvae gradually increase in size. The third instar larvae are white or pale yellow in color and much larger than first instar larvae. They have a pair of oral hooks and posteriorly they have a pair of spiracles. The larvae are apodous but can vigorously wriggle assisted by the spines present on the body. The posterior spiracles are roughly triangular in shape and possess three S-shaped slits.

Pupa

The last larval skin forms the outer covering which later forms a hard puparium. The pupae are 4.5–6.0 mm. They are capsule-like and reddish brown in color.

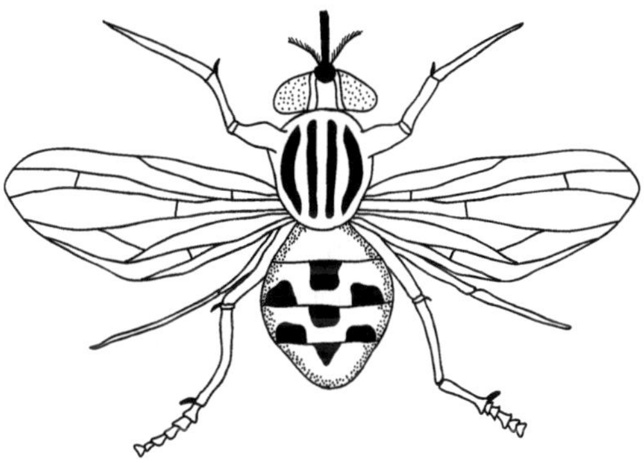

Fig. 7.23 *Stomoxys* sp.

Life Cycle

Mating

Mating occurs after 3–5 days of emergence. The flies start laying eggs after 3–4 days of mating. Both the flies suck blood from different hosts. The blood meal is required for maturation of eggs.

Egg Laying

The flies breed in moist decomposed organic matter. During the lifetime, the flies lay different batches of eggs. Every clutch of egg laying pertains to separate blood meal. Every batch contains 60–130 eggs. A total of about 700–800 eggs are laid by one female fly. Females lay eggs three times throughout its lifetime.

Egg Hatching

Eggs hatch within 2–3 days. Due to extreme environmental situation, hatching period may be prolonged.

Larval Development

Putrefying organic matter is chosen by the females for egg laying. These are manure, hay, etc. After hatching, the larvae come out. The larvae feed on organic substances. The eggs hatch within a day and first stage larvae come out. The larvae feed on the decaying vegetation and develop. The larvae pass on three larval stages. Larval stage lasts for 2–3 weeks.

Pupa Formation

After complete maturity, a cocoon is formed within which the instar remains. The pupa is coarctate type. The last larval skin of larva forms the hard puparium. The pupa is chestnut brown colored. They are about 5–7 mm long. Pupal stage lasts for 1 week or more depending on environmental temperature.

Adult Emergence

The completely matured adults come out from the pupa. They cannot fly immediately. Within 1 h, they start flying.

Importance

Direct Damage

They are obligate bloodsucking flies. Cattle, sheep, goat, horse, etc. are primary hosts affected by the flies. But they also attack humans and dogs during unavailability of these hosts. They are abundantly found in summer or winter season. But usually the flies are available throughout the year. The flies congregate at the leg regions, belly part, etc. Humans are also affected at their leg and knee region. The animals are highly disturbed by constant bloodsucking activity of this fly. The animals try to avoid this disturbance by stamping, shaking the body, shaking the tail, etc. Interestingly the bite is not allergic and painful in human beings as is caused by others. Notable damage occurs if the flies suck blood from the prominent blood vessels of ears. Frank lesions are found at the affected areas. A typical dipping posture is adopted by the animals to skip the attack by the flies. The animals dip their bodies in water keeping the head out which reduces feeding time. Body weight and milk production are reduced.

Indirect Harm

Type of pathogen	Name of species/genera
Virus	Virus of African swine fever West Nile virus Rift Valley virus Virus of equine infectious anemia Capripox virus
Rickettsia	*Anaplasma* spp., *Rickettsia* spp.
Protozoa	*Trypanosoma* spp. (*T. brucei*, *T. evansi*), *Besnoitia* spp.
Helminths	*Dirofilaria* sp. *Habronema* sp. *Hymenolepis carioca*

Habits

The flies are diurnal in feeding habit. They attack during early morning and afternoon, and in cooler climate, they are active during midday. For full engorgement, they do not take much time. They are rapid bloodsuckers and quickly suck blood to get fully engorged within few minutes. They take rest under shade of tree and other objects after complete feeding as found in some other bloodsucking flies. Cattle, sheep, goat, horse, etc. are primary hosts affected by the flies. But they also attack humans and dogs during unavailability of these hosts. They are abundantly found in summer or winter season. But usually the flies are available throughout the year. The stable flies attack different animals, i.e., cattle, sheep, goat, pig, horse, etc. The flies congregate at the leg regions, belly part, etc. Humans are also affected at their leg and knee region.

Control

Control is mostly similar to *Musca* flies. See control of *Musca* fly.

Genus: *Haematobia*

Hosts affected: cattle, buffalo, sheep, horse

Morphology

The flies are small in size, almost half of the size of *Musca* flies. They have palp, proboscis, and arista. The palps are as long as proboscis. The arista is plumose dorsally. Two dark longitudinal stripes are present on the thorax.

Life Cycle

- Eggs are laid by the female flies in the feces of cattle and buffalo.
- Hatching of eggs occur within 24 h.
- The larvae are similar to housefly larvae. The larvae feed on the manure material and grow up for a period of about 4 days.
- Red to reddish brown colored pupa is formed in the heap or soil.
- The adult flies emerge out from the pupa.

Importance

Haematobia exigua (buffalo flies) severely affect buffaloes. They continuously feed on buffaloes. The animals do not feed and drink properly because of constant nuisance created by the flies. The flies transmit *Habronema majus* and *Trypanosoma evansi*. *H. irritans* (horn fly) affect cattle. Horse, sheep, and dogs are occasionally affected. During blood meal, they constantly pierce the skin which causes irritation and lesions. The affected area may emerge as sore which attracts the screw worm flies. These flies transmit one important roundworm, *Stephanofilaria stilesi*.

Control

Traps

Alsynite traps are used for control of the flies.

Repellants

Natural repellents containing citronella are often used on horses.

Biocontrol Agents

Dung beetles of the family Scarabaeidae are regarded as biocontrol agent of *Haematobia exigua*.

Section: Calypterae—Family: Glossinidae

Family Synopsis

This family includes an important genus, *Glossina*. The members of this genus are called tsetse flies. They transmit trypanosomosis in humans and animals. The flies are larviparous. The larvae grow in the uterus of the flies. The control of these flies is mostly based on use of insecticidal agents.

Genus: *Glossina*

Morphology

Adult

Color: The color is dark brown or tan colored.
Size: They are about 1.5 cm long.
Body: The body is tough having hard outer covering.
Body division: The body is divided into three parts, i.e., head, thorax, and abdomen. The thorax has three fused segments.
Head: The head is large and distinct.
Mouthparts: The mouthparts include the labium, labrum epipharynx, and hypopharynx. The mandibles and maxillae are absent.
Proboscis: The proboscis is thin and long. The proboscis is prominent and projected forward and attached with bottom of the head.
Antenna: Antenna is three segmented. The last segment is pea pod shaped. The arista is present on the antenna and has hairs (plumose dorsally).
Thorax: The thorax is large.
Abdomen: The abdomen is not thin but rather it is broad.
Wings: A pair of wings are present. A pair of halteres are also present posterior to the wings. The wings have hatchet shaped discal cell. The hachet shaped cell looks like meat cleaver. The wings remain folded at rest. One wing remains over other.
Legs: Three pairs of medium sized segmented legs are present.

Larva

Third instar larvae are maggot-like.

Pupa

Pupa is small (0.5–1.0 cm). A pair of small dark lobes are present. The outer covering of pupa is hard. This is oblong in shape.

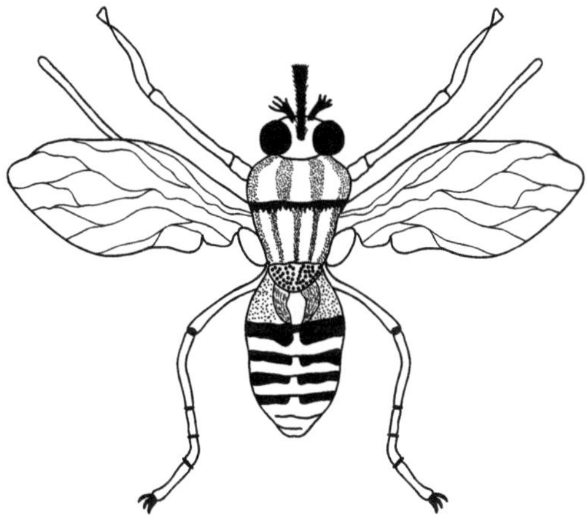

Fig. 7.24 *Glossina* sp.

Life Cycle

Fertilization

The female fertilizes only one egg at a time and retains the fertilized egg in the uterus.

Egg Hatching

Eggs hatch in the uterus itself. The larvae grow in the uterus. The nutrient in the form of uterine milk is provided to the larvae. Up to third larval instar develops in the uterus itself. This is called **adenotrophic** viviparity.

Larval Development

Larva develops in the uterus. Prepupal stages of larvae are laid by the flies. The larva is oval and only 7 mm long. The larvae are negatively phototrophic and wriggle into the soil and start formation of pupa.

Pupal Development

The third larval stage of this fly are laid which are transformed to complete pupal stage within few hours. The larva crawls into the soil and forms a hard puparial case in which the adult stage remains.

Adult Development

Adult flies come out from the pupa.

Habitat

Climate Requirement

For carrying out the complete life cycle of the flies, environmental temperature, humidity, and climate availability of shade are necessarily required. However, minor variation of environmental factors is required by different species. Some species (*G. palpalis*) require almost saturated atmosphere, whereas some others (*G. pallidipes*) are active in moderately dry climate.

Host Finding

The color and smell of the hosts are the cues for finding the hosts. The CO_2 exhaled by the animals attracts the flies. That's why CO_2 is used as bait for controlling many bloodsucking flies. The flies can detect the hosts from long distance because their olfactory system is very strong. They can sense CO_2, acetone, phenol, etc. released by the hosts.

Flight

The flies fly at low height from the ground level. They frequently bite their shadows. The moving objects attract the flies.

Feeding

All species of *Glossina* are bloodsuckers. They suck blood from their suitable hosts. Particular species or group of species have their preference for their hosts. The particular group of flies are found in a particular place if their suitable hosts are available. Some prefer to suck warm blooded animals, whereas some prefer both warm blooded animals and cold blooded animals. The avian species are also affected by these flies. The flies are diurnal in feeding habit. During bloodsucking, some anticlotting factors (anticoagulin) are released by the flies which enhance the flow of the blood, and the flies suck blood easily. An interval of feeding is found in case of these flies. The flies feed every 3 days. After feeding, the flies take rest in the shade of the trees or shade of any object.

Importance

Glossina flies transmit different spp. of *Trypanosoma*, i.e., *Trypanosoma congolense*, *T. brucei*, *T. gambiense*, and *T. rhodesiense*.

Control

Habitat Modification

Since the *Glossina* flies require shaded areas, bush or jungle clearing has significant effect on reduction of *Glossina* flies. But practical implication is difficult because it is labor intensive, time consuming, and highly expensive. The vegetation rapidly comes back again after cleaning and this is a great constraint.

Chemical Control

Insecticide: Application of insecticides is a very important control measure.

Repellants: Application of repellant plays role for controlling *Glossina* flies.

Household management: Medicated screen is useful.

Use of Trap

Different types of traps are used for attracting the flies using olfactory and visual cues. Yeast and sugar are used for production of CO_2, and this CO_2 acts as a bait to attract the flies. The flies are trapped to an insecticide-treated target. "Host decoy trap" can be also used for controlling these flies.

Biological Control

This is an environment-friendly technique. *Bacillus thuringiensis* var *israelensis* can be used for control of *Glossina* flies. Some small flies, i.e., *Mutilla glossinae*, are parasites of *Glossina* flies which can be used for biological control of *Glossina* pupae.

Sterile Insect Technique

This is an environment friendly technique. Sterile insect technique can be applied for production of sterile male *Glossina*. By this technique, sterile flies are produced by irradiation or chemical. This technique does not have any residual effect. The modified organisms do not produce any poisonous or toxic material. In this technique, a large number of flies are produced in large sized insectaries following standard protocol. The newly produced flies are sexually assorted. The male flies are subjected to sterilization. This is done by irradiation or genetic manipulation. However, chemosterilization is another way to produce sterile male. For performing this technique, one should have crucial knowledge about the biology of flies which can be harnessed to develop the perfect protocol. A lot of questions pertain to the development of this technology. The relevant questions are as follows: When are the sterile males be released? How many sterile males are to be released in a particular area? What is the reproductive potential of the flies? Many and many such questions are there. Sterile insect technique is nonintrusive to the environment. It has no adverse effect against the nontarget organisms. Moreover, this is species specific. This technique does not have any residual effect. The modified organisms do not produce any poisonous or toxic material. In totality, this technique is environmentally friendly.

We know that reproductive potential of *Glossina* is less compared to other flies. They do not lay hundreds or thousands of eggs. They only lay one larva in one gonotrophic cycle. In their lifetime, they lay only about 6–12 larvae. Therefore, this technique is not suitable for control of this fly. Study shows that 12,000 sterile flies are required for 1 km^2. Study also shows that about ten sterile males are required for one female fly. It is to be noted that the area should be restricted. For demarcation of a particular zone, geographical information system is utilized. When this technique was first invented, there had many constraints for practical implementation. But day by day, new technology and collating of different technique have made it easy for practical implementation. There may have proven record of eradication and reduction of fly population by using this technique. Many attempts have been successful for controlling agricultural pests. The prerequisite of applying this technique is to reduce the fly population of a particular zone by application of suitable insecticidal agents or by using standard traps. There are certain demerits of this technique. Only one species can be controlled. Multiple species cannot be controlled by using single sterile species. Thus, constraints are related to limited use. Experienced technical persons are required. This technique is very expensive and slow acting. At a time, it is not possible to see the maximum efficacy. This technique is not sustainable. Once the technique has been implemented, it does not continue for long time. Insects are released from jet plane.

Section: Calypterae—Family: Oestridae

Family Synopsis

This family contains many species about 150 in number. The flies are usually hairy. The larval stage is obligatory parasitic. The adults have nonfunctional mouthparts. An important species is *Oestrus ovis* which causes nasal myiasis. This results in loss of production.

Genus: *Oestrus*

Common name: sheep nasal fly, sheep nasal botfly

Morphology

Adult

Color: The flies are dark gray colored. The head and legs are yellowish in color. Small black spots are present on the body which are more prominent in the thorax.

Size: The flies are 12 mm long.

Body: The body is gray colored and scarcely hairy.

Head: The head is nonfunctional and cannot move. Small black pits are present in between the eyes. The flies do not feed during their adult life.

Mouthparts: The mouthparts are rudimentary.

Thorax: The thorax bears tubercles. Light brown hairs are found.

Wings: The wings look glassy.

Legs: The legs are short and yellowish in color.

Larva

The larvae are variable in size. During the developmental process, the larvae increase in size from 1 mm to 20 mm. They are tapering anteriorly and have a flat surface posteriorly. Dark stripes are present on the dorsal aspect of every body segment. Rows of small spines are present on the ventral surface. A pair of hornlike sharp hooks are present at the mouth. The posterior spiracles are found in the cleft of the last larval segment. Every spiracle possesses D-shaped, dark colored stigmal plate.

Pupa

The pupa is coarctated type. The color is dark brown to black.

Fig. 7.25 *Oestrus* sp.

Life Cycle

Fertilization and Egg Hatching

Fertilization and hatching of eggs occur in the fly itself.

Larva Laying

The fly glues the larvae on the mucus material of the nasal cavity. The first larval instars start migrating in the nasal sinus. Second instar is formed after one molting.

Pupa Formation

The larvae come back from the nasal sinus and fall on the ground. They penetrate the soil surface and transform to pupa.

Pathogenesis

Annoyance

The flies try repeatedly to lay larva in the nostril. This creates a great annoyance to the animals. To alleviate the irritation caused, the animals rub their nose on trees, hard objects, or other sheep. They may rub their nose on the ground. These are characteristic symptomatology for guessing *Oestrus* larva infection.

Invasion of Nasal Cavity

The larvae irritate the nasal cavity which results in secretion of viscid mucus material in the nasal passage. The profuse nasal mucus may obstruct the nasal passage, and the affected animals sneeze constantly. This larval infection sets an inflammatory reaction in the nasal wall damaging nasal mucosal layer.

Invasion of Brain

The larvae may proceed to ethmoidal and upper sinuses. The larvae may even erode the skull tissue and reach the brain. This may result in different neurological signs.

Clinical Signs

The clinical signs include vigorous shaking of the head, incoordination of movements, and high stepping gait which resembles the symptoms of GID. That's why this condition is called "false GID."

Dead Larva

The condition becomes severe if the larva is dead. In that situation, the dead larva gets decomposed at the vicinity, and the decomposed material containing the toxic substance is absorbed. This aggravates the condition.

Genus: *Hypoderma*

Common name: heel fly, gad fly, warble fly, cattle grub

Species: *H. bovis*, *H. lineatum*

Morphology

Adult

Color: The flies are black in color. Hairs are yellow or yellow-orange colored.

Size: The flies are about 15 mm long.

Mouthparts: The mouthparts are nonfunctional.

Head: Greenish yellow hairs are present on the head and anterior part of the thorax of *H. bovis*. Yellowish hairs are present on the same region in case of *H. lineatum*.

Abdomen: Light yellow hairs are found at the anterior portion of the abdomen. The posterior portion bears a band of dark and orange yellow hairs.

Body: The body is divided into three parts, i.e., head, thorax, abdomen. Yellow and orange colored hairs are present in the abdomen.

Egg

The eggs are small measuring 1 mm in length. These are white in color. At one side, they possess clasp by which the eggs get attached on the hairs.

Larva and pupa

After hatching, the larvae come out. The color of the larvae changes during development. They start from white to yellow in color which may turn brown or dark. The larvae possess flat tubercles and hairs. The pupal case is black in color.

Life Cycle

Egg Laying

The flies lay their eggs on the hairs of legs, feet, and flank region. However, different species have site preference. It is found that *H. bovis* mostly lays their eggs on the hairs of legs. The fly lays one egg on single hair. *H. lineatum* attach several eggs on single hair. This is the difference of egg laying feature of *H. bovis* and *H. lineatum*.

Hatching

Hatching of eggs occurs on the laying site. Hatching occurs within 4–5 days.

Larva Migration

After hatching, the larvae crawl down the hairs and penetrate the skin. In case of *H. bovis*, the larvae migrate to the connective tissue of the skin. The larvae migrate in the internal organs during autumn and winter. The location where the larvae remain in the winter is called winter resting site. In case of *H. lineatum*, the larvae are found in the submucosa of the esophagus which is the winter resting site of this fly. The larvae of *H. bovis* are found in the spinal canal and epidural fat which is the winter resting site of *H. bovis*.

Warble Stage

The larvae are transformed into second stage which travels toward the subcutaneous connective tissue of the back whereby they transform into third stage larvae during the month of January and February. This is called spring resting site. The larvae cause swelling of the skin which later perforates. The larvae keep their posterior stigmal plates directed outwardly for the purpose of respiration. This is called warble stage.

Pupa and Adult

The larvae wriggles out of the cysts and drop on the ground. Larva is transformed into pupa. Pupal case is black in colour. Adult is released from the pupa.

Importance

Annoyance

The flies cause annoyance to the animals when they approach the animals to lay their eggs. The animals rapidly run away to escape this attack. They may fall down on the fences, water, etc. This is called "gadding." This annoyance causes disturbances in feeding and drinking water resulting in loss of body weight and milk production.

Hide Depreciation

This parasite has importance in livestock economy. The hide quality goes down due to perforations caused by the larvae of the parasite. The larvae migrate in the subcutaneous tissue. They cause swelling of the subcutaneous tissue and create holes there.

Carcass Depreciation

Meat quality also gets deteriorated. A jellylike material is produced in the meat called "bucher's jelly." The deteriorated meat is called "licked beef." The depreciation is caused due to inflammatory swelling of the tissue, formation of tract in tissues, formation of abscess in tissues, etc.

Neurological Disease

There have been reports of *H. lineatum* infection in horses. The first stage larvae may migrate to the brain of horse causing damage to the brain tissue.

Zoonotic Importance

These flies have zoonotic importance. The larvae may cause local warble formation resulting in subdermal migratory tract.

Control

- Manual removal of the larvae by pressing the warbles with fingers
- Use of suction to remove the larvae
- Surgical removal of the larvae using sharp instrument

Section: Calypterae—Family: Calliphoridae

Family Synopsis

The important genus of this family is *Calliphora* and *Lucilia*. This is a large family having more than 1000 species. They are called blowflies. Different species of this family causes cutaneous myiasis. The body of these flies has metallic color. The species of *Calliphora* are called bluebottle flies and *Lucilia* greenbottle flies. The flies are morphologically characterized by presence of hypopleural bristles and bright color with metallic sheen.

This family has two subfamilies:

Subfamily: Calliphorinae
Sarcophaginae

Subfamily: Calliphorinae

The members under this family are bluebottle and greenbottle flies. The larval stage develops in the skin of different animals and causes lesions there which is called cutaneous myiasis.

Genus: *Lucilia*

Species

Two important species under this genus are *L. sericata* and *L. cuprina* which are called "greenbottle fly" as they have green metallic sheen of the thorax and abdomen.

Genus: *Calliphora*

The important species are *C. erythrocephala*, *C. vomitoria*, and *C. grahami*. They are known as bluebottle fly.

Genus: *Phormia*

The member under this genus are called black blowfly. The species is *P. terraenovae*. The flies cause myiasis in sheep.

Genus: *Chrysomya*

The important species under this genus is *C. bezziana*. Other species are *C. albiceps*, *C. micropogon*, and *C. rufifacies*. They are bluish green in color.

Harm Caused

By Primary Blowfly (*Calliphora, Lucilia*)

The primary blowflies lay their eggs on the wool of sheep. The larvae are liberated after hatching and crawl down to the skin. The flies lacerate the skin by their hooks. They also secrete proteolytic enzymes which digest the tissues at the vicinity resulting in frank lesion.

By Secondary Blowfly (*Chrysomya*)

The lesions caused by primary blowflies emit some odor which attracts the secondary blowflies, and the secondary blowflies aggravate the lesion to higher grade. Secondary bacterial infection occurs and the condition becomes more complicated. The wound spreads gradually. The wool of the affected area falls down. Close examination reveals the presence of the larvae on the affected area keeping their posterior part outwardly. Heavy infection may even lead to death of the animals. The myiasis in sheep is called "strike".

Indirect Effect

The lesion caused by the blowflies led the affected animals to lose their condition. They do not feed and drink well. They stand away from the main flock keeping their head down.

Treatment and Control

The affected animals should be treated properly. The wool of the affected area should be clipped, and the larvae should be removed from the affected area. The wound should be dressed properly with suitable insecticidal agents. See the end part of this chapter for treatment of myiasis.

Crutch Removal

The skin of the crutch is frequently soiled with urine which attracts the blowflies. Removal of the skin of the crutch reduces breech strike.

Tail Docking

Skin folds are found at the tail region and this is soiled easily which attracts blowflies. To remove the skin folds at the tail, docking of the tail is done. This is an effective measure to control tail strike.

Selective Breeding

Breeding policy should be directed in such a way that the animals give birth of offsprings having less skin folds.

Proper Disposal of Carcass

The carcass of animal is the main breeding ground of blowflies; therefore, the carcass should be disposed properly.

Subfamily: Sarcophaginae

The important under this genus is *Sarcophaga*. The species is *Sarcophaga dux*. Larvae are reported to be present in skin lesions of different animals.

Subfamily: Sarcophaginae (The members are flesh flies, carrion flies)

Genus: *Sarcophaga* (*S. dux* larvae were reported to be present in the skin lesion of different animals).

Section: Pupipara—Family: Hippoboscidae

Family Synopsis

The members of this family are morphologically characterized by dorsoventrally flattened body, piercing and sucking type of mouthparts, soft leathery indistinctly segmented abdomen, and legs equipped with strong claws. They do not transmit any pathogen to the host but cause marked mechanical damage to their hosts.

Genus: *Hippobosca*

Common name : *Hippobosca equina* is called Forest fly, Horse ked

Morphology

Color: The flies are reddish brown or brown in color.

Size: They are 6–9 mm in size.

Body: The body is tough and scarcely hairy.

Head: The head is flat with piercing type of mouthparts. Head is roughly triangular in shape.

Palps and proboscis: The palpi are short and ensheath the proboscis.

Wings: Roughly subtriangular wings are found. The wings are reddish dark in color. The characteristic feature of wings is that the wing veins crowd toward the anterior border.

Legs: The legs are stout and short in size and are provided with powerful claws.

> **Note**
> *Hippobosca* flies shed their wings when they are on their hosts. But there is exception (*H. equina*). *Hippobosca equina* do not shed their wings while on their hosts and can leave their hosts if disturbed and again attach their hosts subsequently.

Life Cycle

Eggs Hatching

Eggs hatch in the uterus itself.

Larval Development

Larva develops in the uterus. Prepupal stages of larvae are laid by the flies on the soil or humus. About 5–12 larvae are laid. The larvae are yellow in color and globular in shape. A dark spot is found at the posterior pole of the larvae.

Pupal Development

Prepupal stage of larvae (third stage larvae) is laid which is transformed to complete pupal stage within few hours. Their pupae are dark in color.

Adult Development

Adult flies completely develop in the pupa and emerge out from the pupa.

Habitat

The flies are obligatory ectoparasites. They cannot live or perform life cycle without hosts. They suck blood using their piercing type of mouthparts. Very characteristically they are found in the region of hosts where hairs are absent or very less (perineal region, udder, etc.).

Importance

Though the flies do not transmit any pathogen to the hosts, their bites cause direct harm to the hosts. The flies are very much in population during the summer and rainy season. They are not easily disturbed and consistently attack on the perineal and pubic region. Constant irritation and nuisance cause less feed intake which leads to loss of production.

Genus: ***Pseudolynchia***

Species: ***P.*** *canariensis*

Common name: pigeon louse fly or pigeon fly

Hosts: They mostly occur in pigeons. But other birds are also affected.

Morphology

Color: The flies are brown in color.

Size: They are 5–6 mm in size. They are slightly smaller than *Hippobosca*.

Body: The body is tough and scarcely hairy. They are dorsoventrally flattened. They have tough exoskeleton.

Head: The head is found with piercing type of mouthparts. The head is roughly triangular in shape and flat dorsoventrally.

Wings: Roughly subtriangular wings are found. They do not shed off their wings as *Hippobosca* flies when they are on the hosts.

Legs: The legs are stout and short in size. The legs are provided with strong and stout claws, and they can strongly cling to the feathers or hairs.

Life Cycle

Eggs Hatching

Eggs hatch in the uterus itself.

Larval Development

Larva develops in the uterus. The larvae take the nutrition from the milk gland present in the uterus. Three larval instars develop in the uterus. Prepupal stages of larvae are laid by the flies.

Pupal Development

Prepupal stage of larvae is laid which is transformed to complete pupal stage within few hours. The pupae are initially whitish in color and later turn to dark. The pupae are seed shaped dark brown in color.

Adult Emergence

Adult flies come out from the pupa. A specific structure called ptinilium remain consealed on the head. This breaks the pupal case.

Habitat

The flies are obligatory ectoparasites. They cannot live or perform life cycle without hosts. They suck blood using their piercing type of mouthparts. Very characteristically their claws are adapted to cling to the hairs or feathers. They are specifically found in pigeon but they can be found in other birds also. These lice suck blood from the hosts.

Genus: *Melophagus*

Common Name: Sheep Ked

Morphology

Shape: These flies are tick-like and hence they are called tick flies.

Size: They are 4–6 mm long.

Head and mouthparts: The mouthparts are non-functional. The head is not movable.

Body: The body is dorsoventrally flattened. The body is covered by numerous hairs.

Wings: The main characteristic feature is that they do not have any wings.

Legs: The legs are long and provided with powerful claws.

Life Cycle

The complete life cycle of the flies occurs on the hosts. The flies lay the larvae on the skin. At a time, only one larva is laid. A total of 10–15 larvae are laid by the flies. The eggs hatch in the uterus. The larva is retained until it is ready to pupate. The larva remain attached to the milk gland present in the uterus from where the they get nutrition. They develop and get matured after three instar stages. The flies glue the larvae on the wool of the hosts. The larvae are whitish and glistening. The shape is oval. The larvae soon transform to pupa having dark puparium. The pupal stage lasts for 1 month or more depending on the environmental temperature. The adults come out from the pupa.

Importance

Mechanical Damage

The flies take blood from the hosts. A kind of skin reaction (pruritis) is a common feature found as a result of *Melophagus* infection. To alleviate irritation, they rub or scratch the area. As a consequence, the condition is aggravated. Constant irritation leads to interference of feeding and drinking. The body condition of the sheep goes down. They may cause nodule formation which reduces hide quality.

Pathogen Transmission

Virus: bluetongue virus of sheep and goat

Rickettsia: These flies transmit *Anaplasma ovis*, *A. phagocytophilum*, *Borrelia burgdorferi*, and *Bartonella* spp.

Bacteria: *Acinetobacter* sp.

Protozoa: *Trypanosoma malophagium*

Control

- Dipping of animals in water mixed with insecticide.
- Shearing of fleece markedly reduces the parasite because the flies attach the larvae on the wool. Shearing should be repeated.

Genus: Pseudolynchia

The important species under this genus is *P. canariensis* which transmit a haemoprotozoa, *Haemoproteus columbae*.

Order: Coleoptera

The members of this order are called beetle. This is the largest order containing lakhs of species.

Morphology

The typical morphological features of a dipteran insect are found. However, some morphological features are different. The body is divided into three parts, i.e., head, thorax, and abdomen. Two pairs of wings are present. The front pair of wings are very hard which cover the hind pair of wings. The hard pair is "elytra" which is salient identifying character of beetle. The legs are strong and stout. The head is also hard.

Biology

The beetles perform holometabolous type of life cycle having four stages, i.e., egg, larva, pupa, and adult.

Importance

The beetles transmit important pathogens. They are different helminths comprising *Raillietina cesticillus*, *Choanotaenia infundibulum*, *Hymenolepis carioca*, *Gongylonema* spp., *Acuaria* spp., Spirocerca lupi, Ascarops spp. and *Physaloptera* spp.

Subphylum: Chelicerata—Class: Arachnida

The members under this class are ticks, mites, spiders, scorpions, etc. They do not have true mandibles and antennae. They have a sucking pharynx by which they suck blood/tissue fluid. The mouthparts consist of a pair of chelicerae and a pair of pedipalps. The chelicerae of ticks and mites are adapted for piercing and earing the skin. Another highly characterized toothed structure is present called hypostome.

Order: Acarina

This order contains hard and soft ticks, mites. The mouthparts of Acarina consist of a pair of chelicerae, a pair of pedipalps, a toothed structure, and hypostome. The structures are housed on a specific platelike structure called basis capitulum, a part of gnathosoma. Different developmental stages are egg, larva, nymph, and adult. The larval stage has three pairs of legs. The nymphs have four pairs of legs and similar to adults only differ from the adults in having no sex organs. The nymph stages may be single or several. These are protonymph, deutonymph, and triotonymph.

Suborder: Ixodoidea—Family: Ixodidae

Different features of ixodid ticks:

- The ticks have hard chitinous shield called scutum. The shield covers the whole body in case of male and covers a small portion in case of larva, nymph, and adult of female tick.
- The mouthparts are equipped with a capitulum visible from dorsal surface.
- Eyes are present or absent; when present, they are situated at the lateral margin of the scutum.
- In some species, the posterior border has toothlike notched structure (11 in number) called festoons.
- The males may have chitinous plates on the ventral surface.
- Some ticks possess colored enamel-like areas and are called "ornate ticks." Some ticks do not have colored areas and are called "inornate ticks."
- Spiracles are two in number and present behind the fourth coxae of imago.
- Mating of the hard ticks occurs on the host itself.
- The stages of development are egg, larva, nymph, and adult. In hard ticks, only one nymph stage is found.
- The ticks suck blood continuously for days together until their full engorgement.

Fig. 7.26 Engorged tick

General Life Cycle

In accordance with the pattern of life cycle, they are one host tick, two host tick, and three host tick. The ticks lay their eggs in sheltered spots. The female tick lays all their eggs at a time. Thousands of eggs are laid by the female; the number may be 1000–20,000 in accordance with different species. Under suitable environmental condition, the eggs are hatched and the larvae are liberated. The larvae are called seed tick. The larvae have three pairs of legs. The seed ticks climb onto the grass blades or shrubs and wait for getting a suitable host on which the ticks climb and attach by the claws. The description given below provides a comprehensive life cycle in different types of ticks:

One Host Tick

Some ticks comprising *Rhipicephalus* (*Boophilus*) *microplus*, *Rhipicephalus* (*Boophilus*) *annulatus*, etc. are one host ticks. The unfed larvae of these ticks attach to the hosts and pass two ecdyses forming nymph and adult stage, respectively. All three stages suck blood. After full engorgement, the adults are dislodged from the hosts and drop on the ground and lay eggs in sheltered spot.

Two Host Tick

Some ticks comprising *Rhipicephalus evertsi*, *Rhipicephalus bursa*, etc. are two host ticks. The unfed larvae of these ticks attach to the hosts, and after full engorgement, they pass one ecdysis and are transformed to nymph. The nymphs engorge and drop on the ground after full engorgement. The nymph transforms into adult on the ground and the adult ticks seek for getting a host. The adults attach on the hosts and engorge. After full engorgement, the adults are dislodged from the hosts and drop on the ground and lay eggs in sheltered spot.

Three Host Tick

Some ticks comprising *Rhipicephalus appendiculatus*, *Ixodes ricinus*, *Amblyomma hebraeum*, *A. americanum*, etc. are three host ticks. All three stages (larva, nymph, adult) separately suck blood from three different hosts. The unfed larvae of these ticks attach to the hosts, and after full engorgement, they fall on the ground and pass on one molting on the ground and are transformed to nymph later. The nymphs attach on another host and get engorged. After full engorgement, the nymph gets dislodged and falls on the ground. After one molting, the nymphs transform into adult. The adults attach to another host and suck blood from the hosts. After full engorgement, the adults are dislodged from the hosts and drop on the ground and lay eggs in sheltered spot.

Genus: *Rhipicephalus* (*Boophilus*)

This is a genus under the family Ixodidae. The ticks are very common and worldwide in distribution. The ticks are very important arthropods as they transmit many pathogens like *Babesia*, *Anaplasma*, *Borrelia*, *Coxiella*, etc. These are hard ticks. The palps are small and ridged. The hypostome is short. First coaxa is bifid. The basis capitulum is hexagonal. The spiracles are oval or circular. Festoon is absent.

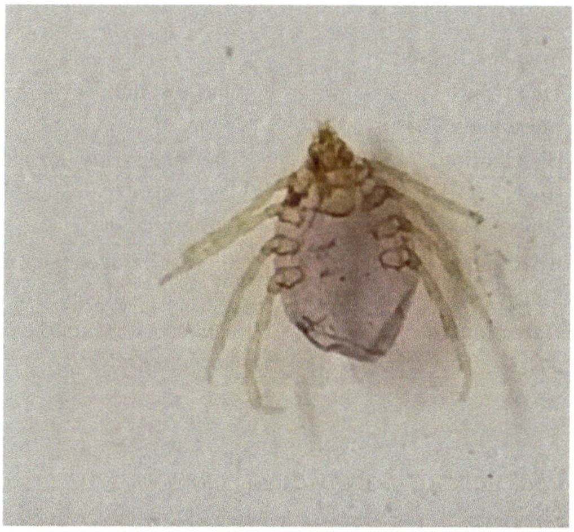

Fig. 7.27 *Rhipicephalus* (*Boophilus*) sp.

Biology

This tick is "one host tick." The female lays more than 2000 eggs. The eggs hatch and the larvae come out. The larvae get ready to attack the host. The larvae feed on the host for few days and then get molted to become the nymph. The nymph feeds on the host like the larvae and gets molted to become the adult. All three stages (larva, nymph, and adult) occur on the same host.

Rhipicephalus (Boophilus) annulatus

Brief description	
Tick type	One host tick
Common name	This is called "North American tick"
Distribution	The ticks are distributed in parts of Africa, USSR, parts of South America, Mexico, the Middle East, the Near East Mediterranean regions, etc.
Hosts	The hosts which are affected by the ticks are cattle, buffalo, Asian water buffalo, Bactrian camel, sheep, and goat
Site	Skin
Disease transmitted	Texas fever, red water fever, piroplasmosis
Brief morphology	The palps are too short with dorsal and lateral ridges. The male has adanal shields. The anal groove is not prominent in male. This may be inconspicuous or absent in female. The festoon is absent
Stages of life cycle	They have all stages of development, i.e., egg, larva, nymph, and adult
Harm caused	The ticks transmit *Babesia bigemina*, *B. bovis*, *Anaplasma ovis*, *A. marginale*, *A. mesaeterum*

Rhipicephalus (Boophilus) calcaratus

Brief description	
Tick type	One host tick
Common name	None
Distribution	They are mostly prevalent in Central Asia, parts of Africa, etc.
Hosts	The hosts which are affected by the ticks are cattle
Site	Skin
Brief morphology	Short and ridged palp, hexagonal basis capitulum, absence of festoon
Stages of life cycle	They have all stages of development, i.e., egg, larva, nymph, and adult
Harm caused	The ticks transmit the pathogens *Babesia berbera*, *Babesia bigemina*, *Anaplasma marginale*, etc.

Rhipicephalus (Boophilus) decoloratus

Brief description	
Tick type	One host tick
Common name	These are also called blue tick, African blue tick
Distribution	The ticks are distributed in sub-Saharan Africa. They are found in Namibia, Botswana, Mozambique, Zimbabwe, Zambia, etc. The ticks are also distributed in West Africa like Senegal, Mali, etc. They are also distributed in some parts of Asia
Hosts	The hosts are cattle, horse, donkey, sheep, goat, etc.
Site	Skin
Brief morphology	Engorged female tick is slaty-blue colored. The legs are pale yellowish in color. The inner margin of first segment of palp bears bristles
Stages of life cycle	They have all stages of development, i.e., egg, larva, nymph, and adult
Harm caused	These ticks transmit the pathogens, i.e., *Babesia bigemina*, *B. ovis*, *Anaplasma marginale*, *Borrelia theileri*, *Anaplasma marginale*, etc.

Rhipicephalus (Boophilus) microplus

Brief description	
Tick type	Hard tick (One host tick)
Common name	The common name of the ticks is tropical cattle tick, Asian blue tick
Distribution	The ticks are originally from Asia. They are distributed in South Africa, East Africa, East Asia, and other parts of Africa. The ticks are also prevalent in South and Central America and Australia
Hosts	The hosts are cattle. But they also affect sheep, goat, and horse. They may affect wild ungulates too
Site	Skin
Brief morphology	The palps are small and ridged. Festoons are absent. The basis capitulum is hexagonal dorsally. Dentition of hypostome is characteristic. The dentition has two columns. Each column has few rows. Each row contains four denticles. Inner margin of first segment of palp is concave and bears no bristle
Stages of life cycle	They have all stages of development, i.e., egg, larva, nymph, and adult
Harm caused	These ticks transmit the pathogens, *Babesia bigemina, B. bovis, Borrelia theileri, Coxiella burnetii, Anaplasma marginale*

Genus: *Rhipicephalus*

Fig. 7.28 *Rhipicephalus* sp

Rhipicephalus appendiculatus

This is a species under the family Ixodidae. This is a three host tick.

Brief description	
Tick type	Three host tick
Common name	The common name of the ticks is brown ear tick
Distribution	The ticks are available in different countries of Africa
Hosts	The hosts are cattle, sheep, goat, and horse. The dogs are also affected by the ticks
Site	The preferable attachment site of the ticks is ear and tail regions
Brief morphology	The ticks are medium sized. The color is reddish brown. The basis capitulum is hexagonal with protruding lateral ends. Palps are small. Festoons are absent.
Stages of life cycle	They have all stages of development, i.e., egg, larva, nymph, and adult
Harm caused	The pathogens transmitted by the ticks are *Babesia bigemina, Theileria mutans, T. parva, Hepatozoon canis, Rickettsia conorii*, virus of Nairobi sheep disease, etc.

Rhipicephalus bursa

Brief description	
Tick type	Hard tick (Two host tick)
Common name	It seems that there is no common name.
Distribution	The species is distributed in Africa, Europe, etc.
Hosts	The hosts are cattle, sheep, horse, dog, man, etc.
Site	Skin
Brief morphology	The basis capitulum has sharp lateral angles. Palps are short. Porose area is oval. The scutum is dark
Stages of life cycle	They have all stages of development, i.e., egg, larva, nymph, and adult
Harm caused	The ticks are important as they transmit several important pathogens like *Babesia ovis*, *B. caballi*, *B. berbera*, *B. equi*, *Theileria ovis*, *T. separata*, *Anaplasma marginale*, *Rickettsia ovina* and *Coxiella burnetii*, Crimean-Congo hemorrhagic fever virus, etc.

Rhipicephalus capensis

Brief description	
Tick type	Hard tick (Three host tick)
Common name	Cape brown tick
Distribution	South Africa
Hosts	The hosts are cattle
Site	Skin
Any special morphology	Long, broad, expanding cervical groove is present.
Stages of life cycle	They have all stages of development, i.e., egg, larva, nymph, and adult
Harm caused	They transmit some important pathogens, i.e., *T. parva*, the causative agent of East Coast fever of cattle and *Anaplasma marginale*

Rhipicephalus evertsi

Brief description	
Tick type	Hard tick (Two host tick)
Common name	The common name of the ticks is red legged tick
Distribution	This species is distributed in different countries of Africa, America, etc.
Hosts	The hosts are a wide range of mammals i.e.cattle,sheep,goat,horse etc.
Site	Skin
Brief morphology	The tick has characteristically bicolor presentation. The scutum is dark brown colored and the legs are reddish orange colored
Stages of life cycle	They have all stages of development, i.e., egg, larva, nymph, and adult
Harm caused	They transmit some important pathogens like *Theileria parva*, the causative agent of East Coast fever; *T. mutans*; *T.equi Babesia caballi*, *B.bigemina*, the causative agent of red water fever, *Anaplasma marginale*

Rhipicephalus sanguineus

Brief description	
Tick type	Three host tick
Common name	The common name of the ticks is Brown dog tick
Distribution	This species is distributed in different countries of Africa but also occurs in other parts of the world
Hosts	Dogs are the common hosts of this tick but may also occur in a wide range of mammals
Site	Skin
Brief morphology	The ticks are 4-5 mm long.They are yellowish brown to reddish brown having dark colored scutum They have hexagonal basis capitulum and festoon
Stages of life cycle	They have all stages of development, i.e., egg, larva, nymph, and adult
Harm caused	The ticks are important as they transmit the important pathogens like *Babesia canis*; *B. vogeli*; *B. caballi*; *B. equi*; *Ehrlichia canis*; *Hepatozoon canis*; different rickettsial spp. (*Rickettsia rickettsia*, *R. conorii*, *Coxiella burnetii*, *Anaplasma marginale*); *Pasteurella tularensis*, the causative agent of East Coast fever; *T. mutans*; *Babesia caballi*, the causative agent of red water fever

Rhipicephalus jeanelli

These are hard ticks under the family Ixodidae which transmit *Theileria parva* in cattle.

Rhipicephalus neavei

The hosts are cattle. The ticks transmit *T. parva*.

Rhipicephalus pulchellus

The ticks affect sheep, zebra, elephant, etc. The common name is zebra tick. The ticks transmit Crimean-Congo hemorrhagic fever virus, Nairobi sheep disease virus, *Theileria taurotragi*, etc.

Genus: *Dermacentor*

The ticks are usually ornate. The festoons are present. The eyes are also present. The palp and hypostome are short. The first coxa is bifid. Fourth coxa is much larger than coxa 1–3.

Dermacentor albipictus

This is a species under the family Ixodidae.

Brief description	
Tick type	Hard tick (one host tick)
Common name	The common name of the ticks is "winter tick" or "moose tick"
Distribution	They are predominantly distributed in North America. Central America,Canada etc.
Hosts	The hosts are moose but the ticks also occur in the cattle, horse, deer, elk, bison, etc.
Site	Skin
Brief morphology	The ticks are reddish brown to gray brown in color. They are medium sized ticks. The basis capitulum is rectangular. Festoons are present. Posterior edge of the tick has folds
Stages of life cycle	They have all stages of development, i.e., egg, larva, nymph, and adult
Harm caused	The ticks are the vector of *Anaplasma* sp.

Dermacentor andersoni

Brief description	
Tick type	Three host tick
Common name	The common name of this tick is Rocky Mountain wood tick
Distribution	These ticks are mostly available in arid area of rocky mountain region. They are also found in different parts of Southwestern Canada, Northwestern United States
Hosts	The hosts of the ticks are cattle, horse, deer, dog, etc. They also affect humans
Site	Skin
Brief morphology	The ticks are brown or reddish brown in color. A dorsal silver-gray ornamentation (shield marking) is present in the female tick. The males are spotted gray and white but do not possess any such shield marking as is found in the female. Festoons are present
Stages of life cycle	They have all stages of development, i.e., egg, larva, nymph, and adult
Harm caused	These ticks are the vectors of *Pasteurella tularensis*, causative agent of tularemia and virus of Rocky Mountain spotted fever. These ticks also transmit *Coxiella burnetii*, *Anaplasma marginale*, etc. The ticks also transmit Colorado tick fever

Dermacentor nitens

Brief description	
Tick type	One host tick
Common name	The common name of the ticks is "tropical horse tick"
Distribution	The ticks are commonly available in Central America, South America, Mexico, etc. They are also distributed in southern regions of Texas and Florida
Hosts	The primary hosts are equids (horse, mule, asses) but it may also affect other hosts, i.e., cattle, sheep, deer, etc.
Site	Skin
Brief morphology	Scutum is inornate. Basis capitulum is rectangular. Eyes are present. Spiracular plate is circular to elliptical. Festoons are present.The shape is oval
Stages of life cycle	They have all stages of development, i.e., egg, larva, nymph, and adult
Harm caused	This is one host tick. The ticks cause equine piroplasmosis caused by *Babesia caballi* and *Theileria equi*

Dermacentor occidentalis

Brief description	
Tick type	Three host tick
Common name	Pacific Coast tick
Distribution	Different parts of California, Western US etc.
Hosts	The common hosts are deer. But they may also occur in cattle, horse, donkey, etc. They also occur in humans
Site	Skin
Stages of life cycle	They have all stages of development, i.e., egg, larva, nymph, and adult
Harm caused	The ticks are the vector of Q fever, tularemia, bovine anaplasmosis, tick paralysis, Rickettsial organisms of spotted fever group have been isolated

Dermacentor reticulatus

Brief description	
Tick type	Three host tick
Common name	Meadow tick, ornate cow tick, ornate dog tick, marsh tick
Distribution	England, Germany, Poland, Hungary, etc. They are also found in USSR
Hosts	Cattle, equines, and dogs are the common hosts. Humans are also affected
Site	Skin
Brief morphology	These are ornate ticks. The scutum is ornate and has reticulate (net like design)
Stages of life cycle	They have all stages of development, i.e., egg, larva, nymph, and adult
Harm caused	These ticks transmit *Babesia canis*, *B. equi* and *B. caballi*, *Francisella tularensis*, *Coxiella burnetii*, *Rickettsia* spp.

Dermacentor nigrolineatus

The hosts are white tailed deer. Common name of the tick is "brown winter tick." But it also occurs in cattle horse.

Dermacentor variabilis

This is a species under the family Ixodidae. The members are called American dog tick or wood tick.They are mostly prevalent in North America. The dogs are the primary hosts.Humans are also affected.The virus of Rocky Mountain Spotted Fever is transmitted.

Genus: *Haemaphysalis*

Morphology

This is a genus under the family Ixodidae. The hosts are cattle, dog, rabbit, etc. These are inornate and brevirostrate type. Palpi and the hypostome are short. Second segment of the palp has lateral projection. The chitinous plates are absent in the ventral surface of the male ticks.

Biology

These are mainly three host ticks. The female lays large number of eggs. After hatching, the larvae come out. The larvae get ready to attack the host. Within a week or more, the larvae are fully engorged. Then the larvae drop on the ground. Within 3 weeks or more, the larvae get molted to become the nymph. The nymphs then attack another host and start feeding. Within a week or more, the nymphs are engorged. After full engorgement, the nymphs fall on the ground. Within about 3 weeks, the nymphs molt to become the adult. Then the adult attacks another host. It is to remember that the durations (engorgement, molting, etc.) vary in different species.

Importance

The pathogens transmitted by the ticks are *Babesia* spp., *Theileria* spp., *Coxiella* spp., *Rickettsia* spp., and *Anaplasma* spp.

Haemaphysalis bispinosa

Brief description	
Tick type	Hard tick
Common name	No name
Distribution	India, Pakistan, Nepal, Sri Lanka, Australia, Indonesia, Malaysia,China etc etc.
Hosts	The hosts are cattle. Other animals, i.e., sheep, goat, etc. They may affect dogs also
Site	Skin
Stages of life cycle	They have all stages of development, i.e., egg, larva, nymph, and adult
Harm caused	This tick transmits *Babesia canis*, *B. gibsoni*, *B. bigemina*, *Anaplasma marginale*, Kyasanur forest disease, and bartonellosis

Haemaphysalis humerosa

Brief description	
Tick type	Hard tick
Common name	The ticks are called bandicoot tick
Distribution	Found in Australia
Hosts	The hosts are bandicoot. They are also found in cattle cattle
Site	Skin
Stages of life cycle	They have all stages of development, i.e., egg, larva, nymph, and adult

Brief description	
Harm caused	The ticks transmit *Coxiella burnetii*, the causative agent of Q fever

Haemaphysalis leachi

Brief description	
Tick type	Three host tick
Common name	The common name of the ticks is yellow dog tick.
Distribution	The ticks are distributed in Asia, Africa, and Australia
Hosts	The common hosts are dogs
Site	Skin
Stages of life cycle	They have all stages of development, i.e., egg, larva, nymph, and adult
Harm caused	The ticks are important as they transmit important pathogens, i.e., *Babesia canis*, the causative agent of babesiosis in dogs. The ticks also transmit *Rickettsia conorii* and *Coxiella burnetii*. The ticks also transmit boutonneuse fever

Haemaphysalis leporispalustris

Brief description	
Tick type	Hard tick (three host tick)
Common name	The common name of the ticks is rabbit tick
Distribution	They are commonly found in the USA
Hosts	The hosts are rabbits
Site	Skin
Brief morphology	The morphology of this tick is similar to other species of this genus. But it is to be noted that the basis capitulum is hexagonal. Basis capitulum is not rectangular. The eyes are absent. The palpi are wider than long
Stages of life cycle	They have all stages of development, i.e., egg, larva, nymph, and adult
Harm caused	The ticks transmit *Coxiella burnetii*, the causative agent of Q fever, and *Pasteurella tularensis*, the causative agent of tularemia.

Haemaphysalis longicornis

Brief description	
Tick type	Three host tick
Common name	They are also called Asian long-horned tick, long-horned tick, bush tick, etc.
Distribution	The ticks are native to East and Central Asia, Australia, New Zealand, etc.
Hosts	They affect a wide range of mammals and birds. Cattle, sheep, goat, horse, dog, and cat are affected by these ticks. Wild animals comprising deer, bear, and fox are also affected by these ticks
Site	Skin
Stages of life cycle	They have all stages of development, i.e., egg, larva, nymph, and adult. They perform sexual reproduction and asexual reproduction by parthenogenesis
Harm caused	The ticks transmit different pathogens, i.e., *Coxiella burnetii*, the causative agent of Q fever, and *Theileria* sp., to cattle. They also transmit different spotted fevers to humans

Haemaphysalis chordeilis
The hosts are birds. This is a common bird tick. The ticks transmit tularemia.

Haemaphysalis bancrofti
This is a species under the family Ixodidae. The ticks occur in cattle and wallaby. They are called wallaby tick.

Genus: *Hyalomma*

The members of this genus are important ticks because these ticks transmit important pathogens. The ticks are inornate but occasionally the ticks may be ornate. The hypostome and the palps are long. The spiracle character is specific. It is comma shaped in the male and triangular in female. Festoons may be present or absent. The pathogens which are transmitted by the ticks are the *Babesia caballi*, *B. equi*, *Theileria dispar*, *T. annulata*, and *T. parva*. *Rickettsia conorii*, *Rickettsia bovis*, and *Coxiella burnetii* are also transmitted by these ticks.

Biology
The eggs are laid by the female ticks. Hatching occurs within 2 months or more. The larvae attack the hosts and engorge within a week. After engorgement, the larvae fall on the ground. Then the larva molts to become nymph within 1–2 weeks. The nymphs attack another host and engorge within 1 week or more. Then the nymphs drop on the ground and molt to become adult after 2 weeks or more. The adults attack another host. Thus, the three host life cycle is completed.

Hyalomma detritum mauretanicum
This is a species under the family Ixodidae.

Hyalomma detritum scupense
This is a species under the family Ixodidae.

Hyalomma dromedarii
This is a species under the family Ixodidae.

Hyalomma excavatum
This is a species under the family Ixodidae.

Hyalomma marginatum
The ticks affect birds and cattle. The common name is Mediterranean tick. The ticks are ditropic as larvae remain on the same host to molt to nymph. Previously it was treated as *Hyalomma marginatum* complex comprising three subspecies, i.e., *H. marginatum marginatum*, *H. marinatum scupense*, and *H. marginatum isaci*.

Hyalomma rufipes
These ticks transmit *Theileria spp.,Babesia spp.*

Genus: *Amblyomma*

These ticks have festoons and eyes. These ticks are longirostrate because they have long palpi and hypostome. Ventral plates are absent in male. They are usually ornate.

Amblyomma americanum

Brief description	
Tick type	Three host tick
Common name	These are called lone star tick, northeastern water tick, turkey tick
Distribution	They are mostly prevalent in North America. They are also found in Eastern United States and Mexico
Hosts	The birds are mostly affected
Site	Skin
Brief morphology	The ticks are longirostrate. The palps and hypostome are long. The common name of the tick is lone star tick because a large white spot is present on the scutum of female
Stages of life cycle	They have all stages of development, i.e., egg, larva, nymph, and adult
Harm caused	They transmit *Coxiella burnetii*, causative agent of Q fever; Rocky Mountain spotted fever caused by *Rickettsia rickettsii*; and tularemia fever caused by *Francisella tularensis*. *Ehrlichia chaffeensis*, *E. ewingii*

Amblyomma hebraeum

Brief description	
Tick type	These are three host tick
Common name	The common name of the tick is bont tick
Distribution	The ticks are predominantly distributed in sub-Saharan Africa
Hosts	Sheep, goat, and cattle are commonly affected by the ticks. But different wild ungulates are also affected. The ticks affect birds too
Site	Skin
Brief morphology	The ticks are longirostrate. The palps and hypostome are long. The male tick has fine black or brown spots on greenish white background. They have dark brown legs with apical yellow bands. The dorsum of the female is dark greenish brown which is punctate and striae
Stages of life cycle	They have all stages of development, i.e., egg, larva, nymph, and adult
Harm caused	They transmit heart water disease caused by *Rickettsia ruminantium* in sheep, goat, and cattle

Amblyomma maculatum

This is a three host tick. They are called Gulf Coast tick.

Amblyomma variegatum

Brief description	
Tick type	Hard tick
Common name	The common name of the tick is variegated tick, tropical bont tick, Antigua gold tick, Senegalese tick
Distribution	The ticks are prevalent in different parts of Africa. This is a three host tick
Hosts	The hosts are mostly the different species of mammals
Site	Skin
Brief morphology	The palps and the hypostome are long. They are longirostrate. The ticks are distinct from other ticks in having bright coloration on the body. Bright yellow gold coloration of the male tick led to give the name Antigua gold tick
Stages of life cycle	They have all stages of development, i.e., egg, larva, nymph, and adult
Harm caused	This tick transmits *Coxiella burnetii*, Nairobi sheep disease, heart water

Genus: *Ixodes*

This is a family under the suborder Ixodoidea. The members coming under the family are called hard ticks. The ticks are oval. The anterior part is somewhat tapering. There is presence of a dorsal shield which is called scutum. The scutum covers the whole body of male and some portion of female. Eyes when present are located at lateral area of scutum. The spiracles are present posterolaterally to the fourth coxae. The anal groove surrounds the anus anteriorly.

Ixodes ricinus

The biologically developmental stages are egg, larva, nymph, and adult.

Brief description	
Tick type	Hard tick (three host tick)
Common name	The common name of the ticks is "castor bean tick" or sheep tick
Distribution	This tick is common in Europe. They are found in some parts of Asia also. They are also distributed in neighboring parts of North Africa, some parts of Russia, etc.
Hosts	The common hosts are dogs. Other hosts are cattle, goat, sheep, horse, cat, etc.
Site	Skin
Brief morphology	These are inornate ticks. They do not have eyes. Festoons are absent Spine of coxa-I overlap coxa -II
Stages of life cycle	They have all stages of development, i.e., egg, larva, nymph, and adult
Harm caused	The ticks transmit several pathogens like *Anaplasma marginale*, *Babesia divergens*, and *B. bovis*. This species also transmits the virus of louping ill and *Staphylococcus aureus*, the causative agent of tick pyemia, and *Coxiella burnetii*

Ixodes angustus

Brief description	
Common name	It seems no common name is available
Distribution	They are mostly found in Canada, North America
Hosts	The hosts are dog. Many other animals and humans are affected. Rodents and squirrels are also affected
Site	Skin
Brief morphology	They have diamond shaped dorsal shield
Stages of life cycle	They have all stages of development, i.e., egg, larva, nymph, and adult
Harm caused	They transmit *Borrelia* sp., a causative agent of Lyme disease. As they rarely suck blood from human hosts, this tick is not considered as principle vector of Lyme disease

Ixodes canisuga

Brief description	
Type of tick	Three host tick
Common name	British Dog tick
Distribution	They are predominantly found in Europe
Hosts	The hosts are dog, cat, sheep, horse, etc.
Site	Skin
Any special morphology	The general morphological features are similar to other spp. The palpi of male are small. The tarsus of first pair of legs are humped. The first coxa does not bear any spine
Stages of life cycle	They have all stages of development, i.e., egg, larva, nymph, and adult
Harm caused	

Ixodes cookei

Brief description	
Common name	These ticks are called Ground hog tick, Woodchuck tickBritish dog tick
Distribution	They are found in North Eastern United states, Canada ,etc. found in Europe.
Hosts	The hosts are ground hog,racoon
Site	Skin
Brief morphology	The general morphological features are similar to other spp. The palps are short. Basis capitulum is rectangular. Festoons are absent
Stages of life cycle	They have all stages of development, i.e., egg, larva, nymph, and adult
Harm caused	The ticks are active in the summer. They transmit Powassan virus, causative agent of Powassan encephalitis. The squirrel, ground hogs, and rodents carry this virus. The ticks get infected when they bite and suck blood from those hosts and transmit the same to humans and animals

Ixodes hexagonus

Brief description	
Common name	These ticks are called hedgehog tick
Distribution	They are predominantly found in Western Europe
Hosts	The hosts are hedgehogs, dog, otter, etc.
Site	Skin
Brief morphology	They are dark brown in color. The shield is heart shaped. Tarsus of first pair of legs are humped. Coxa 1 bears a spine
Stages of life cycle	They have all stages of development, i.e., egg, larva, nymph, and adult
Harm caused	These ticks transmit *Borrelia* sp., causative agent of Lyme disease, to humans and other animals

Ixodes holocyclus

Brief description	
Common name	These are called Australian paralysis tick or simply paralysis tick. Other common names are scrub tick, bluebottle tick, bottle tick
Distribution	This tick is distributed in different parts of Australia
Hosts	The hosts are man, cat, dog, bandicoot
Site	Skin
Brief morphology	They are inornate and nonfestooned. The pigmented eyes are present on the scutum. The first and last pair of legs are clearly darker than two middle pair of legs. The anal groove is complete pear shaped or oval. That's why the name is *Ixodes holocyclus*. Holocyclus means complete circle.
Stages of life cycle	They have all stages of development, i.e., egg, larva, nymph, and adult
Harm caused	They remain firmly attached with the host skin and even may enter the skin. The ticks release neurotoxic material and cause paralysis to both human and animal. They are the vectors of *Coxiella burnetii*

Ixodes pacificus

Brief description	
Tick type	Hard tick
Common name	The ticks are called "California black-legged tick"
Distribution	They are found in western region of America
Hosts	The hosts are cattle, sheep, and horse. The ticks also occur in dog and cat
Site	Skin
Brief morphology	The female has brown-black plate on the anterior back.
Stages of life cycle	They have all stages of development, i.e., egg, larva, nymph, and adult
Harm caused	They transmit *Borrelia* sp., a causative agent of Lyme disease

Ixodes scapularis

Brief description	
Tick type	Hard tick
Common name	The common name of the tick is black-legged tick, shoulder tick, deer tick
Distribution	The ticks are distributed in the USA, Canada, etc.
Hosts	The hosts are cattle, sheep, horse, dog, etc.
Site	Skin
Brief morphology	These are inornate ticks. Festoons are absent. Anal groove is present above the anus
Stages of life cycle	They have all stages of development, i.e., egg, larva, nymph, and adult
Harm caused	They transmit *Borrelia burgdorferi*, causative agent of Lyme disease, *Babesia* sp., *Anaplasma* sp., Powassan virus

Ixodes texanus

They are called racoon tick occurring in racoon.

Ixodes pilosus

They are called bush tick and russet tick. The hosts are horse, cattle, sheep, goat, etc.

Ixodes rubicundus

The hosts are sheep, cattle, and goat. This tick is called Karoo paralysis tick. These ticks are mostly prevalent in South Africa.

Ixodes kingi

The hosts are dogs. The common name of the ticks is rotund tick.

Ixodes muris

The host is mouse. This is also called the “mouse tick.”

Biology of Ixodes

It is a three host tick. In this life cycle, three hosts are required. The female tick lays thousands of eggs. After few weeks, the larvae hatch. Then the larvae attack the host and take blood meal. After full engorgement, the larvae drop on the ground. In the same manner, other two hosts are required for completion of the entire life cycle.

Treatment and Control

See the end part of this chapter for treatment. For control , see the chapter “Trends in Control of Livestock and Poultry Parasites”

Suborder: Ixodoidea—Family: Argasidae

Family Character

- Unlike hard ticks, soft ticks are resistant to drought and desiccation and can live for several years.
- The members of this family are soft and leathery ticks. The body is frequently mammillated which is highly flexible.
- Scutum is absent.
- The gnathosoma or capitulum and mouthparts are located anteriorly on the ventral surface and are not readily visible from dorsal aspect.
- The eyes are usually absent .But rudimentary eyes may be present in some species. If present, it is found in the supracoxal fold.
- The developmental stages are egg, larva, nymph, and adult. Unlike hard ticks, several nymphal stages are found.
- The soft ticks are nocturnal in feeding habit. They suck blood intermittently, not continuously as is found in case of hard ticks.

General Life Cycle Pattern

- The gravid ticks lay eggs in sheltered spots, i.e., cracks, crevices, etc.
- The larvae are liberated after hatching of eggs.
- The larvae attack hosts and feed rapidly and drop to the ground.
- On the ground, the larvae molt to become nymph. Several nymph stages (three to four stages) pass on.
- The adult is emerged out from the last larval stage.
- The adults seek host for feeding. They are engorged and mate on the ground (off host mating).
- The females exhibit multiple gonotrophic cycle. They feed and oviposit frequently.

Some Important Points of Biology of Soft Tick

- Most argasid are nidicolous and can live years together without feeding.
- Immature stages suck blood. But there is exception also. For example, *Ornithodoros savignyi* larvae molt to nymph without taking blood meal.
- The adults feed repeatedly on the host, and the copulation takes place off host, but there is exception. For example, the female of *Otobius* lays eggs autogenically.

Genera
Argas
Otobius
Ornithodoros

Genus: ***Argas***
Species: *A. persicus*
Common name: fowl tick, poultry tick
Distribution: The ticks are worldwide in distribution.

Morphology

- Starved ticks are yellowish-brown in color and engorged ticks slaty-blue colored.
- The size of the ticks is 4–13 mm × 2.5–6 mm.
- Roughly oval shaped body.
- Outer margin of the ticks is sharp.
- Unfed ticks are yellow in color and engorged ticks are dark.
- Due to having translucent cuticle, the dark intestine is visible.

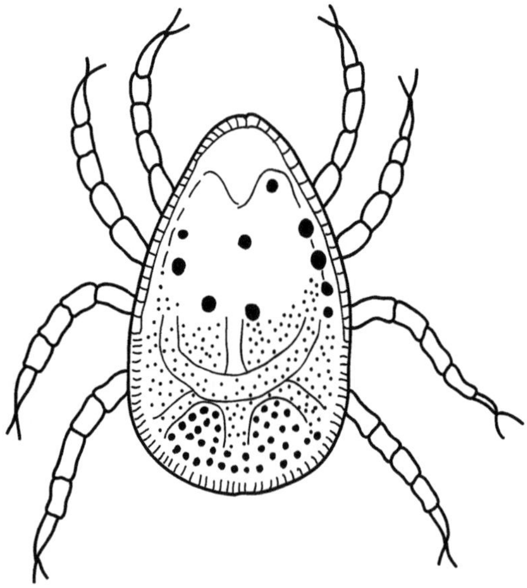

Fig. 7.29 *Argas persicus*

Biology

- The gravid females lay their eggs in sheltered spots, i.e., cracks, crevices, holes, etc. The eggs are laid in batches of 20–100 reaching a total of about 500–900.
- Hatching occurs after a few weeks and the larvae are liberated after hatching.
- The unfed larvae attach to the hosts and start feeding particularly on the body area where there are less feathers like the neck and thigh.

- After full engorgement, the larvae drop on the ground and hide away in search of appropriate dark places and proceed to molting. After molting, they transform into nymph stage. Two nymph stages are usually found in the life cycle of *Argas persicus*. Each nymph stage takes a blood meal and drops on the ground.
- Then the fully developed second nymph attacks the hosts and becomes engorged.
- After full engorgement, the nymph drops on the ground.
- On the ground, the nymph undergoes molting and is transformed into adult.
- The adult continues to feed on the hosts and leaves the hosts and drops on the ground.
- On the ground, the ticks copulate and gravid females are produced.

Importance

The ticks transmit *Anaplasma marginale*, *Aegyptianella pullorum*. They also transmit *Borrelia anserina*, a causative agent of avian spirochetosis severe pathogen of poultry. The ticks attack humans.

Continuous irritation caused by the ticks caused reduction of feed and drinking water intake. The birds lose the condition. The meat and egg production go down.

Genus: ***Otobius***
Species: *O. megnini*
Common name: spinose ear tick
Hosts: cattle, sheep, goat, dog, etc.
Site: The larvae and nymphs remain attached on the ear.

Morphology

- They are bluish gray in color. Legs and mouth parts are pale yellow in color.
- The ticks do not have eyes.
- The ticks have a distinct constriction at the middle part of the body providing the body a fiddle shape.
- The larva is spherical.
- The nymph is covered by many spines.

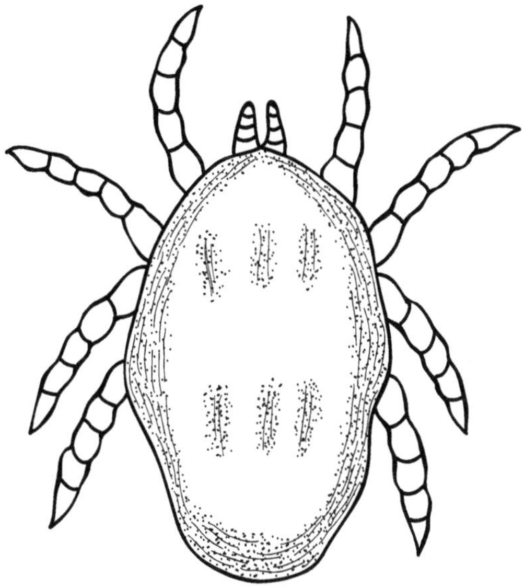

Fig. 7.30 *Otobius* sp. (Nymph stage)

Biology

- The adult ticks lay eggs in the sheltered spots.
- After hatching, the hexapodal larvae are liberated.
- The larvae attack the hosts and reach the ear. The larvae start feeding on the ear.
- After full engorgement, the larvae pass on molting and are transformed into octapodal nymph inside the ear. The ticks undergo two nymphal stages.
- The nymph feeds on the hosts for about 1–7 months and then drops on the ground and takes shelter in cracks and crevices.
- The nymphs then transform into adult.

Note: Only the larvae and nymph are parasitic.

Importance

The ticks do not transmit any pathogen but harm the host by mechanical damage to the ear and eardrum. The nymphs suck blood. They cause irritation and inflammation at the affected site.

Genus: *Ornithodoros*
Species
Ornithodoros moubata
Common name: eyeless tampan of Africa

Morphology

The female is about 10 mm long and the males are 8 mm long. The ticks are leathery and wrinkled. The body is covered by small nodules. Outer border of the ticks has serrated edges. The tick lacks "suture line" at the border of dorsal and ventral surface.

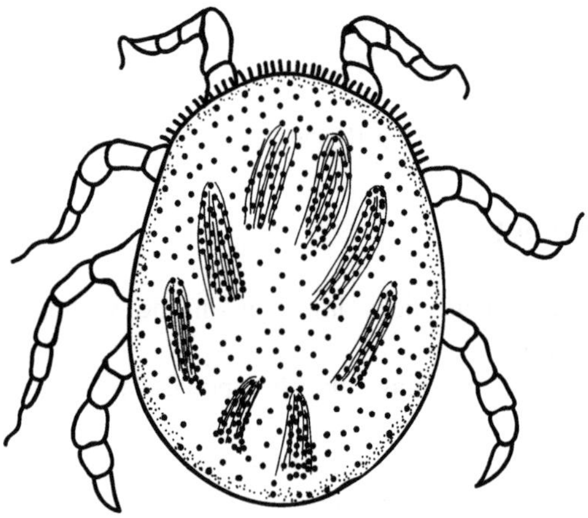

Fig. 7.31 *Ornithodoros* sp.

Biology

- The female tick takes blood meal and then only copulation takes place. The copulation stimulates the ovarian development, and the female becomes gravid later.
- The gravid tick lays the eggs on sand.
- After 2 weeks, the eggs are hatched and the larvae are liberated. The larvae do not feed.
- The larvae pass on one molting after a few days and are transformed into nymph.
- The nymphs attack the hosts and get engorged.
- After engorgement, they drop on the ground and transform into adult. Four nymphal stages are passed for formation of male adult and five nymphal stages for adult female tick.

Importance

O. *moubata* transmits *Borrelia duttoni*. *O. lahorensis* transmits *Anaplasma* spp. They are important reservoir hosts of virus of African swine fever. They are also regarded as vector of *Aegyptienella pullorum of* fowl.

Suborder: Trombidiformes—Family: Trombiculidae

Family Synopsis

The larval stages are called harvest mite, chigger mites, lepte automnale. They are morphologically characterized by large orange, yellow, red colored or dark red colored body, different body segments (gnathosoma, propodosoma, hysterosoma), and densely hairy body giving them a velvety appearance.

Genus: *Trombicula*

The larval stage is called harvest mites, chigger mites, aoutat, lepte automnale etc. The mites are large in size and red, orange, or yellow in color. The body is covered with profuse hairs. The pedipalps are characteristic in having thumb-like nail found on the last segment. They cause black soil mange in sheep and cause "heel bug" in horse. *Trombicula autumnalis* is called harvest mite, aoutat, and lepte automnale in Europe. *Trombicula akamushi* are the vectors of *Rickettsia tsutsugamushi*, a causative agent of scrub typhus or mite typhus.

Family: Demodicidae

Taxonomic rank	
Kingdom	Animalia
Phylum	Arthropoda
Class	Arachnida
Order	Acarina
Suborder	Trombidiformes
Family	Demodicidae
Genus	*Demodex*

Suborder: Trombidiformes—Family: Demodicidae

Genus: *Demodex*

Family Synopsis

These are important pathogenic ectoparasites occurring in different types of mammals. They are morphologically characterized by crocodile-like elongated body having head, thorax, and elongated striated abdomen. There are four pairs of stumpy legs. The mites occur in the hair follicle and sebaceous glands. The life cycle is direct type. The developmental stages are egg, larva, nymph, and adult. They cause demodectic mange or follicular mange.

Morphology

Adult

Shape: They are elongated.

Size: They are 0.25 mm long

Body division: The body of the mite is elongated. This is divided into three parts comprising the head, thorax, and abdomen.

Head: The head is distinct. The mouthparts consist of a pair of chelicerae, a pair of pedipalps, and one unpaired median hypostome.

Abdomen: The abdomen is striated transversely.

Legs: Four pairs of stumpy legs are present.

Egg

The eggs are spindle shaped. The eggs are 85–105um long. The width is 32–56um. Formed larva is found in most of the eggs.

Larva

Protolarva

The body is elongated. Three pairs of legs are present. The distal plates of the appendages bear a single tridentate tubercles.

Deutolarva

These are elongated. Three pairs of legs are present. The basal podal plates of the appendages (3 pairs) possess two tridentate tubercles each.

Nymph

Protonymph

The protonymphs are quite similar to deutolarva. Only difference lies on number of appendages. In deutolarva, three pairs of legs are found whereas the protonymphs possess four pairs. Both these stages possess tridentate tubercles in their leg appendages.

Deutonymph

The deutonymphs are quite similar to adult. But the appearance is more delicate compared to adult. The spine like structures on dorsal capitulum are poorly developed.

Fig. 7.32 *Demodex* sp.

Biology

Fertilization

Fertilization is internal. Mating occurs in the follicle opening.

Egg Deposition

The female mites deposit their eggs in the hair follicles and sebaceous glands. One female deposits about 20–24 eggs.

Hatching

After hatching, the larva comes out from the eggs.

Larva and Nymph Development

There are two larval stages i.e., protolarva and deutolarva. Larva transforms to the nymph. There are two nymphal stages: protonymph and deutonymph.

Adult Development

Along with the sebaceous material, the larvae and nymph come to the outlet of the hair follicle and get matured there.

Predisposing Factor for Demodicosis

The pathogenesis caused by demodectic mange is not well understood. It has been found that immunosuppression due to hyponutrition and other causes results in severe demodicosis. There is genetic predisposition on occurrence of demodicosis. Certain breeds are more susceptible to generalized demodicosis. Long steroid therapy is also a predisposing factor. Different endocrine diseases, i.e., hyperadrenocorticism, hypothyroidism, also predispose generalized demodicosis.

Pathogenesis

All the spp. under the genus are pathogenic. Though the *Demodex* sp. is an inhabitant commensal occurring in the skin, this causes severe disease in dogs mainly in the young animals of low immunity attributed to lack of nutrition. Usually the disease occurs in two forms, i.e., squamous form and pustular form. The squamous form occurs as initial inflammatory reaction along with erythema which furthers up with thickening of the skin, distension of the hair follicle, hyperkeratosis, hyperpigmentation, alopecia, etc. The skin color changes to coppery red which is characteristic features of this mange. Pustular form of secondary bacterial invasion results in severe pyoderma. The pustules get ruptured and discharge pus and exudate material. The pus and exudate get dried to form powdery crusts which is a common characteristic feature. There may be death of the animals due to toxemia.

Pathology

Structural Pathology

As described before, the mites live in hair follicle and sebaceous glands where they multiply rapidly and cause distension of the hair follicle and sebaceous glands and they massively affect the skin.

Histopathology

Histopathological study depicts huge mites in the hair follicle with folliculitis, perifolliculitis, and furunculosis. Marked hyperplasia of the sebaceous gland occurs. Examination of interstitial and perivascular exudate reveals lymphocytes, plasma cells, and macrophages. Follicular hyperkeratosis is also found.

Immunopathology

In canine demodicosis, cell mediated immunity plays the predominant role. Different studies show presence of pro-inflammatory cytokines and suppressor cytokines. Pro-inflammatory cytokines cause inflammation along with recruitment of different inflammatory cells. IL10 and TGF beta are two immunoregulatory cytokines which inhibit the suppression of inflammation and other immune activity. That's why the mites remain for long period in the host.

Molecular Pathology

Soman et al. (2020) worked on molecular reason of pathogenesis. TLR2 receptors are present on the keratinocytes of the skin. These receptors are very important for recognition of the chitin material present in mites. The aforesaid worker demonstrated that the clinical condition of demodicosis is attributed to elevated expression of TLR2 receptors. They also found that the organisms escape the host immunity by inducing elevated expression of genes of IL10 and TGF beta. Their works further depicted that proliferation of the organisms is due to reduced expression of IL6 genes.

Other *Demodex* Species

Demodex bovis	These mites occur in the cattle. These are called "cattle follicle mite"
Demodex caprae	The species cause mange in goats. These are called "goat follicle mite"
Demodex equi	These mites occur in equines
Demodex caviae	The hosts are guinea pigs
Demodex cervi	The hosts are red deers
Demodex folliculorum	The hosts are humans
Demodex ovis	The species cause mange in sheep
Demodex phylloides	These mites occur in pig

Diagnosis and Treatment

Examination of skin scraping is done for identification of this mite. Deep scraping is taken until capillary oozing occurs because tissue material from the hair follicle is required. For treatment, see the end part of this chapter.

Suborder: Trombidiformes—Family: Cheyletidae

Family Synopsis

These are important pathogenic ectoparasites occurring on the skin of different mammals. They are characterized by a waist of body, elongated palp, palpal claws, M-shaped gnathosomal peretremes, bladelike chelicerae, and comblike tarsal appendages.

Genus: *Psorergates*

Psorergates ovis

This is a species under the family Cheyletidae. The hosts are sheep. The disease caused by the mites is also called "Australian itch" occurring in the Republic of South Africa.

Psorergates bos

This mite affects cattle. These mites are cattle itch mite.

Suborder: Sarcoptiformes—Family: Sarcoptidae

Family Synopsis

These are important pathogenic ectoparasites occurring on the skin of different types of mammals, i.e., cattle, sheep, goat, dog, rabbit, etc. They are characterized by globose shape, four pairs of stumpy legs, scaly and spinose body, absence of eyes. The life cycle is direct. The developmental stages are egg, larva, nymph, and adult. They are true burrowing mites and cause sarcoptic mange.

Taxonomic position	
Kingdom	Animalia
Phylum	Arthropoda
Class	Arachnida
Order	Acarina
Suborder	Sarcoptiformes
Family	Sarcoptidae
Genus	*Sarcoptes*

Genus: *Sarcoptes*

The members of this genus occur in humans, cattle, sheep, goat, pig, etc. The seat of predilection of these mites is the skin where they cause mange known as scabies. The shape of the mites is globular. The body bears scales and spines. One important species under this genus is *S. scabiei*.

Morphology

Size: The males are 200–240 um × 150–200 um. The females are 330–600 um × 250–400 um.

Shape: Typically they appear like tortoise. The dorsal surface is convex. The ventral surface is flattened. They are oval in shape.

Eyes: The eyes are absent.

Body: The body is undivided. As others, they do not have cephalothorax division. The minute spines and scales are present over the body. Body covering has typical foldings at the dorsal surface.

***Legs*:** Two pairs of legs are found. The last two pairs of legs do not cross the body margin. The legs are short and bear caruncle (bell shaped suckers). But it is to be noted that all legs do not bear the caruncles. Other legs end in bristles. In the male, the first, second, and fourth pair of legs bear caruncles. In the female, the first and second pair of legs bear caruncles. The pedicels bearing the curuncles are not segmented.

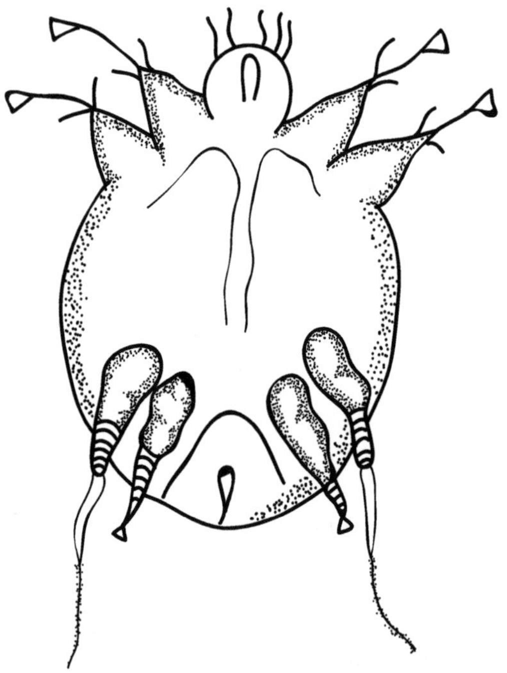

Fig. 7.33 *Sarcoptes scabiei*

Biology

The females penetrate the skin and form tunnels where they lay their eggs. After hatching, the six legged larvae come out. Some larvae come out from the tunnel and reach the surface and get perished.Others burrow the stratum corneum and form moulting pockets. The larvae then transform into the nymphal stages in those moulting pockets. There are two nymphal stages, i.e., protonymph and dutonymph. Then the adults are formed from the last nymph stage. The second pair of legs do not cross the body margin.

Pathogenesis

The fertilized females burrow the skin and form tunnels. In these created tunnels, the females lay eggs. The eggs hatch. After hatching, the larvae are liberated and get attached to the hair follicles. In the process of development, nymph and adults are formed. The infection starts with erythema. Swelling of the skin occurs as a result of allergic reaction due to hypersensitivity I reaction. Papules are formed. Due to allergic reaction, histamine is released and this histamine causes local allergic reaction. This allergic reaction results in itching and scratching. Constant scratching exacerbates the condition. The papules get ruptured and discharge exudate. The skin surface appears ragged and moist with exudate material. Pyoderma commonly occurs as a result of secondary bacterial infection. Secondary fungal infection may also occur. In the course of progression of the disease, the exudate and pus become dried and form crusts. In conclusion, the clinicopathological features occurring in different times are erythema, swelling, thickening of the skin, wrinkle formation, papule formation, hyperkeratosis, acanthosis, exudation, crust formation, etc.

Usual sites of infection in humans are the hands, elbows, groins, umbilicus, back, buttock, penis, etc. Diagnostic clinical symptoms are intense itching. Other features are vesicle formation, papule formation, etc. Weeping eczema is also a feature. Since the organisms affect the skin, dermatitis is the main pathological feature. Dry crust formation, alopecia, wrinkle formation, and thickening of the skin are the pathological features exhibited by the animals infected with *Sarcoptes scabiei.*

Pathology

Anatomical Pathology

Formation of burrow in the skin, erythema, papule formation, pyoderma, swelling, thickening of skin, hyperkeratosis, etc. are some anatomical pathologies.

Immunopathology

There are different ways by which pathology is caused. Two hypersensitivity reactions are very important, i.e., hypersensitivity I and hypersensitivity IV. In hypersensitivity I reaction, the antigen is bound to IgE antibody. In this state, mast cell or basophil binds to the Fc portion of that immunoglobulin. This results in degranulation of mast cell and histamine is released from the mast cells. This histamine causes tissue reaction with allergic manifestation. Type IV hypersensitivity is a delayed type of hypersensitivity which is caused by T cells, macrophages, etc. Langerhans cells take up the parasitic antigen of the parasites. CD4+ T cells and CD8+ T cells are produced. CD4+ T cells release different cytokines which activate the macrophage. The activated macrophages release TNF alpha and IFN gamma. Cytotoxic T cells release different toxic materials and cause tissue reaction. The cytokines recruit different immune cells there. Many eosinophils also aggregate in the affected area. ADCC mediated by eosinophils results in production of different chemicals by the eosinophils.

Molecular Pathology

Bhat et al. (2020) reported numerous differentially expressed genes including pro-inflammatory cytokines IL17A, IL8, IL19, OSM, etc. in CS (crusted scabies) relative to OS (ordinary scabies). They also demonstrated influence of genes associated with immune regulation comprising CD274/PD-L1, IL27, immune signaling (TLR2,TLR8), and antigen presentation (RFX5, HLA5, HLA-DOB) in early host response to CS.

Diagnosis

In addition to analysis of specific clinical signs, diagnosis of the disease is done by examination of skin scraping. The scraped material is digested in 10% KOH, and then the sediment is examined for detection of mites.

Treatment

Benzyl benzoate and ivermectin are some effective drugs. See end part of this chapter.

Genus: *Notoedres*

Species

Notoedres cati

The morphological features are similar to *Sarcoptes*. Dorsal position of the anus is found in *Notoedres* but it is terminal in *Sarcoptes*. Pedicels are not segmented. These are burrowing mites. These mites infect cats. But rodents are also affected. The disease is called notoedric mange or feline scabies. The infection usually starts at the ear which spread to the eyes and gradually captures the whole body. Alopecia, exudation, and dry crust formation are some clinical features.

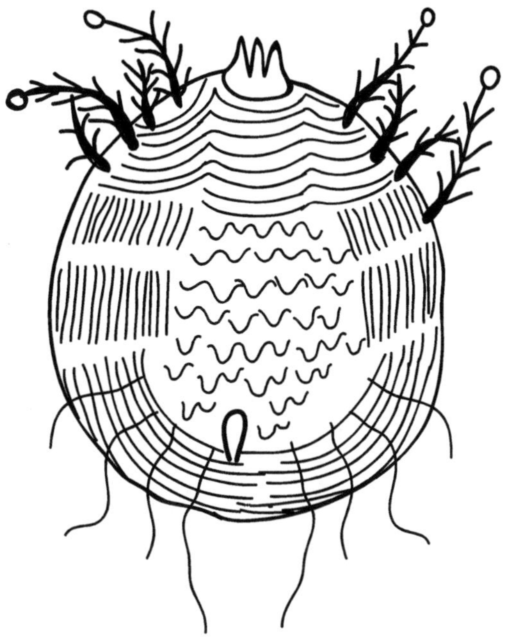

Fig. 7.34 *Notoedres* sp.

Genus: *Cnemidocoptes*

This is a genus under the family Sarcoptidae. The hosts are birds. The condition produced by the mite is called "scaly leg" condition and "depluming itch." They are globose. Scales are present on the body but spines are absent.They are globose. Scales are present on body but spines are absent.

Species

Cnemidocoptes gallinae

This is a species under the family Sarcoptidae. The hosts are birds. The condition caused by the mites is called depluming itch. The mites attack the shaft regions of the feathers and set up inflammation there. Constant irritation occurs at the region, and the birds inflict the condition by picking and spilling the area.

Cnemidocoptes pilae
This species comes under the family Sarcoptidae. These mites are causative agents of scaly leg of budgerigar or parakeet.

Cnemidocoptes jamaicensis
This is a species under the family Sarcoptidae. The hosts are golden thrush. The condition produced by the mite is called "scaly leg".

Cnemidocoptes mutans
This is a species under the family Sarcoptidae. The hosts are fowls. The condition produced by the mite is called scaly leg of poultry. Under the scales, the skin is affected and inflammation starts. The scales are displaced later. The skin becomes thickened and keratinization occurs.

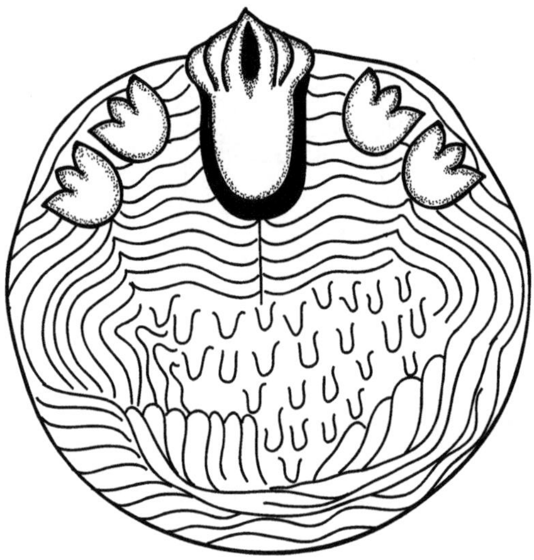

Fig. 7.35 Cnemidocoptes

Suborder: Sarcoptiformes—Family: Psoroptidae

Family Synopsis

These are important pathogenic ectoparasites occurring on the skin of different mammals. They are characterized by oval body. They are nonburrowing mite. The chelicerae are stylet-like. All legs cross the body margin. Bell shaped suckers (caruncles) are present in some or all legs joined by pedicels. The pedicels are segmented in *Psoroptes*, but short or unsegmented in other two genera i.e., *Chorioptes* and *Otodectes*. The life cycle is direct type. The different developmental stages are egg, larva, nymph, and adult. They cause psoroptic mange.

Species of *Psoroptes*

Psoroptes ovis
The members of this species affect sheep. The seat of predilection of the mites is the skin. The condition caused by the mite is called the "sheep scab." The body is oval. Bell-shaped suckers (caruncles) are present in the first, second, and third pair of legs of male, but these are present in the first, second, and fourth pair of legs of female. The pedicels are jointed. The epidermis is affected by the mites and they set up local inflammatory reaction. There occur swelling and itching. Serus exudate comes out from the affected area.

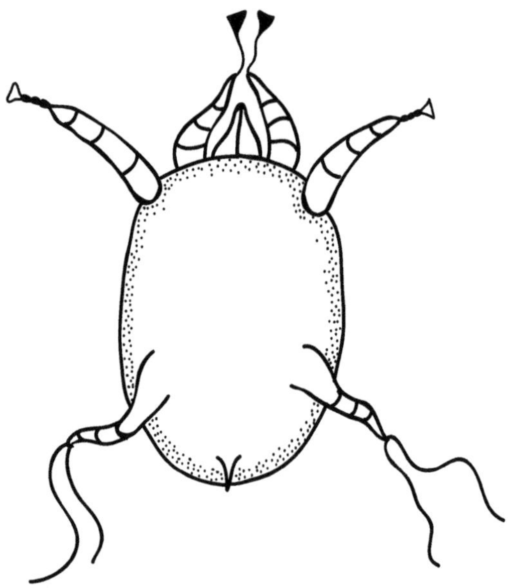

Fig. 7.36 *Psoroptes* ovis

Psoroptes bovis
These mites occur in cattle.

Psoroptes caprae
These mites occur in goat.

Psoroptes cervinus
These mites occur in bighorn sheep.

Psoroptes cuniculi
These mites occur in rabbit, sheep, goat, donkey, and horse.

Psoroptes equi
These mites occur in equines.

Psoroptes hippotus
These mites occur on the ears of horse and mule.

Psoroptes natalensis
These mites occur on cattle, Indian water buffalo, and zebu.

Pathogenesis
The mange caused by the members coming under the genus *Psoroptes* is called psoroptic mange. The species are *P. ovis*, *P. equi*, *P. cuniculi*, *P. natalensis*, and *P. equi*. Of these species, the most important species is *P. ovis* which causes sheep scab. The wool become loose and fall out later on. To alleviate irritation, the sheep scratch and pull out the wool. Serum exudes from the affected area which coagulates and form crusts. The mites cannot survive in the bare patches, and they subsequently migrate to the edge of the lesion and the lesion spread. A large area is affected. *P. cuniculi* affect the ear of rabbit.

Genus: *Chorioptes*

This is a genus which comes under the family Psoroptidae. These are mites which cause chorioptic mange. The parasites resemble the *Psoroptes* sp. Tarsal suckers bear unjointed pedicels. All legs of male are provided with suckers, but in case of female, the first, second, and fourth pair of legs bear suckers. Copulatory tubercles are present in pubescent female. Male bears abdominal lobes.

Chorioptes bovis
This is a species under the genus *Chorioptes*. The hosts are cattle and horse. The conditions caused by the mites are foot mange or itchy legs of horses. On the fetlock regions, a characteristic scab like lesion is produced.

Chorioptes texanus
This is a species under the family Psoroptidae. The seat of predilection of the mites is the ears. The hosts are reindeers, goat, etc.

Genus: *Otodectes*

The important species is *Otodectes cynotis*. They affect dog, cat, fox, etc. They resemble the *Chorioptes*. They have unjointed pedicel. In male, all legs are equipped with suckers, but these are only in the first and second pair of legs of females. The mites cause ear mange. Mange starts at the ear which spread gradually. The mites affect the epidermis. There occurs severe itching. The animals scratch the ears. Secondary bacterial infection occurs. Repeated self-infliction causes development of hematoma. Purulent hemorrhage is caused due to secondary bacterial infection. Perforation of the tympanic membrane may occur. In more advanced stage, the ears droop.

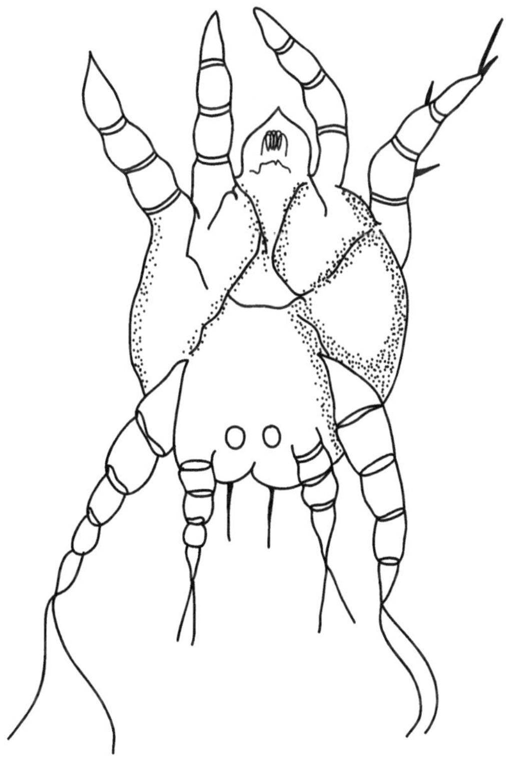

Fig. 7.37 *Otodectes* sp.

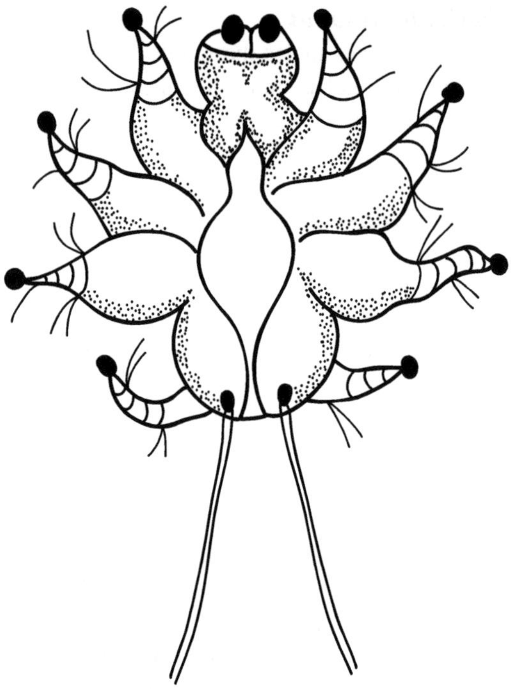

Fig. 7.38 *Psorergates* sp.

Order: Acarina—Suborder: Mesostigmata—Family: Dermanyssidae

Family Synopsis

These are important pathogenic ectoparasites occurring on the skin of different animals. They are serious bloodsuckers and cause marked anemia to the hosts. The mites are noticeably large in size and grayish white in color and become red after engorgement.

Taxonomic rank	
Kingdom	Animalia
Phylum	Arthropoda
Class	Arachnida
Order	Acarina
Family	Dermanyssidae

This is an important family of the suborder Mesostigmata. The common hosts are mostly poultry birds. The *Dermanyssus gallinae* are found to transmit the *Borrelia anserina* causative agent of fowl spirochaetosis. The mites are large in size as these can be found by the unaided eyes.

Genus: *Dermanyssus*

These are poultry mites. The important species is *Dermanyssus gallinae*.

Common name: red mite of poultry, red poultry mite, bird mite, chicken mite
Site: Skin
Hosts: Domestic and wild birds.
Note: Though the common name of the birds is "red mite of poultry," they may also affect mammals including humans.

Morphology

The mites are approximately 1 mm long. The unfed mites are usually grayish white in color. After engorgement, the mites look reddish brown. Characteristically, they have dorsal shield which has truncated posterior end. The legs are notably long. The pair of chelicerae are long and characteristically whip-like.

Life Cycle

Pre Egg Laying

The mites are usually nocturnal feeders. That means they feed the hosts during night and leave the hosts after feeding and hide themselves in cracks and crevices.

Egg Laying

In the cracks and crevices, the engorged mites mate. After mating, the females lay their eggs.

Hatching

Hatching of eggs occurs in the cracks and crevices.

Further Development

Further developmental stages are larva, protonymph, deutonymphs, and adult.

Harm Caused

Direct Harm

The mites affect the skin of birds. The breast and legs are mostly affected. The mites cause irritation which subsequently causes dermal pathology. Hyperpigmentation occurs. They scratch the affected area and self-inflict the areas. This may lead to secondary bacterial infection resulting in pustule formation. Rupture of pustules results in exudation. Later scab is formed.

Indirect Harm

The affected birds may suffer from anemia as the mites suck blood from the hosts. The affected birds take less food and drinking water which results in debility, loss of egg production, loss of body weight, etc. The mites transmit *Borrelia anserina*, causative agent of avian spirochetosis. They also act as vector of eastern and western encephalitis.

Control

- Treatment of the affected birds should be done using different ectoparasiticidal agents including ivermectin.
- Poultry house should be cleaned properly. Inspection of the poultry house should be done whether any cracks and crevices are there.
- Novel control has been demonstrated by different workers using some predatory mites, i.e., *Androlaelaps casalis* and *Hypoaspis miles*.

Genus: *Ornithonyssus*

The mites occurring in the warmer parts of the world is *Ornithonyssus bursa*. These mites are called "tropical fowl mite." These mites attack different types of birds, i.e., fowl, pigeon, sparrow, etc. These mites also attack humans causing pruritis. *Ornithonyssus sylviarum* is called northern mite of fowl. *Ornithonyssus bacoti* occur in rat and is called tropical rat mite.

Treatment Against Tick and Lice

Drug	Concentration	Mode of administration	Dose
Amitraz (liquid)	12.5% w/v	Spray on body	2–4 ml/l in water; Not recomended in horse, cat, pup.
Cypermethrin (liquid)	10% w/v	Spray on body or plunge dip	1 ml/l in water
Deltamethrin (liquid)	1.25% w/v	Spray on body or plunge dip	2 ml/l in water
Malathion (liquid)	50% w/v	Spray on body	10 ml/l in water
Coumaphos (wettable powder)	50% w/v	Spray on body	0.5 g/l in water (for lice); 1–2 g/l in water (for ticks)
Carbaryl (Powder) 5%; (liquid) 10% w/v	5% (powder); 10% w/v (liquid)	Spray/wash	Puppies and kittens below 4 weeks of age not recommended
Ivermectin injection Ivermectin solution Ivermectin tablet Ivermectin pour on	1% w/v in case of Ivermectin injection; 0.8% w/v in case of Ivermection solution; 0.5% w/v in case of Ivermectin pour on	S/C (injection) Oral Oral Pour on	0.2 mg/kg bw 0.2 mg/kg bw 0.2 mg/kg bw 2.5 ml/10 kg bw
Moxidectin injection	1% w/v	S/C or I/M	0.2 mg/kg bw

Treatment Against Mange

Drug	Concentration	Mode of administration	Dose
Amitraz	0.025%- 0.05%%	Dipping of cattle; Not recomended for horse cat, pup	Repeat after 2 weeks
Coumaphos	0.3%	Dipping of cattle and sheep	Repeat after 2 weeks
Ivermectin injection		S/C (injection)	0.2 mg/kg bw (repeat after 10 days in case of cattle, sheep, goat and after 14 days in case of dog)
Doramectin injection		S/C (injection)	0. 2 mg/kg bw for cattle, 0.3 mg/kg bw for sheep and goat

Treatment Against Myiasis

Type of myiasis	Drug	Mode of administration	Dose
General myiasis	Ivermectin injection Ivermectin tablet Ivermectin pour on (0.5% w/v) Moxidectin injection	S/C Oral Pour on S/C or I/M	0.2 mg/kg bw 0.2 mg/kg bw 2.5 mg/kg bw 0.2 mg/kg bw
Nasal myiasis and other myiasis	**Rafoxanide** **Nitroxynil** Ivermectin (as for ticks)	Oral Oral	15–20 mg/kg bw 15–20 mg/kg bw
Intestinal myiasis/ gastric myiasis	Ivermectin and moxidectin (as for ticks) Closantel Coumaphos (0.06%) in water	Oral Hot sponging (40–48 °C) on hairs where eggs are attached	10 mg/kg bw
Cutaneous myiasis	Ivermectin injectioni/m	I/M or S/C	0.2 mg/kg bw

Flea dermatitis: Pyrethrin 1% dipping for dog and cat; Fenthion spot on formulation 15.2 mg/kg bw topically for dog; 6.2 mg/kg bw for cat

Note: All formulations of ivermectin and moxidectin are contraindicated in animals found hypersensitive in previous administration.

Class: Pentastomida—Family: Linguatulidae

Genus: *Linguatula*

Species: *Linguatula serrata*

This species is called "tongue worm." They occur in nasal and respiratory passage of dogs, foxes, etc. They rarely occur in humans. The parasite is tongue-shaped. The body is transversely striated. The dorsal surface is slightly convex and the ventral surface is flattened. They are less pathogenic, but heavy infection may cause different respiratory trouble comprising sneezing, coughing, etc.

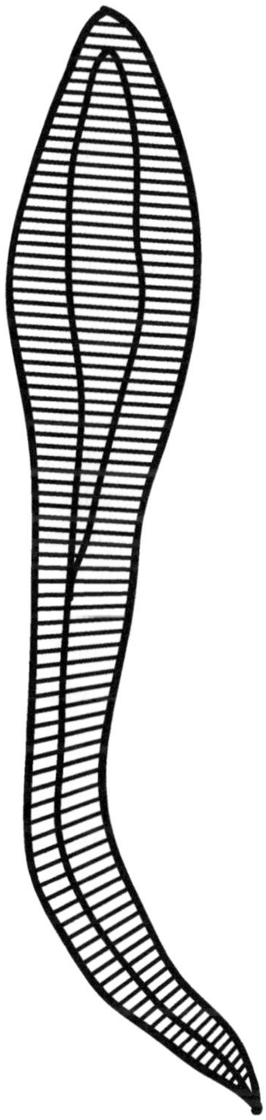

Fig. 7.39 *Linguatula* serrata

Special Part

Zoonosis by Arthropod Parasites

Pulicosis

Causative Agents

The important species is *Pulex irritans* which is a human flea. Dog, pig, and monkey may be affected by this flea also.

Mode of Infection

The organisms are transmitted directly.

Pathological Features

The fleas cause dermatitis which is as a result of hypersensitivity. **Alopecia** may occur.

Diagnosis

Gross and microscopical examination of the parasite is done for diagnosis.

Control

Use of medicated shampoos and combing are two common control steps. Attention should be paid to the surroundings also. The clothing should be cleaned properly. The floor should be cleaned properly.

Gasterophiliasis

Causative Agents

Some important species are *Gasterophilus intestinalis*, *G. haemorrhoidalis*, *G. inermis*, G. *nasalis*, *G. pecorum*, and *G. meridionalis*. *G. intestinalis* and *G. haemorrhoidalis* have zoonotic importance.

Hosts Affected

G. nasalis occurs in horses. *G. pecorum*, *G. nigricornis*, and *G. intestinalis* occur in horses and donkey. *G. inermis* occurs in horses and zebras. *G. meridionalis* occurs in zebra. *The larvae of *Gasterophilus* may also affect the subcutaneous tissue of humans. *Gasterophilus intestinalis* and *G. haemorrhoidalis* are important for humans. The larvae penetrate the skin and cause serpentine linear lesion. The myiasis is called "myiasis linearis."

Mode of Infection

Female flies lay eggs on the face, lips, fetlock regions, etc. The eggs are ingested when the animals lick. Hatching of eggs is assisted by temperature created during licking and mechanical friction. The larvae find their way to reach their relevant site and develop there. Female flies also lay eggs on human skin. The eggs are hatched and the first stage larvae cause creeping myiasis. The larvae can never develop into adult in case of humans.

Pathological Features

The larval stages cause gastric myiasis in equines. They cause craterlike ulcer in the gastric wall. In humans, irritation and allergic reaction occur on the skin.

Diagnosis

The larvae of the flies should be examined properly for identification basing on the morphological characters. Exposure to horse and equids may be suggestive for diagnosis of human infection.

Control

(a) Since the matured larvae come out through the feces, proper disposal of feces is an important control technique. The floor of the animals should be cleaned properly.
(b) A sand paper can be used for rubbing the skin to remove the eggs. This can also be done using a device having a sharp edge.

Hypodermiasis

Causative Agents

The important zoonotic species is *Hypoderma bovis* which is called "northern cattle grub." *H. bovis* normally affect the cattle, but they may also occur in horses, goats, and humans. There have reports of cutaneous myiasis in humans caused by larvae of *H. lineatum*.

Mode of Infection

The *Hypoderma* flies lay eggs on the hairs of the skin. After hatching, the larvae are liberated. Then the liberated larvae penetrate the skin of the hosts.

Pathological Features

In animals, larvae of *H. bovis* cause fat necrosis, inflammation, etc. at the vicinity of the parasitic infection. They may further reach the periosteum and bone causing periostitis and osteomyelitis. Warble stages of the flies cause breathing hole wherein the larvae remain and cause cutaneous swelling and suppurative abscess. In humans, the larvae are found in skin and subcutaneous tissue where they cause lesions.

Diagnosis

The larvae can be identified basing on characteristic morphological features. Funnel-shaped spiracular plate is characteristic identifying feature. Flat tubercles and small spines are found on third stage larvae of flies.

Oestrus sp.

The normal hosts of *Oestrus ovis* is sheep. But these flies may also attack dogs and humans. In both humans and dogs, the larvae can't develop beyond the larval stage. In humans, the larvae are dropped by the female flies in the orbit, mouth, nostril, ear, eye, etc. The larvae cause acute conjunctivitis, throat swelling, and frontal headache in accordance with presence of larvae in different regions.

Dermatobia spp.

Dermatobia hominis are called berne, nuche, and forcel which affect humans. They also affect cattle, dogs, cats, sheep, etc. *Dermatobia hominis* have unique egg depositing behavior. They catch a mosquito, fly, or rarely a tick and lay a batch of eggs. When these mosquitos and flies take blood meal, the eggs are transferred to the hosts. The larvae hatch out and enter through the skin of man, animal. The larvae cause swelling of skin which has a central opening used for breathing of larvae. Skin swellings also occur in the animals.

Old World Flesh Fly

Wohlfahrtia magnifica is called old world flesh fly. They deposit first instar larvae in the external ear of man and also in the lesions of the skin, tongue papillae, sore eyes, etc. The larvae may cause disfigurement of the affected area of man and animal.

Wohlfartia vigil vigil

Wohlfahrtia vigil is called Nearctic flesh fly. They deposit larvae on uninjured skin of humans. Boil-like traumatic myiasis occurs. They also cause myiasis in animals.

Wohlfahrtia vigil opaca

They deposit làrvae on the skin of children and cause furuncular myiasis in man. They also cause myiasis in animals.

Sarcophaga spp.

S. haemorrhoidalis, *S. carnaria*, and *S. fusicauda* usually larviposit on ulcerated human tissues. There have been report of intestinal myiasis in man caused by the larvae of *Sarcophaga*.

Sarcoptosis

Causative Agents

Sarcoptes scabiei affects domestic animals and humans.

Mode of Infection

Direct transmission is the mode of infection.

Pathological Features

Usual sites of infection of humans are the hands, elbows, groins, umbilicus, back, buttock, penis, etc. Diagnostic clinical symptoms are intense itching. Other features are vesicle formation, papule formation, etc. **Weeping eczema** is also a characteristic pathological feature. These mites cause dermatitis. Thickening of the skin, dry powdery crust formation, wrinkle formation, exfoliation of the skin, and alopecia are the pathological features found in dogs.

Diagnosis

In addition to analysis of specific clinical signs, diagnosis of the disease is done by examination of skin scraping. The scraped material is digested in 10% KOH, and then the material is centrifuged and the sediment is examined for detection of mite.

Cheyletiella Infection

Causative Agents

There are different species of *Cheyletiella* which occur in different hosts. *C. yasguri* occurs in dogs. *C. parasitovorax* occurs in rabbit, cat, dog. The mites may infect humans also by direct contact.

Transmission

Transmission occurs by direct contact.

Pathological Features

Diffuse papular lesions occur on the arms, chest, and abdomen of humans. *C. parasitovorax* causes walking dandruff in humans. A dry scaly dermatitis occurs in animals.

Dermanyssus and *Ornithonyssus* Infection

Dermanyssus gallinae are called "red mite" of poultry. *Ornithonyssus bursa* are called "tropical fowl mite." *Ornithonyssus bacoti* is called tropical rat mite. These mites affect humans also.

Tick Infection

Soft Tick Infection

There are different soft ticks which have zoonotic importance. These are *Argas persicus*, *Otobius megnini*, and *Ornithodoros moubata*. *Argas persicus* is a soft tick which affects birds. But these ticks occasionally affect humans and cause painful wounds. The larval and nymph stages of *Otobius megnini* attack the ears of dogs. But they also affect sheep, goats, humans, etc. *Otobius megnini* is called spinose ear tick.

Hard Tick Infection

The hard ticks having zoonotic importance are *Ixodes holocyclus*, *Rhipicephalus (Boophilus) annulatus*, *Amblyomma maculatum*, *Haemaphysalis leporispalustris*, *Dermacentor andersoni*, *D. variabilis*, etc. These ticks affect both man and animals.

Harm Caused

The ticks cause the following harms to humans:

Bite Injury

The ticks pierce the skin and suck blood using the hypostome and chelicerae. This activity may cause bite injury to the skin. In most of the cases, bite is not painful. But the mouthparts of the ticks may be left in the skin if removed carelessly which causes pain.

Tick Paralysis

The toxin from the salivary glands is injected in the host during bloodsucking. This toxin causes acute ascending flaccid paralysis. The ticks which cause tick paralysis are *Ixodes holocyclus*, *Dermacentor andersoni*, and *Amblyomma maculatum*.

Disease Transmission

Different pathogens are transmitted by different ticks. These are Rocky Mountain spotted fever, Colorado tick fever, Crimean-Congo hemorrhagic fever, Kyasanur forest disease, etc.

Diagnosis

Diagnosis can be done by microscopical examination and identification of ticks basing of morphological characters.

Control

- Mechanical removal and destruction of ticks from the skin
- Habitat alternation by destroying their nests
- Application of insecticidal agents in the animal houses
- Sealing of cracks and crevices of the houses

Linguatulosis

Causative Agents

Linguatula serrata is the causative agent which is called dog tongue worm. This parasite affects dogs, cats, foxes, etc. Humans are rarely infected.

Mode of Infection

The parasite lays eggs in the upper respiratory passage of the definitive hosts. The eggs are expelled either with cough material during sneezing or through the feces. The eggs are ingested by intermediate hosts (sheep, goats, etc.). The eggs are hatched and larvae are liberated in the intestine of the intermediate hosts. The larvae penetrate the intestinal wall and enter mesenteric lymph node where the larvae molt to nymph stage. They form cysts in which they lie.The nymphs may break out of cysts and wander in different organs. The humans get the infection by ingestion of raw or undercooked meat of intermediate hosts harboring larva. The larvae are transformed into infective nymph stage. The final hosts get the infection by ingestion of infected muscles of the intermediate hosts harboring the infective stages of the parasites.

Pathological Features

In Animals

Encysted larvae are found in the tissues of intermediate hosts. The larval stages may cause granuloma formation, and they remain in those granulomas. Two types of linguatuliasis occur comprising visceral linguatuliasis and **nasopharyngeal linguatuliasis**. The hatched out larvae migrate in different visceral organs. The larvae induce formation of cysts or granulomas at the affected areas. The definitive hosts get the infection by ingestion of encysted larvae, and they suffer from nasopharyngeal linguatuliasis. The immature stages come out from the cysts, and they crawl up the esophagus and subsequently take the route to reach the nose, pharynx, tongue, etc. resulting in coughing, sneezing, etc. Nasopharyngeal linguatulosis is also called Halzoun syndrome or Marrara. Interestingly the adult parasites remain asymptomatic in dogs.

In Humans

Humans get the infection by ingestion of the infected liver of sheep and goats. The symptoms exhibited by humans are called "halzoun" and "marrara." The symptoms include irritation in the throat which occurs immediately after eating the infected liver of sheep, goats, etc. Different pathological features associated with this infection are congestion and edema formation in the larynx, lip, nose, pharynx, etc. Respiratory symptoms include nasal discharge, sneezing, coughing, dyspnea, etc. GI involvement results in vomiting. Different eye complications include photophobia, lacrimation, exophthalmia, etc.

Diagnosis

Eggs can be detected on examination of nasal discharge and fecal sample. Examination of biopsy material may reveal the nymph stage of the parasite.

Immuno-Arthropodology

There are usually three ways by which immunity against arthropods is displayed. These are as follows:

- Saliva proteins of some arthropods have low molecular weight which do not act as antigen. But these proteins after binding with skin collagen can elicit immune response. Subsequent exposures cause delayed type of hypersensitivity (hypersensitivity IV).
- The salivary antigens bind to epidermal Langerhans cells resulting in cutaneous basophil hypersensitivity.
- Hypersensitivity I also occurs in different arthropod infections.

Immunity against arthropods occurs by innate immunity and adaptive immunity. The dermal layer itself is a protective layer against any ectoparasitic infection. Different specific cells perform specific functions. These are eosinophils, neutrophils, basophils, mast cells, NK cells, macrophages, etc. These cells perform innate immunity. These cells play role in adaptive immunity also. Preformed inflammatory substances, acute phase proteins, enzymes, cytokines, etc. are present in these cells. These substances cause changes in the microenvironment of parasite-infected site. Both innate and adaptive immunity result in the following:

- Reduction of egg production
- Reduction of molting capability
- Reduction of weight of engorged arthropod
- Increasing bloodsucking period
- Death of ectoparasite

It has been found that antibody can be produced against salivary protein of bloodsucking flies. In case of flea bite, the saliva is not capable to elicit immune response, but if the saliva is conjugated with skin collagen, then it can produce immune response. Here, the salivary protein acts as hapten and the skin collagen acts as the carrier. In the skin, there is skin-associated lymphoid tissue (SALT). In case of tick, mite, and other ectoparasitic infestations, there occurs hypersensitivity I reaction. In this hypersensitivity reaction, an allergic reaction is set up due to reexposure of a specific antigen. The Fc part of IgE (antigen-bound) is bound to receptor of mast cells. This receptor is **Fc€R1**. In this state, degranulation of mast cells occurs. The granules release histamine resulting in allergic reaction and tissue injury which affect feeding of tick and other ectoparasites. In case of ectoparasitic infection, hypersensitivity IV reaction also occurs. This is delayed type of hypersensitivity having a lag phase between infection and occurrence of reaction. Th1 cells recognize soluble antigen of ectoparasite and release IL2 and IFN gamma. This IFN gamma activates the macrophage. This activated macrophage releases some cytokines which also cause tissue injury. Eosinophils play important role in ectoparasitic infection. Eosinophil contains cysteinyl leukotrienes and arachidonic acid based inflammatory materials. These cells also contain EBP (eosinophilic basic protein), ECP (eosinophilic cationic protein), and EDN (eosinophil derived neurotoxin). A set of cytokines are also present in the eosinophils which are **IL2**, **IL4**, **IL6**, **IL10**, **IL13**, **IFN gamma**, **TGF beta**. The enzymes present in the eosinophils are catalase, aryl sulfatase, elastase, phospholipase, etc. During ADCC, the eosinophils attach to the Fc part of immunoglobulin (antigen-bound). In this state, the eosinophils release EBP, EDN, ECP, etc. which cause tissue injury. NK cells also have Fc¥RIII receptor and by this receptor the NK cells bind to Fc part of antibody. NK cells also release their granules. These granules contain granzymes, perforins, etc. These substances also cause tissue reaction. Due to tissue reaction, the damaged area becomes unfavorable for feeding of arthropods.

Immunity to Biting Flies

Many biting flies take blood meal from the hosts. During blood meal, the saliva and other substances are injected in the hosts. These injected substances elicit immune response in the hosts, and due to these reasons, it is found that the severity of skin reaction decreases after several bites. This decreased skin reaction is due to development of immunity.

Immunity to Mange Mite

Both cell mediated and humoral immunity occur against mite infection. In the infected area, infiltration of different mononuclear cells including CD8+ T cells has been detected. Presence of CD8+ T cells indicates hypersensitivity IV reaction. This is a cell mediated immunity in which there is no role of antibody. However, it is evident that mite infection may also induce antibody production.

Immunity to Flea

The flea injects saliva in the hosts during bite. The protein component of saliva has low molecular weight. This protein of low molecular weight binds with skin collagen and become immunogenic. This modified form of immunogen elicits a type I hypersensitivity reaction. Later type IV hypersensitivity occurs. Due to this immune reaction, mononuclear cell infiltration comprising CD8+ T cells and macrophages occurs. Type I hypersensitivity reaction results in eosinophilic infiltration. This allergic reaction develops immunological resistance in cats against flea infection. In different vaccination trials, investigators found low flea population and reduced fecundity of the female flea.

Immunity to Myiasis-Causing Larvae

The migrating myiasis-causing larvae release enzymes and other substances which trigger immune response. It is evident that the sheep raises resistance against calliphorine myiasis after several attacks. Larvae of warble flies (*Hypoderma lineatum*, *Hypoderma bovis*) release enzyme and protease during migration in the tissues. This protease (hypodermin A) is immunogenic. Study shows that vaccination with *Hypoderma* protein can protect cattle from this infection.

Immunity to Ticks

There is a long history on investigation of acquired immunity to ticks. The first laboratory investigation of immunity against ticks was made in the year 1913. Bovine resistance to tick infestation includes both innate and acquired immunity. Immunity against ticks occurs in different ways. These comprise of hypersensitivity (immediate and delayed hypersensitivity), complement activation, antibody production, ADCC, cellular reactions, etc. The alternative pathway of complement activation results in acquired resistance. It has been found that tick infection induces increased levels of C3. Increased level of tick specific immunoglobulins has also been found by different investigators. The antigen-antibody complexes are deposited in the tick infested area. Subsequent activation of complement results in lesion formation. Complement activation mediated by classical pathways or alternative way results in generation of anaphylatoxin, degranulation of mast cells and basophils, and concomitant release of eosinophil chemotactic factors, i.e., histamine and vasoactive amines. The different bioreactive molecules released by different effector cells are histamine, leukotriene, prostaglandins, enzymes, etc. which cause cellular inflammation and lesion formation. C3a and C5a complement fragments cause degranulation of basophil and mast cells. C5a also contributes to the cellular influx at bite site. In lesioned areas, the tick cannot suck blood properly. The NK cells are actually analogous to cytotoxic T cells. Therefore, it is clear that all these substances cause obstruction or reduction of feeding of the ectoparasites. The immune substances also damage the physiology of ticks and other ectoparasites. Chemoattractants and vasoactive factors are responsible for leucocytic influx observed at tick attachment site. Hypersensitivity reactions, antibody dependent cellular cytotoxicity, alternative complement pathway, etc. occurs.

Tick Modulation of Host Immunity by Salivary Gland

Tick saliva contains **PGE2**. PGE can inhibit Th1 but cannot inhibit production of Th2 cytokines. In the skin of unsensitized hosts, tick saliva causes degradation of mast cells by breaking the plasma membrane using enzymes present in saliva. Tick saliva also contains substances which can inhibit alternative pathway of complement activation. The tick is capable of inhibition of anaphylatoxin of hosts produced due to complement reaction. Tick saliva reduces the activity of cytokines produced by macrophages. The saliva of tick contains anti-inflammatory and immunosuppressive molecules which reduce host defense. Tick saliva also can suppress function of lymphocytes.

Immunoprophylaxis Against Arthropods

An array of works were carried out in the field of anti-arthropod vaccination. Different types of antigens were used to perform immunoprotection trials comprising crude antigens constituting egg antigen, larval antigen, other somatic antigen of adult, etc. But these vaccines fail to produce considerable immunity due to lack of specificity and immunoprotective potential. Variable protection was reported in the literature on the utilization of larval antigen, but no vaccine has been commercialized using larval antigen. As has been mentioned above, many works have been carried out by different researchers using the antigens derived from the adult parasite, but considerable protection was not achieved. Later on, research work on immunoprotection was pinpointed toward characterization of concealed gut antigen of tick. A satisfactory result was achieved; this gave a good indication of use of these antigens for immunoprotection against tick infection. Researchers were encouraged to produce the vaccine using the tick gut antigen. But it is needless to mention that all proteins present in the tick gut are not immunodominant. Therefore, an array of works were directed toward identification of most immunodominant antigen of gut of tick. SDS-PAGE and Western blot technique were employed for detection of immunoprotective antigens. The efficacy of **Bm 86** and **Bm 91** was assessed, and these are two candidate proteins of *Boophilus microplus* tick gut which showed promising results. The target gene encoding immunodominant protein is taken out from the whole genome, and the gene encoding the specific protein is recombined with the plasmid or other vectors and is inserted into the *Escherichia coli*, *Pichia pastoris*, etc. for expression of the required protein. Then, the specific protein is isolated by specific protein purification protocol. Two vaccines using Bm 86 were commercialized in Australia and Cuba with the name **TickGARD** and **Gavac**, respectively. Efforts were also made by some workers to immunize the animals using the salivary proteins, and these trials displayed some protection.

Studies show identification and characterization of an arthropod derived storage protein, hexamerin. Two such proteins found in *Musca* sp. are Hex L and Hex F. Hex L is synthesized by larval fat bodies and secreted into the hemolymph. This protein provides amino acids and energy during metamorphosis. **Hex F** is synthesized by adult insect fat bodies and similarly secreted in the hemolymph. In females, this protein acts as amino acid reservoir which is used during the final stages of egg formation. Genomic clones containing full length copies of the genes M Hex L1 and Md Hex F1 encoding subunits of larval and adult female hexamerin were isolated and characterized (Moreira et al. 2004).

Molecular Arthropodology

In this segment, different molecular aspects comprising molecular taxonomy, molecular study of biology, molecular pathology, etc. have been described.

Molecular Taxonomy

If we look into the progress of taxonomic study, we will find that many works are being carried out in this dimension which are exploring new molecular elements for classification, mostly the molecular classification. The sophisticated instruments, i.e., SEM (scanning electron microscope), TEM (transmission electron microscope), FTIR (Fourier-transform infrared spectroscopy), AAS (atomic absorption microscopy), etc. are giving more valuable and acceptable data to enrich the taxonomic methodology. Different lineages of DNA and RNA more particularly tRNA and rRNA are being unfurled which are providing substantial knowledge of evolutionary progress of arthropods.

Some Molecular Phylogenetic Study

It is very much difficult to identify the mosquitoes morphologically. In a study of Paredes-Esquivel et al. (2008), two mosquito species, i.e., *Anopheles barbirostris* and *A. campestris*, of Barbirostris serogroup were taken for molecular characterization. The mosquitoes were collected from Thailand and Indonesia. The researchers took Col and ITS2 as molecular markers and performed phylogenetic analysis basing on **Bayesian methods**, **distance measures**, and **maximum parsimony** method which revealed five clades (I–V). Of these clades, clade I to III were of Barbirostris complex. Clade V was of *Anopheles campestris*. Depaquit et al. (2006) carried out phylogenetic *Paraphlebotomus* relationships using ITS2 as marker gene and sequenced nine *Paraphlebotomus* species – *P. alexandri*, *P. andrejevi*, *P. jacusieli*, *P. kazeruni*, *P. mireillae*, *P. mongolensis*, *P. saevus*, *P. sergenti*, and *P. similis* – and two out-group species of the subgenus *Phlebotomus*: *P. papatasi* and *P. duboscqi*. In this study, *Paraphlebotomus alexandri* appeared as the sister group of all other *Paraphlebotomus* sandflies. Morita et al. (2008) carried out one research work on phylogeny of Tabanid flies using nucleotide data from four genes, i.e., mitochondrial Col, nuclear 28S, CAD, and AATS. Phylogeny was reconstructed using Bayesian inference. Their result showed nonmonophyletic Bouvieromyiini and Rhinomyzini as a sister to Tabanidae. They also found that Pangoninae and Tabaninae are monophyletic lineages. Athericidae are strongly supported as the lineages most commonly related to Tabanidae. Thanwisai (2006) carried out a study on genomic characterization of *Simulium* flies basing on the sequences of the second internal transcribed spacer (ITS2) of ribosomal DNA (rDNA) of 40 blackfly species of 4 subgenera from Thailand. In 13 species, intraindividual variation (clonal variation) was found. Two major monophyletic clades were found of which the first monophyletic clade contained the subgenus *Simulium* and *Simulium baimaii* whereas the subgenera *Nevermannia* and *Montisimulium* were placed as the sister group to the subgenus *Gomphostilbia* in the second monophyletic clade.

Some Molecular Study of Biology

Molecular work is done for studying the gene expression level of different organisms. In the present day, there is much possibility to see the mRNA expressional level during the biological development of any organism. RNA seqs. of different organisms are studied, and the possible transcriptomes are analyzed in the processes of their biological development. The developmental progress is correlated with gene expression differences of different developmental stages. The RT qPCR is very useful for this purpose which can quantify the mRNA expressional level. DNA microarray technique is also useful for the identification of the different genes at the different stages of biological development of an organism. These techniques are also used in the field of arthropodology. For example, the basic stages of biological development of arthropods are egg, larva, pupa, and adult. It is presumed that some genes which express in the egg stage will not be found to be expressed in larval stage. In the larval stages, some different genes express as per biological requirement. Similarly, some specific genes which express in larval stage do not express in adult stages.

It has been found that genes express in insect salivary glands encoding the proteins which facilitate feeding. It is very interesting that studies show expression of genes restricted to particular sex (female) during blood feeding stage. This category of genes encodes a wide array of enzymes for taking blood meal and initiating digestion of the blood. These genes encode trypsins, peptidases, etc. which are expressed according to need. The expression of carboxy peptidase is also upregulated which is involved in the digestion of blood meal of flies. Following blood feeding, the specific genes involved in the reproductive cycle are expressed. Some works related to differential expression of genes in different biological developmental state of arthropods have been described:

Rewitz and Gilbert (2008) performed a research work on Halloween genes of *Daphnia pulex* by cytogenetic methodology, i.e., cloning, expression, etc. Expression product was characterized as cytochrome p450 enzymes (CYPs) which are responsible for biosynthesis of arthropod molting hormone, **20-hydroxye-dysone (20E)**. Insects and crustaceans use 20 E as principal molting hormone. Blast gene analysis showed Halloween genes present in *Daphnia* also. This is also present in Lepidoptera, Hymenoptera, Coleoptera, etc. The mRNA level of obp7 gene (**Agam-obp7**) of *Anopheles gambiae* is high in female antennae than proboscis. It is reduced after blood meal which led to conclude that this gene is responsible for host seeking behavior. Harker et al. (2013) reported differential expression of genes in different developmental stages of *Aedes* (early larva, late larva, early pupa, late pupa, adult). They analyzed **custom cDNA microarray analysis**. They found 1551 differentially expressed transcripts. The study displayed significant difference of expression of transcripts in different aforesaid biological stages.

Nayduch et al. (2014) reported differential expression of genes in different stages (teneral sucrose, blood feeding) of female *Culicoides sonorensis*. They constructed a reference transcriptome of *C. sonorensis* containing 19,041 genes. Only less than 7% were expressed during sucrose meal, and up to 52% of genes were expressed during blood meal. This result displayed complex physiological process during hematophagy.

Integrated Pest Management

Integrated pest management means a combined control step which is economic and eco-friendly. There are many ways by which ticks and other acarines can be controlled. This includes surveillance, pasture management, chemotherapy, immunoprophylactic control, biological control, and molecular approaches for reduction of vector competence.

Surveillance

Surveillance is prerequisite for controlling tick infection. Tick population varies in different climatic zones because biology of ticks is dependent on different environmental factors. A surveillance can provide a clear picture of trends of tick infection in different geo-climatological conditions.

Pasture Management

Both hosts and ground are used for completion of tick life cycle. All three (one host, two host, three host) types of ticks lay eggs on the ground. The eggs are hatched in the ground. Other stages, i.e., larva and nymphs, are found on the ground. Therefore, pasture management is very essential to control tick infection effectively. Plowing and burning of pasture can be done.

Chemotherapy

Application of different chemical is done for control of ectoparasite and flies. The insecticidal agents are used by local spraying and aerosol spraying. The acaricidal agents are used by spray, dipping, pour on, etc. Different repellants are also used for control of ticks and flies. These chemicals pertain to two groups, i.e., organochlorine and organophosphorus compounds. The organophosphorus compounds include malathion, parathion, dichlorvos, fenthion, chlorpyriphos, and diazinon. Synthetic pyrethroids include fenvalerate, permethrin, cypermethrin, etc. Organochlorine compounds include DDT, BHC, lindane, chlordane, eldrin, dieldrin, etc. Different nerve agents are sarin, soman, etc.

Immunoprophylaxis

Described earlier

Biological Control

Biological agents include different biopesticides. Though there are many natural enemies of ticks, only a few are evaluated as biocontrol agents (BCA). There are biopesticides, i.e., bacteria, entomopathogenic fungi (*Metarhizium anisopliae*, *Beauveria bassiana*), entomopathogenic nematode, and ixodiphagus wasps. Ixodiphagus wasps are the natural enemies of ticks and are regarded as biocontrol agents of ticks.

Landscape or Vegetation Management

This includes xeriscaping and hardscaping.

Nanotechnological Control

Metal and metal oxides prepared through green synthesis are found to be highly effective against arthropods. Studies show that silver and silver chloride nanoparticles have anti-tick activity. Other plants which are used for synthesis of silver nanoparticles are *Ocimum canum*, *Cissus quadrangularis*, *Musa paradisiaca*, *Manilkara zapota*, etc. As reported, silver nanoparticles are effective against different ticks, i.e., *Rhipicephalus (Boophilus) microplus*, *Hyalomma anatolicum*, *H. isaaci*, *Haemaphysalis bispinosa*, etc. Plant synthesis of silver and gold nanoparticles from different plants is possible. These nanoparticles are effective against different mosquitoes also.

Other Control

See the chapter "Trends in Control of Livestock and Poultry Parasites".

Modulation of Vector Competence to Transmit Parasitic Infections Using Molecular Genetics by Developing Transgenic Vector

Recently novel disease control methods are being developed. Scientists are involved in production of genetically modified mosquitoes. These mosquitoes reduce naturally occurring mosquito population without affecting their genetics. It is found that offsprings from mating of transgenic male mosquitoes and naturally occurring female mosquitoes cannot survive up to adult stage. A number of works were carried out on modulation of vector competence. This is majorly based on strategic application of molecular knowledge to generate pathogen-resistant arthropods. Efforts are being made on isolation and characterization of genes which influence mosquito vector competence. Recently, many molecular technologies have been developed. These are related to molecular characterization of genes which influences vector competence. Introduction of potential anti-pathogen gene is being under research of present trend. This is really one of the novel control strategies of parasite. The working hypothesis in this dimension was to genetically engineer mosquitoes, so that the mosquitoes or other arthropods can develop resistance to parasite development by introduction of parasite-resistant gene in the arthropods. These genetically engineered mosquitoes can drive parasite-resistant gene into wild, susceptible vectors. In this way, in endemic areas, disease transmission can be reduced.

References

Bhat SA, Walton SF, Ventura T, Liu X, McCarthy JS et al (2020) Early immune suppression leads to uncontrolled mite proliferation and potent host inflammatory responses in a porcine model of crusted versus ordinary scabies. PLoS Negl Trop Dis 14(9):e0008601. https://doi.org/10.1371/journal.pntd.0008601

Depaquit J, Ferte H, Leger N, Killick-Kendrick R, Rioux JA et al (2006) Molecular systematics of the phlebotomine sandflies of the subgenus *Paraphlebotomus* (Diptera, Psychodidae, Phlebotomus) based on ITS2 rDNA sequences. Hypo disp spec Gen 128(1–3):177–204

Harker BW, Nehru SK, de Bruyn BS (2013) Stage specific transcription during development of *Aedes aegypti*. BMC Dev Biol 13:29. https://doi.org/10.1186/1471-213x-13-29. PMID: 23875547

Moreira CK, Capurro MDL, Walter M, Pavlova E (2004) Primary characterization and basal promoter activity of two hexamerin genes of Musca domestica. J Insect Sci. 4:2. https://doi.org/10.1093/jis/4.1.2

Morita S, Bayless K, Wiegmann B (2008) Molecular phylogeny of the horse flies: a frame work for renewing tabanid taxonomy. Syst Entomol 41(1):12145. https://doi.org/10.1111/syen.12145

Nayduch D, Lee MB, Saski CA (2014) The reference transcriptome of the adult female biting midge (*Culicoides sonorensis*) and differential gene expression profiling during teneral, blood, and sucrose feeding conditions. PLoS One 9(5):e98123. https://doi.org/10.1371/journal.pone.0098123. PMID: 24866149

Paredes-Esquivel C, Donnelly MJ, Harbach RE, Townson H (2008) A molecular phylogeny of mosquitoes in the *Anopheles barbirostris* Subgroup reveals cryptic species: implications for identification of disease vectors. Mol Phylogenet Evol 50(1):141–151

Rewitz KF, Gilbert LI (2008) Daphnia Halloween genes that encode cytochrome p450s mediating the synthesis of the arthropod molting hormone: evolutionary implications. BMC Evol Biol 8:60. https://doi.org/10.1186/1471-2148-8-60. PMID:18298845

Soman SP, Singh SK, Kumari P, Choudhury S et al. (2020) Quantification of immune-regulatory cytokines and toll like reveptors gene expression in dogs with generalized demodicosis.Vet Par.280:109063. https://doi.org./10.1016/j.vetpar.2020.10906

Thanwisai A, Kuvangkadilok C, Baimai V (2006) Molecular phylogeny of black flies (Diptera: Simuliidae) from Thailand, using ITS-2 rDNA. Genetica 128:177–204. https://doi.org/10.1007/s10709-005-5702-Z. PMID: 17028950

8 Veterinary Protozoology

The protozoa are eukaryotic cells like many other organisms. The term 'Protozoa' means first animal. They have distinct nucleus whereas the prokaryotes do not have distinct nucleus. Like other eukaryotic cells, they have a nucleus, endoplasmic reticulum, mitochondria and a Golgi body. In addition to these, they possess different sub-cellular organelles which perform different funtions.

The locomotion of different protozoa is performed by different locomotory organs. These are flagellum (e.g. trypanosomes), pseudopodia (e.g. *Entamoeba*), cilia (*e.g. Balantidium*). In some protozoa i.e. extracellular stages of Eimeria do not have any locomotory organs. However, they perform locomotion by gliding movement.

Three types of nutrition are found in protozoa. These are holozoic (i.e. food material taken up by pseudopodia or cytostome and digestion of food by digestive enzyme) saprozoic (i.e. liquid food material taken by absorption through general body surface by diffusion) and, holophytic (e.g. having some characteristic of plants and carbohydrate synthesized by chlorophyll carried out in chromatophores in presence of light).

Excretion is usually performed by eliminating the waste product by diffusion through cell membrane. The contractile vacuoles also perform excretion by discharging the waste material through the body surface. In few instances, waste material is eliminated through the anal pore.

The reproduction of protozoa is either asexual or sexual. The asexual reproduction includes binary fission (i.e two daughter cells produced from single parent cell, division being along the longitudinal axis found in trypanomes or some other protozoa and transverse axis found in ciliates), budding (i.e. formation of two or more daughter forms from parent cell by unequal fragmentation of nucleus and cytoplasm), endopolyogeny (i.e. internal budding forming a progeny), endodyogeny (i.e. formation of two daughter cells by internal budding), schizogony (i.e. multiple division of the nucleus of the parent organism and the each nuclear body being surrounded by a part of cytoplasm resulting in a progeny of organisms). Sexual reproduction is performed by syngamy. In this process two gametes are formed (i.e. microgamete and macrogamete). The microgamete is male gamete and the macrogamete is female gamete. These two gametes fuse to form zygote. Conjugation is another process of reproduction in which two organisms pair side by side and exchange their nuclear material. Later the organisms separate and reorganization of nuclear material takes place. Sporogony occurs after syngamy in which the zygote divides into a number of sporozoites. See also the special part of this chapter and Chap. 2 for more information.

S. C. Mandal, *Textbook of Veterinary Parasitology*, https://doi.org/10.1007/978-981-99-2924-5_8

Classification of Protozoa of Veterinary importance

Subkingdom: Protozoa

Phylum: Sarcomastigogophora (Presence of single nucleus, flagella, pseudopodia, or both flagella and pseudopodia serving as locomotory organs; absence of spore formation; syngamy is the major mode of sexual reproduction if sexuality present)

Subphylum: Mastigophora (Reproduction basically by binary fission, trophozoite possessing one or more flagella)

Class: Zoomastigophorea (Flagella number varying from one to several or may be absent in some members; absence of chromatophores; asexual reproduction by binary fission; sexuality in some groups))

Order: Kinetoplastida (Mostly parasitic; one to four flagella found; kinetoplast present near kinetosome having mitochondrial material)

Family: Trypanosomatidae (leaf -shaped,round form; presence of single flagellum, undulating membrane in some groups, blepheroplast, kinetoplast)

Genera

Trypanosoma

Leishmania

Family: Cryptobiidae (Organisms look like trypanosomes but they are biflagellate)

Genus: Cryptobia

Order: Retortamonadida (Two or four flagella found in the organisms; of these, one flagellum turns posteriorly)

Family: Retortamonadidae (Two or four flagella)

Genus

Chilomastix

Cochlosomatidae (Presence of a ventral sucker anteriorly; presence of a parabasal body)

Genera

Cochlosoma

Order: Diplomonadida (Bilaterally symmetrical body; two similar nuclei; cyst formation found)

Family: Hexamitidae

Genera

Giardia

Hexamita

Order: Trichomonadida (Four to six flagella, one is recurrent flagellum attached to undulating membrane; axostyle present

Family: Trichomonadidae (Pyriform in shape; single nucleus; costa and axostyle present; four to six flagella, one recurrent flagellum attached to the undulating membrane.)

Genera: *Tritrichomonas*

Trichomonas

Tetratrichomonas

Pentatrichomonas

Family: Monocercomonadidae (Three to five anterior flagella; a recurrent flagellum usually free; absence of costa)

Genera

Histomonas

Parahistomonas

Order: Retortamonadida (Presence of two or four flagella; one flagellum projecting posteriorly)

Family: **Genus**: Chilomastix

Subphylum: Sarcodina (pseudopodia present; presence of flagella only during developmental stages; asexual reproduction by fission; sexual reproduction usually by flagellate gametes)

Superclass: Rhizopoda

Order: Amoebida (Presence of naked protoplast at the trophic stage, usually unicleate)

Family: Endamoebidae (Multiplication by binary fission; cysts formation common)

Genera

Entamoeba

Dientamoeba

Endamoeba

Iodamoeba

Endolimax

Family: Hartmannellidae (Mainly free living, absence of flagellum)

Genera

Acanthamoeba

Hartmannella

Phylum: Apicomplexa

Class: Sporozoea (Well-developed apical complex; both sexual and asexual reproduction; formation of oocysts; microgametes possessing flagella)
Subclass: Coccidia (Formation of intracellular and dissimilar gamonts by anisogamy)

Order: Eucoccidiida
Suborder: Eimeriina (Independent development of macro- and microgametes; sporozoites found in sporocysts; formation of a large number of microgametocytes; nonmotile zygote)
Family: Eimeriidae (Both sexual and asexual reproduction, formation of schizonts,gamonts, oocysts with nil or variable number of sporocysts,non motile zygote)
Genera
Eimeria
Isospora
Tyzzeria
Wenyonella
Family: *Cryptosporidiidae*
Genus: *Cryptosporidium*
Family: Sarcocystidae
Subfamily: Toxoplasmatinae
Genera
Toxoplasma
Hammondia
Besnoitia
Cystoisospora
Subfamily: Sarcocystinae
Genera
Sarcocystis
Frenkelina
Suborder: Adeleina
Family: Klossiellidae (microgamete without flagella, inactive zygote, typical oocysts not found)
Family: Haemogregarinidae (Ookinete is produced)
Suborder: Haemosporina (There occurs individual development of micro- and macrogamete; motile zygote; absence of conoid; pigment formation in host cells; schizogony in vertebrate and sporogony in invertebrate)
Genera
Haemoproteus
Leucocytozoon
Plasmodium
Subclass: Piroplasmia (Short apical complex; absence of conoid; pigment does not form in the host cell)

Order: Piroplasmida
Family: Babesiidae
Genus: *Babesia*
Family: Theileriidae
Genus
Theileria
Phylum: Ciliophora (Presence of cilia or ciliary organelles at least in one stage of biological development; usually two types of nucleus; reproduction by binary fission)
Class: Kinetofragminophorea (Often presence of cytostome; usual presence of prominent cytopharyngeal apparatus, presence of macro and micro nucleus)

Order: Trichostomatida
Family: Balantidiidae (A vestibulum present near anterior end of the body, body is uniformly ciliated that is long, holotrichous)
Genus: *Balantidium*

Phylum: Sarcomastigophora—Subphylum: Mastigophora—Class: Zoomastigophorea—Order: Kinetoplastida—Family: Trypanosomatidae

Family Synopsis

These are leaf-like protozoan organisms found in the blood and or/ tissues of different vertebrates. They are worldwide in distribution. They are characterized by the presence of a flagellum, an undulating membrane, a kinetoplast, and a blepharoplast. The developmental stages are trypomastigote, epimastigote, promastigote and amastigote. Some species of this family have pathogenic significance causing massive loss in the livestock industry. The gold standard for the diagnosis of this parasitic infection is the examination of blood smears. Control of the parasite mostly relies on chemotherapy.

History

Griffith Evans discovered *Trypanosoma evansi* from the blood of horses and camels in 1880. D. Bruce first described tsetse fly as vector of *Trypanosoma brucei* in the year 1895. Sir D. Bruce discovered *T.gambiense* in the year 1903. He also identified *T.rhodesiense* in the year 1911. Carlos Chagas discovered *Trypanosoma cruzi* in 1909.

Genus: *Trypanosoma*

Morphology

The organisms are leaf-like and flagellated protozoan parasites. One nucleus is present. Posterior to the nucleus, one granule-like structure is present which is called basal granule or blepharoplast. From the blepharoplast the flagellum originates. Posterior to the blepharoplast, a dot-like structure is present which is called kinetoplast. The flagellum is attached to the body of the organism by a membrane which is called an undulating membrane.

Minute Structure

Attempts were made by some researchers to create an ultrathin section of *Trypanosoma*. During this research progress, ultramicrotome was invented. Glass knives were made for cutting ultrathin sections. This microtome is called the Porter and Blum microtome. In the earlier period, by observing the ultrathin section, the kinetoplast was recognized as a vacuole-like structure. The fibrillar structure present in vacuole is called the kinetoplast. The physical connection of the kinetoplast DNA (KDNA) with basal granules was not established earlier. Recently, a set of fibrillar structures connecting the kinetoplast DNA with basal granules has been described. The study shows the presence of protein (P166) between the KDNA and basal granule.

Note

It is thought that *T. evansi* originated from *Trypanosoma brucei brucei*. Gradually, maxicircles of kinetoplast DNA were lost during evolution. The parasites were no longer able to undergo their cycle in *Glossina*.

Important Salivarian Species

Trypanosoma evansi

These are salivarian *Trypanosoma* which come under the anterior station group. They are slender organisms. The nucleus is placed almost centrally. The kinetoplast is subterminal. The flagellum arises from the blepharoplast and reaches to the anterior end, extending beyond the body of the organism, providing a free flagellum. Like *Trypanosoma equiperdum, T. evansi* originated from *T. brucei* after losing a part of kinetoplastic DNA (maxicircle) in the course of their evolutionary pathway. Due to the loss of this kinetoplastic DNA, they cannot develop in the vector and are thus transmitted mechanically. Since they are transmitted mechanically, a wide range of vectors are associated with the transmission of this parasite, such as *Tabanus*, *Stomoxys*, *Haematobia*, *Chrysops*, etc. *T. evansi* is the causative agent of an important disease called as Surra. They can affect a large number of hosts, such as horses, donkeys, mules, cattle, buffalo, sheep, goats, camel dogs, elephants, etc. However disease severity varies among different species of animals.

Trypanosoma equiperdum

These are derivatives of *T. brucei*. In the process of evolution, they are formed by losing kinetoplastic DNA. That is why they cannot undergo biological development in vectors. The organisms are transmitted sexually. No vector is required for their transmission. The hosts are equines. The disease caused by these parasites is called dourine. It is characterized by the development of edema in the vulva, prepuce, and other genital organs. Urticarial plaques are formed under the skin, which are felt like rupee coins inserted under the skin. The circumscribed plaques are popularly known as dollar spots.

Trypanosoma vivax

This species belongs to the subgenus Duttonella. The organisms are transmitted mechanically by *Glossina* flies. This species is predominant in Africa and Latin America. Bovines are the main hosts. Horses are also affected by this species to a lesser extent. This species is characterized by bulbous posterior part, large kinetoplast and short free flagellum. The hosts are cattle, buffalo, camel, horse, sheep, goat etc. They causes mild, moderate, and severe form of disease in cattle depending on different geographic location.

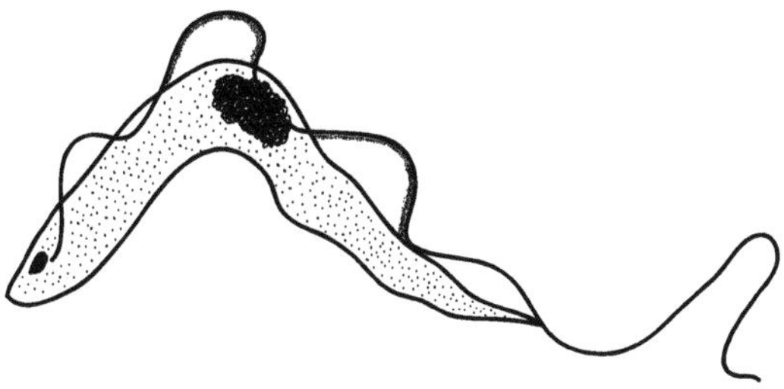

Fig. 8.1 *Trypanosoma vivax*

Trypanosoma brucei

The body is elongated and streamlined with tapering ends. They are polymorphic (slender, stumpy, intermediate form) having subterminal kinetoplast, conspicuous undulating membrane. There are three subspecies of *Trypanosoma brucei*: *T. brucei brucei*, *T. brucei rhodesiense*, and *T. brucei gambiense*. These subspecies are collectively called *T. brucei* complex. The organisms are transmitted by *Glossina* flies. They are transmitted cyclically. The most severe form of disease occurs in equines. Intermittent fever, rapid emaciation, oedema etc. are clinical features.

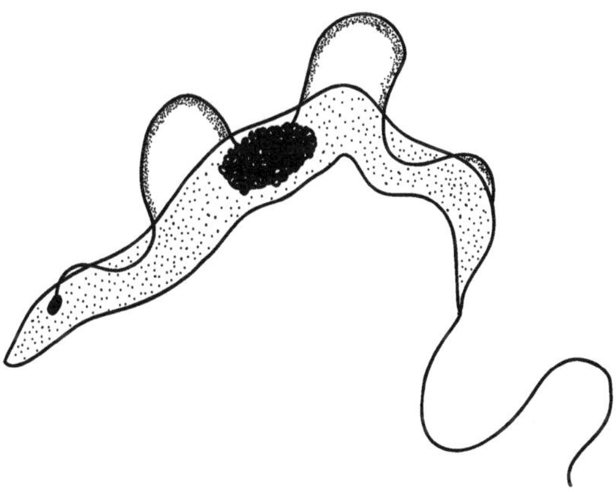

Fig. 8.2 *Trypanosoma brucei*

Trypanosoma congolense

They are small trypanosomes having blunt posterior extremity with marginal kinetoplast present slightly distant from the posterior end. Free flagellum is absent. This species causes Nagana in cattle. Nagana is a 'Zulu' word meaning 'to be in low or depressed spirit'. The disease is characterized by severe normocytic and normochromic anemia, emaciation, cachexia, etc. Other clinical signs include fever, nasal and eye discharge, hind leg paralysis etc. Other animals, such as sheep, goats, horses, camels, and pigs, are also affected. This species is transmitted by *Glossina* flies. *T. congolense* also affects dogs. The clinical signs exhibited by the affected dogs are diarrhea, hemorrhagic diarrhea, vomiting, etc. They may show various nervous disorders such as tremors, seizures, etc. Other signs are intermittent fever, inappetence, loss of body weight, ocular, nasal discharge, etc. The organisms are found in trypomastigote form in the blood of vertebrate hosts. Epimastigote stages occur in the proboscis.

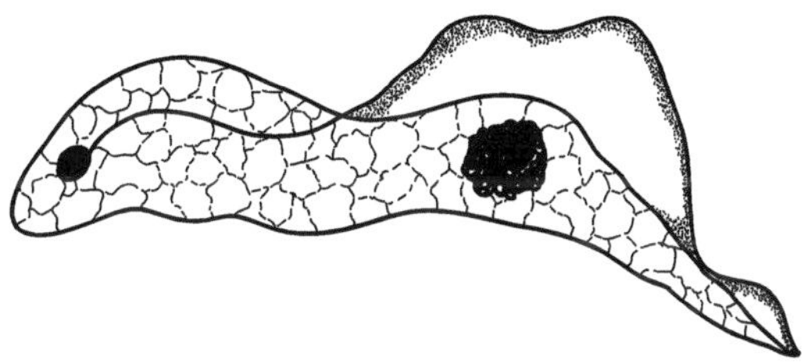

Fig. 8.3 *Trypanosoma congolense*

Trypanosoma equinum

This species comes under the anterior station group of *Trypanosoma*. They are monomorphic. They are large sized trypanosomes similar to *Trypanosoma evansi*. This species originated from *T. brucei*. The equines are mostly affected. Severe disease entity may be found. The disease mostly progress with a chronic course. However, an acute form of this disease is also found. Emaciation and fever are very common clinical signs. Affected animals may die after a few months. The disease caused by *T. equinum* is called "mal de caderus". The animals suffer from lateral recumbency. Weakness, conjunctivitis, keratitis, and formation of urticarial plaques are common clinical features.

Important Stercorarian Species

Trypanosoma cruzi

They are typically "u-" or "c-" shaped organisms appearing in the blood films with Giemsa stain. The organisms are large in size and crescent-shaped. They have free flagella. The nucleus is almost placed centrally. A kinetoplast is placed posterior to the nucleus and it is subterminal. The flagellum originates from the blepharoplast and runs towards the anterior tip and later extends beyond the body with a part of free flagellum. The organisms affect both humans and animals. In human beings, they cause a serious disease called Chagas disease. The vectors of Chagas disease are triatomine bugs (kissing bugs, cone-nosed bugs). Three developmental stages are found during the biological development of this organism: trypomastigote, epimastigote, and amastigote. Trypomastigote and amastigote stages are found in vertebrates. Epimastigote and trypomastigote stages are found in invertebrates. They cause chagas disease in humans.

Note

- Trypanosomes are heterotrophic organisms. The paraxial rod present in the flagella is a common characteristic feature of Kinetoplastida.
- Kinetoplast is a clump of DNA present at one end of a long mitochondrion.

Different Stage of Biological Development

Trypomastigote

This developmental stage is found in both vertebrate and invertebrate hosts. The kinetoplast is present posterior to the nucleus. The undulating membrane is well developed.

Epimastigote

This stage is found in invertebrate hosts. But in few species, this may be found in vertebrate hosts. The kinetoplast is present anterior to the nucleus. The undulating membrane is shortly developed.

Promastigote

This stage is found only in invertebrate hosts. The kinetoplast is present at the anterior tip of the organism. A flagellum is present. The undulating membrane is absent. This stage is found in invertebrate hosts.

Amastigote

This stage is found in both vertebrate and invertebrate hosts. Flagellum is absent in this stage. Instead of a flagellum, a fibril-like structure is present. A kinetoplast is present.

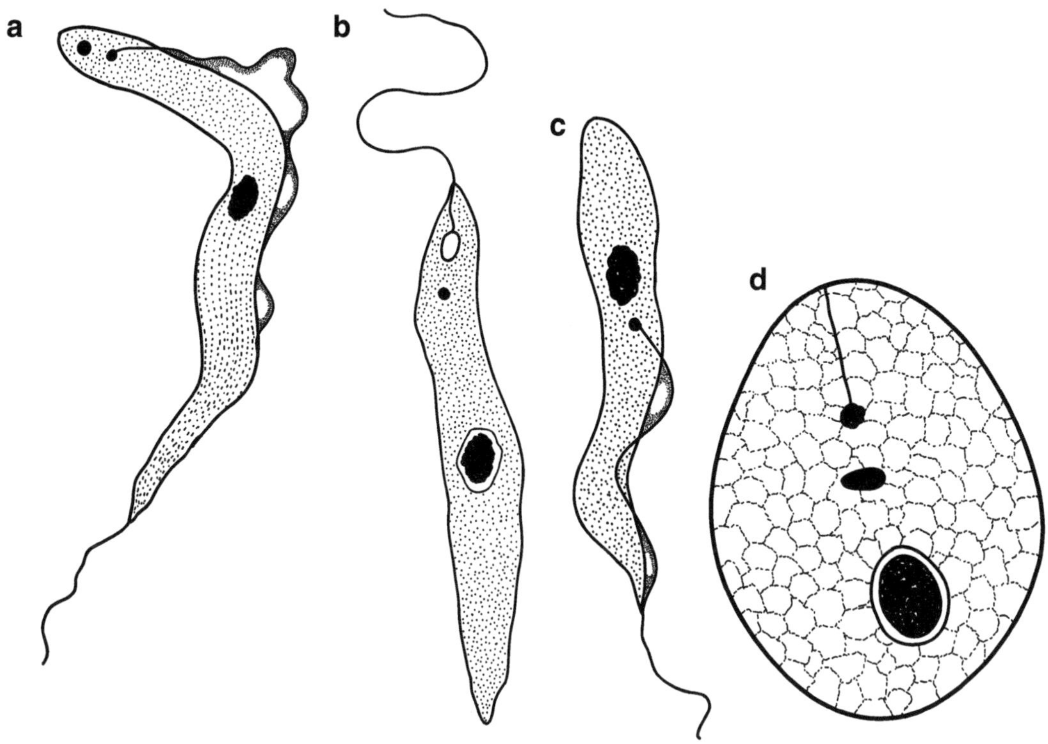

Fig. 8.4 Delopmental stages of members of Trypanosomatidae. (**a**) Trypomastigote (**b**) Promastigote (**c**) Epimastigote (**d**) Amastigote

> **Note**
> - *T. cruzi* induce the phagocyte to be engulfed by receptor mediation, opsonization, etc.
> - The trypanosomes absorb their food materials through their outer membrane. The trypanosomes possess different enzymes.
> - Congenital transmission may occur in the case of *T. cruzi*. The dogs get infection by ingestion of the infected vector.

Biology of Trypanosomes

Transmission of *Trypanosoma*

- Different vectors are involved in the transmission of *Trypanosoma*. Different bloodsucking flies, such as *Tabanus*, *Stomoxys*, *Haematobia*, and *Glossina*, are the vectors of *Trypanosoma*.
- Some trypanosomes (*T. evansi*) may be transmitted to the hosts through the ingestion of infected meat or offals. Dogs and cats may get infection of *T. evansi* by the ingestion of infected fresh meat and offals.
- There are some trypanosomes which are transmitted congenitally.
- Dogs may get infection of *T. cruzi* by the ingestion of infected bugs.
- Vampire bats (vertebrates) also transmit trypanosomes to the hosts. *Trypanosoma evansi* may be transmitted by vampire bat.

Different Steps of Anterior Station Development

- Metacyclic trypanosomes gain entry into the gut of the insect vector when they take their blood meal from the hosts.
- Biological development starts in the gut of the insect vector. The ingested trypomastigotes transform into stumpy form and slender form at the area of the gut allied to the peritrophic membrane.

- The parasites migrate towards the anterior part of the gut and reach the salivary glands via the proventriculus, esophagus, pharynx, and hypopharynx.
- During this migration, they transform into epimastigote stages.
- The epimastigotes further transform to trypomastigotes at the mouth region.
- Metacyclic trypomastigotes are transferred to the mammalian hosts when the infected vectors suck blood from the mammalian hosts.

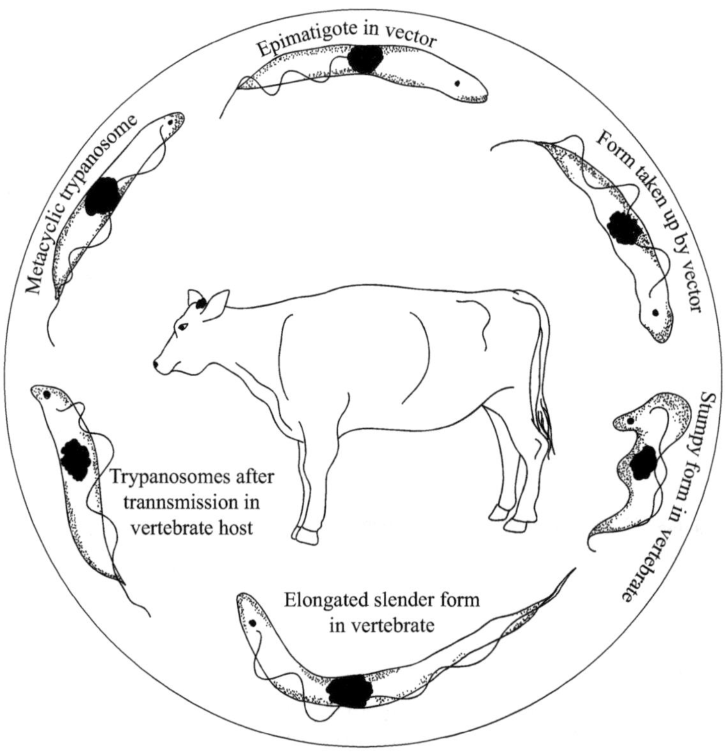

Fig. 8.5 Life cycle of *Trypanosoma* sp.

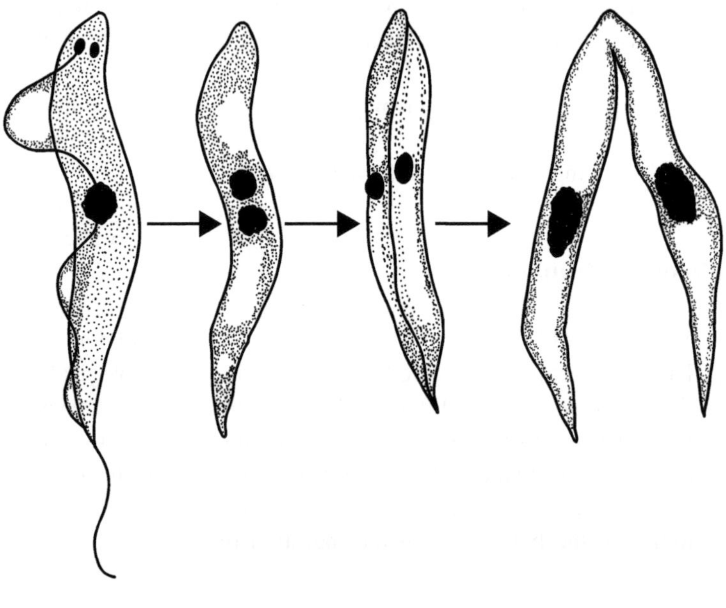

Fig. 8.6 Binary fission of *Trypanosoma* sp.

Steps of Posterior Station Development

- Trypomastigotes gain entry into the gut of the insect vector when they take their blood meal from the hosts.
- Biological development starts in the gut of the insect vector. The ingested trypomastigotes transform to slender epimastigote stages during migration from stomach to the rectum. Some trypomastigotes transform into spherical spheromastigotes.
- Epimastigotes multiply in the gut of the vector and transform to trypomastigote stages. This stage is called metacyclic trypanosomes. These metacyclic trypanosomes are infective stages of the parasites. They are ready to infect vertebrate hosts.
- Mode of infection is typical. The vectors suck a large amount of blood. The extra volume of ingested food exerts pressure on the gut of the vectors which causes defecation of the vectors. After defecation, they bite the mammalian hosts. In this way, the bite area gets contaminated with the feces and the infective stages present in the feces enter the host.
- The organisms are taken up by the macrophages or they invade host cells.
- In the macrophages and other host cells, the organisms transform to round- or oval-shaped amastigote stages. These amastigote stages multiply in the macrophages.
- The amastigotes transform to trypomastigotes.
- The number of organisms increases so much that the macrophages get ruptured and the trypomastigotes come out and mix into the blood.

Different Disease Forms of Trypanosomosis

Surra

In the Indian subcontinent, the disease caused by *T. evansi* in horses is called Surra. Surra is a Hindi word meaning "rotten." In horses, the disease occurs in acute or chronic form. Affected animals die within a few weeks to a few months. The disease is characterized by emaciation, anemia, severe weakness, formation of urticarial plaques, edema, etc. Usually, the plaques are formed in the flank and neck region. Initially, the plaques get ruptured centrally and subsequently, the lesion spreads gradually. The lesion is also found at the mucocutaneous junction. The disease occurs in chronic form in camels, particularly in Asia. The affected animals suffer from intermittent fever, emaciation, anemia, etc. In Sudan, the disease is called gufar. It also occurs in the dogs. The symptoms exhibited by affected dogs are different from other forms. The larynx is severely affected. Pronounced edema of the larynx occurs which creates a roaring sound similar to rabies. The cattle and buffalo are considered as reservoir hosts. The disease manifests in subclinical form. Occasionally, the affected animals may die due to strain variation. The disease can take a serious form if the animals are immunosuppressed due to vaccination, FMD disease, etc.

Nagana

Nagana is a Zulu word. The meaning of "Nagana" is "to be low or depressed spirit". The causative agent is *Trypanosoma congolense*. This disease affects primarily cattle and horses in African countries. Affected animals suffer severely due to massive destruction of RBC and depression of hemopoitic system. The clinical features are fever, anemia, edema, and nasal and eye discharge. Severe anemia results in severe weakness. Paralysis occurs which starts from the hind legs and later the whole body becomes paralysed.

Mal de Caderas

The causative agent of this disease is *Trypanosoma equinum* transmitted by *Glossina* flies. The equines are affected. Weakness, anemia, lateral recumbency, formation of urticarial plaque, etc., are some of the important clinical signs. The disease manifests in three forms:

Acute Stage

The occurrence of the acute form is insidious. In this form, fever, progressive weakness, and emaciation are characteristic clinical features.

Chronic Stage

The fever remains for about 1 day. Thereafter, temperature becomes normal and the animals enter into the chronic stage. Urticarial plaques are formed in the flank and neck region. Along with emaciation, progressive anemia is also found.

Paralytic Stage

The paralytic stage occurs later. The animals sway their legs wide apart. They drag their feet when they try to walk. Later the animals cannot stand without support. If treatment is not provided, death ensues.

African Sleeping Sickness

A number of *Trypanosoma* spp. like *T. brucei, T rhodesiense, T gambiense,* etc., are available in African countries. Of those species, *T. rhodesiense* and *T. gambiense* are very important as these two species cause severe disease in human beings, which is mostly fatal. The important clinical sign is sleeping sickness, also called African sleeping sickness. The nervous system is affected by these species and results in different nervous disorders.

Acute Form of African Sleeping Sickness

T. rhodesiense produces an acute form of African sleeping sickness. The pathological features of infection with *T. rhodesiense* in humans are similar to that of *T. gambiense* infection. The organisms can cross blood-brain barrier and invade the CNS. The disease is characterized by febrile paroxysm, enlargement of lymph node, weakness, myocarditis, weight loss, etc. The symptoms are similar to *T. gambiense* infection but the symptoms develop more rapidly. Death occurs after months and not after years. Diagnosis can be done similarly to the chronic form of African sleeping sickness. Examination of a blood smear and a biopsy of lymph node demonstrate the organisms. The infection is resistant and nonresponsive, hence pentamidine and suramin can be tried.

Chronic Sleeping Sickness

This is caused by *T. gambiense.* The phases are the incubation period, symptomless patent period, acute febrile period, mild chronic period, and sleeping sickness stage. The disease becomes evident at the acute febrile stage which is characterized by an attack of fever followed by nervous disorders. Congenital transmission may occur in the case of *T. cruzi.* Chronic sleeping sickness is initiated when invasion of the nervous system by the pathogens occurs. This stage is characterized by severe headaches, mental dullness, apathy, etc. The patients may become morose or excited. Dullness, depression, incoordination, muscle spasm, trembling, stiffness of the neck, spastic or flaccid paralysis are the associated clinical signs. Diagnosis is done by examination of cerebrospinal fluid, bone marrow, lymph node fluid, etc. The drugs used are pentamidine isethionate, suramin, melarsoprol, etc. Congenital African trypanosomiasis is also found. There are several records of congenital transmission of *Trypanosoma gambiense*. The organisms are found in the newborns from infected mothers.

Dourine

This disease is caused by *Trypanosoma equiperdum*. Dourine is an Arabic word meaning unclean/dirt. Horses and donkeys are mainly affected. Three phases of clinical features are observed. The first phase is the stage of edema occurring particularly in the genitalia. There occurs ulcer on the vaginal and vulval mucosa. The second phase is the stage of urticaria. The plaques are circular and sharply circumscribed which appear as "dollar spots" under the skin. The third phase is the stage of paralysis.

Chagas Disease

This is a disease caused by *Trypanosoma cruzi.* This is transmitted by reduviid bugs. It is also called as American human trypanosomiasis. The organisms are transmitted in a very typical manner. Reduviid bugs feed on lip and eye areas. The metacyclic trypanosomes gain entry through the cut area. The incubation period is 1–2 weeks. Swelling of skin (post of entry area) occurs. The organisms then invade the heart, skeletal muscle, nervous system, etc. The acute form is characterized by facial edema, lymphadenopathy, splenomegaly, anemia, cardiomyopathy, and meningoencephalitis. The chronic form of the disease is characterized by cardiac arrhythmias and different Chagas diseases can be diagnosed in different ways. Examination of blood smears and stained films of lymph node juices may reveal the presence of organisms. Biopsy material of lymph nodes can be examined. Animal inoculation tests can be performed by injecting infected blood. Xenodiagnosis can be done.

Reduviid bugs are allowed to feed on the suspected patient and the organisms are detected in the gut of bugs.

Chagoma

Reduviid bugs (kissing bugs) feed near the eyes and lips. The mode of infection of these organisms is either by the ingestion of the infected bugs or contamination of the wound by the feces of infected bugs. The organisms at first enter the histiocytes wherein they multiply and increase in number. There occurs inflammation and swelling in different areas of the face. These are called as chagoma lesions.

Diagnosis

Symptomatological Diagnosis

Symptomatology cannot provide a confirmed diagnosis. Symptoms are not specific for a confirmatory diagnosis. However, lesions at the mucocutaneous junction, fever, emaciation, anemia, etc., are suggestive of *Trypanosoma* infection. Symptoms of Mal de caderas and dourine are more indicative of this infection.

Parasitological Diagnosis

Diagnosis can be made by examination of the blood smear after staining with different stains.

- Examination of the blood smear is the gold standard method for the diagnosis of *Trypanosoma* infection. Both thick and thin smears should be prepared. Different Romanowsky stains are used. Giemsa and Leishmann stains are generally used. The blood is collected from the tip of the tail or directly from the jugular vein of affected animals.
- For the diagnosis of human sleeping sickness, CSF is examined for the presence of any organisms. At least two smears should be examined.
- Thick concentration method has been described for easy detection by **hematocrit centrifuge technique** and **dark ground buffy coat** technique.
- Anion exchange chromatography technique has also been used for the diagnosis of low-level infection.
- Wet smear technique is also performed using fresh movement of the organism seen under 40× magnification. The examination of wet smear is a rapid and easy test. There is no need for staining or requirement for chemicals. No expert persons are needed to perform this test. However, it is less sensitive compared to the staining technique. Examination of a wet smear is usually done for mass screening of *Trypanosoma* infection in field conditions.
- Animal inoculation and demonstration of the organisms in laboratory animals is another way of diagnosis of trypanosomosis. Subinoculation of *T. brucei* in mice is useful for better demonstration of organisms. However, all trypanosomes may not be established in all laboratory animals. Specific laboratory animals are suitable for specific trypanosomes.

Chemical Tests

Different chemical tests can also be done. These include Stilbamidine test, Jones' nitric acid test, mercuric chloride test, and formol gel test.

Immunodiagnosis

Many works were carried out on immunodiagnosis using different crude and purified antigens of trypanosomes. Recombinant antigens have also been tried. However, to date no immunodiagnostic test is available for field application due to a lack of specificity. Although complement fixation tests, Fluorescent antibody tests and Indirect fluorescent antibody test are sensitive, they lack specificity. These tests do not have practical implications in case of animal infection. However, in some cases, species-specific antigens and species-specific monoclonal antibodies have been described.

- Different recombinant proteins have been used for the diagnosis of trypanosomosis. Sullivan et al. (2014) evaluated the immunodiagnostic potential of sVSG 117 by dual antigen lateral flow test. For this purpose, one more antigen is required. The second antigen used in this study was rISG 65. They reported a sensitivity of 97.3% and a speci-

ficity of 83.3% for the diagnosis of *T. brucei gambiense*.
- It is known that the tandem repeat protein of *Trypanosoma* usually has good antigenic properties. Thuy et al. (2012) reported the immunodiagnostic potential of a tandem repeat protein for the diagnosis of *Trypanosoma evansi*. They reported that the tandem repeat protein of *T. brucei brucei* GM6(TbbGM6) could also be used as a sero-diagnostic antigen for chronic *T.evansi* infection. The CATT card agglutination test is used for the diagnosis of West African trypanosomiasis in humans. After some modification, this test has been found usable for the diagnosis of *T. evansi* infection in water buffalo.
- The use of monoclonal antibodies has opened up many opportunities for specific diagnosis up to the species level in many cases. Monoclonal antibodies can be used for early detection and for the detection of subclinical infection. These antibodies are used to capture the antigen.

Molecular Diagnosis

Molecular identification of trypanosomes up to the species level is possible using suitable markers and PCR techniques, such as multiplex PCR, nested PCR. Different reports are available on the successful use of molecular markers for the accurate identification of *T. evansi*. These are species-specific ITS1, Ro Tat VSG. Studies show that multiplex PCR is expensive and less time consuming. Since we know that there might be multiple species of infection, multiple primers of different species are used simultaneously. Otherwise, single PCR is used several times, which is cumbersome. Nested ITS PCR and SRA-LAMP have been successfully used for the identification of different *Trypanosoma* spp. RFLP-PCR is also used for the diagnosis of *Trypanosoma* infection. In the case of human trypanosomiasis, xenomonitoring of tsetse infection is done to detect enzootic area using high-resolution melt qPCR. For this, species-specific primers are used to bind only with the target gene.

Treatment

Drug	Route of administration	Doses
Diminazene aceturate	I/M	Cattle and buffalo: 7 mg/kg bw Horse: 3.5 mg/kg bw; should be repeated after 4–5 h Camel: 3.5 mg/kg bw Dog: 3.5 m g/kg bw; 5–8 repeated injections at 2–3 weeks intervals
Isometamidium chloride	I/M or S/C	Cattle: 0.5 mg/kg bw Horse: 0.25 mg/kg Repeat after 4-5 hours)

Control

Chemotherapy

Chemotherapy is still the only resort for controlling this disease. However, many other control measures have been demonstrated. But their significance is too low.

Control of Vector

- *Chemical control*: Since various bloodsucking flies act as vectors, the flies should be controlled using different insecticidal agents.
- *Use of trap*: Recently, a study on the olfactory physiology of vectors is advancing, providing knowledge about creating new traps, or modifying the existing traps of vector flies.
- *Endosymbionts:* The use of the knowledge about endosymbionts creates the possibility to develop novel control strategies against vector flies. *Wigglesworthia glossinidia* and *Sodalis* are two endosymbionts found in *Glossina*. *Wigglesworthia* synthesize key vitamin complexes of *Glossina*. This knowledge can be exploited to design control regimens against these flies.
- *Sterile insect technique*: The male flies are subjected to sterilization. This is done by irradiation or genetic manipulation. Chemosterilization is another way to produce sterile males. To perform this technique, one should have crucial

knowledge about the biology of the flies and this knowledge is harnessed to develop perfect protocol. A lot of questions pertain to the development of this technology as follows:

(a) When the sterile males are to be released?
(b) How many sterile males are to be released in a particular area?
(c) What is the reproductive potential of the flies?

We know that the reproductive potential of *Glossina* is less compared to other flies. They do not lay hundreds or thousands of eggs. They lay only one larva in one gonotrophic cycle. In their lifetime, they lay about 14 larvae. That is why this technique has created ample scope to control this fly. Several thousands sterile flies are required for 1 km^2. It is to be noted that the area subject to control should be restricted. For the demarcation of a particular zone, a geographical information system is utilized. When this technique was first invented, it was not easy to control flies. But day by day, this technology has been modified and has gained practical implementation. There are certain demerits of this technique as follows:

- Only one species can be controlled. Multiple species cannot be controlled using a single sterile species. Thus, constraints are related to the limited use of this technique.
- Experienced technical persons are required for the implementation of this technique.
- This technique is very expensive and slow acting.
- At any given time, it is not possible to see the maximum efficacy.
- This technique is not sustainable. Once implemented, it does not continue for a long time.

This technique has many advantages as follows:

(a) It is nonintrusive to the environment.
(b) It has no adverse effect against nontarget organisms.
(c) It is a species-specific technique.
(d) This technique does not have any residual effect.
(e) The modified organisms do not produce any poisonous or toxic material. In totality, this technique is environmentally friendly.
(f) In this technique, a large number of flies are produced in large-sized insectaries following standard protocol.
(g) The newly produced flies are sexually assorted.

Immunological Control

An array of works has been carried out for the development of vaccines using different antigens. The development of a vaccine against *Trypanosoma* is very problematic because of the variable surface glycoproteins present in the trypanosome coat. However, it is notable that trypanosomes also express some nonvariable surface molecules for the assimilation of factors required for their survival. These molecules are actually the targets for the scientists of the present day. After a rigorous search, a number of nonvariable or less variable molecules have been found. Of these molecules, the most important and well-versed antigen molecules are flagellar pocket antigens. This is an invaginated membrane found in the area of origin of the flagellum. Many vital functions are performed by this flagellar pocket, which has drawn the attraction of the researchers for the development of vaccines. For the last two decades, the research works have revealed partial protection by the vaccine using a flagellar pocket. Another important protein subject to immune-protection trial was trans-sialidase. Trans-sialidase is an important membrane-bound enzyme that transfers sialic acid from sialylated glycoconjugate of the host cell to the surface of cells. Other proteins that were used for immune-protection study are actin and tubulin. Several works were carried out on the immunoprotective potential of tubulin, beta-tubulin, and actin, which displayed partial protection. Conclusively, all these aforesaid proteins were not found to be potent vaccine candidates for conferring significant protection.

Breeding Management

A breeding program using trypanotolerant animals is an important step in controlling trypanosomosis. There are different breeds of cattle in Africa, such as N-dama, West African short horn, Boran, Omra Boran (cross between Omra and Boran), N-dama, Baole, etc. It is believed that these animals are tolerant because of their capability to elicit protective immunity. Research study shows that this is an innate trait and heritable to subsequent generations. It has also been found that the trypanotolerant animals are not only resistant to *Trypanosoma* but also to *Babesia*, *Anaplasma* and heartwater disease. Therefore, this has created a good opportunity to control several pathogens simultaneously.

Gene Immunization

Reports are available on the use of naked DNA vaccines in different research trials. However, it seems that there has been no field application of any gene vaccine. DNA plasmid encoding a unique enzyme, transialidase, was used as a naked DNA candidate.

General Pathogenesis of Trypanosomosis

The important clinical features are fever, lesions at the mucocutaneous junction, anemia, emaciation, splenomegaly, etc. The organisms mechanically damage the RBCs, resulting in a significant amount of anemia, which further leads to emaciation in prolonged cases. They play an immunological role in the occurrence of anemia. The spleen and lymph nodes are markedly enlarged. Urticarial plaques are formed. Frank hemorrhages occur at the junction of the mucus and cutaneous area. In some cases, plaques are formed. The plaques get ruptured and later necrosed centrally.

General Clinical Signs of Trypanosomosis

The common clinical signs are as follows:

- Fever
- Anemia
- Emaciation
- Severe depression
- Cachexia
- Lesion at mucocutaneous junction
- Plaque formation
- Edema
- Lacrymation

Genus: *Leishmania*

History

Leishmania donovani was first isolated independently by Leishman and Donovan in 1903. J H Wright discovered *Leishmania tropica* in the year 1903.

Introduction

Leishmaniosis is an important disease caused by different *Leishmania* spp. The organisms are transmitted by bloodsucking flies, such as *Phlebotomus* spp. Two forms of leishmaniosis are found: visceral leishmaniasis and cutaneous leishmaniasis. The important causative agents of visceral leishmaniasis are *Leishmania donovani* and *Leishmania infantum* and cutaneous leishmaniasis is caused by *L. tropica*.

Morphology

Two developmental stages are found during its life cycle: amastigote and promastigote.

These are as follows.

Promastigote

This stage is found only in invertebrate hosts. They are spindle-shaped. The kinetoplast is present at the anterior tip of the organism. A flagellum is present. The undulating membrane is absent.

Amastigote

This stage is found both in vertebrate and invertebrate hosts. Flagellum in the form of fibril (a trace of flagellum), basal granule, and a nucleus are present.

Different Species

Leishmania infantum

This species is mostly similar to *Leishmania donovani*. Many authors opine this species to be subspecies of *Leishmania donovani*. This species is a causative agent of infantile kala-azar. Dog, fox, etc. are reservoir hosts. *Phlebotomus major*, *P. chinensis*, *P. papatasi,* etc. are the vectors of this species. This species is prevalent in North America and the Mediterranean region.

Leishmania major

This species causes ulcerative cutaneous leishmaniasis in both humans and animals. Ulcerative cutaneous leishmaniosis is found in muzzle and feet of dogs infected with this species. Generalized lymphadenopathy and splenomegaly have not been reported in dogs infected with this species.

Leishmania tropica

The specis affect the skin involving the adjacent lymph nodes. This is causative agent of Old world cutaneous leishmaniasis, Oriental sore, Delhi boil, aleppo button. Wide spread dermatitis along with lymphadenopathy and spleenomegaly have been described in older dogs with *L. tropica* infection.

Leishmania donovani

This species causes different types of kala-azar i.e. Indian kala azar, African kala azar, mediterranian kala azar, Chinese kala azar, etc.

Leishmania braziliensis Complex

This complex consists of the following parasites.

Leishmania braziliensis brazilesnsis

This protozoan causes espundia in human beings. The disease is characterized by skin lesions and disfigurement of cartilaginous tissue as a result of erosion. It causes cutaneous leishmaniosis in dogs.

Leishmania braziliensis guyanensis

This protozoan causes pian bois or buba of North and South America. This is characterized by skin ulceration that spontaneously heals and metastases in nasal mucosa may occur in some cases. This parasite may affect dogs.

Life Cycle

The organisms require two hosts: vertebrate hosts and invertebrate hosts (sand flies). The organisms (promastigote stage) are transmitted to the animals through the bite of infected phlebotomine flies. The organisms are engulfed by macrophages and other mononuclear phagocytic cells. In these cells, the organisms multiply by binary fission and transform into amastigote forms.

The following events occur during the life cycle of this protozoan:

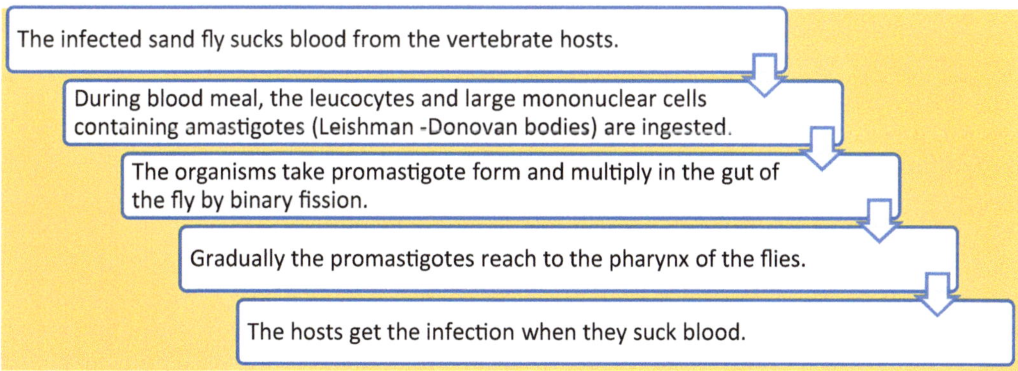

Note

It is commonly known that *Leishmania* sp. performs asexual reproduction. However, Kreutzer et al. (1994) reported evidence of sexual reproduction in *Leishmania*. Akopyants et al. (2009) also reported genetic exchange during cyclical development of *Leishmania* in sand fly.

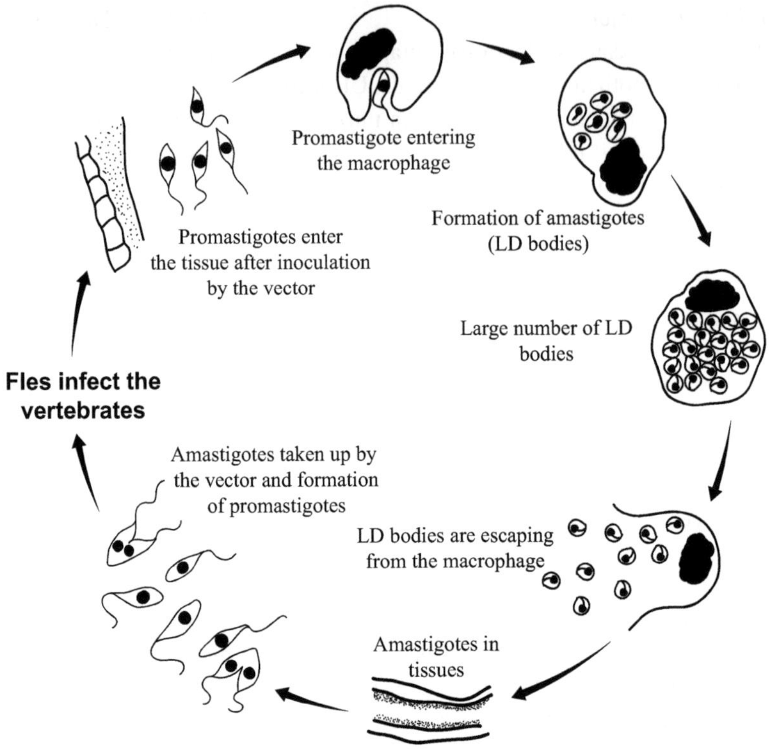

Fig. 8.7 Life cycle of *Leishmania* sp.

Pathogenesis

The promastigote stages are transmitted by the bloodsucking flies during their blood meal. The organisms enter the macrophages and endothelial cells where they multiply and produce a large number of amastigote stages. The parasitized macrophages disseminate the organisms to different parts of the body, most predominantly the liver and spleen. Splenomegaly and hepatomegaly are commonly observed. The disease occurs in two forms as follows.

Visceral Leishmaniosis

The infection caused by *Leishmania* in humans is commonly termed as leishmaniasis. There are three types of visceral leishmaniasis: Indian kala-azar caused by *L. donovani*, Mediterranean kala-azar caused by *L. infantum*, and American kala-azar caused by *Leishmania chagasi*. This disease is also called kala-azar. The liver, spleen, and bone marrow are affected. Splenomegaly and hepatomegaly are common pathological features. The disease starts with fever, followed by headache, abdominal pain, dysentery, and diarrhea. Bleeding from the nostrils and mouth is also common. The dogs may be affected by Leishmania donovani.The pathological features are hepatomegaly, spleenomegaly, lymphadenomegaly

Cutaneous Leishmaniosis

Cutaneous leishmaniosis occurs on the skin, particularly in the area around the eye. This is characterized by the formation of ulcers. The ulcer is of two types: moist and dry. Small papules develop at first, which later on get ruptured and coalesce to form massive lesions. This may be Old World cutaneous leishmaniasis or American cutaneous leishmaniasis occurring in humans. In Old World cutaneous leishmaniasis, the first detectable lesion appears after 3–6 weeks of insect bite. Reddish papules form, become crusted, and transform into ulcers. These ulcers may heal within 2 months or take a longer period of 1 year or more leaving a deeply pigmented depressed scar. American cutaneous leishmaniasis is of two types: metastasizing and non-metastasizing. Espundia is a characteristic metastasizing leishmaniasis affecting nasal septum, nasopharynx, and larynx. Cutaneous leishmaniasis may be localized or diffused. Old world and Middle East cutaneous leishmaniasis are caused by a well-known important species *Leishmania major*. American cutaneous leishmaniasis is caused by *Leishmania mexicana mexicana*, and *Leishmania mexicana amazonensis, L. mexicana panamensis Dogs may suffer from cutaneous leishmaniosis caused by L. tropica, L. mexicana. Hyperkeratitis, ulcerative lesion, alopecia,nodule formation, etc. are some pathological features.*

Mucocutaneous Leishmaniasis

The causative agents are *L. braziliensis, L. panamensis, and Leishmania guyanensis.*

> **Note**
>
> Daulatabad et al. (2015) reported mucocutaneous leishmaniasis in man caused by *Leishmania donovani*.

Clinical Signs

- High fever
- Dullness
- Depression
- *In dogs, the condition appears as viscerocutaneous leishmaniosis. In this case, the following are the clinical signs:
- Fever
- Weakness
- Disinclination to body exercise
- Skin sores, baldness appearing from the head and extending later
- Lethargy
- Inappetence or less appetence
- Bleeding from the nose

The dogs may suffer from cutaneous leishmaniosis. The clinical signs are as follows:

- Ulcer
- Hyperkeratosis
- Depigmentation or dark coloration
- Conspicuous thickening of foot pad and muzzle
- Nodule formation on the skin

Diagnosis

Symptomatological Diagnosis

Symptoms exhibited by humans may be suggestive of this parasitic infection. Characteristic cutaneous lesions may suggest cutaneous leishmaniasis. Fever with splenomegaly and hepatomegaly are also suggestive of kala-azar. However, symptomatology cannot confirm the disease.

Examination of Blood Smear

Diagnosis can be done by the examination of blood smear. Different Romanowsky stains are used for the identification of this protozoan parasite. Giemsa stain and Leishman stain can be used for this purpose.

Biopsy Examination

Biopsy examination is a common practice for the diagnosis of leishmaniasis. Aspiration of minute tissue using fine needle from the spleen, bone marrow, and lymph node is performed. These aspirated materials are smeared and stained or the sections are stained for the detection of the parasites. Although splenic aspiration is more risky than bone marrow aspiration, the splenic aspiration is more sensitive. There may be a rupture of the spleen and uncontrolled hemorrhage may cause the fatal conditions. After staining, the organisms become visible. The amastigote stages are found in monocytes, macrophage, etc. The cytoplasm and nucleus display blue and red color, respectively. Right angle to the red nucleus, a deep red or violet colored rod-like kinetoplast is found.

Imprint cytology can be performed using tissue biopsy material of the liver, spleen and lymph nodes. The cut tissue is placed on the slide and repeated pressure is applied on the flat cut surface to prepare a perfect slide. This slide is stained with Giemsa stain. Repeated searching may be required because the organisms may be distributed unevenly. Histology of tissue section can be performed using Hematoxylin and Eosin stain. Tissue specimens can be stained by fluorescent-dye-tagged antibodies. Fluorescein isothiocyanate isomer-tagged antibody is used for staining tissue sections. Rhodamine B isothiocyanate-conjugated antiserum is also used for the same. Monoclonal antibodies conjugated with fluorescent dyes can also be used to accomplish the same.

Culture of *Leishmania*

Direct demonstration of organisms from specimens may face problems due to insufficient availability of the organisms. Two types of cultural media are used: monophasic and diphasic culture media. The monophasic media are Schneider's insect medium, M199, Grace's media, etc. The diphasic media are Novy-MacNeal-Nicolle medium, Tobie's medium, etc. The aforesaid media are used for the culture of promastigote stages of *Leishmania*. For the culture of amastigote, the tissue culture techniques using monocytes and macrophages as cells are used.

Animal Inoculation

In this method, organisms are demonstrated in laboratory animals after artificial infection with the specimen containing the organisms. Golden hamsters are the most suitable laboratory animals. Common routes of inoculation are intraperitoneal and intrasplenic. Both promastigotes and amastigotes can be used for inoculation. After inoculation, the animals are monitored at weekly intervals for the observation of clinical signs, i.e., cutaneous lesions and hepatosplenomegaly. The amastigotes can be demonstrated in biopsy materials of the spleen, liver, etc., of the laboratory animals. The infected animals may be sacrificed and tissue materials may be examined for the detection of amastigotes.

Immunodiagnosis

Immunodiagnosis is done by antigen detection and antibody detection.

Antigen Detection

Antigen detection is an effective way for the diagnosis of leishmaniasis. This technique is more specific compared to antibody detection. Different specific antigen molecules (72–75 and 123 kDa) are detected in urine. This test displays more than 90% sensitivity for diagnosis of visceral leishmaniasis. Another test for antigen detection is the new latex agglutination test (KTEX). This test displays satisfactory sensitivity and specificity. The test is useful for the early detection of Leishmania antigen of visceral leishmabiasi.

Antibody Detection Test

In the past, nonspecific tests were developed for the detection of immunoglobulins. These include the Chopra antimony test, Napier's formol gel test, and the aldehyde test. These tests are unreliable and lack both sensitivity and specificity. These tests have recently been replaced by reliable tests. A number of immunodiagnostic tests were tried for the diagnosis of leishmaniasis. These include counter immunoelectrophoresis, indirect hemagglutination test, indirect fluorescent antibody test, etc. All these tests are not reliable because they lack specificity. Only IFA is used on a small scale. A modified test DAT has been established. This test is considerably sensitive and also specific. In this test, trypsin-treated promastigotes are used as antigen. This antigen is stained by coomassie brilliant blue. Positive reaction displays agglutination. ELISA has been found to be a reliable immunodiagnosis test. Soluble promastigote antigen is used for performing this test. The suspension of promastigotes is freezed and thawed several times. Then the suspension is centrifuged at 10,000–30,000 × g. The supernatant is used as an antigen. The study shows high sensitivity and specificity of fucose–mannose ligand of 36 kDa glycoprotein found in both amastigote and promastigote. The study also shows encouraging results using the recombinant antigen (rk39). Similar to other protozoan infections, the immunochromatographic test is also usable for rapid detection of VL in field screening. This test is ready to use and easy to perform. In this test, the recombinant antigen rk39 is immobilized on the nitrocellulose membrane.

Molecular Diagnosis

Molecular diagnosis is done using different markers. The following markers are used:

- PCR using kinetoplast DNA as a marker.
- PCR using ribosomal DNA, i.e., ITS1 and ITS2 as a marker.
- PCR using 18SrRNA as a marker.
- PCR using gp63 gene as a marker.

Treatment

Miltefosine, sodium stilbogluconate, amphotericin B are some drugs for treatment of *Leishmania* infection.

Control

The following steps are taken to control leishmaniosis.

General Management

The affected persons/animals should be treated properly. The enzootic areas should be detected. Special attention should be paid to the enzootic areas. Using rapid diagnostic tests based on immunodiagnosis and molecular diagnosis, screening of infection should be done. Screened patients should be treated properly. *Phlebotomus* flies are the vectors of the disease. Control of vector is important for the control of leishmaniasis.

Immunological Control

There have been many trials conducted for the development of a vaccine against leishmaniasis. Most of the vaccines under research were subject to trials against cutaneous leishmaniasis. Whether any vaccine is effective both against cutaneous and visceral leishmaniasis is yet to be unfurled. Whole-killed organisms and defined proteins are used for vaccine trials. Different trials have been conducted taking recombinant antigens as vaccine candidates. The recombinant proteins are surface-expressed glycoproteins such as leishmanolysin (gp63), parasite's surface antigen (PSA2), etc. The level of protection was achieved varied. Although most of the research works were carried out in laboratory animals, the results were encouraging. One more important antigen subject to extensive study is *Leishmania* homologue for receptors of activated c kinase (LACK), which is a common protein expressed during the life cycle of *Leishmania*. Other defined antigens that were subject to different vaccination trials include cysteine proteinase A2. Different membrane proteins of developmental stages have also been used to see the immunoprotective potential of those proteins. These include amastigote membrane protein, i.e., P4. Subcellular fractions were also used for the development of vaccines. These include kinetoplastid membrane protein 11 (KMP 11), paraflagellar rod protein 2 (PFR-2), etc. A series of trials have been conducted on the immunoprotective potential of a fused candidate protein, Leish111f composed of *Leishmania major* stress-inducible protein 1 LmSTI 1, *L. major* homologue of eukaryotic thiol-specific antioxidant (TSA), and *L. braziliensis* elongation and initiation factor (LeIF). Studies show encouraging protection levels in animal models and humans along with report of partial protection against visceral leishmaniasis. Unfortunately, this vaccine candidate did not show any protection in dogs. Studies on the use of salivary proteins from vector (SP15, maxadilan) revealed some amount of protection against *L. major* in mice models.

Looking into the failure of vaccination by killed organisms and the inability to provide long-term immunity by subunit vaccines, another method was developed and work is in progress. This method involves the use of attenuated whole live parasites attributed to the concept of delivering all probable proteins present in the organisms. For this, the organisms are genetically attenuated and a knockout batch of leishmania organisms is developed. For example, genetically attenuated *Leishmania mexicana* were produced by researchers through the deletion of the cysteine proteinase gene. Recent studies show that vaccination of mouse models with dihydrofolate reductase thymidylate synthase (dhfr-ts) knockout parasites provided protection.

Phylum: Sarcomastigophora—Subphylum: Mastigophora—Class: Zoomastigophorea—Order: Trichomonadida—Family: Trichomonadidae

Genus: *Tritrichomonas*

Species: The important species is *Tritrichomonas foetus*.

Hosts: Cattle, buffalo, pig, horse

Site: Genital organs

Disease name: Bovine trichomonosis or bovine trichomonad abortion

Distribution: This parasitic infection is worldwide in distribution

Morphology

These are pear-shaped organisms. They have four flagella, of which three are anterior and one posterior flagellum trails behind. One rod-like structure is present which is called an axostyle. This axostyle slightly protrudes out posteriorly. One well-developed undulating membrane is present. The base of the undulating membrane is called as a costa. A prominent nucleus is present. A sausage-shaped parabasal body is present.

> **Note**
>
> - These protozoan organisms lack any mitochondria. Instead, they contain many membrane-bound electron-dense packets called as hydrogenosomes. The function of hydrogenosome is more like mitochondria which synthesize ATP.
> - *Tritrichomonas* organisms are coated with lipophosphoglycan (LPG), which are shed.
> - Based on some antigenic characters, three serotypes have been described: Brisbane, Belfast, and Manley.

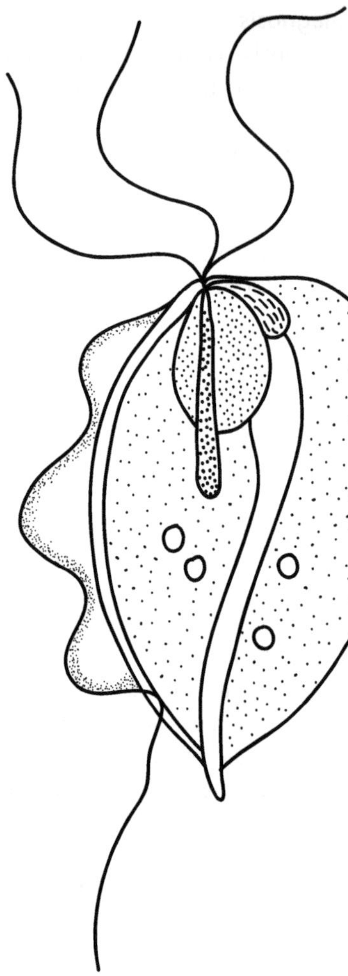

Fig. 8.8 *Tritrichomonas foetus*

Transmission

The organisms are transmitted sexually. Multiplication occurs by binary fission. Artificial insemination, gynecological examination, etc., may be other modes of infection. The organisms cause vaginitis when they remain there. Subsequently, they proceed to the cervix causing cervicitis. Then they reach the uterus where they cause endometritis. If the animals are pregnant, the organisms can affect the fetus.

Pathogenesis

Cows get infected during service by the infected bull or through artificial insemination due to infected semen. The organisms gain entry into the genital tract and multiply by binary fission elsewhere in the genital tract, causing pathological features such as vaginitis, cervicitis, placentitis, endometritis, and pyometra.

After few days of infection, the animals start discharging mucoid material through the vagina. Initial infection may pass unnoticed as there may not be any symptoms. If the animal is pregnant, the fetus is infected, resulting in early abortion. Abortion usually occurs during 8–16 weeks of pregnancy. Irregular heat periods are a common feature in trichomonosis. In some cases, complete elimination of fetus and membrane does not occur, and the retained material is macerated, resulting in purulent endometritis. Despite this parasitic infection, there may be development of the fetus up to full term and normal birth. If the cervix is closed, a closed pyometra may result. In some occasions, mummification of the fetus may also occur. Repeat breeding condition may also result. In some cases, the cervix is closed due to retained corpus luteum, and a closed pyometra develops in this case.

In bulls, the organisms mainly affect the preputial cavity. Initially, the animals feel pain in the infected area and refuse to serve the cows. Painful micturition is another symptom. There is mucous discharge, which may become mucopurulent later on. The symptoms may disappear without any treatment and the animals act as carriers of the organisms.

Clinical Signs

- Mucous or mucopurulent discharge
- Early abortion
- Persistent pyometra after abortion
- Repeat breeding condition

Diagnosis

- Herd history is important for the diagnosis of the disease.
- Constant pyometra after abortion is a diagnostic clue.
- If fresh vaginal discharge is directly examined, the movement of organisms may be visible.
- Vagina should be washed, and vaginal washing can be examined for the detection of the organisms.
- Material from the aborted fetus can be examined for the detection of any organism.
- Allantoic fluid may also be examined.
- Culture of the organisms can be done for better availability for further identification.
- Care should be taken to ensure that the material is not contaminated with urine or any other material. If the material is mixed with lukewarm water, the movement of the organisms may be visible.
- A smear can be prepared and stained following standard staining protocol.

Treatment

Retractor penis muscle is made relaxed by pudendal anaesthesia. Then the penis is washed properly with weak detergent solution. Flavin ointment is introduced in the preputial cavity and the penis is massaged for 15–20 minutes. Parenteral administration of dimetridazole (50 mg/kg bw) daily for five days is effective.

Control

Infected bull should not be allowed for service. Artificial insemination should be performed using semen free from any contamination.

Phylum: Sarcomastigophora—Subphylum: Mastigophora—Class: Zoomastigophorea—Order: Trichomonadida—Family: Monocercomonadidae

Genus

Histomonas

These are flagellated organisms. They occur in the intestine and liver. The organisms occurring in the lumen of caecum have one flagellum. But there may be as many as four flagella. One nucleus is present. The hosts get the infection by the ingestion of this organism along with water and feed. They form colonies in the wall of small intestine and cause enterohepatitis. Different forms of the organisms are found. Invasive forms are found in the early caecal and liver lesions. Invasive forms are 8–17 um long. Vegetative forms are 12–15 um × 12–21 um in size and they are found in the center of the lesion. Resistant forms are 4–11 um in diameter. This form does not produce any cysts.

Histomonas meleagridis

Hosts: The hosts are turkey, pheasants, fowl, guinea fowl, chicken, etc.

Site: Cecum and liver

Disease Name

Infectious enterohepatitis, Blackhead of turkey

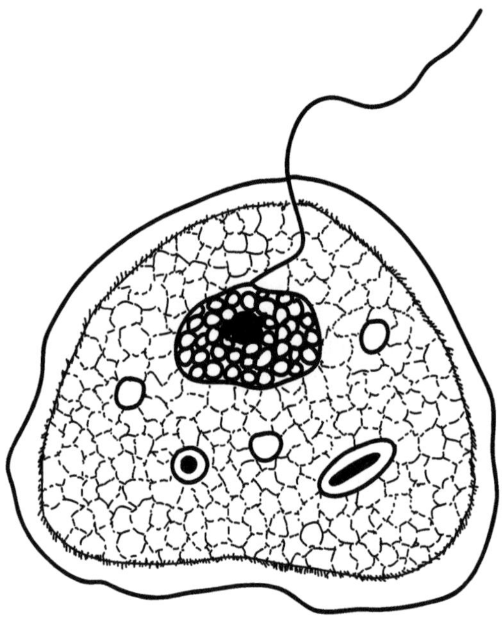

Fig. 8.9 *Histomonas meleagridis*

Life Cycle

- There are several routes of transmission of *Histomonas meleagridis* as follows: The hosts get the infection by the ingestion of infected eggs of *Heterakis gallinarum*. After ingestion, the eggs get hatched and the Histomonads get released. The *Histomonas* organisms enter the epithelial cells of ceca and start forming colony. In the most susceptible birds such as turkey and guinea fowls, the organisms enter the circulation and reach the liver. In the liver cells, they multiply and form colony there.
- Birds may also get the infection by the ingestion of infected earthworm. The earthworms act as transport hosts. They ingest the infected eggs. The eggs hatch inside the body of the earthworm and the larvae are liberated after hatching and remain within the earthworm without any further development. At the same time, the *Histomonas* organisms also remain within the earthworms. The birds may get the infection by the ingestion of infected earthworm.
- The third way of infection is very interesting. The birds may get infection through cloaca by the phenomenon of cloacal drinking. This is a reflexive mechanism where water is taken up by the cloaca. During this activity, the organisms gain entry.

Pathogenesis

Structural Pathology

The organisms cause severe pathology. They cause formation of colonies and induce necrosis in both the intestinal wall and liver. The cause circumscribed depressed ulcer in the liver. The disease caused by this parasite is called blackhead of turkey. The wall of intestine becomes hyperemic and thickened.

Chemical Pathology

Rise in blood sugar and decreases in uric acid levels are some features. Increased levels of aspartate aminotransferase (AST) and lactate dehydrogenase (LDH) are also found.

Symptoms

Different clinical signs are exhibited as follows:

- Weakness.
- Dullness.
- Depression.
- Droopiness.
- Off feeding.

- Cyanotic discoloration occurs. This is usually found in the case of infection of turkey. However, cyanotic discoloration does not occur in chicken.
- Sulfur colored (yellow) feces are another clinical sign. However, in all cases, sulfur colored feces may not be found.
- Loss of body weight.
- Mortality may reach up to 100%.

Diagnosis

Diagnosis is difficult. The typical clinical signs are suggestive of this protozoan parasitic disease. Autopsy of the affected birds reveals the characteristic circumscribed depressed ulcer in liver.

Phylum: Sarcomastigophora—Subphylum: Mastigophora—Class: Zoomastigophorea—Order: Diplomonadida—Family: Hexamitidae

Family Synopsis

The species of this family affect alimentary canal of mammals and birds. The protozoan parasite is characterized by bilaterally symmetrical body with two nuclei and six or eight flagella. The life cycle is direct.

Genus: *Hexamita*

Important Species

Hexamita meleagridis

This protozoan parasite is found in the small intestine of turkey, pea fowl, partridge, quail, etc. It may also be found in the large intestine. It is flagellated and typically pear-shaped. Eight flagella are present. Of these, four are anterior flagella, two are anterolateral flagella, and two are posterior flagella. The organisms multiply by binary fission. The hosts get the infection by the ingestion of cysts contaminated with food and water. The clinical sign include diarrhea. Diagnosis can be done by examination of duodenal and jejunal mucosal scraping. The cysts are resistant.

Genus: *Giardia*

History

Leeuwenhoek discovered Giardia from his own feces in 1681.

Giardia lamblia is an important species occurring in the duodenum and other parts of the small intestine of man. However, this parasite also occurs in monkeys, dogs, cats, etc. In some cases, the parasites are also found in the large intestine. The hosts get the infection by the ingestion of infective cysts. Two other species, *Giardia cati* and *Giardia Bovis*, occur in cats and cattle, respectively. The cysts are resistant to chlorination and ozone treatment.

Morphology

They are pear-shaped organisms having a broad anterior end and a tapering posterior end. The organisms appear like face of monkey. They are dorsally convex and ventrally concave. A large sucking disc is present on the concave ventral side. They have two nuclei and two axostyles. Eight flagella are present, of which two flagella are projected posteriorly. The cysts are ovoid or elliptical in shape containing two or four nuclei and the fibrillar remnant of the trophozoite stage.

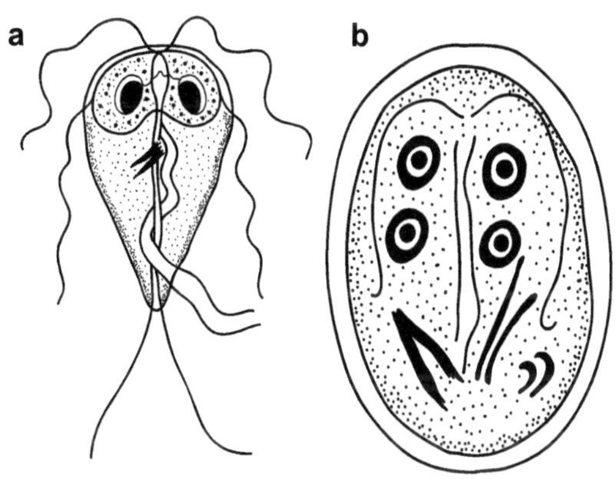

Fig. 8.10 *Giardia* sp. (**a**) Trophozoites (**b**) Cyst

Species

Giardia lamblia
This species occurs in the intestines of humans. This species also affects the monkeys and pigs. The species is worldwide in distribution.

G. canis
This species occurs in the duodenum, jejunum, and upper ileum parts of the intestine. It has been found pathogenic in dogs and causes diarrhea.

G. cati
This species occurs in the small intestine of cats. No pathological significance of this species has been reported.

Giardia muris
This species occurs in the intestines of rodents. The body is pear-shaped, and the median body is roughly round in shape.

Giardia psittaci
This species occurs in the intestines of birds.

G. bovis
This species occurs in the small intestine of ox.

Biology
No intermediate host or vector is required. The life cycle is direct type. The fecal oral route is the mode of infection. The trophozoite stage penetrates the mucosal cells of small intestine and forms colony there. *Giardia* cannot spread via blood circulation.

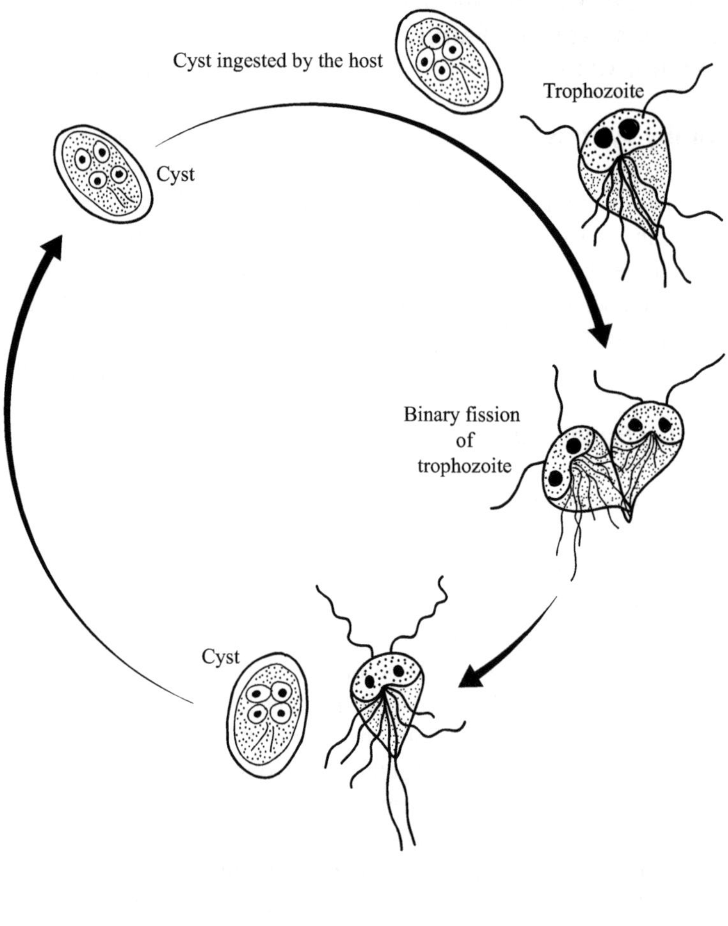

Fig. 8.11 Life cycle of *Giardia* sp.

Clinical Signs

In the animals, this parasitic infection appears to be nonsymptomatic. Occasionally, there may be diarrhea. The diarrhea may be continuous or discontinuous.

Diagnosis

Fecal Examination

Microscopic examination and detection of *Giardia* cysts and trophozoite are the gold standard methods for accurate diagnosis. For this, direct smart technique and concentration technique are used.

- *Direct smear:* Both dry and wet smears are examined for the direct technique. Sensitivity depends on whether direct smear technique or concentration technique is applied. For the identification of trophozoite stage, physiological salt solution is used for mixing with fecal sample. Fecal material can be fixed in sodium acetate–acetic acid formalin (SAF) and stained with Lugol's iodine. In unstained smear, the movement of trophozoite may be visible. However, in stained smear, the movement of trophozoite cannot be seen. For both dry and wet smear techniques, organism may not be readily found on a single prepared slide. One to three slides may be required for examination. It is always advised to examine the sample immediately; otherwise, the trophozoite stages may get deteriorated. To prevent this problem, different preservatives such as 10% buffered formalin and polyvinyl alcohol may be used. If the infection load is low, one should perform concentration technique. Sedimentation and floatation can be used for this.
- *Concentration technique:* When the infection load is low and cysts are not satisfactorily detectable during direct dry and wet smear techniques, a concentration technique should be applied. There are two types of concentration techniques: sedimentation and floatation. The flotation technique is more useful by which the cysts are easily separated. For this purpose, a flotation solution is prepared. A saturated salt solution having a high specific gravity (1.20) is prepared. $NaNO_3$, NaCl, and $ZnSO_4$ are used for this purpose. This technique can be modified with centrifugation. Sedimentation can be performed using formalin ether/formalin ethyl acetate. This chemical is used for the removal of fat droplets. There is a chance of distortion of *Giardia* cysts by $ZnSO_4$. Formalin ether/formalin ethyl acetate does not cause any distortion of the cysts.
- *Kato-Katz:* There exist some reports on the use of Kato-Katz for diagnosis but this technique displays less sensitivity and is not used routinely. Some workers have reported sucrose density gradient centrifugation.

Treatment

Metronidazole, chloroquine, diodoquin, quinacrine are some effective drugs.

Note

Staining of trophozoite and cyst is another step for the perfect diagnosis of this protozoan parasite. By staining, many important morphological details are found. There are different staining methods. Staining may be temporary and permanent. Temporary staining can be done using methylene blue or Lugol's iodine used for wet smear. Permanent staining can be done using iron hematoxylin eosin, Giemsa, etc. Giemsa staining is easy to perform. In Giemsa staining, the flagella and nuclei take on a reddish pink color and the cytoplasm takes blue color. Trichrome is another good stain used for staining fresh as well as PVA-fixed feces.

Entero Test

When microscopical examination of feces is not possible to get any cyst or trophozoite, the string test can be performed. In this test, traces of fluid from duodenal and jejunal area are taken out using lead-weighted capsule containing a highly absorbent nylon string of 90–140 cm in length as per the age of the patient. The patient ingests the capsule. The capsule gets dissolved and the nylon reaches to the target location by peristaltic movement. The string is kept as such for 4 h. Then the string is removed and fluid of the string at bile stained area is taken for microscopical examination and laboratory culture.

Note
- *Giardia* sp. does not have mitochondria. Only mitochondrial remnants called mitosomes are found.
- They also do not have Golgi complex and peroxisomes.
- They have a body which is pear-shaped and a median body which is claw-shaped.
- Acid of stomach activate the process of excystation. However, cysteine proteinase present in the peripheral vesicles contributes to the progression of excystation.
- It is known fact that *Giardia* performs asexual reproduction, but there have been reports of sexual reproduction. The reports are yet to be confirmed.

Phylum: Sarcomastigophora—Subphylum: Sarcodina—Class: Lobosea—Order: Amoebida—Family: Endamoebidae

Genus: *Entamoeba*

The important species is *Entamoeba histolytica*. These parasites occur in large intestine of human beings and primates.

History

Friedrich Losch first described *Entamoeba histolytica* in 1875.

Entamoeba histolytica

Introduction
This organism causes amoebic dysentery in man. Dog, cat, are also susceptible.

Distribution
They are worldwide in distribution.

Hosts
The main host is human. Dogs, cats, monkeys, and pigs also are infected.

Site
The seat of predilection of the organisms is intestine.

Morphology
There are two stages of the organisms: trophozoite and cyst.

Trophozoite
The trophozoite stage is the active form of the organism. Temporary locomotory organs are formed in the trophozoite, which are called as pseudopodia. With these finger-like pseudopodia, the trophozoites can actively move. The cytoplasm of the organism has two main parts: outer ectoplasm and inner endoplasm. Food vacuoles are present.

Cyst Stage
The trophozoites are transformed to cystic stage. The cysts are quadrinucleated. The nucleus of the trophozoite divides into four nuclei. The cysts rarely may contain eight nuclei, as seen in *Entamoeba histolytica*.

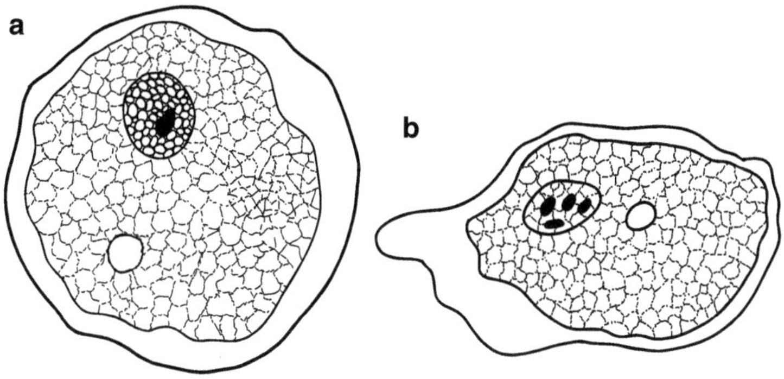

Fig. 8.12 *Entamoeba* sp. (**a**) cyst (**b**) trophozoite

Transmission

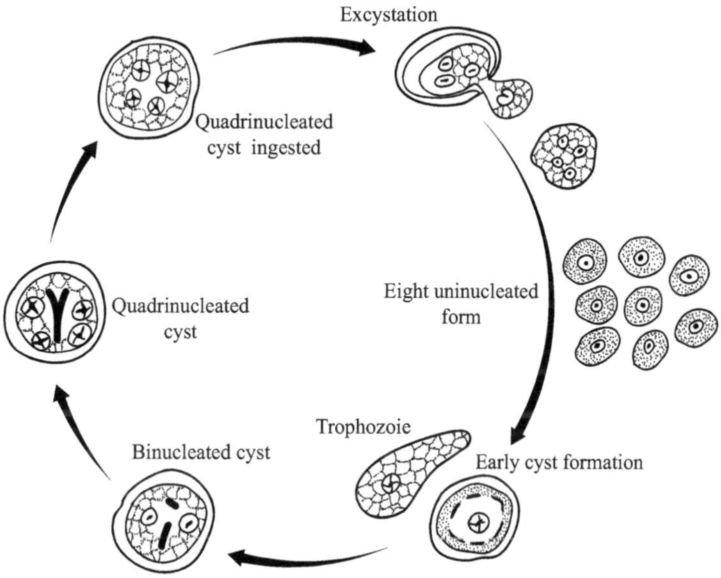

Fig. 8.13 Life cycle of *Entamoeba histolytica*

Pathogenesis and Clinical Signs

In Human

Entamoebiasis is a zoonotic disease. The causative agent of this zoonotic disease is *Entamoeba histolytica*. The intensity of pathogenicity depends on many factors. Infection load, virulence of the organisms, age of the hosts, and presence of bacterial flora are some factors responsible for magnitude of pathogenicity. The infection remains asymptomatic in most of the cases. The disease may occur in both acute and chronic forms. Individuals may be recovered from acute phase, but chronic dysentery can occur. The parasites inhabit the intestines of humans. The trophozoites release proteolytic enzymes that cause lysis of the epithelial layer. In the wall of the intestine, they form compact colonies and destroy the wall of the intestine. This causes formation of flask-shaped ulcer, which has a neck and a dilated distal part. There are two types of lesions: primary or intestinal lesions and secondary or metastatic lesions occurring in the liver, brain, lungs, etc. Metacystic trophozoites penetrate directly into the epithelial layer of mucus membrane, dissolving the protein by histolysin. They reach the submucosal layer where they multiply rapidly and form colonies. An abscess is formed. The abscess gets ruptured, leading to ulcer formation. Ulcers may be generalized, covering the whole wall of colon, or localized involving ileocecal region and sigmoid-rectal region. Ulcer may be a size of a pin's head to an inch in diameter. Microscopical examination may reveal the ulcerative lesions limited to the muscularis mucosa. Amoebic organisms may be found in interglandular spaces. Necrosis and destruction of basement membrane are found in crypts of Lieberkuhn.

The clinical signs are as follows:

- Abdominal discomfort
- Flatulence
- Loose stool with mucus and blood, etc.
- Colic pain
- Steatorrhea

In Dogs and Cats

Transient infection may occur in dogs and cats. The hosts get the infection through fecal-oral route by the ingestion of cysts of the parasites. No pathogenicity is found.

Different Types of Amoebiasis

Chronic Intestinal Amoebiasis

Lesions of chronic intestinal amoebiasis are different from acute amoebic ulcers. Extensive superficial ulcers may be formed. Scarring of intestinal wall, narrowing of intestinal lumen, and adhesions of surrounding viscera are common features. Tumor-like growth of granulation tissues may be formed, which are called as amoeboma.

Hepatic Amoebiasis

Due to metastasis of amoeba organisms in the liver, abscess formation is common in human. Children are seldom affected. The trophozoites of *E. histolytica* are carried to the liver through the hepatic veins from the cecum and colon. These organisms cause cytolysis, leading to the formation of lesions that are initially solid but liquefy later on. Miliary abscesses (minute abscess) are formed initially, which later coalesce to form large-sized abscess. Usually, a single large reddish-brown abscess containing semiliquid consistency is formed. The inner wall of the abscess cavity is ragged. Multiple small abscesses may appear. Three zones are usually formed: central zone of cytolized granular material with no amoebae, an intermediate zone comprising degenerated liver cells, and a peripheral zone comprising congested capillaries and necrosed hepatocytes. The clinical signs of hepatic amoebiasis are as follows:

- Mild evening rise or low remittent temperature
- Pain on right shoulder
- Initial dry cough
- Jaundice occurs sometimes
- Tenderness of liver
- Right-sided chest movement of the patient is restricted or absent

Diagnosis

Examination of fecal sample can be done for detection of trophozoite and/or cyst stage of the parasite. Complement fixation test and ELISA have been described. Molecular diagnosis by PCR, DNA probe etc. are also useful.

Treatment

Metronidazole, di-iodohydroxyquinoline, diloxanide furoate are some effective drugs for treatment.

Cerebral Amoebiasis

Organisms are transported to the brain where they form single or small abscess.

Cutaneous Amoebiasis

This occurs in the skin underlying the visceral amoebic lesion such as areas of pericolic abscess and drainage area of a liver abscess.

Splenic Amoebiasis

Splenic abscess can result from liver abscess.

Pulmonary amoebiasis

The organisms are transported to the lungs from the gut wall via portal circulation. The organisms cause lysis of pulmonary tissues resulting in formation of single or multiple abcess. Secondary pulmonary amebiasis occurs not by transportation through circulation but by adhesion of lung surface with the liver abscess.

Other *Entamoeba* Species

Entamoeba bovis
The hosts are cattle. The seat of predilection of the organisms is the intestine. Cysts are uninucleated.

Entamoeba bubalis
These are protozoan parasites under the Endamoebidae family. They are uninucleated cysts.

Entamoeba coli
These organisms occur in cecum and colon of human beings. The organisms are considered nonpathogenic. Initially, the cysts are uninucleated but total eight nuclei are produced later on.

Entamoeba equi
The organisms occur in the horse. The cysts are quadrinucleated.

Entamoeba gingivalis
This species occurs in the mouth of human beings. The most important thing is that the organisms possibly do not have any cystic stage. Only thc trophozoitc stage is found. Demarcation between endoplasm and ectoplasm is distinctly visible during active movement by the trophozoite. Food vacuoles are present in endoplasm. They are found in gingival tissues. The organisms have also been isolated from tonsil, vaginal, and cervical smears. They remain as commensal in gum and other parts.

Entamoeba gedoelsti
The hosts of the organisms are horses. Cysts formation does not occur.

Entamoeba hartmanni
This species occurs in cecum and colon of dogs, monkeys, and humans. Previously, it was confused with *E. histolytica*. *E. hartmanni* is similar to *E. histolytica* except for size. The trophozoite stages of these organisms do not ingest RBC. The motility of this species is not as vigorous as that of *E. histolytica*. Immature cysts contain one to two nuclei, bar-like chromatoid bodies, and distinct glycogen mass. One to several minute spherical chromatoid bodies are present in mature cyst. The mature cysts contain four nuclei. They are nonpathogenic. Diagnosis is done by the examination of fecal sample (stained with iodine or unstained).

Phylum: Apicomplexa — Class:Sporozoea —Subclass: Coccidia—Order: Eucoccidiida—Suborder: Eimeriina—Family: Eimeriidae

Family Synopsis

The members of this family are mostly the intracellular parasites of the alimentary tract of mammals and birds. The life cycle is direct type exhibiting both asexual and sexual development. Asexual reproduction occurs by schizogony. Endodyogony may be present or absent. Sexual reproduction includes formation of microgamonts and macrogamonts. Zygote formation occurs by the union of micro- and macrogamonts. Oocysts may be without sporocysts or may have several sporocysts, each with one or more sporozoites. Sporogony usually occurs outside the hosts. Many of the species of this family are markedly pathogenic. The gold standard diagnosis is the examination of the feces for the detection of oocysts. Control of this parasite mostly relies on chemotherapy and general management.

Family Eimeriidae

Genera: *Eimeria*

The oocyst contains four sporocysts. Each sporocyst contains two sporozoites.

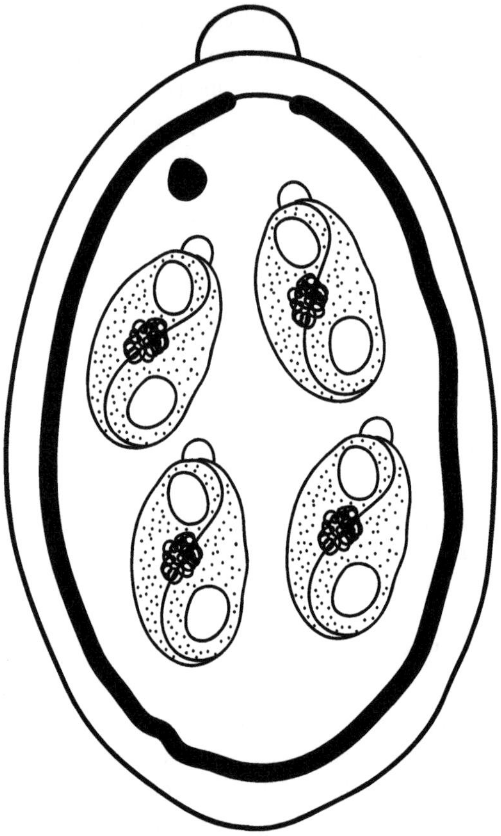

Fig. 8.14 Oocyst of Coccidia (general structure)

Morphology of oocyst

The oocysts are globular, oval, or suboval in shape. They have double wall. The unsporulated oocysts contain only undivided embryonic mass. The sporulated oocyst contains four sporocysts. Each sporocyst contains two fusiform sporozoites. A dot-like structure is found, which is called as polar granule. A micropyle may be present. A dome-like structure is present on the micropyle, which is called as polar cap or micropylar cap. Some residual material is present in the oocyst called as oocystic residual material. Same thing present in the sporocysts are called as sporocystic residual material. A plug-like structure is present at one of the terminal point of sporocyst, which is called as Stieda body.

Life Cycle

It is surprising how different species of *Eimeria* develop in different sites of the gut. It is so because the sporozoites must sense the gut cells. The sporozoites use the amylopectin as energy source during their activity.

The unsporulated oocysts pass out through the feces. In the presence of different environmental factors, the unsporulated oocysts transform to sporulated oocysts. The sporulated oocysts are the infective stages. The hosts get the infection by the ingestion of sporulated oocysts. After gaining entry, the oocysts are affected by different biological factors. Different factors are responsible for release of sporozoites. CO_2, bile, trypsin are some important factors. CO_2 plays major role in rupture of the micropyle and increase the permeability of the wall. Then the enzymes enter the oocysts. The enzyme and bile digest the steida body and a pore is formed through which the sporozoites come out. After hatching, the sporozoites come out and penetrate the epithelial cells of gut. The sporozoites transform to trophozoite. The next stage of development is schizont. The schizont stage gradually increases in size. The schizonts contain many merozoites. Lastly, the schizonts burst and the merozoites come out. Different generations of schizonts are formed. Some of the second-generation merozoites form two types of gamonts: microgametes and macrogametes. Union of two gametes result in the formation of zygote. Plastic granules or wall forming granules form the wall of the oocysts. Then the matured oocysts are extruded out from the tissues and pass out with the feces. Some merozoites may repeat the cycle

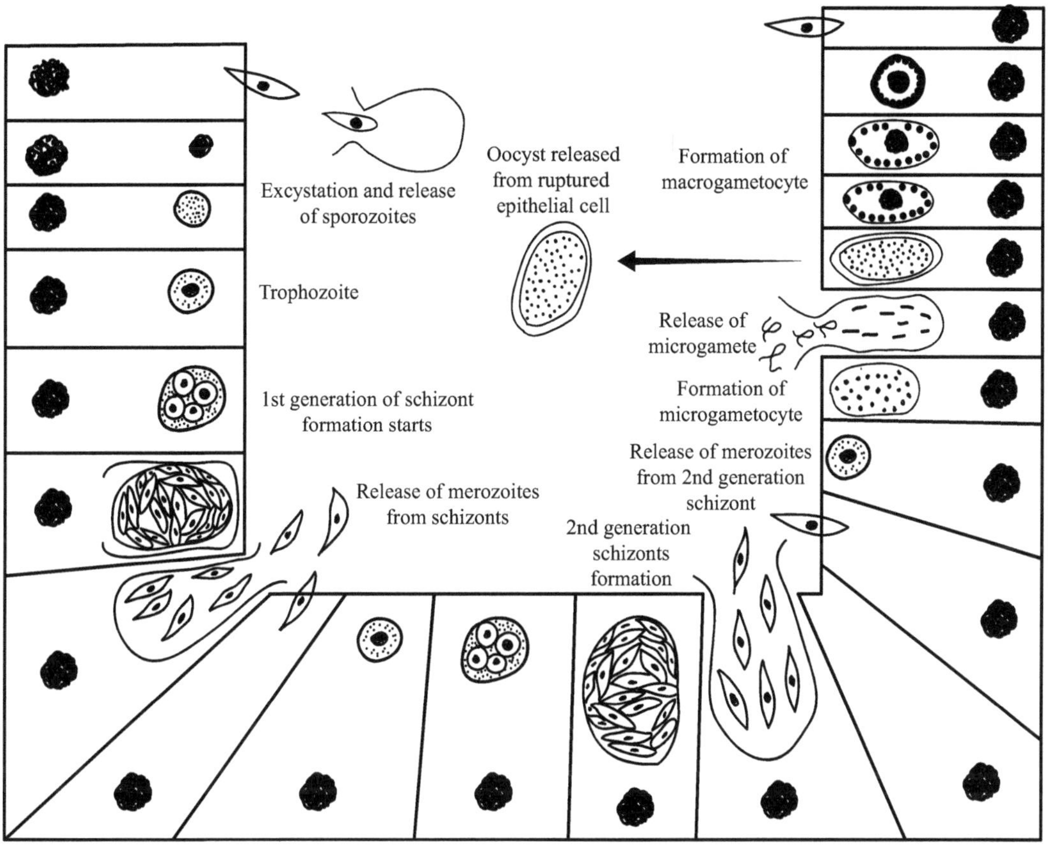

Fig. 8.15 Life cycle of Coccidia

General Pathogenesis

Eimeria tenella causes severe pathology. The young birds are more susceptible. The age group of the birds, falling in high susceptibility stage, is 2–4 weeks. On the first day of infection, there occurs limited or no clinical signs except for slight diarrhea. However, severe pathological features associated with hemorrhagic enteritis occur after 4–5 days of infection. Maximum pathogenesis ensues due to the second generation of schizonts. Colonies of schizonts are formed. The fully formed schizonts burst and the merozoites come out. The gut epithelial cells get denuded. Massive hemorrhage occurs. Blood is drained from the hemorrhagic lesion into the lumen of the intestine, which may result in the ballooning of intestine. Mortality of birds starts. Maximum mortality occurs between between 4th and 6 th day of infection. After recovery a chronic illness starts due to presence persistent cecal core. Other species cause mild, moderate, and severe pathology in accordance with species. Of these species of poultry coccidiosis, *Eimeria necatrix* causes significant pathology. *Eimeria zuernii* causes winter coccidiosis. *E. bovis* also causes significant pathology in cattle.

Clinical Signs

Poultry coccidiosis exhibits the following clinical signs:

- Poultry coccidiosis manifest into mild to severe form
- Simple diarrhea may transform to hemorrhagic diarrhea
- Refuse to take drinking water and food

- Weakness
- Ruffled feather condition
- Droopyness
- High mortality

Bovine coccidiosis exhibits the following clinical signs:

- Hemorrhagic diarrhea
- Tenesmus
- Abdominal pain
- Inappetence
- Loss of production

Diagnosis

Symptomatological Diagnosis

Symptomatology cannot provide a confirmatory diagnosis. However, hemorrhagic enteritis is commonly considered as occurrence of coccidiosis.

Parasitological Diagnosis

Diagnosis can be done by the examination of the fecal sample. The absence of oocysts does not indicate the absence of disease, because after 5–6 days of infection, oocysts discharge starts. If fecal examination is done before that time, oocyst may not be visible. In a flock, a few suspected birds may be sacrificed, and gross lesions in the intestine with different hemorrhagic spots are suspected in the case. Mucosal scraping and feces in the lumen of intestine should be examined. Direct smear examination is done. Under 10×, oocysts may be visible but examination under 40× reveals oocysts more clearly, along with specific structures of the oocyst. We can go for oocyst culture. The unsporulated oocysts are kept in 2.5% solution. The mixed solution is stirred twice daily for aeration. The sporulated oocysts display different minute structures by which the oocysts can be identified more specifically.

Immunodiagnosis

Many works were carried out on immunodiagnosis using different crude antigens and purified antigens. These antigens are oocyst extract (crude antigen), merozoite surface antigen, sporozoite surface antigen, and different recombinant surface antigen. Variable success was achieved in immunodiagnostic tests.

Molecular Diagnosis

Molecular diagnosis is possible using genus-specific or species-specific primers. Different specific primers are available. Different ribosomal DNA, such as ITS1, ITS2, 28SrDNA, etc., are used as molecular markers. Different mitochondrial DNA, such as CO1, ND1, etc., are also used as molecular marker for genomic characterization or identification.

Control

General Management

Affected animals should be treated properly. The litter of the poultry house should be kept dry. Clean utensils should be used. Drinking water and feed should not be contaminated. Proper vaccination schedule should be maintained. For bovine coccidiosis, calves should be isolated within 24 hours of birth. Contaminated feed and drinking water should be avoided. Washing of floor of animal house should be done properly. Over crowding should be avoided. Bedding should be treated with 1.25% sodium hypochlorite.

Chemoprophylaxis

Routine anticoccidial drugs are provided along with the drinking water and food. Long-term application of the drugs results in drug resistance. Two protocols are used to reduce drug resistance. These are **shuttle program,** in which different drugs of different series are used between the batches, or **switch program,** in which the drugs from the same batch are used at different intervals.

Immunological Control

An array of works was carried out for the development of anticoccidial vaccine. Live vaccine, live attenuated vaccine, and subunit vaccine have been tried. See the chapter "Immunoparasitology".

Molecular Vaccine

Molecular vaccination is under research. No molecular vaccine seems not yet available in the market.

Other Genera of Eimeriidae

Wenyonella

The oocyst contains four sporocysts. Each sporocyst contains four sporozoites.

Tyzzeria

The oocyst lacks any sporocyst. Only eight sporozoites are found.

Isospora

The oocysts has two sporocysts. Each sporocyst contains four sporozoites.

Different *Eimeria* Species

Eimeria spp. of Ducks

Eimeria anatis

This species occurs in duck. The oocysts are ovoid and have smooth wall. A distinct micropyle is present. The seat of predilection of the organism is the small intestine. These organisms are nonpathogenic.

Eimeria anseris

This species occurs in goose. The oocyst has a truncated cone at the narrow end. A distinct micropyle is present. Seat of predilection of the organism is the small intestine. This species is less pathogenic, but may cause hemorrhagic enteritis, resulting in the mortality of goose.

Eimeria battakhi

The oocysts are subspherical to ovoid in shape. Micropyle is absent.

Eimeria kotlani

This species occurs in domestic goose. The oocysts are oval. A broad micropyle is present. This parasite causes diarrhea.

Eimeria abramovi

The hosts are wild ducks and geese.

Eimeria boschadis

The hosts are wild ducks and geese. The seat of predilection is kidney.

Eimeria clarkei

This species occurs in wild ducks and geese.

Eimeria koganae

This species occurs in wild ducks and geese.

Eimeria spp. of Domestic Fowl

Eimeria acervulina

This species occurs in domestic fowl. The oocysts are ovoid in shape. A micropyle is present and the polar cap is absent. Stieda body is present but sporocystic residuum is absent. The seat of predilection of the organism is small intestine. This organism causes less pathology. Subacute to chronic intestinal coccidiosis occurs. In the anterior part of the small intestine, the developmental stages are produced. On the surface of the wall of the intestine, pinpoint hemorrhage occurs. Watery and whitish diarrhea are clinical signs.

Eimeria brunetti

The hosts are poultry birds. The seat of predilection of the organism is small intestine, cecum, and cloaca. The parasite is worldwide in distribution. It is the causative agent of rectal coccidiosis, affecting mostly the posterior part of the intestine. The oocysts are ovoid. A micropyle is absent. Both sporocystic residuum and Stieda body are present. The species causes marked pathogenicity. Blood-stained hemorrhagic diarrhea is observed.

Eimeria hagani

The hosts are poultry birds. The oocysts are ovoid in shape. A micropyle is absent. Oocyst wall is smooth. A polar granule is present. The seat of predilection of this species is anterior part of the small intestine. The organisms are less pathogenic. Petechial hemorrhages occur in the wall of the intestine.

Eimeria praecox

The hosts are fowls. The oocysts are ovoid. A micropyle is absent. The seat of predilection of developmental stages of the organism is upper third of the small intestine. The parasite is worldwide in distribution and causes low pathogenicity.

Eimeria necatrix

The hosts are domestic fowl. The oocysts are ovoidal. Both micropyle and polar cap are absent. The seat of predilection of the developmental stages of the organism is the small intestine. These organisms are distributed worldwide. This is an important species out of the lot. In the small intestine, asexual development occurs, while sexual reproduction occurs in the cecum. The species is pathogenic which comes next to *E. tenella*. The organisms cause chronic form of the disease. Hemorrhages occur due to the second generation of schizonts.

Eimeria maxima

The hosts are poultry birds. The oocysts are ovoidal. A micropyle is absent. A polar granule is present. Stieda body is present. The seat of predilection of the organism is the middle third of the small intestine. The parasite is worldwide in distribution. The species is not highly pathogenic. The high load of infection shows disease entity.

Eimeria mitis

The hosts are poultry birds. The oocysts are sub-spherical. Both micropyle and polar cap are absent. The seat of predilection of developmental stages of this species is the anterior half of the small intestine. The posterior part of the small intestine and cecum may also be affected. Polar granule and Stieda body are present. The parasite is worldwide in distribution.

Eimeria mivati

This species of *Eimeria* occur in the small intestine of birds. The seat of predilection is the duodenum to rectum. Oocyst wall is colorless and ellipsoidal to ovoidal. Micropyle is present absent. Sporocystic residuum and Stieda body are present. These organisms are pathogenic. Watery diarrhea occurs. Marked loss of production is found.

Eimeria tenella

The hosts are poultry birds. The seat of predilection of developmental stages of the organism is the cecum. The parasite is worldwide in distribution. It is the causative agent of the cecal coccidiosis. The oocysts are ovoid. The polar granule is present. An indistinct Stieda body is present. Micropyle is absent. This is the most pathogenic Eimeria of birds, which cause great economic loss in the poultry industry.

Turkeys

Eimeria adenoeides

The hosts are domestic turkeys. The oocysts are ellipsoidal. The micropyle may be present or absent. Almost the entire intestine may be affected by this species. This is highly pathogenic species. On fourth day of infection the birds start showing clinical signs. The symptoms include watery diarrhea, ruffle feather, droopingness, etc. Bloody diarrhea may occur. The infection may cause marked mortality.

Eimeria gallopavonis

The hosts are turkeys. The oocysts are ellipsoidal. The micropyle is absent. The seats of predilection of the organism are the lower part of small intestine, rectum, and cecum. They are apparently not pathogenic. Two different forms of second-generation schizonts are produced.

Eimeria meleagrimitis

The hosts are domestic turkeys. The oocysts are sub-spherical. The micropyle is absent. The organisms affect the small intestine. This is a pathogenic species of turkeys, causing up to 100% mortality of the birds.

Eimeria of Cattle and Buffalo

Eimeria alabamensis

The hosts are cattle, water buffalo. The seats of predilections of developmental stages of this species are small intestine, cecum, and colon. The oocyst wall is transparent or may be light yellowish in color and have no micropyle. A parachute-shaped cap is found at the round end of the developing sporocyst. The species is considered nonpathogenic. However, the organisms may cause some amount of pathogenicity in the hosts if the infection load is high.

Eimeria bovis

The hosts are cattle, water buffalo, etc. These are worldwide in distribution. These organisms are markedly pathogenic and cause clinical cases of bovine coccidiosis along with *E. zuernii*. The micropyle is present in the oocysts but micropyle cap is absent. The oocyst residuum is absent. The Globidial schizonts or giant schizonts are produced during biological development. It is true that the schizonts produce marked pathogenicity but maximum pathogenicity is achieved by the gamonts. The feces streaked with blood comes out. There occurs destruction of the crypts of Lieberkuhn and villi occur.

Eimeria cylindrica

The hosts are cattle, buffalo, etc. The shape of the oocyst is cylindrical or subcylindrical. Thin oocyst wall is smooth and colorless. There is no visible micropyle and micropylar cap.

Eimeria ellipsoidalis

The hosts are cattle, water buffalo, etc. The oocysts are ellipsoidal or may be slightly ovoid. Micropyle is absent. Oocystic residuum is absent. Sporocystic residuum is present. Stieda body looks like minute dark knob. Less amount of pathogenicity is produced in this parasitic infection. In young calves, the organisms cause diarrhea.

Eimeria zuernii

The hosts are cattle, Zebu, and water buffalo. The seats of predilection of the organisms are the small intestine, cecum, colon, and rectum. The oocysts are spherical, sub spherical, and ellipsoidal. Both the micropyle and micropylar cap are absent. The species is highly pathogenic and cause clinical cases of bovine coccidiosis along with *E. bovis*. In the acute stage, there occurs hemorrhagic diarrhea. The condition is so intensive that there may be frank blood in the feces.

Eimeria spp. of Goats

Eimeria alijevi

This species occurs in the small and large intestine of goat. The species are prevalent all over the world. The oocysts are ovoid, subspherical, or ellipsoidal. The micropyle is absent or indistinct. The micropylar cap is absent. The Stieda body may be present. The oocystic residuum is absent but the sporocystic residuum is present. They form giant schizonts in the epithelial cells of intestinal wall.

Eimeria arloingi

This species occurs in the intestine of goats. The species is worldwide in distribution. The oocysts are ellipsoidal. The micropyle and the micropylar cap are present. Stieda body is absent. The organisms are pathogenic and cause hyperplasia and pseudo-adenomatous metaplasia of the intestinal villi.

Eimeria christenseni

This species occurs in the small intestine of goats. The oocysts are oval in outline. At the micropylar end, it is flattened slightly. Mound-shaped micropylar cap is prominent. Oocystic residuum is absent but sporocystic residuum is present. Stieda body is present.

Eimeria gilruthi

This species is worldwide in distribution. It occurs in the abomasum but sometimes it may occur in the small intestine. The oocysts look like ellipsoidal. Globidial schizonts are formed which are so large as to be visible by naked eye.

Eimeria ninakohlyakimovae

This species occurs in the small intestine, cecum, and colon of goats. The oocysts are ellipsoidal, ovoid or sub spherical in shape. This is considered to be one of the important species. Both micropyle and polar cap are absent. The oocystic residuum is absent but sporocystic residuum is present.

Eimeria hirci
This species has been reported from the feces of goat. Both micropyle and micropylar cap are present. The oocystic residuum is absent but sporocystic residuum is present.

Eimeria spp. of Sheep

Eimeria ashata
This species occurs in the small intestine of sheep. The oocysts are ellipsoidal to ovoid. Both a micropyle and a dome-shaped polar cap are present. Oocyst wall is yellowish-brown in color. Oocystic residuum is absent but the sporocystic residuum is present. There is evidence of formation of globidial schizont in the small intestine during biological development. Intranuclear development in the experimental infection has been reported.

Eimeria arkhari
This species occurs in the small intestine of sheep. The oocysts are ellipsoidal. Micropyle is absent.

Eimeria crandallis
The oocysts are ellipsoidal or ovoidal. Both micropyle and micropylar cap are present. Oocystic residuum is absent and sporocystic residuum is present.

Eimeria parva
The oocysts are smooth, spherical, or subspherical in shape. The oocysts are light yellow to yellowish green in color. Both micropyle and micropylar caps are absent. Oocystic residuum is absent and sporocystic residuum is present. The organisms occur in sheep and goat.

Eimeria gonzalezi
The oocysts are ovoid or elliptical. The sporozoites have characteristic globules. Both micropyle and micropylar caps are present.

Eimeria ovina
The species occurs in the small intestine of sheep. This species is worldwide in distribution. The oocysts are ellipsoidal to ovoid in shape. The oocyst wall is smooth. Both micropyle and micropylar cap are present. The oocystic residuum is absent. The sporocystic residuum is present. The organisms are apparently nonpathogenic. However, if a large number of organisms affect the hosts, some amount of pathogenicity occurs.

Eimeria pallida
The hosts are sheep and goats. The oocysts appear to be fragile. The oocystic residuum is absent but sporocystic residuum is present. The organisms seem to be nonpathogenic. Both micropyle and micropylar caps are absent.

Eimeria gilruthi
The oocyst is not known. Only they are recognized by the formation of globidial schizonts in the mucosa of abomasum.

Eimeria of Camel

Eimeria cameli
The hosts are camel. The oocysts are characteristically truncated. They are ovoid in shape. During biological development, globidial schizonts are formed. Micropyle is present. Stieda body is absent.

Eimeria rajasthani
The oocysts are ellipsoidal in shape. Stieda body is present. Sporocystic residuum is also present.

Eimeria of Rabbits

Eimeria coecicola
This species occurs in the posterior ilium and cecum of rabbits. The shape of the oocysts is spherical to oval. The oocysts wall is smooth. The micropyle is present. Both oocystic residuum and sporocystic residuum are present.

Eimeria irresidua
This species occurs in the small intestine of rabbits. The shape of the oocysts is ovoidal. Micropyle is present. Oocystic residuum is absent but the sporocystic residuum is present.

Eimeria stiedai
This species occurs in domestic cotton tail rabbits. The micropyle is distinct. The oocysts are

oval to ellipsoidal in shape. The seats of predilection of developmental stages of the species are the liver and bile duct. The species is mostly pathogenic to the young rabbits.

Eimeria of Peasant

Eimeria colchici
The hosts of this species are pheasants. This seat of predilection of the organisms is the small intestine. The oocysts are ellipsoidal.

Eimeria duodenalis
This is a species under the family Eimeriidae. The hosts are peasants. This coccidian spp. are prevalent in Great Britain.

Eimeria spp. of Pigs

Eimeria debliecki
This species occurs in the small intestine of pigs. The shape of the oocysts is ovoid to subspherical. Both micropyle and micropylar cap are absent. The oocystic residuum is absent and sporocystic residuum is present. Stieda body is present. Young pigs suffer most.

Eimeria neodebliecki
This species occurs in pigs. The oocysts are ovoid or ellipsoidal. Both micropyle and micropylar cap arc absent. Oocystic residuum is absent but the sporocystic residuum is present.

Eimeria polita
This species occurs in pigs. The oocysts are ellipsoidal or ovoidal. The oocyst wall is yellowish brown to pinkish brown in color. Micropyle is absent.

Eimeria porci
The oocysts are ovoidal. Micropyle is present and indistinct but the micropylar cap is absent. Oocystic residuum is absent but the sporocystic residuum is present.

Eimeria scabra
The developmental stages are found in the large intestine. It is distributed worldwide. The oocysts are ellipsoidal to ovoidal in shape. This species is pathogenic. There occurs profuse diarrhea.

Eimeria spinosa
This species occurs in the small intestine of pigs. The oocysts are ellipsoidal to ovoid. The brown oocyst wall is covered with spines. Micropyle is absent. Oocystic residuum is absent but the sporocystic residuum is present. Stieda body is present. The young pigs suffer most. Diarrhea and fever occur.

Eimeria suis
This is a species under the family Eimeriidae. The hosts are domestic pigs. The oocysts are ellipsoidal to subspherical. The micropyle is absent. The oocystic residuum is absent but the sporocystic residuum is present.

Eimeria spp. of Horse

Eimeria leuckarti
This species occurs in the small intestine of horses and ass. The oocysts are dark brown in color and ovoid or pyriform in shape. The micropyle is present. The micropylar cap is absent. Giant schizonts are formed during their biological development. The oocystic residuum is absent but the sporocystic residuum is present. This is a pathogenic species.

Eimeria uniungulati
The oocysts are oval to ellipsoidal. The oocysts are orange in color. Both micropyle and micropylar cap are absent. The oocystic residuum is absent but sporocystic residuum is present.

Treatment

In Poultry

Drug	Dose	Application
Amprolium	0.0125–0.025%	In feed
Monensin	0.01–0.0121%	In feed
Zoalene	0.01–0.015%	In feed
Clopidol	0.0125–0.025%	In feed
Nitromide	0.1%	In feed

Cattle and Buffalo

Drug	Dose	Application
Amprolium	10 mg/kg bw daily for 5 consecutive days	In feed (treatment)
Monensin	1 mg/kg bw daily for 5 weeks	In feed (prevention)
Sulfaquinoxaline	13 mg/kg bw daily for 2–3 weeks	In feed (prevention)
Nitrofurazone	15 mg/kg bw daily for a week	Treatment

Sheep and Goat

Drug	Dose	Application
Sulfamezathine	140 mg/kg bw for 3 consecutive days	Orally (for treatment)
Sulfadimidine	500 mg/kg bw	Orally (for treatment)
Lasalocid	2 mg/kg bw for consecutive 20 days	Orally (for prevention)
Monensin	2 mg/kg/day for consecutive 20 days	Orally (for prevention)
Amprolium	10 mg/kg bw for 5 consecutive days	Orally (for treatment)

Phylum: Apicomplexa—Class: Sporozoea—Subclass: Coccidia—Order: Eucoccidiida—Suborder: Eimeriina—Family: Cryptosporididae

Family synopsis

The species of this family affect mostly the alimentary canal of birds and animals. The parasites also affect respiratory system of the hosts. This protozoan parasite has zoonotic importance. Oocysts do not possess sporocysts. Only four free sporozoites are present in the oocyst. Sporogony occurs inside the body. The life cycle is direct type which does not involve any intermediate hosts. The gold standard diagnosis is based on identification of oocysts by a specific stain (Z-N stain). Control of the disease is mainly done by chemotherapy.

History: Tyzzer discovered *Cryptosporidium* in the year 1907.

Genus

Cryptosporidium

Morphology

The oocysts contain four sporozoites. Sporocysts are absent. The oocysts are colorless. Double wall is present. The sporozoites lie parallel.

Cryptosporidium bovis, *Cryptosporidium parvum*, *Cryptosporidium meleagridis*, *Cryptosporidium agni*, etc. are some species of this genus.

Life Cycle

The sporulated oocysts of *Cryptosporidium* contain four sporozoites. The hosts acquire infection by either by ingestion or inhalation. The sporozoites get released from the oocysts after excystation. The parasites do not penetrate the epithelial cells like other Coccidian parasites. The sporozoites remain in the parasitophorous vacuole achieving an "intracellular extracytoplasmic" state. The sporozoites transform into trophozoites. The trophozoites transform into type 1 and type 2 schizonts. Some of the merozoites invade new cells and lead to the formation of either type 1 or type 2 schizonts. However, most of the merozoites initiate sexual multiplication resulting in the formation of microgametes and macrogametes. The microgametes fertilize macrogametes, resulting in the formation of zygotes. The events of life cycle as follows.

- Entrance of oocysts in the hosts
- Excystation of oocysts
- Sporozoites in parasitophorous vacuole
- Formation of trophozoites
- Formation of Type -I and Type-II schizonts
- Merozoites form microgametes and macrogametes

- Microgamete fertilizes macrogamete
- Formation of zygote
- Formation of oocysts

Pathogenesis

Cryptosporidiosis is now an important disease that occurs in young cattle, sheep, and goat. It also affects different gallinaceous birds. The species has zoonotic importance. Human beings get the infection from dairy animals. Enteritis is a common pathological feature, which results in diarrhea. Diffused infiltration of lamina propria occur. The brush border of the epithelial cells in the small intestine is affected. The sporozoites attach to the epithelial cell membrane. These organisms cause damage to the microvilli.

Diagnosis

Diagnosis of the disease is done by different ways:

- Staining and microscopy can be done. Acid-Fast staining of the oocysts can be done, which stain the oocysts red.
- Concentration technique such as Sheather's sugar floatation technique, modified zinc sulphate centrifugal flotation, modified formalin-ethyl acetate concentration technique can be performed for better availability.
- Wet smear examination is done by light microscopy which reveals colorless spherical or slightly ovoid bodies.
- Immunological diagnosis can be done by ELISA.
- Molecular diagnosis can be done by PCR.

> **Note**
>
> Immunomagnetic separation technique is used for the separation of oocysts. Beads are coated with monoclonal antibodies. The oocysts get attached to the beads. The bead-oocyst complex adheres to test tube wall by magnetic force. Suspending fluid is aspirated. The oocysts are detached using acid solution.

Different Species of *Cryptosporidium*

Cryptosporidium agni

The oocysts have four naked sporozoites. They are not enclosed in sporocysts. They affect the microvillous border of epithelial cells of the intestine of sheep. They cause diarrhea and other gastrointestinal pathological features.

Cryptosporidium bovis

The hosts are cattle. Sheep and goat are also affected by this species. The calves are mainly affected. The infection occurs between 3 weeks and 6 months of age in hosts. Severe outbreak may occur in the restricted animals. This is called as winter coccidiosis. The diagnosis is done by detecting oocysts in the feces.

Cryptosporidium meleagridis

The oocysts have four naked sporozoites. They occur on the epithelial cells of the intestine in turkeys. The organisms cause diarrhea and mortality in turkeys.

Cryptosporidium muris

They are found in the house mouse.

Cryptosporidium parvum

This species occurs in cattle, sheep, and pig. Humans are also affected.

Cryptosporidium tyzzeri

This species occurs in the microvilli of epithelial cells of cecum of rodents. Humans are also affected.

Phylum: Apicomplexa — Class: Sporozoea —Subclass: Coccidia—Order: Eucoccidiida—Suborder: Eimeriina—Family: Sarcocystidae

Family Synopsis

The species of this family affect alimentary canal and other organs of mammals and birds. They are worldwide in distribution. Oocysts have two sporocysts. Each sporocyst contains four sporozoites.

The life cycle is of indirect type. Asexual stages occur in intermediate hosts and gametogonous stages in definitive hosts. Cysts or pseudocysts are produced in cells. Toxoplasma is an extremely important protozoan parasite. Some species of Sarcocystis are also very pathogenic. Some parasites of this family cause abortion in the animals.

History

Nicolle and Manceaux discovered *Toxoplasma gondii* in 1909. Frenkel, Miller, Dubey and coworkers pioneered in elucidating the life cycle of *T. gondii*.

Genus: *Sarcocystis*

Morphology

Oocysts

The oocysts of *Sarcocystis* sp. are round. The oocysts may be subglobular. They contain two sporocysts. Each sporocyst contains four elongated sporozoites. They are colorless and have thin walls. Both micropyle and oocystic residuum are absent. Stieda body is also absent.

Sarcocysts

The sarcocysts are formed in the muscle of the intermediate hosts. The size and shape of sarcocysts are variable. They may be large (macro-sarcocysts) or small (micro-sarcocysts). The macrocysts are filamentous, globular, or spindle-shaped and are formed in the skeletal muscles. Sarcocysts are either very long and thin or short and broad. They are found in the parasitophorous vacuole. The sarcocyst has a cyst wall. The thickness of the sarcocyst is variable. The morphology of the sarcocysts needs the description of cryptozoites. Immature sarcocysts contain globular parasites called metrocytes. The metrocytes multiply by the process of endodyogeny. Some of the metrocytes produce banana-shaped zoites called bradyzoites. Some sarcocysts contain a number of compartment divided by septa. The organisms remain in these compartments.

Life Cycle

The events are as follow.

Infection of Definitive Hosts

The definitive hosts get the infection by the ingestion of muscle containing the sarcocysts.

Liberation of Bradyzoites

The muscles get digested and the bradyzoites are liberated in the stomach and intestine. The bradyzoites actively penetrate the intestinal wall.

Gamont Formation

The bradyzoites penetrate the intestinal epithelial cells and form gamonts. Two types of gamonts are formed: macrogamonts and microgamonts. Macrogamonts are uninucleated and oval in shape. They may be round also. Microgamonts are multinucleated and contain several slender microgametes. The formation of gametogonous stages occurs in the goblet cells or enterocytes. Occasionally, gametogonous stages may be formed in the mesenteric lymph nodes.

Oocyst Formation

The union of macro- and microgametes results in the formation of zygote. A wall developed surrounding the zygote and oocysts is formed.

Sporulation

Sporulation of newly formed oocysts occurs in the definitive host itself. The sporulation occurs in the lamina propria. It is to be noted that in the case of *Sarcocystis*, sporulation occurs in the hosts itself, not in the environment.

Oocysts Shedding

Within 7–14 days of infection, the oocysts/sporocysts are shed. Initially, oocyst shedding starts, which is followed by the shedding of sporocysts. The oocyst wall is thin, which ruptures and the sporocysts are released.

Infection of the Intermediate Host

Different herbivorous and omnivorous animals act as the intermediate hosts of *Sarcocystis*. The intermediate hosts get the infection by the ingestion of sporulated oocysts or sporocysts during grazing or drinking water.

Release of Sporozoites

The sporozoites come out from the sporocysts and penetrate the intestinal wall. Then, they enter the mesenteric lymph nodes. The sporozoites are distributed in different internal organs.

Schizont Formation

The sporozoites enter the endothelial cells of small blood vessels of different organs. Schigony starts. The number of generation of schizogony varies with the species involved.

Bradyzoite Formation

The merozoites from second generation schizonts initiate sarcocysts formation in the striated muscles. Initially immature sarcocysts are formed containing the metrocytes (mother cells). The metrocytes multiply by endodyogony. Some metrocytes form banana-shaped zoites called bradyzoites. Bradyzoites are found in the medulla part of sarcocyst whereas the metrocytes are found in the cortex part of the sarcocyst. The metrocytes mature to form bradyzoites. The bradyzoites are banana-shaped and are slow multiplying stages remaining within sarcocyst. The bradyzoites are infective stages.

Pathogenesis

No significant pathology occurs in the definitive hosts. However, the invasive forms may cause vasculitis, myositis, and lymphadenopathy. Transplacental infection occurs in horses.

- *In cattle: Sarcocystis cruzi* is an important species occurring in cattle (intermediate host) and affect the myocardium. Hemorrhages are observed in different skeletal muscles. Low level of pathogenesis occurs in this parasitic infection. The clinical features include anorexia, fever, prostration, loss of body weight, loss of milk production, etc. There may be abortion if the pregnant animals are affected. The animals may recover and enter a chronic phase, which exhibit consistent loss of condition with reduction of body weight, salivation, hair loss, etc.
- *In sheep:* S. *tenella* is an important species of sheep (intermediate host). The similar feature is found in the case of *S. tenella* infection in lambs. The clinical features include fever, anemia, abortion. Nervous disorder, myositis are other clinical features. The animals may even die. *S. arieticanis* transmitted by canids are also important but it is less pathogenic compared to *S. tenella*.
- *In pigs:* Human beings act as the definitive hosts of *Sarcocystis suihominis* and *Sarcocystis hominis*. Nausea, vomiting, abdominal pain, and diarrhea are observed in human beings. The clinicopathological features found in the affected animals are enteritis, myositis, nervous disorders, etc. S. *miescheriana* is not pathogenic to the definitive hosts (dog) but this species cause marked pathogenesis in pig. It causes myositis, dyspnoea, and purpura of skin.
- *In birds*: This protozoan parasite mostly affects the mammals. However, birds may also be affected. Myositis , pneumonia may occur. Several species of *Sarcocystis* affecting birds are *S. wenzeli*, *S. falcatula*.
- *In equines*: Several species occurring in equids are *S. neurona S. betrami, S. equicanis, S. asinus, S. fayeri. All these species are transmitted by dogs. Multifocal lesions in different organs occur in equines. Brain is also affected. S. neurona causes equine protozoal myelonephritis (EPM).*
- *In humans*: Human beings can act as the definitive hosts of the two species *S. hominis* and *S. suihominis* and get the infection by eating undercooked meat harboring the cysts of the parasites. However, human beings may act as the intermediate hosts of *Sarcocystis nesbitti*, a species with reptiles as definitive hosts. Humans get the infection by the ingestion of food contaminated with sporocysts. The

symptoms include fever, headache, etc. Invasive forms cause myositis, vasculitis, etc. Intestinal forms cause diarrhea, abdominal pain, myositis, etc.

Diagnosis

Parasitological Diagnosis

Feces examination: Examination of feces should be done. Sporulation occurs in the host and sporocysts can be detected in the feces. For separation of oocysts, floatation technique is performed using formalin ethyl acetate. There are other density gradient solutions such as zinc sulfate, sodium chloride, cesium chloride, Ficoll-hypaque, Percoll, etc., are superior compared to formalin ethyl acetate. For identification, size of sporocysts and wall character of the oocysts are important features. Not all staining agents stain the wall perfectly, but periodic acid Schiff staining is perfect for staining wall of oocysts. Examination of sarcocyst wall is very important for the identification of *Sarcocystis* up to species level. *Sarcocystis* should be differentially identified from other cyst-forming protozoa such as *Besnoitia*, *Hammondia*, *Frekelia*, *Toxoplasma*, etc.

Biopsy examination: At first, biopsy materials are collected. The collected muscle samples are digested in digestion medium containing pepsin, HCl, and distilled water. The muscle samples are kept in this medium for 24 h at 37 °C. After digestion, the material is centrifuged at 2500 × g for 3 min. The sediment is stained with Giemsa stain and examined under a microscope for detection of sarcocysts.

Immunodiagnosis

Immunodiagnosis is done using different types of antigens. Studies shows that indirect fluorescent antibody test, ELISA, dot ELISA, etc., are useful.

Molecular Diagnosis

Molecular diagnosis is possible by different molecular markers. Different ribosomal DNA, RNA are used. Different mitochondrial DNA, such as ND 1 and Cox1, are used for molecular identification of parasites.

Treatment

There is no effective treatment of chronic sarcocystosis. Pyimethamine and sufadiazine are used for treatment of EPM. Diclazuril and toltrazuril are useful against *S. neurona* infection.

Control

General Management

Since the definitive hosts get the infection by the ingestion of infected meat, the meat should be properly cooked at a minimum temperature of 70 °C temperature for 20 min. Preservation of meet at −10 °C for minimum of 48 h is effective. The animals should be provided uncontaminated food and water. But practically, it is very difficult to do the same.

Immunological Control

Vaccine has not been commercialized. However, studies have shown a certain amount of immunity using different antigens. In trials, different antigens are used. A study has shown beta-irradiated sporocysts conferring protection in mice model. For the analysis of immunity, bradyzoites are also used. Bradyzoite antigens are prepared in different ways. The sarcocysts are collected from the affected muscles. Trypsin digestion is done to remove the attached skeletal muscle. Then the material is homogenized until the color become creamy. The material is centrifuged, the supernatant is discarded, and the pellet is resuspended for use as an antigen. The collected cysts may be washed with saline at pH 7.2 to remove tissue material. The content is then homogenized to rupture the cyst wall. The material is centrifuged at rpm for 10 min. The supernatant is discarded and the pellet is resuspended for use as a vaccine candidate. Different parameters are set up to assess the immunoprotective potential. These parameters are quantity of IgG and IgE. The levels of different cytokines are also determined. These cytokines are usually IFN Gamma, IL5, etc. Different animal models are used. Mice are most suitable animal model used for study. Certain amount of immunity is found by immunization with bradyzoite antigens.

Molecular Control

There is not yet any commercialization of molecular or gene vaccine.

Biological Control

No biological control is available in the field.

Species of Cattle

Sarcocystis bovifelis

The definitive host of this species is the cat, and the intermediate host is cattle. This species is nonpathogenic.

Sarcocystis bovihominis

The definitive host of this species is human and the intermediate host is cattle. This species is nonpathogenic.

Sarcocystis cruzi (*Sarcocystis bovicanis*)

The definitive host of this species is dog and the intermediate host is cattle. This is pathogenic and cause abortion, still birth, etc. Eosinophilic myositis, encephalitis, and enlargement of lymph node are some clinicopathological features. Fever, weakness, loss of body weight, anemia, etc., are some clinical signs.

Species of Sheep

Sarcocystis tenella (*Sarcocystis ovicanis*)

The definitive host of this species is dog and the intermediate host is sheep. This is a pathogenic species. Myositis, myeloencephalitis, edema of spinal cord, and glial nodule are some pathological features. Anemia, fever, paralysis, weakness, and loss of body weight are some clinical signs of this disease in sheep.

Species of Pig

Sarcocystis meischeriana (*Sarcocystis suicanis*)

The definitive hosts are dog, fox, etc., and the intermediate host is the pig. The small sarcocysts occur in tongue, esophagus, diaphragm, etc. Two generations of schizonts are formed in the life cycle. The first-generation schizonts occur in the liver and the second-generation schizonts occur in different organs. The gametogonous stages occur in the definitive hosts affecting lamina propria of jejunum and ileum of dog. This species is not pathogenic to dogs.

Sarcocystis porcifelis

The definitive host is the cat. The intermediate host is the pig. This is a pathogenic species. Myositis and enteritis are clinicopathological features.

Species of Equine

Sarcocystis neurona

This is an important species. The definitive hosts are opossum, dog, etc. The intermediate host is the horse. This is a pathogenic species. Brain is commonly affected. This is causative agent of equine protozoal *myeloencephalitis (EPM). Sarcocystis neurona* causes myeloencephalitis, which is a progressive debilitating disease affecting the CNS of equines. The horse exhibits clinical signs associated with atypical lameness, abnormal respiratory function, and other neurological signs. Ataxia, dysphagia, depression, incoordination are some clinical features. Transuterine infection may occur.

Genus

Genus: *Toxoplasma*

General Description

The only species under this genus is *Toxoplasma gondii*. The cat is the definitive host. The intermediate hosts get the infection by the ingestion of sporulated oocysts. The parasites multiply in the wall of the intestine. There occurs necrosis of the wall of the intestine resulting in enteritis. Cattle suffer from respiratory anomalies including pneumonia, hemorrhagic tracheitis, etc. Renal pathology may occur due to kidney damage. Both nervous and respiratory systems are affected in dogs. Nodules may form in the lungs of dogs.

Gliosis, leptomeningitis, etc., are some associated pathological features observed in animals and humans.

Morphology

The oocysts are globular or subglobular. The oocyst contains two sporocysts; each sporocyst contains four sporozoites. The sporozoites are ellipsoidal in shape. Bradyzoites and tachyzoites are two important biological developmental stages. The bradyzoites are banana-shaped organisms lying in the cyst. These are slowly multiplying stages. The name itself is self explanatory because 'brady' is a Greek word meaning 'slow'. Tachyzoites are crescent-shaped stages. These are rapidly multiplying stages (Tachos=speed in Greek).

Biology

The definitive hosts get the infection by the ingestion of sporulated oocysts, ingestion of the muscles of the intermediate hosts containing bradyzoites, tachyzoites, etc. The intermediate hosts get the infection by the ingestion of sporulated oocysts. In human beings, transplacental infection occurs. Humans also get the infection by the ingestion of infected muscles containing bradyzoites or tachyzoites.

The cats get infection in three ways:

1. By the ingestion of sporulated oocysts.
2. By the ingestion of meat of intermediate hosts containing bradyzoites.
3. By the ingestion of meat of intermediate hosts containing tachyzoites.

Development in the definitive host:

From the bradyzoites

- Ingestion of tissue cysts by cat > Digestion of tissue cyst wall (by enzymes in stomach, intestine) >Release of bradyzoites >Penetration of epithelial cells by bradyzoites > Formation of A-E type of parasitic stages > Merozoites of schizonts of D and E initiate gametogony > Formation of gametocytes in the small intestine >Penetration of macrogamete by microgamete >Formation of zygote > Formation of oocyst wall surrounding the zygote > Rupture of intestinal epithelial cells > Discharge of oocysts in the lumen of intestine > Voiding of oocysts through the feces

From the tachyzoites

- After administration of tachyzoites, formation of asexual stages/enteroepithelial stages is not known.

From the oocysts

- After administration of oocysts, formation of asexual stages/enteroepithelial stage is not known.

Development in the Intermediate Host

From oocysts - Ingestion of oocysts>Rupture of oocysts>Release of sporozoites >Multiplication of sporozoites in the intestinal epithelium and associated lymph node> Formation of tachyzoites> Circulation of tachyzoites through blood and lymph> Formation of cysts in different organs (brain ,liver, cardiac muscle etc.) From bradyzoites - Ingestion of infected meat containing bradyzoites>Lysis of cyst wall> Release of bradyzoites >Transformation of bradyzoites to tachyzoites>Repeated division of tachyzoites and production of large population>Encystment in muscle>Ingestion by cat

Pathological Features

Pathological features are variable. Abortion can occur in human beings and sheep. Abortion, retained placenta, and mummification of fetus occur in ewes. In cattle, the brain and respiratory system are affected mostly. Pneumonia, hemorrhagic tracheitis, and the formation of glial nodules commonly occur in the brain. In cats, enteritis, pneumonia, and nephritis occur. Clinical cases of toxoplasmosis in cat occurs seldom. In dogs, both the lungs and brain are affected, resulting in gliosis in brain, pneumonia, and nodule formation in the lungs. Reports of fatal toxoplasmosis in dog are available. Clinically four categories have been described i.e. general, gastrointestinal, respiratory and neuromuscular.

In humans, the parasites cause diarrhea without blood. The diarrheic material is gruel-like mixed with mucus and fat. Large amount of fat is found in the feces since fat is not properly absorbed. Liver and bile duct may also be affected. The diarrhea may be continuous or discontinuous. Abdominal pain, nausea, and flatulence are some other clinical features. Abortion occurs in humans.

Clinical Signs

The general clinical signs are as follows:

- Diarrhea
- Pneumonic signs
- Incoordination of movement along with other nervous disorders.

Abortion in cats and humans.

Diagnosis

- Examination of fecal sample for the identification of oocysts can be done. For specific identification of oocysts, the oocysts should be cultured for sporulation. The sporulated oocyst has two sporocysts. Each sporocyst contains four sporozoites.
- Biopsy material should be collected from suspected animals. This should be examined after staining for the detection of bradyzoites.
- An easy method can be employed for the detection of bradyzoites. The suspected tissue material is treated with acid-pepsin solution. The material is centrifuged after digestion. The cysts can be observed under stereoscopic microscope like rice grains.
- Animal Inoculation Test–The organisms are isolated from the infected animal. Then the organisms are inoculated (I/P or intracerebrally) into the mice. The virulent organisms cause fatal infection in the inoculated animals after 1–14 days of inoculation. This is the principle of this test.
- Sabin Fieldman Dye Test
- This test is based on immunological reaction. Suspected serum is mixed with *Toxoplasma gondii* parasites. In positive case, the organisms are died by the immunological reaction. The dead organisms do not take up the dye. The unstained organisms are detected.
- Immunodiagnosis

Different immunodiagnostic tests are done for the diagnosis of *Sarcocystis* sp. These include ELISA, IF AT, dot ELISA, Western blot, etc. Usually, the antigen is prepared by the bradyzoites.

Treatment

Sulfadiazine, pyrimethamine, clindamycin, and spiramycin are some effective drugs.

Control

Chemotherapy

Treatment of the affected individuals should be done. Properly cooked meat should be given.

Immunological Control

The bradyzoites are used as vaccine candidate. They are cultured in Vero cells (African Green Monkey Kidney Cells) using RPMI-1640 with 10% fetal calf bovine serum and 5% CO_2.

A series of immunoprotection studies against toxoplasmosis have been conducted pertaining to attenuated vaccine and subunit vaccine.

In the earlier period, attempts were made to produce vaccine using killed organisms. These attempts proved successful for commercial application. Later, attempts were directed towards the use of attenuated or inactivated vaccine.

A recently developed vaccine is **Toxovac** vaccine. This vaccine constitutes live attenuated tachyzoites of S48 strain of *Toxoplasma* for ewes.

Attempts were also made for the development of subunit vaccine. Subunit vaccine does not have the drawbacks of low self-life and reversion of attenuated strain to pathogenic stain, which is the common problem in any live vaccine. But subunit vaccine lacks immunogenicity. To overcome this problem, different adjuvants are used to increase the immunogenicity and to make it more potent. In addition to this, subunit vaccine has its inherent problem of allergic reaction as the subunit peptides are expressed in different host organisms.

Molecular Control

A recent study shows partial protection in mice immunized with TgSOD (*Toxoplasma gondii* super oxide dismutase) gene of *Toxoplasma gondii* by intramuscular inoculation. The study also shows the use of gene of *T. gondii* dense granules protein 6 (GRA6) with partial protection in mice. Other immunogenic proteins are surface antigen (SAG), rhoptry protein (ROP), and microneme protein (MIC).

Genus: *Neospora*

The important species of this genus is *N. caninum*. This species was misdiagnosed as *Toxoplasma gondii* earlier. This species has a wide host range. Dogs and related animals are the definitive hosts. Cattle, sheep, goat, and horse are the main intermediate hosts. Like *Toxoplasma*, the developmental stages of life cycle are tachyzoites, bradyzoites, and oocysts. The dogs probably get the infection by the ingestion of tissue of herbivores harboring the bradyzoites. The herbivores get the infection by the ingestion of sporulated oocysts of this parasite.

A rigorous study was carried out on different aspects of this parasite. It has been tangibly established that this parasite is highly pathogenic to both the cattle and dogs. It causes abortion in cattle. Both transuterine and transcolostral infection occur. The disease appears serious in the dogs. The pups get the infection via transuterine or transplacental mode. The pups suffer from hind limb paralysis. Paralysis of jaws occurs, which results in difficulty in feeding.

Phylum: Apicomplexa—Class: Sporozoea—Subclass: Coccidia—Order: Eucoccidiida—Suborder: Adeleina—Family: Haemogregarinidae

The important genus of this family is *Hepatozoon*. The pathogenic species is *Hepatozoon canis*. The hosts are dog, cat, etc. The schizont stages are formed in spleen, liver, and bone marrow invading the endothelial cells. This parasite is transmitted by the tick, *Rhipicephalus sanguineus*. Sporocysts are formed in the body cavity of the ticks. The dogs become infected by the ingestion of ticks (nymph, adult). The sporozoites are released from the sporocysts and penetrate the wall of the intestine. Then they reach different organs such as liver, spleen, and bone marrow via blood circulation. Schizonts are formed in those organs. Several generations of schizonts occur. Schizonts burst and the merozoites are released and enter polymorphonuclear leucocytes. Gamont stages develop in those cells which are elongated or rectangular forms surrounded by a capsule. The infected cells are taken up by the tick during blood meal. The cells are digested in the gut of tick and the gamonts are liberated. Zygotes are formed which are motile (ookinete). The ookinete penetrates the intestine and hemocoel and transform to oocyst. Later sporocysts are formed. Fever, anemia, and progressive emaciation are some characteristic pathological features.

Phylum: Apicomplexa Class: Sporozoea—Subclass: Coccidia—Order: Eucoccidiida—Suborder: Haemosporina—Family: Plasmodiidae

Family Synopsis

The species of this family affect mostly the RBC and endothelial cells. However, hepatic and other cells are also affected. These protozoan parasites are worldwide in distribution. The life cycle is of indirect type. Schizogony stages occur in vertebrates and sporogony stages occur in invertebrates. The zygote formed by fusion of macrogamete and microgamete is motile. Some species of this family cause severe pathogenicity. Diagnosis is mostly based on the examination of blood smear for the detection of the erythrocytic form of the parasite. Since this protozoan is transmitted by mosquitoes, controlling mosquitoes and chemotherapy are two important resorts for the control of this protozoan.

History

Ronald Ross described the life cycle of avian malarial parasite in mosquito vector in within 1897–1898. Exoerythrocytic stage of human malarial parasite was described by Shortt and Garnham in 1948.

Genus: *Plasmodium*

Introduction

Charles Laveran first discovered *Plasmodium* in the late nineteenth century. The organisms are a global problem, causing malaria in humans and animals. Different species of mosquitoes (*Anopheles*, *Culex*, *Aedes*) are the vectors of malarial pathogens. *Plasmodium vivax*, *Plasmodium falciparum*, and *Plasmodium malariae* affect humans. *P. falciparum* causes mortality in humans. The species occurring in non-human primates are *Plasmodium knowlesi, Plasmodium cynomolgi, Plasmodium simium*, *Plasmodium inui*, *Plasmodium eylesi*, *Plasmodium schwetzi,* etc. Some of these species also occur in humans and have zoonotic importance. As far as the veterinary field is concerned, avian malaria is important. The causative agents are *Plasmodium gallinaceum* and *Plasmodium Juxtanucleare*, which occur in chicken. *Plasmodium relictum* occurs in pigeon. Anemia, pyrexia, and paralysis are common clinical signs in birds. The birds may even die due to this disease.

> **Note**
>
> Black pigment granule of malaria organisms were detected in the blood by Meckel in 1847 and by Virchow in 1849. *Plasmodium malariae* affecting human being was first discovered by CLA Laveran in 1880. Ronald Ross (1897–1898) a renowned and Nobel Laureate first discovered stages of malaria parasites in mosquitoes.

Malaria in Birds

Plasmodium gallinaceum

This species occurs in domestic fowls.

Biology

- The vectors of the parasite are different mosquitoes. The infected mosquitoes transmit sporozoites (infective stage of parasite) to the host while taking blood meal from them.
- Preerythrocytic schizonts form in the macrophages and fibroblasts located at the site of bite.
- Different stages of the parasite develop. First-generation schizonts (pre-erythrocytic schizonts) or cryptozoites are formed.
- First generation schizonts burst and the merozoites are released.
- These merozoites lead to the formation of second-generation schzonts called metacryptozoites.
- Metacryptozoites burst and release merozoites that enter endothelial cells and erythrocytes forming exoerythrocytic schizonts and erythrocytic schizonts respectively.
- Erythrocytic schzonts may initiate further erythrocytic schizogony.
- The merozoites of erythrocytic schizont may form another generation of exoerythrocytic schizonts called phanerozoites.
- Sexual cycle follows after a number of asexual generations. The merozoites enter the RBC.
- Gametocytes form in the RBC, which are taken up by the mosquitoes.
- In the gut of the mosquitoes, subsequent stages form. The microgametocytes undergo exflagellation, resulting in the formation of microgametes.
- The microgametes fertilize macrogametes resulting in the formation of zygotes.
- The zygotes transform into elongated ookinetes.
- The ookinetes penetrate the gut wall and later enter the hemocoel where the oocysts are formed.
- The oocysts undergo sporogony.
- The oocysts rupture and the sporozoites are liberated.
- The sporozoites enter the salivary glands.
- When mosquitoes bite hosts, the sporozoites enter the hosts. Eventually, the sporozoites are produced, which are the infective stages of the parasite.

Fig. 8.16 Life cycle of *Plasmodium gallinaceum*

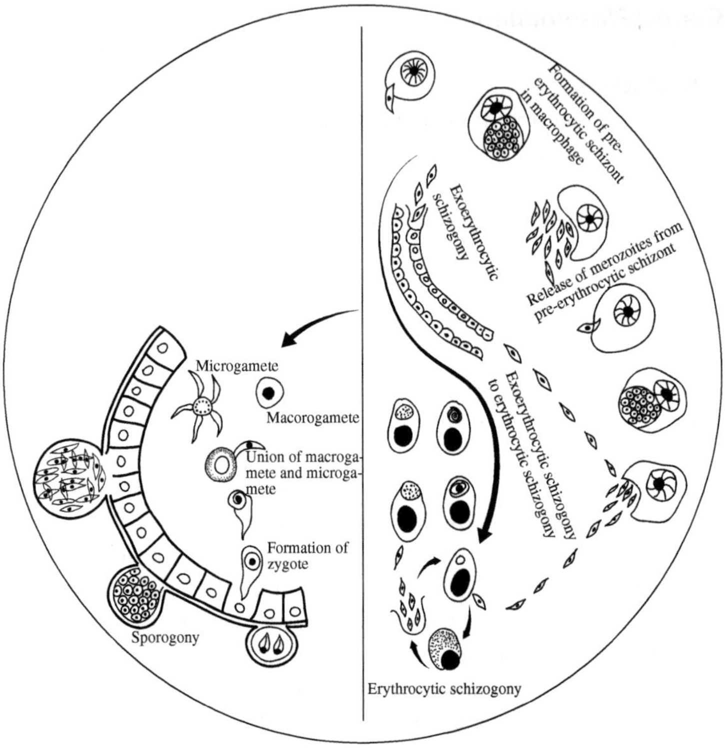

Pathology

Malaria in birds is very severe. Young birds suffer more than the adults. Hepatomegaly and splenomegaly are common pathological features. Glomerular nephritis is also found. Anemia and emaciation are frequently observed and the bird's condition progressively deteriorates, leading to death. Mortality rate is around 80%.

Malaria in Humans

Different species cause malaria in humans. Four important species affecting human beings are *Plasmodium falciparum*, *P. vivax*, *Plasmodium ovale*, and *P. malariae*. P. vivax and P. falciparum are common. *P. falciparum* is a causative agent of severe malaria in humans.

Biology of the Parasite

- The vectors of the parasite are different species of Anopheles mosquitoes. Infected mosquitoes transmit sporozoites (infective stage of parasite) to the host while taking blood meal from them.
- Exoerythrocytic schizonts form in the liver.
- Merozoites from exoerythrocytic schizonts enter the circulation and affect the erythrocytes or these may initiate further exoerythrocytic schizogony.
- Erythrocytic schizogony starts.
- Merozoites from erythrocytic schizonts enter another RBC and schizogonic cycle continues.
- Some merozoites enter the RBC, resulting in the formation of micro- and macrogamonts.
- Further development occurs in the female Anopheles mosquitoes when they take blood meals.
- In the outer surface of gut of the mosquitoes, subsequent stages form. The microgametocytes undergo exflagellation resulting in the formation of microgametes.
- The microgametes fertilize macrogamete, resulting in the formation of zygotes.
- The zygotes transform into elongated ookinetes.
- The ookinetes penetrate the gut wall and later enter the hemocoel, where oocysts are formed.
- The oocysts undergo sporogony.

- The oocysts rupture and the sporozoites are liberated.
- The sporozoites enter the salivary glands.
- When mosquitoes bite hosts, the sporozoites enter the hosts. Eventually, the sporozoites are produced, which are the infective stages of the parasite, as mentioned earlier.

Pathology

Hepatomegaly and splenomegaly are the common pathological features. Headache, fever, joint pain, back pain, and severe weakness are the common clinical features.

Diagnosis

Examination of blood smears after Romanowsky staining may reveal the presence of parasitic stages. This method is the gold standard method for the detection of erythrocytic stages of this protozoan. Immunological tests include counter immunoelectrophoresis (CIEP), indirect hemagglutination test (IHT), indirect fluorescent antibody test (IFAT), latex agglutination test (LAT), and enzyme linked immunosorbant assay (ELISA). Works are also in progress on molecular diagnosis. Different types of PCR have been tried for the diagnosis of malaria, such as RFLP PCR, RT qPCR, etc.

Treatment

Qunine sulfate, pyrimethamine, and mefloquine atovaquone are effective drugs.

Control

Control of the disease could be achieved through proper treatment, control of the vectors, etc. Persons should be protected from mosquito bites.

Different Types of Malaria

Autochthonous Malaria

Malaria infection contracted locally is called as autochthonous malaria.

Imported Malaria

Malaria brought in from outside areas.

Introduced Malaria

Malaria cases that have been proved locally occurred but are derived from imported malaria.

Indigenous Malaria

This is malaria which is natural in a particular area.

Sporadic Malaria

Autochthonous malaria which occurs in few number and scatteredly are called as sporadic malaria.

Induced Malaria

Malaria which is inoculated (blood transfusion, etc.).

Genus

Haemoproteus

The important species in pigeons is *Haemoproteus columbae*. This species also occur in other birds. Other two species are *Haemoproteus sacharovi* and *Haemoproteus meleagridis* occurring in doves and turkeys, respectively. *H. columbae* is transmitted to birds by vectors such as Hippoboscid flies during bloodsucking. The organisms cause avian malaria. Different pathological features occur in the liver and spleen.

Haemoproteus columbae

Site: Gamonts are found in blood.

Hosts: Turkey, quail

Morphology

Gamonts: The gametocytes are found in the RBC. The organisms have typical morphological features. A distinct granular pink-colored nucleus is visible. The organism occupies most of the area of cytoplasm of RBC. A number of pigment granules (hemozoin) are found in the organisms. The number of pigments is used in taxonomy for speciation of this parasite. The organisms are halter-shaped encircling the nucleus with certain displacement of the nucleus. The angle of displacement is also taken into consideration for speciation. A key difference between *Plasmodium* and *Haemoproteus* is that in the case of *Haemoproteus*, only gametocytes are formed in the erythrocyte, whereas no other stages are formed as in the case of *Plasmodium*. In Giemsa-stained blood smears, both gametocytes can be differentiated. The nucleus of macrogametocytes is more compact and cytoplasm is denser. In microgametocytes, the nucleus is not compact. Pigment granules are evenly distributed in macrogametocytes whereas pigment granules are clustered at the polar region in the case of microgametocytes. The pigment granules are comparatively larger in size compared to *Plasmodium*. Reddish-purple volutin granules may be present in some *Haemoproteus* organisms. These granules are frequently confused as schizont stages of *Haemoproteus* spp. Complete expulsion of host cell nucleus may occur.

Schizonts: Schizont stages are formed in different organs.

Life Cycle

The hippoboscid flies (*Pseudolynchia canariensis*) transmit this parasite to the hosts. Hippoboscid flies take up infected RBC containing the gamonts during blood meal. Zygotes are formed in the flies. The flies transmit sporozoites while taking blood meals from the hosts. The following events occur in this life cycle:

Infection of the Vector

The vectors get the infection when they suck blood from the definitive hosts.

Development in Vector

In the gut, ingested blood is digested and the gametocytes released. The sexual reproduction occurs in the gut of the vector. Exflagellation of the microgamete occurs in the gut of the vector. The microgamete fertilizes the macrogamete. The male and female gametes unite to form zygotes. The zygote is motile (ookinete) and reach to the outer surface of midgut. The zygote transforms into oocyst. The oocysts undergo sporogony and sporozoites form. These sporozoites are infective stages of the parasite. Sporogony occurs in the salivary gland of Hippoboscid fly (*Pseudolynchia canariensis*). Sporozoites are formed in oocysts. From the oocysts, sporozoites come out and remain ready to infect the definitive hosts.

Infection of the Definitive Hosts

The vector transmits sporozoites when the vectors take blood meal from the hosts.

Development in the Vertebrate Hosts

After infection, the sporozoites enter the endothelial cells of different organs, such as lungs, liver, spleen, kidney, etc. In the endothelial cells, early schizonts are formed. The early schizonts contain a number of bodies, each containing a nucleus. These bodies are called cytomeres. Each cytomere starts multiplication and division of nucleus by the process of multiple fission. Ultimately, each cytomere produces many merozoites. After the rupture of cytomeres, the merozoites are released. The merozoites enter RBC. In the RBC, male and female gamonts are formed.

Pathogenesis

This parasite does not produce marked pathogenesis, but anemia, weakness, and lameness are some clinical features. Extraerythrocytic forms have been described and are found in liver, lung, etc. The endothelial cells of different organs are affected by the parasites. Congestion and hemorrhage occur in these organs. In severe condition, mortality of birds may occur. Sometimes, mixed infection of *Plasmodium* and *Haemoproteus* occur.

Diagnosis

Symptomatological Diagnosis

It is not possible to diagnose the case only by observation of symptoms because no specific symptom is found.

Parasitological Diagnosis

Parasitological diagnosis is the gold standard. The gamont stages are found in the erythrocyte stages of parasite. The stages can clearly be identified. The organisms are typically halter shaped. Other features are described under the heading of morphology. Sometimes, there might be confusion with *Plasmodium* in RBC. To eliminate this confusion, a differential morphological analysis should be conducted. The pigment granules are larger in size in case of *Haemoproteus* compared to *Plasmodium*. spp. In case of *Plasmodium*, both schizonts and gamonts are found in erythrocytes. But in case of *Haemoproteus* infection, only gamonts stages are found in erythrocytes. It becomes difficult, if mixed infection occurs.

Other Species

Haemoproteus nettionis

This species occurs in ducks. The affected hosts exhibit lameness, respiratory signs, etc.

H. meleagridis

This species occurs in turkeys.

H. sacharovi

This species occur in pigeon and dove. They cause less pathogenicity.

Genus: *Leucocytozoon*

There are several species of this genus. These are Leucoocytozoon simondi, L. caulleryi, L. smithi etc. L. simondi occurs in wild ducks and geese. L. caulleryi occur in domestic chicken. L. smithi occurs in domestic and wild turkey. Young birds suffer more compared to adults. The clinical signs include listlessness, inappetence, greenish diarrhea, anemia etc.

Phylum: Apicomplexa—Class: Sporozoea—Subclass: Piroplasmia—Order: Piroplasmida—Family: Babesiidae

Family Synopsis

These are mostly pear-shaped protozoan organisms found in the blood of different vertebrates. They are worldwide in distribution. The protozoans are characterized by pyriform, oval, round pleomorphic shape, short apical complex, endodyogeny, endopolygeny, fission in erythrocytes. Some species of this family have pathogenic significance causing massive loss in the livestock industry. Gold standard diagnosis of this parasitic infection is the examination of blood smears. Control of the parasite mostly relies on chemotherapy.

Discovery

Victor Babes discovered *Babesia* species in 1888 and the genus name was given in his honour. Theobald and Kilbourne demonstrated transmission of Babesia through tick vector in 1893.

Family: Babesiidae

Genus: *Babesia*

Morphology

The organisms are typically pear-shaped, with narrow ends that form acute angle. However, this typical shape and arrangement are not found in all members of this genus.

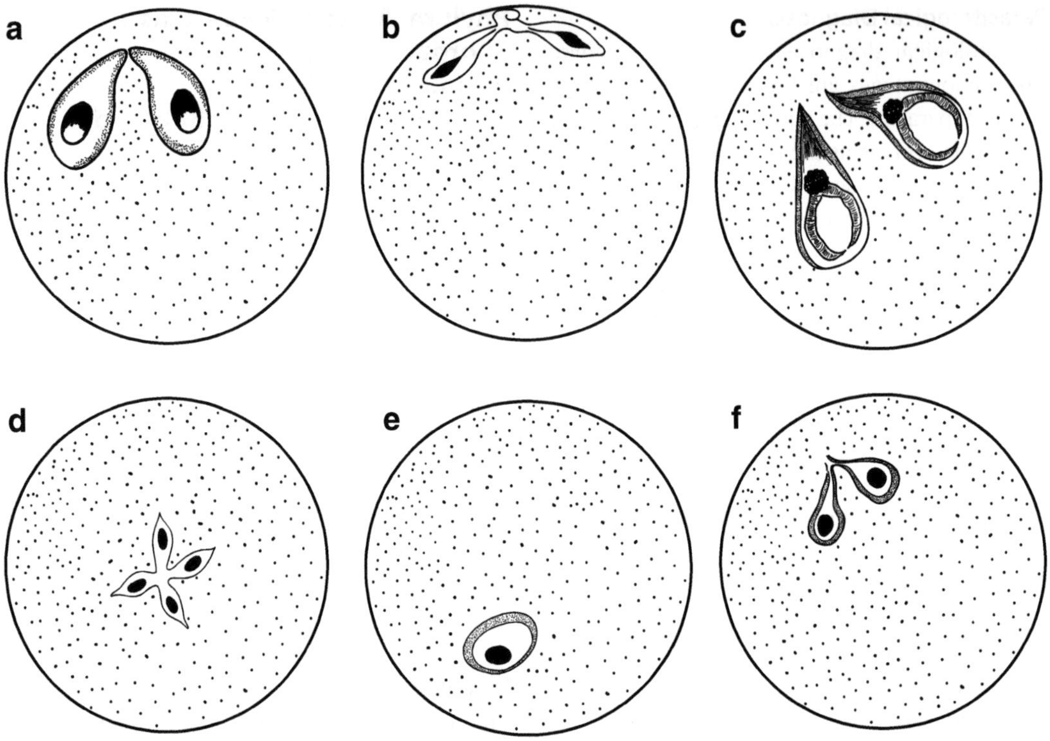

Fig. 8.17 Different forms of *Babesia* sp.

Life Cycle

The hosts get the infection when infected tick bites and transmits sporozoies. These sporozoites attack RBCs. In the RBCs, the organisms multiply by various asexual phenomena, such as endodyogeny, endopolygeny, schizogony, etc. The merozoites leave the RBC and affect new RBCs. Biological development of this organism starts in the vector when it takes blood meal from the host. The RBCs get digested and the organisms become free in the gut of the vector. Zygote formation occurs in the tick. These zygotes are called as vermicules. The club-shaped vermicules reach the malpigian tubules, hemocoel, ovary, etc., of the ticks. The organisms enter the ovary of the ticks and multiply by the process of multiple fission and later pass through the eggs. This is called **transovarian transmission**. The vermicules enter the gut cells of the larvae and multiply there. In the gut cells of larvae, the organisms form fission bodies. The larvae molt to become nymphs, and the organisms are readily available in the salivary glands of the nymphs for transmitting infection to the hosts. In a similar way the adult ticks also transmit infection.This pattern is called "**stage to stage transmission**". Transovarian transmission occurs by one host tick and stage to satge transmission occurs by two or three host tick.

The following events occur during life cycle of this protozoan:

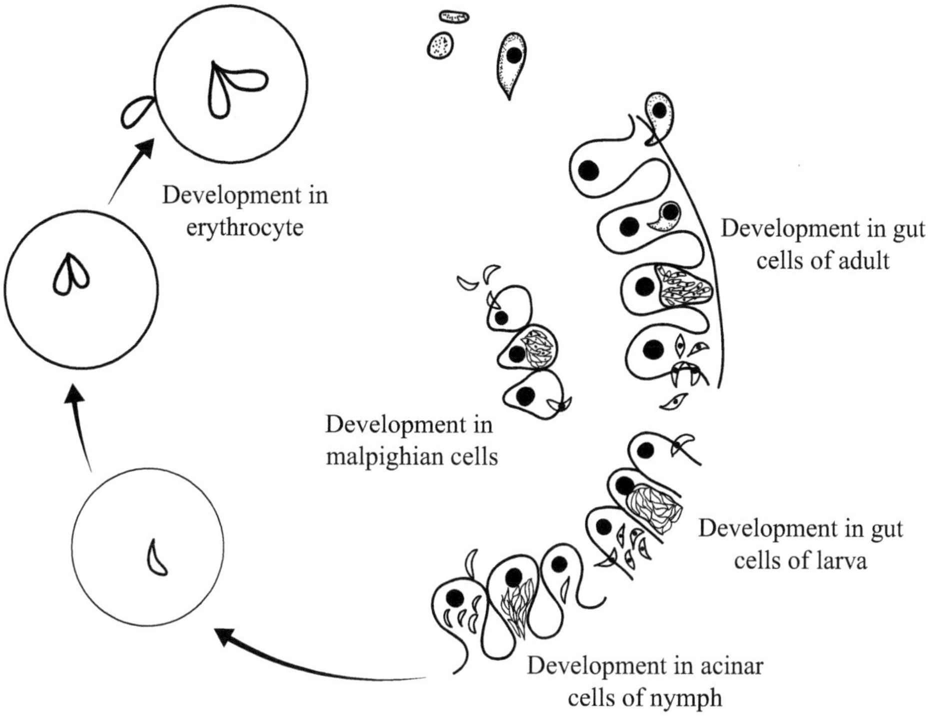

Fig. 8.18 Life cycle of *Babesia* sp.

> **Note**
> *Babesia* organisms may be transmitted through the bites of the infected dogs.

Symptoms

Pathogenesis and Clinical Signs

The pathological features generally observed in *Babesia* infection are anemia, splenomegaly, hepatomegaly, neuropathology, etc. The older animals are more susceptible in comparison to young animals because of inverse age resistance. **High rise of temperature, anemia, and coffee colored urine are very common clinical features.** The disease manifests in different animals in different intensity of pathogenicity. ***B. canis* and *B. bovis*, *B. divergens*, *B. microti* infection exhibit neurological signs.**

Cattle

In cattle, *Babesia bigemina* and *B. bovis* are the main causative agents. Fever, anemia, and hemoglobinuria (coffee-colored urine) are common clinicopathological features observed in those animals. Incoordination of movement is exhibited due to nervous form of the disease in *B. bovis* infection. Cerebral form of the disease also occurs. *B. bigemina* infects cattle and is associated with diseases known as Texas fever, red water fever, and cattle tick fever. The disease may have fatal sequelae. Hemoglobinuria is the common clinical sign found in this disease. High fever is another clinical sign which may last for weeks or more in some cases. The animals stop feeding, drinking water, and may enter into a course of chronic phase and the animals become debilitated, emaciated, and weak. The magnitude of pathogenicity due to *Babesia major* is less than that due to *B. bovis* and *B. bigemina*. Clinical signs are mild hemoglobinuria, fever, anemia, etc. The clinical signs are similar to that exhibited by *B. bigemina*.

Sheep and Goat

Babesia motasi and *B. ovis* are common in sheep and goats. Variable clinical features (subclinical or clinical) occur in these animals. The main clinical signs exhibited due to *Babesia motasi* infection are hemoglobinuria, fever, weakness, and off feeding. Mild form of the disease occurs due to *B. ovis* infection.

Dogs

B. canis occurs in dogs. Fever, anemia, and jaundice are the clinicopathological features due to this infection. The animals become weak and look dull and depressed. Feed intake and water intake is reduced. The clinical symptoms exhibited due to *Babesia gibsoni* infection are anorexia, dullness, hemoglobinuria, anemia, and fever. Jaundice may occur but it is not so frequent. The disease entity runs acute and chronic state. The central nervous system may be affected causing paralysis, incoordination of movement, epilepsy, etc.

Equines

In case of *Babesia caballi* infection, posterior paralysis is common as the central nervous is affected by this parasite. Hemoglobinuria rarely occurs. Fever and anemia are common. Severe form of disease has also been recorded. *Babesia equi* is considered to be very pathogenic compared to *B. caballi*. There is evidence of mixed infection of *B. caballi* and *B. equi*. *B. caballi* causes acute, subacute, and chronic form of the disease. Fever, anemia, icterus are common clinical signs.

Pig

The main clinical signs exhibited due to *Babesia tautmanii* infection are hemoglobinuria, fever, anemia, edema, and weakness. Similar clinical features are found due to *B. perroncitoi infection.*

Diagnosis

History-Based Diagnosis

History-based diagnosis is never a confirmatory diagnosis but this may be suggestive of diagnosis that needs to be confirmed by other techniques. Detection of enzootic area and availability of *Boophilus* ticks is very important. Attention should be paid to the animal to determine whether any ticks are present on the skin. If present, they should be examined properly through microscopic identification.

Symptomatological Diagnosis

Symptomatology cannot provide confirmatory diagnosis. Symptoms are not specific in babesiosis. However, fever, anemia, and hemoglobinuria may be suggestive of babesiosis.

Differential Diagnosis

Differential diagnosis is very important and is done after critical analysis of the clinical signs. Babesiosis should be differentiated from trypanosomosis, theileriosis, anaplasmosis, leptospirosis, postparturient hemoglobinuria, and bacillary hemoglobinuria. The exhibited neurological clinical signs of babesiosis should be differentiated from rabies, toxicosis, etc.

Postmortem Diagnosis

Postmortem diagnosis is conducted through a series of systematic activities. Different smears are prepared from lungs, liver, kidney, brain, etc. Smears should also be prepared from the blood vessels of body extremities. The smears are stained using standard staining protocol.

Examination of Smears

Diagnosis can be done by examination of the blood smear. Different Romanowsky stains are used for the identification of this protozoan parasite. Staining by Giemsa and Leishman stain can be done for this purpose. Since decades, examination of blood smears for the detection of intraerythrocytic stages of *Babesia* have been the gold standard method. This method is useful even up for species identification in many cases. This method is very simple, inexpensive, and less time consuming. However, expert microscopists are required for the detection of organisms, because various confusing things appear in the microscopic field, such as artifacts, stain deposits, etc. Direct detection of organisms in blood smear is mostly possible during the acute stage of infection when high parasitemia prevails. However, in chronic stage of infection, this is of no or very less use. In chronic stage of infection, examination of thick blood smear is important. In this technique, a drop of blood is placed on the slide. The drop of blood is air-dried, fixed with methanol, and stained with standard staining protocol. Source of blood is very important. If the animal is suspected for *B. bovis* or *B. canis* infection, capillary blood from ear vein is useful because this organism adheres to the endothelial cells. On the other hand, when the case is suspected for *B. bigemina* and *B. gibsoni* infection, peripheral blood from jugular vein is useful. Postmortem diagnosis is possible through the examination of the smear of brain tissue when it is suspected for *B. bovis* infection. Brain tissue is taken from the grey matter of cerebral cortex and placed on a slide and smeared using another slide. Drying, fixation, staining, etc., are done following standard protocol.

Since the ticks are the vectors of *Babesia*, identification of organisms can be done in the hemolymph of tick. After engorgement, the organisms get released and remain free in the gut for a short period. Sexual reproduction commences and zygotes are produced. These stages penetrate the midgut and transform into motile kinete, which enters the hemolymph and migrate to different organs. For this test, using small scissor, one leg of female tick is cut and a drop of hemolymph is placed on the slide. This is air-dried and stained following a standard staining protocol.

Animal Inoculation Test

This is a method which is used sometimes when the organisms are not detectable in the chronic stages of infection due to low parasitemia. There may be absence of prominent clinical signs. The clinical signs may not include hemoglobinuria. In this test, suspected material is inoculated into susceptible animals and the inoculated animal is monitored and observed for any signs. In this test, suspected blood (500 ml) is inoculated into susceptible animals. A sleenectomized calf is used as the recipient animal.

General Immunodiagnostic Test

Many works have been carried out on immunodiagnosis using different crude antigen and purified antigen. Variable success was achieved in immunodiagnostic tests. IFAT is an effective test. In this test, antibody of the affected animal is detected using *Babesia* antigen. The antigen of *Babesia* is prepared by culture of intraerythrocytic merozoites. This antigen is smeared on the slide. The slides are wrapped and kept at −20 °C. In this state, the slides can be used for several years. The slides are thawed at first and then air-dried. Then the slides are treated with acetone. The antibody of the affected animal is the primary antibody. The anti-bovine IgG is the secondary antibody that is conjugated with fluorochrome. Fluoresin isothiocyanate is used as fluorochrome. Several steps of incubations and washings are done. After completion of these steps, the fluorochrome is detected by fluorescence microscope. This IFAT is good for sensitivity and specificity. ELISA is a good test that provides both specificity and sensitivity. Different recombinant proteins have been used for the diagnosis of babesiosis. Tebele et al. (2000) worked on immunodiagnostic potential of *Babesia bigemina*. P200 is a merozoite protein used for immunodiagnosis of *Babesia bigemina*. However, the main constraint of using this antigen is variation in antigen quality from batch to batch. The researchers produced recombinant P200 bacterially. They found potential of this recombinant merozoite antigen for immunodiagnostic study. Zou et al. (2016) reported promising serodiagnostic potential of two recombinant antigens: recombinant *Babesia canis* secreted antigen (rBcSA 1) and recombinant *Babesia canis* merozoite surface antigen (rBcMSA-1), which could be used for indirect ELISA. Immunochromatographic test (ICT) can be performed for the diagnosis of *Babesia canis*.

Immunochromatographic Test

This is a reliable test for the diagnosis of babesiosis. In this technique, a paper strip or nitrocellulose paper and two antibodies are used. One antibody (capture antibody) is immobilized on the paper. Another antibody is the detection antibody labeled with a labeling agent. The sample (antigen) is placed in the sample pad area. The antigen forms a complex with the labeled antibody. **This complex move on and reach previously fixed capture antibody and binds to it in positive case**. A colored line is formed. That is the indication of positive reaction.

Antigens used for immunochromatographic test
• *B. bovis*—Merozoite surface antigen 2 (MSA2) • *B. bigemina*—Rhoptry associated protein (RAP-1) • *B. gibsoni*—P50 • *B. caballi*—48 kDa merozoite rhoptry protein

Molecular Diagnosis

Molecular diagnosis is possible using genus-specific or species-specific primers. Different specific primers are available. Different ribosomal DNA, such as ITS1, ITS2, etc., are used as molecular markers. Different mitochondrial DNA, such as CO1, ND1, etc., are also used as molecular markers for genomic characterization or identification. Various molecular methods applied for the diagnosis of babesiosis are as follows:

- Loop-mediated isothermal amplification
- In this method, four primers are designed in such a way that they can amplify six different sequences of the same DNA target. Additional pair of primers (loop primer) enhances the production of DNA due to increased PCR reaction. In this method, a large amount of DNA is produced, which can be analyzed by direct visualization of an intercalating fluorochrome like SYBR green.
- Real-time PCR
- Reverse line blot hybridization: Use of DNA probe for the diagnosis of babesiosis is in progress but practical implementation is lacking so far. This technique is useful for detecting multiple genera, species, and strains simultaneously. In a miniblottter, specific oligonucleotides are applied in an aligned format. The PCR amplified product is labeled with biotin. This biotin-labeled PCR product is then hybridized perpendicularly with the oligonucleotides so that every oligonucleotide gets exposed to every PCR-amplified product. The membrane is incubated in streptavidin-peroxidase conjugate, and the emitted light (chemiluminescence) is detected by an imaging system.

Treatment of Babesiosis

Drugs	Mode of administration	Dose
Diminazine aceturate	I/M or S/C	3–5 mg/kg bw (for cattle, buffalo, sheep, and goat); 5 mg/kg (for horse)
Imidocarb dihydrochloride	S/C	1.2 mg/kg bw (for cattle and buffalo); 2.4 mg/kg in horse (should be repeated after 24 h)
Phenamidine isethionate	S/C	12 mg/kg bw (for cattle, buffalo, and horse); 15 mg/kg bw (for dog)

Control

General Management

Affected animals should be treated properly.

Control of Vector

Ticks act as the vector of this parasite. Control of tick using different acaricidal agents should be done. See tick control described elsewhere.

Chemoimmunization

In the earlier period, experiments were carried out using infection and treatment method. The infected blood was transfused into a healthy animal and that animal was treated using anti-Babesial drug. Field application of this method is also done.

Immunological Control

An array of works was carried out for the development of anti-Babesial vaccine. Live vaccine, live attenuated vaccine, and subunit vaccine have been tried.

Molecular Vaccine

There seems no molecular vaccine has been commercialized.

Important Species of *Babesia*

The species occurring in the cattle are *Babesia bigemina, B. bovis*, *Babesia divergens, Babesia major*, *Babesia berbera,* etc. The species occurring in the goat are *B. motasi*, *B. ovis*, *and B. taylori*; in sheep are *B. motasi*, *B. ovis*, and *Babesia Foliata*; in pigs are *B. trautmanni*, *B. perroncitoi,* etc.; in dogs are *Babesia canis, B. gibsoni, Babesia vogeli;* in cats is *Babesia Felis*; in horse are *B. equi and B. caballi.*

Different Babesia Species

Babesia spp. of Cattle

Babesia argentina

This is an important species of the genus *Babesia* found in cattle. The animals suffer from fever, anemia, jaundice, etc.

Babesia berbera

This is a species of the genus *Babesia* found in cattle. Fever, anemia, jaundice are the symtoms.

Babesia bigemina

Introduction	This species cause cattle tick fever, red water fever, and piroplasmosis
Hosts	The hosts are all bovines, water buffalo, etc.
Site	They are found within RBC
Distribution	This protozoan parasite is prevalent in tropical and subtropical countries
Salient morphology	The organisms are large sized, pear-shaped. The narrow end of the organisms forms an acute angle. Pear shape of the organisms is common. But oval irregular forms are also found
Transmission	The ticks that act as the vectors of this protozoan parasite are *Rhipicephalus* (*Boophilus*) *annulatus*, *R.* (*B.*) *decolaratus*, *R.* (*B.*) *microplus*, *Rhipicephalus bursa*, *R. evertsi*, and *Haemaphysalis punctata*
Pathogenesis	Young animals suffer less. More prominent form of the disease occurs in exotic breeds. The main clinical signs are hemoglobinuria, fever, weakness, off feeding, etc. Cerebral form of the disease also occurs. The affected animals may die. The disease starts with high rise of fever (41.7 °C). High parasitemia is found during early phase of the disease. Fever is followed by other symptoms. The animals stop feeding and drinking water. Depression, absence of rumination, salivation, dryness of muzzle, loss of milk production, etc., are other associated symptoms. The affected animals may enter into a course of chronic phase and become debilitated, emaciated, and weak

Babesia bovis

Introduction	The organisms are small hemoprotozoan parasites
Hosts	The hosts are all bovines
Site	They are found within RBC
Distribution	They are prevalent in different parts of Europe, Asia, Africa, and America
Salient morphology	The organisms are small piroplasms. The shape of the organisms is pyriform and round. They may be irregular in shape
Transmission	The ticks that act as the vectors are *Rhipicephalus* (*Boophilus*) *microplus*, *R. bursa*, *Ixodes ricinus,* etc.
Pathogenesis	Young animals suffer less. The main clinical signs are fever, anorexia, weakness, off feeding, incoordination of movement, etc. The disease starts with high fever, which is followed by hemoglobinuria after about 10 days of infection. Cerebral form of the disease also occurs. Most of the field outbreak in Australia is caused by *B. bovis*

Babesia divergens

Introduction	The organisms occur in the RBC in divergent form
Hosts	The hosts are all bovines
Site	They are found within RBC
Distribution	They are prevalent in different parts of Europe (Northern and Central Europe)
Salient morphology	The organisms are small piroplasms. They are even smaller than *B. bovis*. They are easily identified because of their divergent form lying superficially on the RBC. Circular form of the organisms is also found
Transmission	The ticks that act as the vectors are *Ixodes ricinus*
Pathogenesis	The clinical signs include fever, anorexia, haemoglobinuria, weakness, off feeding, incoordination of movement, etc. The severity is less compared to *B. bigemina*

Babesia major

Introduction	This is a species under the family Babesiidae. The organisms are hemoprotozoan parasites
Hosts	The hosts are cattle
Site	RBC
Distribution	They are prevalent in North and West Africa, South America, Southern Europe, etc.
Salient morphology	The size of the organism is smaller than *B. bigemina* and found centrally in the RBC. The organisms look like *B. bigemina* except that these are smaller in size. The organisms are found in the centre of the RBC. But location of the parasite in RBC may vary.
Transmission	*Haemaphysalis punctata is major vector. Rhipicephalus (Boophilus) calcaratus is another vector.*
Pathogenesis	The clinical symptoms exhibited in the disease are not severe. The symptoms include less fever, mild hemoglobinuria, and slight anemia

Babesia spp. of Equines

Babesia caballi

Introduction	This species occur in equines. These are large form of organisms similar to *B. bigemina.*
Hosts	They occur in the horses, donkey, mule, etc.
Site	They are found within RBC
Distribution	The organisms are distributed in different parts of Asia, Europe, Africa, etc. They are also prevalent in America.
Salient morphology	The organisms are large in size and are mainly pyriform. However, they may be round or ovoid. The organisms resemble *B. bigemina.*
Transmission	Different *Dermacentor* spp., such as *Dermacentor marginatus*, *D. nitens*, *Dermacentor reticulatus*, *Hyalomma* spp., *Rhipicephalus* spp., act as the vectors
Pathogenesis	In the case of *B. caballi* infection in horses, posterior paralysis is common due to disturbances of the central nervous system. The young foals suffer from circling movement, incoordination, restlessness, etc. Hemoglobinuria rarely occurs. Fever and anemia are common clinical features. Severe form of disease has also been recorded. The disease run in acute, subacute, or chronic course. Mixed infection of *B. caballi* and *B. equi* may occur

Babesia equi

Introduction	These are small form of equine *Babesia*
Hosts	They occur in horses, mule, donkeys, and zebra
Site	They are found within RBC
Distribution	Worldwide in distribution
Salient morphology	The shape of the organisms is pyriform, round, amoeboid, etc. The organisms are small in size and usually appear in "Maltese cross" form arranged by four daughter organisms
Transmission	Different *Dermacentor* spp., such as *D. marginatus*, *D. nitens*, *D. reticulatus*, *Hyalomma* spp., and *Rhipicephalus* spp., act as vectors. *Hyalomma* spp. and *Rhipicephalus* spp. are also associated with transmission of this *Babesia* species.
Pathogenesis	The disease starts with high rise of temperature (41.7 °C), which may sustain for few days to weeks. Other symptoms appear subsequently, which include weakness, depression, icterus, hemoglobinuria, etc. The severity of the disease is more than *B. caballi* infection. Posterior paralysis found in *B. caballi* infection is usually absent in *B. equi* infection

Babesia spp. of Dogs

Babesia canis

Introduction	This is a large form of canine *Babesia*
Hosts	The hosts are dogs
Site	They are found in the RBC
Distribution	They are prevalent in different parts of Europe, America, Asia, Africa, etc. This species is prevalent in most of the continents.
Salient morphology	The size of the organism is large with a clear vacuole in the cytoplasm. Pyriform is the characteristic morphological feature. An acute angle is formed by the slender ends of two organisms. However, there may be variable shapes of the organisms, such as amoeboid and ring forms. Multiple organisms in RBC may be found.
Transmission	The ticks that act as vectors are *Rhipicephalus sanguineus*, *Dermacentor marginatus Hyalomma plumbeum plumbeum*, and *Haemaphysalis leachi*
Pathogenesis	The young animals suffer less in general cases of babesiosis. However, in the case of *B. canis* infection, young pups also suffer as that of adult dogs. The clinical symptoms exhibited in the disease are anorexia, dullness, depression, disinclination to movement, hemoglobinuria, anemia, jaundice, etc. Splenomegaly is marked. Both feces and urine may be yellowish in color. The affected animals may die. Some atypical symptoms like bronchitis, pneumonia, ascites, etc., may be observed

Babesia gibsoni

Introduction	This is a small form canine *Babesia* species
Hosts	The hosts are dogs
Site	RBC
Distribution	They are prevalent in different parts of Asia including China, India, Sri Lanka, Japan. They are also prevalent in North Africa, and some parts of Turkistan.
Salient morphology	The size of the organism is smaller than *B. canis*. Pyriform shape is not usually found. Variable shapes of the species are signet ring like, oval, annular, circular, etc.
Transmission	The pathogens are transmitted by different ticks, such as *Rhipicephalus sanguineus, Haemaphysalis bispinosa, and H. longicornis.*
Pathogenesis	The clinical symptoms exhibited in the disease are fever, anorexia, dullness, depression, disinclination to movement, hemoglobinuria, anemia, etc. Jaundice may occur but is not so common. The disease entity runs in a chronic state but not acute as is found in case of *B. canis*

Babesia vogeli

Introduction	This is a large form of canine *Babesia* species
Hosts	The hosts are dogs
Site	RBC
Distribution	They are prevalent in Southern Asia and North Africa, North, America, South America, and Central America.
Salient morphology	Morphologically the organisms are similar to *B. canis* but they are large in size. The organisms may be synonym of *B. canis*.
Transmission	*Rhipicephalus sanguineus*
Pathogenesis	The clinical symptoms exhibited in the disease are fever, anorexia, dullness, depression, disinclination to movement, hemoglobinuria, anemia, etc.

Babesia spp. of Cat

Babesia felis

Introduction	This is feline *Babesia sp.*
Hosts	The hosts are cats and other felines
Site	The organisms are found in RBC
Distribution	Asia, America, and Africa
Salient morphology	The size of the organism is small. The organisms also form Maltese cross arranged by four small daughter organisms. They are round, oval and ring -shaped.
Transmission	Life cycle seems not reported
Pathogenesis	The common clinical signs are weakness, lethargy, inappetence, etc. the occasional clinical features are fever and jaundice. There may be splenomegaly, pulmonary edema, and renal failure

Babesia spp. of Sheep and Goat

Babesia motasi

Introduction	This is a large form of *Babesia* causing ovine and caprine babesiosis
Hosts	The hosts are sheep and goat
Site	RBC
Distribution	They are prevalent in different parts of Europe, Asia, and Africa
Salient morphology	The organisms are large pear-shaped. The narrow ends form an acute angle
Transmission	The ticks that act as the vectors are *Dermacentor silvarum*, *D. marginatus*, *Rhipicephalus bursa*, and *Haemaphysalis punctata*
Pathogenesis	The disease occurs in acute or chronic form. The main clinical signs are fever, weakness, and off feeding. In acute stage, the clinical features include fever, anemia, hemoglobinuria, etc. The clinical signs are similar to the clinical signs exhibited by *B. bigemina*. In chronic form of the disease, no typical clinical feature is found

Babesia ovis

Introduction	This is a small form of *Babesia* causing ovine and caprine babesiosis
Hosts	The hosts are sheep and goat
Site	RBC
Distribution	They are mostly prevalent in tropical and subtropical countries. They are also found in Eastern Europe. and Southern Europe.
Salient morphology	The organisms are mostly round occurring in the margin of of RBC. Pyriform organisms are rarely found. The organisms are much smaller in comparison to *B. motasi.*
Transmission	*Rhipicephalus bursa* and *R. sanguineus* are the vectors.
Pathogenesis	The disease is expressed in acute and chronic form. The acute form of the disease includes fever, jaundice, anemia and hemoglobinuria, etc. in the chronic form of the disease, no marked symptom is found

Babesia foliata

This species occurs in sheep. The organism is similar to *Babesia ovis*. They occur in the central part of the RBC. They do not produce much pathogenesis.

Babesia taylori

This species occurs in goat. The organisms are small. The shape is ovoid or round. 8–16 organisms may be found in one RBC.

Babesia spp. of Pigs

Babesia perroncitoi

Introduction	This is small form of *Babesia* causing swine babesiosis
Hosts	The hosts are pigs
Site	RBC
Distribution	They are prevalent in tropical and subtropical countries. They are also reported from Sudan, Vietnam, and Sardinia
Salient morphology	The organisms are round in shape and vacuolated. Oval and pyriform shapes are also found. The merozoites usually occur singly. More than one merozoite may also be found
Transmission	As per reports, *Rhipicephalus sanguineus, R. (Boophilus) microplus* and *Dermacentor reticulatus* may be the vectors
Pathogenesis	The main clinical signs are hemoglobinuria, fever, anemia, edema, weakness, and off feeding

Babesia trautmanni

Introduction	This is an important swine *Babesia*
Hosts	The hosts are pigs
Site	RBC
Distribution	They are prevalent in different parts of Europe, Russia, and Africa. They are also found in Asia.
Salient morphology	The organisms are elongated narrow occurring in the RBC. The shape of the organisms may also be ring-like or oval
Transmission	The ticks that act as the vectors are *R. (B.) decolaratus, Dermacentor reticulatus,* etc.
Pathogenesis	The clinical signs are hemoglobinuria, fever, anemia, and edema, weakness

Phylum: Apicomplexa —Class: Sporozoea —Subclass: Piroplasmia—Order: Piroplasmida—Family: Theileriidae

Family Synopsis

These are protozoan organisms found in the blood of different vertebrates. They are worldwide in distribution. The protozoans are characterized by pyriform, oval, round pleomorphic shape, short apical complex, schizogony, or binary fission. Ixodid ticks are vectors of these parasites. Ixodid ticks are vectors of these parasites. Some species of this family have pathogenic significance causing massive loss in the livestock industry. Gold standard diagnosis of this parasitic infection is examination of blood smears. Control of the parasite mostly relies on chemotherapy.

History

The genus name *Theileria* was given in honour of Sir Arnold Theiler (early 1900s).

Genus: *Theileria*

Different Species

The common species of this genus are *Theileria annulata* (causative agent of bovine tropical theileriosis), *T. parva* (causative agent of East coast fever), *Theileria lawrenci* (causative agent of corridor disease), *Theileria hirci*, *T. ovis*, *Theileria camelensis,* etc.

Morphology

In the RBC, the organisms appear as dot, comma, rod, and ring-shaped structures. In lymphocytes, schizonts are found. The organisms in the schizont stage are like small granules.

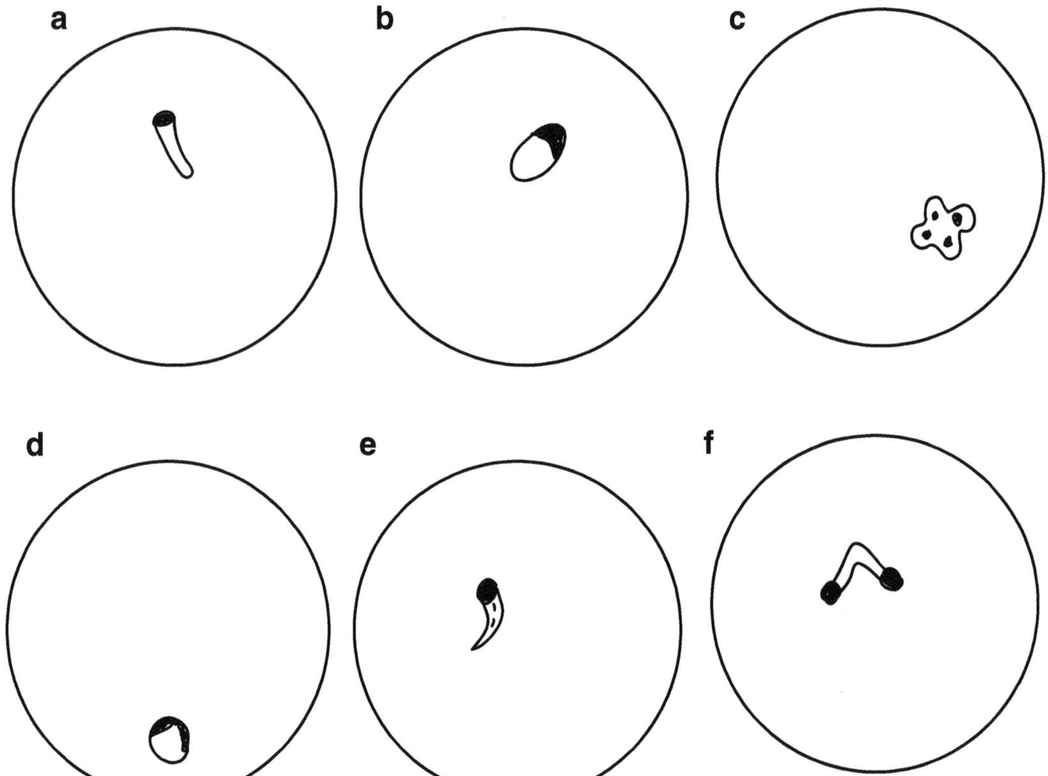

Fig. 8.19 Different forms of *Theileria* sp.

Biology

Animals get the infection when infected ticks bite the hosts. During bloodsucking, the sporozoites enter the host. The organisms enter the lymph node and subsequently enter the lymphocytes. In the lymphocytes, macroschizonts and microschizonts are formed. Microschizonts are also found in reticuloendothelial cells, macrophages etc. Micromerozoites are released from the microschizonts. The merozoites then invade the RBC.

The events found in the life cycle of *Theileria* are as follows:

- The ticks bite the hosts.
- The sporozoites gain entry into the hosts through tick bites.
- The sporozoites enter lymphocytes. The entrance and establishment in host cell includes different steps. At first the parasite recognize the host cells. The next stage is formation of junction between host cell and parasitic stage. A "zipping process" causes internalization of the parasite. In this stage, the parasite is still surrounded by host cell membrane. Disssolution of host cell membrane occurs and the parasite become free from host cell membrane and gain entry. Later formation of microtubule network associated with the parasite is found.
- The parasites cause lymphoproliferation. A synchronous division of both parasite and lymphocytes occurs.
- Macroschizonts and microschizonts are formed. Clonal expansion of schizonts occur. Multinucleated macroschizonts are called Koch's blue bodies.
- Micromerozoites are released from the microschizonts.
- The micromerozoites enter the erythrocytes. Simultaneous occurrence of microschizonts and erythrocytic stages indicate that the erythrocytic stages are derived from the micromerozoites.
- The infected RBCs are ingested when the ticks take a blood meal from the hosts.
- The RBCs are digested in the gut of the tick.
- The organisms are released in the gut.
- Formation of micro- and macrogametes occurs in the gut of tick.
- The gametes unite to form a zygote.
- The zygotes become elongated and transform into kinete stage and pass to the hemolymph from the gut.
- The kinetes reach the acinar cells of the salivary glands.
- Sporogony occurs in the acinar cells.
- Sporonts are formed.
- Subsequent development (infective sporozoites) occurs after moulting to the next stages (nymph, adult).

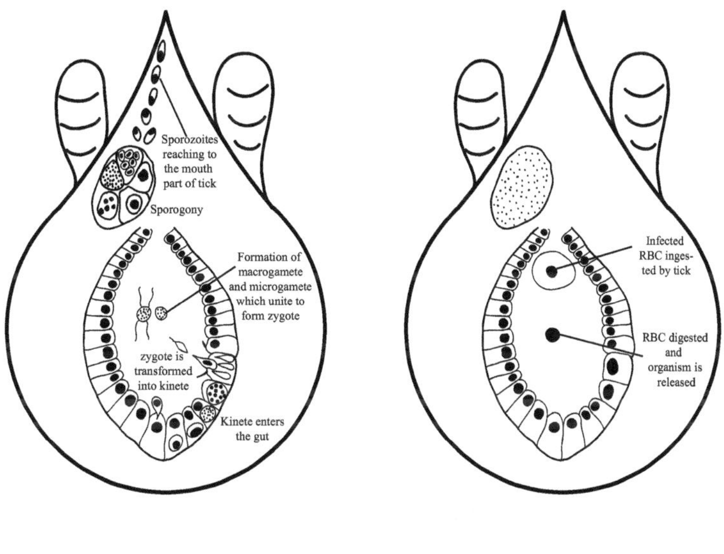

Fig. 8.20 Life cycle of *Theileria* sp. (In tick)

> **Note**
> - Attachment of sporozoites to the host's cell surface is required to generate critical signal which triggers a series of events within both host and parasite resulting development *Theileria*. *T. annulata*, *T. orientalis*, and *T. hirci* show transplacental infection in sheep.

Pathogenesis

The sporozoites enter the lymphocytes. The parasites induce lymphoproliferation. As a result, hyperplasia of lymph nodes occurs. The lymph nodes get enlarged in size. The merozoites come out from the lymphocytes and affect RBCs.

The disease caused by *T. annulata* may run in acute, subacute, and chronic phases. The acute phase of the disease runs only for 3–4 days. The subacute and chronic phases of the disease take a course of 3 weeks or so. The initial onset of the disease is characterized by high fever (40.5–41.5 °C), marked enlargement of lymph nodes, respiratory symptoms, nasal discharge, etc. Gradually, the febrile condition declines. Anemia, jaundice, and severe weakness are found in this stage. The animals may die. A typical "**Punched out ulcer**" in abomasum is found in cattle affected by *Theileria*, particularly *T.annulata*.

A cerebral form of the disease may occur in the case of *T. annulata* infection. The animals show different nervous disorders.

Hematological parameters show a low level of hemoglobin, a low level of TLC, and low PCV. Low levels of serum calcium and protein were also observed.

The disease caused by *T. parva* is similar to caused by *T. annulata*. This causes high mortality. The disease commences with fever (40–41.7 °C) and the animals may turn to death or recovery may occur. A few days after fever, the animals exhibit clinical signs comprising off feeding, dullness, depressioin, nasal dischrge, swelling of lymph nodes and eyelids. Hemorrhagic diarrhea may occur. Calves suffer from subacute form of the disease. Mild form of the disease is associated with less fever and the affected animals may recover. Turning sickness is a condition associated with *T. parva* and *T. mutans infection. Circling movement and abduction of hind legs are observed. T. lawrenci causes "Corridor disease" in cattle and buffalo which is a fatal disease and the mortality reach upto 80% or more. T. mutans is less pathogenic and mild fever is observed. T. hirci is markedly pathogenic in sheep and goat. Fever, nasal discharge, swelling of lymph node, transitory hemoglobinuria are some clinical features found in this infection. T. ovis is less pathogenic.*

Clinical Signs

- High fever with rise of temperature (40–41. 5 °C) is a prominent clinical sign
- Dullness
- Depression
- Off feeding
- Lacrimation
- Hemorrhagic diarrhea
- Turning sickness is exhibited by circling movements due to the occurrence of necrosis of the brain, which is a typical sign found in *T. parva* and *T. mutans* infections.

> **Note**
> - *Theileria buffeli/Theileria orientalis cause oriental theileriosis, also called as Theileria-associated bovine anemia* (TABA). *T. annulata* causes tropical theileriosis, Mediterranean coast fever, and Mediterranean theileriosis.
> - *T. parva* causes East Coast Fever
> - *T. hirci* causes ovine malignant or small ruminant malignant theileriosis.

Diagnosis

History-Based Diagnosis

The hard history and prevalence of *Hyalomma* ticks in the endemic zone are supportive diagnostic elements.

Symptomatological Diagnosis

Symptomatology cannot provide a confirmatory diagnosis because the exhibited symptoms are not specific. However, clinical signs comprising fever and enlargement of prescapular lymph node might be suggestive of *Theileria* infection.

Parasitological Diagnosis

(a) Examination of blood smear

Diagnosis can be done by the examination of the blood smear. Different Romanowsky stains are used for the identification of this protozoan parasite. Giemsa staining and Leishman staining can be used for this purpose.

(b) Examination of biopsy material

Lymph node biopsy is done for the detection of *Theileria* schizonts. *Theileria* schizonts are formed in the lymphocytes. The prescapular lymph node is increased in size because of this. Sometimes, the detection of the piroplasmic stages is not sufficient for a confirmative diagnosis of infection. In that case, a biopsy is done. Some amount of PBS is injected into the lymph node. The pressure created by the PBS enables us to take out the lymph material into the syringe. The smears are then prepared and stained. The stained slides are examined and Koch's blue bodies are detected.

Immunodiagnosis

A series of immunodiagnostic tests have been tried to diagnose cases of theileriosis. These tests are the complement fixation test (CFT), hemagglutination inhibition test, and indirect fluorescent antibody test (IFA). Many works have been carried out on immunodiagnosis using different crude antigens and purified antigens. Variable success was achieved in immunodiagnostic tests. Different recombinant proteins have also been used for the diagnosis of theileriosis. Immunodiagnostic potential of recombinant *Theileria annulata* merozoite surface antigen (Tams 1) for the diagnosis of *Theileria annulata was reported*. The immunodiagnostic potential of this protein was evaluated by ELISA. Bakheit et al. (2004) reported the immunodiagnostic potential of an antigen of *Theileria annulata* (TaSP). The result displayed 99.1% sensitivity and 90.47% specificity in ELISA. Kumar et al. (2013) worked on the immunodiagnostic potential of a recombinant antigen of *Theileria equi*. The recombinant antigen, EMA2 (extraerythrocytic merozoite antigen of *Theileria equi*) were expressed in *E. coli* with GST tag. The immunodiagnostic potential of this protein along with it's suitability for sero prevalence was reported by them. Kumar et al. (2015) developed ELISA based on EMA1 recombinant antigen and reported it's suitability in serodiagnostic prevalence of *Theileria equi*.

Molecular Diagnosis

Molecular diagnosis is possible using genus-specific or species-specific primers. Different specific primers are available. Different ribosomal DNA, such as ITS1, ITS2, 28SrDNA, etc., are used as molecular markers. Different mitochondrial DNA, such as CO 1, ND1, etc., are also used as molecular markers for genomic characterization or identification.

Treatment: Parvaquone, buparvaquone, menoctone, oxytetracycline,etc. are some effective drugs.

Control

General Management

Affected animals should be treated properly.

Control of Vector

Ticks act as the vector of this parasite. Control of ticks by using different acaricidal agents should be done. See the chapter "Veterinary Entomology and Acarology" for control of ticks.

Chemoimmunization

In the earlier period, experiments were carried out by infection and treatment methods. Infected blood was transfused to a healthy animal and that animal was treated with anti-theilerial drugs. This technique is also used in the field for control of theileriosis.

Immunological Control

An array of works was carried out for the development of an anti-Theilerial vaccine. Live vaccine, live attenuated vaccine, and subunit vaccine were used in different trials. One of the successful vaccines is Roxavac-T, which is a schizont culture vaccine. Efforts were made to use recombinant antigens to develop vaccines. A sporozoite surface protein (p67) conferred satisfactory protection. After the invention of the cell culture technique, the concept of culturing theileriosis-infected cells encouraged many scientists to produce live vaccines. In this technique, schizont-infected lymphocytes are cultured following standard protocol. Several passages are completed so that the organism becomes avirulent. The cultured cells are cryopreserved for future use. At the time of vaccination, the cryopreserved ampule is thawed and inoculated into the target animals. The vaccine has been successful for *T. annulata*. However, a high dose is required for *T. parva* vaccination, which is a costly proposition. This was a great constraint on *T. parva* vaccination. The constraints of vaccination against *T. parva* still prevailed, which imposed an urge to develop a vaccine against *T. parva*. Hence, the concept of the use of live sporozoites emerged as an alternative option to cell culture vaccine. In this technique, the infected ticks are homogenized, and this homogenized tick becomes the source of live sporozoites. The prepared material is cryopreserved for future use. This technique also did not produce satisfactory results because a low dose of sporozoites can also produce severe reactions in inoculated animals. Further modification of this technique was required and "inoculation and simultaneous treatment" with oxytetracycline was started. This infection and treatment technique resulted in satisfactory success for conferring protection. A cocktail vaccine known as the Muguga cocktail containing the isolates of *Theileria parva* was developed. The vaccine contains three strains of *T. parva*. These strains are Muguga, Kiambu, and Serengeti- transformed. This vaccination technique resulted in a broad range of protection. The production and application of this vaccine are coupled with different tedious methods comprising establishment of infection of cattle with different isolates, standardization of amount of different isolates, cryopreservation of vaccines in liquid nitrogen, maintenance of cold chain, etc. All these led to limited use of this vaccine.

Treatment

Chlortetracycline, buparvaquone, halofuginone, and menoctone are some important drugs for treatment of theileriosis.

Genus: *Cytoxzoon*

These protozoan organisms are closely similar to the *Theileria* sp. Schizogony stage of these members does not occur in the lymphocytes. This occurs in erythrocytes. There are several species under this genus like *Cytoxzoon felis, Cytauxzoon strepsicerosi*, *Cytauxzoon sylvicaprae*, and *Cytauxzoon taurotragi*. Schizogony stage of these members do not occur in lymphocytes. This occurs in macrophages, endothelial cells etc. *C. felis* occurs in felines and is identified by it's signet ring appearance. Maltes cross form is also found. This highly fatal in cat. Fever, anorexia, icterus are some clinical features.

Cytauxzoon strepsicerosi

This is a species under the family Theileriidae and occurs in Kudu. The common clinical signs exhibited are anemia, icterus, fever, anorexia, etc.

Cytauxzoon sylvicaprae

This is a species under the family Theileriidae. The hosts are duiker. The organisms are prevalent in South Africa. The common clinical signs exhibited are anemia, icterus, etc.

Cytauxzoon taurotragi

This is a species under the family Theileriidae. The hosts are yearling eland. The organisms (erythrocytic forms) are similar to *Theileria parva*. Schizonts occur in the lungs, liver, etc.

Phylum: Ciliophora—Class: Kinetofragminophorea—Family: Balantididae

Genus: *Balantidium*

The important species is *Balantidium coli*.

Site

This species is found in the intestine of pigs and man.

Morphology

Trophozoite

This parasite has two stages i.e., trophozoite and cyst. The trophozoite stage is oval, subcylindrical, and pear in shape. The trophozoites possess two nuclei: one is large in size (macronucleus) and the other is small in size (micronucleus). The macronucleus is kidney-shaped. The micronucleus is smaller than the macronucleus and located nearer to the concave side of the macronucleus. The body is ciliated and possesses a mouth (cytostome) and an excretory organ (cytopyge). There are two contractile vacuoles. One is terminal and the other is near the centre.

Cyst

The cysts are formed either in the gut or in the environment. They are oval or spherical in shape and have a double membrane. The color of the cysts is yellowish or greenish.

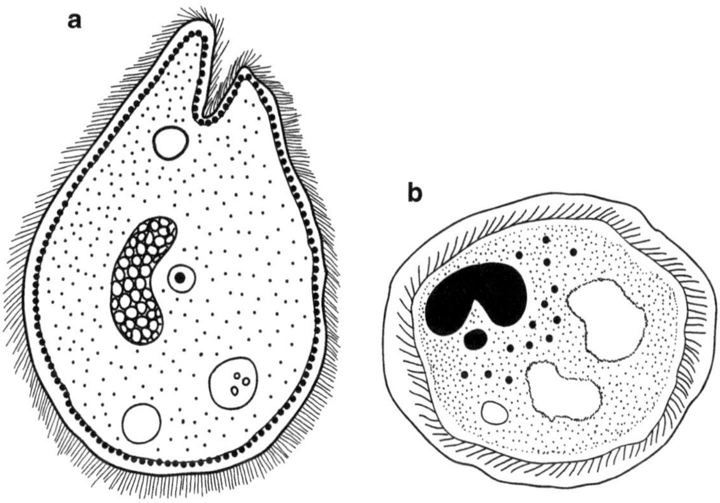

Fig. 8.21 *Balantidium coli* (**a**) trophozoite (**b**) cyst

Pathogenesis

The organisms remain as commensals. However, in some situations of stress condition in the animals, the organisms flare up and cause some amount of pathogenicity.

Diagnosis

Diagnosis can be done in different ways, such as microscopy, staining, and detection of antibodies. Sheather's sugar floatation technique, modified zinc sulfate centrifugal floatation, and modified formalin–ethyl acetate concentration technique can be used.

The organisms inhabit the gut of humans and pigs. In pigs, they do not cause pathogenesis. Dysentery is caused by these organisms in humans. Infection in human occurs by the ingestion of cysts. Trophozoites get released from the cysts and enter the submucosal layer of the colon where they multiply and form colonies. Simple inflammation of the colon may lead to ulceration. Hemorrhagic diarrhea may occur.

Treatment

Nitazoxanide and azithromycin are effective drugs against this parasitic infection. Carbasone, didoquin, oxytetracycline, metronidazole are some other drugs for treatment.

Control

Personal hygiene is very important. Filtrated water should be consumed. People should avoid contaminated ponds or water reservoirs, etc.

Rickettsia

Anaplasmosis

Some rickettsial organisms closely resemble protozoa. *Anaplasma* and *Ehrlichia* are two important rickettsial organisms. *Anaplasma* are very small and occur in the RBCs. The important species of this genus are *A. marginale*, found at the margin of the RBC, and *A. centrale*, found in the center of the RBC. The hosts are cattle and related ruminants. *A. ovis* occurs primarily in sheep, goat, deer etc. The organisms are transmitted by different ticks and bloodsucking flies. Anaplasmosis is characterized by a high rise of temperature, anorexia, dullness, and depression, which may even lead to the death of the animals. Different ticks transmit *Anaplasma* spp.

Ehrlichiosis

The organisms are coccoid or ellipsoidal and are found in circulating leucocytes. They appear singly or as morula. This is a type of rickettsia. The species of the genus *Ehrlichia* are *E. bovis*, *E. canis,* and *E. equi*. *E. canis* is the causative agent of tropical pancytopenia in dogs. These rickettsial organisms occur in the leucocytes and appear as morulae of small bodies. Animals suffering exhibit high rise of temperature, anorexia, etc. The superficial lymph nodes may be enlarged. A high dose of tetracycline is very much effective.

Special Part

Protozoa are unicellular eukaryotic organisms. Some protozoans form colonies. These are called colonial protozoa. The colony is formed by similar organisms. If some flagellates are isolated from a colony, they can undergo further fission and produce a daughter colony. Two types of colonies are observed: spheroid and discoid. In some colonies, the individual organisms are arranged in a branching pattern. This is called an arboroid colony. Stalks are characteristics of an arboroid colony.

The protozoa may have pseudopodia. Four major types of pseudopodia are found: filopodia, lobopodia, myxopodia, and axopodia.

Filopodia

Filopodia are slender and tubular structures that taper to a pointed tip. They can branch. The branchings may be anastomosed. Circulating granules are absent in filopodia.

Lobopodia

Lobopodia are broad pseudopodia with rounded tips. This type of pseudopodia is characteristic of amoeba. The slender lobopodia are hyaline in appearance. However, the large lobopodia possess a clear ectoplasm and granular endoplasm.

Myxopodia

Myxopodia are thin filamentous structures that can branch. This branching pattern leads to the formation of a complex network that acts as a food trap.

Axopodia

Axopodia are thin pseudopodia. They have many microtubules surrounded by ectoplasm. They are also called actinopodia. These are radiating structures that arise singly from apparently spherical organisms.

Flagella

These are slender thread-like structures that act as locomotory organs for many protozoa, bacteria, etc. There are different types of flagella. These include *Stichonematic flagellum*, *Pantonematic flagellum*, *Acronematic flagellum*, *Pantacronematic flagellum*, etc.

- *Stichonematic flagellum*: In the stichonematic type of flagellum, a single row of filaments is found at one side of the flagellar sheath. Mastigonemes are found in one row along one lateral side of flagellum.

- *Pantonematic flagellum*: In this type of flagella, two or more rows of mastigonemes are found at both sides of flagellum.
- *Pantacronematic flagellum*: In the pantacronematic type of flagella, both a terminal filament and one or more rows of mastigonemes are present.
- *Acronematic flagellum*: In this type of flagella, mastigonemes are absent. Only a terminal filament is present. This is also called "lash" flagellum.
- *Trailing flagellum:* This flagellum arises anteriorly but trails posteriorly.

Mastigonemes

These are specific structures found in some flagella. The sheath of some flagella possessess lateral filaments called mastigonemes. The mastigonemes may be arranged in one or more rows.

Axostyle

An axostyle is a specific structure of microtubules that arises from the base of the flagellum. There are variations of axostyles in protozoan organisms. This may be slender (*Monocercomonoides pilleata*). There may be multiple axostyles (*Snyderella tabogae*). Axostyle may possess inclusion bodies (*Tritrichomonas augusta*).

Costa

This is a prominent striated fibrous cytoskeleton structure found in some protozoa. This is the base of the undulating membrane.

Cresta

This is a roughly triangular membranous structure that seems to be a homologue of Costa. The anterior end is broad and joined to the blepharoplast.

Aciculum

This is a needle-like structure found opposite to the kinetoplast and reaches the anterior flagella in some protozoa (*Cryptobia helices*).

Parabasal Apparatus

In many flagellates, parabasal apparatus is present. These parabasal bodies have taxonomic importance. The function of the parabasal apparatus is not known yet. They form a part of the mastigont and are variable in size and shape. Parabasal apparatuses are of many varieties, found in different flagellates. This is also equally variable in the parasitic flagellates. In some uninucleates (i.e., Trichomonadidae), a simple elongated parabasal body is present. In certain flagellates, the parabasal body is found in the form of a coil around the axostyle. It may be branched in some protozoans.

Cilia

Structurally, cilia and flagella are the same, although cilia are shorter and present more in number. The structural components of cilia are axonemes ending in a basal granule and a sheath. An argentophilic layer covers the axial filament. The simple cilium is a locomotory organ. However, many cilia get fused to form different compound organelles which have been formed by fusion of longitudinal and transverse rows in many species. Such organelles are called membranelles, undulating membranes, and cirri. The undulating membrane is formed by the fusion of one or more longitudinal rows of cilia. The membranelles appear as triangular flaps formed by two or more transverse rows of cilia. Cirri are formed by the tufts of cilia.

Fibrillar System

The fibrilar system found in protozoa is very typical. The basal granules of each longitudinal row of cilia are united together by fibrils. In some species of protozoa, it is found that the longitudinal and transverse fibrils are joined to the secondary basal granules. The same is also connected to superficial primary basal granules by rhizoplasts. In some species, the fibrils are joined specifically and form a complex fibrillar system.

Silver Line System (Ciliates)

Scientist Klein demonstrated the silver line system in many ciliate protozoa. The name has been given based on his technique. In this technique, reduced silver gets deposited on the superficial structure of the organisms.

Cytostome

Ingestion of solid food by some protozoa is performed through a specific mouth opening called cytostome. The cytostome often opens into the cytopharynx. In ciliates, another structure associated with the cytostome is the peristome, which surrounds the cytostome. In ciliates, the peristomal area may be provided with an undulating membrane, a row of membranelle, etc.

Contractile Vacuole

In many protozoa, contractile vacuoles are present. Interestingly, the position, number, shape, and accessory structures of vacuoles are widely variable in different protozoa. In amoeba, the position of vacuoles changes in accordance with their movements. These small vacuoles get fused to form large vacuoles. In some protozoa, vacuoles are absent. Instead of vacuoles, an apparently permanent contractile tube is present.

It is very interesting how the protozoa sink in the water and then come up from the depth. Vacuoles play a role in this regard. Vacuoles are present in the outer foamy cytoplasmic layer and help some protozoa remain at a particular depth of water. The sinking of some protozoa is somewhat characteristic which is mediated by the collapse of vacuole and retraction of pseudopodia. This increases the specific gravity of the protozoa and the protozoa sink. When the organisms are to rise again, the new vacuoles develop.

Chromatophores

Chromatophores are found in many phytoflagellates. They have variable colors, numbers, and sizes. The chromatophores may be H-shaped or cup-shaped. In H-shaped structure, two large lobes are connected by a connecting structure. The chromatophores may be flattened in some flagellates.

Pyrenoids

These are some specific structures that are associated with the chromatophores of protozoa and perform the synthesis of carbohydrates. They appear as solid bodies or aggregates of granules found in autotrophic protozoa. These organelles have a role in the synthesis of polysaccharides.

Chondriosomes

This is another name for mitochondria.

Mitochondria

Mitochondria are present in many protozoa. The shape of mitochondria is variable. They may be short, rod-like, granular, filamentous, etc. The mitochondria are distributed randomly in the cytoplasm in most of the protozoa. However, mitochondria may be aggregated in specific areas, such as around food vacuoles. Cytochrome oxidase enzyme in mitochondria play an important role in cellular respiration specially electron transport chain. The presence of succinate dehydrogenase and some other enzymes has been found in protozoan mitochondria. The presence of these enzymes indicates that mitochondria are involved in oxidation.

Pigment of Protozoa

In phytoflagellate organisms, chromatophores are present. The pigments of chromatophores vary among different groups. Other pigments are also present apart from chromatophores. These pigments are red, violet, yellow, green, brown, and blue in colour. They are found in the cytoplasm of many protozoa. Blue-green granules usually disappear after a certain duration of starvation. Pink pigment also disappears due to exposure to sunlight. However, they regenerate in the darkness. By chromatographic technique, certain pigments have been isolated. These pigments are xanthophylls (dinoxanthin, diadinoxanthin, neodinoxanthin), chlorophyll, etc. The exact function of most protozoan pigments is unknown. It is needless to mention that chlorophylls have a role in photosynthesis. The red pigment has a role in reflecting the light from the red end of the spectrum (e.g., red granule present in *Euglena rubra*). Carotenoid pigments are also involved in the absorption of blue-green light of low intensity.

Glycosylphosphatidyl Inositol (GPI)

GPI is a lipid anchor present on many cell surfaces, which is formed during post-translational modification of proteins with glycolipid in eukaryotes. These act as anchors for different receptors, enzymes, transport, proteins, etc. They are assembled in the endoplasmic reticulum and attached to phosphatidyl inositol lipid. The core of GPI has phosphatidyl inositol, glycan comprising three mannose, and one glucosamine. A terminal phosphoethanolamine is attached to the newly formed carboxyl terminus of the protein by amide. GPI anchors and glycoinositolphosphatidyl phospholipids are profusely present in the membranes of parasitic protozoa. Recent studies have revealed many functions of these structures related to the immunostimulatory and regulatory activity of immunity. The bioactivity of GPI anchors is associated with their ability to elicit the synthesis of proinflammatory cytokines and nitric oxide by host macrophage.

Immuno-Protozoology

Immunity to Different Protozoa

Immunity to *Trypanosoma*

Trypanosomes have evolved to produce a dense surface coat of glycoprotein (variant surface glycoprotein), which is shed frequently. At any given time, 10^7 molecules of a VSG are produced. When the hosts further trigger to mount immunity against trypanosomes, the parasites produce a new surface coat. It is very interesting that **new VSGs are generated through programmed DNA rearrangement phenomena in trypanosomes**. This phenomenon is not yet known completely. *Trypanosoma* have a unique characteristic to frequently change their surface antigen as a result of changes in their surface protein. They have a repertoire of VSG genes in a series at the telomeric region. In that region, a particular VSG gene gets activated and expresses a VSG protein. Whenever required, the earlier VSG gene becomes silent, and a new VSG gene gets activated. It is not clearly understood how the silencing of VSG genes occurs. Except for *Trypanosoma cruzi*, all trypanosomes are extracellular. In the case of extracellular trypanosomes, antibodies are produced against this antigen and this antigen binds to the antibody. This antigen-antibody complex attracts the complements. The complements cause a cascade of reactions which are mostly enzyme-mediated and the specific products of the complement reaction clear the pathogen. Complement reactions occur in three pathways: classical pathway, alternative pathway, and lectin pathway. In all these pathways, the initial reactions are different but the three reactions eventually produce a membrane attack complex (MAC). The MAC releases perforin and granzyme, which are toxic to the pathogens. These toxic materials disrupt the membrane of the pathogen. Perforin causes pore formation in the membrane of the pathogen, and through these pores, granzyme enters and kills the pathogen. But trypanosomes soon change their surface protein and escape the host immune response.

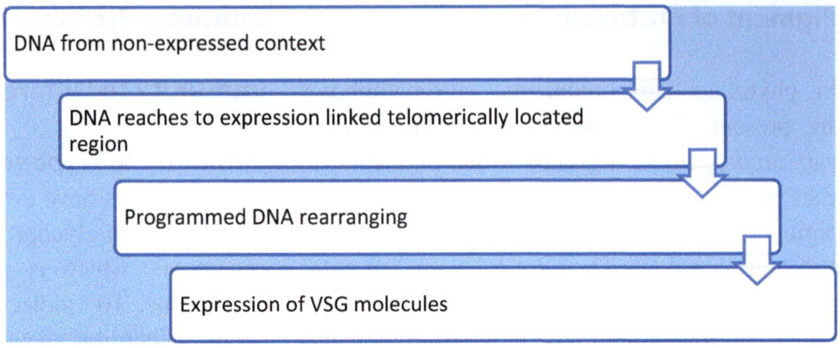

Fig. 8.22 Expression of VSG molecules

Immunity to *Theileria*

As is found in certain viruses, *Theileria* organisms induce antigenic changes on surfaces of lymphocytes, and against this changed antigen, the host mounts cytotoxic T cells. *Theileria* organisms possess a unique character: synchronous division of the protozoan stage with the host cell occurs once it has entered the lymphocytes. The organisms transform the lymphocytes and the lymphocytes start proliferation. Since the organisms remain within the cells, they fail to get access to the host's immune response. In the peripheral blood of *Theileria*-infected animals, cell-mediated cytotoxicity activity has been reported. **Studies show the detection of macrophage inhibitory factor (MIF) in peripheral blood lymphocytes (PBLS) from calves experimentally infected with *Theileria annulata*.** These lymphocytes produce a factor that inhibits macrophage migration, indicating CMI response.

Immunity to *Babesia*

When ticks infect hosts, the sporozoites enter the hosts. Initially, the sporozoites are found free in the blood plasma. At this stage, the sporozoites get neutralized by antibodies. Gradually, the organisms enter the RBCs. Different cells of the innate immune system play a role in restricting the organisms. Macrophages release nitric oxide, TNF alpha, and reactive oxygen species (ROS). NK cells produce IFN gamma. However, it is not clearly known whether these substances cause restriction of intraerythrocytic development. **Th1 cells also produce IFN gamma. This might degrade the intraerythrocytic stages, though it has not been proved.** The humoral immune response provides partial protection to the hosts against *Babesia* infection. Antibodies are more effective against free parasites than intraerythrocytic stages. Evidence indicates that IFN gamma could be directly toxic to the intracellular stages.

Immunity to Coccidia

The developmental stages of coccidian parasites face a physical barrier at first, and then the organisms are engulfed by macrophages. This is an innate immune activity. Apart from macrophages, natural killer cells and dendritic cells also perform innate immunity. Pathogens have conserved molecular patterns. These are called pathogen-associated molecular patterns (PAMPs), which are recognized by pattern recognition receptors (PRRs) present on immune effector cells. In the developmental stage of many *Eimeria* spp., **profilin** has been found to be a major TLR ligand. Another type of immunity is adaptive immunity. Adaptive immunity develops in *Eimeria* infection, which is specific to a particular antigen. There are different types of cells that perform antigen presentation. M cells, dendritic cells, and B cells are responsible for presenting antigens to T cells. In case of intracellular pathogen, IL12 switches the Th1 cells. Cytotoxic T cells (CTL) also get activated. Both schizonts and sexual stages are killed by CTL. In the case of the extracellular pathogen, IL4 leads to a Th2 response.

CD4 + helper cells and CD8+ cytotoxic T cells (CTL) are two major T cell subsets that play a role in coccidian immunity. **Lymphocytes are present in gut-associated lymphoid tissue (GALT) at the region of Peyer's patches of the small intestine, cecal tonsil, etc.** In the intestine, specialized modified epithelial cells called M cells are present. These M cells take up the antigen present in the lumen and transport them to other immune cells. Extracellular antigens taken up by the macrophages is processed and presented to T helper cells. Th1 and Th2 are T helper cells that are produced depending on cytokines activated on T cells. IL 12 switches T cells to Th1 cells. The Th2 cells are produced due to the action of IL 2 and IL 4 cytokines. Th1 cells produce IFN gamma. Previously, it was thought that IFN gamma plays a major role in coccidia immunity. However, different immune events occur to elicit immunity. Different cytokines are produced by Th2 cells, which act on the B cells. IL 4 activates B cells. The B cells transform them into plasma cells. IL 5 activates the plasma cells, and these plasma cells produce antibodies. Different cytokines also act on the cytotoxic T cells (Tc). These Tc cells recognize the infected cells and kill them. CTL-mediated immunity is specific to coccidia immunity.

Immunopathology in Protozoan Diseases

In *Trypanosomosis*

Various immunological events cause immunopathology and these immunopathological responses are due to innate immunity and adaptive immunity. In *Trypanosoma* infection, fever is a common sign. Macrophages release pyrogens, which is transported to the brain and alter the temperature-controlling zone. **TNF alpha is responsible for the cachexia condition observed in trypanosomosis**. Anemia is another pathological feature. Apart from mechanical destruction of RBC by trypanosomes, antigen-antibody complexes adhere to the surface of RBC, which are later engulfed by phagocytes. Edema formation is another pathological consequence of *Trypanosoma* infection. Immune complexes activate prekallikrein, resulting in the production of kininogenase, which acts on kininogen and produces kinin. Kinin increases the permeability of vasculature resulting in extravasation of intravascular fluid in the tissue space. Trypanosomes are important parasites occurring in the blood. They cause severe anemia. Lesions occur at the mucocutaneous junctions. Urticarial plaques and edema are also observed.

In *Theileriosis*

In *Theileria* infection, sporozoites invade the lymphocytes where they get transformed to schizonts, which transform the infected cells and induce a state of uncontrolled proliferation. *Theileria* schizonts induce to produce different antiapoptotic proteins, which inhibit the activity of the **P53** protein. This results in massive lymphoproliferation. This massive lymphoproliferation results in the swelling of the lymph nodes. In *Theileria* infection, marked pathology occurs and the affected animals may even die. Lymphoid hyperplasia is a common clinicopathological feature. Petecheal hemorrhages occur in the GI tract, and ulceration of the abomasal wall is characteristic, termed as a "'punched out" ulcer.

In *Babesiosis*

In *Babesia* infection, pathogenesis occurs in different ways. Apart from the mechanical destruction of RBC by *Babesia* spp., opsonization of RBCs due to adherence by antigen–antibody complexes attracts phagocytes. As a consequence, RBCs get phagocytosed. Antigen–antibody complexes also gets deposited in the spleen, kidney, lymph nodes, etc., where they cause tissue damage. *Babesia* infection causes a mild form of the disease in sheep and goats. However, *Babesia bigemina* and *B. bovis* are very important for causing marked pathology in cattle. They cause fever, anemia, hemoglobinuria, etc. In dogs, the clinicopathological features include fever, anemia, jaundice, etc.

In *Eimeriosis*

In *Eimeria* infection, the intestinal wall is affected by the organisms. The sporozoites enter the epithelial cells and form schizonts during the biological developmental process. The affected cells get ruptured and denuded. As a normal immunological phenomenon of the gut, immune cells aggregate at the affected area and cause tissue damage. Pathogenesis depends on the location of parasites, species or strain of parasites, stage of parasites, etc. Since *Eimeria* spp. occur in the gut, general enteritis and hemorrhagic enteritis are some pathological features. In *Eimeria* infection,both cell mediated and humoral immunity occur. Infected epithelial cells release pro-inflammatory cytokines which cause marked inflammation. The antigen is recognised by Th4+ cells which produce different cytokines. Th1 cells produce interferon -gamma which activate the macrophage .The macrophages produce nitric oxide for killing the parasite. At the same time this nitric oxide cause tissue damage .Cytotoxic T cells also release different toxic substances for killing parasite. These toxic substances also cause tissue damage.The immunity is species -specific and long lasting. Anibody produced due to humoral immunity provide protection,but it is less prominent than CMI.

Immunodiagnosis

Some important tests are described below.

Indirect Fluorescent Antibody Test

This test is based on antigen–antibody reaction. The basic principle is like ELISA. The protozoa-infected erythrocytes are cultured or they can be collected from infected animals after the removal of other cells. Smears are prepared on the glass slides, dried, and washed with acetone to remove the hemoglobin. Then, a series of incubation and washing is done. After acetone washing, the suspected antibody (primary antibody) is added onto the smear. After incubation, the slide is washed again. The secondary antibody is added. The secondary antibody is labeled with fluorochrome. After treatment with the secondary antibody, the slide is washed again. Finally, the slides are dried and examined under fluorochrome microscope.

Immunochromatographic Test

In this technique, a paper strip or nitrocellulose paper is used. Two antibodies are used here. One antibody (capture antibody) is immobilized on the paper. Another antibody is the detection antibody labeled with a labeling agent. The sample (antigen) is placed in the sample pad area. The antigen forms a complex with the labeled antibody. **This complex moves on and reaches to previously fixed capture antibody and binds to it in a positive case**. A colored line is formed. That is the indication of a positive reaction.

Other immunodiagnostic test: See chapter "Parasitological techniques".

Some Immunodiagnostic Study

Zou et al. (2016) reported the promising serodiagnostic potential of two recombinant antigens: recombinant *Babesia canis* secreted antigen (**rBcSA 1**) and recombinant *Babesia canis* merozoite surface antigen (**rBcMSA1).** These two antigens showed encouraging results in indirect ELISA and immunochromatographic test (ICT) for the diagnosis of *Babesia canis*. Tebele et al. (2000) worked on the immunodiagnostic potential of *Babesia bigemina* antigen. **P200** is a merozoite protein used for immunodiagnosis of *Babesia bigemina*. However, the main constraint of using this antigen is the variation in antigen quality from batch to batch. The researchers produced recombinant P200 bacterially. They found the potential of this recombinant merozoite antigen for immunodiagnostic studies.

Kumar et al. (2013, 2015) worked on the immunodiagnostic potential of two recombinant antigens of *Theileria equi*. The recombinant antigens, namely, **EMA 1** and **EMA2** (extraerythrocytic merozoite antigen of *Theileria equi*), were expressed in *E. coli* with GST tag. The proteins displayed encouraging results. Immunodiagnostic potential of recombinant merozoite surface antigen (**Tams 1**) for the diagnosis of *Theileria annulata* was reported. The immunodiagnostic potential of this protein was evaluated by ELISA. Bakheit et al. (2004) reported the immunodiagnostic potential of an antigen of *Theileria annulata* (**TaSP**) for the immunodiagnosis of this parasitic infection. The result displayed 99.1% sensitivity and 90.47% specificity in ELISA.

Sullivan et al. (2014) evaluated the immunodiagnostic potential of **sVSG 117** by dual antigen lateral flow test. For this purpose, one more antigen was required. The second antigen used in this study was **rISG 65**. They reported a sensitivity of 97.3% and a specificity of 83.3% for the diagnosis of *T. brucei gambiense*. It is known that the tandem repeat protein of *Trypanosoma* usually has good antigenic properties. Thuy et al. (2012) reported the immunodiagnostic potential of a tandem repeat protein for the diagnosis of *Trypanosoma evansi*. They reported that tandem repeat protein of *T. brucei brucei* **GM6 (TbbGM6)** could be used for the immunodiagnosis of *T. evansi*.

Immunoprotection Studies

Table: Study of immunoprotection of different protozoan infections using different antigens

Name of antigen	Animals	Name of parasites	Finding	Researchers
EtIMP1(Recombinant *Eimeria tenella* immune mapped protein-1)	Birds (poultry)	*Eimeria tenella*	67% reduction in replication of parasites when compared to unimmunized control birds	Kundu et al. (2017)
Recombinant *Eimeria tenella* gametocyte 22 (**Etgam22**)	Eimeria tenella	*Eimeria tenella*	There was a significant increase in anti-EtGam22 IgY antibody, serum IFN-Y, IL-2, TGF-beta, and IL-4 levels in the vaccinated group of chickens	Rafiqi et al. (2019)
Recombinant *Eimeria tenella* **paraflagellar rod protein (TePFR1** and **TePFR 2)**	Mice	*Trypanosoma evansi*	Extended survival and control of parasitemia following lethal challenge.	Maharana et al. (2020)
Recombinant multi-antigen merozoites surface antigen vaccine against *Babesia bovis* (**MSA2a1, MSA2b, MSA2c**)	Bovine	*Babesia bovis*	Exhibited immunoprotection potential characterized by evasion-inhibitory antibody production, elicit IFNgama secreting cells	Gimenez et al. (2016)
Attenuated cell line of Theileria annulata and sporozoite antigen (Spag 1)	Calves	*Theileria annulata*	A combination of Spag 1 and attenuated cell line of *T. annulata* conferred the best protection (synergistic effect) when compared with immunization with a single antigen	Darghouth et al. (2006)

Advances in the Control of Protozoan Diseases

Naked DNA Vaccine Against Protozoa

The term naked DNA vaccine was changed to nucleic acid vaccine at a meeting organized by the World Health Organization in Geneva in 1994. The principle related to this protocol is the introduction of genes through virus vectors. Different viral vectors are used for this purpose. These include **vaccinia, swine pox, capri pox, feline pox, herpes,** etc. The naked DNA vaccine was accidentally discovered while studying with liposomal delivery system of DNA for immunization. The researchers observed that naked DNA is more efficient than encapsulated liposomes. This opened up a new era of naked DNA vaccines.

Different methods of inoculation of naked DNA vaccine (nucleic acid vaccine) are as follows:

1. Cationic liposomes
2. Gene gun
3. Pneumatic jet

Composition of Vaccine

The naked DNA vaccine is composed of three components as follows:

(a) Plasmid
(b) A strong viral promoter flanking the target gene
(c) Polyadenylation transcription termination sequence

Once the complex containing the aforesaid components enters the cell, the viral promoter starts transcription of the gene. After complete transcription, the polyadenylation transcription termination sequence terminates the synthesis. Then the transcribed mRNA binds to the ribosome and starts translation of protein synthesis. Promoters such as SV40, cytomegalovirus (CMV), RSV virus, etc., are very useful and widely used. The plasmids are isolated from bacteria, *Escherichia coli,* and these plasmids are used for the development of the complex of naked DNA vaccine. The target gene is inserted into the plasmid along with promoter gene sequences. If there is no promoter gene, no translation occurs.

Some Information on DNA Vaccination

Plasmodium

In the late 1998s, several DNA-based vaccine candidates, namely, **HEP17, pfCSP, and pfSSP2,** were developed. A modified vaccinia virus was used for transferring the naked DNA construct. CD8+ response was achieved in these trials.

Toxoplasma

A study shows the development of a gene vaccine against *Babesia* sp. A recombinant plasmid was constructed with the gene of P30 surface antigen of tachyzoites of *Toxoplasma* within a tissue plasminogen activator leader sequence flanked by a CMV promoter. The experiment was carried out in mice. A strong antibody response was found.

Trypanosoma

Trans-sialidase (TS) is an important enzyme present in many parasites. Several works were carried out on the use of this enzyme as a vaccine candidate. The enzyme is used for the survival of the parasites in the hosts and the virulence of this parasite depends on this enzyme. Study shows that a DNA construct made using the plasmid and a gene encoding the catalytic domain of TS elicited a good antibody response upon injection. Trypomastigote surface antigen (**TSA-1**) of *T. cruzi* also generates a strong CTL response after inoculation of the plasmid containing the gene of that respective protein.

Recombinant Antigen

The recombinant antigens of *Babesia* are merozoite surface antigen, apical membrane antigen, rhoptry associated protein etc. The recombinant antigens of *Theileria* are merozoite surface antigen, sporozoie surface antigen. Different recombinant antigens of *Eimeria* are as following:

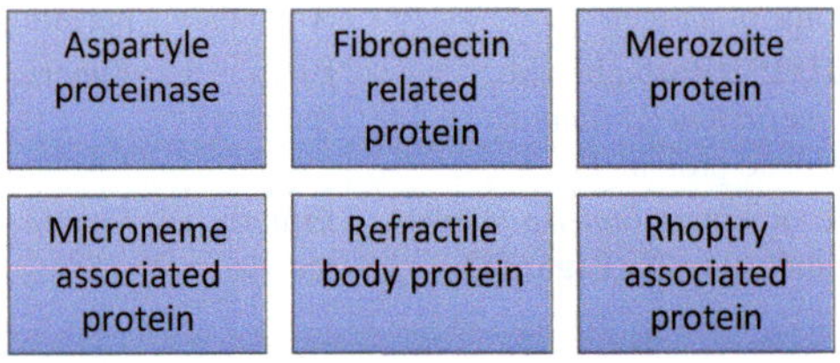

Fig. 8.23 Different recombinant antigens of *Eimeria*

Different Vaccines Against Protozoan Parasites

Coccivac B

This is a live vaccine produced by various workers in the earlier period of the 1940s. The vaccine consisted of different species of coccidian parasites, such as *E. acervulina*, *E. mivati*, *E. maxima. E. tenella,* etc.

Coccivac D

This is also a live vaccine containing sporulated occysts of different species of *Eimeria.* These species include *Eimeria acervulina*, *E. necatrix*, *E. mivati*, *E. hagani*, *E. tenella*, *E. praecox*, *E. brunetti,* and *E. maxima.* The dosage of the parasites is calibrated in such a way that the vaccine once introduced to the birds, at least four life cycles of the parasites would be completed. The vaccine is formulated as freeze-dried pellets.

Nobilis COX ATM

The basic difference between this vaccine and the previous vaccine is that the previous vaccines constitute the strains of the parasite that had never been exposed to any anticoccidial agents, and, obviously, were readily susceptible to anticoccidial agents. However, this vaccine consists of the oocysts that are tolerant to anticoccidial agents. An advantage of this vaccine is that it remains active even in the presence of ionophorus compounds up to 3–4 weeks of age. The vaccine consists of the parasitic species, such as, *Eimeria tenella, E. acervulina,* and *E. maxima.*

Eimeriavax 4m

This is a live precautious *Eimeria* vaccine. It consists of viable oocysts of *E. acervulina, E. maxima, E. tenella,* and *E. necatrix.* This vaccine is administered via eye drops. This vaccine is typically for layer and breeder birds.But this can also be used for broiler birds.

Paracox-8

This is also a live attenuated vaccine consisting of oocysts of eight precocious lines of coccidian species. This vaccine is administered orally for the active immunization of chickens against different *Eimeria* spp. It consists of *Eimeria mitis HP, E. praecox HP, E. brunetti HP, E acervulina HP, E. necatrix HP, E. tenella HP, E. maxima MFP, E.maxima CP.* The vaccine is administered in the drinking water to chickens between the ages of 5 and 9 days inclusive.

Immucox

This is also a live vaccine containing oocysts of several species of *Eimeria.* It is used to protect healthy chicken broilers and Roasters.

Livacox

This is a live attenuated vaccine administered for poultry immunity. The vaccine consists of LIVACOX T for broilers and LIVACOX Q for breeders and layer pullets only. Livacox Q can develop immunity 10–14 days after vaccination. This vaccine consists of four *Eimeria* spp.: *E. tenella, E. maxima, E. acervulina,* and *E. necatrix.* Livacox T consists of three *Eimeria* spp.: *E. tenella, E. maxima,* and *E. acervulina.*

Cox Abic

The first subunit vaccine commercialized for poultry is Cox Abic. The vaccine is prepared by the macrogametocytic antigen of *Eimeria maxima.*

Anaplaz

This is the first commercially available vaccine for cattle against anaplasmosis. This vaccine consists of the organisms, such as *Anaplasma marginale,* harvested from infected cattle. The organisms are killed and processed. The vaccine is administered in cattle against anaplasmosis.

Giardiavax

Killed cultured trophozoites of *Giardia lamblia* are used for the preparation of this vaccine for dogs. It is given in the subcutaneous route to healthy dogs at the age of 8 weeks or older. The second dose is given within 2–4 weeks after the first dose.

Nobivac Piro

This vaccine is prepared from the supernatant part of in vitro cultured parasite antigen. The antigen is prepared from *Babesia canis* and *B. rossi.* The first dose is given at the age of 6 months and the booster dose is given 3–6 weeks after the first dose of vaccination. The vaccine is given at subcutaneous route in dogs against babesiosis.

Bovilis Neoguard

Killed tachyzoite of *Neospora caninum* is used for preparation of this vaccine with SPUR adjuvant. The vaccine is administered in cattle against neosporosis.

Toxovac

This is a live attenuated anti-*Toxoplasma* vaccine. It contains the live attenuated S48 strain of *Toxoplasma* administered to sheep. The dose is given intramuscularly into the anterior half of the neck.

Nanotherapy

Development of Nanodrug

Very limited works have been carried out on the application of nanotechnology in the field of veterinary parasitology; but very recently, a speedy research work has been directed towards nanoparasitology with sufficient funds.

1. Enhancement of the efficacy of different drugs with the help of nanoparticles has been proved.
2. Use of different nanovehicles has proved to reduce the toxicity of antiparasitic drugs.
3. Nanoparticles and nanovehicles are used to increase the bioavailability of the drug for a long period. Thus, the duration of drug activity can be increased.
4. Development of smart antiparasitic drugs is possible.
5. The active gold-coated glass particles commonly known as nanoshells can open a new window to target the parasite without affecting/reducing host tissue damage. Ample works were carried out and many present works are already in progress to envisage the activity of nanoshells. The nanoshells are being used as biosensors in the study of host-parasite interactions.

Delivery of drugs using nanocapsules confers high efficacy against different parasitic pathogens. Reports indicate good efficacy against *L. donovani* both in vitro and in vivo using **mannose grafted chitosan nanocapsules (MnosCNc)** containing AmB. Nanocapsules can be produced from different materials, such as chitosan, sodium alginate, etc. Quinapyramine sulfate encapsulated within nanocapsules shows stronger in vivo efficacy against *T. evansi* than the drug alone. The side effects of quinapyramine are high but when this drug is administered in nanocapsules, a negligible amount of side effects is found. **Polymeric micelles** are another type of nanomaterial, which appears to be a good nanodelivery system, that has garnered considerable attention as a potential drug carrier. Micelles form spontaneously in aqueous solutions and are composed of amphipathic copolymers. Polymeric micelles serve as an effective treatment for *Leishmania* infection. Drug releasing potential of the nanoparticles depend on the binding forces between the drug molecule and the nanoparticle. The efficacy of gold nanoparticles against *Leishmania* sp. has been reported. Several works were carried out on antiparasitic efficacy of silver nanoparticles. In the study of Abbasalipourkabir et al. (2016), the **solid lipid nanoparticles (SLNs)** were used for encapsulation of nitazoxanide (NTX), which is a new antiparasitic and antiprotozoal drug delivery system. Lipid nanodiscs are a potentially useful delivery system for the treatment of intrahistiocyte pathogen. It is composed of a phospholipid bilayer surrounded by the amphiphilic motifs of apolipoproteins. The study shows the use of nanodiscs for the delivery of anti-*Leishmania* drugs. Delivery of drugs using nanocapsules confers high efficacy against different parasitic pathogens. Reports indicate good efficacy against *L. donovani* both in vitro and in vivo using mannose grafted chitosan nanocapsules (MnosCNc) containing AmB. Metallic nanoparticles include inorganic nanoparticles, such as gold, silver, magnesium, titanium, and zinc, etc. Drugs are adsorbed or covalently bound to the surface of these particles. These drug-bound particles remain as colloidal suspension in liquid. Drug releasing potential from the nano particles depend on the binding forces between the drug molecule and the nanoparticle. The efficacy of gold nanoparticles against *Schistosoma* sp., *Leishmania* sp., etc., has been reported. Several works were carried out on antiparasitic efficacy of silver nanoparticles.

Molecular Protozoology

Molecular Study of Biology

The molecular study of biology depicts variations in gene expression during the biological development of parasites. Different stages of parasites perform different functions. Therefore, different functional proteins are required in different biological stages of development. For example, function of tissue invading parasite is entirely different from that of parasites in the stationary phase. Tissue-invading parasites require different enzymes for penetration. Therefore, genes encoding those proteins are transcribed before/during this activity, so that the parasites can perform their specific functions during different developmental stages.

The gene expressions can be analyzed from RNA seq.data. **RNA seq.data analysis** is important for the determination of mRNA expression. Genes are transcribed and spliced in eukaryotic cells, resulting in the production of mature mRNA transcripts. In this technique, the first step is isolation of mRNA. Then, the isolated mRNA is digested into fragments, and these fragments are transformed into ds-cDNA following standard protocol. Using high-throughput, short read sequencing method, the ds-cDNA is sequenced. These sequences are aligned with a reference genome to reconstruct which genome regions were being transcribed. By using RNA-seq. data study and **KEGG analysis**, it has been found that number of gene expression and category of gene expression of merozoite (second) and merozoite (third) are different in some protozoa.

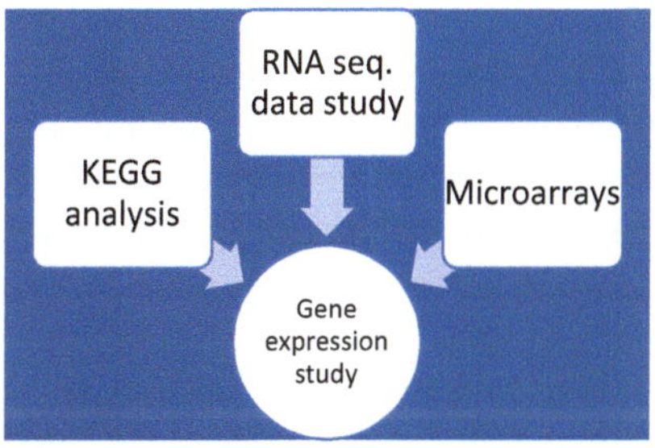

Fig. 8.24 Gene expression study

Difference of gene expression in protozoan parasite has been described by different researchers. Some genes are specifically expressed in different biological events. For example, HAP2 gene is expressed during fertilization of some protozoan parasites. Hussein et al. (2017) documented that **hap2** gene was transcribed during development of *Babesia bovis* within tick midgut. But transcription of this gene did not occur during piroplasmic stage. Therefore, it is clear that some genes are expressed in particular stage/stages of development.

In another study, Camacho-Nuez et al. (2017) found that in *B. bigemina*, Hap2 expression occurred during the formation of sexual stages and not in blood stages. For this study, DNA was isolated from bovine infected with *B. bigemina*. The Hap2 gene was amplified, cloned, and sequenced. Subsequently, the expression of hap2 in the host and vector tissue was analyzed using different techniques including semi-quantitive Rt-PCR, Western blot analysis, immunofluorescence study, etc. These studies revealed that HAP2 was expressed both in sexual stages induced in vitro and in infected ticks. However, no expression was detected in asexual erythrocytic stage of *Babesia*.

Martins et al. (2015) studied the role of a **surface membrane protein (TcSMP)** in *Trypanosoma cruzi*. This is a membrane-spanning protein located at the cellular surface and is released into the extracellular milieu and this TcSMP appears to be the key elements, typical of surface proteins (N-terminal signal peptide or signal anchor). These researchers demonstrated that TcSMP proteins are capable of inhibiting entry of metacyclic trypomastigote into host cells. They also reported that TcSMP binds to mammalian cells capable of triggering Ca^{2+} signaling and lysosome exocytosis, which are responsible for parasitophorous vacuole biogenesis. The researchers concluded that TcSMP may play a role in the invasion of host cell.

KEGG Analysis

The full form of KEGG is Koyoto Encyclopedia of Genes and Genomes. KEGG constitutes the data set of genome, protein, chemical, disease, etc. KEGG pathway maps are designed based on experimental output in organism. The ortholog status of one organism is utilized for other organisms. The data of a particular organism are used to interpret dataset of other organisms.

Molecular Taxonomy

Like other parasites, there has been radical progress in taxonomic study of protozoa. In the earlier period, taxonomy was based on morphological features mainly. Later on, biochemical and immunological characterization of different organisms were done. Molecular characterization of different protozoan parasites has been a trend for identification and placement in molecular phylogenetic tree. Several works have been carried out on molecular identification and characterization of parasites. **ITS1, ITS2, CO1, ND1, and 5, 5S, 5.8S, 18S** are used as genetic markers for molecular characterization of protozoa.

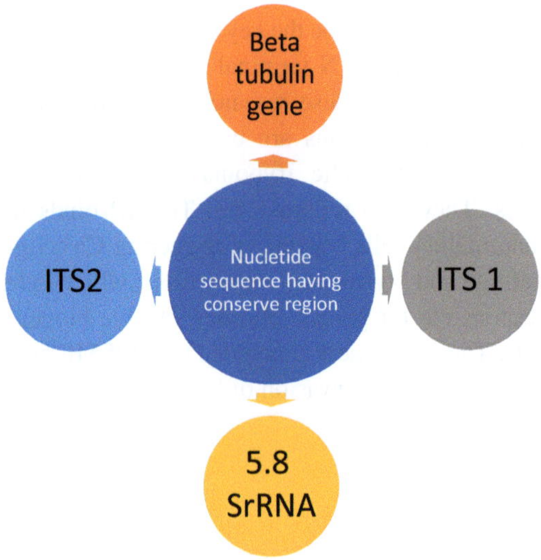

Fig. 8.25 Conserve nucleotide sequence

Some Molecular Phylogenetic Study

Villareal et al. (2013) sequenced complete ITS rDNA including 5.8S of Philipine isolates of *Trypanosoma evansi*. The sequence data displayed 99%–100% similarity between Philipine isolates and known *T. evansi* isolate available in Gene Bank. George et al. (2015) carried out molecular and phylogenetic analysis of *Theileria annulata* from India using 18SrRNA as a molecular marker. Their results showed novel genotypes of *T. annulata* strains from India. Phylogenetic study revealed closed proximity of 13 strains to a strain of China. One Indian strain was similar to *Theileria* sp. from buffalo in South Africa. Molecular phylogenetic studies were also carried out for different *Babesia* spp. The studies used different molecular markers, such as ITS1, ITS2, 18SrRNA, etc. The works depict homogeinity and heterogeneity within the species of different isolates of different areas.

Molecular Pathology by Protozoa

Cytokines are very important in precipitating any pathology. These are small peptide proteins produced by different immune cells. These vital proteins work in different ways by creating an orchestration amongst the immune cells. Thus, the immune cells can respond and cross talk each other. Inflammation is beneficial to the host, but if imbalance of release of cytokines occurs, pathogenesis ensues. Cytokines, as a rule, differentiate T cells, such as Th1 and Th2 cells. There is a dichotomy between Th1 cells and Th2 cells. But if dysregulation of cytokines occurs between these two types of cells or different regulatory cells and substances, the overactive and autoreactive cells are not cleared up. This leads to pathological events, which is called as immunopathology. There are different immunological reasons for the occurrence of pathology in the tissues. These include ADCC, CMI, and nitric oxide produced by macrophages, toxins released by different effector cells, respiratory burst, etc.

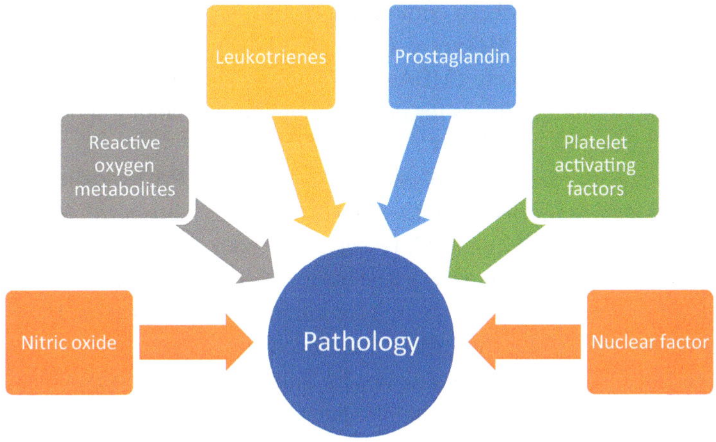

Fig. 8.26 Different pathological factors

In the earlier period, nonspecific innate immunity was considered as the sole immunity. Later, toll-like receptors (TLRs) were discovered. These receptors are present in different specific immune cells. **TLRs have diverse expression patterns that recognize pathogen-associated molecular patterns (PAMPs) of different protozoan pathogens**. Each TLR recognizes a specific PAMP. A specific signal is generated when a PAMP binds to specific TLR. This specific signal results in the release of cytokines from specific immune cells. TLR-mediated recognition creates a link between innate and adaptive immunity because it has been observed that TLR recognition causes maturation of dendritic and other immune cells and T helper cell response occurs.

Protozoan Zoonosis

Cryptosporidiasis

Causative Agents

The main pathogens responsible for causing human cryptosporidiosis are *C. parvum* and *C. hominis*. However, *C. meleagridis*, *C. canis*, *C. felis*, and *C. muris* also occur in humans.

Mode of Infection

The hosts get the infection through fecal oral route. The infective stages of this protozoan are sporulated oocysts. There are reports of autoinfection in the hosts.

Pathological Features

Enteritis is a common pathological feature that results in diarrhea. The brush border of the epithelial cells in the intestine is markedly affected. Intestinal cells are massively damaged, which may lead to edema, ulceration, etc. The marked lesions are found in the jejunum and ileum. Affected animals and human beings exhibit symptoms such as diarrhea, anorexia, dehydration, and abdominal discomfort.

Diagnosis

Symtomatological Diagnosis

The exhibited clinical signs are not specific for confirmation of this parasitic infection.

Parasitological Diagnosis

Parasitological diagnosis is done through microscopical examination of feces for the detection of oocysts by staining. Examination of fecal sample after modified **ZN staining reveals red colored oocysts.** Various concentration methods for the examination of fecal sample can be used. These include Sheather's sugar floatation technique, modified formalin–ethyl acetate concentration technique, modified zinc sulfate centrifugal floatation technique, etc.

Immunological Diagnosis

Many works have been carried out using crude and purified antigens. However, no immunodiagnostic technique has been commercialized so far for field application. Moss et al. (2004) performed a multiplex bead assay (MBA) to analyze oral fluid and serum specimens for immunoglobulin G (IgG) antibodies specific to *Cryptosporidium parvum* antigens coupled to polysterene beads. The recombinant antigens were *C. parvum* 17- and 27-kDa antigens (r17 and r27, respectively) both linked with glutathione-S-transferase (GST) fusion proteins. This was assessed for immunodiagnostic potential. The result concluded that MBA using sera or oral fluid collected from children may be an option for immunodiagnosis and epidemiological study.

Molecular Diagnosis

Molecular diagnosis is possible using different molecular markers. Different ribosomal DNA, ribosomal RNA, mitochondrial DNA could be tried as molecular markers.

Control

General Control

Proper treatment should be done using the specific drug available so far. Personal hygiene should be maintained. Additionally, food and drinking water should be free from contamination.

Immunological Control

Different crude and defined antigens have been used in different research trials but no vaccine has yet been commercialized so far.

Molecular Control

So far, no naked DNA vaccine or gene vaccine is available.

Isosporiasis

Causative Agents

The important species is *Isospora beilli* occurring in human and primates.

Mode of Infection

The hosts get the infection through fecal-oral route by the ingestion of sporulated oocysts containing sporocysts.

Pathological Features

Isospora beilli causes diarrhea in humans. Hemorrhagic diarrhea, abdominal tenderness, etc., are some commonly observed clinical signs in humans. Diarrhea may occur in primates.

Diagnosis

Examination of fecal sample may reveal the presence of oocysts. The oocysts have two sporocysts, each containing four sporozoites.

Control

Food and drinking water should not be contaminated with oocysts. Proper washing of the utensils should be done.

Acanthamoebiasis

Causative Agents

Acanthamoeba castellani and *Acanthamoeba polyphaga* are two important zoonotic species predominantly occurring in humans. However, these parasites may also occur in dogs, thus having zoonotic significance.

Transmission

These are amoebic organisms available in soil, water, air, etc., having a ubiquitous distribution. These organisms are opportunistic protozoans rarely affecting humans. The route of infection of *Acanthamoeba* spp. in humans includes skin, wounds, lesions, etc. *Acanthamoeba* keratitis frequently occurs in contact lens wearers. Also, the organisms may enter the human body through the nasal orifice. Autochonous infection occurs in dogs, which get the infection by the ingestion of the parasitic stage. The dogs get infection also through inhalation.

Pathological Features

In Humans

In humans, the organisms may enter through the nasal orifice and enter alveolar blood vessels. They then cross the blood–brain barrier and affect the nervous system. Granulomatous amoebic encephalitis occurs. Acanthomoeba keratitis may be characterized by corneal ulcer. This may result in blindness.

In Animals

The organisms affect the liver, lung, heart, etc., of dogs, where they cause hemorrhagic foci. Muscle stiffness, breathing difficulty, eye disease, seizures are some clinical features.

Diagnosis

Examination of cerebrospinal fluid can be done, and motile amoeba can be detected during examination. Molecular diagnosis can be done by PCR using specific molecular marker. Cerebrospinal fluid is used as the sample source of parasite.

Control

Proper treatment, maintenance of hygiene, etc., can reduce the chance of this parasitic infection.

Dientamoebiasis

Hosts Affected

Dientamoeba fragilis affects the cecum and colon of humans and monkeys.

Mode of Infection

Direct fecal oral route is considered as the route of infection of *Dientamoeba fragilis*. Howerver, exact route is not known.

Pathological Features

The organisms seem to be nonpathogenic.

Diagnosis

Examination of fecal sample reveals the binucleated trophozoite stage of the organisms after staining with trichrome stain.

Control

Proper hygiene is important. Care should be taken so that food and drinking water do not become contaminated.

Entamoebiasis

Causative Agents

The important species are *Entamoeba histolytica*. These parasites occur in the large intestine of humans and primates. However, transient infection may occur in dogs and cats.

Mode of Infection

The hosts get the infection through fecal oral route by the ingestion of cysts of the parasites.

Pathogenesis and Clinical Signs

In Humans

The intensity of pathogenicity depends on many factors. Infection load, virulence of the organisms, age of the hosts, and the presence of bacterial flora are some factors responsible for the magnitude of pathogenicity. The infection remains asymptomatic in most cases. The disease may occur in both acute and chronc phase. The individuals may recover from acute phase, but chronic dysentery occurs later. They act as the cyst-passers remaining asymptomtic. In general, the symptoms include abdominal discomfort, flatulence, loose stool with mucus and blood, etc. There may be ulcer and carcinoma in the large intestine. Apendicitis may also be caused by this parasitic infection. Human beings are mainly affected by amoebic liver abcess. The clinical signs comprise mild evening rise in temperature or low remittent temperature, pain in the right shoulder, and initial dry cough. Jaundice sometimes occurs. Tenderness of the liver is another clinical finding. Right-sided chest movement of the patient is restricted or absent. Pulmonary amoebiasis occurs if the organisms are transported to the lung from the gut wall via portal circulation. The organisms cause lysis of pulmonary tissues, resulting in the formation of single or multiple abscesses. Secondary pulmonary amoebiasis occurs not by transportation through circulation but by adhesion of lung surface with the liver abscess. It is the extension of amoebic liver abscess up to the lung. Human beings may also suffer from cerebral amoebiasis, splenic amoebiasis, etc.

In Animals

The dogs probably get the infection from human beings. The infected dogs are asymptomatic, but an outbreak of dysentery was reported in cattle.

Diagnosis

Symtomatological Diagnosis

The clinical signs may not confirm the case of *Entamoeba* infection, but the symptoms may be suggestive of this infection.

Parasitological Diagnosis

(a) Feces should be examined properly, which may reveal the cysts of the parasites. Gross examination of feces reveals **foul smelling, copious, and dark red feces**. Microscopical examination reveals trophozoite and quadrinucleated cysts. In the case of fresh feces, wet mount by saline may reveal the movement of trophozoite. Six to seven samples should be examined for better sensitivity.
(b) Concentration technique by formalin ethyl acetate sedimentation method can be used.
(c) For observing the internal structure appreciably, the material can be stained with trichrome or iron hematoxylin.
(d) Histology can be done by examining biopsy material from colonic mucosa after staining with **H&E or PAS stains**.
(e) Culture of organisms can be done using different culture media, such as Boek and Drbohlav egg serum containing Locke's solution, Balamuth's medium, Diamond's medium, Robinson's medium, etc.

Immunological Diagnosis

(a) Different immunological tests, such as ELISA, IHA, and IFA, are used for extraintestinal amoebiasis. ELISA is done for the detection of **170 kDa lectin** antigen.
(b) Coproantigen can be detected by ELISA.
(c) An immunochromatographic test using specific antigen can be done. These tests are commercially available. Immunochromatographic test is done for the detection of **29-kDa** *E. histolytica* surface protein.

Molecular Diagnosis

Molecular diagnosis is possible using various molecular markers. Different ribosomal DNA, ribosomal RNA, and mitochondrial DNA could be tried as molecular markers. **Multiplex real-time PCR using 18SrRNA** have been described.

Imaging Method

A colonoscopy can be done to find a collar button or flask-shaped ulcer.

Control

General Control

Affected persons should be treated properly. Good sanitary measures and proper sewage disposal should be done.

Immunological Control

Different crude and defined antigens have been tried, but no vaccine has yet been commercialized.

Molecular Control

It seems no naked DNA vaccine or gene vaccine is available.

Iodamoebiasis

Causative Agents

Iodamoeba butschelii is an important species. This parasite affects the lower digestive tract of humans, pigs, monkeys, etc.

Pathogenesis

These parasites are nonpathogenic.

Mode of Infection

The route of infection is the fecal-oral route. The hosts get the infection by the ingestion of cysts of the parasites.

Diagnosis

Examination of fecal samples may reveal the presence of uninucleated cysts of the parasites.

Control

Maintenance of hygiene should be done. Pure food and drinking water should be consumed.

Giardiasis

Causative Agent

The causative agent is *Giardia lamblia.*

Hosts

Giardia lamblia affects the duodenum and other parts of the small intestine of the humans. However, this parasite also occurs in monkeys, dogs, cats, birds, etc. In some cases, the parasites are also found in the large intestine.

Mode of Infection

The infection is transmitted through the fecal-oral route. The cysts are resistant to chlorination and ozone treatment. The hosts get the infection by the ingestion of the cystic stage of the parasite.

Pathological Features

In Humans

In human beings, the parasites cause diarrhea without blood. The diarrheic material is gruel-like mixed with mucus and fat. A large amount of fat is found in the feces since fat is not properly absorbed. The liver and bile duct may also be affected. The diarrhea may be continuous or discontinuous. Abdominal pain, nausea, and flatulence are some other clinical features.

In Animals

In the animals, this parasitic infection appears to be nonsymptomatic. Occasionally, there may be diarrhea.

Diagnosis

(a) Examination of fecal samples may reveal the cyst stage of the parasites. Like amoebiasis, RBC and pus cells are not found. Concentration techniques such as **formalin ether sedimentation and zinc sulfate floatation techniques** are used.
(b) **Duodenal aspirate** can be examined for the detection of cysts.
(c) **A string test** can be done. This is simple procedure. A thread is attached to a capsule. The patient is asked to swallow the capsule. The capsule get dissolved in the stomach. The thread reaches the duodenum due to peristaltic movement and the weight attached to it. Then the thread is removed and the attached material is placed in the saline, which is examined for the detection of the trophozoite stage of the parasite.
(d) The coproantigen is detected by ELISA. For this purpose, monoclonal antibodies against cyst wall protein are used. This test is available and found reliable.
(e) Rapid immunochromatographic test, commercially named "Tribe Parasite Panel," is available for simultaneous diagnosis of *Giardia*, *Cryptosporidium*, and *Entamoeba histolytica*.
(f) ELISA test is also available for the detection of antibodies. However, the problem is that the test cannot differentiate between recent and old infections.
(g) *Giardia* can be cultured using different media. Of these culture media, Diamond's medium is commonly used.

Control

Affected individuals should be treated properly. Hygiene and sanitation should be maintained properly. Care should be taken not to contaminate food and drinking water.

Balantidiasis

Causative Agents

The causative agent is *Balantidium coli* occurring in the large intestine of humans and pigs.

Mode of Infection

Infection occurs through oral route by the ingestion of cysts of the parasite.

Pathogenesis

In Humans

These organisms remain as commensals and do not cause any pathology. They may occasionally flare up and cause enteritis. Enteritis may turn to superficial or deep ulceration in the chronic stage of infection. Bloody or watery mucoid diarrhea is another clinical feature. Humans exhibit symptoms comprising concurrent nausea, epigastric pain, vomiting, and intestinal colic.

In Animals

The organisms are found in the pigs as commensals. However, occasionally the organisms may invade the intestinal wall and cause superficial or deep ulcers. There may be simple or hemorrhagic enteritis.

Diagnosis

Examination of fecal samples may reveal the presence of cysts of the organisms. The infection is common in the pigs. Incidence is less in human beings compared to pigs. Fresh fecal samples should be examined as the organisms are easily destroyed.

Control

Feces should be disposed of properly. Drinking water should be free from contamination. Contact with the infected person should be avoided.

Plasmodiasis

Hosts Affected

The species occurring in nonhuman primates are *P. knowlesi, P. cynomolgi, P. simium*. These species also occur in humans.

Mode of Infection

The vectors of the parasite are different mosquitoes. The infected mosquitoes transmit sporozoites (infective stage of the parasite) to the host while taking blood meal from them. Different stages of the parasite are developed. Gametocytes are formed in the RBC, which are taken up by the mosquitoes. In the mosquitoes, subsequent stages are formed and eventually the sporozoites are produced, which are the infective stages of the parasite as mentioned earlier.

Pathological Features

In Humans

Hepatomegaly and splenomegaly are the common pathological features. Headache, fever, joint pain, back pain, and severe weakness are the common clinical features. *P. cynomolgi* causes tertian malaria.

In Animals

In the case of *P. knowlesi*, quotidian malaria occurs. Cerebral involvement is not found. Renal and respiratory involvement are found. *P. cynomolgi* causes tertian malaria in monkey.

Diagnosis

Examination of blood smears after Romanowsky staining may reveal the parasitic stages. For *P. knowlesi*, blood smear examination reveals different stages of the parasite. Late trophozoites are band forms and look like *P. malariae*. Morphologically, it is difficult to distinguish from other species. Early trophozoite stages present as ring form, accole form, etc., which are similar to *P. falciparum*. Specific nested PCR is available.

Control

Since mosquitoes are the vectors, vector control by insecticides should be done. Affected persons should be treated properly. Mosquito nets should be used.

Sarcocystosis

Causative Agents

The important zoonotic spp. are *Sarcocystis hominis*, *S. porcihominis*, *S. bovihominis*, *S. nesbiti*, and *S. lindemanni*.

Mode of Infection

For the completion of the life cycle of this parasite, one intermediate host is required. The sporulated oocysts or sporocysts are expelled from the hosts. The sporulated oocysts/sporocysts are the infective stages. After ingestion, the sporozoites are liberated and develop into schizonts in the respective organs. Sarcocysts are formed in different skeletal muscles. The final hosts get the infection by the ingestion of infected muscles of intermediate hosts containing the cysts. Bradyzoites are formed in the cysts. Gametogony and sporogony occur in the intestines of the definitive hosts.

Pathological Features

In Humans

Human beings can act as the definitive hosts of the species such as *S. hominis*, *S. suihominis,* etc., and they get the infection by eating undercooked meat harboring the cysts of the parasites. However, human beings may act as the intermediate hosts of *S. nesbitti*, a species with reptiles as the definitive hosts. Humans get the infection by the ingestion of food contaminated with sporocysts. The symptoms include fever, headache, etc. Invasive forms include myositis, vasculitis, etc. Pathological features due to invasive forms include diarrhea, abdominal pain, etc. Myositis and myonecrosis also occur.

In Animals

Eosinophilic myositis. Muscle tremors, weight loss, dyspnea, etc., are found.

Diagnosis

Symtomatological Diagnosis

The exhibited clinical signs are nonspecific, which cannot confirm the disease.

Parasitological Diagnosis

(a) Feces should be examined properly, which may reveal the presence of the sporocysts of the parasite.
(b) Examination of biopsy material stained with hematoxylin-eosin or some other stains may reveal the sarcocysts containing bradyzoites.
(c) **Image-based diagnosis** by X-ray may reveal the presence of the parasitic cysts. However this is not always reliable.

Immunological Diagnosis

Many works have been carried out using crude and purified antigens. However, no immunodiagnostic technique has been commercialized so far for field application.

Molecular Diagnosis

Molecular diagnosis is possible using different molecular markers. Different ribosomal DNA, ribosomal RNA, and mitochondrial DNA could be used as molecular markers.

Control

General Control

Uncooked or undercooked meat should not be consumed.

Immunological Control

Different crude and defined antigens have been used in different immunoprotection trials, but no vaccine has yet been commercialized.

Molecular Control

Works were also carried out on gene immunization. However, so far no naked DNA vaccine or gene vaccine is available.

Toxoplasmosis

Causative Agents

The causative agent is *Toxoplasma gondii.*

Mode of Infection

The definitive hosts get the infection by the ingestion of sporulated oocysts or by the ingestion of the muscles of the intermediate hosts containing bradyzoites, tachyzoites, etc. The intermediate hosts get the infection by the ingestion of sporulated oocysts. In human beings transplacental infection occurs. Human beings also get the infection by the ingestion of infected muscles containing bradyzoites or tachyzoites.

Pathological Features

In Humans
Pathological features are variable. Abortion is caused in human beings. Severe form of disease is found in congenitally infected children. Encephalitis occurs in immunosuppressed humans. Fever, malaise, muscle pain, head ache, drowsiness, disorientation are common clinical features.

In Animals
Abortion, retained placenta, and mummification of the fetus occur in ewes. In cattle, the brain and respiratory system are affected mostly. Pneumonia, hemorrhagic tracheitis, and formation of glial nodules commonly occur in the brain. In cats, enteritis, pneumonia, and nephritis occur. In dogs, lungs and brainsare affected, resulting in gliosis in the brain, pneumonia, and nodule formation in the lung.

Diagnosis

Symtomatological Diagnosis
The exhibited clinical signs are not so specific for the confirmation of this parasitic infection.

Parasitological Diagnosis
(a) Feces should be examined properly, which may reveal the presence of the eggs of the parasite.
(b) Image-based diagnosis by CT scan analysis can reveal the presence of parasitic cysts.

The organisms are detected in lymph nodes, brain, liver, etc. **Comma-shaped** tachyzoites are detected in blood, body fluid, etc.
(c) Tissue cyst containing distinct bradyzoites is detected by biopsy tissue examination.

Animal Inoculation
T. gondii-suspected clinical sample is identified and inoculated intraperitoneally in laboratory mice. Organisms can be detected in spleen aspirate and peritoneal fluid.

Culture
Human fibroblast cell lines, such as vero, HeLa, etc., are used for the culture of *Toxoplasma gondii.*

Immunological Diagnosis
(a) Antibody detection by ELISA for the detection of IgG, IgM, and IgE is done.
(b) Antigen detection cannot be done successfully because the antigen does not have sensitivity.
(c) Direct fluorescent antibody test is done following standard protocol.
(d) In human beings, TORCH (*Toxoplasma*, rubella, cytomegalovirus, herpes) screen is performed for the detection of IgG and IgM antibodies against those mentioned antibodies during pregnancy.

Molecular Diagnosis
Molecular diagnosis is possible using different molecular markers. Different ribosomal DNA, ribosomal RNA, and mitochondrial DNA could be used as molecular markers.

Control

General Control
(a) Proper treatment should be administered using specific drugs available so far.
(b) Proper disposal of cat feces should be done.
(c) Routine examination for the detection of *Toxoplasma* infection in cats should be done.
(d) Contact with infected cats should be avoided.

Immunological Control
Different crude and defined antigens have been used in various immunoprotection trials. A live vaccine (Toxovac) has been developed in the UK for ewes and this vaccine is mostly used to provide protection during pregnancy.

Molecular Control
Works were also carried out on gene immunization. However, so far no naked DNA vaccine or gene vaccine is available.

Trypanosomiasis

Causative Agents

The important zoonotic species are *Trypanosoma gambiense*, *T. rhodesiense*, *T. cruzi*, etc. *Trypanosoma rhodesiense* has been recovered from game animals, cattle, etc.

Transmission: *Trypanosoma rhodesiense* and *T. gambiense* are transmitted by *Glossina* flies. *T. cruzi* is transmitted by Reduviid bugs.

Pathological Features

In Humans

T. gambiense and *T. rhodesiense* causes sleeping sickness. *T. cruzi* causes Chagas disease. *T. rhodesiense* causes swelling of lymph nodes particularly at the neck, groin, and limbs. This condition is called as Winterbottom's signs. There may be fever. Nervous system is less affected as compared to that in *T gambiense* infection. The affected persons may die within a few months if left untreated. In the case of *T. gambiense* infection, initially lymph nodes are affected. The initial clinicopathological features include headache, irregular fever, joint pain, etc. Gradually the parasites affect the nervous system. Leptomeningitis occurs. The clinical symptoms include dullness, depression, incoordination of movements, paralysis, and convulsion. Continuous sleepiness starts.

The acute form of Chagas disease is characterized by a high fever for about 2 weeks with swelling of the body. The face becomes swollen. Marked edema of eyelids occurs, which is called as Romana's sign. Enlargement of the liver and spleen occurs. Leucocytosis and lymphocytosis occur in early Chagas disease. Every organ may be affected by the organisms. Destruction of cells or organs results in functional disability and death. "Mega" disease condition occurs, which is characterized by megaesophagus, megacolon, megaureter, etc. Heart may be affected and myocardial cells are parasitized. The chronic phase starts later on when organisms disappear from the peripheral blood. Cardiomegaly occurs. Myocarditis is found. **"Lipochagomas"** may occur due to the invasion of organisms in adipose tissue.

In Animals

Infection of animals (cattle, sheep, horses, dogs, etc.) with *T. gambiense* was reported, but more works are required to establish the zoonotic significance. Several wild animals serve as the reservoir hosts of *T. rhodesiense*. This species is pathogenic to sheep, goats, horses, cattle, dogs, etc. *T. cruzi* may affect dog. Spleenomegaly is a common pathological feature. Anemia has been observed.

Diagnosis

Symtomatological Diagnosis

The exhibited clinical signs are not confirmatory for diagnosis. However, different pathological features as described earlier may be suggestive of this parasitic infection.

Parasitological Diagnosis

Detection of organisms in blood and lymph during the acute phase, and detection of organisms in CSF during the chronic phase, is a gold standard technique.

Immunological Diagnosis

Many works have been carried out using crude and purified antigens. Several immunodiagnostic tests, such as ELISA, IFA, CFT, Western blot, are tried for diagnosis.

Molecular Diagnosis

Molecular diagnosis is possible using different molecular markers, i.e., ribosomal DNA, ribosomal RNA, and mitochondrial DNA could be used as molecular markers.

Control

Since the sand flies are the vectors, control of sand flies using insecticidal agents is very important. Several crude and defined antigens have been used but no vaccine has yet been commercialized. Works were also carried out on gene immunization. However, so far no naked DNA vaccine or gene vaccine is available.

Babesiosis

Causative Agent

Babesia microti is an important zoonotic species of *Babesia*.

Mode of Infection

The pathogen is transmitted by a hard tick *Ixodes scapularis.*

Pathogenesis/Clinical Signs

Weakness, malaise, and fever are some clinical signs. Subsequently, the affected person exhibits a chill feeling and sweating (a characteristic of malaria signs). In severe form occurring in immunodeficient individuals, severe anemia, severe respiratory distress, and cardiac anomalies comprising congestive heart failure are observed. There have been reports of *B. bovis* and *B. divergens* infection in spleenectomized persons. This infection results anemia, pyrexia, haemoglobinuria, and jaundice etc.

Diagnosis

Examination of blood smear should be done, which can reveal the organisms.

Endolimacosis

Hosts Affected

Endolimax nana is a parasite of humans and monkeys and occurs in the hindgut.

Mode of Infection

The hosts get the infection through the fecal-oral route by the ingestion of cysts of the parasites.

Pathological Features

This parasite seems to be nonpathogenic. However, intermittent or chronic diarrhea may occur.

Diagnosis

Examination of the fecal samples may reveal the presence of parasitic cysts.

Though this is not a protozoan parasite, but considering it's relevance, the parasite is described here.

In-Vitro Culture of Protozoan Parasites

Entamoeba histolytica

There are various culture media for the culture of *Entamoeba histolytica*. It is urgently required to culture *E. histolytica* for immunological and molecular research work. The names of the cultures that could be used are Boek and Drbohlav's medium, Balamuth's medium, Diamond's medium, and Cleveland's medium. Two types of culture media are used for the culture of *Entamoeba histolytica*: xenic and axenic. Xenic culture media contains unknown flora And include Balamuth's medium, Modified Boek and Drbohlav's egg diphasic medium, TYSGM-9, Jones's medium are some xenic culture media. Axenic cultures do not contain any unknown flora except the organisms to be cultured. Some media for culture of luminal parasitic protists have been mentioned in the table with several information from the review article of Ahmed (2014)

Name of medium	Type	Composition
Balamuth's medium	Xenic liquid	Phosphate buffer containing monobasic and tribasic potassium phosphate, egg yolk, and whole liver concentrate solution
Boek and Drbohlav's medium	Xenic medium	Locke's solution, human serum inactivated, eggs, rice powder, and antibiotics (penicillin, streptomycin, acryflavin)
Cleveland's medium	Monoexenic diphasic	Thioglycollate, penicillin-inhibited streptobacilli bacteria
Diamond's medium	Axenic liquid	Bovine serum inactivated, yeast extract, trypticase, and minerals
Diamond's medium modified	Axenic liquid	Casein peptone, maltose, cysteine L hydrochloride, yeast extract, streptomycin, etc.

Balantidium coli

Many efforts have been made to culture *Balantidium coli*. This parasite can be cultured in Boek and Drbohlav's medium. The culture of this parasite is very easy. Homemade culture media, comprising water, fecal matter, and serum, can also be used for the culture of this parasite. These are commonly available simple ingredients that could be used for the culture of *Balantidium coli*. Three other xenic media, TYSGM9, LES, and Pavlova media, are useful for the culture of *Balantidium coli*.

Tritrichomonas foetus

Tritrichomonas fetus is cultured in Diamond's medium at 37 °C. The culture media should be routinely monitored for organisms every day for up to 5 days. Claussen's medium is also used for the culture of this protozoan.

Trichomonas vaginalis

This parasite can be cultured in Diamond's medium, Cleveland's medium, and Linstead's medium. Two culture media for culture of vaginal *Trichomonas* organisms for clinical diagnosis are important: Both Kupferberg's STS and Diamond's medium (modified) are commercially available. A study by Gelbert et al. (1989) showed that Diamond's medium modified was better as compared to Kupferberg's (STS) medium. In the case of Kupferberg's medium, a lag phase had to pass before the exponential growth. But in the case of Diamond's medium, there was no lag phase and it exhibited long exponential growth.

Giardia intestinalis

This parasite can be cultured in Diamond's medium. A 20% fresh human serum can be used for culture of *Giardia intestinalis*. The addition of bovine or porcine bile to this medium dramatically enhances the growth with a marked reduction in generation time. PEHPS medium is also a homemade and economic culture medium used for the culture of *Giardia lamblia*. TY1-S-33 is a good medium for *Giardia* culture. But this is expensive.

Table: Name of the media in which free-living protozoa can be cultured (several information from review of Ahmed 2014)

Name of medium	Type of medium	Composition
PYG medium	Axenic liquid	PYG glucose, peptone, yeast extract, etc.
Modified PYG medium	Axenic liquid	Trypticase peptone, peptone, yeast extract, beef extract, glucose, etc.
Nonnutrient agar (NNA)	Monoxenic solid	Page's saline -non nutrient agar, *Escherichia coli*/*Enterobacter* spp. overlay
DGM-21 A	Defined medium	Amino acids, salts, and vitamins
Tissue culture	Tissue culture monolayer	Monkey kidney and rat glioma
TY 1-S-33 medium	Axenic	Trypticase, yeast extract, dextrose NaCl, K2HPO4, etc.

Culture of Hemoprotozoa

Trypanosoma spp.

A new protocol using a biphasic medium composed of **Novy, McNeal, and Nicolle** (NNN) blood agar medium as a solid phase and liver infusion, brain heart infusion, and tryptose (LIBHIT) medium as a liquid phase was employed to culture *Trypanosoma* spp. (Lemos et al. 2013). After 4 days of inoculation of the trypanosomes in the culture tubes containing biphasic mediumcomposed of NNN and LIBHIT, it was found that trypomastigotes had differentiated into epimastigotes and amastigotes. Diphasic medium containing defibrinated blood is commonly used for the culture of *Trypanosoma cruzi.*

Leishmania spp.

Different culture media comprising NNN, Schneider's Drosophila medium, M-199 with 10% FCS etc., are used for the culture of promastigote stages *Leishmania* spp. Macrophage cell line, Grace's insect medium supplimented with calf serum 20% are used for culture of amastigote of *Leishmania*. However, promastigotes can also be cultured in Grace's Insect Medium.

Different culture media for the culture of different stages of hemophlagellates are presented in the following table (several information from Ahmed 2014)

Name of media	Type	Constituents	Stages
NNN medium	Diphasic	Salt agar with rabbit blood	Promastigote/trypomastigote of *Leishmania* sp. and *Trypanosoma* sp.
Schneider's Drosophila medium	Liquid	Inactivated 30% fetal calf serum	Promastigote of *Leishmania* sp.
Macrophage cell line	Tissue-based cultivation	Mouse macrophage cell line	Amastigotes of *Leishmania* sp.
Grace's insect medium	Axenic	Lactalbumin hydrolysate, yeastolate, L glutamine and intestinal homogenate of *Triatoma*	Amastigote of *Leishmania* sp.
M-199 + 10% fetal calf serum	Cell culture medium	Amino acids, vitamins, salts, glucose, tween, cholesterol, and purines, etc.	Promastigotes of *Leishmania* sp.

Babesia spp.

There are many media in which *Babesia* parasites can be cultured. In the following table, different cultures have been mentioned [several information from a review of Schuster (2002), Levy and Ristic (1980), and Adaszek and Winiarczyk (2011)].

Name of species	Medium	Type of serum/constituent/condition
Theileria spp.	RPMI-1640	Glucose, sodium chloride, sodium bicarbonate, disodium phosphate, potassium chloride, magnesium sulfate, and calcium nitrate, etc.
B. divergens	HEPES-buffered RPMI-1640	Human (10)
B. bovis	HEPES-buffered medium 199	Bovine (40)
B. caballi	HEPES-buffered HL-1	Serum-free plus lipid-rich supplements
B. equi	HL-1	Fetal calf (20) plus serum factors
B. gibsoni	HL-1	Dog (20)
B. bigemina	A-DMEM, A-RPMI, A-DMEM/F12, and M-199	Bovine serum
B. canis canis	RPMI-1640 with 40% canine serum in 5% CO2 in air atmosphere, subsequent subcultures into HL-1 with 40% canine serum or EMEM with 40% fetal bovine serum	
B. bovis	MASP (microaerophilous stationary phase culture), HEPES-buffered medium-199	Bovine (40)

Theileria spp.

Theileria-infected lymphoid cells can be cultured in different culture media. These are Leibovitz L-15, DMEM, RPMI-1640, Eagle MEM, M-199, McCoy, etc.

Emerging and Re-Emerging Protozoan Parasites

Cryptosporidium Infection

Cryptosporidium is an important protozoan parasite occurring in both animals and humans. Therefore, this parasite has zoonotic significance. The members of this genus come under the family Cryptosporididae. There are different species of *Cryptosporidium,* such as *C. bovis, C. parvum,* etc. *C. parvum* occurs in cattle. This species also affects human beings. The organisms are homoxenous and are transmitted through the fecal-oral route. They have high dissemination potential because they are transmitted directly from one host to another host. Another reason is that the sporulated oocysts come out through the feces. This parasite is transmitted to hosts through contaminated food and drinking water. The parasites cause enteritis, resulting in diarrhea and leading to significant morbidity in affected hosts. Human beings pick up the infection by consuming contaminated food and drinking water containing sporulated oocyst of *Cryptosporidium.* There are many predisposing factors responsible for the outbreak of the disease. A large outbreak of cryptosporidiasis occurred in Milwaukee where many people in the city were affected and it appeared as a great threat. Water drained from surrounding cattle farms was the major reason for dissemination of the parasitic infection throughout the city by contaminating the city's water supply system.

de Graf et al. (1999) in a review article described cryptosporidiosis in avians as an emerging disease associated with respiratory pathology in chicken and other galliform birds, and intestinal problems in turkeys and quails. The disease in cattle was neglected until the early 1980s (reviewed by de Graf et al. 1999). But outbreak was reported in cattle later in different times. Initially, *Cryptosporidia* infection was not considered as pathogenic as in many cases, the organisms were found along with viral and bacterial infections in animals. However, later, technology developed to accurately identify *Cryptosporidium* species, and many clinical cases were found to be caused by *Cryptosporidium* alone. This disease is recognized as an emerging disease in human beings.

Leishmania Infection

Leishmaniasis is one of the important emerging parasitic diseases. This is an intracellular protozoan parasite. It is transmitted by vectors. Bloodsucking *Phlebotomus* flies act as the vectors or transmitters of this protozoan parasite. Previously, leishmaniasis was thought to be controlled, but more attention should be paid to this disease for the problem of reemergence. More and more research works are directed to control the disease, but in spite of enormous efforts exerted by the researchers, the problem is still prevailing. The reemergence of this parasitic infection has been a big issue. The interaction between humans and sandflies has increased for different reasons. Deforestation, new human settlement, and industrial development are some important predisposing factors. Ecological changes due to different reasons have increased the interaction between humans and sandflies.

In the Old World (the Eastern hemisphere), leishmaniasis is prevalent in parts of Asia, Africa, and Southern Europe. In the New World (the Western hemisphere), this is found in Central America, South America, Mexico, etc. Different literates indicate a sharp rise in cases of imported leishmaniasis in developed, non-endemic countries, associated with military operations, international tourism, and the influx of immigrants from endemic countries (Pavli and Maltezou 2010). Popular Mediterranean destinations are emerging as the main areas of acquisition of visceral leishmaniasis for European travelors (Pavli and Maltezou 2010). The influx of leishmaniasis in non-endemic areas is increasing. In Australia, leishmaniasis was rarely reported. However, Stark et al. (2008) reported 20 cases (17 cutaneous, 2 visceral, and 11 post kaka-azar dermal leishmanoid) imported into Australia.

Chagas Disease

This is caused by an important protozoan parasite, *Trypanosoma cruzi*. This disease is highly prevalent in Latin America. However, the disease is emerging in other countries, particularly in AIDS patients as they are highly immunocompromised. This disease is a zoonotic disease. Many millions of people infected with *T. cruzi* commonly reside in the endemic region of Latin America. Geospatial data from 2002 to 2011 demonstrated that Chagas disease has existed in countries outside of Latin America, namely, Australia, Japan, Europe, etc. (Reviewed, Liu and Zhou 2015). There are many reasons for occurrence of Chagas disease in non-endemic areas. The increasing levels of migration, important epidemiological changes are the predisposing factors Review of Liu and Zhou (2015) revealed diagnosis of Chagas disease in Canada, US, Australia, Japan, and Europe.

Plasmodium Infection

Plasmodium parasites are emerging with a consistent flow. In Southeast Asian countries, *Plasmodium knowlesi* has emerged as a major cause of malaria, although it was not the major cause of malaria there earlier. Emergence as well as resurgence of malaria continues to be evident worldwide, much of it is due to the spread of drug-resistant parasites and insecticide-resistant vectors. Guggisberg et al. (2018) in their review article reported *Plasmodium odoecoili* infection in one-fourth of white-tailed deer population in North America.

Trypanosoma Infection

Different unusual cases of *T. evansi* infection in a human being are reported with history of febrile episode. On and off for a period of 5 months. A systematic review was done by Aregawi et al. (2019) on the global distribution ,host range,and prevalence of *Trypanosoma evansi*. They analyzed 272 references published between 1906 and 2017 and found *T. evansi* reported from 48 countries largely confined to Africa and Asia, with 77% ($n = 48$) of countries with natural *T. evansi* infection. They also found reports of *T. evansi* infection from non-endemic countries, such as Sohuth America (seven countries) and Europe (four countries). *T. evansi* is not endemic there but was imported with infected animals.

Occurrences *of T. evansi* and *T. brucei brucei* infection in human beings are being observed. There are several reports that some animal trypanosomes are emerging as human trypanosomes.

Toxoplasma Infection

Toxoplasma gondii is an important protozoan parasite with a global distribution. Definitive hosts get the infection by the ingestion of either undercooked meat or food contaminated with feces from infected cats. In immunocompetent humans, the infection remains latent, but when immunosuppression occurs, the infection flares up with severe neurological consequences. In areas where this parasitic infection was not found, the cases are being recorded in those areas due to many reasons. These comprise rapid urbanization, global warming, migration of people and animals, natural calamities, etc. All these cause the fragmentation of animal habitats. Thus, new foci (new zones) are created for the new prevalence of *Toxoplasma* infection.

Acanthamoeba Infection

This is one more newly emerging parasitic infection that causes chronic ulceration of the cornea and progressive disseminated disease. The hosts get infection through contact with contaminated water or soil (James 1997).

Babesia Infection

When the emerging status of *Babesia* infection arises, we are concerned with incidences in humans as well as animals. Increased incidences of human babesiasis have been recorded. *Babesia microti* is a zoonotic parasite transmitted to human beings from rodents. In the northeastern United States, *Babesia microti* is a species of human infection. However, in accordance with the recent reports, the infection is spreading. As compared to animal babesiosis, human babesiasis has exponentially increased in the United States. Human cases of babesiasis caused by *B.microti* have been reported from all along the northeastern seaboard and inland comprising Maine to Maryland (reviewed, Vannier et al., 2015). *B. divergens*-like organisms have been reported from Missouri, Kentucky, and Washington (Reviewed, Vannier et al., 2015). Human babesial infections have been reported in temperate regions, such as Europe, France, etc. However, recently, cases have also been occurring in other parts of the world, including Asian countries like India, Taiwan, and China. *B. gibsoni* is a species of dog. In the earlier decades, very few cases of *Babesia gibsoni* infections were reported. However, cases of *B. gibsoni* are gradually increasing in different countries. In the last few decades, many reports on *B. gibsoni* infection have been recorded in different countries, such as Asia, Europe, the Middle East, and many parts of America. *B. gibsoni* infection has increased in North America. Jain et al. (2017) reported a number of cases of *B. gibsoni* infection in Kerala. It is considered that the UK is free from autochthonous infection of canine babesiosis. However, clusters of canine babesiosis have been reported. In the winter months of 2015/16, a cluster of canine babesiosis occurred there. de Marco et al. (2017) confirmed several cases of *B. canis* infection in dogs from Harlow, Essex.

Theileria Infection

Reports on the emerging status of *Theileria* infection in animals are scanty. *Theileria orientalis* is normally prevalent in Australia and the infected animals do not exhibit clinical signs of the disease. However, as reviewed by Kamau et al. (2011), an outbreak of a clinical form of *T. orientalis* in New South Wales was reported. Clinical signs included anemia, depression, fever, hemoglobinuria, etc. This happened due to the introduction of cattle from inland areas to New South Wales. Usually, *T. orientalis* is asymptomatic in the United States. However, virulent form of (*Torientalis Ikeda*) has been reported by Oakes et al. (2019). They identified the species of *T. orientalis* (Ikeda genotype 2) by molecular techniques using rDNA as a marker. The animals infected with *T. orientalis* Ikeda 2 exhibited clinical signs. This genotype was not reported earlier in North America but they reported this genotype as emerging infectious disease with potential for major animal health and economic impact.

Protozoa of Wild Animals

Parasites	Wild hosts	Site
Trypanosoma uniforme	Antelope, bush buck, impala, duiker etc.	Circulatory system
T. congolense	Antelope, zebra, elephant, and warthhog	Circulatory system
T. simiae	Camel and warthog	Circulatory system
T. brucei	Rabbit, rodent, mice, bushbuck, warthogs, waterbuck etc.	Circulatory system
T. evansi	Camel, water buffalo, deer, wild pig, capybara and elephant	Circulatory system
T. lewsi	Rat	Circulatory system
T. nabiasi	Rabbit	Circulatory system
T. rangeli	Opossum, apes, and monkey	Circulatory system
T. cruzi	Fox, ferret, opossum, squirrel, armadillos, racoons, apes and monkey etc.	Different organs (heart, liver, lung, kidney)
Leishmania donovani	Rodent, fox	Different organs (liver spleen)
L. infantum	Jackal, fox, wolf, and porcupine as reservoirs	Different organs spleen, liver, bone marrow, kidney, etc.)
L. aethiopica	Rock hyraxes	Primarily skin
L. major	Hedge hog, giant gerbil	Primarily skin
L. mexicana mexicana	Pocket mouse, tree rat, cotton rat, and vesper rat as reservoir hosts	Primarily skin
L. brasiliense brasiliense	Forest rodent as reservoir hosts	Primarily skin
L. peruviana	Rodent as reservoir hosts	Different organs
Trichomonas gallinae	Mourning dove, hawk, Turkey, finches, raptor and golden eagle	GI tract
Theileria spp.	*Theileria parva* in African buffalo, T. *annulata* in water buffalo	Circulatory system
Babesia spp.	*Babesia bigemina* in water buffalo, *B. bovis* in roe deer, *B. caballi* in donkey, mule, *B. trautmani* in wart hog, bush pig, *B. canis* in wolve, jackal etc	Circulatory system

Babesia Infection

Babesia odocoilei infects white-tailed deer. This os is found in Virginia, Texas, New Mexico, Oulahoma, and Florida. It is transmitted by deer tick (*Ixodes scapularis*). *Babesia odocoilei* also affects elk, reindeer, caribou, desert big horn sheep, etc. Mortality and clinical signs are rare. The pathogens are mainly maintained by two factors: vectors and animal reservoirs. *B. vulpes* commonly infects some species of wild canids. Many *Babesia* spp. in wild animals are distributed across Asia, Africa, and America. *Babesia lotori* occurs in raccoons.

Trypanosoma Infection

African trypanosomes, *T. brucei*, have been reported in lions and gorillas. *T. brucei* and *T. rhodesiense* have been reported in antelopes. *T. evansi* has been reported in tapir, deer capybara. *T. lewisi* occurs in rats. Tigers are infected with *T. evansi*. The animals exhibit clinical signs comprising fever, panting, anorexia, convulsion, and sudden death.

Leishmania Infection

Leishmania major has been reported in gorilla (*Gorilla gorilla gorilla*). Promastigotes and amastigotes were detected in the feces of gorillas.

L. infantum infection has been detected in primates and wild carnivores. Various feline animals, rodents, and marsupials are also affected by *L. infantum.* Donkeys and camels are also affected by *L. donovani.* This species also occurs in rodents.

Coccidia Infection

Eimeria southwelli occurs in cow nose rays. *E. falciformis* is a pathogenic coccidian parasite occurring in the epithelial cells of the large intestine of mice.

Cryptosporidium Infection

Wild birds are affected by *Cryptosporidium baileyi*, *C. galli*, and *C. bayleri*. These spp. affect the respiratory system. Clinical disease with mortality is associated with *C. galli.*

Other Protozoan and Rickettsial spp.

Anaplasma phagocytophilum infects rodents and shrews. They act as reservoirs. *Histomonas meleagridis* occurs in turkeys, guinea fowl, pheasants, and peafowl. *Parahistomonas wenrichi* occurs in pheasants. *Tritrichomonas foetus* is found in deer. *Trichomonas gallinae* occurs in turkeys, morning doves, hawks, and golden eagles. *Trichomonas tenax* occurs in monkeys. *Trichomitus wenyoni* occurs in the cecum of hamsters and rhesus monkeys. *Tetratrichomonus gallinarum* occurs in turkeys, guinea fowl, quail, pheasants, and patridges. *Tetratrichomonas microti* occurs in rodents and hamsters. *Chilomastix mesnili* occurs in primates.

References

Abbasalipourkabir R, Fallah M, Sedighi F, Maghsood AH, Javid S (2016) Nanoencapsulation of nitazoxanide in solid lipid nanoparticles as a new delivery system and in vitro release study. J Bio Sci 16(4):120–127. https://doi.org/10.3923/jbs2016.120.127

Adaszek L, Winiarczyk S (2011) In vitro cultivation of *Babesia canis canis* parasites isolated from dogs in Poland. Parasitol Res 108(5):1303–1307. https://doi.org/10.1007/s00436-010-2181-7. PMID: 21127906

Ahmed NH (2014) Cultivation of parasites. Trop Parasitol 4(2):80–89. https://doi.org/10.4103/2229-5070.138534. PMID: 25250227

Akopyant NS, Kimlin N, Secundino N, Pratik R et al (2009) Demonstration of genetic exchange during cyclical development of Leishmania in the sand fly vector. Science 324(5924):265–8. https://doi.org/10.1126/science.1169464. PMID: 19359589; PMCID: PMC2729066

Aregawi WG, Agga GE, Abdi RD et al (2019) Systematic review and meta- analysis on the global distribution, host range, and prevalence of *Trypanosoma evansi*. Parasit Vectors 12:67. https://doi.org/10.1186/s13071-019-3311-4

Bakheit MA, Schnittger L, Salih DA, Boguslawski K et al (2004) Application of the recombinant *Theileria annulata* surface protein in an indirect ELISA for the diagnosis of tropical theileriosis. Parasitol Res 92(4):299–302. https://doi.org/10.1007/s00436-003-1055-7. PMID:14722760

Camacho-Nuez M, Hernandez-Silva DJ, Castaneda-Ortiz EJ, Paredes-Martinez ME et al (2017) Hap 2, a novel gene in *Babesia bigemina* is expressed in tick stages, and specific antibodies block zygote formation. Parasit Vectors 10(1):568. https://doi.org/10.1186/s13071--017-2510-0. PMID:29132437

Darghouth MA, Boulter NR, Gharbi M, Sassi L et al (2006) Vaccination of calves with an attenuated cell line of *Theileria annulata* and the sporozoite antigen SPAG-1 produces a synergistic effect. Vet Parasitol 142(1–2):54–62. https://doi.org/10.1016/j.vetpar.2006.06.010. PMID:16870344

Daulatabad D, Singal A, Dhawan A, Padhi D, Sharma S (2015) Mucocutaneous leishmaniasis caused by Leishmania donovani infection in an Indian man. Int J Dermatol 54(6):680–4. https://doi.org/101111/ijd.12748. PMID: 25557311

de Graf DC, Vanopdenbosch E, Ortega-Mora LM, Abbassi H, Peeters JE (1999) A review of the importance of cryptosporidiosis in farm animals. Int J Parasitol 29(8):1269–1287. http://doi.org/10.1016/S0020-7519(99)00076-4

De Marco MDMF, Hernadez-Trina LM, Phipps LP, Hansford K et al (2017) Emergence of Babesia canis in southern England. Parasit vectors 10(1):241. https://doi.org/10.1186/s13071-017-2178-5. PMID: 28514953; PMCID: PMC54364

Gelbert SM, Thomason JL, Osypowski PJ, James JA, Hamilton PR (1989) Comparison of Diamond's medium modified and Kufferberg medium for detection of Trichomonas vaginalis. J Clin Microbiol 27(5):1095–6. http://doi.org/10.1128/jcm.27.5.1095-1096.1989. PMID: 2787332; PMCID: PMC267490

George N, Bhandari V, Reddy DP et al (2015) Molecular and phylogenetic analysis revealed new genotypes of *Theileria annulata* parasites from India. Parasit Vectors 8:468. https://doi.org/10.1186/s13071-015-1075-Z

Gimenez AM, Francoso KS, Ersching J, Icimoto MY et al (2016) A recombinant multi-antigen vaccine formulation containing *Babesia bovis* merozoite surface antigens MSA-2a1, MSA-2b and MSA-2c elicits invasion -inhibitory antibodies and IFN gamma producing cells. Parasit Vectors 9(1):577. https://doi.org/10.1186/s13071-016-1862-1. PMID:27842609

Guggisberg AM, Sayler KA, Wisely SM, Odom John AR (2018) Natural history of *Plasmodium odocoilei* malaria infection in farmed white-tailed deer. msphere 3(2):e00067–e00018. https://doi.org/10.1128/msphere.00067-18. PMID:29669881

Hussein HE, Bastos RG, Schneider DA, Johnson WC et al (2017) The *Babesia bovis* hap2 gene is not required for blood stage replication, but expressed upon in vitro sexual stage induction. PLoS Negl Trop Dis 11(10):e0005965. https://doi.org/10.1371/journal.pntd.0005965. PMID:28985216

Jain KJ, Lakshmanan B, Syamala K, Praveena JE et al (2017) High prevalence of small *Babesia* species in canines of Kerala, South India. Vet World 10(11):1319–1323. https://doi.org/10.14202/vetworld.2017.1319-1323. PMID:29263592

James JL (1997) Emerging parasitic infections. FEMS Immunol Med Microbiol 18(4):313–317. https://doi.org/10.1111/j.1574-695X.1997.tb01061.x

Kamau J, de Vos AJ, Playford M, Salim B et al (2011) Emergence of new types of *Theileria orientalis* in Australian cattle and possible cause of theileriosis outbreaks. Parasit Vectors 4:22. https://doi.org/10.1186/1756-3305-4-22. PMID: 21338493

Kreutzer RD, Yemma JJ, Grogl M, Tesh RB, Martin TI (1994) Evidence of sexual reproduction in the protozoanparasite Leishmania (Kinetoplastida:Trypano somatidae) 51(3):301–307. https://doi.org/10.4269/ajtmh.1994.51.301

Kumar S, Kumar R, Gupta AK, Yadav SC, Goyal SK, Khurana SK, Singh RK (2013) Development of EMA-2recombinant antigen based enzyme-linked immunosorbent assay for seroprevalence studies of Theileria equi infection in Indian equine population. Vet Parasitol 198(1–2):10–7. https://doi.org/10.1016/J.Vetpar.2013.08.030. PMID: 24070779

Kumar S, Rakha NK, Goyal L, Goel P et al (2015) Diagnostic application of recombinant equine merozoite surface antigen-1 in ELISA for detection of *Theileria equi* specific antibodies. Jpn J Vet Res 63(3):129–137. PMID:26563032

Kundu K, Garg R, Kumar S, Mandal M et al (2017) Humoral and cytokine response elicited during immunisation with recombinant immune mapped protein 1(EtIMP-1) and oocysts of *Eimeria tenella.* Vet Parasitol 244:44–53. https://doi.org/10.1016/j.vetpar.2017.07.025. PMID:28917316

Lemos M, Souza CS, da Costa SC, Souto-Padron T (2013) Isolation and in vitro culture of trypanosomes from *Leptodactylus ocellatus* from the Atlantic Forest in a new experimental culture medium. J Parasitol 99(1):164–167. https://doi.org/10.1645/GE-2949.1. PMID: 22694360

Levy MG, Ristic M (1980) *Babesia bovis*: continuous cultivation in a microaerophilous stationary phase culture. Science 207:1218–1220

Liu Q, Zhou XN (2015) Preventing the transmission of American trypanosomiasis and it's spread into non-endemic countries. Infect Dis Poverty 4:60. https://doi.org/10.1186/s40249-015-0092-7. PMID:26715535

Maharana BR, Sudhakar NR, Jawalagatti V, Saravanan BC et al (2020) Evaluation of the immunoprotective potential of recombinant paraflagellar Rod proteins of *Trypanosoma evansi* in mice. Vacc (Basel) 8(1):84. https://doi.org/10.3390/vaccines8010084. PMID:32059486

Martins NO, Souza RT, Cordero EM, Maldonado DC et al (2015) Molecular characterization of a novel family of *Trypanosoma cruzi* surface membrane proteins (TcSMP) involved in mammalian host cell invasion. PLoS Negl Trop Dis 9(11):e0004216. https://doi.org/10.1371/journal.pntd.0004216. PMID:26565791

Moss DM, Montgomery JM, Newland SV, Priest JW, Lammie PJ (2004) Detection of Cryptosporidium antibodies in sera and oral fluids using multiplex bead assay. J Parasitol 90(2):397–404. https://doi.org/10.1645/GE-3267. PMID: 15165066

Oakes VJ, Yabsley MJ, Schwartz D, LeRoith T, Bissett C et al (2019) *Theileria orientalis* Ikeda genotype in cattle, Virginia, USA. Emerg Infect Dis 25(9):1653–1659. https://doi.org/10.3201/eid2509.190088

Pavli A, Maltezou HC (2010) Leishmaniasis, an emerging infection in travellers. Int J Infect Dis 14(12):e1032–e1039. https://doi.org/10.1016/j.ijid.2010.06.019. PMID:20952234

Rafiqi SI, Garg R, Ram H, Reena KK et al (2019) Immunoprophylactic evaluation of recombinant gametocyte 22 antigen of *Eimeria tenella* in broiler chickens. Parasitol Res 118(3):945–953. https://doi.org/10.1007/s00436-018-06198-2. PMID:30637470

Schuster FL (2002) Cultivation of *Babesia* and *Babesia*-like blood parasites: agents of an emerging zoonotic disease. Clin Microbiol Rev 15(3):365–373. https://doi.org/10.1128/CMR.15.3.365-373.2002. PMID:12097245

Stark D, van Hal S, Lee R, Marriot D et al (2008) Leishmaniasis, an emerging imported infection: report of 20 cases from Australia. J Travel Med 15(5):351–354. https://doi.org/10.1111/J.1708--8305.2008.00223.x. PMID:19006509

Sullivan L, Fleming J, Sastry L, Mehlert A et al (2014) Identification of sVSG117 as an immunodiagnostic antigen and evaluation of a dual- antigen larval flow test for the diagnosis of human African trypanosomiasis. PLoS Negl Trop Dis 8(7):e2976. https://doi.org/10.1371/journal.pntd.0002976. PMID:25033401

Tebele N, Skilton RA, Katende J, Wells CW et al (2000) Cloning, characterization, and expression of a 200-kilodalton diagnostic antigen of *Babesia bigemina.* J Clin Microbiol 38(6):2240–2247. https://doi.org/10.1128/38.6.2240-2247. PMID:10834983

Thuy NT, Goto Y, Lun ZR, Kawazu S, Inoue N (2012) Tandem repeat protein as potential diagnostic antigen for *Trypanosoma evansi* infection. Parasitol Res 110(2):733–739. https://doi.org/10.1007/s00436-011-2632-9. PMID:21927872

Vannier EG, Direk-Wasser MA, Mamoun CB, Krause P (2015) Babesiosis. Infect Dis Clin North Am 29(2)357–37. https://doi.org/10.1016/j.ide.2015.02.008

Villareal MV, Mingala CN, Rivera WL (2013) Molecular characterization of *Trypanosoma evansi* isolates from water buffaloes (*Bubalis Bubalis*) in The Philippines. Acta Parasitol 58(1):6–12. https://doi.org/10.2478/s11686-013-0110-5. PMID:23377905

Zou M, Cao S, Luo Y, Liu M et al (2016) Molecular identification and antigenic characterization of a merozoite surface antigen and a secreted antigen of *Babesia canis* (BcMSA1 and BcSA1). Parasit Vectors 9:257. https://doi.org/10.1186/s13071-016-1518-1. PMID:27141812

Immunoparasitology 9

Important Immune Cells

Neutrophils

Neutrophils are very important cells of the myeloid system. These are major effector cells and perform immunological roles by phagocytosis, antibody-dependent cellular cytotoxicity (ADCC), respiratory burst, etc. They contain two types of granules: primary granules containing the major enzyme myeloperoxidase and lysozymes. Elastase, acid hydrolases, etc. are some other enzymes found in the primary granules of the neutrophils. The secondary granules possess collagenase and lysozyme. The foreign organism cannot get attached to the surface of the neutrophil because a **zeta potential** occurs due to the negative charge of both the surfaces of the neutrophil and the foreign organisms. The neutrophils become able to bind with the organisms when it is neutralized by a positively charged antibody molecule or complement that is adsorbed on the surface of the neutrophils. This is called as opsonization.

Eosinophils

These cells are produced in the bone marrow. These are the second major polymorphonuclear granulocytes. Eosinophils perform many roles in the destruction of large organisms. The small organisms are cleared up by phagocytosis but extracellular destruction is required in the case of large organisms by the eosinophils. The granules of eosinophils contain acid phosphatase and peroxidase. The peroxidase of eosinophils is far more strong as compared to the neutrophil's peroxidase. They can bind to the organisms when the organisms are coated with antibodies or complement. Eosinophils can perform antibody-dependent cellular cytotoxicity. The eosinophils have an FcR receptor that is specific for binding to the Fc part of the antibody. The antibody molecules bind to the antigen. In this status, the eosinophils bind to the Fc part of the antibody and the eosinophils release their toxic proteins. Eosinophils contain leukotrienes, arachidonic acid-based inflammatory mediators. Eosinophils also contain major basic protein (MBP), eosinophil peroxidase (EPO), eosinophilic cationic protein (ECP), and eosinophil-derived neurotoxin (EDN). Eosinophils also possess a plethora of cytokines, such as IL2, IL4, IL6, IL10, IL13, interferon gamma, transforming growth factor etc. Eosinophils provide an immediate innate immune response and thereby regulate the microenvironment at the site of parasite infection. Eosinophil contains some other important enzymes, such as **phosphatase, catalase, aryl sulfatase,** etc. All these cause the degradation of the extracellular matrix. The eosinophils are produced in the hemopoitic system that are mobilized into the circulation and ultimately reach the

S. C. Mandal, *Textbook of Veterinary Parasitology*, https://doi.org/10.1007/978-981-99-2924-5_9

inflamed tissue where the parasites are under the control of humoral eosinophil stimulation factors (ESF). The eosinophils get adhered to the surface of the parasites that are already coated with antibodies or complement. The eosinophils in this situation release neurotoxins and proteolytic enzymes that affect the tegument of the parasite.

Mast Cells

Mast cells are also called as mastocytes. They originate from the myeloid stem cells. There are two types of mast cells. These are mucosal mast cells (MMC) and connective tissue mast cells (CTMC). MMCs are mostly found in intestinal parasitic infections. These are found in the lamina propria and intestinal epithelial tissues. Degranulation of mast cells is very important in immunity. Mast cells possess an Fc epsilon R1 receptor. In normal phenomena, the Fab part of the antibody binds to the antigen. The Fc part of the antibody links to the mast cell through the Fc epsilon R1 receptor. When this linkage is completed, signal transduction occurs. As an initial step of signal transduction, tyrosine kinases are activated. The activated tyrosine kinases activate phospholipase C. This results in the production of diacylglycerol and inositol triphosphate as a result of hydrolysis of membrane phospholipid. The kinases cause phosphorylation of myosin and the granules move to the cell surface and get fused with the plasma membrane. Then the contents of the granules are released. These granules release histamine, leukotrienes, etc.

Basophils

Basophils are granulocytic cells. The basophils reach the tissues under the influence of different cytokines released by lymphocytes and other cells. The granules contain prostaglandin, leukotriene, histamine, etc. The basophils also have Fc receptors that bind to the Fc part of the antibody molecules. In a similar way as found in the case of mast cells, basophils degranulate and produce different proteases, amines, and cytokines. The major cytokines present in the mast cells are IL4, IL5, IL13, TSLp, etc.

Macrophages

These are mononuclear phagocytic cells that perform a great role in immunity. These are called as Kupffer cells when present in liver sinusoids, alveolar macrophages in the lung, histiocytes in connective tissues and microglia in the brain. The stem cells differentiate into promonocytes that further differentiate to form the monocytes under the influence of the colony stimulating factors. The macrophages perform diversified functions. When foreign pathogens enter or invade any tissue, neutrophils come first at that site using chemotaxis. The neutrophils perform their roles using phagocytosis. The dying neutrophils release some enzymes (elastase and collagenase) that act as the chemotactic factors to attract monocytes. Catalase is a vital enzyme present in the macrophages. Macrophages can perform respiratory burst, but it is less potent as compared to neutrophils. They possess nitric oxide synthetase and produce nitric oxide that kills pathogens. To perform any immune function, macrophages are to be activated. Th1 cells release IFN gamma, which activates macrophages. If the inflammatory condition persists for a long period, a large number of macrophages gather around the foreign body that gives the appearance as of an epithelium. These are called as epithelioid cells. The epithelioid cells may fuse with each other and form the multinucleated giant cells. Many important cytokines are produced by macrophages. The macrophages are antigen-presenting cells and present exogenous antigens to T cells.

NK Cells

These are natural killer cells that are a type of lymphocytes derived from the same progenitor. These are analogous to T cytotoxic cells. NK cells are typically different from other immune cells as they can recognize damaged cells without the presence of antibodies or MHC molecules. They are early activators. Both innate and adaptive immunity are performed by NK cells. They possess small granules containing perforins and granzymes. Perforins cause pore formation in the target cell through which granzymes enter and kill the cell by apoptosis or osmotic lysis. NK cells secrete different important cytokines. These are TNF alpha, TGF beta, IL10, IFN gamma, IL12, GM-CSF, etc. Antibody-dependent cell-mediated cytotoxicity is also performed by NK cells. NK cells have FcgammaRIII receptors that can recognize the Fc part of the antibody already charged with antigen. This results in the release of cytolytic granules and apoptosis.

Immunoglobulins

Immunoglobulins are a class of proteins found in the serum and are capable of binding with the epitopes of antigens. These are secreted by plasma cells by the action of different cytokines. Different cytokines act on the B cells that subsequently transform into plasma cells. In most of the cases, B cells produce a clone of the homologous cells category that differentiate into plasma cells. These plasma cells produce monoclonal antibodies. The immunoglobulins consist of two chains: light and heavy chains. These light and heavy chains are attached by a hinge region. This is called the **hinge** region because the Fab region is movable and can swing freely around the center of the molecule as if they are hinged.

Different Types of Immunoglobulins

There are five types of immunoglobulins: IgG, IgA, IgM, IgD, and IgE. Among these five immunoglobulins, IgG is found in the highest concentration in blood. The molecular weight of this immunoglobulin is **150** kDa. Since the size of these molecules is small, they can pass through the blood vessels more easily as compared to other immunoglobulins. IgM is found in blood in the second highest concentration. IgM plays a significant role in immunity. It possesses five subunits of 180 kDa molecular weight each. It is the immunoglobulin that has the highest number of subunits having **five antigen-binding sites**. Understandably, the functional ability of IgM is more as compared to IgG. IgM cannot easily pass through in the tissue space because of its large size. IgA immunoglobulin is normally found as a dimer. The two molecules are joined by a J **chain**. IgA has a great role in parasitic immunity and is produced in the intestinal wall of mucosa-associated lymphoid tissue (MALT). IgA is also produced by the plasma cells located in skin-associated lymphoid tissue (SALT). IgA bound to the secretory component is called as secretory IgA. Due to the presence of a secretory component, IgA is not digested in the GI tract. The immunoglobulins are produced abundantly in the respiratory tract, GI tract, urogenital tract, etc. where the mucosal layer is present and these immunoglobulins provide immunological protection against any infection. IgE is found in very low quantities. The molecular weight of IgE is approximately 200 kDa. IgE has a great role in type 1 hypersensitivity reaction. The Fc region of IgE binds to basophils or mast cells when the antigen is captured by the antibody molecule. The effector cells (mast cells and basophils) release prostaglandin, leukotriene, histamine, etc. IgD is found in a very low concentration. Its exact function is not known.

Important Immune Phenomena

Antibody-dependent Cellular Cytotoxicity

Tissue-invading helminths are immature larval stages. There are different mechanisms by which their migration is inhibited. The larval stages of hookworms, *Toxocara* and *Fasciola* are tissue-invading parasitic stages. Antibody-dependent cellular cytotoxicity (ADCC) is one of the most important immunological phenomena which restrict tissue-invading parasites. This phenomenon requires effector cells and antibodies. The effector cells are eosinophils, neutrophils, macrophages, platelets, etc. The antibody components involved in ADCC are IgE, IgG, and IgA. The effector cells have receptors for the Fc portion of the antibodies. The parasitic organisms captured by antibodies are destroyed by the toxic materials released by the effector cells carrying receptors for the Fc part of the antibody. When these cells are activated by the fixation of the antibodies with the captured pathogen, the effector cells release toxic substances to the pathogens. For example, these toxic substances of eosinophils are major basic proteins, eosinophil cationic proteins, eosinophil-derived neurotoxins, reactive nitrogen intermediates, etc. The eosinophils contain many other substances like cysteinyl leukotrienes, eosinophilic peroxidase, aryl sulfatase, elastase, and arachidonic acid-based inflammatory mediators. Eosinophils possess a plethora of cytokines also, such as IL2, IL4, IL6, IL10, IL13, IFN gamma, TNF alpha, TGF beta. All these substances affect the parasite and change the microenvironment of parasite penetration. IL5 has a role in the maturation, migration and activation of eosinophils.

Complement Activation

Complements are a series of proteins present in blood. The complement is activated either by the pathogen or by the pathogen-bound antibody. After activation, a cascade of reactions occur leading to the production of an active component, such as membrane attack complex (MAC). The so-produced active component affects the pathogen.

Nitric Oxide Killing

Macrophages possess nitric oxide synthetase and this enzyme oxidizes the guanidino nitrogen of L-arginine resulting in the production of nitric oxide and citrulline. The nitric oxide is very toxic to the pathogen.

Respiratory Burst

The respiratory burst is an important reaction and is one of the important immune phenomena. It occurs in neutrophils, macrophages, etc. It results in the rapid production of reactive oxygen species (ROS), hydrogen peroxide, etc., which kill the internalized pathogens.

Different steps of respiratory bursts:

- The NADPH oxidase enzyme present in the immune cells converts NADPH into $NADP^+$ that results in the release of electrons.
- The oxygen molecule accepts this donated electron and transforms into a superoxide anion.
- Under the influence of superoxide dismutase, two superoxide anions react with H^+ ions and form hydrogen peroxide and oxygen.
- Hydrogen peroxide reacts with Cl^- and forms H_2O and OCl^- under the influence of the myeloperoxidase enzyme.
- OCl^- ultimately reacts with H_2O and hypochlorous acid (HOCl) is produced by this reaction. This hypochlorous acid is highly toxic to the pathogens.

Types of Parasitic Antigens

A parasitic antigen is a substance which is derived from the parasite and it can elicit an immune response in the host. Both crude and purified antigens are used for immunoprotection. Crude antigens are somatic antigens (prepared by the homogenization of the whole parasite/parasitic stages). Excretory and secretory materials of the parasite are also used for immunological studies; they are called excretory–secretory antigens. Purified antigens are also called defined antigens. This is a specific protein used for immunoprotection and immunodiagnosis. Defined antigens are either purified native antigens or specific antigens prepared by genetic manipulation (recombinant DNA technology). Parasitic antigens are complex in nature. The body of a parasite is formed of different tissues. The crude protein of a parasite contains a number of proteins.

Different Types of Antigens

The parasitic antigens can be categorized into two categories: crude antigen and defined antigen, as described below.

1. Crude Antigens

The crude antigen is composed of many proteins. This antigen is prepared from the whole parasite or part of parasite/parasitic stages. No protein is eliminated from the whole crude antigen. The crude antigen is prepared from the following materials:

(a) **Whole parasite/whole parasitic stage**
  (i) Whole adult parasite
  (ii) Whole larval parasite
  (iii) Eggs
(b) **Part of parasite**
  Somatic antigens are prepared from any organ or part of the parasite.
  Examples:
  (i) Antigen prepared from the salivary glands of parasites (tick salivary glands).
  (ii) The gut of the tick is used as an antigen.
  (iii) Gut antigen of *Haemonchus.*
  (iv) Flagellar antigen of trypanosomes.
(c) **Excretory secretory material/moulting fluid**

Excretory–secretory materials of parasites, particularly of the helminthic parasites, are used as antigens. Moulting fluid released during the moulting of the parasite is also used as an antigen.

2. Defined antigens

Defined antigens are the specific antigens. Proteins such as enzyme, beta tubulin, etc. are used as antigens. These are also called specific antigens.

(a) ***Defined antigens of native source*****:** The specific protein molecules are isolated directly from a crude protein of the parasite. Many such molecules are used as antigens. These are enzymes (cathepsin L, cathepsin B, glutathione S-transferases, cysteine proteinase, etc.) and specific proteins (beta tubulin, calmodulin, paramyosin, etc.). A particular protein from a crude mass of the parasitic protein is called a purified antigen. The protein is isolated from the crude protein of parasites by different protein-purification protocols, such as gel exclusion chromatography, ion exchange chromatography, affinity chromatography, etc. In the recent past, several defined proteins have been isolated from *Fasciola* spp., *Schistosoma* spp., and other flukes. From *Fasciola* parasites, glutathione S-transferase, cathepsin L, FABP, and hemoglobin have been isolated. Their immunodiagnostic and immunoprotection performances have been tested. **Sm 26 and Sm 28 proteins** have been isolated from *Schistosoma* sp. The immunopotential of these antigens has also been assessed. Cysteine proteinases have been isolated from *Fasciola* spp., *Schistosoma* spp., *Paragonimus* spp., Paramphistomes, etc. In most of the cases, column chromatography has been found an ideal technique for the isolation of a particular protein.

(b) ***Defined antigens prepared by a recombinant technique*****:** The defined antigens as described earlier can be expressed in the laboratory by a specific protocol. For this purpose, the gene or nucleotide sequence coding for a particular protein is identified. This particular gene is isolated from the whole genome sequence. Restriction endonucleases are used for cutting the nucleotide sequence to get the desired/target gene. The target gene is isolated and incorporated into a plasmid vector. The recombinant plasmid is inserted into the host cells (*Escherichia coli*, yeast cells). The protein is expressed in the host cells. That target protein is isolated from the host cells following standard protocol and used as an antigen.

Characterization of Parasitic Antigens

Characterization of any parasitic antigen can be done as follows:

(a) Immunological characterization
(b) Biochemical characterization
(c) Molecular characterization

Before the characterization of any parasitic antigen, the particular antigen is isolated. Isolation of particular parasitic proteins is done using different protein-purification protocols. In most of the cases, different chromatographic techniques (gel filtration, immuno-affinity, ion exchange, gas–liquid, enzyme-substrate chromatography, etc.) are used for the isolation of a particular parasitic protein. The isolated protein is used as an antigen. Hyperimmune sera are generated by immunizing animals using the antigen. The Western blot technique is performed to see whether the protein is immunodominant or not. Apart from Western blot, many immunological tests are also done, such as **ELISA (different types), CFT, LFT, CIE, DID,** etc. The isolated protein can be characterized biochemically also. Its function and role in the body can be unfurled. Molecular characterization of a particular protein is also done. The amino acid sequence of a particular protein can be determined. The nucleotide sequence can be deduced from the amino acid sequence.

Types of Immunity

Humoral Immunity

Humoral immunity occurs in the case of exogenous antigens. This exogenous antigen is taken up by different antigen-presenting cells, such as macrophage, B cells, and dendritic cells. MHC-II molecules present in antigen-presenting cells are responsible for handing over the processed antigen to T cells. The antigen is internalized into the endocytic vesicle. It is hydrolyzed by lysosome-associated enzymes (lipase, protease, hydrolase, etc.). The protein of the antigen is degraded to small fragments (oligopeptide). The MHC-II molecules bind to peptide fragments. MHC molecules have two polypeptide chains. The antigen is assembled in between these two peptide chains of the MHC. This antigen is then carried to the surface of the APC and presented to T cells. Inactivated CD4+ T cells recognize the antigen using TCR receptor complex. This TCR complex is composed of two glycoprotein heterodimer (alpha and beta) polypeptide chains, one CD3 molecule and one CD4 molecule. The TCR complex of CD4+ T cell comes in contact with the APC. When the T-cell receptor comes in contact with the antigen, signal transduction occurs. The inactive CD4+ T cells get activated to form Th1 and Th2 cells. T-helper cells (Th1 and Th2) are very important because they secrete different cytokines. Th1 cells release IL2, IFN gamma, TNF alpha, TNF beta, etc. Th2 cells release IL4, IL5, IL6, IL13, etc. All these cytokines are functionally connected to mount an immune response. Some cytokines act on B cells (B lymphocytes). IL4 activates B cells. B cells get transformed into plasma cells. Some B cells remain as memory cells.

Cell-mediated Immunity

Cell-mediated immunity occurs if the antigen is endogenous. The antigen is degraded by a protease enzyme present in the proteasome. The endogenous antigen remains in the cytosol. Small fragments of polypeptides are produced after degradation. The processed antigen fragments are bound to MHC-I molecules. The antigen is presented to CD8+ T cells. CD8+ T cells also have a TCR complex (T-cell receptor complex) that contains two heterodimeric polymeric chains, one CD3 molecule and one CD8 molecule. Inactive CD8+ T cells get transformed into T cytoxic cells. T cytotoxic cells are also called as Tc cells. T cytotoxic cells contain different toxic materials, such as granzyme, perforin, etc. The toxic substances kill the organisms.

Hypersensitivity

There are four common hypersensitivity reactions:

Hypersensitivity-I

Hypersensitivity-I is an immediate type of immunological reaction in which an allergic reaction is set up due to reexposure to a specific antigen. IgE antibodies bind to antigens. The Fc part of antigen-bound IgE is linked to the receptor (FcR) of effector cells (mast cells, basophils). In this state, degranulation of mast cells occurs. A number of chemical mediators are liberated from the mast cells comprising histamine, leukotrienes, prostaglandin, etc. These substances cause a hypersensitivity reaction. The hypersensitivity-I reaction occurs in *Haemonchus* infection and *Trichostrongylus* infection in which IgE and mast cells play a role (self-cure phenomena). In **self-cure phenomena,** IgE and mast cells are the prerequisites. The Fab part of the antibody binds antigen. The Fc part of this antigen-bound antibody links with mast cells through the receptor Fc epsilon RI present on mast cells. In this state, signal transduction occurs. As an initial step of signal transduction, tyrosine kinases are activated. The activated tyrosine kinases activate phospholipase C. This results in the production of diacylglycerol and inositol triphosphate through hydrolysis of membrane phospholipid. The kinases cause phosphorylation of myosin and the granules of the mast cells move to the cell surface and get fused with the plasma membrane. Then, the contents of the granules are released. These granules release histamine, leukotrienes, etc. This results in hyperplasia of goblet cells. These increased number of goblet cells produce a huge amount of mucous. The histamine increases the peristaltic movement of the GI tract. All these ultimately result in the expulsion of the parasites. Hypersensitivity-I also occurs in other parasitic allergic reactions, such as those caused by hookworm, *Strongyloides*, *Schistosoma*, etc. Different arthropod infections also result in allergic reactions. These are Queensland itch or sweet itch and mange caused by different mites like *Sarcoptes* and *Demodex*.

Hypersensitivity-II

This is an antibody-dependent immune phenomenon in which a specific antibody binds to the respective antigens (self-antigen or foreign antigens present on host cells) resulting in tissue damage or cell lysis. This cell injury is caused by direct phagocytosis as a result of opsonization, ADCC, and complement-mediated lysis. IgG antibodies bind with FcRIII present on NK cells and macrophages which results in the release of granzyme and perforin ultimately causing cell death by apoptosis. Complement-mediated lysis occurs in different ways. Via the classical pathway, IgG and IgM activate the complement leading to the production of C3b that induces phagocytosis. Complement activation also results in the formation of a membrane attack complex (MAC) that forms pores in the cell membrane. Hypersensitivity-II occurs in many parasitic infections. ADCC and complement activation commonly occur during the invasion of the tissues by larval parasites. Anemia caused by different hemoparasites, granuloma formation in schistosoma infection, and lesions in Chagas disease are the consequences of the hypersensitivity-II reaction.

Hypersensitivity-III

Hypersensitivity-III occurs as a result of the accumulation of antigen–antibody complexes in the tissues. The antigen–antibody complex is not cleared adequately by innate immune cells. Thus, deposition of immune complex occurs that sets up an inflammatory reaction due to the aggregation of leucocytes. Immune complex-mediated hypersensitivity results in nephritis in many parasitic diseases. These are *Plasmodium*, *Leishmania*, *Trypanosoma*, etc.

Hypersensitivity-IV

This is a delayed type of hypersensitivity and it has a lag phase for its occurrence. Type IV hypersensitivity occurs in three ways:

1. Th1 cells recognize soluble antigens and release IFN gamma. The IFN gamma activates macrophages. Macrophages release different cytokines that cause tissue injury.
2. Th2 cells release IL5 cytokines that activate and recruit eosinophils and these eosinophils release different tissue-damaging materials.
3. Cytotoxic T cells directly damage the cells by releasing different toxic substances. An example of parasitic infections in which delayed hypersensitivity occurs is chronic schistosomosis due to granuloma formation surrounding the eggs. Delayed hypersensitivity reaction also occurs in cases of Chagas disease, toxoplasmosis, skin reaction due to *Ascaris* infection, etc.

Regulation of the Immune Response

It is very interesting that a series of interleukins are produced due to any infection and a harmony of correlation amongst the cytokines is found. This picture is also found in parasitic infections. IL10 and TGF beta play a crucial role in reducing the severity of acute disease and allowing survival of the host. A subset of T cells is produced which regulates the immune reaction. This subset of T cells is T-regulatory (Treg cells) cells. Treg cells produce IL10 and TGF beta. These two cytokines are anti-inflammatory cytokines. These cytokines inhibit the Th2 response. In this way, these cytokines downregulate immune activity.

Evasion of Immunity

Parasites can escape the host's immune attack by different phenomena. The following are some ways by which the parasites escape the host's immune attack.

- Trypanosomes frequently change the antigenic character of surface proteins. Since the organisms are able to change their antigens frequently, the hosts fail to mount a strong immunity against this parasitic infection.
- Many helminthic parasites adsorb the host's antigen on their surface. That's why, the hosts do not mount any immunity as the parasites remain hidden within the host's antigen and in this way, the parasites remain safe. This phenomenon of immune evasion is called as molecular mimicry.
- Some parasites release different enzymes (i.e., cathepsin L protease) that cause a reduction of lymphoproliferation. Immunity is ineffective due to a lack of lymphocytes that are the prime cells of immunity. These enzymes also cleave the immunoglobulins of hosts.
- *Plasmodium* parasites hide themselves in hepatocytes to escape the attack of cytotoxic cells.
- Many helminthic pathogens (e.g., *Toxocara* sp.) shed their outer coat when attached to effector immune cells.
- *Toxoplasma* organisms release some substances that inhibit the fusion of phagosome and lysosome and thus remain protected from attack of lysosomal enzymes.
- Juvenile flukes adsorb IgM on their surface. Since the parasites remain hidden within immunoglobulins, eosinophils and other immune effector cells do not produce any Fc receptor causing nonoccurrence of ADCC or other types of immune reaction.

Immunomodulation and Its Use

Immunomodulation is the modulation of the immune system that occurs naturally or it may be induced artificially.

Natural Immunomodulation

In natural conditions, to reduce the immune reaction, T-regulatory cells are produced. Treg cells produce important cytokines such as IL10 and TGF beta which play a role in the regulation of immunity.

Human-induced Immunomodulation

There are many examples of human-induced immunomodulation. These are mostly of immunotherapies that can induce, enhance, prevent, and regulate immune reactions. Immunotherapy is also called as biological therapy. These are of two types as follows:

- **Activation immunotherapy**
  In this immunotherapy, immune substances or immune cells are infused to stimulate immunity. Vaccination of animals with different

parasitic antigens is mostly done as an immunomodulation strategy.

- **Suppression immunity**
  In this immunotherapy, cytostatic or immunosuppressive drugs are given to reduce the immune reaction. In the field of veterinary parasitology, the use of immunosuppressive drugs for suppression of immunity seems less. However, in human medicine, this is done for the treatment of autoimmune diseases.

Immune Response to Helminths, Arthropods, and Protozoa

Immune Response to Helminths

A non-immunological defense mechanism is also found in the case of helminthic infection as that of other parasitic infections. There occur variations of susceptibility of the hosts to the parasites. Breed, sex, species, etc. are the factors which influence the variation of susceptibility along with pathological entities. Helminthic immunity is diversified. There are many reasons for this diversity of the immunity:

(a) The size and shape of the helminthic parasites are much varied. This variation indirectly influences the host's immunity.
(b) The biochemical composition of the helminthic parasites is varied.
(c) There are stage variations of the helminthic parasites during their biological development. The different stages of a parasite do not have equal virulence. The virulence directly or indirectly influences host immunity.
(d) There are many routes of entry of the helminthic parasites that may be oral, transcutaneous, transcolostral, congenital, venereal, etc. The route of infection of helminthic parasites significantly influences host immunity.
(e) The outer morphological features of parasites are also variable. The immune system first recognizes the outer surface of the parasite. Therefore, the interface created by the outer covering of a parasite and the host's skin is very important. We know that the cuticular organization of the helminthic parasite is variable. So immunity patterns must vary due to variation of the helminthic cuticle.
(f) The infective stages of helminthic parasites are also variable. The infective stages may be egg, larvae, etc. The stage of the helminthic parasite also influences the host's immunity.
(g) The location of the helminthic parasite is variable. This may be the gut, liver, lung, brain, and many other organs. Innate immunity varies in different organs.
(h) The category of helminthic parasites varies (cestodes, trematodes, nematodes, flukes, etc.). Therefore, it is obvious that immunity would vary in accordance with the nature of the parasite.

All types of immunoglobulins are produced in helminthic infection, such as IgG, IgD, IgA, IgM, IgE, etc. However, IgE has a great role in helminthic immunity. IgE is also called reagent antibody secreted from MALT, the local mucosa-associated lymphoid tissue. The hypersensitivity-I reaction occurs. The role of IgE antibodies is markedly found in the case of the self-cure phenomenon. In this phenomenon, the reaction between helminthic antigens and mast cell-bound IgE cause the release of vasoactive amines (i.e., histamine) as a consequence of degranulation of mast cells, basophils, etc. This causes high contraction of intestinal musculature. At the same time, histamine increases the permeability of blood vessels and causes the efflux of large amounts of body fluid. Hyperplasia of goblet cells occurs. All these result in the expulsion of the parasites. Another phenomenon which triggers helminthic immunity is antibody- dependent cellular toxicity (ADCC). Cytotoxic T cells (Tc) release toxic substances that kill the parasites. It is needless to mention that strong immunity occurs due to helminthic infection. In spite of that, the helminthic parasites easily survive in the hosts. What is the reason? It is that the parasites are able to modulate the host's immunity. This modulation of host immunity occurs in such a direction that the parasites can survive, creating a chronic status of the disease.

Immune Response to Arthropods

Basic immunological phenomena are the same but some minute variations are taken into consideration. Fleas release saliva that contains proteins of low molecular weight acting as haptens that bind to the collagen material present in the skin. This, in turn, causes a local Type IV hypersensitivity reaction. As a consequence, infiltration of different mononuclear cells occurs in the localized area that is a local allergic reaction. In some arthropod infections, Type I hypersensitivity occurs. A reaction between antigen and mast cell-bound IgE occurs. The activated mast cells get degranulated and release vasoactive amines such as histamine that causes local inflammation manifested as an allergic reaction. Both cell-mediated as well as humoral immunity occur in the case of tick infection. In mite infection, Type IV hypersensitivity occurs as is found in case of flea bites.

Immunity to Protozoa

Non-immunological defense mechanisms play a role in resisting any parasitic infection. There are many examples of non-immunological immunity. N-Dama cattle are good examples of this. They are trypanotolerant animals and are genetically resistant to *Trypanosoma* organisms. Continuous genetic selection of the hosts against the organisms resulted in this resistance. Protozoa can initiate both humoral and cell-mediated immunity. Though all protozoan parasites fail to elicit good immune responses, many protozoans are able to confer significant immune response. Antibody-dependent cellular cytotoxicity and complement-mediated lysis are two major phenomena for inhibiting protozoan infection. In some cases, local antibodies play a great role in this regard.

(a) The immunity occurring in *Babesia* infection is interesting, wherein infected erythrocytes incorporate antigens on their membrane and the antibody molecules adhere to the surface of the RBC. This process is called as opsonization; thus, they are cleared along with RBCs by phagocytic cells.

(b) T cytotoxic cells display cell-mediated immunity in the case of the endogenous antigen present within the cells. The cytotoxic T cells release toxic substances that kill the organisms. Cytotoxic T cells kill the infected lymphoblasts in *Theileria* infection.

(c) In the case of *Toxoplasma* infection both cell-mediated and humoral immune responses have been found. Usually, the sensitized T lymphocytes release cytokines that activates the fusion of the lysosome–phagosome system in the macrophage and after the fusion of the phagosome with the lysosome, lytic enzymes are released that kill the organisms. But, *Toxoplasma* sp. has enormous potential to evade host immunity. They inhibit the fusion of the phagosome and lysosome. The organisms can survive in these cells.

(d) In *Tritrichomonas* infection, the release of local antibody (IgE) occurs that binds to the antigen. In this situation, the Fc receptor present on the basophil and mast cells bind to the Fc part of the antibody. Then, degranulation of mast cells or basophils occurs. These granules release histamine, prostaglandin, and leukotriene. This initiates a Type I hypersensitivity response and this hypersensitivity reaction increases the permeability of blood vessels by the action of histamine; and IgG can easily pass through the blood vessels as the permeability has been increased and reaches the site of infection and kills the organisms.

(See also Chap. 8 for detailed description)

Immunodiagnosis

Different immunodiagnostic tests are performed for the diagnosis of parasitic infections. Agar gel precipitation test, counter immuno-electrophoresis, Bentonite flocculation test, latex agglutination test, complement fixation test, immunofluorescence test, indirect hemagglutination test, and radioimmune assay are some different immunodiagnostic tests. But, recent immunodiagnosis has been pinpointed toward more specific tests like ELISA, indirect ELISA, sandwich ELISA, micro-ELISA, dot blot ELISA, dipstick ELISA, etc. Varieties of modifications are being incorporated into ELISA as per requirement because simple ELISA is not able to give good results in some specific cases. In the earlier decades, crude antigens prepared from the adult, larva, and eggs were used for immunodiagnostic studies. In those cases, positive results were found obviously but the tests were not specific and were precipitating the chances of cross reactions. To overcome this problem, work on using different purified proteins started. Many defined antigens have been isolated from flukes, such as glutathione S-transferase, FABP, cathepsin L, hemoglobin, etc. Of these proteins, cathepsin L has been proved to have good diagnostic value. The work has not been stopped here. These immunodominant proteins are expressed *in vitro* by recombinant DNA technology to generate large amounts of antigen at a time. In the similar way, many specific proteins have been isolated and are also procured by recombinant DNA technology for other parasites.

Some immunodiagnostic tests have been described as follows:

Indirect Fluorescent Antibody Tests

In an indirect fluorescent antibody test, a reaction occurs between a particulate antigen and an immune serum. Antiglobulins are labeled with fluorescin dye (e.g., fluorescein isothiocyanate or rhodamine). In UV light, these fluorochromes are capable of emitting visible light. Usually, corpuscular antigens of parasites are used. Frozen sections of helminth or smear of intact protozoa is mounted on the glass slide. Fixation is done by controlled heat or chemicals like methanol, formalin, glutaraldehyde, acetone, etc. Then, the antigen is treated with sera. Slides are washed. In positive cases, the antibody reacts and binds to the surface antigen of the parasite and forms an antigen–antibody complex. The slides are washed again and fluorochrome-labeled antiglobulin is added and incubated again. The slides are further washed and examined under a fluorescent microscope. This test is sensitive as well as specific. This test can be done for the diagnosis of *Toxoplasma*, *Babesia*, and many other parasites.

Indirect Hemagglutination Test

This test is very common, simple, and widely used. This test is sensitive. It is simply an agglutination reaction between coated antigen- and serum. Sheep RBCs are used here. Several compounds are used for the attachment of sheep RBCs with parasitic antigens. These are formaldehyde, glutaraldehyde, tannic acid, etc. The antigen-coated RBCs are kept in tubes or microtiter plates and mixed with a fixed amount or serially diluted sera of the suspected animal. In positive cases, erythrocytes form a "net pattern" at the button of tubes or wells. In negative cases, erythrocytes form simply a button. This test is done for the diagnosis of different parasitic infections, such as schistosomosis, fasciolosis, etc. (See Chap. 13 for other immunodiagnostic tests)

Vaccines

The first step to developing an immunological control is to find out a potent antigen for immunization of the hosts. The immunogens found out in the earlier decades were mostly crude antigens that were used in different research trials. The crude antigen failed to confer significant protection against parasitic infection. Later, standard methods were developed to isolate the defined protein by different purification techniques. It is very difficult to produce a good vaccine against the parasites due to the complex nature of the protein and frequent variation of the protein. The scientists are involved in producing a potent vaccine. But, till date, a very limited number of vaccines are available in the field of parasitology.

Live Vaccines

Many works were carried out on the use of live vaccines against different parasitic infections. In this technique, the animals are inoculated with the organisms collected from the infected animals after culturing or propagating. In most of the cases, the result was not encouraging because of many drawbacks. The main complications include manifestation of the disease from the inoculated organisms due to an imbalance of the dose of the inoculum and the variation of the susceptibility. This type of vaccination strategy was applied in many animals against different protozoan diseases like coccidiosis, babesiosis, trypanosomiasis, and theileriosis. Efforts were made to produce a potent vaccine against coccidia using live oocysts. **Coccivac** was produced using eight species of *Eimeria* and was commercially introduced in the field. Later, a number of anticoccidial vaccines were developed based on similar technology; they are Immucox, Livacox, etc. Subcutaneous inoculation of infected (*Babesia bigemina*) blood from carrier animals to susceptible calves and treatment with an anti-babesial drug (Diminazine) is an established practice to provide immunological control for babesiosis.

Live Attenuated Vaccines

There have been many efforts regarding the use of attenuated vaccines against different parasitic infections. Taking the drawbacks of live vaccine into account, the concept of attenuated vaccines emerged. Many vaccination trials were carried out using live attenuated organisms. The organisms are collected from the infected animals and cultured for propagation. Attenuation of the parasite is done to reduce its virulence. Radiation, chemical treatment, and genetic manipulation of the parasites are done to attenuate them. The attempt of lungworm vaccination was done using irradiated third-stage larvae. The vaccines against lungworms (*Dictyocaulus filaria*, *Dictyocaulus viviparus*) consisted of two doses of 1000 irradiated larvae at 1-month intervals. Anti-hookworm vaccination using X-irradiated infective third-stage larvae of *Ancylostoma caninum* was started. An effective vaccine was also produced against *Oesophagostomum*, *Bunostomum*, *Gaigeria*, etc. using irradiated larvae. Live attenuated vaccine candidates were produced by different scientists against coccidiosis, babesiosis, theileriosis, trypanosomosis, and toxoplasmosis. A potent cell-culture vaccine containing irradiated schizonts is used in the field against theileriosis in the name of Rakshavac T. Attenuated anticoccidial vaccines are produced using three attenuation protocols: gamma irradiation, embryo adaptation, and development of precocious strains. Precocious strains of *Eimeria* species are developed by repeated in vitro passage of virulent parasites. The early maturing and less pathogenic oocysts are selected.

Killed Vaccine

In this type of vaccine, killed parasites are used as vaccine candidates. Bovilis Neoguard is a vaccine against *Neospora caninum* that is prepared using killed tachyzoites of *Neospora caninum*. Anaplaz and Giardiavax are two other killed vaccines against anaplasmosis and giardiasis.

Recombinant Vaccines

A particular parasitic protein can be produced in the laboratory by recombinant DNA technology. For this purpose, the target gene responsible for the expression of a particular protein is taken out from the whole genome of the parasite using different restriction enzymes. The gene is amplified by PCR. The PCR product contains multiple copies of the target gene. This target gene is inserted in the plasmid. The plasmid has a complementary site where the target gene can be ligated. This gene is attached to the complementary site by the ligase enzyme. Now, the plasmid is transfected in the host organisms. Two important organisms used for protein expression are *E. coli*, *Pichia pastoris*. The desired protein is expressed in these cells. Then, the target protein is isolated using a specific isolation protocol. The same is utilized as an antigen for immunoprotection studies or immunodiagnostic studies. In the process of preparation of recombinant proteins, many hurdles are faced by researchers. The main problem frequently encountered is protein folding. The post-translated protein does not have the same folding. However, at the present time, efforts are being made to develop different techniques for the perfect folding of proteins.

A set of recombinant proteins from many parasites has been produced. Some isoenzymes like glutathione S-transferases, cathepsin-L, FABP, and many other proteins of different parasites have already been expressed by recombinant DNA technology. These recombinant proteins exhibited satisfactory protection against fasciolosis. Recombinant GST proteins were used in different research trials of immunoprotection against schistosomosis (*Schistosoma japonicum*, *Schistosoma bovis*, *Schistosoma mansoni*, etc.). The study shows satisfactory protection against *Taenia ovis* using different recombinant GST proteins: GST 16, GST 18, GST-45W, and GST-45B/X. One recombinant *Echinococcus granulosus* protein (EG-95) conferred a high degree of protection. Recombinant vaccines have been tried against *Rhipicephalus (Boophilus) microplus* (vector of *Babesia* species), *Theileria parva*, *Babesia bovis,* and many other parasite infections.

Subunit Vaccine

Subunit vaccines is a type of vaccine that contain specific part/parts of pathogen i.e., particular protein, peptide etc. Different subunit vaccines against *Rhipicephalus* (*Boophilus*) are TickGARD, Gavac etc. Barbervax is another one subunit vaccine against haemonchosis that contains two specific gut proteins i.e., H11 and H-Gal-GP.

Nucleic Acid Vaccines

The term naked DNA vaccine was changed to nucleic acid vaccine in a meeting organized by the World Health Organization in Geneva in 1994. The principle related to this protocol is the introduction of genes through viral vectors. Different viral vectors are used for this purpose. These are vaccinia, swine pox, capri pox, feline pox, herpes, etc. The naked DNA vaccine was accidentally discovered while studying a liposomal delivery system of DNA for immunization. The researchers observed that naked DNA is more efficient than encapsulated liposomes. This provides a new era of naked DNA vaccines.

Different methods of inoculation of naked DNA vaccines:

1. Cationic liposomes
2. Gene gun
3. Pneumatic jet

Composition

The naked DNA vaccine is composed of three components as follows:

1. Plasmid
2. A strong viral promoter flanking the target gene
3. Polyadenylation transcription termination sequence

Once the complex containing the aforesaid components enters a cell, the viral promoter starts transcription of the gene. After complete transcription, the **polyadenylation transcription termination sequence** terminates the synthesis. Then, the transcribed mRNA binds to the ribosome and starts translation of proteins. **SV 40 promoter, CMV (cytomegalovirus), RSV virus, etc.** are very useful promoters and are widely used. The plasmids are isolated from bacteria, *E. coli* and these plasmids are used for the development of the complex of naked DNA vaccine. The target gene is inserted in the plasmid along with promoter gene sequences. The promoter gene is essential for translation.

Different Vaccines and Their Candidates

Category	Name of vaccine	Vaccine candidate	Animals
Live virulent strain	Coccivac B	*Eimeria acervulina, E. tenella, E. maxima, E. mivati*	Broiler birds
	Coccivac D	*E. acervulina, E. mivati, E. necatrix, E. hagani, E. tenella, E. brunetti, E. maxima, E. praecox*	Chicken
	Inovocox	*E. tenella, E. acervulina, E. maxima (two strains)*	Chicken
	Inovocox EM1	*E. tenella, E. acervulina, E. maxima (one strain)*	Chicken
	Immucox C1	*E. acervulina, E. tenella, E. necatrix, E. maxima*	Chicken
	Immucox C2	*E. tenella, E. acervulina, E. brunetti, E. maxima, E. necatrix*	Broiler birds
	Advent	*E. maxima, E. tenella, E. acervulina*	Chicken
	Nobilis COX-ATM	*E. tenella, E. acervulina, E. maxima*	Layers/breeders
Live attenuated strain	Paracox 5	*E. acervulina, E. maxima, E. tenella, E. mitis*	Broiler birds
	Paracox 8	*E. mitis, E. acervulina, E. brunetti, E. necatrix, E. tenella, E. maxima, E. praecox*	Broiler birds
	Livacox T	*E. acervulina, E. maxima, E. tenella*	Broiler birds
	Livacox Q	*E. tenella, E. acervulina, E. maxima, E. necatrix*	Layer and pullet
	Eimeriavax	*E. tenella, E. necatrix, E. maxima, E. acervulina*	Broiler birds
	Raxavac T	Live attenuated schizonts of *Theileria annulata*	Cattle
	Dictol	Irradiated attenuated larvae (third stage) of *Dictyocaulus viviparus*	Cattle
	Difil	Irradiated attenuated larvae (third stage) of *Dictyocaulus filaria*	Sheep
	Toxovac	Live attenuated S48 strain of *Toxoplasma* sp.	Sheep
Killed vaccine	Neospora Bovilis Neoguard	Killed tachyzoites of *Neospora caninum*	Pregnant cows, heifer
	Anaplaz	Killed *Anaplasma marginale*	Cattle
	GiardiaVax	Killed cultured trophozoites of Giardia lamblia	Dog
Subunit vaccine	TickGARD	Recombinant protein (Bm86) of consealed gut protein of *Rhipicephalus* (*Boophilus*) *microplus*	Cattle
	GAVAC	Recombinant protein (Bm86) of consealed gut protein of *Rhipicephalus* (*Boophilus*) *microplus*	Cattle
	Gam22	Recombinant *E. tenella* protein	Poultry birds
	Barbervax	Recombinant vaccine containing gut membrane proteins (Hƚ1and H-gal-GP)	Sheep
Transgenic parasite vaccine	AMA 1	*E. maxima, E. tenella*	Poultry birds
	IMP1	*E. tenella, E. maxima*	Poultry birds

Note: The above list is the information of vaccines. Some vaccines are not available at present

Immunopathology

Scientists of the present day do not stop only at the traditional pathological study of parasitic diseases. However, they are rapidly proceeding to understand the exact molecular mechanism of pathology. Immunopathology has taken a great dimension and innumerable research works have been done to unfurl the role of different immune cells and immune substances. These are T cells, B cells, complement, immunoglobulin, cytokines, etc. There are balanced activities of these immune cells and immune substances. The biochemical substance (cytokines) that is released by any cell, is obviously influenced by the genetic map of that cell. Hence, the molecular pathology is more significant. Similarly, many events that happened due to different parasitic infections can be unfurled and analyzed. However, we will have to wait a much more time for the same. The initial work is already in progress. In many research studies, cytokine therapy, and immune cell therapy have been started. At present, we can describe the inflammation in a more precise way than simply describing the rubor, dolor, calor and tumor.

Remote Sensing and Geographic Information System in Parasitology

10

Remote sensing is a recently developed technology by which information about the earth's surface, atmosphere, etc. can be gathered from a distance through satellite or sensor-based aircraft. The sensors collect the data and the data is deduced into a comprehensive form. In the recent days remote sensing is applied in different field of science including biological science. This is very useful for envisaging the occurrences of parasitic diseases based on image analysis of geoclimatological condition of a large area.

Satellite and Imaging Scanner System

Satellite remote sensing contains

(a) Diverse spatial resolution
(b) Temporal resolution
(c) Spectral resolution

Spatial Resolution

This specifies the pixel size of a satellite image of the earth's surface .

Temporal Resolution

The temporal resolution specifies revising the frequency of a satellite sensor for a particular location.

Spectral Resolution

The number and position of bands in the electromagnetic spectrum is taken into consideration for the collection of reflected radiance. The resolution derived is called as spectral resolution. This is the capability of the sensor to detect different wavelengths.

What is a Spectral Signature?

A spectral signature indicates the differentiation of spectral lines as a result of variation in the emittance of a material which is reflected as a variation in the wavelength. The interaction between incidental EM wavelength and material of that particular section of the electromagnetic spectrum is critically examined. The concept of spectral signature lies in the differentiation of spectral lines. The objects on the earth can

S. C. Mandal, *Textbook of Veterinary Parasitology*, https://doi.org/10.1007/978-981-99-2924-5_10

absorb, reflect, transmit, and even can emit electromagnetic energy. Modern technology has enabled us to develop special digits or sensors which are fairly capable of measuring different types of electromagnetic energy. Thus, we can measure the changes occurring on earth or in the earth's atmosphere by calibrating spectral signatures. There are different types of light, such as visible, infrared, etc. The reflectance of this light from different objects is different. For example, the reflectance of light from water, forest, sand, soil, etc. is individually different. The reflectance of these objects is plotted and compared.

Impression of Satellite Imaging

Three types of satellite images: visible, infrared and water vapor, are found. Satellite imagery are interpreted crucially and attention is paid to seeing different band patterns as created by different objects. Size, shape, location, texture tone, shadow, etc. of different elements on the earth and earth's atmosphere are observed. There are three types of satellite images. These are true color, infrared, and water vapor. The main obstacle in satellite image detection is radiographic artefacts. These artefacts are to be corrected or eliminated to get the perfect image. Image patterns are different. A satellite can cover approximately 185 km of landscape. Straight-line images indicate human-made road, canal, etc. Volcanos appear as circular outlines. Mountains appear as wrinkles and mountain ranges as wavy lines.

Digital Image Processing

The processing of images is a very important aspect. An unprocessed image is a general image which may not provide the details of a specific field. Digital image processing is needed to focus on a particular subfield according to our needs. In this context, it is to be mentioned that there are two types of image processing: analog image processing and digital image processing. Digital image processing is better in many aspects over analog image processing. For image processing, computer algorithms are used to process digital images. In digital image processing, a much wider range of algorithms can be applied to the input data.

Fundamentals of GIS

This is a special type of computer-based information system which is used to capture, store, retrieve, process, visualize and analyse geospatial data related to the Earth's surface. For a GIS system, two types of data are used: raster data and vector data.

Components of GIS

1. **Hardware**

   Hardware is the computer by which GIS is analyzed. It possesses main components such as monitor along with accessories, CPU, RAM, etc. RAM should have not less than 8 GB, but it is better to have 16 GB memory.
2. **Software**

   There are several software tools which are used to serve the purpose. Software is composed of different tools, methodologies, etc. There are some tools which can convert physical data into digital data. The digital data are stored. These stored data can be retrieved and analyzed whenever necessitated. The software is used from the centralized computer server to desktop.
3. **Data**

   Data are the collected information. The information is stored in different codes. After decoding, the actual data are observed. In GIS, different types of data are used, such as vector data, raster data, images, etc.
4. **Method**

   Different methods are applied to convert physical data into digital data. There are different methods for the analysis of data.
5. **People**

   They are persons who work for GIS and are the main component of any GIS. The persons specialized in the field of GIS work there. They create different plans according to the need. Based on the need, different methods are developed according to the accumulated data pattern.

6. **Network**

   This is the communication of data. This is a computer algorithm which represents interconnected lines. These interconnected lines are also called as edges.

How Many Types of Data are Utilized?

There are three types of data:

(a) Spatial data
(b) Nonspatial data
(c) Metadata

Spatial Data

This is a set of data that is directly or indirectly related to location on the surface of the earth. These are also called as geospatial data. Spatial data are of two types: raster data and vector data. Raster data is composed of grid cells. The columns and rows represent the data. A geographic area is divided into many small cells of images. These are actually matrices of square areas. In the case of vector data, X and Y coordinates are used to locate lines, dots, polygons, or any other area. This has multiple dimensions.

Nonspatial Data

This set of data does not describe any location or object. These are nonphysical data, such as number, character, etc. The data is placed in tables or spreadsheets. This data is independent of geometric considerations. This has a single dimension.

Metadata

These are not the active data. These are actually the passive information about one data. That is why metadata is called "data of data". For example, suppose 70% of students passed in one subject in a certain school in a particular year. This will be called active data. Metadata includes file name, date of entry, etc., of the data.

Why Do We Need GIS?

GIS stands for Geographical Information System. Visual inspection of a particular location is difficult because any instrument and/or technical persons have to reach the place physically which is tedious and cumbersome job to be present in uncomfortable areas. But a distant imaging system is very comprehensive and convenient and can be inspected from a particular location to visualize distant places based on accumulated signals collected from that distant place. Moreover, GIS can cover a large area with many elements such as roads, forests, dams, water bodies, atmosphere, etc. The following can be done by GIS:

1. We can analyse geographically referred data and can generate new information from that data using different analytical methods.
2. We can simulate, compare, and restructure the data collected from different sources.
3. By GIS we can locate or identify any object in a particular area. We can know what changes are going on in a particular location. Data is generated at different times that are critically analysed to determine the changes in that location.
4. There are many specific applications of GIS in the fields of mapping, science, industry, climate, health management, transportation, archeology, telecommunication, etc.

Raster Data Representation

Rasters are digital aerial photographs or imagery which is digitized in the form of grids or matrices of cells or pixels. These matrices of cells or pixels are arranged into columns and rows. Each cell of the matrix bears a specific value which provides information about something.

1. **Continuous Data**

   Continuous data represents a phenomenon but not an object. Discrete or thematic data represent objects but continuous data represent phenomena like temperature, air, etc.
2. **Thematic Data**

   This is also known as discrete or discontinuous data. This data is mainly related to objects. This has a starting point and an end point. This can be easily identified by analyzing its boundaries. Some good examples are trees, lakes, etc.

   Use of GIS in the Field of Parasitology

   There are immense applications of GIS in the field of parasitology. The locations of niches of the parasites as well as the breeding areas of vectors are identified. The expansion or reduction of these breeding areas can also be determined in relation to different environmental factors at different time intervals.

Difference Between Raster Data and Vector Data

Raster data consists of a matrix of squares containing many features within this limit. Different names have been given to this matrix of squares such as pixels, cells or grids. Vector data actually uses two coordinates: X and Y, to define particular locations of points, lines, polylines, polygons, etc.

Data Input, Editing, Analysis and Modeling

GIS uses mainly two types of data as follows:

1. Raster data (spatial)
2. Vector data (spatial)

The description has been given earlier.

Digitizing

This is a method by which geographic data or any other data, such as text, sound, and signal, is converted into digital format.

There are two types of digitizing:

1. Manual digitizing

   In this technique, a magnetic pen is used to create different features as per need and the same is input into the computer.
2. Automatic scanning

   The scanner automatically scans the spatial data.

*GIS Output as a Map

GIS has four components as follows:

(i) Data input

Data generated by collected signals.

(ii) Database management

Collected data are stored for a long time.

(iii) Analysis

Data can be analyzed whenever required.

(iv) Output

Output may be as follows:

(a) Maps
(b) Charts
(c) Table
(d) Reports
(e) Statistics

Mainly GIS output is a map. A map can show a variety of data such as streets, forests, buildings, water bodies, etc. Google maps is an example.

Types of Maps

Category Map

In this type of map, data of different categories are represented in different shades.

Quantity Map

The density of a particular category is represented in a quantity map. It is color coded which uses different shades of same color indicating different quantities of something. For example, the highest population is represented by red color shade and the lowest population by white shade.

Cluster Map

The density of a particular category in a particular area is represented in this map. This type of map combines color, shape, labels to cluster densely packed points of data.

Case Study by GIS

To study a GIS pattern which correlates with a particular parasitic disease, the following components are taken into consideration:

1. Host component
2. Parasite component
3. Vector/intermediate host component
4. Environment component

For a case study of any parasitic disease the following steps are done:

1. **Selection of** a **problem**
   The first step is to select a problem or determine a purpose. This may be as follows:
   (a) Forecasting any parasitic disease.
   (b) Determination of the spread of parasitic disease.
   (c) Determination of intermediate host/vector population.
   (d) Determination of prevalence of zoonotic diseases.
   (e) Determination of risk factors for any parasitic diseases.
   (f) Determination of status of previous parasitic diseases.
   (g) Identification of new foci of parasitic diseases.
2. **Retrieving geospatial data related to the problem**
   (a) Data of a particular location.
   (b) Data of a particular time.
3. Identification of statistical software of GIS system.
4. Development of problem-related graphs, tables, etc.
5. Analysis of geospatial data related to environmental factors, such as temperature, rainfall, humidity, etc. by using different statistical software.

Molecular Parasitology with Diagnostics and Vaccine

11

Basics of Molecular Biology

In this section, the prokaryotic and eukaryotic types of cells are discussed in brief because we should know about the different components of cells. A parasitologist, aspiring to molecular knowledge of veterinary parasitology, should have a proteomic and genomic background. In the present day, every parasitologist should have a background in biochemistry because today's parasitology deals with immunoparasitology and molecular parasitology simultaneously. Therefore, some basic knowledge of biochemistry is obviously required to understand the molecular trend of research in veterinary parasitology. There are two types of cells. These are as follows:

Prokaryotes

Prokaryotic cells are the simplest living cells having simple cellular mechanisms. They are abundantly available in different environmental niches. The prokaryotic cell is enclosed by a plasma membrane consisting of two lipid bilayers within which protein molecules are present which perform different functions for entry and exit of different small molecules. A rigid cell wall is present in most of the prokaryotes. This protects the cells from any shrinking or swelling in the environment. The cytoplasm of the cell contains a single circular chromosome found in the nucleoid and the same is attached to the membrane. Plasmids having limited genetic information are also found in the cytoplasm. Others are RNA, ribosomes, etc. Some proteins are found attached to the plasma membrane but distinct subcellular organelles are absent in prokaryotic cells. The outer wall may carry pili used for the attachment to other cells. Flagella may be present which helps in locomotion by their rotating movement. The cellular components are the cell wall, plasma membrane, cytoplasm, nucleoids, plasmids, and ribosomes.

Eukaryotes

The characteristic features of eukaryotes are membrane-bound organelles. Every organelle is identically different in structure and function. They are surrounded by a plasma membrane. Plants, fungi, and protists have rigid cell walls. Apart from the presence of organelles, an array of protein fibers called the cytoskeleton is found. These fibers include microtubules made up of tubulin and microfilaments made of actin. These cells contain plasma membrane, nuclei, cytoplasm, mitochondria, Golgi apparatus, ribosomes, microbodies, etc. The eukaryotic nucleus is membrane-bound containing multiple chromosomes carrying vast genetic information. Each chromosome carries a single DNA molecule. The

S. C. Mandal, *Textbook of Veterinary Parasitology*, https://doi.org/10.1007/978-981-99-2924-5_11

nuclear envelope made up of a lipid double membrane bears many pores which act as passageways for different molecules. In the nucleus, transcription of RNA occurs; and the processed RNA molecules pass into the cytoplasm where translation of protein takes place. The nucleus contains specific bodies called nucleoli in which r-RNA is synthesized.

Mitochondria and Chloroplasts

The cells contain an important organelle called mitochondria which is called the powerhouse of the cell. Formation of adenosine 5′ triphosphate (ATP) occurs in the mitochondria. One cell may contain 1000–2000 mitochondria. The mitochondria have a smooth outer membrane. The inner membrane is convoluted and forms typical protrusions called as cristae. Small circular DNA molecules, ribosomes and mitochondrial-specific RNA are found in the mitochondria. Some proteins are also produced in mitochondria. The chloroplast present in plants is comparable to mitochondria.

Endoplasmic Reticulum

This is a massive membrane system present in the cytoplasm and continues to the nuclear envelope. There are two types of endoplasmic reticulum: smooth endoplasmic reticulum and rough endoplasmic reticulum (RER). The smooth endoplasmic reticulum possesses different enzymes performing detoxification of foreign material. RER is so-called because ribosomes are attached to the endoplasmic reticulum. The lipids and proteins synthesized on the RER are transported to the Golgi apparatus.

Microbodies

Lysosomes, peroxisomes, and glyoxysomes are different microbodies. Lysosomes are important microbodies in which many important enzymes are found. These enzymes play important roles by degrading protein, lipid, carbohydrate, etc. Peroxisomes perform a great role in which free radicles, H_2O_2, etc. are destroyed. Likewise, in plants, glyoxysomes are specialized plant peroxisomes which carry out reactions of the glyoxylate cycle.

Carbohydrates

Sugar has different numbers of carbons. These are trioses (three carbons), tetroses (four carbons), pentoses (five carbons), and hexoses (six carbons). Pentoses are found in animal cells. A reduced form of pentose, such as deoxyribose, is found in DNA. Hexoses are the main ingredients of nutrients. They are aldose or ketose in accordance with the presence of aldehyde or keto group. It is to be kept in mind that all hexoses are aldoses (glucose, galactose, and mannose) except fructose which is a ketose having a keto group. The most important hexose is glucose. It is the prime source of energy and is a cellular fuel. Fructose is a fruit sugar and is the sweetest sugar. It has a ketonic group. Benedict's test and Fehling's test are based on this property for the detection of glucose in the urine. Galactose is a major component of milk sugar. Galactolipids are found in plant leaves. Mannose is of aldohexose series of carbohydrate. Mannitol is an alcohol of mannose found commonly in brown algae. The general formula is CH_2O. Aldose has aldehyde groups. Ketose has ketone groups. Oligosaccharides are formed by the condensation of 2–10 monosaccharide molecules. Disaccharides are formed by two molecules of monosaccharides. One molecule of H_2O is released in the formation of saccharides. One molecule of H_2O is released in the formation of the 1–4 glycosidic bond. Sucrose, maltose and lactose are common examples of disaccharides. Sucrose is especially made up of glucose and fructose found in sugar cane, sugar beet, etc. Free aldehyde or ketone groups are absent in sucrose. Maltose is a which gives two molecules of glucose on hydrolysis. Maltose is produced commercially by the hydrolysis of starch by the enzyme diastase. Lactose is available in milk.

Lactose does not occur in nature except as a product of the mammary glands. Cellobiose has two molecules of beta-D glucose; it cannot be spilt by mammalian enzymes. Raffinose is a trisaccharide having galactose, glucose, and fructose molecules.

Lipids

Lipids are insoluble biomolecules. But these are soluble in nonpolar organic solvents, such as chloroform and benzene. Several examples of lipids are fatty acids, waxes, glycolipids, phospholipids, steroids, etc. A fatty acid is a long hydrocarbon chain. The chain is saturated or unsaturated. Saturated fatty acids have single carbon–carbon bonds whereas unsaturated fatty acids have one or more double bonds in the hydrocarbon chain. The common formula of saturated fatty acids is $Ch_3(Ch_2)_nCOOH$. A fatty acid having only one double bond is called as monounsaturated; examples include palmitoleic acid and oleic acid. If a fatty acid has more than one double bond such as two in linoleic acid, three in linolenic acid and four in arachidonic acid it is called as polyunsaturated acid. Unsaturated fatty acids are more abundant in nature as compared to saturated fatty acids. An example of the most common monounsaturated fatty acid is oleic acid. Double bonds are absent in saturated fatty acids. Some polyunsaturated fatty acids are called as essential fatty acids which are present with food. Examples are arachidonic,' linoleic, and linolenic acids.

Proteins

Proteins are polymers of amino acids. There are some amino acids which have charged side chains, such as arginine, histidine, lysine, glutamic acid, and aspartic acid. There are some amino acids which have polar uncharged side chains. These are asparagine, glutamine, cysteine, serine, and threonine. There are some amino acids which have nonpolar aliphatic side chains, such leucine, isoleucine, valine, glycine, alanine, methionine, and proline. There are some amino acids which have aromatic side chains, such as tryptophan, tyrosine, and phenylalanine. There are some amino acids which have polar uncharged side chains. These are asparagine, glutamine, cysteine, serine, and threonine.

Basic Structure of Amino Acids

- The basic structure of amino acids includes an alpha carbon linked to a carboxyl group, a primary amino group, a proton, and a specific side chain.
- The side chain (R) differs in different amino acids.
- The alpha carbon has four different groups except glycine.
- The amino acids behave as both acid and base and thus are called amphoteric in aqueous solution.
- Different amino acids have different properties depending on the shape, size, charge, and chemical reactivity.
- There are some nonstandard amino acids, such as 5 hydroxy lysine, 4-hydroxyproline, etc.

Basic Introduction to Amino Acids

Glutamine: This is one amino acid which is an amide derivative of glutamic acid.

Asparagine: This is one amino acid which is an amide derivative of aspartic acid.

Threonine: This is one amino acid which has a hydroxyl group.

Serine: This is one amino acid which has a hydroxyl group.

Cysteine: This amino acid has a thiol group (sulfhydryl group) which oxidizes to cysteine.

Methionine: This is one amino acid. It possesses one sulfur atom in a thio ether link within its alkyl side chain.

Valine: This is one amino acid having a hydrophobic alkyl group for the side chain by which it exhibits hydrophobic interactions in protein structures.

Alanine: This is one amino acid having a hydrophobic alkyl group for the side chain by which it exhibits hydrophobic interactions in protein structures.

Leucine: This is one amino acid having a hydrophobic alkyl group for the side chain by which it can exhibit hydrophobic interactions in protein structures.

Isoleucine: This is one amino acid having a hydrophobic alkyl group for the side chain by which it can exhibit hydrophobic interactions in protein structures.

Tyrosine: This is one amino acid which has a bulky hydrophobic side chain. The UV absorption property of this amino acid is because of its aromatic structure.

Tryptophan: This is one amino acid which has a bulky hydrophobic side chain. The UV absorption property of this amino acid is because of its aromatic structure.

Phenylalanine: This is one amino acid which has a bulky hydrophobic side chain. The UV absorption property of this amino acid is because of its aromatic structure.

There are two types of proteins. These are as follows:

Fibrous Proteins

Fibrous proteins are those proteins which have a high axial ratio (length:width). Example: keratin and silk fibroin are found in hair and wool.

Globular Proteins

The protein molecules are folded compactly forming a spherical shape in water. (Example: different enzymes).

Glycoproteins

Glycoprotein has bound nonprotein material (carbohydrate).

Lipoproteins

This protein has bound nonprotein material (lipid).

Definition of Peptides

Dipeptide: This is formed by the linkage of two amino acids by a peptide bond.

Polypeptide: Many amino acid residues are linked (by peptide bond) to form polypeptide. The structural backbone of the polypeptide is formed by the addition of a repeating sequence of alpha carbon. The side chains of amino acids provide the functionality of a protein.

Structure of Proteins

Primary Structure of Proteins

- The primary structure is formed by the covalent linking of the alpha-carboxyl group of one amino acid with the alpha-amino group of the next amino acid by an amide bond.
- This amide bond is also known as a peptide bond. Polypeptides are formed by many amino acids linked by peptide bonds.
- The polypeptides are directional with N-terminus and C-terminus.
- One end of the polypeptide has an alpha amino group while the other end has a free alpha–carboxyl group.
- The sequence of amino acids from the N-terminus to the C-terminus of a polypeptide is called its primary structure.

Secondary Structure of Proteins

- Polypeptides can form folds and form a number of regular structures by hydrogen bonding.
- In this way, the backbone of polypeptide forms a helix structure (L-helix) with 3.6 amino acid residues per turn.

- Each peptide's N-H group is hydrogen bonded to the C=O group three residues away. Example: The alpha-helical structure is found in globular and fibrous proteins.
- Another secondary structure is the formation of a Beta-sheet by placing the side chain (R) alternatively above and below the sheet.
- N-H and C=O groups of one section of polypeptide are hydrogen bonded to the complementary groups of another section.
- The sheet may be parallel when these sections run in the same direction (N-terminus to C-terminus) or it may be antiparallel when the sections alternate (C terminus to N terminus).
- Different sections of polypeptide chains are arranged side by side and give a sheet structure with the side chains (R) projecting alternately above and below the sheet.

Tertiary Structure of Proteins

It is a structure formed by the alpha-helix, beta-sheet, loops and other minor secondary structures. It is an inherent property of protein. Specific proteins get folded into a tertiary structure. Misfolding of the tertiary structure is prevented by some other proteins called chaperones. The tertiary structure is a three dimensional structure that is formed by specific folding in such a way that the hydrophilic side chains are located mainly on the exterior while hydrophobic amino acids are buried in the interior. This provides overall stability to the three-dimensional structure. The whole setup is further stabilized by different non-covalent interactions between side chains comprising hydrogen bonding, van der Waal's forces, electrostatic salt bridges, etc. Apart from non-covalent interaction, two cysteine residues form a covalent bond.

Quaternary Structure of Proteins

The quaternary structure is formed by stabilization of two or more polypeptide chains (subunits). These subunits may be identical or different. For example, hemoglobin has two alpha-globin and two beta-globin chains.

Function of Proteins

- Enzyme protein performs many biochemical reactions.
- Hemoglobin transport O_2 to RBC.
- Transferrin transports iron to the liver and lipoprotein transports dietary fat to the blood. Body fluid, muscle, and skin are structurally composed of protein.
- Protein is essential for growth.
- Some proteins are receptor proteins which perform signaling roles. The receptors transmit signals.
- Cytokines are proteins and they play a great role in immunological reactions.

Structure of Nucleic Acids

There are four heterocyclic bases of DNA: purine—adenine and guanine; pyrimidine—thymine and cytosine. Bases covalently bonded to pentose sugar molecules are called nucleosides.

- Ribose + Base = Ribonucleoside or nucleoside in the case of RNA.
- 2′-deoxyribose + base = 2′-deoxyribonucleoside or deoxynucleoside in case of DNA.

Nucleotides

Phosphate group binds to sugar of a nucleoside and the resultant molecule is called nucleotide.

Deoxyribonucleic Acid

- The bases of nucleic acid are covalently joined to a pentose sugar. Pentose sugar is a component of nucleotides.
- Ribose sugar is found in RNA whereas 2′-deoxyribose is found in DNA. The pentose sugar is attached to the 1′ position (N-1) of pyrimidine and 9′ position (N-9) of purines.

- These sugars attach to the bases by glycosidic bonds.
- DNA forms polymers by the joining of deoxyribonucleotides. RNA forms polymers by the joining of ribonucleotides.
- The phosphate group joins with 5′ hydroxyl of one ribose and 3′ hydroxyl of another ribose. In this way, a chain of nucleotides is formed. This kind of link or bond is called as a phosphodiester bond.
- Two chains characteristically wind around each other and result in a right-handed double helix.
- Two strands are joined by non-covalent hydrogen bonding. Two strands are directed antiparallel in terms of the 5′–3′ direction.
- The striking feature of these two strands is that adenine pairs with thymine. Guanine pairs with cytosine, thus creating purine–pyrimidine pairs.
- The sequence of one strand indicates the sequence of the other strand, i.e., if A (adenine) is in one strand, T would exactly be at the opposite strand.
- The planar bases remain in the centre of the helix and the negatively charged sugar-phosphate backbone remains outside the helix.
- The two strands run antiparallel and around 10 base pairs are found per turn.
- RNA is a single-stranded molecule and it does not form any helical structure but it forms a globular structure and local helical structures are formed due to intramolecular hydrogen bonding.

DNA Tertiary Structure

In the tertiary structure of DNA, large-scale folding in linear polymer occurs. A three-dimensional shape is found. Handedness (right or left), number of base pairs per turn, length of the helix turn and size difference between major and minor grooves indicate different areas.

Different Types of DNA

A-DNA: The main characteristic feature of A-DNA is that it has wide and shallow minor grooves making it more accessible to protein. On the other hand, it has deep and narrow major grooves which is not easily accessible to the protein. The base pairs are tilted relative to the axis of the helix.

B-DNA: Unlike A-DNA, B-DNA has a wide and deep major groove making it more accessible to proteins; however, the minor groove is narrow and deep. The base pairs of B-DNA are almost perpendicular to the axis of the helix.

Z-DNA: It seems to be a left-handed double helix. It has a more narrow and elongated helix as compared to A-DNA and B-DNA.

DNA Damage

- DNA damage occurs in different areas of DNA molecules.
- In highly acidic conditions, the nucleic acid is hydrolysed and get separated into bases, ribose or deoxyribose and phosphate.
- Alkali also changes the tautomeric state of the bases.
- A high concentration of formaldehyde and urea causes denaturation of both DNA and RNA by disrupting hydrogen bonding.
- High temperature also causes denaturation of DNA.
- High pH also causes denaturation of DNA as well as RNA resulting in cleavage of the phosphodiester backbone.

Density Gradient Centrifugation

- DNA is mixed with high molecular weight salt (i.e., cesium chloride, CsCL) and centrifuged at very high speed. DNA samples migrate to a sharp band at the position in the gradient corresponding to their own buoyant density.

- This is called as equilibrium density gradient centrifugation or isopycnic centrifugation.

Spectroscopic Properties/Thermal Properties

- Aromatic bases of nucleic acids can absorb UV light with a lambda max of 260 nm. The concentration of nucleic acid can be measured at the absorbance of 260 nm.
- Purity of double-stranded DNA sample can be determined by measuring the ratio of A260/A280.
- It is pure if the value is 1.8. This value increases when there is RNA contamination.

Functions of DNA

1. This is a storehouse of hereditary information. Three types of genes are mainly found: structural genes (cistron), regulator genes, and operator genes.
2. DNA can perform replication which provides a duplicate copy at the time of cell division, which is called autocatalytic function.
3. Different types of RNAs are synthesized from template DNA.
4. Different types of proteins are synthesized by protein expression machinery.
5. Mutations occur in DNA due to sudden inheritable variation.
6. During crossing over (meiosis), variations occur due to a mixture of paternal and maternal genes.
7. To correct heritable disease, gene therapy is possible.
8. Different developmental stages occur due to signals operated by DNA.
9. DNA fingerprinting is possible for studying the hypervariable minisatellite DNA sequences.
10. A specific target gene in pathogens is made silent using a complementary nucleotide sequence(gene silencing technique).

Histone Proteins

A nucleosome or nu-body is a subunit of chromatin formed by histones wrapped in DNA. The hydrophobic regions remain towards the centre and positively charged regions remain outwardly that attracts the negatively charged phosphate-containing DNA. The length and thickness of nucleosomes are 10 nm (100 A) and 5–5.7 nm (50–57 A), respectively. A DNA of 200 bp forms 1.75 turns over a nucleosome. Under the electronic microscope, it appears as beads of string. Two types of chromatins are found: euchromatin and heterochromatin. Euchromatin is lightly stained and loosely packed for which it is transcriptionally active. However, the heterochromatin is densely packed and transcriptionally inactive. Nonhistone chromosomal proteins are also used for chromatin packing. Histones are small protein molecules. There are four families of core histones: H2A, H2B, H3, and H4. H1 is another kind of protein family which has different characteristics. Histone is a highly positive charge protein which enables them to strongly bind with negatively charged DNA molecules. A histone octamer forms a characteristically wedge-shaped disk, surrounding which the DNA molecule wraps. The histone molecules protect the DNA from the effect of micrococcal nucleases. They remain packed in the nucleus. A DNA–protein complex is formed for packaging in the nucleus and is known as chromatin.

DNA Replication

Replication starts at a particular region of the DNA. This region has a particular nucleotide sequence. This is called as an autonomous replication sequence or ARS. This is also called as origin of replication (Ori). In prokaryotes, a single Ori is found and in eukaryotes several Ori is found. A replicating unit is called a replicon. The DNA of eukaryotes possesses several replicons whereas the DNA of prokaryotes acts as a single replicon. Replication may be unidirected or bidirected. Four types of deoxynucleotides: dAMP, dGMP, dCMP, and dTMP transform to dATP, dGTP, dCTP, and dTTP (triphosphate). The reaction between monophosphate and H_3PO_4 results in the production of triphosphate in the presence of phosphorylase. The Ori complex recognizes Ori which attracts enzymes and helicase. Helicase unbinds the DNA helix by breaking H bonds. The double strand gets separated into a single strand and these strands are stabilized by single-stranded binding protein (SSBP). Then, there occurs the release of Topoisomer I and II. A "Y"-shaped replication fork is produced. An RNA primer is synthesized by DNA-dependent RNA polymerase or primase. The RNA primer acts as the 5′ end of the new strand and provides area (3-OH group) for the joining of DNA nucleotides. Separated new strands act as templates. Complementary phosphorylated nucleotides bind to the new strand. DNA strands at the replication fork function as templates and the nitrogen bases attract phosphorylated nucleotides. Polymerase III of prokaryotes and polymerase delta of eukaryotes, energy (ATP) and Mg^{2+} cause the formation of phosphodiester bonds between successive nucleotides. The DNA polymerase accurately attaches nucleotides but sometimes the wrong nucleotide is attached and the same is removed and replaced by the correct nucleotide by the action of different enzymes. The phenomenon of DNA replication is based on the principle that each strand of DNA double helix carries the same information. Therefore, each parental strand gets separated and each separated part acts as a template to which a new complementary daughter strand is formed by the normal base pairing rules (A with T; G with C). Different enzymes perform their individual activity as follows:

- DNA polymerase-I performs proofreading and repair.
- DNA polymerase-II performs special repair.
- DNA polymerase alpha performs replication. DNA polymerase alpha causes initiation of replication.
- DNA polymerase beta performs DNA repair. DNA polymerase gamma performs mitochondrial DNA replication and repair. DNA polymerase delta catalyze replication and participate in DNA repair.
- DNA polymerase epsilon participate in DNA replication.

Individual (newly made) double-stranded molecules pass to the daughter cells during cell division. The substrates required for DNA synthesis are dNTPs (deoxynucleoside 5′ triphosphates) such as dATP, dCTP, dTTP, and dGTP. The piece of DNA which replicates as a single unit is called a replicon. This could be called a unit of replicating DNA.

Parasitic Genomics

Genomics, by definition, is the study of the structure and function of genes. A genome is a complete set of DNA containing all of its genes. Therefore, parasitic genomics indicates the study of the structure and function of parasitic genomes. Genomics is not the study of a particular gene; rather, it encompasses the study of whole aspects of a set of genes. At present, three genomes of parasites are of importance. These are **nuclear DNA, mitochondrial DNA, and ribosomal DNA**. Many studies are being carried out on the molecular characterization of different parasites. The research workers are paying keen attention towards the use of robust molecular markers which can be helpful for the identification of parasites up to certain taxonomic levels (genus, species, strain). Parasitic genomics includes the detailed study of different DNA and RNAs. It includes sequencing of DNA and RNA by standard protocols. The analysis of sequence is done using different standard software/tools. Analysis of the nucleotide sequence helps in molecular phylogeny, molecular epidemiology, molecular-based evolutionary study, etc. Once a complete nucleotide sequence of a parasite is known, then protein profiles of that parasite can be derived.

DNA and RNA Technology

The important technologies which are frequently used in the field of veterinary parasitology are described below.

Isolation of DNA

For carrying out any DNA-based molecular work, the prerequisite is the procurement of DNA. Molecular identification of parasites by polymerase chain reaction (PCR) is the most reliable technique, for which DNA isolation from parasites is essential. These are the following steps for the isolation of DNA.

Step-1
(Lysis)
The first step of DNA isolation is the lysis of cells. This can be done in two ways: mechanically and chemically. Mechanically, this can be done by homogenization, sonication. Chemically, it is done by treating with detergent and the enzyme proteinase K. Chemical isolation is widely used.

Step-2
(Precipitation)
The next step is precipitation. The material is treated with a concentrated salt solution. After lysis, DNA becomes free from the nucleus but it still remains mixed with cellular debris materials. Alcohol is added to it. DNA is not water soluble and it is precipitated out of the aqueous solution.

Step-3
(Purification)
The next step is the purification of DNA. The material is centrifuged. The debris material is clumped by centrifugation and DNA is separated from the clump of debris. Now, the DNA has to be separated out from the reagent, salt, etc. For this, the material is subject to ice-cold ethanol extraction or isopropyl alcohol extraction. The same could be done also by phenol–chloroform precipitation. After this treatment, further centrifugation is done and DNA comes to the supernatant.

Isolation of RNA

- Lysis of cells is done by using a buffer containing sodium dodecyl sulfate (SDS), urea, phenol, guanidium isothiocyanate, etc.
- Denaturation of DNA is done by DNAase. Denaturation of protein is done by proteinase K.
- Cellular components are separated by centrifugation after adding chloroform. The material gets separated into two phases: organic phase and the aqueous phase. RNA is present in the aqueous phase.
- Recovery of RNA is done using isopropyl alcohol.

Isolation of Plasmid

- Plasmids are naturally occurring extrachromosomal circular DNA molecules 2–200 kb in size.
- These are found in many numbers (a few hundred) in *E. coli.*
- There are specific biotechnological methods by which plasmid DNA can be isolated. At first, *E. coli* (sample) is taken and inoculated into a culture broth.
- After overnight growth, the organisms enter into a stationary phase. The cells are centrifuged and cell pellets are formed.
- The cell pellet is treated with a buffer solution containing lysozyme which can digest the cell wall.
- Then, a cell lysis solution containing SDS in NaOH is added.
- This solution causes the denaturation of protein and DNA causes the hydrolysis of RNA.
- Potassium acetate is used to neutralize the solution and it precipitates denatured protein and chromosomal DNA.
- The material is further centrifuged and the supernatant contains plasmid.

- Then, the pure plasmid is isolated from the lysate by phenol extraction.

Methods of Electrophoresis of DNA

There have been many studies on the electrophoresis of parasitic DNA. This is one of the basic techniques used for carrying out different molecular works in the field of parasitology. Whether DNA has been amplified or not is generally confirmed by electrophoresis. DNA bands appear in the gel. The gel is stained with ethidium bromide (ETBR) which is mixed at the time of gel preparation. If the gel is placed in the gel documentation system, the bands are visible. The appearance of bands proves that amplification has occurred and that parasite-specific primer has worked for the amplification of specific DNA fragments. Electrophoresis is done for other purposes also. After electrophoresis, a particular band can be isolated. The particular DNA fragments are eluted from the gel and attached to the plasmid following standard protocol. This is very important for carrying out recombinant DNA technology.

Electrophoresis

- DNA can migrate in an electrophoretic field.
- Agarose gel electrophoresis is done for the migration of DNA. Agarose is a polysaccharide which forms a solid gel between 0.5% and 2% (w/v) concentration.

Fig. 11.1 Gel electrophoresis of DNA

- DNA fragments run in the agarose gel in the electrophoretic field in the presence of a specific buffer.
- Since the DNA is negatively charged it can move towards the positively charged field.
- DNA of varying sizes can be separated by agarose electrophoresis using different concentrations of agarose gel (1%, 1.5% w/v).
- DNA is stained by adding ethidium bromide to the gel. If the gel is placed under UV light, the bands of DNA are detected.
- If the digested and undigested plasmid DNA are run in different lanes along with the markers, we will get different band patterns of digested and undigested plasmids.

DNA Cloning

DNA cloning is a very important technique which is frequently used in the field of veterinary parasitology. This technique is a prerequisite for the production of recombinant proteins. The basics of DNA cloning are described hereby:

- The gene responsible for the expression of a particular protein is taken from a large genome following standard protocol.
- The fragment of the genome is attached to the complementary site of the plasmid (vector).
- The target gene attached to the plasmid and the resultant molecule is called as recombinant DNA.
- This can replicate independently without the original genome in another host species. A number of genetically identical organisms or clones containing the recombinant DNA are produced as a result of propagation of host organisms (*E. coli*).
- Manipulation of very large genomes is done in different host organisms (*Saccharomyces cerevisiae*).
- *E. coli* are prokaryotic cells which are used for the insertion of a recombinant DNA construct in these cells for expression. ***Pitchia pastoris*** are eukaryotic cells which are also used for the expression of recombinant parasitic proteins.

Development of a Genomic Library

Many works have already been carried out and many works are in progress for the development of genomic libraries of different parasites. The development of a genomic library is important for having the data of sequences of different genes present in the parasites. The basic technique is almost the same for parasites and other organisms.

Important Points of a Genomic Library

- A genomic library is a collection of different genes of parasites.
- The DNA sequence of a particular parasite is collected following standard protocol.
- The whole genomic DNA is subjected to many fragments using different restriction endonucleases.
- An individual DNA fragment is inserted into the plasmid to form recombinant DNA.
- The recombinant DNA containing the fragment is inserted into the host organism and cloned.
- There are two types of gene libraries: genomic library and cDNA library. If the library is developed by considering genomic DNA it is called as the genomic library and by considering cDNA it is called as the **cDNA library**.
- The whole genome is at first isolated from an organism. There are different protocols to extract the genomic DNA from these organisms.
- The common extraction procedures are phase extraction or protease digestion.
- These two methods are applied to remove protein, lipid and other unwanted macromolecules.
- The genomic DNA extracted in this way is a long chain of several hundred or more kilobases.
- This genomic DNA is cut into a number of fragments by physical shearing or by restriction endonucleases.
- Restriction enzymes result in the fragmentation of DNA molecules.
- Complete digestion of genomic DNA using restriction enzymes is done. The restriction enzyme is chosen as per need.
- If the genome size is small (5 kb), then the library could be developed using the plasmid.
- If the genomic size is larger, **phage lambda, yeast artificial chromosome (YAC), and bacterial artificial chromosome (BAC)** are used.

Genomic libraries have already been constructed for many parasites. Some names of the parasites, whose genomic libraries have been constructed, are *Ancylostoma ceylanicum*, *A. caninum*, *Leishmania donovani*, *Babesia microti*, *Plasmodium falciparum*, etc.

Gene Expression

If we want to know about gene expression we must know transcription. Transcription has been described as follows:

Transcription

One strand of a DNA double helix acts as a template and from this template DNA strand, an mRNA chain is generated. Three major events occur during transcription. These events are as follows:

(a) Initiation
(b) Elongation
(c) Termination

At first, the DNA is unzipped due to the disruption of hydrogen bonds by the respective enzyme (RNA polymerase). RNA polymerase is a very important enzyme which reads the DNA strand from the 3′ to the 5′ ends and synthesizes mRNA from the 5′ to the 3′ ends. The structure of RNA is similar to DNA but, in RNA, uracil takes the place of thymine. The single-stranded mRNA leaves the nucleus through the nucleopores and ultimately migrates to the cytoplasm. In the case of prokaryotic cells, mRNA is formed directly without any requirement of transcriptional modification. Whereas, in eukaryotic cells, primary transcripts are subjected to **capping with 7-methyl-guanosine and tailing with a poly-A tail**. The final mRNA is produced after the splicing of introns (noncoding parts). The part of the DNA strand which takes part in transcription is called the transcription unit. This has a promoter region at the beginning and a terminator region at the end. A structural gene remains in between the promoter and the terminator region. The promoter region occurs at the 5′ end of the template DNA and the terminator region occurs at the 3′ region of the template DNA. The promoter has different sites which attach transcription factors and RNA polymerase.

Different RNA Polymerases

RNA polymerase-I	• This enzyme is responsible for the synthesis of ribosomal RNA except for 5S
RNA polymerase-II	• This enzyme is responsible for the synthesis of mRNA, SnRNA
RNA polymerase-III	• This enzyme is responsible for the synthesis of tRNA, 5srRNA, and SnRNA

Prokaryotic RNA polymerase has five subunits. Of these, the sigma factor is very important which recognizes the start signal of the promoter region. The transcription has several steps which are as follows:

Activation of Ribonucleotides

Four types of nucleotides are available in the nucleoplasm. Before incorporation, the nucleotides are activated by a specific enzyme and converted to triphosphate. The core enzyme of RNA polymerase travels along the template DNA strand with the help of a specific protein factor called NUsA protein.

Initiation

RNA polymerase reaches the promoter region which is recognized by the sigma factor. After the attachment of RNA polymerase with the promoter, unzipping of the template DNA starts.

Base Pairing

The activated and phosphorylated ribonucleotides attach to opposite complementary nitrogen bases of the template strand. A pyrophosphate radical is released and hydrolysis occurs. The energy released due to the unzipping of DNA is utilized for chain formation by establishing the bands. ATP = AMP + 2Pi + energy

Between ribonucleotides, a phosphodiester bond is formed with the help of energy, core enzymes and Mg^{2+}.

Elongation

Elongation of RNA occurs till the enzyme reaches the terminator. RNA polymerase gets separated by the rho factor and Nus G.

mRNA Processing

Before the onset of protein synthesis, the primary transcript has to be processed into a mature transcript called messenger RNA (mRNA). Processing of mRNA includes:

(a) Both ends are protected from degradation by enzymes. The ends of the primary transcript are susceptible to degradative enzymes such as exonucleases. A CAP is added to the 5′ end of the primary transcript. One unusual linkage is formed due to capping. The exonucleases fail to recognize CAP. Thus, this end is protected. But, the question is that CAP is formed at one end; then, how is the whole transcript protected? We know that exonucleases can only can work from an end. Since the end is capped, the whole transcript is protected. Similarly, the 3′ end is protected by the addition of a poly-A tail by the enzyme poly-A polymerase.

(b) The most interesting thing is that the noncoding intron sequences are removed through a process called as **splicing**. This mRNA is now mature and comes out through nucleopores.

Translation

Protein synthesis occurs in the ribosomes. The ribosomes have two subunits: small and large. In accordance with the trinucleotide genetic code, a specific polypeptide chain is produced. **mRNA acts as a template for the accurate synthesis of specific polypeptides**. The following events occur during translation:

(a) **Activation**

The amino acid is attached to the respective tRNA by the carboxyl group of the amino acid and the 3′ -OH of tRNA by an ester bond. This is called charged tRNA.

(b) **Initiation**

During this event, the small subunit binds to the 5′ end of mRNA with the help of initiation factors (IF).

(c) **Elongation**

Elongation occurs when the next aminoacyl tRNA, along with GTP and elongation factors, binds to the ribosome.

(d) When the ribosome faces a stop codon (**UAA, UAG and UGA**), then the tRNA cannot recognize the ribosome. The releasing factors cause the termination of the polypeptide chain.

This is a process by which new proteins are produced in the cell as per the requirement of the body. The events occur during protein biosynthesis is summarized below:

1. Formation of messenger RNA (mRNA)
2. Aminoacylation of transfer RNA (tRNA)
3. Transport of amino acids
4. Formation of polypeptide chains
5. Post-translation modification of proteins

Regulation of Gene Expression

There are two types of genes: housekeeping and regulatory genes. Housekeeping genes express themselves continuously in all cells of the body because these proteins are required consistently for the continuous biochemical reaction process. Housekeeping genes are also called as constitutive genes. Other types of genes are luxury genes or non-constitutive genes or regulatory genes. The activity of non-constitutive genes is regulated. Two types of genes are expressed. These are positive and negative types of genes. In positive genes, expression continues till their activity is suppressed by repressors. This activity is called as repressive regulation. Repression occurs due to the action of the product of a regulatory gene. Negative genes remain non-expressed until they are induced to do so. This type regulation is called as inducible regulation.

In the field of parasitology, many works have been carried out using RT-q PCR. The RT-q PCR is mostly done to see the expression level of a gene by quantifying mRNA. For RT-q PCR, many housekeeping genes are used as reference genes. Glyceraldehyde-3-phosphate dehydrogenase (GAPDH) and ubiquitin are some housekeeping genes which are used as reference genes for performing RT-q PCR.

Recombinant Protein Production

A particular parasitic protein can be produced in the laboratory by recombinant DNA technology. For this purpose, the target gene responsible for the expression of a particular protein is taken out from the whole genome of the parasite by using different restriction enzymes. The gene is amplified by PCR. The PCR product contains multiple copies of the target genes. This gene is inserted in plasmids. The plasmids have a complementary site where the target gene can be recombined. This gene is attached to the complimentary site by the ligase enzyme. Now, the plasmid is transfected in the host cells. Two important organisms such as *Escherichia coli*, *Pichia pastoris,* or other yeast cells act as host cells for the expression of the protein. The desired protein is expressed in these cells. Then, the target protein is isolated by a specific isolation protocol. The same is utilized as an antigen for immunoprotection studies or immunodiagnostic studies. In the process of preparation of recombinant protein, many hurdles are faced by the researchers. The main problem, frequently encountered, is protein folding. The post-translated protein does not have the exact folding as that of the natural protein. However, at the present time, efforts are being made to develop several techniques to have a proper folding of a protein. **Some isoenzymes and other specific proteins of parasites such as glutathione S-transferases, cathepsin-L, cathepsin-B, FABP, etc. have already been expressed by recombinant DNA technology**.

Hybridoma Technique

This is a technique by which antibody-producing B cells are fused with myeloma cells and the fused cells are called as hybridoma. These cells are capable of the production of identical antibodies (monoclonal) for an infinite period as myeloma cells are immortal. The following are the steps:

Step-1
Immunization of mice with target antigen
Step-2
Collection of splenocytes from immunized mice
Step-3
Isolation of B cells from splenocytes
Step-4
Fusion of B cells with myeloma cells (hybridoma)
Step-5
Dilution of hybridoma cells

Method in Detail

First, the mice are inoculated with the target antigen by a standard immunization protocol constituting several injections of antigens at suitable time intervals so that antibodies can be produced against that antigen. Then, the mice are sacrificed. The B cells are isolated as mentioned earlier. These B cells are fused with myeloma cells by the method of electrofusion, keeping these cells in an electric field. Fusion of cells can also be done by a chemical method using polyethylene glycol. The fused cells are incubated in a specific medium, **HAT medium (hypoxanthine, aminopterin, and thymidine)**. The aminopterine compound present in the HAT medium blocks the pathway of nucleotide synthesis. As a result, the unfused myeloma cells die because they are not able to perform nucleotide synthesis as they lack HGPRT which is responsible for carrying out nucleotide synthesis by a salvage pathway. However, hybridoma cells can survive as they have HGPRT and they can perform nucleotide synthesis by the salvage pathway. B cells die automatically after their lifespan. Thus, hybridoma cells capable of producing specific antibodies are selected and are allowed to grow in culture. The monoclonal antibodies are collected from the culture supernatant.

Applications

There are many uses of monoclonal antibodies:

- Monoclonal antibodies are used for providing passive immunity to animals that can give instant immunoprotection. There are many infective diseases, for which no chemotherapy is available. In that case, a monoclonal antibody can be used.
- Monoclonal antibodies can be used to produce vaccine candidates. Immuno-affinity chromatography columns can be made using monoclonal antibodies. The corresponding antigens can be prepared. Defined antigens can be isolated by this purification protocol using monoclonal antibodies.
- In different research programs, monoclonal antibodies are used. To know the function of any cytokine, the anti-cytokine monoclonal antibody is used to neutralize the action of the cytokine. By this method, the function of any cytokine can be determined. The function and nature of other specific proteins can also be determined.

Molecular Diagnosis

Among the new molecular diagnostic techniques, PCR is the most important and widely used method. This technique was developed to amplify extremely minute amounts of DNA to a large amount. Application of PCR methodology is not only used for the detection and identification of the parasite, but is also used in the fields of virology and bacteriology. There is a limitation of the use of this technique in the field of parasitology because it is not possible in many cases to collect parasitic specimens from the hosts. In the case of hemoprotozoan parasites, it is easy to collect blood. Some hemoprotozoan parasites remain in the RBC, lymphocytes, monocytes, or any other cell of the blood; or they may be present extracellularly in the blood. So if we collect the blood, we can easily get the parasites along with the cells or in blood constituents. However, it is not possible to collect the parasites which remain in the internal vital organs like lungs, liver, heart, etc. However, techniques are available to isolate the DNA from the parasites by taking out biopsy material from the suspected organ. For example, *Fasciola* parasites live in the liver and bile duct. However, we can develop the technique to collect the biopsy material from the liver wherein the juvenile flukes migrate. But to adopt the technique is very cumbersome. A general PCR is of no use for the specific identification of parasites. A specific primer is required for the identification of parasites. Different PCR techniques are used in the field of veterinary parasitology. These are allele-specific PCR, PCR-RAPD, RFLP-PCR, RT-qPCR, multiplex PCR, and nested PCR.

Biotechnological Tools in the Diagnosis of Parasitic Diseases

Diagnosis

Polymerase Chain Reaction

Among the new molecular diagnostic techniques, polymerase chain reaction (PCR) is the most important and widely used method. This technique was developed to amplify extremely minute amounts of DNA to a large amount. Application of PCR methodology is not only used for detection and identification of the parasite, but this technique is also used in the field of virology and bacteriology. There is a limitation of the use of this technique in the field of parasitology because it is not possible in many cases to collect the parasitic specimen from the hosts. In the case of hemoprotozoan parasites, it is easy to collect blood. Some hemoprotozoan parasites remain in the RBC, lymphocytes, monocytes, or any other cell of blood; or they may be present extracellularly in the blood. So, if we collect the blood, we can easily get the parasites along with the cells or in blood constituents. However, it is difficult to collect the parasites that remain in the internal vital organs like the lung, liver, heart, etc. However, techniques are available to isolate the DNA from the parasites by taking out biopsy material from the suspected organ. For example, *Fasciola* parasites live in the liver and bile duct. However, we can develop the technique to collect the biopsy material from the liver wherein the juvenile flukes migrate. But, to adopt the technique is very cumbersome. A general PCR is of no use for specific identification of parasites. A specific primer is required for the identification of parasites.

Restriction Fragment Length Polymorphism

Restriction endonucleases are used to make the fragments of the DNA in question and differences in homologous DNA sequences can be detected by the presence of fragments of different lengths after digestion of the DNA samples. RFLP is subject to a molecular marker technique and is specific to a single clone, restriction enzyme combination. For this, an RFLP probe is made and labeled that hybridizes with one or more fragments of the digested DNA samples after they were separated by gel electrophoresis that reveals unique blotting pattern characteristics to a particular genotype at a specific locus. Different short, single or low-copy genomic DNA or cDNA clones are specifically used as an RFLP probe. In this context, it is to be mentioned that most RFLP markers are codominant. That means both alleles in heterozygous samples can be detected. RFLP probes are frequently used for **genome mapping, forensics, hereditary diseases, paternity tests, etc**. the development of the RFLP probe is carried out as follows:

- Total DNA is digested using restriction endonucleases.
- The so-formed digested DNA is size-fractionated by running through a preparative agarose gel. The fragments ranging from 500 to 2000 bp are excised and eluted. The fragments are amplified or cloned into plasmid vectors (e.g., pUC18). Amplification or cloning are done generate sufficient copies of probe.
- Southern blot is used to check the inserts that can be probed with total sheared DNA.

PCR-RFLP

In this technique, PCR amplicons are digested with endonucleases. The digested DNA fragments are electrophoretically separated. The specific band pattern is analyzed. Since the PCR is done, a very small amount of DNA is required. And there is no need for a hybridization step.

Examples

- A study shows the genotyping of *Toxoplasma gondii* by multilocus PCR-RFLP markers. A set of markers was used for genotyping *Toxoplasma gondii*.
- A study shows a comparison of PCR-RFLP and Genescan-based genotyping for analyzing infection dynamics of *Plasmodium falciparum*. It was found that the Genescan method is more precise and can achieve a higher throughput.
- A study shows evaluation of PCR-RFLP assay of ITS genes for direct identification of *Leishmania* species.

Nested PCR

This is a modification of PCR that is intended to reduce nonspecific binding in products as a result of amplification of unexpected primer binding sites. In this PCR technique, both sensitivity and specificity are enhanced. Two primer sets are used here. The first set of primers is designed in such a way that it can anneal to sequences upstream from the second set of primers. The amplicons produced in the first PCR reaction are used as a template for the second PCR reaction.

Example

A study shows nested PCR-RFLP for the genotyping of *Cryptosporidium* parasites isolated from calves and children.

Multiplex PCR

Multiplex PCR is used for multiple targets in a single PCR experiment. Different target DNA sequences are amplified simultaneously using multiple primer pairs in a restriction mixture. It should be kept in mind that amplicon sizes of different genes such as their base pair length should be different so that distinct bands can be visualized by gel-electrophoresis. In this PCR, a single template or multiple templates are used. Multiplex PCR is used in high-throughput single-nucleotide polymorphism (SNP) genotyping, mutation analysis, template quantification, gene deletion analysis, RNA detection, forensic analysis, etc. In this type of PCR, different DNA templates or different regions of a gene are amplified. Primer designing is very essential for this PCR; many points that include primer length, melting temperature, GC content, etc., should be taken into consideration. The steps of multiplex PCR are as follows:

- The PCR reaction tube is taken.
- The required amount of Master Mix or dNTPs is poured.
- A multiplex buffer containing $MgCl_2$, DMSO, KCl, etc. is added.
- Sets of primers are added simultaneously to the reaction.
- Dye is added to the reaction.
- Template DNA is added (either different DNA templates or different regions of a gene).
- Nuclease-free water is added to make the final volume of the reaction.
- A PCR machine is run.

The study shows an array of research works on this type of PCR using different parasites such as *Trypanosoma* spp., *Leishmania* spp., *Babesia* spp., hookworms, filarial worms, etc.

Random Amplified Polymorphic DNA (RAPD)-PCR

This is also called arbitrary priming PCR (AP-PCR). In this type of PCR, segments of DNA are amplified at random. For this purpose, several arbitrary short primers are developed (8–12 nucleotides). In normal PCR, the known DNA templates are amplified using specific primers. In RAPD-PCR, the sequence to be amplified is unknown as arbitrary primers are used. There are several constraints while performing RAPD-PCR. High expertise and skill are required to get perfect results. There should be no DNA contamination because very short-length primers are used in this PCR and there is a chance of nonspecific binding of the contaminated DNA by the arbitrary primers. The low reproducibility of this PCR is another constraint. A study discusses many research works on the use of RAPD-PCR for the determination of polymorphisms of different parasites such as *Trypanosoma* spp., *Leishmania* spp., *Babesia* spp., *Theileria* spp., hookworms, etc. The PCR product is run in the gel and the band patterns of similar parasites is compared. In this PCR, polymorphism of the same species can be detected.

Cleavase Length Fragment Polymorphism

Cleavage length fragment polymorphism is a recent molecular method. This is a subtyping system based on exploiting the property of an enzyme cleavase. This enzyme recognize junctions between single and double stranded regions of DNA which is formed after denaturation and cooling. Sequence-dependent secondary structures are formed that contain folded hairpin-like structures. The enzyme that is used for cleaving of 5′ of the hairpin loop at the junction is unique to that strand of DNA. The CLFP pattern is detected.

Nucleic Acid Probes

A nucleic acid probe is a labeled, short sequence of nucleotide that is used to detect a target sequence. The short sequence of the nucleotide binds to a region of the target sequence. For that, there should be homology between the short sequence and the region of the target sequence to which the short sequence binds. There are several ways by which the short sequence can be labeled. These are radiolabeling, chemiluminescence labeling, fluorescence labeling, etc. The general steps are as follows:

- Isolation of total genomic DNA
- Digestion of total genomic DNA by restriction enzymes
- Separation of DNA fragments by gel electrophoresis
- Formation of different DNA bands on agarose gel
- Transfer of DNA bands onto a nitrocellulose membrane following standard protocol (blotting)
- Putting the nitrocellulose membrane in a solution containing a radiolabeled DNA probe
- Hybridization of the DNA probe complementary to the DNA fragment occurs

Many works were carried out on the use of nucleic acid probes. The rDNA probe is used for the identification of *Bebesia* spp., *Leishmania* spp., *Trypanosoma* spp., etc. Species-specific DNA probes are used for the identification of different *Trypanosoma* spp., *Eimeria* spp., etc. DNA probes are also used for the identification of helminthic parasites, such as *Fasciola* spp., *Ancylostoma* spp., etc. Works were carried out on the use of DNA probes for the detection of the larval stage of the trematode parasite in snails. The nucleic acid probe is also used for epidemiologicalstudies. See the Chap. 13 "Parasitological Techniques" for other diagnostic methods.

Phylogeny

Taxonomic study in the earlier period was mostly based on gross and microscopic morphological features of one organism. The term till today is used to describe mostly the outer and inner structural features of any organism. In the case of bacteria and viruses, the placement of these organisms in a phylogenetic tree is mainly based on the outer and inner structural features, biochemical characteristics, physiological characteristics, and genetic characteristics. But, in the field of parasitology, the parasites have been placed in the phylogenetic tree mostly based on the internal and external structures. But, the characterization of parasites was done to a lesser extent on the study of biochemical properties, physiological properties, genetic properties, antigenic characters, etc. But, the time has come to pay attention to more and more minute structures for the proper placement of the parasites. In the earlier period, before the invention of the microscope, visual analysis of the outer and inner structure was the sole resort for the identification and study of the parasites. At that time, the gross morphometric data and different structures were the routine practice for the study of any parasite. There have been many records of tapeworm and roundworms in the ancient period. The records tell us about the gross morphological features. At that time, the technique of staining of the parasite was not known to the people. The color, length of the parasite, number of segments, etc. were taken into account for morphological study. In the later period, the discovery of lens was made by Leeuwenhoek that later emerged as a microscope. That was a great landmark of the age, not only in the field of parasitology, but also in other fields of all biological sciences. The microscope did wonders at that time. There have been enormous studies on microscopy-based morphology of different organisms.

In recent times, phylogeny has taken on a new dimension that has crossed the biochemical and immunological characterization of parasites; rather, it has sharply entered into the field of molecular characterization and identification of parasites. See other chapters.

Expression of Antigen and Antibody Fragments Useful as Diagnostic Reagents and Vaccines

For the expression of antigens or antibody molecules, different steps of molecular techniques have to be carried out:

Important Points

- Our purpose of recombination is to insert a gene encoding a specific protein into the plasmid (vector). For this, the DNA fragment of interest is incorporated into the plasmid.
- DNA ligase enzymes are used for this purpose. Ligase can seal the broken phosphodiester bond.

Transformation of Recombinant Molecules

- The recombinant molecules are now transferred into a host organism. It is well known that *E. coli* are good host organisms.
- A specific technique is used for the introduction of a vector into the host organism. *E coli* cells treated with Ca^{2+} easily take up DNA.
- The cells after the above treatment are able to take up foreign DNA and these are called as transformed cells.
- The cells, after transformation, are subjected to heat shock at 42 °C for 1–2 min that allows the cells to recover from the unusual condition of the transformation process. Then, the cells are allowed to grow in a growth medium and then spread on an agar plate until a single colony develops.

Transfection of Recombinants in Eukaryotic Host Cells

- There are different ways by which DNA could be transfected into eukaryotic cells. These are injection, bombardment with solid particles, digestion of the cell wall, or precipitation of DNA onto the cell wall. The following are the different processes for the transfection of DNA.
- This transfection is not as easy as that of prokaryotic cells.
- The principle of transformation lies on minor digestion of the cell wall using degrading enzymes to make the protoplast fragile.
- It should be done in such a controlled manner that, despite the creation of a fragile protoplast, there would not be any harm to the cell.
- After the withdrawal of enzymes the cell walls become repaired later.
- The cells which have no cell wall are subjected to precipitation of DNA onto their surface with calcium phosphate.
- DNA uptake by the cells can be increased by high voltage that causes the formation of transient pores.
- This process is called as electroporation. DNA can be microinjected by fine glass pipettes. A gene gun is also used for the introduction of DNA.

Selection of Competent Cells

- There is a standard method to identify the *E. coli* that has taken up the recombinant plasmid.
- When the *E. coli* are allowed to grow in an agar plate, thousands or millions of colonies are formed.
- It is a fact that all the colonies do not contain the recombinant plasmid molecules. An antibiotic-resistant gene is provided in the vector. Example: the beta-lactamase gene confers resistance against ampicillin.
- Therefore, if the plasmid along with the beta-lactamase gene is used for recombination, then the transformed cells will survive and grow in the medium containing ampicillin since it has ampicillin-resistant genes.
- In the media, we will get a single colony.

Polymerase Chain Reaction

The polymerase chain reaction is done to amplify minute amounts of DNA to large amounts. This is done for many purposes. One of the most important uses of the polymerase chain reaction is diagnosis. The following are notable points of PCR:

- The primers are extended by DNA polymerase.
- The steps are denaturation, primer annealing and, subsequently, polymerization.
- If 95 °C (60 s) heat is applied, the double-stranded DNA gets separated into two templates.
- Each separated strand is called as template DNA. Then, the temperature is reduced to 55 °C. At this temperature, primers are annealed to template DNA.
- After annealing, the temperature is further increased to 72 °C for 60–90 s for optimal polymerization.
- dNTPs and Mg^{2+} are provided for polymerization. The cycle is repeated several times to achieve a number of copies of the target DNA.

Different Types of PCR Used in the Field of Parasitology

PCR is pre-requisite to carry out different molecular parasitic techniques. In most of the molecular techniques, different PCR techniques are performed. DNA cloning, DNA library construction, production of recombinant protein -all such techniques require PCR as a very important tool.

Types of Immune Response

See the previous Chap. 9 'Immunoparasitology'.

Novel and Other Antigens

In the earlier period, crude antigen was used for different vaccination trials. Later, different specific parasitic proteins having a vital role in the parasites are being used for vaccination trials. Day by day, different, novel antigens are being discovered:

- Different vital enzymes such as glutathione-S-transferase, cathepsin L, cathepsin B, etc. are promising antigens.
- Different other vital proteins (nonenzyme) such as beta-tubulin, calmodulin, etc. are important antigens.
- Salivary proteins are used for vaccination trials against tick infestation.
- Different gut proteins of *Haemonchus* sp., such as **H11 and HgalGP**, are promising antigens.
- Different surface proteins, such as larval surface protein of helminthic parasites, merozoite surface protein, sporozoite surface proteins, etc., are used for vaccination trials.
- Different endosymbionts of parasites are used for vaccination trials.

Molecular Vaccine

Vectored Parasitic Vaccines

The vectored parasitic vaccine has been a novel technique for delivering any candidate protein of the vaccine through a vector. This technique provides the advantage of inducing a cell-mediated immune response. In this technique, different viruses, such as pox virus, adenovirus, etc., are used. The gene encoding genetic instruction of a particular antigen of a parasite is stitched to the virus vector's genome. This genetically engineered vectored vaccine is administered in the hosts. There have been trials on the production of vectored parasitic vaccines, such as *Plasmodium* sp., *Theileria* sp., etc.

Recombinant Vaccine

In this technique, the gene fraction encoding a particular immunodominant protein is incorpo-

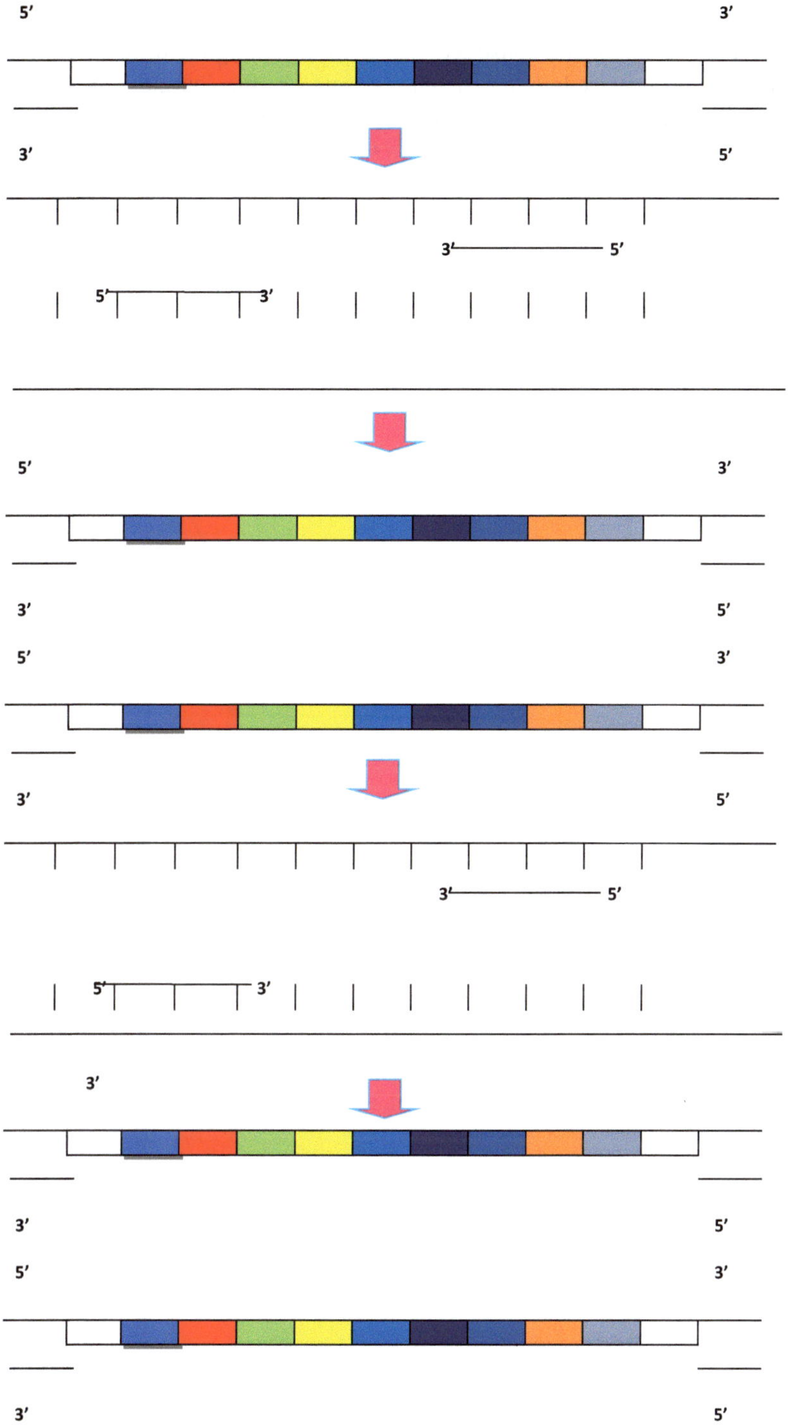

Fig. 11.2 Schematic representation displaying of PCR events

rated into a plasmid. This is called as recombinant construct. This recombinant construct is inserted into expression host cells (*Escherichia coli*, *Pichia pastoris,* etc.). The targeted protein is expressed in these specific cells. The protein is then isolated and used as the vaccine. A number of trials have been conducted in the field of parasitology using recombinant vaccines. Recombinant vaccines have been tried against *Boophilus microplus* (vector of *Babesia* sp., *Theileria parva*, *Babesia bovis,* and many other parasites). See the Chap. 9 "Immunoparasitology" for other types of vaccines.

DNA Sequencing

- The earlier chemical method of DNA sequencing is due to Maxam and Gilbert.
- In this method, base-specific cleavage of DNA is done to produce different fragments of different sizes.
- One end of the DNA (either 5′ or 3′ end) is labeled by radioactive phosphate. For this, modification of the bases is done first using specific chemicals.
- Dimethyl sulfate is used to methylate at G bases. Piperidine is used to break the sugar–phosphate backbone.
- Hydrazine and formic acids are two approaches for attacking the different bases. Formic acid attacks the purine base (A and G) for depurination.
- Hydrazine is also used to attack pyrimidines (C+T) for hydrolysis.
- The treated material **(G, A+G, C+T, C)** is allowed to run into the four lanes. Sequencing is done by the analysis of band patterns.

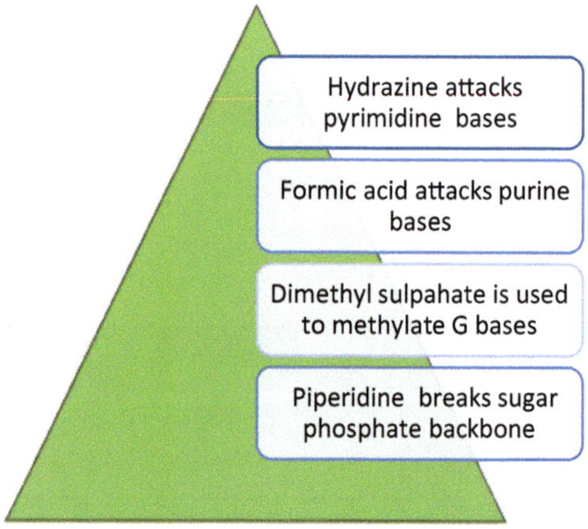

Fig. 11.3 Schematically displayed chemicals that break the bases of nucletide

12 Trends in the Control of Livestock and Poultry Parasites

Conventional and Novel Methods of the Control of Helminths

There are different ways to control helminthic parasites. These are as follows:

Proper Nutrition and Health Management

The health of the animals should be maintained properly with proper nutrition. Nutrition is very important which boosts the animals' immunity. The nutrients should be chosen with antioxidants. Biological reactive species are very detrimental to any biological system and reduce the general immunity of animals. The animals become very much prone to infection due to poor health conditions. The animals should be provided with optimum vitamins and minerals. Proper nutrition restores the immunity power of an animal.

Silage Feeding

Silage feeding to the animals can be done when the pasture is highly infected with parasite/parasitic stages. When the grass or other vegetation is converted to silage, the eggs, larvae and metacercariae are killed.

Control of Intermediate Hosts

A number of intermediate hosts play a role in the transmission of different helminthic parasites. These are snails, slugs, arthropods (fly, mosquitoes, grasshopper, cockroaches, earthworms). Different methods for control of snails are as follows:

Mechanical

(a) *Collection and killing*: This is a simple method. The snails are collected from the ponds, canals, ditches, etc. by hand and/or fishing net. The collected snails are killed mechanically by burning, application of hot water, etc.
(b) *Habitat alteration*: The water can be drained off from the water-logged area by making channels.
(c) *Removal of vegetation*: The submerged vegetation is the food of snails. If the submerged vegetation is removed, the snails will die from being deprived of food.

Chemical

(a) Different chemical compounds have molluscicidal effects.
(b) Different phytochemicals have molluscicidal effects.

S. C. Mandal, *Textbook of Veterinary Parasitology*, https://doi.org/10.1007/978-981-99-2924-5_12

The earthworms are controlled by ploughing the soil. The mosquitoes and flies are controlled in different ways. The most effective control method is the application of insecticidal agents.

Rotational Grazing

For the purpose of rotational grazing, a particular grazing land (pasture) is divided into different paddocks. The animals are allowed to graze rotationally in different plots alternatively. In this way, any parasitic stages on a particular paddock die due to lack of hosts when this paddock of the pasture is in the resting phase. Rotational grazing can be done using different species.

(a) Rotational grazing with the same species.
(b) Rotational grazing with different species.

Bioclimatograph

Bioclimatograph represents the correlation between the climatological variables and occurrences of living organism. Bioclimatograph can be created for determining the relationship between the parasite and the prevailing climatological condition. The graph also depicts the deposition of parasite/parasitic stages in different areas.

Feces Disposal

The feces are the source of different parasitic stages, such as eggs, larvae, etc. If the feces are disposed of properly, the chance of occurrence of parasitic disease would be less. If the feces are disposed of in a particular pit, the deposited feces get fermented and the heat produced due to fermentation can kill the developmental stages of parasites.

Use of Feces as Biofertilizer

The feces could be used as biofertilizer.

Reproduction Timing

Lactation or parturition of the animals causes a reduction in their immunity. So the timing of reproduction of the animals can be adjusted in such a way that the parturition would happen in a season favorable for their food and nutrition.

Pasture Management

Pasture becomes contaminated with different stages of parasites (egg, larva, etc.) after the grazing of animals. Management of pasture can be done as follows:

(a) The pasture should be ploughed intermittently. In this way, the deposited parasites/parasitic stages are buried and killed. An additional advantage of pasture ploughing is that it can control the earthworms also.
(b) The pasture can be burnt which can kill the parasite/parasitic stages.

Bedding Management

There is every chance of contamination of the bedding of animals by feces, urine, nasal discharge, etc. The eggs and larvae are voided with these materials. Therefore, the bedding should be replaced frequently.

Management of Animal House

The animal house should have some basic requirements for controlling parasitic disease. These are as follows:

(a) The floor of an animal house should have no cracks and crevices and it should easy to clean. The floor should be cleaned properly with KOH, hot water or steam water.
(b) The designing of the house should be done in such a way that sufficient sunlight and air can enter the house.

(c) The windows and door should be perfectly closed.

can be done by pulse release, use of urea molasses block, etc.

Drinking Management

The animals should be provided with under water/tap water for drinking. They should not be allowed to drink pond or river water because this water may be contaminated with metacercariae or other infective stages of parasites.

Stocking Rate

The stocking rate of animals should be maintained. The number of animals should be proportionate to the space of the animal's house. There should be no overcrowding. Overcrowding can increase animal-to-animal contact. Moreover, overcrowding reduces the immunity of the animals.

Zero Grazing

The animals are stall-fed and not allowed in the pasture when there is a high grade of infection in the pasture.

Segregation/Isolation

The severely affected animals should be maintained separately so that they cannot spread infection to the healthy stock.

Proper Use of Anthelmintic

The anthelmintics should be used at a proper dose and interval which reduces the chance of drug resistance.

Long-term Medication

There are different ways by which long-term medication is done. The long-term medication

Manger and Water Troughs

The manger and water troughs should have a standard height so that they cannot be easily contaminated with feces, urine, etc.

Weaning

Young animals are highly susceptible to parasitic infection. For this, they should be separately housed and grazed.

Biological Control

There are different nematophagous or nematode-trapping fungi which are used for the biological control of parasites. These fungi include *Duddingtonia flagrans* and *Arthrobotrys oligospora*. See also different biological control techniques in other section of this chapter.

Dose and Move

The animals are dewormed and allowed in the pasture.

Meteorological Forecasting

Meteorological forecasting is done based on climatological data from several years. Rainfall, temperature, and humidity are taken into consideration for forecasting parasitic infection.

Remote Sensing by Satellite Image Analysis

Remote sensing has become a trend for providing information about the chance of parasitic infection. Satellite image analysis is done. Image of large area indicates water stagnation, plantation, etc. which indicates the probable occurrence of parasitic infection. For more details, refer to Chap. 10.

Breeding Policy

Resistant animals can be used for the development of farms. Different breeds are resistant to some helminthic parasites. For example, Red Maasai, Finn Dorset, and Garole sheep are resistant to *Haemonchus* infection. These animals can be used for breeding purposes.

Anthelmintics and Their Mode of Action

Name of anthelmintic group	Mode of action
Benzimidazole/ pro-benzimidazole	The drugs inhibit microtubule polymerization thus causing prevention of glucose uptake
Imidazothiazole/ tetrahydropyrimidine	Depolarizing neuromuscular blocking agent. The drugs act at synaptic and extra-synaptic nicotinic acetylecholine receptors of parasitic muscle cells and cause paralysis of worms.
Avermectin	The drugs bind with glutamate -gated chloride ion channels which increase permeability of cell membrane to chloride ions. This causes hyperpolarization of cells resulting in paralysis and death of parasite. This drug causes the release and binding of GABA leading to the inhibition of signal transduction that results in paralysis
Salicylanilides/ substituted phenols	The drug causes uncoupling of oxidative phosphorylation in mitochondria. This interferes with ATP production
Organophosphates	The drugs are cholinesterase inhibitors. This leads to neuromuscular paralysis of worms
Piperazine	This drug acts by anticholinergic action. The drug shows GABA agonistic action and depresses the action of acetyl-choline. This leads to the paralysis of worms

Anti-nematodal Drugs

- Benzimidazoles/pro-benzimidazoles: Albenzazole, fenbendazole, cambendazole, mebendazole, oxfendazole, parbendazole, oxibendazole, thiophanate, and netobimin
- Piperazine: Piperazine salts and diethylcarbamazine
- Imidazothiazoles/tetrahydropyrimidines: Tetramisole, levamisole, morantel, and pyrantel
- Avermectin: Ivermectin, moxidectin
- Organophosphate: Dichlorvos, trichlorfon, etc.
- Salicylanilides/substituted phenols: Closantel and nitroscanate

Anti-trematodal Drugs

- Salicylanilides/substituted phenols: Rafoxanide, noxyclozanide, nitroxynil
- Benzimidazole: Triclabendazole, Albendazole

Anticestodal Drug

- Salicynilide/substituted phenol: Niclosamide
- Others: Praziquantel, Bunamidine

Characteristics of Ideal Anthelmintics

1. The anthelmintic should be non-toxic and should not produce any side effects.
2. The drug should be biocompatible. The chemical should be well tolerated by the animals.
3. The drug should be excreted easily. There should be no residual effect of the drug in the tissues of the treated animals.
4. Less amount should be needed for clearing any parasitic infection.
5. Parenteral administration could be possible.

6. An anthelmintic drug should be non-corrosive.
7. An anthelmintic drug should not develop frequent resistance.

Anthelmintic Resistance

There are many reports on anthelmintic resistance. If we pay slight attention to the literature, we will be able to find that many reports on drug resistance of benzimidazole compounds. It is needless to mention that there are also many reports on ivermectin resistance. It is really a serious threat as ivermectin is the only good and broad-spectrum anthemintic. Research work should be intensified so that the drug resistance could be avoided. The following are the reasons for drug resistance:

1. **Frequent use of drug**
2. **Use of same group of drugs repeatedly**
3. **Sub-dosing**

Researchers are still in search of potential mechanisms of drug resistance. The following general mechanisms have been identified so far. All may not be related to anthelmintic resistance.

1. Conversion or transformation of an effective drug into an ineffective compound.
2. Development of alternative biochemical pathways to bypass inhibited reactions.
3. Alteration or modification of the drug-binding site
4. Increase in efflux or decrease in influx of the drug.

   The drugs come out from the cells. The entry of the drug into the cells is inhibited.
5. Production of drug-sensitive enzymes in higher level.
6. To decrease the production of effective molecules from the drug.

The following are the predisposing factors for the occurrence of anthelmintic resistance.

1. Genetic mutation of the organisms as a result of abuse of anthelmintics.
2. Genetic mutation results in alteration of gene expression.
3. Differential selection of resistant individuals due to the creation of selection pressure.

Spectrum of Activity

The spectrum of activity of an anthelmintic is very important. It indicates which drug should be chosen to control more than one parasite. For example, some drugs have narrow spectrum, while some have a broad spectrum. Ivermectin is a broad-spectrum drug as it is effective against both roundworms and ectoparasites. On the other hand, antimony compounds are effective against schistosomiasis and a narrow range of other parasites. The spectrum of a drug is due to the chemical nature of the compound. If a compound inhibits enzymes present in many classes of parasites, then this compound would obviously have a broad range of efficacy.

Integrated Control Method

The integrated control method is a simulation of different control methods. It has been observed that a single control strategy is not sufficient to significantly control parasitic infections. That is why different control strategies should be simulated in such a way that a significant level of control could be achieved. For the control of helminthic parasites, general management of parasites (pasture management, grazing management, etc.), chemotherapy, and immunological control can be employed. Biological control using nematode-trapping fungi and nematophagous fungi is in the research process. Commercial application of this biological control has not been realized. Immunological control using a potent vaccine could be undertaken.

Immunological Control of Helminths

Many works were carried out to control helminthic parasites. Earlier attempts were made to develop vaccines using attenuated parasites.

Dictol and Difil were produced to immunize animals against lung-worm infection. Vaccines were developed against hook-worm infection using third stage larvae. Later on, a series of works have been carried out using purified antigens. Native purified antigens were used for the development of vaccines. In the recent past, attention has been pinpointed to the use of recombinant antigens. A particular parasitic protein can be produced in the laboratory by recombinant DNA technology. For this purpose, the target gene responsible for the expression of a particular protein is taken out from the whole genome of the parasite by using different restriction enzymes. The part of a gene is amplified by PCR. The PCR product contains multiple copies of the target gene. This part of a gene is inserted in a plasmid. The plasmid has a complimentary site where the target gene can be ligated. This gene is attached to the complementary site by the ligase enzyme. Now the plasmid is transfected into the host cells, two important expression cells are used: ***Escherichia coli*** and ***Pichia pastoris***. The desired protein is expressed in the cells. Then the target protein is isolated using a specific isolation protocol. The same is utilized as an antigen for immunoprotection studies or immunodiagnostic studies. In the process of preparation of recombinant protein, many hurdles are faced by the workers. The main problem, frequently encountered is protein folding. The post-translated protein does not have the exact folding as the native protein. However, at present time, efforts are being made to develop several techniques to have a proper folding of a protein. Some isoenzymes like glutathione S-transferases, cathepsin- L, fatty acid-binding protein (FABP), etc. have already been expressed by recombinant DNA technology.

Immunoprotection Against Trematodes

Fatty (Acid-Binding Protein)

This is an important protein found in many organisms. This protein has an affinity to fatty acid. The trematodes cannot synthesize sterol in their own body. A fatty acid is required for the lipid layer of cells. Therefore, the flukes are dependent on lipid material from their hosts. This protein plays a role in this regard. Looking into the importance of this protein, an array of research works were carried out for the development of a vaccine using native purified FABP and recombinant FABP using a series of adjuvants comprising Freund's adjuvant, Quil A, montenide 80, etc. A certain level of efficacy was found which was reflected by the reduction of fluke number and fluke weight in a vaccination trial using native FABP (Estuningsih et al. 1997). In this study, they used recombinant FABP at a dose of 600 μg for three consecutive doses at 3-weeks interval. Kumar et al. (2012) also carried out work on the immunopotential capability of FABP and found a certain level of protection. Rabia Aly et al. (2012) found heterologous protection against *Schistosoma mansoni* using native FABP from *Fasciola gigantica.*

Cysteine Proteases

This is a class of thiol enzymes. A number of enzymes are found in this class. The important enzymes having both diagnostic and immunoprotection potential are cathepsin L, cathepsin B, etc. **Cysteine proteases** are present in a wide range of organisms, ranging from lower group (bacteria, protozoa, rickettsia, etc.) to higher group of organisms (mammals, reptiles, etc.). This class of enzymes performs different biological functions. Cysteine proteases of parasites also perform different vital functions comprising digestion, tissue invasion, immune evasion, etc. Several works were carried out on immunodiagnosis and immunoprotection for parasitic diseases. Wijffels et al. (1994) carried out immunoprotection studies against *Fasciola* sp. using cysteine proteinase as a vaccine candidate and got satisfactory result.

Glutathione (S-transferase)

Glutathione S-transferases (GSTs) are a family of isoenzymes which perform detoxification and excretion of xenobiotic agents (Kumar et al. 2016). GSTs occur as multiple functional enzymes in most life forms comprising aerobic eukaryotes and prokaryotes. These enzymes are essential for the life of most organisms. The hypothesis is that anti-GST antibodies can neutralize the function of GST in the parasite and the parasite will be damaged/harmed. Based on this hypothesis, a series of works were carried out to envisage the immunoprotection potential of GST against *Fasciola* parasite in different animals (Estuningsih et al. 1997; Sexton et al. 1994; Morrison et al. 1996; Paykari et al. 2002). The native form of GST is isolated usually by glutathione–agarose affinity chromatography technique. Different adjuvants were used to potentiate the GST protein for conferring good immunoprotection. In different studies, a number of adjuvants, i.e., Quil A, montenide 80, DEAE, Freund' s adjuvant, etc. were used in different combinations in different experimental works. In these works, animal models used were mice, rabbits, sheep, cattle, buffalo, etc. Efficacy parameters measured were antibody titre, number of flukes recovered, weight of flukes, EPG, etc. Encouraging results were recorded in these experimental works. In some trials, efficacy reached 70% and above. Works were also carried out to produce recombinant GSTs. This protein was expressed in *E. coli* bacteria and the same recombinant protein was used as a vaccine candidate along with different adjuvants as mentioned earlier. The immunoprotective capability of this enzyme was found against other flukes also.

Cathepsin

Cathepsin is a vital enzyme found in many life forms. This enzyme belongs to the cysteine proteinase enzyme family. Cathepsin L and cathepsin B are important. Cathepsin L has different forms, such as cathepsin L1, cathepsin L2, etc. Immunoprotective potential of this protein was assessed against different flukes with encouraging results.

Hemoglobin

Hemoglobin of *Fasciola* parasite was tested for evaluating its immunoprotection potential. Several works were carried out on its immunoprotection efficacy. The studies show a low protection level as compared to GST and cathepsin. Dalton et al. (1996) carried out immunoprotection studies using hemoglobin as a vaccine candidate.

Immunoprotection Against Cestodes

For immunoprotection purposes, different types of antigens are used as vaccine candidates.

Crude Antigen

(a) **Whole Somatic Antigen**

In the earlier decades, vaccination works were carried out in a very simple way. Somatic antigen was prepared by homoginization of the whole worm or a part of the worm. After centrifugation, the supernatant part was used for vaccination trials. This was inoculated into the animals for vaccination trial. Some works were also carried out using antigens derived from cyst fluid. It is to be noted that all these proteins are complex. These protein moieties contain many proteins. Multiple proteins are less potent as compared to specific proteins. The efficacy of vaccination using complex protein was not up to the satisfactory level. But these works have significance as these were the pioneer works in this direction.

(b) **Excretory Secretory Antigen**

Works were also carried out using excretory–secretory antigens. The adult parasites are placed in culture media **(RPMI-1640)** with antibiotics. It is to be noted that the parasites should be cultured strictly in aseptic conditions. The cultured material is centrifuged. After centrifugation, the supernatant material is used for vaccination trial. The metacestodes are collected from affected animals aseptically. These are cultured in culture media following standard protocol. The cultured material is centrifuged and the supernatant is taken. The supernatant material is dialyzed to make it concentrated.

Vaccination Using Defined Antigen

Defined antigens are specific proteins which are used as vaccine candidates. The specific proteins can be isolated in native form or can be produced by recombinant DNA technology. Isolation of specific native proteins can be done using different column chromatographic techniques. These are gel filtration chromatography, ion exchange chromatography, immuno-affinity chromatography, etc. Before proceeding to isolate any specific protein, it is a prerequisite to identify or detect the specific protein. For this, Western blot technique is used. For this purpose, the crude protein is fractionated by SDS-PAGE technique. These fractionated proteins are transferred to the nitrocellulose membrane. The immunodominant protein is detected in the Western blot using hyperimmune sera as the primary antibody, secondary antibody conjugated with enzyme and finally treated with substrate. Defined antigens can be produced by recombinant DNA technology. The gene encoding a specific protein is attached to the plasmid. This recombinant plasmid is inserted in the *E. coli* or yeast cells for the expression of the protein.

Name of antigen	Animal	Finding	Researchers
Recombinant antigen of *Echinococcus granulosus* (rEgmyofilin) was used for inoculation subcutaneously	Sheep	After completion of vaccination, challenge infection was given with eggs of *E. granulosus*. Number of cysts was reduced in the vaccinated animals	Zhu et al. (2016)
Recombinant antigen of *Echinococcus granulosus* (rEg29)	Sheep	After completion of vaccination, challenge infection was given with eggs of *E. granulosus*. 94.5% immunoprotection was achieved. Significantly increased levels of IgG, IgG1, IgG2, IL2, IL4, and IFN gamma were also found	Wang et al. (2016)
Two recombinant proteins, i.e., Tm16 and Tm18, were expressed in *E. coli* as fusion protein with GST. This recombinant protein was used with the adjuvant Quil A	Sheep	Immunoprotection of sheep was achieved against this protein *of Taenia multiceps*	Gauci et al. (2008)

Immunoprotection Against Nematodes

Different crude and defined antigens have been used for the development of vaccines. Difil and Dictol were produced against lung-worms using irradiated third stage larvae. Both these vaccines were discontinued due to different constraints. The following table displays different research works on immunoprotection using recombinant antigens (see Chap. 9):

Different recombinant antigens used for immunoprotection against nematodes

Name of antigen	Animals	Finding	Researchers
Intranasal immunization of As 16 (recombinant *Ascaris suum* 16 coupled with cholera toxin)	Mice	Studies show immunoprotection values of recombinant As16 against *Ascaris suum*	Tsuji et al. (2001)
The recombinant antigen rAs 16 was expressed in yeast cells and the same was formulated with ISA 720 or alumn-induced	Mice	The formulation displayed a significant reduction (36.7%) of larvae. The same recombinant protein was also formulated with an adjuvant, cholera toxin B (CTB) which displayed 99.6% protection against *A. suum* eggs in mice	Wei et al. (2017)

Different recombinant antigens used for immunoprotection against nematodes			
Name of antigen	Animals	Finding	Researchers
Recombinant proteins of HcARF1 (rHcARF1) of *Haemonchus contortus*. This ADP-ribosylation factors (ARFs) are Ras-related small GTP ase family which are involved in vesicular trafficking regulations. The gene of *H.contortus* small GTP ase ADP-ribosylation was cloned and recombinant protein was expressed.		This protein modulated cytokine production profile by increasing IL4, IL10, and IL17. At the same time IFN gamma production was decreased significantly. The study further displayed NO production. All these indicated immunomodulatory effect of CMI by the aforesaid recombinant protein	Gadahi et al. (2017)
Recombinant protein 26/23 (rHcp 26/23)	Sheep	Conferred protection in sheep with 92.2% egg count reduction and 59.9% worm (*H. contortus*) burden reduction	Kandil et al. (2017)
H11, H110D, H-gal-GP (recent hidden antigen)		Promising results had been found using different hidden gut antigens, i.e., H11, H110D, H-gal-GP (recent hidden antigen)	Review of Newton (1995)
Work on immunoprotection potential of a recombinant antigen, rHc23. They vaccinated lambs with 100 μg rHc23/dose+bacterial immunostimulant (B1) (LPS from *Escherichia* coli+propionibacterium acnes extract). Trickle infection was performed using two doses (6X,1000 infective or 6X,2000L3).	Lambs	This vaccine schedule could reduce the epg and worm burden	Gonzalez-Sanchez et al. (2019)

Biological Control

Biological control has been an interesting strategy to control helminthic parasites. There are many advantages of biological control. These are as follows:

- Biological control is ecofriendly. There is no chance of environmental damage due to the use of toxic drugs.
- The technique is economic.
- There is no requirement for high-graded technology for the preparation and delivery of biological agents.
- There is no chance of development of resistance.

A series of works is carried out on the use of different biocontrol agents, such as birds, rodents, insects, fungi, bacteria, etc. In the field of helminthology, the use of different fungi is providing promising results. Though there are several drawbacks in developing the technology of using fungi as biocontrol agents, rigorous research may bring this technique for field use. There are different categories of fungi according to their physiological, ecological, and behavioral nature The following are the categories of fungi:

Predaceous Fungi

The predaceous fungi form different structures on their mycelium:

- **Adhesive knobs**
- **Ring structures**
- **Network structure**

The fungi can trap parasites using those aforesaid structures. Several trials have been conducted using the predaceous fungi, such as *Arthrobotrys oligospora*, *Duddingtonia flagrans*, *Pochonia* sp. The motility of the parasites influences the predaceous activity of fungi. In the case of *Arthrobotrys oligospora*, one constraint is faced by the researchers. The mycelium and conidia of this fungi are destroyed in the GI environment. *Duddingtonia flagrans* are better in this regard. They produce thick-walled spores (chlamydospores). These chlamydospores are used for controlling parasites. Many trials have been conducted on the use of these fungi for the control of nematode parasites. The chlamydospores are mixed with the feed of the animals. Then the feces are cultured and larvae count is done to assess their efficacy. Significant reduction in larvae has been found in different trials when compared to the uninfected control group. Satisfactory reduction in free-living larvae was also found in different plot trials. The study showed a satisfactory reduction in larvae in feces when animals are fed with fungal spores for 3–4 months.

Endoparasitic Fungi

Endoparasitic fungi are obligatory parasites of nematodes. They produce sticky spores which adhere to the parasite's cuticle and enter into the body of the parasite and destroy it enzymatically. An example of endoparasitic fungi is *Harposporium anguillulae*. The use of these fungi significantly reduces the larvae count of *Haemonchus contortus* after fecal culture.

Egg Parasitic Fungi

Some fungi are parasitic on the eggs of parasites. Examples are *Verticillium chlamydosporium*, and *Paecilomyces lilacinus*. These fungi secrete enzymes and destroy the eggs of parasites (*Ascaridia galli*, *Parascaris equorum*).

Formulation of Deworming Schedule

The deworming schedule is determined based on the following factors:

1. Location
2. Climate of the location
3. Animals (species, breed, sex, etc.)
4. Animal population

The deworming schedule varies in accordance with the above factors. Following are some common deworming schedules:

Dog, Cat, and Kitten

1. Deworming every week until the age of 3 months starting from the age of 2 weeks
2. 3–6 months of age—deworming could be done once a month
3. After 6 months—heartworm preventive medication should be done.

Cattle

1. Start deworming from the first week of age (a single dose of piperazine adipate—10 mg can be given).
2. Then deworming should be done every month for first 6 months.
3. Then deworming should be done once in 3 months.

Horse

1. The horse should be dewormed every 6-month intervals.

Conventional and Novel Control Strategies against Protozoan Diseases

Chemotherapy

Antiprotozoal drugs and their mode of action

1. Ionophores compounds—Salinomycin sodium, monensin, and lasalocid

   Mechanism of action: They cause cytotoxicity by disrupting the membrane potential by conducting ions through the lipid membrane in the absence of a protein pore.
2. Hydroxyquinolines—Buquinolate, decoquinate

   Mechanism of action: They cause the inhibition of RNA synthesis.
3. Thiamine analogues—Amprolium

   Mechanism of action: These are thiamine antagonist which bind with thiamin pyrophosphokinase thus inhibiting the formation of active form of thiamine.
4. Sulfur compounds—Sulfaquinoxaline and sulfadimidine

   Mechanism of action: Sulfur drugs usually inhibit the synthesis of dihydrofolic acid by competing with para-aminobenzoic acid (PABA) for binding to dihydropteroate synthetase, ultimately resulting in a decrease in the synthesis of nucleotides and DNA.

5. Diloxanide
   Mechanism of action: Unknown
6. Metronidazole
   Mechanism of action: This disrupts DNA and inhibits nucleic acid synthesis.
7. Menoctone
   Mechanism of action: This is ubiquinone antagonist.
8. Imidocarb
   Mechanism of action: Not clearly known
9. Diminazine
   Mechanism of action: This causes the inhibition of S-adenosyl-L-methionine decarboxylase. This enzyme plays a role in the synthesis of polyamines. Diminazine also binds to kDNA and disrupt kDNA.
10. Buparvaquone
    Mechanism of action: It blocks the respiratory chain.
11. Pentamidine
    Mechanism of action: Not clearly known. It is thought that it causes the inhibition of DNA synthesis.
12. Quinine derivative
    Mechanism of action: This is thought to inhibit heme polymerization.
13. Quinapyramine
    Mechanism of action: It causes kinetoplastic DNA condensation.
14. Isometamidium compounds
    Mechanism of action: Kinctoplastic topoisomerase type II is selectively inhibited.
15. **Suramin**
    Mechanism of action: It distrots intracellular membrane in lysosomes and inhibits gycerophosphate dehydrogenase which is the cause of trypanocidal activity.

Integrated Control Method

There is ample scope for integrated control of protozoal disease. A number of antiprotozoal vaccines are available. However, this number of vaccines is not available in the case of helminthic diseases. Therefore, general management, chemotherapy, and vaccination can be done as simulation strategies for the integrated control of protozoan parasites. Though in different sources, biological control has been mentioned to be an integrated control strategy, this control strategy still now has not hatched any practical implication. For the control of parasitic disease, the type of life cycle (indirect or direct life cycle) matters. If the protozoan parasite does not have any vector or intermediate hosts, then the question of control of intermediate hosts or vectors does not arise. If the protozoan parasite is vector transmitted, as is found in trypanosomosis, theileriosis, babesiosis, etc., control of flies and ticks must be included along with chemotherapy/treatment and general management of the animals or birds. It has been observed that a single control strategy is not sufficient to significantly control protozoan parasites. That is why different control strategies should be simulated in such a way that a significant level of control could be achieved. Different control strategies have been described.

Resistance Against Antiprotozoal Drugs

These are some factors which cause resistance.

Loss of Enzyme

In some cases, drug resistance occurs due to a change in the structure of the enzyme. In the case of visceral leishmaniasis, it has been found that the pentavalent antimonials are converted to trivalent antimonials that are the active principle. Some enzymes are responsible for this conversion. Loss of reductase enzyme occurs that results in resistance.

Drug Efflux

In some cases, it has been evident that P-glycoproteins and homologous multidrug transporters contain a phosphorylated linker sequence which might control drug efflux. P-glycoprotein increases the drugs efflux and cause drug resistance. Efflux-mediated drug resistance was reported while using anti-Leishmanial drug.

Alteration in Drug Target

An amino acid is substituted in the enzyme dihydrofolate reductase and thymidylate synthase which cause drug resistance found in malarial treatment.

Loss of Drug Activity

The drug activity is lost in some protozoa. In the case of metronidazole, this type of drug resistance occurs.

Role of Specific Protein

P-glycoprotein causes drug resistance found in *Plasmodium falciparum.* P-glycoprotein (pgpA) in *Leishmania* causes resistance of antimonial compound.

Immunological Control

See earlier Chaps. 8 and 9.

Molecular Control

See Chaps. 8 and 11.

Conventional and Novel Methods of Control of Arthropods

Chemotherapy/Insecticidal Agents

There are many insecticidal agents. These are eldrin, dieldrin, crufomate, crotoxyphos, coumaphos, fenthion, malathion, dichlorvos, permethrin, etc. These insecticidal agents are used for killing insects (mosquitoes, blood-sucking flies). Some chemicals are used as ectoparasiticidal agents. These agents kill the ectoparasites (ticks, mites, lice, fleas, etc.). Some examples are permethrin, cypermethrin, deltamethrin, diazinon, malathion, carbaryl, ivermectin, moxidectin etc.

Insecticide Resistance

Workers are still in search to find out a potential mechanisms of drug resistance. The following reasons have been identified so far.

1. Enhanced detoxification rate of insecticidal agent
2. Alteration in target site

Insecticidal resistance occurs due to increased enzyme activity for the detoxification of insecticidal agents. Gene upregulation of enzymes occurs that results in the shortening of half-life of insecticidal agents. The enzymes causing rapid detoxification are oxidative enzymes, hydrolysis enzymes, glutathione S-transferase, etc. Oxidative enzymes are cytochrome P450-dependant monooxygenase. Insecticides become less toxic after oxidation. Hydrolysis causes total detoxification of insecticide. Insecticides target sites in the nervous system of the insect, tick or mite. Common target sites are sodium channel, acetyl cholinesterase, acetyl choline receptor, etc. All these are proteins have several subunits. Due to genetic alterations, changes occur in amino acids of these proteins. Thus, they become insensitive to these insecticidal agents.

Integrated Control Method

The integrated control method is a simulation of different control methods. It has been observed that a single control strategy is not sufficient to significantly control parasites. That is why different control strategies should be simulated in such a way that a significant level of control could be achieved. For the control of arthropod parasites, the following are the possible ways:

1. Application of insecticidal agents/ chemotherapy
2. Immunological control
3. Biological control
4. Genetic control

Chemotherapy

Described earlier

Immunological Control of Arthropods

An array of works was carried out in the field of anti-tick vaccination. Different attempts were taken as research projects for the development of vaccine against tick infestation. Different categories of antigens comprising crude antigens such as eggs antigen, larval antigen, adult somatic antigen, etc. were used to perform immunoprotection trials. But these vaccines fail to produce considerable immunity due to a lack of specificity and immunoprotective potential. Variable protection was reported in the literature on the utilization of larval antigen but no vaccine has been commercialized using larval antigen. As has been mentioned above many works have been carried out by different workers using antigens derived from adult parasites but mentionable protection was not achieved. Later, research work on immunoprotection was directed towards the characterization of the concealed gut antigens of ticks. A satisfactory result, however, was achieved providing a good indication of immunoprotection using novel antigens from the gut. Researchers were encouraged to produce the vaccine by using the tick gut antigen. But it is needless to mention that all proteins present in the tick gut are not immunodominant. Therefore, an array of works was directed towards the identification of the most immunodominant antigens of the gut of ticks. Western blot techniques were employed for the detection of immunoprotective antigens. The efficacy of Bm 86 and Bm 91, the two candidate proteins of *Boophilus microplus* tick gut, were screened out, which showed promising results. Thanks to recombinant DNA technology, a large amount of protein can be produced. The target gene encoding the immunodominant protein is taken out from the whole genome and the gene is recombined with the plasmid or other vectors and inserted into the *Escherichia coli*, *Pichia pastoris*, etc. for the expression of the required protein. The specific protein is isolated by different specific protein purification protocols. Two vaccines using Bm 86 were commercialized in Australia and Cuba with the name "TickGARD" and "Gavac," respectively. Efforts were also employed by some workers to immunize the animals by using salivary proteins and get considerable protection.

Biological Control of Arthropod

Different biocontrol agents are used for the control of mosquitoes. These are as follows:

1. **Predatory fish**
   Examples
   (a) Mosquito fish (*Gambusia affinis*)
   (b) Cyprinid fish (carps)
   (c) Killifish
2. Fly
   (a) Dragonfly larvae feed on larvae of mosquitoes.
   (b) Adult dragonflies feed on adult mosquitoes.
3. Predator mosquitoes
   Example: Toxorhynchites
4. Fungi
   Example: *Metarhizium* sp.
5. Bacteria
   Bacillus thuringiensis israelensis (BTI) is used as biocontrol agents. They interfere with the digestive system of larvae of arthropods. These bacteria can be dispersed by hand or by helicopter.

Sterile Male Technique

This is a unique technique by which artificial mutation is induced in insects by radiation to disrupt the DNA. These insects are released in the wild. The insect population is reduced in this way. See Chap. 7 (Family: Glossinidae).

Breeding Policy

Some breeds of animals are resistant to tick infestation. These breeds can be used for farming.

Novel Control Strategies against Parasitic Infection

Naked DNA Vaccine or Gene Vaccine

See previous description

Nanotechnology

Nanotechnology literally means a technology which is performed on a nanoscale. This science has already contributed many interesting and surprising nanosystems in many spheres of mankind. The use of physical, chemical or biological systems at nanoscales is the mainstay of nanotechnology. This surprising technology is flourishing in many scientific fields. Of these different fields, the field of nanomedicine has also emerged with splendiferous facets comprising the development of smart drug and the development of novel vaccines. All these particularly involve the identification of precise targets, i.e., cells and receptors related to specific clinical conditions. The efficacy of the drug can be increased to many folds. Very limited works have yet been carried out in the field of application of nanotechnology in parasitology. However, at presen,t many works are being carried out in the field of veterinary parasitology.

History and Journey of Nanotechnology

The *Fantastic Voyage* (1966) was the first movie in which the seed of nanotechnology remained hidden. This was a famous movie in which nanotechnology was very interestingly applied in medicine and introduced to the large public that was the seed of today's nanomedicine. In this film, the life of a scientist, Jan Benes, was saved from severe blood clotting by an expedition using the miniaturized submarine *Proteus*, measuring to the scale of 1000 nm. The film bagged an extraordinary Oscar award for best special effects. This laid the foundation for nanomedicine. This science has been blessed with the invention of sophisticated instruments, such as scanning tunneling microscope, scanning electron microscope, transmission electron microscope, Fourier transmission electron microscope, etc. Atomic force microscope has also been invented which can scan atomic levels and generate image with extreme detail. In different scientific fields, multidirectional research on nanotechnology started and made an explosion of knowledge in this field of science. Nanotechnology in medicine emerged as a natural consequence of events. Scientists from all branches welcomed this science and the field of parasitology remain silent. Many works have also been carried out in this field.

Different Devices

Lipid systems: Amongst the nanoparticulate systems, lipid particles are used on a large scale. Since their composition is similar to plasma membrane lipids and cholesterol in human beings and other vertebrate animals, they appear to have a good biocompatibility and enormous potential to create a very good drug delivery system. Lipid nanoparticles have low toxicity as compared to polymeric particles. Solid lipid nanoparticles (SLNs) and nanostructured lipid carriers (NLCs) can maintain high as well as constant levels of drugs in the blood (constant plasma levels). Intravenous, intramuscular, and subcutaneous are the parental routes of administration of lipid nanoparticles. These nanoparticles are produced in different ways comprising double emulsion: water-oil-water or single emulsion: oil-water. These nanoparticles can also be prepared by another method, i.e., solvent diffusion method with ethanol and anti-solvent water. These nanoparticles are manufactured from solid or liquid lipids and stabilized by emulsifiers. Lipid nanoparticles are good as they can incorporate both hydrophilic and hydrophobic drugs.

Lipid-based systems are as follows:

1. Micelles
2. Liposomes
3. Lipid nanodiscs

4. Solid lipid nanoparticles (SLNs)
5. Nanospheres

Micelles: Micelles are formed by the amphipathic nature of fatty acid. They have hydrophilic and hydrophobic parts. Micelles are a good category of drug delivery system and the same has been evaluated by researchers applying in vitro and in vivo techniques. Micelles can potentially enhance the activity of antiparasitic drugs. Studies show the use of micelles as drug delivery devices for antitrypanosomal, anti-leishmanial, and some other antiparasitic drugs.

Liposomes: Liposomes are vesicular structures within which the target substance can be incorporated. They possess a lipid bilayer and a hollow sphere encompassing an aqueous phase. As they have both aqueous and lipid compartments, it is possible to encapsulate drug molecules or substances in either the aqueous compartment or the lipid bilayer compartment. Studies have shown the use of liposome system for antiparasitic drug, such as anti-leishmanial drugs, anti-trypanosomal drugs, and others.

Nanodiscs: Lipid nanodiscs are potentially useful delivery systems for the treatment of intrahistiocyte pathogens. These are composed of a phospholipid bilayer surrounded by the amphipathic motifs of apolipoproteins. Studies have shown the use of nanodiscs for the delivery of anti-*Leishmania* drugs.

Solid lipid nanoparticles (SLNs): SLNs are of different types of structures, such as SLN, nanostructured lipid carrier (NLC), and lipid-conjugates. SLN contain lipids. Since lipids are normal biochemical compounds of animal cells, they are biocompatible and biodegradable substances. They are solid and rigid providing good protection for the drug incorporated. SLNs have some disadvantages, such as low drug loading efficiency because of the restricted allosteric space. SLNs are typically produced using a high-pressure homogenization technique. Research studies have shown the efficient use of SLN for the delivery of antiparasitic drugs, such as praziquantel, albendazole, etc. with higher efficacy. Abbasalipourkabir et al. (2016) found efficacy in the nanocapsulation of nitazoxanide in solid lipid nanoparticles as a new drug delivery system and in vitro release study.

Polymeric systems: Polymeric systems consist of polymeric nanocapsules and nanospheres. These are solid particles in the colloidal form.

Nanocapsules: Delivery of drugs using nanocapsules confers high efficacy against different parasitic pathogens. Reports indicate good efficacy against *Leishmania donovani both in vitro* and in vivo using mannose-grafted chitosan nanocapsules (MnosCNc) containing AmB. Nanocapsules can be produced from different materials, such as chitosan, sodium alginate, mannitol, etc. Quinapyramine sulfate encapsulated within nanocapsules shows stronger in vivo efficacy against *Trypanosoma evansi* than the drug alone. The side effects of quinapyramine are high but if this drug is administered in nanocapsule, a negligible amount of side effects is found.

Polymeric micelles: Polymeric micelles are another type of nanomaterial which appears to be a good nanodelivery system that has garnered considerable attention as a potential drug carrier. Micelles form spontaneously in aqueous solutions and are composed of amphiphilic copolymers. They are used for the effective drug delivery for treatment of *Leishmania* infection.

Metalic nanoparticles: Metallic nanoparticles include inorganic nanoparticles, such as gold, silver, magnesium, titanium, zinc, etc. Drugs are adsorbed or covalently bound to the surface of these particles. These drug-bound particles remain as colloidal suspensions in liquid.

Drug-releasing potential of the nanoparticles depends on the binding forces between the drug molecule and the nanoparticle. The efficacy of gold nanoparticles against *Schistosoma* sp., *Leishmania* sp., etc. was reported. Several works were carried out on the antiparasitic efficacy of silver nanoparticles.

Different fungi are used for green synthesis of different nanoparticles. These are *Duddingtonia flagrans* and *Trichoderma harzianum.* Silver and zinc nanoparticles can be synthesized by plant synthesis protocol using different fungi. Curcumin-nisin nanoparticles against *Fasciola* species were assessed.

Anti-arthropodic Efficacy of Nanoparticles

Metal and metal oxides prepared through green synthesis have been found to be highly effective against arthropods. Studies show that silver and silver chloride nanoparticles have anti-tick activity. Other plants which are used for the synthesis of silver nanoparticles are *Ocimum canum*, *Cissus quadrangularis*, *Musa paradisiacal* (L), *Manilkara zapota,* etc. These plants are also used for synthesis of zinc nanoparticles. As reported, silver nanoparticles are effective against different ticks, such as *Rhipicephalus* (*Boophilus*) *microplus*, *Hyalomma anatolicum*, *Hyalomma isaaci*, *Haemaphysalis bispinosa,* etc. Plant synthesis of silver and gold nanoparticles from different plants is possible. These nanoparticles are effective against different mosquitoes also.

Nanoparticle-based Vaccine

Different nanoparticles are used to potentiate vaccine efficacy. Nanoparticles are used as adjuvants for different vaccine candidate.

Development of Nanodrug

Very limited works have been carried out on the application of nanotechnology in the field of veterinary parasitology but very recently a speedy research work has been directed towards nanoparasitology with sufficient funds. The following are the scopes for the application of nanotechnology in the field of veterinary science.

- Enhancement of the efficacy of different drugs with the help of nanoparticles has been proved.
- Use of different nanovehicles has proved to reduce the toxicity of antiparasitic drugs.
- Use of nanoparticles and nanovehicles to increase the bioavailability of the drug for a long period. Thus, the duration of drug activity can be increased.
- Development of smart antiparasitic drugs can be possible.
- The active gold-coated glass particles commonly known as gold nanoshells can open a new window to target the parasite without affecting/reducing host tissue damage. Several works were carried out and many present works are already in progress to envisage the activity of nanoshells. The nanoshells are being used as biosensors in the study of host-parasite interaction.
- Use of nanovehicles can reduce drug resistance of antiparasitic drugs. Pentavalent antimony sodium is the drug of choice for the treatment of cutaneous leishmaniasis. But the main problem of this drug is occurrence of increasing resistance against these compounds. Liposomes are very important for production of nanovehicles. Nanovehicles are produced by mixing phospholipid with water. Use of nanovehicles for delivery of the aforesaid drug has created opportunity to reduce drug resistance.

Development of Nanovaccine

(a) Nanoparticles substitute the plasmid vector for the introduction of recombinant DNA in host cells (*E. coli*) for the production of recombinant antigens which may be used for the development of vaccines.
(b) The knowledge of nanotechnology is also providing different concepts of generating modest adjuvants to potentiate antiparasitic vaccines.
(c) It is very interesting that it is possible to convert the character of antigens by using nano-

technology. As per desire, one antigen could be used for humoral or cell-mediated immune response.

(d) Use of nanoadjuvant can reduce the number of booster doses of antiparasitic vaccines.

Some Nanotherapeutic Research

In a study by Abbasatipourkabir et al. (2016), the solid lipid nanoparticles (SLNs) were used for the encapsulation of nitazoxanide (NTX) which is a new antiparasitic and antiprotozoal drug delivery system. Studies show that praziquantel-loaded hydrogenated castor oil solid lipid nanoparticle has enhanced efficacy. PZQ-loaded hydrogenated castor oil solid lipid nanoparticle (PZQ-HCO-SLN) suspension is prepared by hot homogenization and ultrasonication method. Silver, chitosan, and curcumin nanoparticles have anti-Giardial activity.

Kar et al. (2014) performed in vitro testing of the efficacy of gold nanoparticles against the cestode parasite exhibited paralysis times of 2.67, 2.1, and 1.47 h and death times of 3.22, 2.83, and 2.55 h for dosages of 0.25, 0.5, and 1.0 mg/ml, respectively. The most efficacious dose was determined as 1.0 mg/ml, wherein the onset of paralysis of parasites occurred after 1.47 h and death after 2.55 h of treatment. In different trials, the enzyme activity of cestodes was also assayed to see the efficacy of nanoparticles. In the treated worms, there were marked changes in the ultrastructural morphology, such as disorganization of the tegumental architecture showing major damages in the form of wrinkles, lesions and ruptures on the surface tegument of the parasite. The clustered appearance of microtriches was found all over the body surface. The AcPase activity in *Raillietina* sp. paralyzed with the test materials (gold nanoparticles and genistein) was diminished by 29.16% and 34.72%, respectively. The study of Kar et al. (2014) also showed that the gold nanoparticles (6–18 nm diameter) produced by bioreduction of chloroauric acid using mycelia-free culture filtrate have a detrimental effect on the ultrastructure of ***Raillietina*** sp. The disorientation of microtriches, spines, and scales was noted along with distortion and disfigurement of the tegument, breakage, and sloughing off of the tegumental surface structures.

Biologically synthesized AgNPs against *Haemonchus contortus* were also reported. AgNPs using *Ziziphus jujuba* leaf extract displayed anthelmintic activity against *H. contortus*. The green synthesis of nanoparticles using microbes such as actinomycetes, bacteria, and fungi is undoubtedly easy and less time-consuming methods for getting novel nanomolecules. It is needless to mention that fungi can be grown easily in the laboratory as well as on a large scale as mycelia can withstand extreme conditions of the environment. Different fungi are used for the production of different nanoparticles. These fungi are *Fusarium*, *Aspergillus*, *Penicillium* sp., etc. The use of different nanoparticles using different fungi is different.

References

Abbasalipourkabir R, Fallah M, Maghsood AH, Javid S (2016) Nanocapsulation of nitazoxanide in solid lipid nanoparticles as a new drug delivery system and in vitro release study. J Biol Sci 16(4):120–127. https://doi.org/10.3923/jbs.2016.120.127

Dalton JP, McGonigle S, Rolph TP, Andrews SJ (1996) Induction of protective immunity in cattle against infection with Fasciola hepatica by vaccination with cathepsin L proteinases and with hemoglobin. Infect Immun 64(12):5066–5074. https://doi.org/10.1128/IAI.64.12.596

Estuningsih SE, Smooker PM, Wiedosari E, Widjajanti S, Vaiano S et al (1997) Evaluation of antigens of *Fasciola gigantica* as vaccines against tropical fasciolosis in cattle. Int J Parasitol 27(11):1419–1428. https://doi.org/10.1016/s0020-7519(97)00096-9. PMID:9421734

Gadahi JA, Eshan M, Wang S, Zhang Z, et al. (2017) Recombinant protein of *Haemonchus contortus* small GTP ase ADP-ribosylation factor 1(HcARF1) modulate the mediated immune response in vitro. Oncotarget 8(68):112211–112221. doi:https://doi.org/10.18632/oncotarget.22662 . PMID:29

Gauci C, Vural G, Oncel T, Varcasia A et al (2008) Vaccination with recombinant oncosphere antigens reduces the susceptibility of sheep to infection with *Taenia* multiceps. Int J Parasitol 38(8–9):1041–1050. https://doi.org/10.1016/j.ijpara.2007.11.006. PMID:18160069

Gonzalez-Sanchez ME, Ndombasi-Bokuy M, Cuquerella M, Alunda J (2019) Immunization with recombinant rHc23 partly protects lambs against trickle infections by *Haemonchus contortus*. BMC Vet Res 15(1):333. https://doi.org/10.1186/s12917-019-2084-5

Kandil OM, Abdelrahman KA, Shalaby HA, Hendawy SHM (2017) Evaluation of a crude larval protein and recombinant somatic protein 26/23(rHcp26/23) immunization against Haemonchus contortusin sheep. Vet World 10(7):758–763. https://doi.org/10.14202/vetworld.2017.758-763. PMID:28831218. PUBMED

Kar PK, Murmu S, Saha S, Tandon V et al (2014) Anthelmintic efficacy of gold nanoparticle derived from phytopathogenic fungus, *Nigrospora oryzae*. PLoS One 9(1):e84693. https://doi.org/10.1371/journal.pone.0084693

Kumar N, Varghese A, Gaurav N, Dinesh C, Samanta S et al (2012) Vaccination of buffaloes with Fasciola gigantica recombinant glutathione s-transferase and fatty acid binding protein. Prasitol Res 110:419–426. https://doi.org/800.gl/pkppmm

Kumar N, Maurya PM, Solanki JB, Mishra AK (2016) Immunoprophylactic control strategy for tropical fasciolosis: a possibility. J Vaccines Immunol 2(1):036–040. https://doi.org/10.17352/jvi.000019

Morrison CA, Colin T, Sexton JL, Bowen F, Wicker J et al (1996) Protection of cattle against Fasciola hepatica infection by vaccination with glutathione S-transferase. Vaccine 14(17–18):1603–1612. https://doi.org/10.1016/s0264-410x(96)00147-8. PMID:9032888

Newton SE (1995) Progress on vaccination against Haemonchus contortus. Int J Parasitol 25(11):1281–1289. https://doi.org/10.1016/0020-7519(95)00065-a. PMID:8635880. PUBMED

Paykari H, Dalini A, Madani R (2002) Immunization of sheep against *Fasciola gigantica* with glutathione S-transferase. Vet Parasitol 105(2):153–159. https://doi.org/10.1016/s0304-4017(01)00546-5. PMID:11900929

Rabia Aly I, Diab M, AM EI-A, Hendawy M, Kadry S (2012) Fasciola gigantica Fatty binding protein (FABP)as a prophylactic agent against Schistosoma mansoni infection in CD1mice. Korean JParasitol 50(1):37–43. https://doi.org/10.3347/kjp.2012.50.1.37. PMID22451732; PMCID: PMC3309049

Sexton JL, Wilce MC, Colin T, Wijffels GL et al (1994) Vaccination of sheep against Fasciola hepatica with glutathione s-transferase: identification and mapping of antibody epitopes on a three dimensional model of the antigen. J Immunol 15 152(4):1861–1872. PMID:7509829. PUBMED

Tsuji N, Suzuki K, Kasuga-Aoki H, Matsumoto Y et al (2001) Intranasal immunization with recombinant *Ascaris suum* 14-kilodalton antigen coupled with cholera toxin B subunit induces protective immunity to *A. suum* infection in mice. Infect Immun 69(12):7285–7292. https://doi.org/10.1128/IAI.69.12.7285-7292.2001. PMID:11705899

Wang H, Li Z, Gao F, Zhao J et al (2016) Immunoprotection of recombinant Eg.p29 against Echinococcus granulosus in sheep. Vet Res Commun 40(2):73–79. https://doi.org/10.1007/s11259-016-9656-7. PMID:27094043

Wei J, Versteeg L, Liu Z et al (2017) Yeast -expressed recombinant As16 protects mice against Ascaris suum infection through induction of a Th2 skewed immune response. PLOS Negl Trop Dis 11(7):e0005769. https://doi.org/10.1371/journal.pntd.0005769. PMID:28708895. PUBMED

Wijffels GL, Salvatore L, Dosen M, Waddington J et al (1994) Vaccination of sheep with purified cysteine proteinases of Fasciola hepatica decreases worm fecundity. Exp Parasitol 78(2):132–148. https://doi.org/10.1006/expr.1994.1014. PMID:8119370

Zhu M, Gao F, Li Z, Wang X, Wang Z, Zhao J, Zhao W (2016) Immunoprotection of recombinant Eg.myophilin against Echinococcus granulosus infection in sheep. Exp Ther Med 12(3):1585–1590. https://doi.org/10.3892/etm.2016.3501. PMID:27602078. PMCID:4998173

13 Parasitological Techniques

The different parasitological techniques include collection and preservation of entire parasite and / or parasitic stages. It also furthers up identification of parasites or parasitic stages following conventional techniques,immunological techniques, molecular tecniques. A collective information has been given in this chapter which comprises fecal examination using different techniques, biopsy examination technique, blood smear examination for identification of haemoprotozoa and other parasites. This chapter also encompasses different immunodiagnostic techniques i.e. ELISA, Indirect Fluorescent Antibody Test, Indirect Haemagglutination Test, Compliment Fixation Test, etc. The important molecular techniques have also been described in this chapter. The techniques described in this chapter is the overview. The different techniques and data (amount of chemical, amount of solution,duration of time in different steps of methods mentioned in this chapter may vary. One should standardize every technique:

- **General parasitological techniques**
  The general techniques usually include
  (a) Collection of parasites and parasitic stages
  (b) Preservation of parasites and parasitic stages
  (c) Permanent mounting of parasites and parasitic stages
- **Immunoparasitological techniques**
  This technique generally includes
  (a) Preparation of parasitic antigen
  (b) Use of parasitic antigen for immunodiagnosis and immunoprotection
- **Molecular parasitological techniques**
  Molecular techniques include molecular diagnosis, other molecular techniques related to vaccine production, etc.

General Parasitological Techniques

Collection of Helminths

Purpose

The helminthes are collected for different purposes, such as for the preservation of intact worms, larvae, and eggs in the museum, practical laboratory hall, preparation of antigens, in vitro culture, etc.

Specimens

Whole Adult Specimen

(a) Whole specimens are collected for preservation in the laboratory as gross specimens or for mounting on slide.
(b) Whole specimen is also collected for the purpose of preparation of whole somatic antigen.

S. C. Mandal, *Textbook of Veterinary Parasitology*, https://doi.org/10.1007/978-981-99-2924-5_13

(c) They are also collected for culture in the laboratory.

Part of Specimens

Part of the specimens are collected for the purpose of permanent preservation in the museum, preparation of permanent mounted slides, preparation of antigens, extraction of genetic material, etc.

Eggs and Larvae

Eggs are collected for laboratory preservation for future demonstration to the students or other persons concerned, preparation of egg antigen, extraction of genetic material, etc.

Requirements During Collection

- Beakers
- Petri dishes
- Sample containers
- Sieves of different gauzes
- Phosphate buffer
- Normal water
- Distilled water
- Ice box with ice

Time of Collection

The time of parasite collection is very important. Many tapeworms are very small in size. The scolex of tapeworm is very small. Clear sunlight is essential for the detection of these small worms or parts of the worms. Therefore, cloudy climate, dim light, evening, and night are not ideal times.

Collection Method

The collection of worms is usually done from the slaughtered animals of a slaughterhouse. Since the predilection site of tapeworms is mostly the small intestine, the small intestine part is opened. Easily detectable tapeworms are collected at first. The scolex of the tapeworms remains attached to the wall of the intestine. The scolices are to be removed very gently so that the scolex part is not detached from the worm. The small tapeworms that are not easily detectable can also be collected in different ways. For this, the intestinal content is taken out and the content of the intestine is mixed with water to dilute the content. The diluted content is sieved using a sieve of different pore sizes. The collected tapeworms are thoroughly washed with water. The worms subject to permanent preservation are placed in 10% formalin. The worms or part of worms subject to permanent slides are placed in between slides. The slides are then bound by thread or rubber bands on either side. The pressed worms are placed in 10% formalin. The worms subject to culture are placed in the container having culture media. The worms subject to antigen preparation are placed in the ice box.

For the collection of trematodes, the respective organs are taken out. For collection of liver flukes, the affected liver are sliced by a knife. Then, the pieces are squeezed to remove the flukes. The bile duct is opened and the worms are collected from the bile duct. The amphistomes are collected from the rumen and reticulum. Sometimes, innumerable amphistomes are found attached to the ruminal or reticular wall. The worms are washed thoroughly and placed in the container having 10% formalin. In a similar way as described in the case of cestodes, the trematodes are placed in different containers.

For a collection of nematodes, the worms are also collected in a similar way as described earlier.

Fecal Examination

Purpose

Fecal samples are collected from different types of hosts. Fecal examination is done for the diagnosis of different parasitic infections. Diagnosis of gastrointestinal (GI) parasitic infection is possible by fecal examination. Intact whole parasites, segments, eggs, and larvae are passed with the feces. Detection of the parasite and parasitic stages is done for the diagnosis of parasitic infection.

Requirements

Top cork plastic specimen collecting container, polyethylene bag, and plastic bags.

Method of Collection

- Collect uncontaminated feces. The feces should not be contaminated with urine, dust, grass, grains, etc.
- Utmost effort should be made to collect fresh feces. For the collection of feces from large animals, direct per-rectal collection of feces using gloves should be done. In the case of small animals, such as sheep or goats, feces are collected by inserting the index finger using gloves. A gentle massage is required to relax the sphincter muscle.
- Feces can be collected from the ground also, if the animal is found defecating. In this condition, there will be no problem of detection of animals.
- The containers are labeled perfectly with the mentioning of date of collection, time of collection, location, animal species, sex, and age.

Preservation of Feces

The feces is preserved in 10% formalin or 70% alcohol. If it is not possible to examine the feces immediately, it may be kept in the refrigerator at 4 °C.

Shipment of Feces

If the feces are subject to examination in laboratory at distant places, they should be packed perfectly. For that, two containers are used for a single sample: one for keeping feces in 10% formalin (1:3, feces:formalin) and the other for placing feces in PVA solution required for trichrome staining, if suspected. The corks of the containers should be tightened. These containers should be further wrapped with paraffin tape. The pack should be labeled properly.

The collection containers should be labeled properly with full identification of the animals, date and time of collection, etc. Feces can be collected from the ground if the animal is found defecating. In that situation, the upper part of the feces should be taken. The material adjacent to the ground should not be taken because of the risk of contamination. The feces should not be contaminated with dust, debris, urine, hair, etc. that create confusion during examination.

It is better to collect the feces directly from the suspected animals following a per-rectal collection technique by wearing clean gloves. In the case of small animals like sheep, goats, etc., a moistened index finger is inserted and massaged gently until the sphincter muscle is properly relaxed. To collect feces from birds, it is advised to keep the single bird in a single cage fitted with a clean metal or plastic tray for collecting feces.

If it is needed to examine the whole herd, fecal samples are collected from some representative ailing animals or suspected animals.

Examination of Feces

Gross examination of feces is very important because it provides many clues for diagnosis. The color and consistency should be noted. The feces should be examined for whether any blood streaks or mucous material are present. It is to be further noted whether the feces is completely formed. Completely formed feces are semisolid and cylindrical.

1. **Direct smear technique:** This is easy, convenient, less time consuming, and nonexpensive. A pinch amount of feces is placed on a cleaned slide. Up to 1–2 drops of distilled water are added to the feces. A clean stick is used to make the smear. The smear should be carefully prepared so that no air bubble is created. The smear should be semitransparent and it should not be hazy.
2. **Concentration techniques**

Sedimentation Method

Purpose

This technique is done when the load of parasitic infection is less and the eggs, cysts, and oocysts are not easily found during the examination of the fecal smear.

Requirement

- Feces
- Tea strainer or double-layer gauze
- Distilled water
- Conical flask
- Beaker

Procedure

- Take approximately 10 g of feces in a glass container.
- Add 100 ml of water to the feces.
- Mix the material thoroughly.
- Filter the diluted feces using a tea strainer or double-layer cloth
- Place the filtrate in a conical flask for 5 min.
- Decant the supernatant carefully without disturbing the material at the bottom.
- Repeat the process for 8–10 times.
- Then, examine the sediment to find any parasitic egg or parasitic stages.

Observation

Eggs and parasitic stages are found during examination.

Direct Floatation Method

Purpose

This technique is done when the load of parasitic infection is less and the eggs, cysts, and oocysts are not easily found during examination of fecal smear.

Principle

The eggs that float upwards if these are placed in a saturated salt or sugar solution. Here, the difference of specific gravity of eggs, cysts, oocysts, and the flotation solution is exploited. The specific gravity of eggs, cysts, etc., is less as compared to the specific gravity of the floatation solution.

Requirements

Chemical Items

- Salt (NaCl)
- Sugar
- Distilled water

Other Items

- Feces
- Beakers
- Gauze for filtration
- A narrow container

Procedure

- Take a glass container.
- Pour 10 ml of floatation solution (saturated sodium chloride or sugar solution in general and saturated Zinc Chloride or Zinc Sulphate for trematode eggs).
- Place 5 g of feces in the floatation solution.
- Mix it properly with a glass rod.
- Filter it with a gauze.
- Take the filtrate in a narrow glass container.
- Keep it for 20–30 min undisturbed.
- Place 2–3 drops from the upper surface of the floatation solution on the slide.
- Examine under a microscope.

Observation

Eggs or any other parasitic stages may be found.

Centrifugal Floatation Method

Purpose

This technique is done in low concentrations of eggs or any other parasitic stages in feces. The eggs, cysts and oocysts are not easily found during the examination of the fecal smear.

Principle

As with that of the sedimentation method, here also, the difference of specific gravity of eggs, cysts, oocysts, and floatation solution is exploited. The specific gravity of eggs, cysts, etc. is less as compared to the specific gravity of the floatation solution, and these float upwards and reach the upper surface of the floatation solution. The combination of centrifugation and floatation increase the sensitivity of this method.

Requirements

Chemical Items

- Salt (NaCl)
- Sugar
- Distilled water

Other Items

- Feces
- Beakers
- Gauze for filtration
- Centrifuge tube

Method

- Take a glass container.
- Pour some amount of floatation solution (10 ml) into it.

- Place 5 g of feces in floatation solution.
- Mix it properly with a glass rod.
- Filter it with a gauze.
- Take the filtrate in a centrifuge tube.
- Centrifuge for 5 min at 1200 rpm.

The procedure is same as above. The filtrate is kept in a centrifuge tube and centrifuged at 1200 rpm for 5 min. After centrifugation, a glass rod is touched to the upper surface of the fluid and one to two drops of liquid is taken out and placed on a glass slide. A cover slip is placed on the fluid and examined under a microscope. Or, in another way, a disposable pipette is gently placed under the surface touching the wall of the tube and fresh floatation solution is added to form a meniscus at the top. A coverslip is placed on the top and allowed to stand for 10 min. Then, the cover slip is placed on the slide and examined under the microscope.

Observation: Eggs or any other parasitic stages may be found.

Willis Technique

Purpose
The purpose of this technique is similar to the other two techniques described before. This is indeed direct floatation method.

Principle
As with that of sedimentation method, here also, the difference of specific gravity of eggs, cysts, oocysts, and floatation solution is exploited. The specific gravity of eggs, cysts, etc. is less as compared to the specific gravity of the floatation solution and these float upwards and reach to the upper surface of the floatation solution.

Requirements

Chemical Items
- Salt (NaCl)
- Distilled water

Other Items
- Feces
- Beakers
- Cheese cloth

Method
- Take a glass container.
- Pour some amount of floatation solution (10 ml) into it.
- Place 5 g of feces in the floatation solution.
- Mix it properly with a glass rod.
- Filter it with a gauze.
- Take the filtrate in a narrow glass cylinder and fill it up to the top and a clean cover slip is slid away in such way that there will be no air bubble in between the slide and cover slip. Keep it for 20-30 minutes.
- Quickly remove the cover slip and place on a slide and examine under microscope.

Observation
Eggs or any other parasitic stages may be found.

Stoll's Dilution Technique

Purpose
The purpose of this technique is to determine eggs per gram of feces (EPG).

Principle
The number of eggs is determined by counting the number of eggs in a sample of feces. The number of eggs in a sample of feces represents the number of eggs in a gram of feces by specific calculation.

Requirement
- 3 g of feces
- N/10 NaOH solution
- Test tube graduated at least for 45 ml
- A few glass beads

Method
- Take a test tube graduated at least for 45 ml.
- Pour 45 ml N/10 NaOH in the test tube.
- Add 3 g of feces in the NaOH solution.
- Stir the material uniformly by using a few glass beads.
- Place 0.15 ml of the material on the slide and place a cover slip on it.

- Examine it after 5 min under microscope and count the number of eggs.

Observation and Calculation

The eggs or parasitic stages are identified and counted. The EPG is calculated as follows:

$$\text{EPG} = \text{Number of eggs counted} \times 100$$

McMaster Method

Purpose

The purpose of this technique is to determine eggs per gram (EPG) using a McMaster slide.

Principle

The McMaster slide is a specific slide made up of two slides separated by transversely placed strips of glass. On either side of the slide, 1 cm^2 is marked. The height is 0.15 mm in between two slides. Therefore, if the area is filled with fecal material in solution, the volume is 0.15 ml.

Requirements

- Feces
- Floatation solution
- Glass beaker
- McMaster slide

Procedure

- Take 2 g of feces and soak it in 30 ml of water and wait until the feces is sufficiently soft.
- Add 30 ml of saturated salt solutions in it.
- Stir it properly with a glass rod.
- Take the sample in a Pasteur pipette and load the sample in the McMaster slide filling all the spaces.
- Count the eggs within the ruled area of chamber.

Observation and Calculation

The eggs are identified and counted. The number of eggs within the ruled area of a chamber multiplied by 200 represents the number of eggs per gram of original feces sample for both the chambers. **Calculation**: Eggs Per Gram = Number of eggs in a ruled area of a chamber × 200

Culture of Feces

Fecal culture is done in different ways. This can be done by using a Petri dish or test tube.

Petri Dish Method

Purpose

In vitro culture of feces for the collection of larvae.

Principle

The eggs of many nematodes are passed with the feces. The eggs hatch in the environment in the presence of suitable environmental factors. In the fecal culture technique, the same environmental factors are provided for the hatching of the eggs.

Requirements

- One large Petri dish and one small Petri dish
- Feces
- Blotting paper
- Charcoal powder

Method

- Place a small-sized Petri dish on a large-sized Petri dish upside down.
- Mix some amount of feces with charcoal powder uniformly and keep the material ready.
- Place a moistened filter paper on the small Petri dish. Some amount of feces is smeared on the small Petri dish.
- Pour water into the large Petri dish to more than half of its volume.
- Connect the smeared feces with water by using a blotting paper and keep this arrangement in room temperature usually for seven days.
- Examine the water of the lower Petri dish for the detection of larvae.

Observation

The water of the large Petri dish is examined under a microscope at 10 X. The movement of larvae is clearly detectable.

Test Tube Method (Harada Mori Method)

Purpose
In vitro culture of feces for the collection of larvae.

Principle
Water is a very important factor for the hatching of nematode eggs. In this technique, the smeared feces are arranged in such a way that the feces come into contact with water.

Requirements
- One large test tube
- Glass slide
- Blotting paper
- Charcoal powder

Procedure
- Wrap the glass slide with blotting paper.
- Some amount of feces is smeared over one side of the slide.
- Take some amount of water in the test tube.
- Place the slide smeared with feces in a slanting position.
- Pour some amount of water gently to fill one-third of the smeared area.

Observation
The water of the test tube is examined under a microscope at 10×. The movement of larvae is clearly detectable.

Collection of Larvae

Baermann Technique

Purpose
By this technique, larvae from feces or any other sample are collected.

Principle
Time duration and moisture are two important factors for the hatching of eggs. This condition is maintained in the Baermann apparatus.

Requirements
- Baermann apparatus: This apparatus constitutes a stand with a holder and a funnel fitted with a rubber tube. The terminal part of the tube is fitted with a clamp or clip. The funnel is filled up with water. A cheesecloth is placed on a wire sheet placed on the funnel.
- Centrifuge tube
- Water

Procedure
- Fill up the funnel with warm (40 °C) water.
- Place 10 g or more amount feces on cheesecloth.
- Keep this apparatus in this state overnight.
- Open the clump or clip and collect 10 ml of fluid from the rubber tube.
- Centrifuge the fluid.
- Examine the material of the bottom part of the test tube for detection of larvae.

What Happens?
The larvae present in the feces move out and start passing through the cheesecloth and proceed downward.

Observation
Examine the material of the bottom part of the test tube for detection of larvae.

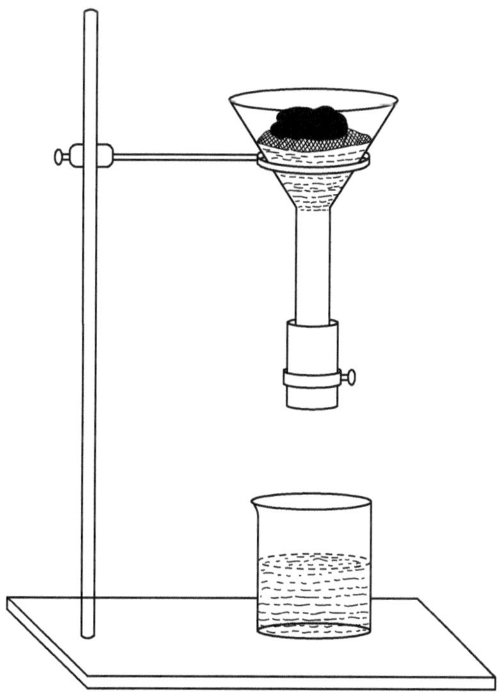

Fig. 13.1 Baermann apparatus

Preparation of Permanent Slides of Cestodes and Trematodes

Purpose

Permanent slides are prepared for the identification of the parasites. The prepared slide can be kept for long periods (many years) for demonstration to the students or other concerned persons.

Requirements

- AFA (alcohol-formalin-acetic acid)
- 30%, 50%, and 70% grade alcohol
- Borax carmine
- Acid alcohol solution
- Xylol
- Canada balsam
- Slides
- Coverslip

Procedure

- The whole parasite or part of the parasite is placed in between two slides and the specimen is flattened gently by pressing.
- The slide is bound by a thread on either side so that the slides are not reached.
- The overlapped slides are placed in AFA used for fixation for 20 min.
- Then, the bound thread is opened and the parasites are removed carefully avoiding any rupture.
- The specimen is passed serially in different graded alcohols (30%, 50%, and 70%) for 20 min each.
- Thorough washing of the specimen is done with 70% alcohol in a rectangular cavity block.
- The specimen is then transferred in alcoholic borax carmine for 1 h or more depending on the size of the parasites.
- Then, the specimen is washed in 70% alcohol twice for 20 min in every step separately.
- Then, the specimen is serially passed through 80%, 90% and absolute alcohol one by one for a duration of 20 min for every step in separate containers.
- Transfer the specimen in Xylol-1 and Xylol-2 in separate containers for washing.
- Transfer the specimen to clove oil for 30 min.
- Place in Xylol for washing.
- Mount the specimen in Canada balsam.

Techniques of Protozoal Examination

The methods mentioned here are only the overview. One should standardize the methods along with data in their laboratories for getting accurate results.

Examination of Blood Smear

Examination of the blood smear is the gold standard method for the identification of hemoprotozoan parasites. A wet smear is examined in some cases. In the case of trypanosoma infection, the examination of a wet smear is done. Although no morphological features are observed by the wet smear technique, the technique provides clues for diagnosis. Giemsa staining and Leishman staining are commonly done. Other stains are methylene blue and Wright's stains. It may be required to examine more than one slide. Both thick and thin smears are prepared, because in some cases, parasitemia is too low. In those cases, a wet smear would be useful. The method is as follows:

Leishman Staining

Purpose

The purpose of this staining is to detect hemoprotozoan infection.

Procedure

- Air-dry the smeared slides.
- Pour some Leishman stain on the smeared slides in such a way that it will cover the smear but not overflood it.
- Allow the arrangement for 2–3 min
- Pour an equal amount of distilled water on the smear.
- Allow it to stand for 10 min for staining.
- Remove the slides from the staining rack and decant the stain.

- Wash immediately.
- Air-dry the smear.
- The slide is ready for examination.

Observation

The slides are examined for the detection of hemoprotozoan infection. The microscopical examination is done under 40× and 100×.

Giemsa Staining

Purpose

The purpose of this staining is to detect hemoprotozoan infection.

Procedure

- Air-dry the smeared slides.
- Prepare the fresh working solution by mixing the stain with deionized distilled water (1:9).
- Pour some of the stain on the smeared slides in such a way that it will cover the smear but not overflooded it.
- Allow the slides undisturbed for 30 min. The slides should be monitored so that the slides will not dry.
- Remove the slides from the staining rack and decant the stain.
- Wash immediately.
- Air-dry the smear.
- The slide is ready for examination.

Observation

The slides are examined for the detection of hemoprotozoan infection. Microscopical examination is done under 40× and 100×.

Lymphnode Biopsy Technique

Purpose

This technique is used to detect *Theileria* infection in lymphocytes.

Background Information

During the life cycle of *Theileria*, the sporozoites are transmitted by ticks to the hosts. The sporozoites enter the lymphocytes. Using the lymph node biopsy technique, the biopsy material is used for the preparation of smears. The organisms are found in the lymphocytes.

Requirements

A syringe with a needle, phosphate buffered saline (PBS), and clean grease-free slide

Procedure

- Restrain the suspected animal.
- Locate the prescapular lymph node and hold it by the left hand.
- Inject 5–10 ml of PBS into the lymph node.
- Massage the lymph node.
- Take out lymph material from the lymph node by a syringe.
- Prepare the smear using the lymph material.

Observation

The smears are stained by Giemsa stain. Schizonts (Koch's blue bodies) are detected.

Modified Ziehl-Neelsen Method

Purpose

The purpose of this staining is to stain *Cryptosporidium* by the modified Ziehl–Neelsen method.

Requirements

- Suspected feces
- Carbol fuchsin
- Methanol
- Acid water solution (1% HCl in methanol)
- Malachite green (0.4%)

Procedure

- Prepare the fecal smear.
- Dry the fecal smear.
- Fix the smear with methanol for 3 min.
- After fixation, pour some amount of Carbol Fuchsin, and allow it to stand for a duration of 15–20 min.
- Remove the slides from the staining rack.
- Decant the Carbol Fuchsin and wash the smear with distilled water.
- The next step is done for the decolonization of Carbol Fuchsin. Pour 1% of HCl in methanol and keep the slide for 20–30 s.

- Remove the slides and wash them with distilled water.
- Counterstain the smear with Malachite Green (0.4%) for 30–60 s.
- Remove the slides, decant the stain, and wash off.
- Air dry the slides.
- The stained slides are washed with graded alcohol and are cleared by xylene in 3–4 changes. The stained slide is mounted with synthetic resin.

Trichrome Staining

Purpose

Trichrome staining is done for staining of fecal smear.

Requirement

- Schaudinn's solution
- 70% ethanol
- Trichrome stain
- Ethanol
- Xylene

Procedure

- Prepare the fecal smear on the grease-free slide.
- Place some amount of Schaudinn's solution in a staining jar and place the slides for 30 min for fixation.
- Place the slide in 70% ethanol and allow for a period of 5 min.
- Keep the slide in a solution containing 70% ethanol with iodine.
- Place the slide in 70% ethanol for 5 min.
- Place the slide in 70% ethanol for 3 min in a separate container.
- Stain the slide using Trichrome stain for 7–10 min.
- Then, place the slide in acid-alcohol solution (90% ethanol and 1% acetic acid by volume) for 1–3 s.
- Use two separate containers having absolute ethanol and place the slide separately in each container for 3 min.
- Pass the slides in two separate xylenes for 5–10 min each.

Culture of Oocysts

Purpose

We should know the purpose of oocyst culture. These are as follows:

(a) *Specific identification*
Oocysts are cultured for specific identification. In unsporulated oocysts, different identifying morphological features are not found. But in the sporulated oocysts, sporocysts, sporozoites, polar granules, etc. are clearly visible and these features can be used for the identification of oocysts.

(b) *Artificial infection*
Sporulated oocysts are the infective stage of coccidian parasites. Therefore, sporulated oocysts are necessary for artificial infection to the hosts.

Requirements

- Petri dish
- Distilled water
- 2.5% potassium dichromate solution

Procedure

- Take a Petri dish.
- Place some amount of feces subject to culture.
- Pour 2.5% potassium dichromate solution.
- Mix it properly.
- Stir the material twice daily.

Observation

The oocysts are examined under a microscope after 3–4 days under 40×. The sporulated oocysts are found.

Direct Smear Examination for Protozoa

Direct smear technique reveals the presence of oocysts, cysts of *Entamoeba*, *Giardia*, *Balantidium*, etc. Follow the technique of preparation of direct smear as mentioned earlier. Add one drop of Lugol's iodine for better visibility for *Entamoeba* or *Giardia*.

Concentration Technique

The methods of floatation are done for detection of protozoan parasitic stages (oocyst, oocyst, etc.) as described earlier.

Practical Part of Arthropods

Collection and Preservation of Arthropods

Purpose

The arthropods are collected for keeping in the Parasitology Museum for demonstration to the students or other persons concerned. They are also collected for the development of arthropod colonies for research purposes. These are collected for direct preparation of antigens or extraction of genetic material.

Requirements

- For fly collection—Fly-catching net
- For lice collection—Fine toothed comb can be useful
- For mite collection—Blunt forceps are used to scrape infected skin.

Collection Method

For the collection of flies, fly-catching nets are used. One should search for a swarm of flies and by sudden effort, the flies should to be collected; otherwise, they will be dispersed. Different baited traps are useful for the collection of flies. Other traps including sweeping traps and light traps can also be used for catching flies. For ground-dwelling beetles or other ground-dwelling insects, different pan traps are used. For a collection of lice, the highly infected area should be searched out. If the lice or nits are found attached to the hairs, furs, etc., the affected hairs should be clipped out by scizures. Otherwise, a mixture of BHC and sand (1:4) is applied on the affected area. The lice will fall down. They are collected carefully. The ticks are usually collected manually by hand. They are found in some particular location, such as ears and in between the hooves. The ticks are held by hands first and then removed from the affected site by gentle pulling outwardly in such a way that the mouth apertures remain intact. This is because, for identification of the ticks, mouth apertures are important. Initially, some damage may occur to the ticks but gradually one would be trained to collect the ticks without losing any body parts. For a collection of mites, the affected area is identified. By using a blunt forceps, the affected skin is scraped.

Preservation of Arthropods

The collected flies, lice and ticks are kept in a cyanide bottle containing a freshly prepared bed of potassium cyanide and plaster of Paris mixture on the floor with a few sheets of blotting papers over it. Pieces of paper are placed in bottles so that the arthropods cannot immediately cling together. Collected flies, lice, ticks, and mites are preserved in 70% alcohol or 10% formalin. The fleas are preserved only in 70% alcohol. Formalin is not used for the preservation of fleas.

Mounting of Arthropods

Some arthropods, such as flies and beetles, are mounted dry in the entomology box. The entomological box is a wooden box containing a bed of wax with naphthalene. The arthropods are pinned down to the wax bed vertically. Thin and long entomological pins are used for this purpose. The labels of respective arthropods are also pinned up. A few naphthalene balls are placed there so that the arthropods are not destroyed by insects.

Preparation of Permanent Slides

Purpose

Permanently mounted slides are prepared for identification of the arthropods and to keep the slides in the museum for demonstration to the students or other concerned persons.

Requirements

Before proceeding to prepare permanently mounted slides, the required items should be kept ready. These are distilled water, graded alcohol (50%, 70%, and 90% absolute), 10% KOH solution, Canada balsam or DPX, slides, and coverslip.

Method

- The different body parts, viz., legs, heads, wings, etc., are passed in alcohol and then transferred to xylene for clearing. Then, these specimens are mounted on slides. The arthropod egg and nits are also mounted in similar ways.
- The ticks are pierced by pins in three to four areas so that KOH solution can enter the body and digest the inner tissue of the arthropods. They are washed properly with distilled water. The washed arthropods are passed in the graded alcohol for dehydration. Then, these are transferred to xylene for clearing. The cleared specimen is mounted by Canada balsam.

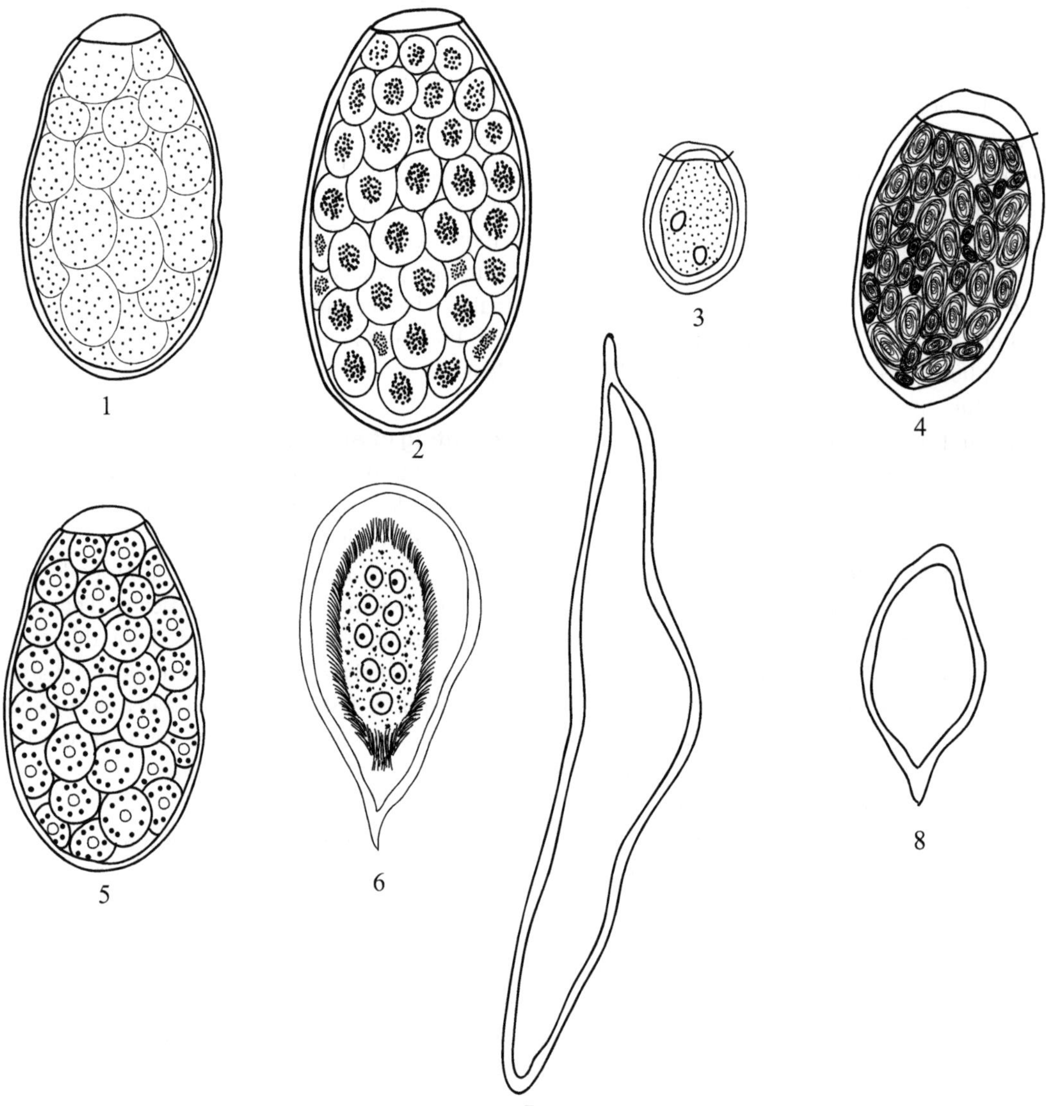

Fig. 13.2 Eggs of trematodes trematodes (1. *Fasciola* spp. 2. Amphistome 3. *Dicrocoelium* spp. 4. *Fascioloides* sp. 5. *Fasciolopsis* sp. 6. *Schistosoma incognitum* 7. *Schistosoma nasalis* 8. *Schistosoma indicum*

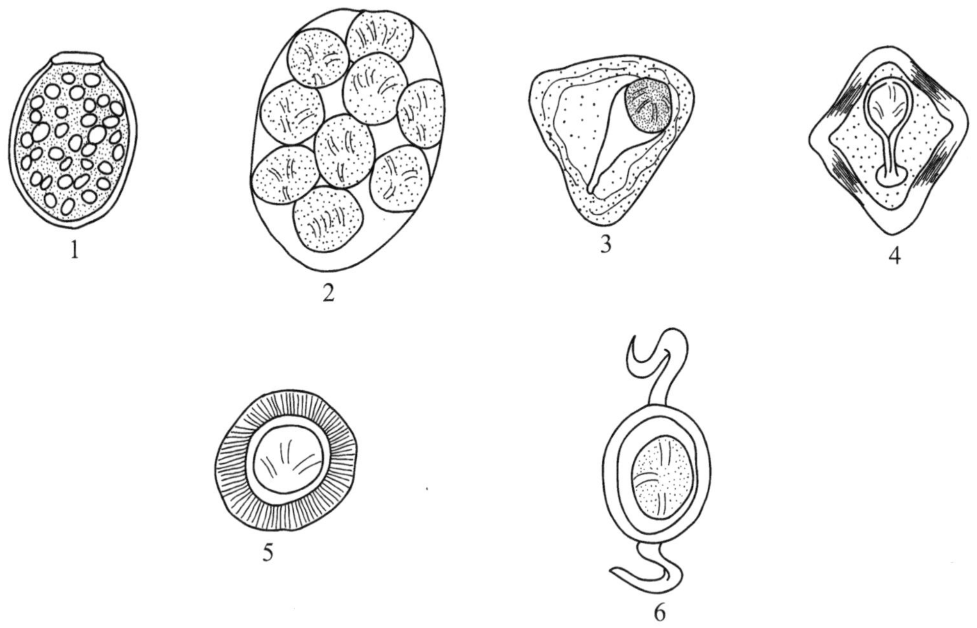

Fig. 13.3 Eggs of cestodes (1. *Diphyllobothrium* spp. 2 *Dipylidium* spp.; egg capsule containing eggs 3. *Moniezia expansa* 4. *Moniezia benedeni* 5. *Taenia* spp. 6. *Choanotaenia infundibulum*)

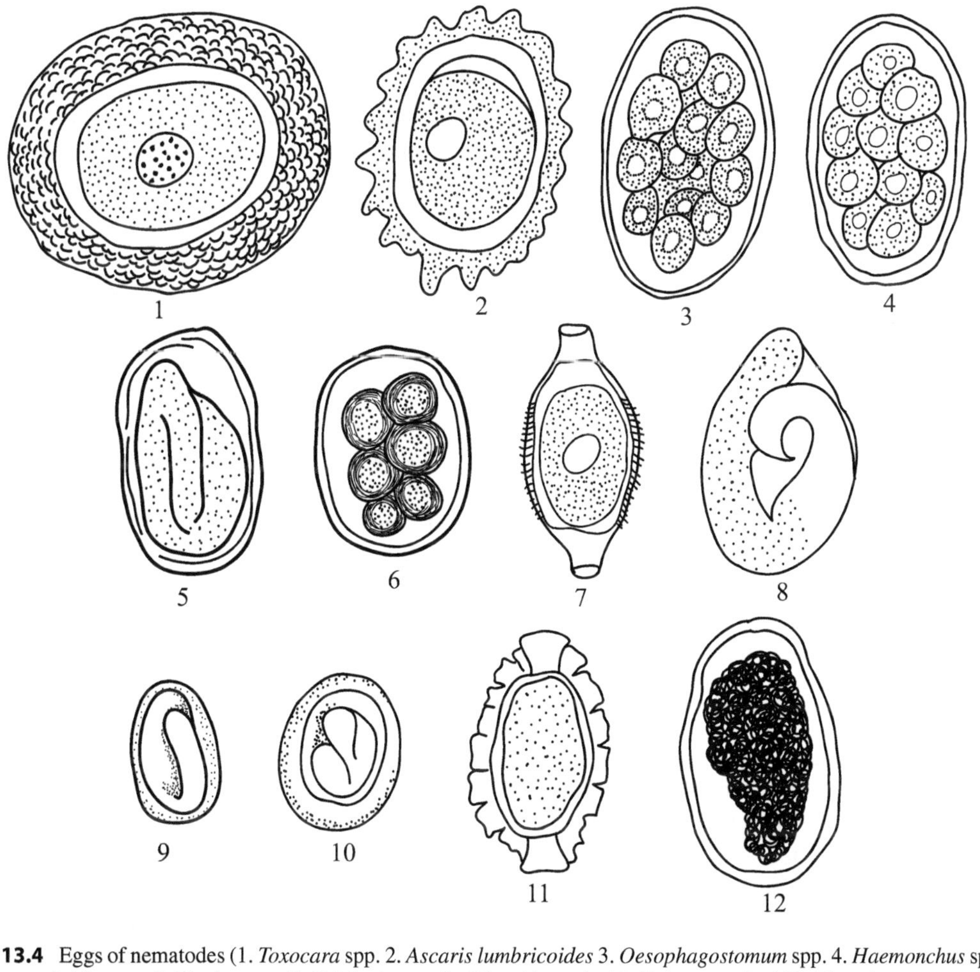

Fig. 13.4 Eggs of nematodes (1. *Toxocara* spp. 2. *Ascaris lumbricoides* 3. *Oesophagostomum* spp. 4. *Haemonchus* spp. 5. *Strongyloides* spp. 6. Hook worm 7. *Trichuris* spp. 8. *Filaroides osleri* 9. *Spirocerca lupi* 10. *Physaloptera* spp. 11. *Dioctophyma renale* 12. *Marshalagia* sp.

Immunoparasitological Techniques

Theoretical Background

Humoral Immunity

Humoral immunity occurs in the case of exogenous antigens. This exogenous antigen is taken up by different antigen-presenting cells, such as macrophages, B cells, and dendritic cells. Major histocompatibility complex (MHC) II molecules present in antigen-presenting cells are responsible for handing over the processed antigen to T cells. The antigen is internalized into the endocytic vesicle. The antigen is hydrolysed by lysosome-associated enzymes (lipase, protease, hydrolase, etc.). The protein of the antigen is degraded to small fragments (oligopeptides). The MHC-II molecules bind to these peptide fragments. These peptide fragments are taken up by MHC molecules. MHC molecules have two polypeptide chains. The antigen is assembled in between these two peptide chains of MHC. This antigen is then carried to the surface of antigen presenting cells (APC) cells and presented to T cells. Inactivated CD4+ T cells recognize the antigen by the T-cell receptor (TCR) complex. CD4+ T cells contain the TCR complex. This TCR complex is composed of two glycoprotein heterodimers (alpha and beta) polypeptide chains, one CD3 molecule and one CD4 molecule. The TCR complex of CD4+ T cell comes in contact with the APC. When the TCR comes in contact with an antigen, signal transduction occurs. The inactive CD4+ T cells get activated to Th1 cells and Th2 cells. T-helper cells (Th1 and Th2) are very important because they secrete different cytokines. Th1 cells release interleukin-2 (IL2), interferon (IFN) gamma, tumour necrosis factor (TNF) alpha, TNF beta, etc. Th2 cells release IL4, IL5, IL6, IL13, etc. All these cytokines are functionally connected to mount an immune response. Some cytokines act on B cells (B lymphocytes). IL4 activates B cells. B cells get transformed into plasma cells. Some B cells remain as memory cells.

Cell-mediated Immunity

Cell-mediated immunity occurs if the antigen is endogenous. The antigen is degraded by a protease enzyme present in the proteasome. The endogenous antigen remains in the cytosol. Small fragments of polypeptides are produced after degradation. The processed antigen fragments are bound to MHC I molecules. The antigen is presented to CD8+ T cells. CD8+ T cells also have TCR complex that contains two heterodimeric polymeric chains, CD3 molecule, and CD8 molecule. Inactive CD8+ T cells get transformed into T cytotoxic cells. T cytotoxic cells are also called as Tc cells. T cytotoxic cells contain different toxic materials, such as granzyme, perforin, etc. The toxic substance kills the organisms.

Hypersensitivity

There are four common hypersensitivity reactions:

Hypersensitivity-1

Hypersensitivity-1 is an immediate type of reaction in which an allergic reaction is set up due to re-exposure to a specific antigen. The IgE antibody binds to the antigen. The Fc part of IgE is bound to the receptor (FcR) of effector cells (mast cells and basophils). In this state, degranulation of mast cells occurs. A number of chemical mediators are liberated from the mast cells comprising histamine, leukotrienes, prostaglandin, etc. These substances cause the hypersensitivity reaction. The hypersensitivity-1 reaction occurs in *Haemonchus* infection and *Trichostrongylus* infection in which IgE and mast cells play a role (self-cure phenomena).

Hypersensitivity II

This is an antibody-dependent immune phenomenon in which a specific antibody binds to a respective antigen resulting in tissue damage or cell lysis. Cell injury occurs if the antigen is present on the cell surface and is mediated by direct phagocytosis as a result of opsonization, antibody-dependent cell-mediated cytotoxicity (ADCC) and complement-mediated lysis. IgG antibodies bind with Fc¥RIII present on NK cells and macrophages and results in the release of granzyme and perforin ultimately causing cell death by apoptosis (ADCC). Complement-mediated lysis occurs in different ways. Via the classical pathway, IgG, and IgM activate complements leading to the production of C3b that mediates phagocytosis. Complement activation also results in the formation of the membrane attack complex (MAC) that forms pores in the cell membrane.

Hypersensitivity-III

Hypersensitivity-III occurs as a result of the accumulation of antigen–antibody complex in the tissues. The antigen–antibody complex is not cleared adequately by innate immune cells. Thus, deposition of the immune complex occurs which sets up an inflammatory reaction due to the aggregation of leucocytes.

Hypersensitivity-IV

This is a delayed type of hypersensitivity and it has a lag phase for its occurrence. Type IV hypersensitivity occurs in three ways:

1. Th1 cells recognize soluble antigens and release IFN gamma. The IFN gamma activates macrophages. Macrophages release different cytokines that cause tissue injury.
2. Th2 cells release IL5 cytokines that activate and recruit eosinophils and these eosinophils release different tissue-damaging materials.
3. Cytotoxic T cells directly damage the cells by releasing different toxic substances. An example of parasitic infections in which delayed hypersensitivity occurs is chronic schistosomiasis due to granuloma formation surrounding the eggs. Delayed hypersensitivity reaction also occurs in cases of Chagas disease, toxoplasmosis, skin reaction due to *Ascaris* infection, etc.

Immune complex-mediated hypersensitivity is also found in *Plasmodium* infections (*Plasmodium vivax*, *P. falciparum*). As a result of this reaction, nephritis occurs.

Types of Parasitic Antigens

Parasitic antigens are substances that are derived from parasites and they can elicit an immune response in the host. Both crude and purified antigens are used for immunoprotection. Crude antigens are somatic antigens (prepared by the homogenisation of the whole parasite/parasitic stages). Excretory and secretory materials of the parasite are also used for immunological studies and are called as excretory–secretory antigens. Purified antigens are also called as defined antigens. These are specific proteins used for immunoprotection and immunodiagnosis. Defined antigens are either purified native antigens or specific antigens prepared by genetic manipulation (recombinant DNA technology). Parasitic antigens are complex in nature. The body of a parasite is formed of different tissues. The crude protein of a parasite contains a number of proteins.

Different Types of Parasitic Antigens

The parasitic antigens can be described in two categories:

1. Crude Antigens

Crude antigens are composed of many proteins. These antigens are prepared from the whole parasite or part of the parasite/parasitic stages. No protein is eliminated from the whole crude antigen. The crude antigen is prepared from the following materials:

(a) **Whole parasite/whole parasitic stage**
  (i) Whole adult parasite
  (ii) Whole larval parasite
  (iii) Eggs

(b) **Part of parasite**
  Somatic antigens are prepared from any organ or part of the parasite.
  Examples:
  (i) Antigen prepared from salivary glands of parasites (tick salivary glands).
  (ii) The gut of the tick is used as an antigen
  (iii) Fluid of bladder worm (i.e., hydatid cysts)

(c) **Excretory-secretory materials/moulting fluid etc.**
  Different excretory-secretory materials, particularly of helminthic parasites are used as antigens. Moulting fluid, particularly of helminthic and arthropod parasites are used as antigens.

Excretory secretory materials of parasites, particularly of the helminthic parasites, are used as antigens. Moulting fluid released during the moulting of parasites is also used as an antigen.

2. Defined antigens

Defined antigens are specific antigens. Proteins, such as enzyme, beta tubulin, etc., are used as antigens. These are also called as specific antigens.

a. **Defined antigens of native sources**
   The specific protein molecules are isolated directly from crude proteins of parasites. Many such molecules are used as antigens. These are enzymes (cathepsin L, cathepsin B, glutathione-S-transferases, cysteine proteinase, etc.) and specific proteins (beta tubulin, calmodulin, paramyosin, etc.).
   A particular protein from a crude mass of the parasitic protein is called as a purified antigen. The protein is isolated from the crude protein derived from the whole parasite by different protein-purification protocols, such as gel exclusion chromatography, ion exchange chromatography, affinity chromatography, etc. In the recent past, several target proteins have been isolated from *Fasciola* spp., *Schistosoma* spp., and other flukes. From *Fasciola* parasites, glutathione-S-transferase, cathepsin L, fatty acid binding protein (FABP), and hemoglobin have been isolated. Their immunodiagnostic and immunoprotection performances have been tested. **Sm26 and Sm28 proteins** have been isolated from *Schistosoma* sp. The immunity performance of these antigens has been tested. Cysteine proteinases have been isolated from *Fasciola* spp., *Schistosoma* spp. *Paragonimus* spp., Paramphistomes, etc. In most cases, column chromatography is an ideal technique for the isolation of a particular protein.
b. **Defined antigens prepared by recombinant technique**
   The defined antigens, as described earlier, can be expressed in the laboratory by specific protocols. For this purpose, the gene or nucleotide sequence coding for a particular protein is identified. This particular gene is isolated from the whole genome sequence. Restriction endonucleases are used for cutting the whole nucleotide sequence to get the desired/target gene. The target gene is isolated and incorporated into a plasmid vector. The recombinant plasmid is inserted into host cells (*Escherichia coli* and yeast cells). The protein is expressed in these host cells. That target protein is isolated from the host cells and is used as an antigen.

Characterisation of Parasitic Antigens

Characterisation of any parasitic antigen can be done as follows:

(a) Immunological characterisation
(b) Biochemical characterisation
(c) Molecular characterisation

Before the characterisation of any parasitic antigen, the particular antigen is isolated. Isolation of particular parasitic proteins is done by different protein-purification protocols. In most cases, different chromatographic techniques (gel filtration, immuno-affinity, ion exchange, gas–liquid, enzyme–substrate chromatography, etc.) are done. The isolated protein is used as an antigen. Hyperimmune sera are generated by immunising animals using the antigen. The Western-blot technique is performed to see whether the protein is immunodominant or not. Apart from Western blot, many immunological tests are also done. The tests are enzyme-linked immunosorbent assay (**ELISA,** different types), complement fixation test **(CFT), CIE, DID,** etc. The isolated protein can be characterised biochemically also. Its function and role in the body can be unfurled. Molecular characterisation of a particular protein is also done. The amino acid sequence of a particular protein can be determined. The nucleotide sequence can also be deduced from the amino acid sequence.

Preparation of Parasitic Antigens

Steps of somatic antigen preparation

- Collect helminthic parasites (cestodes, trematodes, nematodes).

- Wash the worms thoroughly with PBS solution.
- Keep the washed worms in fresh PBS.
- Add some antiprotease.
- Bring to the laboratory in an ice box.
- Keep some amount of PBS in a glass tube usable for homogenisation.
- Homogenize by applying several thrust
- Centrifuge the material at 10,000 g at 4 °C. The supernatant is used as antigen.

Immunodiagnosis

(All the methods described here are only the overviews. One should vividly standardize the methods along with data given here in their laboratories for getting accurate result)

Precipitin Test

- Take a capillary tube.
- Take an antigen in the capillary tube (0.1 ml).
- Take test serum in the same capillary tube.
- Seal the capillary tube with plasticine.
- Keep the tube for 10 min.
- Precipitin material is seen at the interface of antigen and antibody.

Immunodiffusion Test

A very common test is the double immunodiffusion test.

The steps are as follows:

- Glass slides are taken.
- The slides are washed and cleaned with alcohol to make them free of grease.
- The cleaned slides are placed on the laboratory table.
- 0.1% agar is prepared in a beaker.
- Precoating is done by placing the slide in a hot precoating solution and then removing it.
- The precoated slides are kept slanting and allowed for drying for 10 min.
- In another beaker 1% agar solution is prepared.
- Then, on the precoated slides, 3 ml of agar solution is overlayed.
- The slides are kept until the agar is solidified.
- The wells are made by gel puncture.
- Usually, one central well is made first. Then, several wells are made surrounding the central well.
- Some amount of serum is added to the central well. Antigen is added in the peripheral well.
- In the positive case, a precipitin arc is formed.

Bentonite Flocculation Test

- Take a few bentonite particles.
- Coat these particles with antigens.
- Take a few drops of antigen-coated bentonite particles on slides.
- Add diluted serum to this.
- Keep the slide for 30 minutes.
- Observe the floccules under a microscope.
- Formation of floccules is the positive reaction.

Indirect Hemagglutination Test

It is simply the agglutination of antigen-coated erythrocytes by serum activity. Sheep red blood cells (RBCs) are used here. Several compounds are used for the attachment of sheep RBC to parasitic antigens. These are formaldehyde, glutaraldehyde, etc. The antigen-coated RBCs are kept in tubes or microtitre plates and mixed with the diluted sera of animals. In positive cases, erythrocytes form a net pattern at the button of tubes or wells. In negative cases, erythrocytes form simply the button.

Preparation of Sheep RBC

- Collect sheep blood.
- Mix the blood with Alsever's solution at a 10:12 ratio.
- Incubate the blood for 3 days at 4 °C.
- Treat the RBCs with tannic acid.
- The surface of the RBC is coated with antigens.

Steps of Test

- Add 0.05 ml of 1% test serum in the wells of hemagglutination plates (both test and control).
- Then, add 0.025 ml of serum and from the 12th well, go on transferring and discarding.
- Add 0.025 ml of sensitized sheep RBC to every test well.
- Add unsensitized RBC in the control well.
- Mix the material in the well.
- Incubate for 1 h at 37 °C and for 1–3 h at room temperature.
- A mat or carpet formation is seen in a positive reaction.

Intradermal Test

- 0.1 ml of antigen is injected intradermally.
- At the adjacent area, the same amount of PBS is injected which will act as control.
- Wait for 24–48 h. The reaction starts with thickening of the skin. The reaction may persist for a few days.

Complement Fixation Test

Principle

A mixture of antigen (known), unknown patient's serum and complement is incubated. Sheep RBCs are added to the mixture. In positive cases, there will be no lysis of sheep RBC. In positive cases, antigen and antibody reactions occur. Due to this reaction, a complement is utilized for fixation. Since there is no free complement, the binding of complement with sheep RBC (coated with sheep antibody) will not occur. Lysis of RBC does not occur.

Requirements

Chemicals and Biological Agents

Washed sheep RBC suspension (5%), guinea pig complement free from antibodies against the present antigen, hemolysin (rabbit anti-sheep RBC), barbital buffer diluent.

Equipment

Water bath, microtiter plates, pipettes, color standard for accessing hemolysis, barbital buffer diluent, and a centrifuge adapter for microtiter plates.

Method

- Serum is prepared from the collected blood
- Serum is inactivated by heating at 56 °C for 30 minutes.
- Incubate a mixture of known antigens (0.1 ml), inactivated patient's serum (0.1 ml) and a small amount of complement (0.1 ml).
- Add hemolysin-coated RBC.

Interpretation

(a) If there is no lysis of sheep RBC, it indicates the presence of antibodies in the serum (positive result).

(b) Lysis of RBC indicates the absence of antibodies in the serum (negative result).

ELISA

This is an immunodiagnostic test that is sensitive. This immunodiagnostic test is done for the diagnosis of many parasitic diseases. This test is also done for the detection of antigens by sandwich ELISA. In the case of sandwich ELISA, the plates are coated with antibodies instead of antigens.

Different Steps of ELISA

- ELISA plates (99-well) are taken.
- First, the wells are coated with parasitic antigen antigen (antigen mixed with coating buffer at a concentration of 1–10 μg/ml). Coating buffer is 0.05 M Carbonate – bicarbonate buffer (pH 9.6). Coating is done using 100 μl coating buffer for each well. Incubation is done for overnight at 4 °C or 1–2 hours at 37 °C.
- The antigen solution is removed. The wells are washed with washing buffer. The washing buffer is prepared by mixing PBS with 0.05 M Tween 20.
- The wells are blocked by a blocking buffer. Blocking buffer is 1% Bovine Serum Albumin

in PBS. 300 μl of blocking buffer is used for each well. Incubation is done for 1–2 hours at room temperature.
- Then, the blocking buffer is removed.
- Then, the wells are washed again by washing buffer.
- Then, primary antibody is added. Usually 100 μl of primary antibody solution is added. Incubation is done for overnight at 4 °C or 1–2 hours at room temperature.
- Then, the primary antibody solution is removed.
- The wells are washed by washing buffer.
- Then, a secondary antigen, conjugated with an enzyme (Horseradish peroxidase or Alkaline phosphatase) is added. Conjugate solution is diluted in diluent (PBS with blocking agent) at a ratio of 1:1000 to 1:10,000. Usually 100 μl of enzyme -conjugated secondary antibody solution is added. Incubation is done for 1–2 hours at room temperature.
- Then, the secondary antibody conjugated with the enzyme is removed.
- The wells are washed with washing buffer.
- Substrate is added (APS in case of Alkaline phosphatase and TMB in case of HRPO).
- Enzyme–substrate color reaction occurs.
- A stop solution (50 μl) containing 0.5M–2.0M Sulfuric acid is added. The antibody titre is determined by an ELISA reader.

Double-Antibody Sandwich ELISA
- ELISA plates (99-well) are taken.
- First, the wells are coated with antiserum.
- The antiserum solution is removed.
- The wells are blocked by a blocking buffer.
- Then, the blocking buffer is removed.
- Then, the wells are washed again by a washing buffer.
- Then antigen is added.
- Then, the unbound antigen solution is removed.
- The wells are washed by a washing buffer.
- Again, an antiserum solution is added. Washing is done using washing buffer.
- Secondary antibody conjugated with enzyme is added.
- The Secondary antibody conjugated with the enzyme is removed.
- The wells are washed with a washing buffer.
- A substrate is added.
- An enzyme–substrate color reaction occurs.
- The antibody titre is determined by an ELISA reader.

Dot ELISA
- Take a piece of nitrocellulose membrane.
- Charge antigen solution on the nitrocellulose membrane dropwise keeping distance in between the drops.
- The rest of the area is blocked by a blocking buffer.
- Then, the blocking buffer is removed.
- Then, the membrane is washed again by a washing buffer.
- Then, a primary antibody is added.
- Then, the primary antibody solution is removed.
- The membrane is washed by a washing buffer.
- Then, a secondary antigen conjugated with an enzyme is added.
- Then, the secondary antibody conjugated with the enzyme is removed.
- The membrane is washed with the washing buffer.
- A substrate is added.
- An enzyme–substrate color reaction occurs.

Dipstick ELISA
- Take a narrow piece of nitrocellulose membrane.
- Charge a drop of antigen solution at one side of the membrane.
- Dip the antigen-spotted area in a blocking buffer.
- Dip the antigen-spotted area in a washing buffer.
- Dip the antigen-spotted area in a primary antibody solution.
- Dip the antigen-spotted area in the washing buffer.
- Dip the antigen-spotted area in a conjugate solution.

- Dip the antigen-spotted area in the washing buffer.
- Dip the antigen-spotted area in substrate solution.
- Observe the color development.

Indirect Fluorescent Antibody Test

Here, a reaction occurs between a particulate antigen and an immune serum. Antiglobulins are labeled with fluorescing dye (e.g., fluorescein isothiocyanate or Rhodamine). In UV light, these fluorochromes are capable of emitting visible light. In general, corpuscular antigens of parasites are taken. Frozen sections of helminth or smear of intact protozoa are mounted on a glass slide. Fixation is done by controlled heat or chemicals like methanol, formalin, glutaraldehyde, acetone, etc. Then, the antigen is treated with sera. Slides are washed. In positive cases, the antibody reacts and binds to the surface antigen of the parasite and forms an antigen–antibody complex. The slides are washed again and flurochrome-labeled antiglobulin is added and incubated again. The slides are further washed and examined using a fluorescent microscope. This test is sensitive as well as specific. This test can be done for the diagnosis of infections caused by *Toxoplasma*, *Babesia*, and many other parasites.

Different steps are as follows:

- Frozen sections of helminth or smears of intact protozoa are mounted on a glass slide.
- Fixation is done by controlled heat or chemicals like methanol, formalin, glutaraldehyde, acetone, etc. Acetone is used for fixation of frozen tissue. The slide is kept in −20 °C for 10m.
- The slides are washed with PBS.
- The slides are treated with primary antibody (1:50 to 1:1000 depending on standardization). The slides are kept in room temperature for 30–60m.
- The slides are washed again with washing buffer (PBS with 0.05% Tween 20. 3–5 washings are done, each for 5–10m.
- The slides are treated with secondary antibody tagged with fluorescein isothiocynate.The concentration of secondary antibody is same as primary antibody.
- The slides are further washed and examined using a fluorescent microscope.

Hybridoma Technique

This is a technique by which antibody-producing B cells are fused with myeloma cells and the fused cells are called as hybridoma capable of producing identical antibodies (monoclonal) for infinite periods as the myeloma cells are immortal. The following are the steps:

Step-1
Immunization of mice with target antigen
Step-2
Collection of splenocytes from immunized mice
Step-3
Isolation of B cells from splenocytes s
Step-4
Fusion of B cells with myeloma cells (hybridoma)
Step-5
Dilution of hybridoma cells

Method in Detail

First, the mice are inoculated with the target antigen by a standard immunization protocol constituting several injections of antigens with suitable time intervals so that antibodies can be produced against that antigen. Then, the mice are sacrificed. B cells are isolated as mentioned earlier. These B cells are fused with myeloma cells by the method of electrofusion keeping these cells in an electric field. Fusion of cells can also be done employing a chemical method using polyethylene glycol. The fused cells are incubated in a specific medium, HAT medium (hypoxanthine, aminopterin, thymidine). The aminopterin compound present in the HAT medium blocks the pathway of nucleotide synthesis. As a result, the unfused myeloma cells die because they are not able to perform nucleotide synthesis as they lack h**ypoxanthine-guanine phosphoribosyltransferase** (HGPRT) responsible for carrying out nucleotide synthesis by a salvage pathway. However, hybridoma cells can survive as they have HGPRT and they can perform nucleotide synthesis by the salvage pathway. B cells die automatically after their lifespan ends. Hybridoma cells producing specific antibodies are selected and are allowed to grow in culture. The monoclonal antibodies are collected from the culture supernatant.

Application

There are many uses of monoclonal antibodies:

- Monoclonal antibodies are used for providing passive immunity to animals. It gives instant immunoprotection to the animals. There are many infective diseases, for which no chemotherapy is available. In that case, a monoclonal antibody can be used.
- Monoclonal antibodies can be used to produce vaccine candidates. An immuno-affinity chromatography column can be made by using monoclonal antibodies. The corresponding antigens can be prepared. Defined antigens can be isolated by this purification protocol using monoclonal antibodies.
- In different research programs, monoclonal antibodies are used. To know the function of any cytokine, an anti-cytokine monoclonal antibody is used to neutralize the action of the cytokine. By this method, the function of any cytokine can be determined. The function and nature of other specific proteins can also be determined.

Techniques of Molecular Parasitology

Some Theoretical Background

Primary Structure of Proteins

- The primary structure is formed by the covalent linking of the alpha–carboxyl group of one amino acid with the alpha–amino group of the next amino acid by an amide bond.
- This amide bond is also known as a peptide bond. Polypeptides are formed by many amino acids linked by peptide bonds.
- The polypeptides are directional having an N-terminus and a C-terminus.
- One end of the polypeptide has an alpha-amino group while the other end has a free alpha-carboxyl group.
- The sequence of amino acids from the N-terminus to the C-terminus of a polypeptide is called its primary structure.

Secondary Structure of Proteins

- The polypeptides can form folds and form a number of regular structures by hydrogen bonding.
- In this way, the backbone of a polypeptide forms a helix structure (α-helix) with 3.6 amino acid residues per turn.
- Each peptide's N-H group is hydrogen bonded to the C=O group three residues away. Example: The α helical structure is found in globular and fibrous proteins.
- Another secondary structure is the formation of a β-sheet with placing the side chain (R) alternatively above and below the sheet.
- N-H and C=O groups of one section of a polypeptide are hydrogen bonded to complementary groups of another section.
- The sheet may be parallel when these sections run in the same direction (N-terminus to C-terminus) or it may be antiparallel when the sections alternate (C-terminus to N-terminus).
- Different sections of polypeptide chains are arranged side by side and give a sheet structure with the side chains (R) projecting alternately above and below the sheet.

Tertiary Structure of Proteins

- It is a structure formed by the α-helix, β-sheet, loops, and other minor secondary structures. It is an inherent property of proteins.
- The specific proteins get folded into a tertiary structure.
- Misfolding of the tertiary structure is prevented by some other proteins that are called as chaperones.
- Tertiary structure is a three dimensional structure formed by specific folding in such a way that the hydrophilic side chains are located

mainly on the exterior while hydrophobic amino acids are buried in the interior.

- This provides overall stability to the three-dimensional structure.
- The whole setup is further stabilized by different non-covalent interactions between side chains comprising hydrogen bonding, van der Waals forces, electrostatic salt bridges, etc.
- Apart from non-covalent interactions, two cysteine residues form a covalent bond.

Quaternary Structure of Proteins

- The quaternary structure is formed by the stabilization of two or more polypeptide chains (subunits).
- These subunits may be identical or different. For example, hemoglobin has two α- globin and two β-globin chains.

Structure of Nucleic Acids

- There are four heterocyclic bases of DNA.
  (a) Purine—adenine (A) and guanine (G)
  (b) Pyrimidine—thymine and cytosine.
- A base (i.e., adenine, guanine, etc.) covalently bonded to a pentose sugar molecule is called a nucleoside.
- Ribose + Base = Ribonucleoside or nucleoside in the case of RNA. 2′Deoxyribose + Base = 2′-deoxyribonucleoside or deoxynucleoside in case of DNA.
- Nucleotides—Nucleosides along with bound one or more phosphate groups are called nucleotides.

Deoxyribonucleic Acid

- The bases of nucleic acid are covalently joined to a pentose sugar. Pentose sugar is a component of nucleotides.
- Ribose sugar is found in RNA whereas 2′-deoxyribose is found in DNA. The pentose sugar is attached to the 1 position (N-1) of pyrimidines and 9–position (N-9) of purines.
- These sugars attach to the bases by glycosylic or glycosidic bonds.
- DNA forms polymers by joining deoxyribonucleotides. RNA forms polymers by joining ribonucleotides.
- The phosphate group joins with the 5′ hydroxyl of one ribose and the 3′ hydroxyl of another ribose. In this way, a chain of nucleotides is formed. This kind of link or bond is called as phosphodiester bond.

DNA—Double Helical Structure

- Two chains characteristically wound around each other result in a right-handed double helix.
- The two strands are joined by non-covalent hydrogen bonding. The two strands are directed antiparallel in terms of the 5′–3′ direction.
- The striking feature of these two strands is that adenine pairs with thymine. Guanine pairs with cytosine thus creating a purine–pyrimidine pairs.
- The sequence of one strand indicates the sequence of the other strand, i.e., if A (adenine) remains in one strand, T would be exactly in the opposite strand.
- The planar bases remain in the centre of the helix and the negatively charged sugar-phosphate backbone remains outside the helix.
- The two strands run antiparallel and around 10 base pairs remain per turn.
- RNA is a single-stranded molecule and it does not form any helical structure but it forms a globular structure and local helical structures are formed due to intramolecular hydrogen bonding.

DNA Tertiary Structure

In the tertiary structure of DNA, large-scale folding in linear polymer occurs. A three-dimensional shape is found. Handedness (right or left), number of base pairs per turn, length of the helix turn and size difference between major and minor grooves are different features detected.

DNA and RNA Protocols

(All the techniques described here are only the overviews. One should vividly standardize these techniques along with data in their laboratories for getting accurate result)

Isolation of DNA

For carrying out any DNA-based molecular work, the prerequisite is the procurement of DNA. If we want to identify any parasite, we use polymerase chain reaction (PCR). PCR is also done and it seems to be the prerequisite for many other purposes, such as gene sequencing, DNA cloning, recombinant DNA technologies, etc. To perform PCR, the isolation of DNA from the parasite is very important. These are the following steps for the isolation of DNA:

Step-1

Lysis

The first step of DNA isolation is the lysis of cells. This can be done in two ways: mechanically and chemically. Mechanically, this could be done by homogenization, sonication, etc. Chemically, it is done by treating with detergent and enzyme proteinase K.

Step-2

Precipitation

The next step is precipitation. The material is treated with a concentrated salt solution. After lysis, DNA becomes free from the nucleus but it remains mixed with cellular debris materials. For this, alcohol is added to it. DNA is not water soluble and it is precipitated out of the aqueous solution.

Step-3

Purification

The next step is the purification of DNA. The material is centrifuged. The debris material is clumped by centrifugation and DNA is separated from the clump of debris. Now, the DNA has to be separated from the reagent, salt, etc. For this, the material is subject to ice-cold ethanol or isopropyl alcohol extraction. The same could be done using phenol–chloroform precipitation. After this treatment, further centrifugation is done and DNA comes to the supernatant.

Isolation of RNA

- Cells are lysed by using a buffer containing sodium dodecylsulphate (SDS), urea, phenol, guanidium isothiocyanate, etc.
- Denaturation of DNA is done by DNAase. Denaturation of protein is done by proteinase K.
- Cellular components are separated off by centrifugation after adding chloroform. The material gets separated into two phases: organic phase and aqueous phase. RNA is present in the aqueous phase.
- Recovery of RNA is done by using isopropyl alcohol.

Isolation of Plasmid

- Plasmids are naturally occurring extrachromosomal circular DNA molecules (2–200 kb) in size.
- These are found in many numbers (a few hundred) in *E. coli*.
- There are specific biotechnological methods by which plasmid DNA can be isolated. At first, *E. coli* (a sample) inoculated into a culture broth.
- After overnight growth, the organisms enter a stationary phase. The cells are centrifuged and cell pellets are formed.
- The cell pellet is at first subjected to a buffer solution containing lysozyme that can digest the cell wall.
- Then, cell lysis solution containing SDS in NaOH is added.
- This solution causes denaturation of proteins, DNA, and hydrolysis of RNA.
- Potassium acetate is used to neutralize the solution and it precipitates denatured protein and chromosomal DNA.
- The material is further centrifuged and the supernatant contains plasmid.
- Then, the pure plasmid is isolated from the lysate by phenol extraction.

Methods of Electrophoresis of DNA

There have been many studies on the electrophoresis of parasitic DNA. This is one of the basic techniques used for carrying out different molecular works in the field of parasitology. Now, the question is in which case, is electrophoresis done? For the diagnosis of some parasitic diseases, PCR is done. Whether DNA has been amplified or not, it is confirmed by electrophoresis. The DNA bands appear in the gel. The gel is stained with ETBR that is mixed at the time of gel preparation. If the gel is placed in the gel documentation system, the bands are visible. The appearance of bands proves that amplification has occurred that means the parasite-specific primer has worked for the amplification of specific DNA fragments. Electrophoresis is done for other purposes also. After electrophoresis, a particular band can be isolated. The particular DNA fragments are eluted from the gel and attached to the plasmid, following standard protocol. This is very important for the production of recombinant antigens of parasites.

Principle

- DNA fragments run in the agarose gel in the electrophoretic field in the presence of a specific buffer.
- Since the DNA is negatively charged it can move towards the positively charged field.
- DNA of small size and large sizes can be separated by agarose electrophoresis providing different concentrations of agarose gel (1%, 1.5% w/v).
- DNA is stained by adding ETBR to the gel. If the gel is placed under UV light, the bands of DNA are detected by the illumination of UV light.
- If the digested and undigested plasmid DNA are run in different lanes along with the markers, we will get different band patterns of digested and undigested plasmids.

Requirements

Electrophoresis tray, 1% (or as per requirement) agar solution, ETBR solution, power supplies, electrodes, well-forming combs, etc.

Steps of Electrophoresis

- The concentration of agarose gels varies. Usually 0.5%–2% solution is prepared (w/v) in an Erlenmeyer flask. This solution is prepared with a running buffer. The common buffers used for running gel are Tris/Borate/EDTA (TBE) [45 mM Tris-borate, 1 mM ethylene diamine tetraacetic acid (EDTA)] and Tris/acetate/EDTA (TAE) (40 mM Tris-acetate, 1 mM EDTA).
- This is heated in the microwave to make the agarose solution.
- Now, add ETBR solution (0.5 µg/ml) to the agarose solution. Otherwise, the gel can be stained after the completion of electrophoresis in the running buffer after adding ETBR.
- Agarose should be kept in an incubator at 65 °C.
- Place the gel running tray in the gel-casting apparatus. Alternatively, the open edges of the tray are wrapped by tape to make a mold.
- Place a comb of appropriate size in the tray.
- Pour the agarose into the tray.
- Allow the arrangement at room temperature for solidification.
- Place the tray in the gel box. Pour the running buffer into the tray so that it can cover the gel.
- Remove the comb.
- Prepare the DNA sample with loading dye (6× containing 0.25% bromophenol blue, 0.25% xylene cyanol, 30% glycerol).
- Load the sample gently.
- Connect the lids to the power supply.
- Run the gel (1–5 V/cm.) unless the dye reaches the defined p
- Turn off the power supply after completion of gel run.
- Remove the gel tray and drain off the extra running buffer.
- Remove the gel from the tray.
- The gel is examined under the gel documentation system.
- Bands are visible.

DNA Cloning

DNA cloning is a very important technique that is frequently used in the field of veterinary parasitology. This technique is a prerequisite for the production of recombinant proteins. The basics of DNA cloning are described hereby:

- **Restriction enzyme digestion**: The gene responsible for the expression of a protein is taken from a large genome following standard protocol (restriction enzyme digestion, PCR amplification, gel electrophoresis).

- **Ligation of DNA fragment**: The fragment of the genome (DNA fragment) is ligated to the complementary site of the plasmid (vector). Ligation reaction mixture contains 10-100ng vector,insert (3:1 to 1:1 molar ratio), 1-2 units T4 DNA Ligase, Ligase buffer as per manufacturer's instruction. This whole substance is incubated at 16 °C for 16-24 hours.
- **Transformation of host cell**: The target gene attached to the plasmid is called as recombinant DNA. Bacterial or yeast cells are made competent to take up the recombinant DNA. The cells are treated with CaCl2 solution (usually 50–100 mM). Electroporation with brief electric pulse can be done for creating temporary pores in the cell membrane for DNA uptake. Heat shock from 0 °C to 42 °C is also another technique which increases the membrane permeability.
- **Selection of transformants**: The transformed cells are plated on solid selective medium containing antibiotics or other selective agents (LB Agar 15 gm/L, Ampicillin 50–100 μg/ml). The bacteria taken up the plasmid containing antibiotic resistant gene will grow. Other transformed cells will be died by the action of the antibiotics.
- **Confirmation of clones**: Confirmation of clones is done by DNA sequencing, restriction enzyme digestion etc.

Method in detail

- Cut donor DNA segment using restriction endonuclease. Alternatively, complementary DNA can be prepared using mRNA and reverse transcriptase enzymes.
- Insert the target DNA
- Selection of a vector is very important. It should replicate in the host cell. Plasmid, yeast artificial chromosome (YAC), bacterial artificial chromosome (BAC), etc. are used as expression vectors.
- The plasmid vector is cut using the same RE enzyme that was used for preparation of donor DNA fragment.
- Base pairing of donor DNA fragment and plasmid vector is done by DNA ligase enzyme.
- The recombinant vector is transferred in the host cell (*Escherichia coli*).
- The recombinant host cell is grown in culture media.
- Both recombinant and nonrecombinant cells grow in the media. Therefore, the colonies of recombinant host cells are isolated. For this purpose, the marker gene of the plasmid vector is used. For example, the PBR 322 plasmid vector contains ampicillin- and tetracycline-resistant gene.

Recombinant Protein Production

A particular parasitic protein can be produced in the laboratory by recombinant DNA technology. For this purpose, the target gene responsible for the expression of a particular protein is taken out from the whole genome of the parasite by using different restriction enzymes. The gene is amplified by PCR. The PCR product contains multiple copies of the target genes. This gene is inserted in plasmids. The plasmid has a complementary site where the target gene can be recombined. This gene is attached to the complementary site by the ligase enzyme. Now, the plasmid is transfected in the host cells. Two important organisms such as *E. coli* and *Pichia pastoris* or other yeast cells act as host cells for the expression of the protein. The desired protein is expressed in these cells. Then, the target protein is isolated by a specific isolation protocol. The same is utilized as an antigen for immunoprotection studies or immunodiagnostic studies. In the process of the preparation of recombinant protein, many hurdles are faced by researchers. The main problem, frequently encountered, is protein folding. The post-translated protein does not have the exact folding as that of the natural protein. However, at the present time, efforts are being made to develop several techniques to have a proper folding of a protein. **Some isoenzymes and other specific proteins of parasites, such as glutathione-S-transferases, cathepsin-L, cathepsin-B, FABP, etc., have already been expressed by recombinant DNA technology.**

Polymerase Chain Reaction

Preparation of Primers

- Primers can be procured if commercially available.
- Otherwise, primers can be designed using different software available (Primer3, Primer-BLAST, Oligo, Geneious).
- The designed nucleotide sequence of primer is used for production.

Stepwise Procedure of PCR

- Prepare the PCR mixture (PCR master mixture commercially available also)
- The PCR volume of 50 μl contains 25 μl of 2× Top Taq DNA polymerase, 20 pmol of each primer (1 μl each), 6 μl of template DNA, 0.2 mM of each deoxy nucleotide triphosphate (dNTP) and nuclease-free water variable.
- The steps are as following:
- 94-98 °C for 30 s (Denaturation).
- 53 °C for 45 s (Annealing).
- Followed by 72 °C for 1-2 min (Extension).

Electrophoretic Separation of DNA

- Arrange the Mini Horizontal Electrophoresis apparatus
- Prepare 1% TBE agarose gels.
- Add ETBR (0.5 μg/ml of gel).
- Mix it gently in the liquid state of the agar.
- The open side of the electrophoretic apparatus is closed by paraffin tape.
- We can calibrate the expected volume of agar so that the head of combs can slightly penetrate the agar. This can be done by using distilled water.
- Then, approximately 25 ml of ETBR-treated agar is poured into the apparatus.
- Then, the comb is placed at one side to make well.
- After complete solidification, the combs are gently removed.
- Then, TBE buffer is poured.
- Then, DNA sample present in the loading buffer is loaded in the well and run the gel.

MIX
Papier aus verantwortungsvollen Quellen
Paper from responsible sources
FSC® C105338

If you have any concerns about our products,
you can contact us on
ProductSafety@springernature.com

In case Publisher is established outside the EU,
the EU authorized representative is:
Springer Nature Customer Service Center GmbH
Europaplatz 3, 69115 Heidelberg, Germany

Printed by Libri Plureos GmbH
in Hamburg, Germany